Properties, Definitions, and Formulas

Properties of Real Numbers

$a + b = b + a, ab = ba$	**Commutative Properties**
$(a + b) + c = a + (b + c), (ab)c = a(bc)$	**Associative Properties**
$a + 0 = a, 1 \cdot a = a$	**Identity Properties**
$a + (-a) = 0, a \cdot \dfrac{1}{a} = 1$	**Inverse Properties**
$a \cdot 0 = 0$	**Multiplication Property of 0**
$a(b + c) = ab + ac$	**Distributive Property**
$-(a - b) = b - a$	**Opposite of a Difference**
$a(-1) = -1a = -a$	**Multiplication Property of -1**
$-\dfrac{a}{b} = \dfrac{-a}{b} = \dfrac{a}{-b}$	**Signs in Quotients**

Order of Operations

1. Operations within grouping symbols
2. Exponents
3. Multiplication and division
4. Addition and subtraction

Absolute Value

$$|a| = \begin{cases} a, & \text{if } a \geq 0 \\ -a, & \text{if } a < 0 \end{cases}$$

Slope Formula

$$m = \frac{y_2 - y_1}{x_2 - x_1}$$

Equation of Line

Slope-Intercept Form: $y = mx + b$

Point-Slope Form: $y - y_1 = m(x - x_1)$

Exponents

$$b^m \cdot b^n = b^{m+n} \qquad (b^m)^n = b^{mn}$$

$$(ab)^n = a^n b^n \qquad b^0 = 1$$

$$b^{-n} = \frac{1}{b^n} \qquad \frac{b^m}{b^n} = b^{m-n}$$

$$\left(\frac{a}{b}\right)^n = \frac{a^n}{b^n} \qquad \left(\frac{a}{b}\right)^{-n} = \frac{b^n}{a^n}$$

Radicals

$$\sqrt[n]{a}\,\sqrt[n]{b} = \sqrt[n]{ab}$$

$$\frac{\sqrt[n]{a}}{\sqrt[n]{b}} = \sqrt[n]{\frac{a}{b}}$$

$$b^{m/n} = \sqrt[n]{b^m}$$

Factoring Patterns

$$A^2 - B^2 = (A + B)(A - B)$$

$$A^2 + 2AB + B^2 = (A + B)^2$$

$$A^2 - 2AB + B^2 = (A - B)^2$$

$$A^3 + B^3 = (A + B)(A^2 - AB + B^2)$$

$$A^3 - B^3 = (A - B)(A^2 + AB + B^2)$$

Fractions and Rational Expressions

$$\frac{a}{c} + \frac{b}{c} = \frac{a + b}{c},$$

$$\frac{a}{c} - \frac{b}{c} = \frac{a - b}{c}$$

$$\frac{ac}{bc} = \frac{a}{b}$$

$$\frac{a}{b} \cdot \frac{c}{d} = \frac{ac}{bd},$$

$$\frac{a}{b} \div \frac{c}{d} = \frac{a}{b} \cdot \frac{d}{c}$$

$$\frac{a}{b} = \frac{c}{d} \text{ if and only if } ad = bc$$

Zero Factor Property

If $ab = 0$ then $a = 0$ or $b = 0$.

Quadratic Formula

$$x = \frac{-b \pm \sqrt{b^2 - 4ac}}{2a}$$

Elementary Algebra

Elementary Algebra

Discovery and Visualization

Third Edition

Elaine Hubbard

Kennesaw State University

Ronald D. Robinson

Houghton Mifflin Company   Boston • New York

Editor-in-Chief: Jack Shira
Editorial Associate: Marika Hoe
Senior Production/Design Coordinator: Jodi O'Rourke
Senior Manufacturing Coordinator: Marie Barnes
Marketing Manager: Ben Rivera

Printed in the U.S.A.

Library of Congress Control Number: 2001133278

Student Text ISBN: 0-618-22386-X
Instructor's Annotated Edition ISBN: 0-618-22387-8

3456789-DOW-06 05

Contents

Chapter 4   Properties of Lines   208

Chapter 5   Systems of Linear Equations   272

Chapter 6   Exponents and Polynomials   320

Preface

Every learning theory emphasizes the value of widening the sensory spectrum. To accomplish this in teaching mathematics, we believe that one must give students a visual connection that allows students to "see" mathematics in context.

In the third edition of *Elementary Algebra,* we continue to use our extensive teaching experience to create a balance between traditional approaches and the use of a graphing calculator. We have developed an approach to teaching with a graphing calculator that works successfully for us, for our colleagues, and for students.

As in our previous editions, much of the presentation begins with the visualization of concepts. This gives the student a logical basis for definitions and allows the student to discover basic rules and principles. However, our overall goal is to place considerable emphasis on multiple approaches as an instructional strategy.

Approaches

Graphical Approaches   In this edition, we have retained the many advantages of the graphing calculator as an instructional tool. The use of graphing calculator technology provides the student with concrete, visual experiences that lead toward the generalizations and formal structure of mathematics in a natural way.

Numerical Approaches   We have greatly increased the use of numerical approaches in this edition. Calculator tables and hand-drawn tables are introduced early in the text. Such tables have a wide range of applicability—from evaluation of expressions to the detection of extraneous solutions. Tables often provide a vital link between graphical informality and symbolic formality.

Symbolic Approaches   The specificity of the graphical and numerical approaches provides a foundation for symbolic representation, the essence of mathematics. Thus, although the pedagogical devices of graphs and tables often lead the way in the exposition, the focus and goals of the content are mathematical structures, concepts, and procedures.

Verbal Approaches   We believe that verbal skill and mental discipline go hand in hand. Thus, many of the features new to this edition are designed to assist students in thinking and communicating in the language of mathematics.

These approaches are most effective when they are used in mutually reinforcing ways with all other techniques. Whenever appropriate, we present topics with a visual (graphical) and intuitive (numerical) foundation and progress to general (symbolic) conclusions. Through such verbal features as Speaking the Language, Writing Exercises, and Data Analysis, we synthesize all of the approaches into a meaningful whole. In our own classroom experiences, we have found that collaboration is a natural outcome.

First We See, Then We Do

We view the graphing calculator as an instructional tool for promoting a greater understanding of mathematical concepts. The graphing calculator gives us a pathway from visualization to discovery, but the confirmation of our observations is through algebra, which receives the primary emphasis throughout this text. In short, first we see the concept, then we do the algebra.

A variety of calculator models are used in instructional programs, and the extent to which calculators are required and used also varies widely. We believe that the TI-83 Plus is representative, and our presentation is based on that model.

1. The feature "Calculator Key Words" is found throughout the text, but especially in the first half of the book. One or more calculator operations are described at their first point of use. The T1-83 Plus keystrokes are given for a specific example.
2. The separate *Graphing Calculator Keystroke Guide* contains even more information about the TI-83 Plus, as well as keystroke information for the TI-86.

Content

In this third edition, the overall content has been modified only slightly. Two of the most significant changes are as follows.

1. Some of the introductory material has been moved to the point at which it is first used. For example, concepts and notation originally introduced in Chapter 1 but not needed until later chapters are now introduced where and when they are needed. As a result, Chapter 1, especially Section 1.1, has been streamlined somewhat.
2. We realize that the new placement of calculator keystroke information within the text might lead to the perception that this is a "calculator book." Although that information is designed for the convenience of calculator users—instructors and/or students—our book can be used very effectively without any references to or use of a graphing calculator.

Examples and Exercises

Immodestly, we believe that our examples and exercise sets are major strengths of our book. Our intent has always been to provide a pool of varied activities from which instructors can draw and to rely on the instructors to create a syllabus that reflects the goals of their curricula.

1. Exercises labeled as "Modeling with Real Data" have been updated or replaced as needed in order to provide the most current data available.
2. Exercises formerly labeled as "Group Projects" are now titled "Data Analysis" to reflect their actual intent. These exercises have also been updated or replaced as needed.
3. Exercises formerly labeled "Looking Ahead" are now entitled "Warm-Up Skills" and placed in the applicable chapter openers. These exercises review essential math skills that are needed for the current chapter.
4. The topic "thread" found in each individual chapter, which begins with the chapter opener application and is expanded on in a Data Analysis feature, has been updated or replaced as needed.

Art

Much of the art has been revised.

1. Although art involving calculator screens can be treated as generic, all such art has been redesigned to make it consistent with the display of a TI-83 Plus.
2. Graphs that were formerly labeled with "bubbles" are now in color with corresponding color-coded labels.
3. Graphs of inequalities, including systems of inequalities, have been improved to help the student see the solution regions.
4. Additional new art has been included to provide more visual assistance to the student.

The goals of this book are to provide an effective instructional framework for the classroom and to engage students in a better understanding of the nature of mathematics in their personal and career lives. We earnestly hope that our work will promote achievement and success for all.

Supplements

For Students

HM³ CD-ROM Tutorial (Student Version)   (0-618-22392-4) Every algebraic topic in the text is supported by this state-of-the-art tutorial software featuring algorithmically generated exercises and quizzes, as well as animated solution steps, selected video clips, and lessons/problems presented in a colorful, lively manner.

Graphing Calculator Keystroke Guide   (0-618-22388-6) Contains keystroke information and examples for the TI-83 Plus and the TI-86.

Student Solutions Manual   (0-618-22389-4) Contains worked-out solutions for odd-numbered problems from the student text.

Math Study Skills Workbook by Paul D. Nolting   (0-395-98225-1) This workbook is designed to help reinforce skills and minimize frustration in any math class, lab, or study skills course. It offers a wealth of proven study tips and sound advice on note taking, time management, and reducing math anxiety. In addition, numerous self-assessment opportunities enable you to track your own progress.

Lecture Videos   (0-618-22396-7) These comprehensive lecture videos presented by Dana Mosley explain and reinforce the algebraic concepts for each chapter section in the textbook. In addition, students will see how to use their graphing calculator to solve selected problems from the text.

SMARTHINKING™ Live Online Tutoring   Houghton Mifflin has partnered with SMARTHINKING to provide an easy-to-use and effective online tutoring service. A dynamic **Whiteboard** and **Graphing Calculator Function** enables students and e-structors to collaborate easily. SMARTHINKING offers three levels of service:

- **Text-Specific Tutoring** provides real-time, one-on-one instruction with a specially qualified e-structor.
- **Questions Any Time** allows students to submit questions to the tutor outside the scheduled hours and receive a reply within 24 hours.
- **Independent Study Resources** connect students with around-the-clock access to additional educational services, including interactive web sites, diagnostic tests and Frequently Asked Questions posed to SMARTHINKING e-structors.

For Instructors

Instructor's Annotated Edition   (0-618-22387-8) This special version of the text contains teaching tips in the text margins and provides answers to the odd- and even-numbered exercises for the end-of-section activities, review problems, Chapter Tests, Cumulative Tests, and Selected Features including Warm-Up Skills, Think About It, and Speaking the Language.

Instructor's Solutions Manual   (0-618-22390-8) Contains worked-out solutions for all problems in the text.

HM³ CD-ROM Tutorial (Instructor Network Version)   In addition to the student version features, this software offers instructors a classroom management system to track student performance. To request a copy, please contact your Houghton Mifflin representative or e-mail us at college_math@hmco.com.

Printed Test Bank   (0-618-22391-6) Provides a variety of test items for each chapter of the text.

HM Testing   (0-618-22393-2) This new computerized test generator database is designed to produce an unlimited number of tests for each chapter of the text, including cumulative tests and final exams. This hybrid CD-ROM, which works on both Microsoft Windows and Macintosh platforms, contains numerous algorithms as well as online testing and gradebook functions.

Web Site and PowerPoint Slides   A book-specific web site features a number of resources, including PowerPoint slides for easy classroom presentations. Go to *http://math.college.hmco.com/instructors,* and follow the Developmental Math links to the Hubbard/Robinson, *Elementary Algebra: Discovery and Visualization,* 3rd Edition site.

Acknowledgments

We would like to thank the following colleagues who reviewed the third edition manuscript and made many helpful suggestions:

Maria Calzada	*Loyola University, New Orleans*
Joan A. Carr	*National Technical Institute for the Deaf*
Peggy Clifton	*Redlands Community College*
James E. Coleman	*Baltimore City Community College*
Ray E. Collings	*Georgia Perimeter College at Clarkston*
Linda Crabtree	*Metropolitan Community College*
Vincent A. Daniele	*National Technical Institute for the Deaf*
George DeRise	*Thomas Nelson Community College*
Dennis C. Ebersole	*Northampton Community College*
Robert Farinelli	*Community College of Allegheny County*
Dale S. Felkins	*Arkansas Tech University*
Catherine Griffin	*Lansing Community College*
Celeste Hernandez	*Richland College*
Glenn Hunt	*Riverside Community College*
Vijay S. Joshi	*Virginia Intermont College*
Jeff A. Koleno	*Lorain County Community College*
Lee J. McEwan	*Ohio State University*
Timothy McLendon	*East Central College*
Barbara D. Sehr	*Indiana University, Kokomo*

We are grateful to Helen Medley, who took great care to ensure accuracy of the text and answers. We also thank Merrill Peterson for his many contributions to the production of this book.

Algebra Basics, Equations, and Inequalities

Overall, the television ratings for the Democratic and Republican national conventions have declined since 1968. The data in the bar graph can be modeled by **algebraic expressions.** Such expressions can be used to write **equations** and **inequalities** that we can use to compare ratings and to describe trends. (For more on this real-data problem, see Exercises 89–92 at the end of Section 2.7.)

(Source: Neilson Media Research.)

Chapter Snapshot

This chapter opens with a discussion of algebraic expressions and how to evaluate them algebraically and graphically. We then discuss equations whose solutions we estimate graphically. By applying the properties of equations, we learn how to obtain exact solutions algebraically. We conclude with graphing and algebraic methods for solving inequalities.

Warm-Up Skills

The following exercises review concepts and skills that you will need in Chapter 2.

In Exercises 1 and 2, use the Distributive Property to perform the indicated operation.

1. Multiply $3(x - 5)$.
2. Factor $4x + 4y$.
3. Use an associative property to multiply $-4(3y)$.

In Exercises 4 and 5, evaluate the given expression.

4. $2(-3)^2 - 4(-3)(2) - 5^2$
5. $7 - 2(5 - 1)$
6. Graph the set $\left\{-4, \frac{5}{3}, 0, \sqrt{25}\right\}$ on a number line.

In Exercises 7–9, insert $<$, $>$, or $=$ to make the statement true.

7. $2(-2) + 5 \quad\underline{}\quad 1 - (-2)$
8. $\frac{1}{3}(-12) - 16 \quad\underline{}\quad 2(-12) + 4$
9. $2(-1) + 1 \quad\underline{}\quad \dfrac{3(-1) - 1}{2}$
10. Explain why $x - 10$ can be written as $-10 + x$.

In Exercises 11 and 12, use the given property to complete the statement.

11. Multiplication Property of -1: $-a = \underline{} \cdot a$
12. Associative Property of Multiplication:
$$6\left[\frac{5}{6}(y + 3)\right] = \underline{}\,(y + 3)$$

71

Opening Features

Chapter Opener

Each chapter begins with a short introduction to a real data application, which is then expanded on in a Data Analysis feature. The chapter opener also includes the new feature Warm-Up Skills, a list of 10 to 12 review exercises that focus on essential math skills. A section-by-section table of contents and a helpful Chapter Snapshot briefly summarize the topics that will be covered in the chapter.

Section Openers

Each section begins with a list of subsection titles that briefly outlines the material found within that section. Various technology 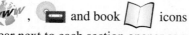, and book icons appear next to each section opener as a reminder that these resources are available for students. For a complete description of resources, please see the Supplements section of the Preface.

---

72   CHAPTER 2   Algebra Basics, Equations, and Inequalities

2.1 Algebraic Expressions

Evaluating Algebraic Expressions • Terms and Coefficients • Equivalent Expressions • Combining Like Terms • Removing Grouping Symbols • Simplifying Algebraic Expressions

Evaluating Algebraic Expressions

In Chapter 1 we described a numerical expression as any combination of numbers and operations. A numerical expression may also contain grouping symbols to indicate the order of operations.

Similarly, an **algebraic expression** is any combination of numbers, variables, grouping symbols, and operations. Each of the following is an algebraic expression.

$$x^2 + 3x - 4 \qquad \frac{a - 4}{5} \qquad 3y + 2(y - 4) \qquad 2x - 3y$$

A numerical expression has a specific value that can be obtained by performing the indicated operations. An algebraic expression does not have a specific value until the variables in the expression have been replaced with numbers.

Evaluating Algebraic Expressions
1. Replace every occurrence of each variable with the given value for that variable.
2. Evaluate the resulting numerical expression.

EXAMPLE 1   Evaluating Algebraic Expressions with One Variable

Evaluate the given algebraic expression for the given value of the variable.
(a) $2x - 7$; $x = 4$ (b) $t^2 - 3t + 1$; $t = 5$
(c) $-x^2$; $x = -3$

LEARNING TIP
An important part of evaluating an expression is applying the Order of Operations correctly.

Solution
(a) $2x - 7 = 2(4) - 7$ Replace x with 4.
$ = 8 - 7$ Evaluate the numerical expression.
$ = 1$

(b) $t^2 - 3t + 1 = 5^2 - 3(5) + 1$ Replace t with 5.
$ = 25 - 15 + 1$ Evaluate the numerical expression.
$ = 11$

(c) $-x^2 = -1 \cdot x^2$ Multiplication Property of -1
$ = -1 \cdot (-3)^2$ Replace x with -3.
$ = -1 \cdot 9$ Square -3; then multiply.
$ = -9$

Modeling with Real Data

In accordance with the current standards of professional organizations such as AMATYC and NCTM, an abundance of examples and exercises are provided that teach students how to organize and interpret data, how to model it with mathematical functions, and how to use such models to extrapolate and predict outcomes. Sourced data from a variety of subjects are used in numerous examples, groups of modeling exercises, and Data Analysis exercises.

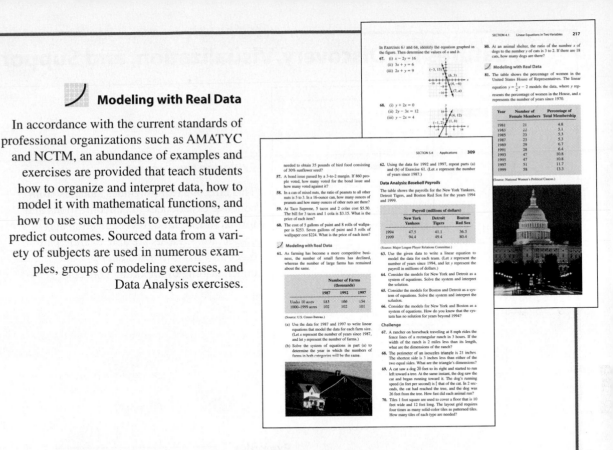

Emphasis on Real Data and Real Life

Real-Life Applications

The majority of application problems are written in the context of real-life knowledge and experience. Some sections are devoted exclusively to examples and exercises involving real-life applications, and most other sections contain dedicated blocks of such problems. According to national teaching standards, students gain a better understanding of the practical nature of mathematics when the concepts and topics are connected to real-world experiences and situations.

Features for Discovery, Visualization, and Support

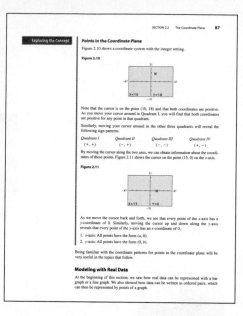

Exploring the Concept

This effective feature guides students from concrete experiences to generalizations and formal rules. Through visualization and conjecture, students are able to anticipate outcomes and arrive at a deeper understanding of important concepts.

Examples

All sections contain numerous, titled examples, many with multiple parts graded by difficulty. These examples illustrate concepts, procedures, and techniques, and they reinforce the reasoning and critical thinking needed for problem solving. Detailed solutions include helpful comments that justify the steps taken and explain their purpose.

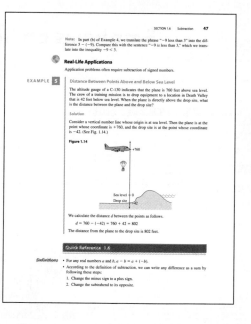

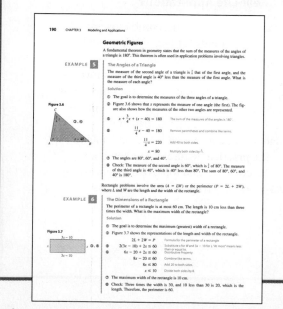

Graphs

Both traditional and calculator graphs are used throughout the exposition and exercises to assist students in visualizing concepts. All calculator displays are now based on the TI-83 Plus due to the popularity and functional design of this graphing calculator model.

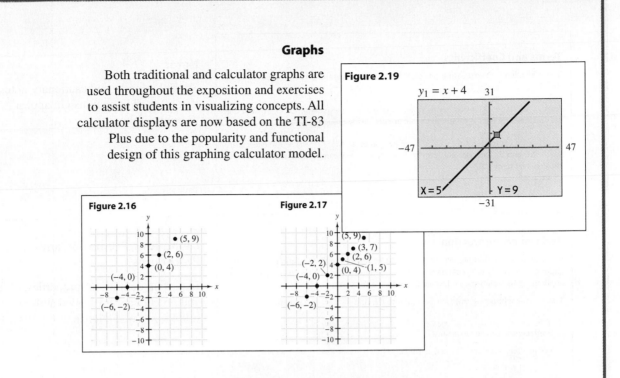

Figure 2.19

$y_1 = x + 4$

Figure 2.16

Figure 2.17

Calculator Key Words

Calculator Key Words explain certain calculator functions when they are first introduced in the text. Key Words reference the TI-83 Plus model. In addition, the separate *Graphing Calculator Keystroke Guide* provides detailed step-by-step instructions and examples for all Key Words for both the TI-83 Plus and TI-86 models.

We can also use a calculator to evaluate an expression. For now we will limit our discussion to expressions with one variable, and we will consider only one of several possible methods.

Keys to the Calculator

Store

We assign a value to a variable by *storing* the value in the variable.

Evaluate

You can use your calculator to evaluate an expression for a given (stored) value of the variable. To let $x = 3$ and evaluate $x^2 - 5$, press

3 [STO→] [X] [ENTER]

[X] [^] 2 [−] 5 [ENTER]

3→X
                    3
X^2−5
                    4

Terms and Coefficients

The addends of an expression are called **terms.** A **constant term** is a term with no variable.

Note: We can think of the terms of an expression as the parts of the expression that are separated by plus or minus signs that are not inside grouping symbols.

Because the expression $6x^2 - 8x + 7$ can be written as $6x^2 + (-8x) + 7$, the terms are $6x^2$, $-8x$, and 7. The constant term is 7. Because $x + y$ can be written as $x + y + 0$, the constant term is 0.

Notes

Special remarks and cautionary notes that offer additional insight appear throughout the text.

Performing Subtraction

For simple subtraction problems, you should be able to apply the definition mentally and not have to rely on a calculator. (When you do use a calculator, you do not need to apply the definition because the calculator does it for you.)

The rule for subtracting fractions is similar to the rule for adding fractions.

Subtraction Rule for Fractions

For any real numbers a, b, and c, where $c \neq 0$,

$$\frac{a}{c} - \frac{b}{c} = \frac{a - b}{c}$$

In words, subtract the numerators and retain the common denominator.

Definitions, Properties, and Procedures

Important definitions, properties, and procedures are shaded and titled for easy reference.

Learning Tip

Every section has at least one Learning Tip in the margin that offers students helpful strategies and alternative ways of thinking about concepts.

Think About It

Located in the margin of each section, Think About It is a question or series of questions that requires critical thinking and reasoning in order to broaden and extend concepts. These questions are designed to spark students' imagination and interest.

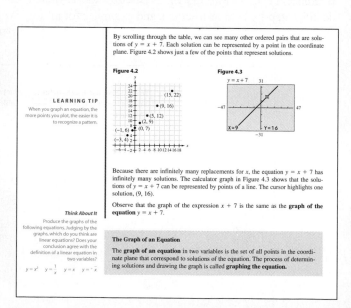

By scrolling through the table, we can see many other ordered pairs that are solutions of $y = x + 7$. Each solution can be represented by a point in the coordinate plane. Figure 4.2 shows just a few of the points that represent solutions.

Figure 4.2

Figure 4.3

LEARNING TIP
When you graph an equation, the more points you plot, the easier it is to recognize a pattern.

Because there are infinitely many replacements for x, the equation $y = x + 7$ has infinitely many solutions. The calculator graph in Figure 4.3 shows that the solutions of $y = x + 7$ can be represented by points of a line. The cursor highlights one solution, (9, 16).

Observe that the graph of the expression $x + 7$ is the same as the **graph of the equation** $y = x + 7$.

Think About It
Produce the graphs of the following equations. Judging by the graphs, which do you think are linear equations? Does your conclusion agree with the definition of a linear equation in two variables?

$y = x^2$ $y = \frac{1}{x}$ $y = x$ $y = ^-x$

The Graph of an Equation

The **graph of an equation** in two variables is the set of all points in the coordinate plane that correspond to solutions of the equation. The process of determining solutions and drawing the graph is called **graphing the equation.**

End-of-Section Features

A typical section ends with two features—Quick Reference and Speaking the Language.

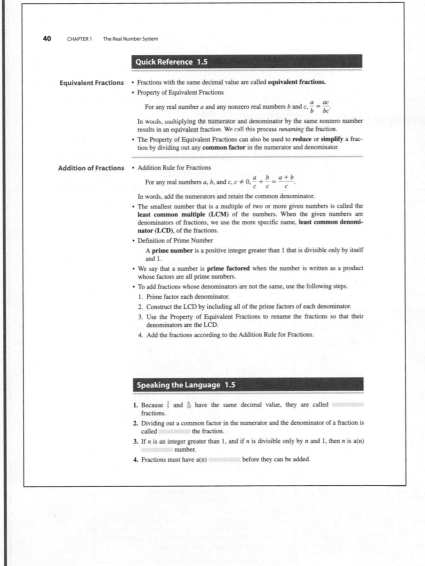

Quick Reference 1.5

Equivalent Fractions
- Fractions with the same decimal value are called **equivalent fractions.**
- Property of Equivalent Fractions

  For any real number a and any nonzero real numbers b and c, $\dfrac{a}{b} = \dfrac{ac}{bc}$.

  In words, multiplying the numerator and denominator by the same nonzero number results in an equivalent fraction. We call this process *renaming* the fraction.
- The Property of Equivalent Fractions can also be used to **reduce** or **simplify** a fraction by dividing out any **common factor** in the numerator and denominator.

Addition of Fractions
- Addition Rule for Fractions

  For any real numbers a, b, and c, $c \neq 0$, $\dfrac{a}{c} + \dfrac{b}{c} = \dfrac{a+b}{c}$.

  In words, add the numerators and retain the common denominator.
- The smallest number that is a multiple of two or more given numbers is called the **least common multiple (LCM)** of the numbers. When the given numbers are denominators of fractions, we use the more specific name, **least common denominator (LCD)**, of the fractions.
- Definition of Prime Number

  A **prime number** is a positive integer greater than 1 that is divisible only by itself and 1.
- We say that a number is **prime factored** when the number is written as a product whose factors are all prime numbers.
- To add fractions whose denominators are not the same, use the following steps.
  1. Prime factor each denominator.
  2. Construct the LCD by including all of the prime factors of each denominator.
  3. Use the Property of Equivalent Fractions to rename the fractions so that their denominators are the LCD.
  4. Add the fractions according to the Addition Rule for Fractions.

Speaking the Language 1.5

1. Because $\frac{1}{5}$ and $\frac{6}{10}$ have the same decimal value, they are called _____ fractions.
2. Dividing out a common factor in the numerator and the denominator of a fraction is called _____ the fraction.
3. If n is an integer greater than 1, and if n is divisible only by n and 1, then n is a(n) _____ number.
4. Fractions must have a(n) _____ before they can be added.

Quick Reference

Quick Reference appears at the end of all sections except those dealing exclusively with applications. This handy reference and review tool provides detailed summaries of the important rules, definitions, properties, and procedures grouped by subsection.

Speaking the Language

Speaking the Language appears at the end of all sections except those dealing exclusively with applications. This feature helps students to think and communicate in the language of mathematics by reinforcing vocabulary and contextual meanings.

Exercises

A typical exercise set in each section includes exercises from each of the following groups: Concepts and Skills (including Writing 🖉 and Concept Extension CE), Real-Life Applications, Modeling with Real Data, Data Analysis, and Challenge. Certain exercises in the Instructor's Annotated Edition contain a graphing calculator icon ▦ indicating that a graphing calculator is recommended.

Concepts and Skills

Most exercise sets begin with the basic skills and concepts discussed within the text. These include Writing Exercises, designed to help students gain confidence in their ability to communicate, and Concept Extension exercises, which go slightly beyond the text examples.

Data Analysis

The Data Analysis exercises appear in many of the end-of-section exercise sets. Students may work independently or in groups to solve these problems by analyzing and interpreting real-life data.

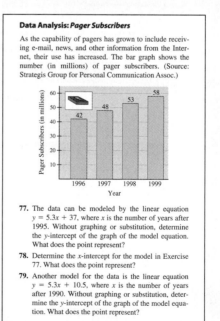

Data Analysis: Pager Subscribers

As the capability of pagers has grown to include receiving e-mail, news, and other information from the Internet, their use has increased. The bar graph shows the number (in millions) of pager subscribers. (Source: Strategis Group for Personal Communication Assoc.)

77. The data can be modeled by the linear equation $y = 5.3x + 37$, where x is the number of years after 1995. Without graphing or substitution, determine the y-intercept of the graph of the model equation. What does the point represent?

78. Determine the x-intercept for the model in Exercise 77. What does the point represent?

79. Another model for the data is the linear equation $y = 5.3x + 10.5$, where x is the number of years after 1990. Without graphing or substitution, determine the y-intercept of the graph of the model equation. What does the point represent?

Challenge

In Exercises 77–80, the coordinates of the endpoints of a line segment are given. Determine the coordinates of the midpoint of the segment.

77. $(-2, 3), (6, 3)$ 78. $(-5, -2), (-1, -2)$

79. $(-4, 1), (-4, 3)$ 80. $(3, -3), (3, 7)$

81. One side of a rectangle has vertices $(-2, -3)$ and $(7, -3)$. Another side is 8 units long. Point P is a third vertex in Quadrant I. Determine the coordinates of point P.

82. A square whose sides are 8 units long is drawn with its center at the origin. What are the coordinates of the four vertices?

Challenge

These problems appear at the end of most exercise sets and offer more challenging work than the standard and Concept Extension problems.

End-of-Chapter Features

At the end of each chapter, these features appear in the following order: Chapter Review Exercises, Chapter Test, and Cumulative Test (at the end of selected chapters).

Chapter Review Exercises

Each chapter ends with a set of review exercises. These exercises include helpful section references that direct students to the appropriate sections for review. The answers to the odd-numbered review exercises are included at the back of the student text.

Chapter Test

A Chapter Test follows each chapter review. The answers to all test questions, with the appropriate section references, are included at the back of the student text.

Cumulative Test

A Cumulative Test appears at the end of Chapters 3, 5, 8, and 10. The answers to all test questions, with section references, are included at the back of the student text.

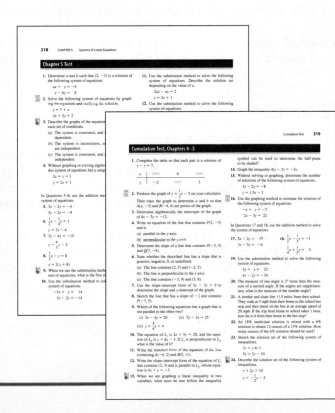

How To Use This Book

You paid a lot of money for this book, so how can you use it most effectively? Part of the answer depends on your instructor's approach to teaching the course. For example, are you taking a traditional course, with classes meeting 3–5 times per week? Is your course primarily independent study, with tutors and lab assistance available? Programs of study come in many flavors. How you use this textbook will depend on how you are expected to participate in your own learning.

Another factor is the role, if any, of a graphing calculator. For example, your instructor might plan to use a graphing calculator just to make classroom presentations. Or are you required or encouraged to use one? Or perhaps your instructor wants you to ignore the textbook's references to a graphing calculator. A clear understanding of your instructor's expectations is a major factor in how you use this book.

As instructors with many years of classroom teaching experience, we know a few things about how to help you reach your goals, and we wrote this book with those things in mind. Here are a few tips.

Reading the Text

"Ask about this!"

Many students feel that they can't read a math textbook. However, if you go about it the right way, you definitely can. Reading mathematics is not the same as reading a novel. You will need to go slowly, pay attention to detail, and make notes about questions you need to ask.

Some students prefer to read material before it has been presented. Others prefer to do their reading afterwards. Choose the style that works best for you, but make a resolution to read the text!

Explorations and Examples

Many sections of the book contain **Explorations.** An exploration logically takes you from something that you know to something that you can conclude. As you work through an exploration, your main task is to observe. When you get to the end, you should feel that you have seen enough evidence to suggest a general concept or rule. Work the exploration with us. If you do, you will be persuaded rather than just taking our word for it.

Your book has hundreds of **Examples** that give you guided practice with concepts and skills. It is not enough simply to read the examples. You should actually work the examples with paper and pencil at hand. See how far you can go on your own before you need to peek at the worked-out solution.

Calculator Key Words

If you are expected to use a graphing calculator, you will want to pay special attention to **Keys to the Calculator.** This feature describes certain operations that you can perform with a calculator, and it gives the keystrokes for the TI-83 Plus model. If you are using a TI-86, use the **Key Word** to refer to the keystrokes in your separate *Graphing Calculator Keystroke Guide.*

Thinking and Learning

Throughout the book you will find two features in the margin. As the name implies, **Learning Tips** give you some other ways of thinking about a concept and some approaches that you might find helpful. Pay attention to these and try them.

The feature called **Think About It** has questions that challenge you to expand on a related topic. Have fun with them, but if you feel too pushed, don't worry about them.

Expressing Yourself

Every subject has its own vocabulary. Learning to think, speak, and write in the language of mathematics is essential to a reasonable understanding of the material.

Nearly every section has a feature called **Speaking the Language** and exercises that ask for a written answer. We recommend that you try these even if they are not assigned. Even if all you do is look at the answers, you will have learned something important.

Applying Yourself

Your instructor will undoubtedly provide you with a syllabus that includes homework assignments. This syllabus is a major hint at what your instructor expects you to know.

No one can learn to play a violin by watching someone else play a violin. The exercises that we provide give you a chance to pick up the fiddle. If you are serious about succeeding, go beyond the assignments. We give the answers to the odd-numbered exercises so that you can gauge how you are doing right away.

Don't be afraid of making mistakes—learn from them. Above all, stay current. If you fall behind, you may have great difficulty recovering and catching up.

Refreshing Yourself

At the end of nearly every section, we include a **Quick Reference,** which is a summary of the key concepts and rules of the section. One excellent use of this feature is in your preparation for exams. Work through the summary and check off those things that you feel confident that you know. If you are unsure about a topic, go back and review, work examples, and work related exercises.

Every chapter concludes with a set of **Review Exercises** and a **Chapter Test.** Some chapters also have a **Cumulative Test** that includes material from preceding chapters. Naturally, we can't guarantee that these exercises and test questions will exactly match the questions on your exam, but the review and the test-taking practice will be very valuable to you.

Elementary Algebra

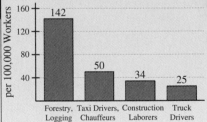

(Source: Bureau of Labor Statistics.)

The accompanying bar chart shows the four occupations with the highest fatality rates per 100,000 workers.

*The data in the bar chart can be written in the form of **rational numbers,** which are **real numbers** of a specific form. Operations can be performed with these numbers to draw conclusions about the hazards of these occupations. For more on this real data problem, see Exercises 79–82 at the end of Section 1.5.*

The Real Number System

Chapter Snapshot

We begin our study with an examination of the structure, order, and properties of the real number system. We then turn to the rules and methods for adding, subtracting, multiplying, and dividing real numbers. These basic skills are essential to your success in this and all future courses in mathematics.

Warm-Up Skills

A feature called Warm-Up Skills appears at the beginning of every chapter. In this feature, we present 10–12 brief exercises that review certain mathematical skills discussed in previous chapters. You will need to understand these essential skills before learning the concepts presented in the new chapter. If you feel shaky about a particular exercise, we suggest you review the relevant concept or skill by going through the prior chapter material.

Because Chapter 1 has no preceding math topics to review, we have developed a list of important questions that will help you get organized for success and prepare for your upcoming course. To view this list of questions, please visit our web site http://math.college.hmco.com/students and follow the links for Elementary Algebra, 3e.

1.1 The Real Numbers

The Integers • The Rational Numbers • The Real Numbers • Order of the Real Numbers

The Integers

It is reasonable to believe that the earliest use of numbers was for counting. We call the numbers 1, 2, 3, . . . the **counting numbers** or **natural numbers.**

In mathematics it is convenient to organize a collection of objects, such as numbers, into a **set** and to name the set for easy reference. For example, we can write the set N of natural numbers as $N = \{1, 2, 3, \ldots\}$. Braces are used to enclose the set of numbers, and the three dots indicate that the numbers continue without end. The numbers in the set are called the **elements** of the set.

Adding just one additional element 0 to set N results in a new set W, which is called the set of **whole numbers.**

$$W = \{0, 1, 2, 3, \ldots\}$$

Note that every natural number is also a whole number.

In everyday life we encounter numbers that describe measurements that are less than 0, such as a temperature that is below zero or a bank account that is overdrawn, or that represent decreases or losses, such as a decline in stock value or a loss of yardage in a football game.

To provide for such numbers, we expand the set W of whole numbers into the set

$$J = \{\ldots, -3, -2, -1, 0, 1, 2, 3, \ldots\}$$

We call set J the set of **integers.** Note that the natural numbers and the whole numbers are also integers.

A visual way to represent the integers is a **number line.** To draw a number line, we select any point of a line and associate it with the number 0. This point is called the **origin.** Then we associate the remaining integers with points that are to the left and right of the origin and that are spaced one unit apart. We call this distance the **unit distance.** (See Fig. 1.1.) Numbers to the right of the origin are called **positive numbers.** Numbers to the left of the origin are called **negative numbers** and they are identified with the symbol $-$. The number 0 is neither positive nor negative.

Figure 1.1

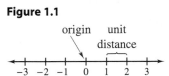

Note: For emphasis, we sometimes identify positive numbers with the symbol $+$, but usually this symbol is omitted. We read $+9$ as *positive* 9, not plus 9. Similarly, the symbol -6 is read *negative* 6, not minus 6.

⚷ Keys to the Calculator

Negative

A special key is used to enter negative numbers. To enter -5, press

[(−)] 5 [ENTER]

Note: The keystrokes that are presented throughout this book are for the TI-83 Plus calculator model. If you are using a TI-86, use the key words (for example, "Negative") to refer to your Calculator Supplement.

The Rational Numbers

For positive integers, the basic operations of addition, subtraction, multiplication, and division are familiar to us from arithmetic. In particular, division can be indicated in the form of a fraction such as $\frac{3}{5}$ or $\frac{12}{7}$. The number above the fraction bar is called the **numerator,** and the number below the fraction bar is called the **denominator.**

A special kind of fraction is one in which the numerator and denominator are both integers. Such fractions are called **rational numbers.**

> **Definition of a Rational Number**
>
> A **rational number** is a number that can be written in the form $\frac{p}{q}$, where p and q are integers and q is not 0.

Note: We use the letter Q to represent the set of rational numbers.

A letter (or any other symbol) used to represent an unknown number is called a **variable.** In the definition of a rational number, the numerator is represented by the variable p, and p can be replaced with any integer. The denominator is represented by the variable q, and q can be replaced with any integer except 0.

According to the definition, each of the following is a rational number.

$$\frac{5}{8}, \quad \frac{-10}{3}, \quad \frac{7}{1}, \quad \frac{0}{2}, \quad \frac{-9}{-16}$$

Figure 1.2

From arithmetic, we know that if p is a positive integer, then p can be written as the rational number $\frac{p}{1}$. Later we will see that this is also true if p is a negative integer. In short, every integer is also a rational number. Figure 1.2 shows the relationship among the sets N, W, J, and Q.

Every rational number $\frac{p}{q}$ has a decimal name that can be determined by dividing p by q.

| Rational numbers |
| Integers |
| Whole numbers |
| Natural numbers |

> 🔑 **Keys to the Calculator**
>
> **Divide**
>
> Although the ÷ key is used for division, the slash symbol / is displayed on the screen.
>
> To perform $31 \div 250$, press:
>
> **31** ÷ **250** ENTER

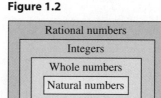

```
31/250
           .124
```

EXAMPLE 1   **Decimal Names for Rational Numbers**

Use your calculator to determine the decimal names for the following rational numbers.

(a) $\frac{5}{8}$ (b) $\frac{1}{3}$ (c) $\frac{8}{11}$

Solution

(a) $\dfrac{5}{8} = 5 \div 8 = 0.6250000\ldots$

(b) $\dfrac{1}{3} = 1 \div 3 = 0.333333\ldots$

(c) $\dfrac{8}{11} = 8 \div 11 = 0.72727272\ldots$

In Example 1, the decimal in part (a) is usually written simply as 0.625. We call this a **terminating decimal** because only zeros follow the 0.625. In parts (b) and (c), the decimal names have repeating patterns and are called **repeating decimals.** A convenient way to represent a repeating decimal is with a line over the repeating block. Thus, $\frac{1}{3} = 0.\overline{3}$ and $\frac{8}{11} = 0.\overline{72}$.

🔑 **Keys to the Calculator**

Mode

You can set the number of decimal places that you want your calculator to display. The calculator will round your answer to the number of decimal places that you select.

```
Normal Sci   Eng
Float 0123456789
Radian Degree
Func Par Pol Seq
Connected Dot
Sequential Simul
Real a+bi, re^θi
Full Horiz G-T
```

1. Press ⬚MODE⬚ to display the screen shown in the figure.
2. Use your down arrow key to move to the second line.
3. Using your left and right arrow keys, move to the number of decimal places that you want to display.
4. Press ⬚ENTER⬚.

If you select "Float," the calculator will display up to 10 digits. If the next nondisplayed digit is 5 or greater, the last displayed digit will be rounded up. For example, 0.555555 . . . will be displayed as 0.5555555556.

The Real Numbers

Not all decimal numbers are terminating or repeating. For example, the number π (the ratio of a circle's circumference to its diameter) has the decimal name 3.141592654 . . . . Because this decimal does not terminate and has no block of repeating digits, it is an example of a **nonterminating, nonrepeating decimal.** Such numbers are called **irrational numbers.**

Irrational numbers are often the result of taking the square root of a number. Although $\sqrt{9}$ is the rational number 3 (because $3^2 = 9$), $\sqrt{5}$ is the irrational number 2.236067977 . . . . It can be shown that every rational number has either a terminating or a repeating decimal name. Thus the set of rational numbers and the set of irrational numbers are entirely different sets with no elements in common.

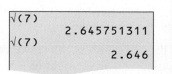

Keys to the Calculator

Square Root

To calculate $\sqrt{7}$, press

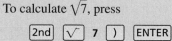

The figure shows the result in the float mode and rounded to three decimal places.

Think About It

Do you think that the decimal number 0.101101110 ... has a pattern? Is it a rational number?

If we join together the set of rational numbers and the set of irrational numbers, the combined set is called the set **R** of **real numbers.** (See Fig. 1.3.) We can describe the real numbers as all numbers that have decimal names.

Figure 1.3   The Real Numbers

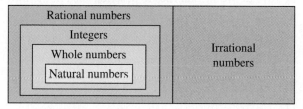

A number line can be used to represent all real numbers. Each number associated with a point of the number line is called the **coordinate** of the point. Highlighting points whose coordinates are given is called **plotting** the points, and the resulting picture is called a **graph.**

EXAMPLE   2

Classifying and Graphing Real Numbers

Consider the set $A = \left\{ -\dfrac{7}{2}, -2, 0, 1.5, \sqrt{10}, 5 \right\}$.

(a) Identify the sets to which the numbers in set A belong. When necessary, use your calculator to determine the decimal name of the number.

(b) Graph set A on a number line.

Solution

(a)

	$-\dfrac{7}{2}$	-2	0	1.5	$\sqrt{10}$	5
Natural number						✓
Whole number			✓			✓
Integer		✓	✓			✓
Rational number	✓	✓	✓	✓		✓
Irrational number					✓	
Real number	✓	✓	✓	✓	✓	✓

(b) **Figure 1.4**

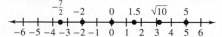

Order of the Real Numbers

We say that two numbers are **equal** if they are associated with the same point of the number line. We use the symbol $=$ to indicate equality. For example, $\frac{1}{2} = 0.5$. To indicate that two numbers are not equal, we use the symbol $\neq$. For example, $-3 \neq 3$.

We can use a number line to visualize the **order** of the real numbers, which we can indicate with the **inequality** symbols $<$ and $>$. Because -2 is to the right of -5 on the line, we say that -2 *is greater than* -5 and write $-2 > -5$. Equivalently, -5 is to the left of -2. Thus we say that -5 *is less than* -2 and write $-5 < -2$.

EXAMPLE **3**

Writing Inequalities

Write two inequalities to describe each of the following.

(a) 10 is less than 12.   (b) 1 is greater than -1.

(c) a is less than b.   (d) $0.5 \neq 0.\overline{5}$

Solution

(a) $10 < 12$ and $12 > 10$ (b) $1 > -1$ and $-1 < 1$

(c) $a < b$ and $b > a$ (d) $0.5 < 0.\overline{5}$ and $0.\overline{5} > 0.5$

EXAMPLE **4**

Order of Real Numbers

Insert $<$, $>$, or $=$ to make the statement true.

(a) -3 ▨ -4 (b) -2.5 ▨ $-\dfrac{5}{2}$

(c) $\dfrac{1}{6}$ ▨ $\dfrac{1}{5}$ (d) $\dfrac{1}{4}$ ▨ $0.\overline{25}$

LEARNING TIP

To help you determine the order of two numbers, visualize their location on a number line. Because -1 lies to the right of -2, $-1 > -2$.

Solution

(a) $-3 > -4$

(b) $-2.5 = -\dfrac{5}{2}$

(c) $\dfrac{1}{6} = 0.1\overline{6}$ and $\dfrac{1}{5} = 0.2$,   so $\dfrac{1}{6} < \dfrac{1}{5}$

(d) $\dfrac{1}{4} = 0.25$ and $0.\overline{25} = 0.252525\ldots$,   so $\dfrac{1}{4} < 0.\overline{25}$

To refer to numbers that are *less than or equal to* a given number, we use the symbol $\leq$. Similarly, we use the symbol $\geq$ to indicate numbers that are *greater than or equal to* a given number. Each of the following statements is true.

$$-2 \leq 5 \qquad -7 \leq -7 \qquad -1 \geq -3 \qquad 0 \geq 0$$

EXAMPLE 5

Inequalities and Sets

Suppose that n represents an integer and w represents a whole number. Write the set that is described by each inequality.

(a) $n > -4$ (b) $n \geq -4$

(c) $w \leq 3$ (d) $w \leq 0$

Solution

(a) $\{-3, -2, -1, 0, 1, 2, \ldots\}$ The symbol $>$ means that -4 is not included.

(b) $\{-4, -3, -2, -1, 0, 1, 2, \ldots\}$ The symbol $\geq$ means that -4 is included.

(c) $\{0, 1, 2, 3\}$

(d) $\{0\}$ The symbol $\leq$ means that 0 is included.

Quick Reference 1.1

The Integers

- A **set** is a collection of objects called the **elements** of the set.
- The set N of **counting numbers** or **natural numbers** is
    $$N = \{1, 2, 3, 4, \ldots\}.$$
- The set W of **whole numbers** is $W = \{0, 1, 2, 3, \ldots\}$.
- The set J of **integers** is $\{\ldots, -3, -2, -1, 0, 1, 2, \ldots\}$.
- A **number line** is a line whose points are associated with numbers. The point associated with 0 is the **origin.** The other integers are spaced at **unit distances** to the left and right of the origin.
- Numbers to the right of the origin are **positive numbers,** which are sometimes indicated with the symbol $+$. Numbers to the left of the origin are **negative numbers,** which are indicated with the symbol $-$. The number 0 is neither positive nor negative.

The Rational Numbers

- In a fraction, the number above the fraction bar (division symbol) is the **numerator,** and the number below the fraction bar is the **denominator.**
- A **variable** is a letter or other symbol that is used to represent an unknown number.
- A **rational number** is a number that can be written in the form $\frac{p}{q}$, where p and q are integers and q is not 0.
- Every rational number has a decimal name that is either a **terminating** or a **repeating decimal.** Terminating decimals have a finite number of nonzero digits. Repeating decimals have repeating patterns of digits that can be indicated with a line above the repeating block.

The Real Numbers

- Numbers with decimal names that are nonterminating and nonrepeating are **irrational numbers.**
- If we join together the set of rational numbers and the set of irrational numbers, the combined set is called the set **R** of **real numbers.** We can describe the real numbers as all numbers that have decimal names.

- Each number associated with a point of the number line is called the **coordinate** of the point. Highlighting points whose coordinates are given is called **plotting** the points, and the resulting picture is called a **graph.**

Order of the Real Numbers

- Two numbers are **equal** if they are associated with the same point of the number line. The symbol = means *is equal to;* the symbol ≠ means *is not equal to.*
- The **inequality** symbols < (*is less than*) and > (*is greater than*) are used to indicate the **order** of real numbers. We also use the symbols ≤ (*is less than or equal to*) and ≥ (*is greater than or equal to*).

Speaking the Language 1.1

1. The set $\{\ldots, -3, -2, -1, 0, 1, 2, 3, \ldots\}$ is called the set of ▭.

2. If p and q are integers and $q \neq 0$, then a number of the form $\dfrac{p}{q}$ is called a(n) ▭.

3. The decimal name of an irrational number is neither ▭ nor ▭.

4. If 3 is associated with point A of a number line, then 3 is called the ▭ of point A.

Exercises 1.1

Concepts and Skills

1. How does the set of whole numbers differ from the set of natural numbers?

2. What are two ways in which the symbol + is used? Give an example of each.

In Exercises 3–12, write the described set of numbers.

3. Whole numbers that are less than 4

4. Integers that are greater than -3

5. Integers that are less than 4

6. Whole numbers that are greater than -3

7. Even natural numbers between 7 and 15

8. Odd natural numbers between 6 and 16

9. Whole numbers that are rational numbers

10. Whole numbers that are not natural numbers

11. The four largest integers that are less than or equal to -5

12. The three smallest integers that are greater than or equal to -2

13. How do we know that every integer is also a rational number?

14. How do irrational numbers differ from rational numbers?

In Exercises 15–24, determine whether the statement is true or false.

15. Every whole number is a rational number.

16. Every rational number is a real number.

17. The number 0 is both a whole number and an integer.

18. The number 1 is both a rational number and a real number.

19. Positive numbers and natural numbers are the same.

20. Positive numbers and rational numbers are the same.

21. Negative integers are not rational.

22. Some real numbers are not rational.

23. All real numbers are either rational or irrational.

24. No rational number is irrational.

In Exercises 25–28, describe in words numbers that satisfy the given conditions.

25. Whole numbers that are not natural numbers

26. Rational numbers that are not integers

27. Integers that are not negative

28. Integers that are not whole numbers

▼ **29.** Explain the difference between 0.35 and $0.\overline{35}$.

▼ **30.** Describe how to determine the decimal name of a rational number.

In Exercises 31–36, use your calculator to determine the decimal name for the given rational number. State whether the decimal is terminating or repeating.

31. $\dfrac{7}{5}$ **32.** $\dfrac{4}{3}$

33. $\dfrac{5}{6}$ **34.** $\dfrac{11}{20}$

35. $-\dfrac{5}{11}$ **36.** $-\dfrac{23}{50}$

In Exercises 37–42, list all the numbers in the given set that are members of the following sets. Use your calculator to write decimal names when necessary.

(a) Natural numbers

(b) Whole numbers

(c) Integers

(d) Rational numbers

(e) Irrational numbers

37. $\left\{ -3, -\dfrac{2}{3}, 0, 5, \dfrac{35}{5} \right\}$

38. $\left\{ -5.2, -1, \dfrac{6}{6}, 9 \right\}$

39. $\left\{ -\sqrt{7}, 1, \sqrt{9}, \dfrac{4}{3} \right\}$

40. $\left\{ -3.7, -\sqrt{4}, 0, \pi, \dfrac{24}{3} \right\}$

41. $\left\{ -2\dfrac{1}{3}, 532, -0.35, \dfrac{18}{6}, \sqrt{5}, 1.\overline{14} \right\}$

42. $\left\{ -\dfrac{5}{3}, 0, \dfrac{12}{3}, -\pi, 3.794, -10 \right\}$

In Exercises 43–50, draw a number line graph of the given set.

43. $\{ -1, 3, 8 \}$

44. $\{ -4, 0, 1, 2 \}$

45. $\left\{ \dfrac{3}{4}, -\dfrac{5}{2}, 0, 3.5 \right\}$

46. $\left\{ -\dfrac{2}{3}, 1.5, -3, 2\dfrac{1}{4} \right\}$

47. All natural numbers n such that $n < 5$

48. All integers n such that n is negative and $n \geq -4$

49. $\left\{ -5, -2.2, \dfrac{3}{10}, \sqrt{5}, 3.\overline{25} \right\}$

50. $\left\{ -\dfrac{18}{5}, -1, 0.75, \pi, \sqrt{17}, 5.\overline{3} \right\}$

In Exercises 51–60, write two inequalities to describe the given information.

51. 5 is greater than 0.

52. 7 is less than 10.

53. -12 is greater than or equal to -20.

54. 30 is greater than or equal to 30.

55. 24 exceeds -17.

56. -29 is more than -30.

57. A number n is at least -5.

58. A number c is at most 7.

59. A number y is no more than 4.

60. A number x is no less than -9.

In Exercises 61–68, determine whether the inequality is true or false.

61. $-2 > 1$ **62.** $-26 < 19$

63. $-8 \leq -8$ **64.** $14 \geq 14$

65. $0 \leq 0$ **66.** $\dfrac{57}{239} \geq 0$

67. $\dfrac{5}{6} > \dfrac{6}{7}$ **68.** $-0.001 < 0$

In Exercises 69–78, insert $<$, $>$, or $=$ to make the statement true.

69. -9 ▮▮▮ -6 **70.** -1 ▮▮▮ -7

71. $\dfrac{5}{4}$ ▮▮▮ 1.25 **72.** 0.625 ▮▮▮ $\dfrac{5}{8}$

73. $\dfrac{3}{4}$ ▅▅▅ $-\dfrac{8}{5}$

74. $-\dfrac{5}{3}$ ▅▅▅ $\dfrac{3}{5}$

75. $-\dfrac{1}{4}$ ▅▅▅ $-\dfrac{1}{5}$

76. $-\dfrac{9}{7}$ ▅▅▅ $-\dfrac{7}{9}$

77. $\dfrac{22}{7}$ ▅▅▅ π

78. 3.14 ▅▅▅ π

In Exercises 79–84, assume that w represents a whole number and that n represents an integer. Write the set described by the given inequality.

79. $n < -6$

80. $n \geq -3$

81. $w < 2$

82. $w \leq 5$

83. $w \leq 0$

84. $n \leq 2$

Modeling with Real Data

85. The table shows the average prices for three popular holiday trees.

Typical Prices for 7-Foot Trees	
Fraser fir	$45
Leyland cypress	$35
Virginia pine	$24

(Source: National Christmas Tree Growers Association.)

What is the appropriate symbol ($<$, $>$, or $=$) to compare the revenue from selling 7 million Fraser fir trees and that from selling 9 million Leyland cypress trees?

86. Refer to the data in Exercise 85. Suppose that of the 38 million trees sold in 2000, 10 million were Fraser fir, 20 million were Leyland cypress, and 8 million were Virginia pine. If you plotted the revenue from each on a number line, which would be

(a) to the left of the other two?

(b) to the right of the other two?

(c) between the other two?

Data Analysis: *Fatal Home Fires*

Winter months have the highest percentages of fatal home fires.

Months with the Highest Percentages of Fatal Home Fires	
November	9.1%
December	13.1%
January	13.3%
February	10.8%
March	9.6%

(Source: National Fire Protection Association.)

87. Compare the percentages for each pair of consecutive months by using $<$ or $>$.

88. In which months do more than $\frac{1}{10}$ of the fatal home fires occur?

89. What 2-month period accounts for more than $\frac{1}{4}$ of the fatal fires?

90. What is the total percentage of fatal home fires for December, January, and February?

Challenge

91. The symbol $\in$ is used to represent the phrase *is an element of.* Thus $5 \in W$ means that 5 is an element of the set of whole numbers. Determine whether each of the following statements is true or false.

(a) $0 \in N$

(b) $3 \in N$ and $3 \in W$

(c) $7 \in \{7\}$

(d) $-0.5 \in J$

92. The inequality $2 < 5 < 9$ means that 2 is less than 5 and 5 is less than 9. Suppose the numbers a and b are to the left of the number c on the number line. Explain why $a < b < c$ is not necessarily true.

In Exercises 93–96, write the given decimal as a simplified rational number.

93. 1.8

94. 0.26

95. $0.\overline{6}$

96. $0.\overline{3}$

97. (a) Write the decimal form for $\frac{1}{9}$ and $\frac{2}{9}$.

(b) Based on your results in part (a), what is your conjecture about the decimal forms for $\frac{3}{9}$ and $\frac{4}{9}$?

(c) What rational number has the decimal form $0.\overline{7}$?

1.2 | Operations with Real Numbers

The Basic Operations • Exponents and Square Roots • Opposites and Absolute Value •
The Order of Operations

The Basic Operations

The basic operations with real numbers are addition, subtraction, multiplication, and division. Simple calculations are performed with the familiar arithmetic facts.

> 🔑 **Keys to the Calculator**
>
> **Add**
>
> Use the $\boxed{+}$ key for addition.
>
> **Subtract**
>
> Be sure to use the $\boxed{-}$ (minus) key for subtraction rather than the $\boxed{(-)}$ (negative) key.
>
> **Multiply**
>
> Although the $\boxed{\times}$ key is used for multiplication, the symbol $*$ is displayed on the screen.

Figure 1.5

```
7+5
              12
4-0
               4
3*8
              24
```

The following information refers to Fig. 1.5:

$7 + 5 = 12$:   The **sum** of 7 and 5 is 12. The 7 and 5 are called **addends.**

$4 - 0 = 4$:   The **difference** of 4 and 0 is 4. We call 4 the **minuend** and 0 the **subtrahend.**

$3 \cdot 8 = 24$:   The **product** of 3 and 8 is 24. The 3 and 8 are called **factors.**
    Other ways to represent this product are $3(8)$, $(3) \cdot 8$, and $(3)(8)$.

Note: The multiplication symbol $\times$ is used in arithmetic, but to avoid confusing that symbol with the letter x, we use the symbol $\cdot$ in algebra.

For a division such as $20 \div 2 = 10$, we say that the **quotient** of 20 and 2 is 10. We call 20 the **dividend** and 2 the **divisor.** Other ways to represent this quotient are $20/2$ and $\frac{20}{2}$.

For differences and quotients, the operations are indicated in the order in which the numbers are given. Thus the difference of x and y is $x - y$, and the quotient of b and a is $b \div a$.

Exponents and Square Roots

We frequently have occasion to multiply a number by itself two or more times. A compact way to write such products is to use an **exponent.**

$$3^{\overset{\text{exponent}}{5}} = \underbrace{3 \cdot 3 \cdot 3 \cdot 3 \cdot 3}_{\text{5 factors}} = 243 \qquad 4^{\overset{\text{exponent}}{2}} = \underbrace{4 \cdot 4}_{\text{2 factors}} = 16$$

base ⌐   base ⌐

The numbers 3^5 and 4^2 are called **exponential expressions.** We call the number being multiplied by itself the **base.** The exponent indicates the number of factors of the base. Thus 3^5 means 5 factors of 3, and 4^2 means 2 factors of 4.

Exponential expression	*Read as*
7^2	7 to the second power or 7 squared
4^3	4 to the third power or 4 cubed
6^8	6 to the eighth power

EXAMPLE **1**

Writing Products as Exponential Expressions

Write each of the following as an exponential expression.

(a) $9 \cdot 9 \cdot 9 \cdot 9 = 9^4$ 4 factors of 9

(b) $y \cdot y \cdot y \cdot y \cdot y = y^5$ 5 factors of y

(c) $8 = 8^1$ 1 factor of 8

0—⌐ **Keys to the Calculator**

Exponent

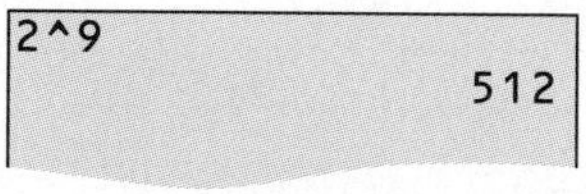

```
2^9
          512
```

The $\boxed{\wedge}$ key is used to raise a number or expression to a power. To calculate 2^9, press:

2 $\boxed{\wedge}$ **9** $\boxed{\text{ENTER}}$

EXAMPLE **2**

Determining the Value of an Exponential Expression

Perform the indicated operations. Use your calculator for parts (e) and (f).

(a) 7^2 (b) 3^3

(c) 0^{17} (d) 1^{36}

(e) 8^5 (f) 27^2

Figure 1.6

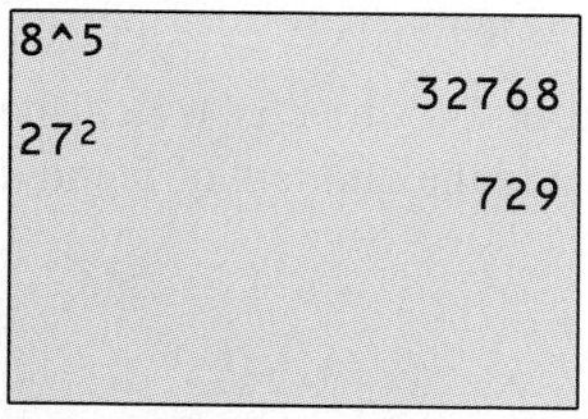

```
8^5
             32768
27²
               729
```

Solution

(a) $7^2 = 7 \cdot 7 = 49$

(b) $3^3 = 3 \cdot 3 \cdot 3 = 27$

(c) 0^{17} means 17 factors of 0. The product is 0.

(d) 1^{36} means 36 factors of 1. The product is 1.

(e) and (f) See Fig. 1.6.

Related to the operation of squaring a number is the operation of taking the **square root** of a number. The symbol $\sqrt{}$ is used to indicate this operation.

We say that $\sqrt{36} = 6$ because $613^2 = 36$, $\sqrt{100} = 10$ because $10^2 = 100$, and $\sqrt{1} = 1$ because $1^2 = 1$. Note that the *square* of 9 is 81, and a *square root* of 9 is 3.

EXAMPLE **3**

Determining the Value of a Square Root

Perform the indicated operations. Use your calculator for parts (d) and (e).

(a) $\sqrt{25}$ (b) $\sqrt{64}$ (c) $\sqrt{0}$

(d) $\sqrt{324}$ (e) $\sqrt{2601}$

Think About It

Square roots of numbers (other than perfect squares) are usually unfamiliar. Describe an easy way to estimate the value of a square root such as $\sqrt{50}$ without your calculator.

Solution

(a) $\sqrt{25} = 5$ because $5^2 = 25$.

(b) $\sqrt{64} = 8$ because $8^2 = 64$.

(c) $\sqrt{0} = 0$ because $0^2 = 0$.

(d) and (e) See Fig. 1.7.

Figure 1.7

```
√(324)
                    18
√(2601)
                    51
```

Opposites and Absolute Value

Figure 1.8

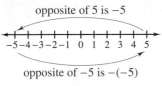

opposite of 5 is -5

opposite of -5 is $-(-5)$

We use the word **opposites** to refer to two numbers that are the same distance from 0 on the number line. The symbol $-$ is used to indicate the opposite of a number. For example, the opposite of 5 is indicated by -5, and the opposite of -5 is indicated by $-(-5)$.

We can see from the number line in Fig. 1.8 that the opposite of -5 is 5, and so we can write $-(-5) = 5$. We generalize this observation in the following property.

Property of the Opposite of the Opposite of a Number

For any real number x, $-(-x) = x$.

In words, the opposite of the opposite of a number is the number.

In general, we can say that the opposite of a positive number is a negative number, and the opposite of a negative number is a positive number. For completeness, we agree that the opposite of 0 is 0.

Note: It is unfortunate that the symbol $-$ is used for three different purposes.

(a) *Minus*, as in $8 - 2$ ("eight minus two")

(b) *Negative*, as in the number -7 ("negative 7")

(c) *Opposite*, as in $-x$ ("opposite of x")

It is particularly important that you read the symbol $-x$ as "the opposite of x," not "negative x," because $-x$ does not necessarily represent a negative number.

The **absolute value** of a number is its distance from 0 on the number line. Symbolically, absolute value is represented by vertical lines on either side of the number. Because distance is not negative, the absolute value of a number is always a nonnegative number.

Number	*Distance from 0*	*Absolute Value*
4	4 units	$|4| = 4$
-8	8 units	$|-8| = 8$
0	0 units	$|0| = 0$

Definition of Absolute Value

For any real number a,

$$|a| = \begin{cases} a & \text{if } a \geq 0 \\ -a & \text{if } a < 0 \end{cases}$$

In words, if a number is positive or 0, then its absolute value is simply the number. However, if a number is negative, then its absolute value is the opposite of the number.

🔑 **Keys to the Calculator**

Absolute Value

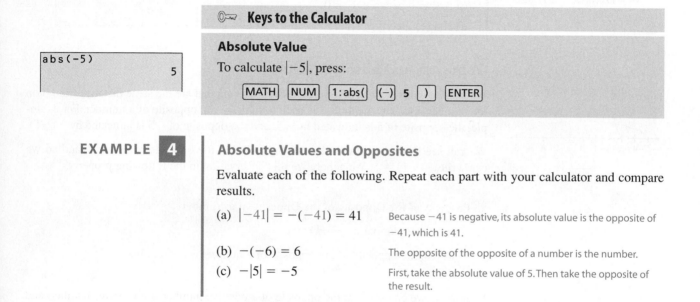

abs(-5)
                    5

To calculate $|-5|$, press:

[MATH] [NUM] [1: abs(] [(−)] **5** [ )] [ENTER]

EXAMPLE 4   **Absolute Values and Opposites**

Evaluate each of the following. Repeat each part with your calculator and compare results.

(a) $|-41| = -(-41) = 41$ Because −41 is negative, its absolute value is the opposite of −41, which is 41.

(b) $-(-6) = 6$ The opposite of the opposite of a number is the number.

(c) $-|5| = -5$ First, take the absolute value of 5. Then take the opposite of the result.

The Order of Operations

Any combination of numbers and operations is called a **numerical expression.** The following are examples.

$$7 + 3(2) \qquad \frac{5^2 - 6}{2} \qquad \sqrt{25 - 9} + |-8|$$

Performing the indicated operations in a numerical expression is called **evaluating the expression.**

Consider the expression $3 + 2 \cdot 8$. If we add first, the result is

$$3 + 2 \cdot 8 = 5 \cdot 8 = 40$$

If we multiply first, the result is

$$3 + 2 \cdot 8 = 3 + 16 = 19$$

To avoid such ambiguity, we must agree on the order in which operations are to be performed. In particular, we will rank multiplication ahead of addition. Thus the correct result is 19.

A numerical expression may also include **grouping symbols** to indicate which operations are to be performed first. For example, $(12 - 8) - 1 = 4 - 1 = 3$, whereas $12 - (8 - 1) = 12 - 7 = 5$. The most common grouping symbols are parentheses, but brackets [ ] and braces { } are also used. Moreover, square root symbols, fraction bars, and absolute value symbols also serve as grouping symbols. When we evaluate a numerical expression, we agree to perform operations in the following order.

The Order of Operations

1. Begin by performing all operations within grouping symbols. If there is more than one operation, use the order in steps 2–4.
2. Perform all operations involving exponents and square roots.
3. Perform all operations involving multiplication and division from left to right.
4. Perform all operations involving addition and subtraction from left to right.

If there are grouping symbols within other grouping symbols, we begin with the innermost grouping symbols and work outward.

Note: The Order of Operations is not the result of how calculators work. Rather, calculators are designed to follow the Order of Operations.

EXAMPLE 5

Evaluating Numerical Expressions

Evaluate each of the following expressions. Repeat each part with your calculator and compare results.

(a) $18 - 5 + 7 = 13 + 7 = 20$ Operate from left to right.

(b) $18 - (5 + 7) = 18 - 12 = 6$ Begin inside the grouping symbols.

(c) $20 \div 4 \cdot 5 = 5 \cdot 5 - 25$ Operate from left to right.

(d) $20 \div (4 \cdot 5) = 20 \div 20 = 1$ Begin inside the grouping symbols.

(e) $\sqrt{16 + 9} = \sqrt{25} = 5$ Begin with the sum under the square root (grouping) symbol.

(f) $\dfrac{2 + 6}{2} = \dfrac{8}{2} = 4$ Begin with the sum above the fraction bar (grouping symbol).

(g) $|5 + 9| = |14| = 14$ Begin with the sum in the absolute value (grouping) symbol.

EXAMPLE 6

Evaluating Numerical Expressions with the Order of Operations

LEARNING TIP

A big part of success in algebra is just being careful. When you evaluate expressions using the Order of Operations, write the complete expression at each step.

Evaluate each numerical expression. Use your calculator to check your results.

(a) $20 - 3^2 = 20 - 9 = 11$ Exponent first

(b) $\sqrt{16 \cdot 2 \div 8} = \sqrt{32 \div 8}$ Begin under the square root (grouping) symbol.

   $= \sqrt{4}$ Multiply and divide from left to right.

   $= 2$

(c) $(12 + 3) \cdot 2^3 - 7 = 15 \cdot 2^3 - 7$ Grouping symbols first

$\qquad\qquad\qquad\quad = 15 \cdot 8 - 7$ Then the exponent

$\qquad\qquad\qquad\quad = 120 - 7$ Then multiplication

$\qquad\qquad\qquad\quad = 113$ Finally, subtraction

(d) $\dfrac{18}{20 - 9 \cdot 2} = \dfrac{18}{20 - 18}$ Begin below the fraction bar. Multiplication first

$\qquad\qquad\quad = \dfrac{18}{2}$ Then subtraction

$\qquad\qquad\quad = 9$ Finally, division

(e) $17 - [2 + 3(4 + 1)] = 17 - [2 + 3(5)]$ Innermost grouping symbols first

$\qquad\qquad\qquad\quad = 17 - [2 + 15]$ Work inside the brackets.

$\qquad\qquad\qquad\quad = 17 - 17$ Multiplication, then addition

$\qquad\qquad\qquad\quad = 0$

Quick Reference 1.2

The Basic Operations
- An indicated addition $a + b$ is called a **sum.** The numbers a and b are **addends.**
- An indicated subtraction $a - b$ is called a **difference.** The number a is the **minuend,** and the number b is the **subtrahend.**
- An indicated multiplication $a \cdot b$ is called a **product.** The numbers a and b are called **factors.**
- An indicated division $a \div b$ is called a **quotient.** The number a is the **dividend,** and the number b is the **divisor.**

Exponents and Square Roots
- The **exponential expression** b^n indicates that the **base** b is to be multiplied by itself n times, where n is the **exponent.**
- The operation of taking the **square root** of a number is indicated with the symbol $\sqrt{\;}$.

Opposites and Absolute Value
- Two numbers that are the same distance from 0 on the number line are called **opposites.** The opposite of a number x is represented by $-x$, which is read "the opposite of x."
- Property of the Opposite of the Opposite of a Number

    For any real number x, $-(-x) = x$.
- The **absolute value** of a number is its distance from 0 on the number line. The absolute value of a number x is represented by $|x|$.
- Definition of Absolute Value

    For any real number a,

    $$|a| = \begin{cases} a & \text{if} & a \geq 0 \\ -a & \text{if} & a < 0 \end{cases}$$

The Order of Operations
- A **numerical expression** is any combination of numbers and operations. Performing the indicated operations is called **evaluating the expression.**

- **Grouping symbols** in a numerical expression indicate the operations that are to be performed first. Parentheses, brackets, braces, square root symbols, fraction bars, and absolute value symbols are all grouping symbols.

- The following Order of Operations is an agreement as to the order in which operations are to be performed.

  1. Begin by performing all operations within grouping symbols. If there is more than one operation, use the order in steps 2–4.

  2. Perform all operations involving exponents and square roots.

  3. Perform all operations involving multiplication and division from left to right.

  4. Perform all operations involving addition and subtraction from left to right.

- If there are grouping symbols within other grouping symbols, begin with the innermost grouping symbols and work outward.

Speaking the Language 1.2

1. In the product xy, y is called a(n) ▧▧▧▧▧, whereas in the quotient $\dfrac{x}{y}$, y is called the ▧▧▧▧▧.

2. In the expression x^3, we call x the ▧▧▧▧▧ and 3 the ▧▧▧▧▧.

3. The symbol $-y$ represents the ▧▧▧▧▧ of y, whereas the symbol $|y|$ represents the ▧▧▧▧▧ of y.

4. In the expression $2[x + (y - 3)]$, the brackets and the parentheses are called ▧▧▧▧▧.

Exercises 1.2

Concepts and Skills

1. What is the difference between addends and factors?

2. Which of the following is the correct way to write the difference of x and 5? Explain.

  (i) $x - 5$ (ii) $5 - x$

In Exercises 3–10, match the word in column B with the number described in column A.

Column A	Column B
3. 6 in the quotient $12 \div 6$	(a) base
4. 13 in the sum $0 + 13$	(b) minuend
5. 8 in the difference $8 - 7$	(c) divisor
6. 4 in the product $7(4)$	(d) addend
7. 3 in the exponential expression 5^3	(e) factor
8. 12 in the quotient $\dfrac{12}{4}$	(f) subtrahend
9. 1 in the difference $6 - 1$	(g) exponent
10. 2 in the exponential expression 2^5	(h) dividend

11. Write the product of 2 and 5 in three different ways.

12. Write the quotient of 15 and 3 in three different ways.

13. What is the purpose of a natural number exponent?

14. How do we know that $\sqrt{16} = 4$?

In Exercises 15–18, write the given exponential expression as a product.

15. 6^4 **16.** 0^3

17. t^5 **18.** n^2

In Exercises 19–24, write the given product in exponential form.

19. $5 \cdot 5 \cdot 5 \cdot 5$ **20.** $8 \cdot 8 \cdot 8 \cdot 8 \cdot 8 \cdot 8$

21. $yyyyy$ **22.** aaa

23. 14 factors of x **24.** 9 factors of y

In Exercises 25–32, evaluate the given exponential expression. Use your calculator only when necessary.

25. 8^2 **26.** 2^3 **27.** 2^5 **28.** 3^4

29. 2^{10} **30.** 3^8 **31.** 15^5 **32.** 12^6

In Exercises 33–38, determine the square root. Use your calculator only when necessary.

33. $\sqrt{81}$ **34.** $\sqrt{16}$ **35.** $\sqrt{49}$

36. $\sqrt{9}$ **37.** $\sqrt{676}$ **38.** $\sqrt{1369}$

39. What are three ways in which the symbol $-$ is used? Give an example of each.

40. What is a way of describing two numbers that are opposites?

In Exercises 41–44, evaluate the expression.

41. $|-13|$ **42.** $-|8|$

43. $-|-8|$ **44.** $|-(-3)|$

In Exercises 45–50, write the opposite and the absolute value of the given number.

45. 7   **46.** -3 **47.** -2

48. 9   **49.** $-\dfrac{5}{8}$ **50.** 5.2

In Exercises 51–54, graph the given number, its absolute value, and its opposite.

51. -6 **52.** 5   **53.** 7   **54.** -8

55. What number(s) have an absolute value of 8?

56. What number(s) have an absolute value of 0?

57. What number(s) are three units from 0?

58. What number(s) are five units from 0?

59. What number(s) are four units from -2?

60. What number(s) are two units from 5?

In Exercises 61–68, insert $<$, $>$, or $=$ to make the statement true.

61. $|-3|$ ▒▒▒ -3 **62.** $-|-1|$ ▒▒▒ $-(-1)$

63. $-|7|$ ▒▒▒ -7 **64.** $|-10|$ ▒▒▒ $|10|$

65. $-(-4)$ ▒▒▒ 4 **66.** $|-3|$ ▒▒▒ $|2|$

67. $-|-6|$ ▒▒▒ $|-10|$ **68.** 6 ▒▒▒ $|-7|$

In Exercises 69–86, evaluate the numerical expression.

69. $7 + 3(2)$ **70.** $19 - (5)(3)$

71. $2 + 3^2$ **72.** $(2 + 3)^2$

73. $8(5) - 3 \cdot 7$ **74.** $5(6) - 5^2$

75. $4(3 - 1)^2$ **76.** $(9 - 7)^2 + 2(4)$

77. $3^2 - \sqrt{3^2}$ **78.** $4(2) + \sqrt{3 + 1}$

79. $\sqrt{3^2 - 4(1)(2)}$ **80.** $\sqrt{6^2 - 4(1)(5)}$

81. $17 - 3(2^2 + 1)$ **82.** $2 + 5(2 + 5)$

83. $7 - 2[5 - (6 - 4)]$

84. $6 + [5(7 - 6) + 3]$

85. $\dfrac{8^2 - 5 \cdot 3}{5(10) - 1}$ **86.** $\dfrac{4 \cdot 2 - 7}{11 - 2 \cdot 3}$

In Exercises 87–90, write a numerical expression and use your calculator to evaluate it.

87. The sum of 59 and 370

88. The difference of 253 and 185

89. The product of 87 and 236

90. The quotient of 554,022 and 63

In Exercises 91–96, evaluate the numerical expression.

91. $|-4| - |4|$ **92.** $|6| + |-7|$

93. $|-12| - |-10|$ **94.** $|-13| + |-15|$

95. $\dfrac{|-8|}{|-2|}$ **96.** $\dfrac{10}{|-5|}$

Modeling with Real Data

97. In 1999, Ken Griffey, Jr., averaged 1 home run for every 12.625 at bats, whereas Mark McGwire averaged 1 home run for every 8.015 at bats. Griffey had 606 at bats and McGwire had 521 at bats. (Source: CNNSI.) How many home runs did each hit?

98. In a 1999 survey of 1013 African-Americans, 89% said that they were happy with their lives. This was up from 56% in 1992. (Source: *African-American Monitor*.) Of those surveyed, how many more said that they were happy in 1999 than in 1992?

Data Analysis: *Election Information Sources*

The following table shows how voters in a recent presidential election obtained information about the candidates.

Where Voters Obtained Information	Percentage
Presidential debates	45
Newspaper stories	32
TV news	30
Newspaper editorials	23
News magazines	21
TV analysis after debate	18
TV political talk shows	18

(Source: Media Studies Center.)

99. Which categories were sources of information for more than $\frac{1}{4}$ of the voters?

100. Which categories were sources of information for less than $\frac{1}{5}$ of the voters?

101. Two categories describe information from newspapers: stories, 32%; and editorials, 23%. Because $32 + 23 = 55$, should we conclude that more than half of the voters obtained information from newspapers?

102. What percentage of the voters did *not* obtain information from presidential debates?

Challenge

103. Suppose $\sqrt{x} = a$, where x is a whole number. What do we know about a^2?

104. If n is any natural number, what are the values of 0^n and 1^n?

In Exercises 105–108, insert grouping symbols so that the numerical expression is equal to the given value.

105. $7 \cdot 2 + 3 \cdot 1$;   35

106. $3 \cdot 5 - 2^2$;   27

107. $4^2 - 1 \cdot 3$;   45

108. $\sqrt{3 \cdot 13 - 1}$;   6

1.3 Properties of the Real Numbers

The Commutative Properties • The Associative Properties • Properties Involving 0 and 1 • Additive and Multiplicative Inverses • The Distributive Property

Physical objects have certain identifiable characteristics such as shape, weight, and color. We call such characteristics *properties* of the object.

The real numbers also have certain observable characteristics that we can state as general rules called **properties.** We have already stated one such property in Section 1.2.

> **Property of the Opposite of the Opposite of a Number**
>
> For any real number x, $-(-x) = x$.

In this section, we will consider some other fundamental properties of real numbers.

The Commutative Properties

Some properties are self-evident from our experience with arithmetic. For example, we know that numbers can be added or multiplied in any order and the results will be the same.

$$5 + 3 = 3 + 5 = 8 \qquad 7(2) = 2(7) = 14$$

The Commutative Properties of Addition and Multiplication

For any real numbers a and b,

1. $a + b = b + a$ Commutative Property of Addition
2. $ab = ba$ Commutative Property of Multiplication

Note: Subtraction and division are *not* commutative operations. For example, $12 \div 6 = 2$, but $6 \div 12 = 0.5$. In Section 1.6, we will see that $8 - 5 \neq 5 - 8$.

The Associative Properties

Although grouping symbols indicate the operations that are to be performed first, groupings in sums and products do not affect the results.

$$(3 + 5) + 12 = 8 + 12 = 20 \qquad (4 \cdot 2) \cdot 7 = 8 \cdot 7 = 56$$
$$3 + (5 + 12) = 3 + 17 = 20 \qquad 4 \cdot (2 \cdot 7) = 4 \cdot 14 = 56$$

The Associative Properties of Addition and Multiplication

For any real numbers a, b, and c,

1. $(a + b) + c = a + (b + c)$ Associative Property of Addition
2. $(ab)c = a(bc)$ Associative Property of Multiplication

Note: Subtraction and division are *not* associative operations.

$$(20 - 15) - 2 = 3, \quad \text{but} \quad 20 - (15 - 2) = 7$$
$$(32 \div 16) \div 2 = 1, \quad \text{but} \quad 32 \div (16 \div 2) = 4$$

Properties Involving 0 and 1

The real numbers 0 and 1 play special roles in addition and multiplication, respectively. When 0 is added to any number, the result is the original number. We call 0 the **additive identity.** When any number is multiplied by 1, the result is the original number. We call 1 the **multiplicative identity.** These observations are summarized by the following properties.

The Additive and Multiplicative Identity Properties

For any real number a,

1. $a + 0 = 0 + a = a$ Additive Identity Property
2. $a \cdot 1 = 1 \cdot a = a$ Multiplicative Identity Property

The following summary states other properties of 0 and 1.

> **Properties Involving 0 and 1**
>
> For any real number a and any natural number n,
>
> 1. $\dfrac{a}{1} = a$ 2. $\dfrac{a}{a} = 1, a \neq 0$ 3. $\dfrac{0}{a} = 0, a \neq 0$
>
> 4. $0^n = 0$ 5. $1^n = 1$

Additive and Multiplicative Inverses

The more formal name for *opposite* is **additive inverse.** The following property states the relationship between a number and its additive inverse.

> **Property of Additive Inverses**
>
> For any real number a, there is a unique real number $-a$ such that
> $a + (-a) = -a + a = 0.$

Note: The word *unique* means *exactly one* or *one and only one.*

The Property of Additive Inverses guarantees that every real number has a unique additive inverse and that their sum is the additive identity 0.

The **reciprocal** of a number is found by interchanging its numerator and its denominator.

Number	$\dfrac{3}{8}$	6	-2	$\dfrac{1}{3}$	$-\dfrac{5}{7}$
Reciprocal	$\dfrac{8}{3}$	$\dfrac{1}{6}$	$\dfrac{1}{-2}$	3	$-\dfrac{7}{5}$

The number 0 does not have a reciprocal.

Note: To determine the reciprocal of an integer, we can write the integer as a fraction: $6 = \frac{6}{1}$, and so the reciprocal is $\frac{1}{6}$.

The more formal name for *reciprocal* is **multiplicative inverse.** The following property states the relationship between a number and its multiplicative inverse.

> **Property of Multiplicative Inverses**
>
> For any nonzero real number a, there is a unique real number $\dfrac{1}{a}$ such that
>
> $$a \cdot \dfrac{1}{a} = \dfrac{1}{a} \cdot a = 1$$

The Property of Multiplicative Inverses guarantees that every nonzero real number has a unique multiplicative inverse and that their product is the multiplicative identity 1.

EXAMPLE **1** | **Identifying Properties of Real Numbers**

Name the property that is illustrated.

(a) $3 + (x + 5) = 3 + (5 + x)$ Commutative Property of Addition

(b) $7x \cdot 1 = 7x$ Multiplicative Identity Property

(c) $-2a + 2a = 0$ Property of Additive Inverses

(d) $3 \cdot \left(\dfrac{1}{3}x\right) = \left(3 \cdot \dfrac{1}{3}\right)x$ Associative Property of Multiplication

(e) $(x + 8) + 0 = x + 8$ Additive Identity Property

EXAMPLE **2** | **Using the Properties to Rewrite Expressions**

Use the given property to complete the statement.

(a) Commutative Property of Multiplication: $5 \cdot (x + 2) = $ ▒▒▒▒▒▒▒ .

(b) Associative Property of Addition: $3 + (5 + x) = $ ▒▒▒▒▒▒▒ .

(c) Property of Multiplicative Inverses: $4x \cdot \dfrac{1}{4x} = $ ▒▒▒▒▒▒▒ , $x \neq 0$.

Solution

(a) $5 \cdot (x + 2) = (x + 2) \cdot 5$

(b) $3 + (5 + x) = (3 + 5) + x$

(c) $4x \cdot \dfrac{1}{4x} = 1,\, x \neq 0$

The Distributive Property

To evaluate $4(5 + 3)$, we can follow the Order of Operations by adding inside the grouping symbols first and then multiplying by the outside factor 4.

$$4(5 + 3) = 4(8) = 32$$

We obtain the same result if we multiply each number within the parentheses by the outside factor 4 and then add the results.

$$4(5) + 4(3) = 20 + 12 = 32$$

This suggests the following important property.

The Distributive Property

For any real numbers a, b, and c,

1. $a(b + c) = ab + ac$

2. $ab + ac = a(b + c)$

More precisely, we call this property the Distributive Property of Multiplication over Addition. The property can also be stated for subtraction.

1. $a(b - c) = ab - ac$

2. $ab - ac = a(b - c)$

We refer to both versions simply as the Distributive Property. In the first forms, we *multiply* to convert an indicated product $a(b + c)$ into an indicated sum $ab + ac$. In the second forms, we *factor* to convert an indicated sum $ab + ac$ into an indicated product $a(b + c)$.

EXAMPLE

Using the Distributive Property to Multiply and Factor

(a) Multiply $3(x + 7)$.   (b) Multiply $9(2 - c)$.

(c) Factor $3x + 3y$.   (d) Factor $5a - 10$.

LEARNING TIP

The Distributive Property plays a significant role in algebra. Make sure you understand how to use the property to factor as well as to multiply.

Solution

(a) $3(x + 7) = 3 \cdot x + 3 \cdot 7 = 3x + 21$

(b) $9(2 - c) = 9 \cdot 2 - 9 \cdot c = 18 - 9c$

(c) $3x + 3y = 3 \cdot x + 3 \cdot y = 3(x + y)$

(d) $5a - 10 = 5 \cdot a - 5 \cdot 2 = 5(a - 2)$

EXAMPLE

Using the Distributive Property to Simplify Arithmetic

Use the Distributive Property to rewrite the given numerical expression so that the computation is easy to perform mentally.

(a) $8 \cdot 41 + 2 \cdot 41$ (b) $37 \cdot 101$

Think About It

The Distributive Property can be applied to evaluate $9(3 - 1)$ and to simplify $9(a - 1)$. Why is it essential in one and optional in the other?

Solution

(a) $8 \cdot 41 + 2 \cdot 41 = 41(8 + 2) = 41 \cdot 10 = 410$

(b) $37 \cdot 101 = 37(100 + 1) = 37 \cdot 100 + 37 \cdot 1 - 3700 + 37 = 3737$

Example 4 shows how a property of real numbers might be used to evaluate a numerical expression. However, the primary algebraic use of the properties is for describing the structure of the set of real numbers and for deriving new rules. These basic properties play an important role in all of our future study of algebra.

Quick Reference 1.3

The Commutative Properties

- The Commutative Property of Addition
    For any real numbers a and b, $a + b = b + a$.
- The Commutative Property of Multiplication
    For any real numbers a and b, $ab = ba$.
- Subtraction and division are not commutative.

The Associative Properties

- The Associative Property of Addition
  For any real numbers a, b, and c, $(a + b) + c = a + (b + c)$.
- The Associative Property of Multiplication
  For any real numbers a, b, and c, $(ab)c = a(bc)$.
- Subtraction and division are not associative.

Properties Involving 0 and 1

- In addition the **additive identity** is 0. In multiplication the **multiplicative identity** is 1.
- The Additive Identity Property
  For any real number a, $a + 0 = 0 + a = a$.
- The Multiplicative Identity Property
  For any real number a, $a \cdot 1 = 1 \cdot a = a$.
- Other properties involving 0 and 1:
  For any real number a and any natural number n,

  1. $\dfrac{a}{1} = a$ 2. $\dfrac{a}{a} = 1, a \neq 0$

  3. $\dfrac{0}{a} = 0, a \neq 0$ 4. $0^n = 0$

  5. $1^n = 1$

Additive and Multiplicative Inverses

- The words *opposite* and *reciprocal* have the more formal names **additive inverse** and **multiplicative inverse,** respectively.
- Property of Additive Inverses
  For any real number a, there is a unique real number $-a$ such that
  $$a + (-a) = -a + a = 0$$
- Property of Multiplicative Inverses
  For any nonzero real number a, there is a unique real number $\dfrac{1}{a}$ such that
  $$a \cdot \frac{1}{a} = \frac{1}{a} \cdot a = 1$$

The Distributive Property

- The Distributive Property
  For any real numbers a, b, and c,
  1. $a(b + c) = ab + ac$
  2. $ab + ac = a(b + c)$
- The property can also be stated for subtraction.
  1. $a(b - c) = ab - ac$
  2. $ab - ac = a(b - c)$
- In the first forms of the Distributive Property, we *multiply* to convert an indicated product $a(b + c)$ into an indicated sum $ab + ac$. In the second forms, we *factor* to convert an indicated sum $ab + ac$ into an indicated product $a(b + c)$.

Speaking the Language 1.3

1. The ▨▨▨▨ properties allow us to group addends and factors.

2. The product of a number and its ▨▨▨▨ is the multiplicative identity.

3. We can use the ▨▨▨▨ Property to remove parentheses in the expression $4(c - 3)$.

4. Writing an indicated sum as an indicated product is called ▨▨▨▨ the expression.

Exercises 1.3

Concepts and Skills

1. In your own words, what is the Commutative Property of Addition? the Commutative Property of Multiplication?

2. In your own words, what is the Associative Property of Addition? the Associative Property of Multiplication?

In Exercises 3–6, rewrite the given numerical expression so that the computation is easier to perform mentally. Then evaluate the resulting expression. State the property that justifies the change.

3. $(73 + 92) + 8$

4. $2 + (48 + 15)$

5. $5 \cdot (4 \cdot 7)$

6. $(13 \cdot 25) \cdot 4$

In Exercises 7–16, name the property that is illustrated.

7. $4 + (7 + 2) = (4 + 7) + 2$

8. $a + 6 = 6 + a$

9. $3x + (-3x) = 0$

10. $8(x + 4) = 8x + 32$

11. $0 + y = y$

12. $2(5z) = (2 \cdot 5)z$

13. $3d + 6 = 3(d + 2)$

14. $\dfrac{1}{2a} \cdot 2a = 1, a \neq 0$

15. $n \cdot 7 = 7n$

16. $x = 1x$

17. In connection with addition, what do we call 0?

18. In connection with multiplication, what do we call 1?

In Exercises 19–28, use the given property to complete the statement.

19. Multiplicative Identity Property: $c = $ ▨▨▨▨ .

20. Associative Property of Addition: $(x + y) + 3 = $ ▨▨▨▨ .

21. Property of Additive Inverses: $-7x + $ ▨▨▨▨ $= 0$.

22. Commutative Property of Multiplication: $x(y + 4) = $ ▨▨▨▨ .

23. Distributive Property: $5x + 5y = $ ▨▨▨▨ .

24. Associative Property of Multiplication: $3(7a) = $ ▨▨▨▨ .

25. Property of Multiplicative Inverses:

▨▨▨▨ $\cdot \dfrac{1}{x} = 1, x \neq 0$.

26. Additive Identity Property: $0 + (b + 4) = $ ▨▨▨▨ .

27. Commutative Property of Addition: $4(6 + x) = $ ▨▨▨▨ .

28. Distributive Property: $4(6 + x) = $ ▨▨▨▨ .

In Exercises 29–32, name the property that justifies each step.

29. $(2a + 7) + 3a = 2a + (7 + 3a)$ (a) ▨▨▨▨

$= 2a + (3a + 7)$ (b) ▨▨▨▨

$= (2a + 3a) + 7$ (c) ▨▨▨▨

$= a(2 + 3) + 7$ (d) ▨▨▨▨

$= (2 + 3)a + 7$ (e) ▨▨▨▨

$= 5a + 7$

30. $-x + (3 + x) = -x + (x + 3)$ (a) ▨▨▨▨

$= (-x + x) + 3$ (b) ▨▨▨▨

$= 0 + 3$ (c) ▨▨▨▨

$= 3$ (d) ▨▨▨▨

31. $(3x)(4x) = (3 \cdot x \cdot 4 \cdot x)$ (a) ▨▨▨▨

$= (3 \cdot 4 \cdot x \cdot x)$ (b) ▨▨▨▨

$= (3 \cdot 4) \cdot (x \cdot x)$ (c) ▨▨▨▨

$= 12x^2$

32. $2\left[\dfrac{1}{2}(a + x)\right] = \left(2 \cdot \dfrac{1}{2}\right)(a + x)$ (a) ▨▨▨▨

$= 1(a + x)$ (b) ▨▨▨▨

$= a + x$ (c) ▨▨▨▨

In Exercises 33–38, indicate which property in column B justifies each statement in column A. Assume that x and a are whole numbers and n is a natural number.

Column A	Column B
33. $(5 \cdot 0)^{17} = 0$	(a) $\dfrac{a}{1} = a$
34. $\dfrac{x + 3}{1} = x + 3$	(b) $\dfrac{a}{a} = 1,\ a \neq 0$
35. $(5 - 4)^{12} = 1$	(c) $\dfrac{0}{a} = 0,\ a \neq 0$
36. $\dfrac{0}{x + 4} = 0$	(d) $0^n = 0$
37. $\dfrac{x + 12}{x + 12} = 1$	(e) $1^n = 1$
38. $\dfrac{n - n}{n} = 0$	

39. Suppose that you need to multiply a number by 11. Use the Distributive Property to explain why multiplying that number by 10 and then adding the number to the product produces the desired result.

40. In your own words, what is the Additive Identity Property? the Multiplicative Identity Property?

In Exercises 41–48, use the Distributive Property to multiply.

41. $2(x + 3)$ **42.** $5(a - 4)$

43. $4(y - 3)$ **44.** $6(5 + x)$

45. $x(2 + y)$ **46.** $c(a + b)$

47. $a(x - 5)$ **48.** $k(4 - x)$

In Exercises 49–58, use the Distributive Property to factor.

49. $4x + 4y$ **50.** $2a - 2b$

51. $7y - 7z$ **52.** $12a + 12y$

53. $ax + ay$ **54.** $kx - ky$

55. $3x + 6$ **56.** $10x - 20$

57. $8a - 16$ **58.** $5x + 15$

In Exercises 59–66, use the Distributive Property to rewrite the given numerical expression so that the computation is easy to perform mentally.

59. $15 \cdot 102$ **60.** $25 \cdot 1001$

61. $98 \cdot 7$ **62.** $61 \cdot 50$

63. $98 \cdot 17 + 2 \cdot 17$ **64.** $23 \cdot 7 + 23 \cdot 3$

65. $31 \cdot 13 - 31 \cdot 3$ **66.** $27 \cdot 5 - 7 \cdot 5$

🌐 Real-Life Applications

67. At a small college, 140 women and 132 men applied for parking permits that cost $8 each. Use the Distributive Property to write two numerical expressions that represent the total paid to the college for parking permits.

68. At a bake sale, pies and cakes both cost $6 each. A person buys 4 pies and 3 cakes. Use the Distributive Property to write two numerical expressions that represent the total cost of the sale.

✎ Modeling with Real Data

69. The table shows the results of a survey to determine how often American families eat breakfast together.

American Families Eat Breakfast Together . . .	
Weekends only	49%
1–2 weekdays	12%
3–4 weekdays	—
Never	12%
Daily	11%
Don't know	1%

(Source: General Mills.)

(a) What percentage of the families eat together on 3 or 4 weekdays?

(b) What percentage of the families eat together on 1–4 weekdays?

70. Children of ages 9–13 play video games an average of 1.4 hours per day. (Source: *Sports Illustrated.*) Approximately how many equivalent days per year does the average child play video games?

Challenge

In Exercises 71–74, use the Distributive Property to rewrite the given numerical expression so that the computation is easy to perform mentally.

71. $73 \cdot 5 + 73 \cdot 3 + 73 \cdot 2$

72. $9 \cdot 97 + 9 \cdot 6 - 9 \cdot 3$

73. $15 \cdot 98$

74. $18(1.5)$

In Exercises 75–78, fill in the blank and state the property that justifies your answer.

75. $\frac{3}{4}(8x) = ($ ⬛⬛⬛ $)x$

76. $2x + 7x = ($ ⬛⬛⬛ $)x$

77. $-2 \cdot 5 + (-2)y = ($ ⬛⬛⬛ $)(5 + y)$

78. $(x + 5)(x + 2) = (x + 2)($ ⬛⬛⬛ $)$

1.4 | Addition

Addends with Like Signs • Addends with Unlike Signs • Real-Life Applications

Addends with Like Signs

The methods for adding two real numbers differ depending on the signs of the addends. We begin with sums in which the addends have the same signs. Because we know how to add numbers that are positive, we consider the sum of two negative numbers.

Exploring the Concept

Sums of Negative Numbers

Figure 1.9 shows three sums in which the addends are negative numbers.

Figure 1.9

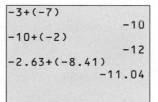

```
-3+(-7)
                -10
-10+(-2)
                -12
-2.63+(-8.41)
             -11.04
```

In each case, the result is negative, and the absolute value of the result is the sum of the absolute values of the addends. These observations can be proved to be true for the sum of any two negative numbers.

The following summary states the addition rules for addends with like signs.

> **Addition Rules for Addends with Like Signs**
> 1. The sign of the sum is the same as the signs of the addends.
> 2. The absolute value of the sum is the sum of the absolute values of the addends.

EXAMPLE  1

Adding Numbers with Like Signs

Determine each sum without your calculator. Then use your calculator and compare your results.

LEARNING TIP

Note the use of parentheses. If the second addend is negative, it is usually enclosed in parentheses.

(a) $-9 + (-7) = -16$

(b) $-100 + (-1) = -101$

(c) $-3.7 + (-2.5) = -6.2$

Addends with Unlike Signs

We now turn to sums in which the addends have unlike signs.

Exploring the Concept

Sums of Numbers with Unlike Signs

Figure 1.10 shows four sums in which one addend is positive and the other addend is negative.

Figure 1.10

```
-4+7
                3
4+(-7)
              -3
-9+3
              -6
9+(-3)
                6
```

Observe that the addend with the larger absolute value appears to dictate the sign of the result. Also, we can obtain the absolute value of the result by subtracting the absolute values of the addends.

Again, the rules that are suggested by this experiment can be proved, but we accept them as given.

> **Addition Rules for Addends with Unlike Signs**
> 1. The sign of the sum is the same as the sign of the addend with the larger absolute value.
> 2. The absolute value of the sum is the difference of the absolute values of the addends.

Note: For addends with unlike signs, when we subtract the absolute values of the addends, we subtract the smaller absolute value from the larger absolute value.

EXAMPLE 2

Adding Addends with Unlike Signs

Determine each sum without your calculator. Then use your calculator and compare your results.

(a) $15 + (-5) = 10$ (b) $-20 + 14 = -6$

(c) $-12 + 12 = 0$ (d) $4.3 + (-7.5) = -3.2$

EXAMPLE 3

Using the Addition Rules to Calculate Sums

Perform the addition.

(a) $-8 + 5 = -3$

(b) $-3 + (-12) = -15$

(c) $15 + (-6) = 9$

(d) $-4 + 11 = 7$

(e) $12 + (-13) = -1$

(f) $-584 + 379 = -205$

(g) $-18.97 + (-34.5) = -53.47$

When a sum contains three or more addends, grouping symbols may be used to indicate the order of operations. Also, the addends of a sum can be reordered and grouped however we wish.

EXAMPLE 4

Sums with More than Two Addends

Perform the addition.

(a) $-4 + 7 + 9 = 3 + 9 = 12$ No parentheses, so add from left to right.

(b) $(-2 + 5) + [6 + (-9)] = 3 + (-3)$ Add inside the grouping symbols first.

$= 0$ Property of Additive Inverses

(c) $2 + (-5) + (-7) + 8 + (-10)$

$= 2 + 8 + (-5) + (-7) + (-10)$ Commutative Property of Addition

$= [2 + 8] + [(-5) + (-7) + (-10)]$ Associative Property of Addition

$= 10 + (-22)$

$= -12$

Note: Because addends with like signs are easy to add, mentally grouping positive and negative addends makes the sum easier to determine.

EXAMPLE 5

Translating into Sums

In each part, write the described sum and perform the addition.

(a) 6 plus -9

(b) the sum of -5, -3, and 8

(c) 5 more than -2

(d) 7 added to -16

Solution

(a) $6 + (-9) = -3$

(b) $-5 + (-3) + 8 = 0$

(c) $-2 + 5 = 3$

(d) $-16 + 7 = -9$

Think About It

Consider the following sum.
$1 + (-2) + 3 + (-4) + 5 + (-6) + \cdots$
What is the sum if there are 50 addends? How would you represent the sum if there are n addends, where n is even?

Note: In part (c) of Example 5, we translate the phrase "5 more than -2" into the sum $-2 + 5$. Compare this with the sentence "5 *is* more than -2," which we translate into the inequality $5 > -2$.

Real-Life Applications

We need to be able to add signed numbers in order to solve applications such as the following.

EXAMPLE 6

Calculating a Depth Below Sea Level

A marine biologist for the National Oceanographic Institute descended 20 feet below the ocean surface to gather data on ocean plant life. After gathering the data, she descended another 50 feet to collect more information. Finally, she descended another 70 feet for the deepest sample. From there, she ascended 30 feet for a final sample before returning to the surface. The depth changes are given by -20, -50, -70, and $+30$ feet. How deep was the biologist when she collected her deepest sample? How deep was she when she collected her last sample?

Solution

The deepest point is the sum of the depths at which she collected the first three samples.

$$-20 + (-50) + (-70) = -140$$

We interpret the result as 140 feet below the ocean surface. The location of the final sample is the sum of all of the biologist's depth changes.

$$-20 + (-50) + (-70) + 30 = -110$$

We interpret the result as 110 feet below the ocean surface.

Quick Reference 1.4

Addends with Like Signs

• If a and b are real numbers with like signs, then $a + b$ is determined as follows.

1. The sign of the sum is the same as the signs of the addends.

2. The absolute value of the sum is the sum of the absolute values of the addends.

Addends with Unlike Signs

• If a and b are real numbers with unlike signs, then $a + b$ is determined as follows.

1. The sign of the sum is the same as the sign of the addend with the larger absolute value.

2. The absolute value of the sum is the difference of the absolute values of the addends.

• When a sum consists of three or more addends, grouping symbols may be used to indicate the order of operations.

• Grouping positive and negative addends together makes the sum easier to determine.

Speaking the Language 1.4

1. If x and y are both negative, then the sign of the sum of x and y is _____.

2. If x and y have unlike signs, then the sign of the sum of x and y is the same as the sign of the addend with the larger _____.

3. If a sum of three or more addends is written without grouping symbols, then we add from _____ to _____.

4. If a and b have unlike signs, then the absolute value of the sum is determined by _____ the absolute values of a and b.

Exercises 1.4

Concepts and Skills

1. Explain how to determine the sum of two numbers that have like signs.

2. Explain how to determine the sum of two numbers that have unlike signs.

In Exercises 3–34, determine the sum.

3. $8 + (-5)$

4. $9 + (-4)$

5. $-6 + 2$

6. $-7 + 4$

7. $-3 + (-7)$

8. $-5 + (-10)$

9. $-6 + (-2)$

10. $-9 + (-5)$

11. $-16 + 16$

12. $-14 + 14$

13. $-3 + 5$

14. $-7 + 5$

15. $12 + (-4)$

16. $13 + (-7)$

17. $-6 + 14$

18. $-14 + 8$

19. $-7 + 0$

20. $-9 + 0$

21. $4 + 16$

22. $5 + 18$

23. $0 + (-55)$

24. $0 + (-22)$

25. $-9 + 17$

26. $14 + (-20)$

27. $-12 + 12$

28. $17 + (-10)$

29. $-11 + (-11)$

30. $-13 + (-8)$

31. $100 + (-101)$

32. $-50 + (-5)$

33. $-14 + 3$

34. $36 + (-9)$

In Exercises 35–44, determine the sum.

35. $-14 + (-2) + (-1)$ **36.** $5 + (-8) + (-12)$

37. $-3 + [6 + (-9)]$ **38.** $-16 + [3 + (-7)]$

39. $6 + (-7) + 2$

40. $-20 + 6 + (-1)$

41. $-12 + (-2) + (-5) + 10$

42. $15 + (-1) + 2 + (-6)$

43. $-6 + 8 + [(-5) + 3]$

44. $(-2) + [(-5) + (-3)] + 9$

In Exercises 45–50, predict the sign of the sum and use your calculator to find the sum.

45. $-2.3 + (-4.5)$ **46.** $2.8 + (-4.6)$

47. $-45.84 + (-5.73)$ **48.** $-0.35 + 78.3$

49. $24.47 + (-34.78)$ **50.** $-975.3 + (-537.37)$

In Exercises 51–70, perform the indicated operations.

51. $17 + (-8)$ **52.** $-36 + (-9)$

53. $6 + (-8) + (-2)$ **54.** $-5 + 10 + (-7)$

55. $-6 + |-8|$ **56.** $|4| + |-3|$

57. $187.24 + (-187.24)$ **58.** $72.72 + (-73.73)$

59. $-50 + 5$ **60.** $-8 + (-9)$

61. $-10 + (-2) + 15$ **62.** $18 + 3 + (-10)$

63. $0 + (-18.3)$

64. $17.459 + 0$

65. $-12 + (-3) + 10 + (-2)$

66. $3 + (-8) + (-6) + 11$

67. $|-7| + (-7)$ **68.** $|-5| + 2$

69. $-0.003 + (-0.219)$ **70.** $-0.5 + 0.3$

71. When a sum has many addends, we may wish to add the positive addends first and then add the negative addends. What properties allow us to reorder and regroup the addends?

72. What are two names that are given to pairs of numbers whose sum is 0?

In Exercises 73–76, state the property of real numbers that justifies the step.

73. $-3 + 11 + (-4)$

$\quad = -3 + (-4) + 11$ (a) ▬▬▬

$\quad = [-3 + (-4)] + 11$ (b) ▬▬▬

$\quad = -7 + 11$

$\quad = 4$

74. $6 + (-11 + 8) + 3$

$\quad = 6 + (-11) + (8 + 3)$ (a) ▬▬▬

$\quad = 6 + (-11 + 11)$ (b) ▬▬▬

$\quad = 6 + 0$ (c) ▬▬▬

$\quad = 6$ (d) ▬▬▬

75. $3x + y + (-3x)$

$\quad = 3x + (-3x) + y$ (a) ▬▬▬

$\quad = [3x + (-3x)] + y$ (b) ▬▬▬

$\quad = 0 + y$ (c) ▬▬▬

$\quad = y$ (d) ▬▬▬

76. $a + (-b) + b + (-a)$

$\quad = a + (-a) + (-b) + b$ (a) ▬▬▬

$\quad = [a + (-a)] + [(-b) + b]$ (b) ▬▬▬

$\quad = 0 + 0$ (c) ▬▬▬

$\quad = 0$ (d) ▬▬▬

In Exercises 77–86, perform the indicated operations.

77. $-8 + 3^2$

78. $-12 + 5(4) + (-8)$

79. $-4 + 5(-4 + 5)$ **80.** $-(-7 + 5)$

81. $\dfrac{-7 + 11}{-6 + 2(5)}$ **82.** $\dfrac{12 + (-6)}{4(3) - 2(5)}$

83. $-14 + 30 - 2^3$ **84.** $[10 + (-8)][3^3 - 5^2]$

85. $-6 + \sqrt{4^2 + (-12)}$ **86.** $|-19 + 3^2| + (-8)$

In Exercises 87–92, write a numerical expression and then evaluate it.

87. What is the sum of -964 and 351?

88. What is 11 more than -6?

89. What is -9 increased by 5?

90. What is the sum of -7, 3, and -10?

91. If 12 is added to the sum of -5 and 2, what is the result?

92. If -10 is increased by the sum of -5 and 2, what is the result?

In Exercises 93–96, determine the unknown number.

93. The sum of what number and 2 is -5?

94. The sum of -3 and what number is 7?

95. Two of three addends are -1 and 4. If the sum of the three numbers is -9, what is the third addend?

96. If the sum of three numbers is zero, and two of the numbers are 2 and -2, what is the third number?

Real-Life Applications

97. In their first possession of the football, a team had a loss of 9 yards on the first down, a gain of 4 yards on the second down, and a gain of 7 yards on the third down. What was the total gain or loss of yardage in these three downs?

98. Suppose that after you write a tuition check for $765, your checkbook balance is −$362.94. Before dropping the check in the fee collection box, you deposit a student loan check for $536.50. What is your checkbook balance after the deposit and the tuition payment?

99. A wrestler's weight was 216 pounds when he went on a diet. His change in weight for each of the seven subsequent weeks was as follows.

Week	Weight (lb)
1	5 (Loss)
2	1 (Gain)
3	2 (Loss)
4	1 (Loss)
5	4 (Gain)
6	3 (Loss)
7	2 (Gain)

What was the wrestler's weight after the seven weeks?

100. During a 1-week period in International Falls, Minnesota, the daily high temperatures (in degrees Fahrenheit) were $-10°$, $17°$, $5°$, $-9°$, $0°$, $6°$, and $4°$. What was the average high temperature for the week? (*Hint*: The **average** or **mean** of n numbers is found by adding the n numbers and dividing the sum by n.)

Challenge

In Exercises 101–104, let a represent a negative number and b represent a positive number. Determine whether the given sum is positive or negative.

101. $|a| + b$

102. $(-a) + b$

103. $a + (-b)$

104. $-b + b + a$

105. Suppose a person takes one step backward and then two steps forward. Repeating this pattern, how many steps must the person take to be 27 steps from the starting point?

106. (a) Calculate $1 + 2 + (-3) + (-4)$.

(b) Calculate $1 + 2 + (-3) + (-4) + 5 + 6 + (-7) + (-8)$.

(c) Consider a sum that follows the same pattern as in part (b), with the last two addends being -47 and -48. What is your conjecture about the value of the expression?

1.5 Addition with Rational Numbers

Equivalent Fractions • Addition of Fractions

Equivalent Fractions

Many different fractions can have the same decimal value. For example, consider the fraction $\frac{1}{4} = 0.25$. In the following fractions, we multiply the numerator and denominator of $\frac{1}{4}$ by 2, 3, 5, and 12.

$$\frac{2 \cdot 1}{2 \cdot 4} = \frac{2}{8} = 0.25 \qquad \frac{3 \cdot 1}{3 \cdot 4} = \frac{3}{12} = 0.25$$

$$\frac{5 \cdot 1}{5 \cdot 4} = \frac{5}{20} = 0.25 \qquad \frac{12 \cdot 1}{12 \cdot 4} = \frac{12}{48} = 0.25$$

Because all the resulting fractions have the same decimal value, we say that they are **equivalent** fractions.

Property of Equivalent Fractions

For any real number a and any nonzero real numbers b and c,

$$\frac{a}{b} = \frac{ac}{bc}$$

In words, multiplying the numerator and the denominator by the same number (except 0) results in an equivalent fraction.

Note: Multiplying the numerator and the denominator of a fraction by the same nonzero number is equivalent to multiplying the fraction by 1.

$$\frac{a}{b} = \frac{a}{b} \cdot 1 = \frac{a}{b} \cdot \frac{c}{c} = \frac{ac}{bc}$$

Multiplying the fraction by 1 does not change its value.

One way in which the Property of Equivalent Fractions is used is to write a given fraction with a different denominator. We call this process *renaming* the fraction.

EXAMPLE 1

Renaming Fractions

(a) Rename $\frac{1}{9}$ so that the denominator is 27.

(b) Rename $-\frac{3}{4}$ so that the denominator is 8.

Solution

(a) $\frac{1}{9} = \frac{1 \cdot 3}{9 \cdot 3} = \frac{3}{27}$ Because $9 \cdot 3 = 27$, multiply the numerator and the denominator by 3.

(b) $-\frac{3}{4} = -\frac{3 \cdot 2}{4 \cdot 2} = -\frac{6}{8}$ Multiply the numerator and the denominator by 2.

We can also use the Property of Equivalent Fractions to rename a fraction so that its denominator is smaller.

$$\frac{6}{15} = \frac{3 \cdot 2}{3 \cdot 5} = \frac{3}{3} \cdot \frac{2}{5} = 1 \cdot \frac{2}{5} = \frac{2}{5}$$

Note that both the numerator and the denominator contain the factor 3. Thus we say that 3 is a **common factor.** Any factor that is common to the numerator and the denominator can be divided out, and the resulting fraction is equivalent to the original fraction. We call this process **simplifying** or **reducing** the fraction.

EXAMPLE 2

Simplifying Fractions

If possible, simplify each fraction.

(a) $\frac{8}{20}$ (b) $-\frac{7}{21}$ (c) $-\frac{26}{100}$ (d) $\frac{9}{8}$

LEARNING TIP

In algebra, as in arithmetic, we always reduce fractions. However, we usually do not write an improper fraction as a mixed number. For instance, we reduce $\frac{12}{8}$ to $\frac{3}{2}$, but we usually do not write the result as the mixed number $1\frac{1}{2}$.

Solution

(a) $\dfrac{8}{20} = \dfrac{4 \cdot 2}{4 \cdot 5} = \dfrac{2}{5}$ Divide out the common factor 4.

(b) $-\dfrac{7}{21} = -\dfrac{7 \cdot 1}{7 \cdot 3} = -\dfrac{1}{3}$ Divide out the common factor 7.

(c) $-\dfrac{26}{100} = -\dfrac{2 \cdot 13}{2 \cdot 50} = -\dfrac{13}{50}$ Divide out the common factor 2.

(d) $\dfrac{9}{8} = \dfrac{3 \cdot 3}{2 \cdot 2 \cdot 2}$

Because 9 and 8 have no common factor, the fraction cannot be simplified.

Addition of Fractions

For the sum $\frac{3}{5} + \frac{4}{5}$, the result is found by adding the numerators and retaining the common denominator: $\frac{3}{5} + \frac{4}{5} = \frac{7}{5}$. The following is the general rule for adding fractions.

> **Addition Rule for Fractions**
>
> **For any real numbers a, b, and c, where $c \neq 0$,**
>
> $$\frac{a}{c} + \frac{b}{c} = \frac{a+b}{c}$$

Note: When performing operations with fractions, we should simplify the result whenever possible.

EXAMPLE 3

Adding Fractions with the Same Denominators

Perform the indicated operation.

(a) $\dfrac{3}{7} + \dfrac{1}{7}$ (b) $\dfrac{-1}{18} + \dfrac{7}{18}$ (c) $\dfrac{7}{12} + \dfrac{-5}{12}$

Solution

(a) $\dfrac{3}{7} + \dfrac{1}{7} = \dfrac{3+1}{7} = \dfrac{4}{7}$

(b) $\dfrac{-1}{18} + \dfrac{7}{18} = \dfrac{-1+7}{18} = \dfrac{6}{18} = \dfrac{2 \cdot 3 \cdot 1}{2 \cdot 3 \cdot 3} = \dfrac{1}{3}$

(c) $\dfrac{7}{12} + \dfrac{-5}{12} = \dfrac{7+(-5)}{12} = \dfrac{2}{12} = \dfrac{2 \cdot 1}{2 \cdot 2 \cdot 3} = \dfrac{1}{6}$

To perform addition with fractions that do not have the same denominators, we must first rename the fractions so that they do have the same denominators. Although it is not essential, it is convenient for this common denominator to be the smallest number possible.

The smallest number that is a multiple of two or more given numbers is called the **least common multiple (LCM)** of the numbers. When the given numbers are denominators of fractions, we use the more specific name, **least common denominator (LCD),** of the fractions.

For simple fractions, the LCD is often apparent.

$$\frac{1}{2}, \frac{2}{3} \quad \text{The LCD is 6.} \qquad\qquad \frac{3}{4}, \frac{5}{8} \quad \text{The LCD is 8.}$$

For other fractions, such as $\frac{5}{12}$ and $\frac{7}{30}$, a numerical routine may be needed to determine the LCD. This routine includes the use of **prime numbers.**

> **Definition of a Prime Number**
>
> A **prime number** is a positive integer greater than 1 that is divisible only by itself and 1.

Note: We say that a is *divisible* by b if the remainder of $a \div b$ is 0.

The first ten prime numbers are 2, 3, 5, 7, 11, 13, 17, 19, 23, 29. We say that a number is **prime factored** when the number is written as a product whose factors are all prime numbers.

Exploring the Concept

Sums of Fractions with Unlike Denominators

Consider the following sum.

$$\frac{5}{12} + \frac{7}{30}$$

If we prime factor the denominators, we obtain

$$\frac{5}{12} + \frac{7}{30} = \frac{5}{2 \cdot 2 \cdot 3} + \frac{7}{2 \cdot 3 \cdot 5}$$

The LCD must include the prime factors of the first denominator, 12.

LCD (so far): $2 \cdot 2 \cdot 3$

The LCD must also include the prime factors of the second denominator, 30. However, two of these factors, 2 and 3, are already in the LCD, so we only need to include the factor 5.

LCD: $2 \cdot \overbrace{2 \cdot 3 \cdot 5}$

(with 12 over the first part and 30 under the last part)

Now we use the Property of Equivalent Fractions to rename the fractions with $2 \cdot 2 \cdot 3 \cdot 5$ for both denominators.

$$\frac{5}{12} + \frac{7}{30} = \frac{5}{2 \cdot 2 \cdot 3} + \frac{7}{2 \cdot 3 \cdot 5}$$

$$= \frac{5 \cdot 5}{2 \cdot 2 \cdot 3 \cdot 5} + \frac{7 \cdot 2}{2 \cdot 2 \cdot 3 \cdot 5} \qquad \text{Property of Equivalent Fractions}$$

$$= \frac{25}{60} + \frac{14}{60}$$

$$= \frac{39}{60} \qquad \text{Addition Rule for Fractions}$$

$$= \frac{3 \cdot 13}{3 \cdot 20} \qquad \text{Factor the numerator and denominator.}$$

$$= \frac{13}{20} \qquad \text{Divide out the common factor 3.}$$

EXAMPLE 4

Adding Fractions with Different Denominators

Perform the indicated operations.

(a) $\dfrac{3}{5} + \dfrac{2}{7}$ (b) $\dfrac{-3}{4} + \dfrac{1}{2}$ (c) $\dfrac{1}{8} + \dfrac{5}{12}$

Think About It

How do you enter a mixed number such as $3\frac{5}{8}$ in a calculator? How do you change a mixed number such as $3\frac{5}{8}$ to an improper fraction?

Solution

(a) $\dfrac{3}{5} + \dfrac{2}{7} = \dfrac{3 \cdot 7}{5 \cdot 7} + \dfrac{5 \cdot 2}{5 \cdot 7} = \dfrac{21}{35} + \dfrac{10}{35} = \dfrac{31}{35}$

(b) $\dfrac{-3}{4} + \dfrac{1}{2} = \dfrac{-3}{2 \cdot 2} + \dfrac{1}{2} = \dfrac{-3}{2 \cdot 2} + \dfrac{1 \cdot 2}{2 \cdot 2} = \dfrac{-3}{4} + \dfrac{2}{4} = \dfrac{-1}{4}$

(c) $\dfrac{1}{8} + \dfrac{5}{12} = \dfrac{1}{2 \cdot 2 \cdot 2} + \dfrac{5}{2 \cdot 2 \cdot 3}$

$$= \frac{1 \cdot 3}{2 \cdot 2 \cdot 2 \cdot 3} + \frac{2 \cdot 5}{2 \cdot 2 \cdot 2 \cdot 3}$$

$$= \frac{3}{24} + \frac{10}{24}$$

$$= \frac{13}{24}$$

🔑 **Keys to the Calculator**

Fraction

You can use your calculator to report results in simplified fraction form. Here are the keystrokes for simplifying $\frac{8}{20}$ [Example 2(a)] and for adding $\frac{3}{5} + \frac{2}{7}$ [Example 4(a)].

8 ÷ 20 [MATH] [1:Frac] [ENTER]

3 ÷ 5 + 2 ÷ 7 [MATH] [1:Frac] [ENTER]

```
8/20▶Frac
                    2/5
3/5+2/7▶Frac
                  31/35
```

Quick Reference 1.5

Equivalent Fractions

- Fractions with the same decimal value are called **equivalent fractions.**
- Property of Equivalent Fractions

  For any real number a and any nonzero real numbers b and c, $\dfrac{a}{b} = \dfrac{ac}{bc}$.

  In words, multiplying the numerator and denominator by the same nonzero number results in an equivalent fraction. We call this process *renaming* the fraction.

- The Property of Equivalent Fractions can also be used to **reduce** or **simplify** a fraction by dividing out any **common factor** in the numerator and denominator.

Addition of Fractions

- Addition Rule for Fractions

  For any real numbers a, b, and c, $c \neq 0$, $\dfrac{a}{c} + \dfrac{b}{c} = \dfrac{a+b}{c}$.

  In words, add the numerators and retain the common denominator.

- The smallest number that is a multiple of two or more given numbers is called the **least common multiple (LCM)** of the numbers. When the given numbers are denominators of fractions, we use the more specific name, **least common denominator (LCD)**, of the fractions.

- Definition of Prime Number

  A **prime number** is a positive integer greater than 1 that is divisible only by itself and 1.

- We say that a number is **prime factored** when the number is written as a product whose factors are all prime numbers.

- To add fractions whose denominators are not the same, use the following steps.

  1. Prime factor each denominator.

  2. Construct the LCD by including all of the prime factors of each denominator.

  3. Use the Property of Equivalent Fractions to rename the fractions so that their denominators are the LCD.

  4. Add the fractions according to the Addition Rule for Fractions.

Speaking the Language 1.5

1. Because $\frac{3}{5}$ and $\frac{6}{10}$ have the same decimal value, they are called _____ fractions.

2. Dividing out a common factor in the numerator and the denominator of a fraction is called _____ the fraction.

3. If n is an integer greater than 1, and if n is divisible only by n and 1, then n is a(n) _____ number.

4. Fractions must have a(n) _____ before they can be added.

Exercises 1.5

Concepts and Skills

1. What do we mean by equivalent fractions?

2. Describe the method for renaming a fraction. What property is used in this process?

In Exercises 3–6, rename the fraction so that its denominator is each of the given numbers.

3. $\dfrac{1}{6}$; 18, 30, 42

4. $\dfrac{3}{5}$; 10, 35, 50

5. $-\dfrac{2}{7}$; 14, 35, 56

6. $-\dfrac{1}{3}$; 12, 21, 27

In Exercises 7–16, determine the numerator.

7. $\dfrac{3}{5} = \dfrac{\blacksquare}{15}$

8. $\dfrac{1}{3} = \dfrac{\blacksquare}{9}$

9. $\dfrac{7}{4} = \dfrac{\blacksquare}{12}$

10. $-\dfrac{5}{2} = -\dfrac{\blacksquare}{24}$

11. $-\dfrac{5}{6} = -\dfrac{\blacksquare}{30}$

12. $-\dfrac{2}{7} = -\dfrac{\blacksquare}{77}$

13. $-6 = -\dfrac{\blacksquare}{2}$

14. $8 = \dfrac{\blacksquare}{7}$

15. $\dfrac{7}{36} = \dfrac{\blacksquare}{144}$

16. $-\dfrac{2}{35} = -\dfrac{\blacksquare}{140}$

17. Under what condition can a fraction be reduced or simplified?

18. Which of the following is an example of reducing a fraction? Explain.

(i) $\dfrac{12}{9} = \dfrac{4}{3}$ (ii) $\dfrac{4}{3} = 1\dfrac{1}{3}$

In Exercises 19–28, reduce the given fraction to its lowest terms.

19. $\dfrac{4}{6}$

20. $\dfrac{9}{12}$

21. $\dfrac{6}{48}$

22. $\dfrac{9}{45}$

23. $-\dfrac{30}{12}$

24. $-\dfrac{49}{21}$

25. $\dfrac{12}{35}$

26. $-\dfrac{16}{27}$

27. $-\dfrac{72}{16}$

28. $\dfrac{24}{18}$

29. For the sum $\frac{1}{2} + \frac{1}{4}$, is 8 a common denominator? Is 8 the least common denominator? Explain.

30. Describe the LCD of two fractions if their denominators have no prime factors in common.

In Exercises 31–38, add the fractions.

31. $\dfrac{2}{5} + \dfrac{7}{5}$

32. $\dfrac{4}{3} + \dfrac{7}{3}$

33. $\dfrac{5}{8} + \dfrac{3}{8}$

34. $\dfrac{3}{2} + \dfrac{7}{2}$

35. $\dfrac{-3}{7} + \dfrac{5}{7}$

36. $\dfrac{-8}{5} + \dfrac{11}{5}$

37. $\dfrac{5}{4} + \dfrac{-3}{4}$

38. $\dfrac{13}{9} + \dfrac{-1}{9}$

In Exercises 39–50, add.

39. $\dfrac{1}{2} + \dfrac{1}{3}$

40. $\dfrac{1}{5} + \dfrac{1}{3}$

41. $\dfrac{2}{5} + \dfrac{1}{10}$

42. $\dfrac{5}{8} + \dfrac{3}{4}$

43. $3 + \dfrac{-2}{7}$

44. $\dfrac{-9}{2} + 5$

45. $\dfrac{5}{9} + \dfrac{7}{12}$

46. $\dfrac{9}{20} + \dfrac{3}{8}$

47. $\dfrac{4}{15} + \dfrac{5}{18}$

48. $\dfrac{1}{12} + \dfrac{3}{16}$

49. $\dfrac{5}{7} + \dfrac{-3}{14}$

50. $\dfrac{-8}{15} + \dfrac{2}{3}$

In Exercises 51–56, add.

51. $\dfrac{5}{3} + \dfrac{-2}{3} + \dfrac{-1}{3}$

52. $\dfrac{-3}{8} + \dfrac{7}{8} + \dfrac{-1}{8}$

53. $\dfrac{-2}{5} + \dfrac{3}{5} + \dfrac{1}{2}$

54. $\dfrac{4}{3} + \dfrac{3}{4} + \dfrac{-1}{3}$

55. $\dfrac{1}{8} + \dfrac{1}{4} + \dfrac{1}{2}$

56. $\dfrac{2}{3} + \dfrac{1}{6} + \dfrac{1}{2}$

In Exercises 57–60, perform the indicated operations.

57. $2\dfrac{1}{4} + 3\dfrac{1}{3}$

58. $1\dfrac{1}{5} + 2\dfrac{2}{3}$

59. $1\dfrac{5}{6} + 4\dfrac{1}{2}$

60. $3\dfrac{5}{16} + 2\dfrac{7}{8}$

In Exercises 61–70, add.

61. $\dfrac{-1}{3} + \dfrac{-4}{3}$

62. $\dfrac{-3}{5} + \dfrac{-6}{5}$

63. $\dfrac{-7}{8} + \dfrac{3}{2}$

64. $\dfrac{9}{5} + \dfrac{-3}{10}$

65. $4 + \dfrac{3}{7}$

66. $\dfrac{2}{3} + 6$

67. $\dfrac{2}{5} + \dfrac{1}{3}$

68. $\dfrac{1}{2} + \dfrac{1}{3} + \dfrac{1}{4}$

69. $\dfrac{1}{2} + \dfrac{1}{4} + \dfrac{1}{8} + \dfrac{1}{16}$

70. $\dfrac{1}{3} + \dfrac{1}{9} + \dfrac{-1}{27}$

In Exercises 71–74, write a numerical expression for the described operation and then perform the operation.

71. Find the sum of $1\dfrac{1}{8}$ and $\dfrac{3}{8}$.

72. What is the sum if the addends are $\dfrac{8}{3}$ and $\dfrac{5}{12}$?

73. What is the sum if the addends are $2\dfrac{2}{3}$ and $\dfrac{-5}{6}$?

74. Find the sum of $\dfrac{-7}{13}$ and $\dfrac{-1}{13}$.

Real-Life Applications

75. If a stock was purchased at $25\frac{3}{8}$ and then sold after its value had increased by $\frac{3}{4}$, what was the stock's selling price?

76. The finished width of a 2 × 4 board is 1.5 inches. If a $\frac{3}{16}$-inch trim strip is fastened to the board, what is its total width?

Modeling with Real Data

77. In a survey, people were asked what they do with the giblets from their holiday turkey. The responses, along with the fraction of those people who gave the response, are shown in the table.

What People Do with Turkey Giblets	
Soup/gravy	$\dfrac{41}{100}$
Throw away	$\dfrac{1}{5}$
Eat	$\dfrac{7}{50}$
Feed to pet	$\dfrac{2}{25}$
Other	$\dfrac{1}{25}$
"The what?"	—

(Source: *USA Today*.)

(a) What fraction of the people surveyed use the giblets for personal consumption?

(b) What fraction of those surveyed did not know what giblets were?

78. Write the fractions in Exercise 77 as decimals.

(a) Which response was given by the greatest number of people surveyed? by the least number?

(b) What response ranked third?

Data Analysis: *Dangerous Occupations*

The accompanying bar chart shows the four occupations with the highest fatality rates per 100,000 workers. (Source: Bureau of Labor Statistics.)

Figure for 79–82

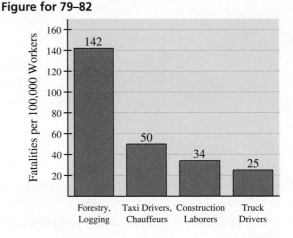

79. The data in the bar chart can be written as ratios (fractions), such as $\frac{142}{100,000}$ for forestry and logging. Write the data for the other three categories as fractions and simplify all four fractions.

80. Why is it easier to make comparisons with the original data (fatality rates per 100,000 workers) than with the simplified fractions obtained in Exercise 79? State two ways to compare fractions when their denominators are not the same.

81. Add the four fractions in Exercise 79. What is your interpretation of the result?

82. The national average for all occupations is 5 fatalities per 100,000 workers. How many times the national average is the rate for each occupation in the table?

Challenge

In Exercises 83 and 84, perform the indicated operation. Assume all variables represent positive numbers.

83. $\dfrac{3}{a} + \dfrac{7}{a}$ **84.** $\dfrac{5}{x} + \dfrac{-2}{x}$

85. Which of the following is a common denominator

for $\dfrac{a}{b}$ and $\dfrac{c}{d}$?

(i) $b + d$ (ii) bd (iii) ac (iv) $a + c$

Select values for the variables to show that the common denominator is not necessarily the LCD.

86. Write six fractions whose numerators are the first six natural numbers and whose denominators are the first six even natural numbers. Then show that the sum is 3.

1.6 | Subtraction

Definitions • Performing Subtraction • Real-Life Applications

Definitions

Recall that in the expression $8 - 5$, the number 8 is called the **minuend,** and the number 5 is called the **subtrahend.** Both the expression and the value of the expression are called the **difference.**

Exploring the Concept

The Relationship Between Subtraction and Addition

Figures 1.11 and 1.12 suggest that subtraction and addition are closely related.

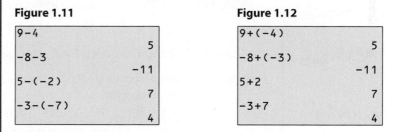

Figure 1.11

$9-4$	
	5
$-8-3$	
	-11
$5-(-2)$	
	7
$-3-(-7)$	
	4

Figure 1.12

$9+(-4)$	
	5
$-8+(-3)$	
	-11
$5+2$	
	7
$-3+7$	
	4

Comparing each difference in Figure 1.11 to the corresponding sum in Figure 1.12, we can make two observations.

1. The results are the same.
2. The minus sign is changed to a plus sign, and the subtrahend is changed to its opposite.

These observations suggest the following definition of subtraction.

Think About It

To illustrate $5 - 2$ on a number line, we locate 5 and then move 2 units to the left. How do you think we could illustrate $4 - (-5)$?

Definition of Subtraction

For any real numbers a and b, $a - b = a + (-b)$.

This definition provides the method for performing subtraction.

> **Performing Subtraction with Real Numbers**
>
> To subtract any two real numbers,
>
> 1. change the minus sign to a plus sign, and
> 2. change the subtrahend to its opposite.
>
> Then evaluate the sum.

An obvious benefit of converting a difference into a sum is that we already know the rules of addition.

Performing Subtraction

For simple subtraction problems, you should be able to apply the definition mentally and not have to rely on a calculator. (When you do use a calculator, you do not need to apply the definition because the calculator does it for you.)

The rule for subtracting fractions is similar to the rule for adding fractions.

> **Subtraction Rule for Fractions**
>
> For any real numbers a, b, and c, where $c \neq 0$,
>
> $$\frac{a}{c} - \frac{b}{c} = \frac{a - b}{c}$$
>
> In words, subtract the numerators and retain the common denominator.

EXAMPLE **1**   **Performing Subtraction with Real Numbers**

Perform the indicated operations. Use your calculator in parts (g) and (h) and to verify the results in all other parts.

(a) $5 - 12$ (b) $-20 - 4$

(c) $16 - (-5)$ (d) $-7 - (-3)$

(e) $0 - (-8)$ (f) $\dfrac{1}{3} - \dfrac{5}{2}$

(g) $-2.47 - (-6.91)$ (h) $-538 - 297$

LEARNING TIP

Initially you should write the addition step. Omitting that step often leads to errors and confusion.

Solution

(a) $5 - 12 = 5 + (-12) = -7$

(b) $-20 - 4 = -20 + (-4) = -24$

(c) $16 - (-5) = 16 + 5 = 21$

(d) $-7 - (-3) = -7 + 3 = -4$

Figure 1.13

```
-2.47-(-6.91)
              4.44
-538-297
          -835.00
```

(e) $0 - (-8) = 0 + 8 = 8$

(f) $\dfrac{1}{3} - \dfrac{5}{2} = \dfrac{2}{6} - \dfrac{15}{6}$ The LCD is 6.

$\qquad = \dfrac{2 - 15}{6}$ Subtract numerators, retain the LCD.

$\qquad = \dfrac{2 + (-15)}{6}$ Definition of subtraction

$\qquad = \dfrac{-13}{6}$

(g) and (h) See Figure 1.13.

We now investigate the effect of reversing the order of the numbers in a difference.

Exploring the Concept

Reversing the Order of Subtraction

Compare the following pairs of differences.

$\qquad 9 - 5 = 4 \qquad\qquad\qquad 5 - 9 = -4$

$\qquad -3 - (-8) = 5 \qquad\qquad -8 - (-3) = -5$

$\qquad -7 - 10 = -17 \qquad\qquad 10 - (-7) = 17$

Note that in each pair the results are *opposites* (or *additive inverses*). Symbolically, we say that $a - b$ and $b - a$ are opposites; that is, $-(a - b) = b - a$ and $-(b - a) = a - b$.

Property of the Opposite of a Difference

For any real numbers a and b, $-(a - b) = b - a$.

As we will see in our later work, this is an important rule for simplifying expressions and for reversing the order of subtraction when we want to do so.

EXAMPLE 2

Using the Property of the Opposite of a Difference

Use the Property of the Opposite of a Difference to rewrite each expression without grouping symbols. In part (a), evaluate the expression.

(a) $-(3 - 11)$

(b) $-(5 - x)$

Solution

(a) $-(3 - 11) = 11 - 3 = 8$

(b) $-(5 - x) = x - 5$

EXAMPLE 3

Performing Mixed Operations

Perform the indicated operations.

(a) $12 - (3 - 7)$ Parentheses first

$= 12 - [3 + (-7)]$ Definition of subtraction

$= 12 - (-4)$

$= 12 + 4$ Definition of subtraction

$= 16$

(b) $4 - (5 + 9) - 7$

$= 4 - 14 - 7$ Parentheses first

$= 4 + (-14) + (-7)$ Definition of subtraction

$= -17$

(c) $(6 - 9) - (-8 + 2)$

$= -3 - (-6)$ Parentheses first

$= -3 + 6$ Definition of subtraction

$= 3$

(d) $5 - [4 - (1 - 2)]$ Innermost grouping symbols first

$= 5 - [4 - (-1)]$

$= 5 - [4 + 1]$ Definition of subtraction

$= 5 - 5$ Brackets first

$= 0$

EXAMPLE 4

Translating into Differences

In each part, write the described difference and perform the subtraction.

(a) -6 minus -5

(b) -9 less than 3

(c) $\dfrac{2}{3}$ decreased by $-\dfrac{8}{7}$

(d) the difference between 7 and -1

Solution

(a) $-6 - (-5) = -1$

(b) $3 - (-9) = 12$ Subtract -9 from 3.

(c) $\dfrac{2}{3} - \left(-\dfrac{8}{7}\right) = \dfrac{2}{3} + \dfrac{8}{7}$ Definition of subtraction

$= \dfrac{2 \cdot 7}{3 \cdot 7} + \dfrac{8 \cdot 3}{7 \cdot 3}$

$= \dfrac{14}{21} + \dfrac{24}{21} = \dfrac{38}{21}$

(d) $7 - (-1) = 8$ The first given number is the minuend, and the second given number is the subtrahend.

Note: In part (b) of Example 4, we translate the phrase "-9 less than 3" into the difference $3 - (-9)$. Compare this with the sentence "-9 *is* less than 3," which we translate into the inequality $-9 < 3$.

Real-Life Applications

Application problems often require subtraction of signed numbers.

Distance Between Points Above and Below Sea Level

The altitude gauge of a C-130 indicates that the plane is 760 feet above sea level. The crew of a training mission is to drop equipment to a location in Death Valley that is 42 feet below sea level. When the plane is directly above the drop site, what is the distance between the plane and the drop site?

Solution

Consider a vertical number line whose origin is at sea level. Then the plane is at the point whose coordinate is $+760$, and the drop site is at the point whose coordinate is -42. (See Fig. 1.14.)

Figure 1.14

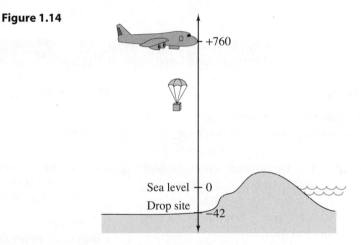

We calculate the distance d between the points as follows.

$$d = 760 - (-42) = 760 + 42 = 802$$

The distance from the plane to the drop site is 802 feet.

Quick Reference 1.6

Definitions

• For any real numbers a and b, $a - b = a + (-b)$.

• According to the definition of subtraction, we can write any difference as a sum by following these steps:

1. Change the minus sign to a plus sign.

2. Change the subtrahend to its opposite.

Performing Subtraction

- Subtraction Rule for Fractions

    For any real numbers a, b, and c, where $c \neq 0$, $\dfrac{a}{c} - \dfrac{b}{c} = \dfrac{a-b}{c}$.

    In words, subtract the numerators and retain the common denominator.

- Property of the Opposite of a Difference

    For any real numbers a and b, $-(a - b) = b - a$.

Speaking the Language 1.6

1. We can write any difference as a(n) ▨▨▨▨.
2. The expressions $x - c$ and $c - x$ are ▨▨▨▨.
3. To subtract fractions that have a common denominator, we subtract only the ▨▨▨▨.
4. The words *less than* refer to a(n) ▨▨▨▨, whereas the words *is less than* refer to a(n) ▨▨▨▨.

Exercises 1.6

Concepts and Skills

1. What two changes must be made to the expression $-5 - 8$ to convert it into a sum with the same value?

2. In the expression $2 + 8$, we call both the 2 and the 8 addends. However, in the expression $2 - 8$, we give the 2 and the 8 different names. What are the names? Why is it that 2 and 8 can have the same names in $2 + 8$ but must have different names in the expression $2 - 8$?

In Exercises 3–26, determine the difference.

3. $4 - 15$
4. $3 - 17$
5. $-7 - 3$
6. $-6 - 5$
7. $9 - (-4)$
8. $8 - (-7)$
9. $-2 - (-8)$
10. $-4 - (-12)$
11. $-5 - 8$
12. $-12 - 9$
13. $0 - (-52)$
14. $0 - 25$
15. $13 - (-5)$
16. $15 - (-8)$
17. $-12 - 12$
18. $12 - (-12)$
19. $-6 - (-3)$
20. $-10 - (-5)$
21. $24 - (-24)$
22. $5 - (-24)$
23. $-4 - 16$
24. $-17 - 9$
25. $-8 - (-5)$
26. $-17 - 24$

27. Interpret $-(a - b) = b - a$ in your own words.

28. Suppose you enter X − Y in your calculator and the displayed result is −4. What will you obtain if you enter Y − X? Why?

In Exercises 29–34, use the Property of the Opposite of a Difference to rewrite each expression without grouping symbols. Evaluate all expressions that do not contain a variable.

29. $-(-5 - 7)$
30. $-(4 - 8)$
31. $-[4 - (-6)]$
32. $-[-8 - (-1)]$
33. $-(a - 9)$
34. $-(3 - c)$

In Exercises 35–44, perform the indicated operations.

35. $15 - 20 - 5$
36. $-3 - 5 - 7$
37. $10 - (6 - 9)$
38. $-5 - (-2 + 12)$
39. $-8 - 10 - (-3) + 7$
40. $18 - (-2) - 25 + 8$
41. $-3 + (-2) - (-1 + 10)$

42. $6 + 2 - (5 - 12)$

43. $16 - [(-3) + 10 - (-9)]$

44. $-7 + [3 - (-11) - 13]$

In Exercises 45–48, translate the given description into a numerical expression and evaluate it.

45. What is the difference between -17 and -18?

46. What is -3 decreased by 4?

47. What is 9 less than -1?

48. Determine the difference between $3\frac{2}{9}$ and $1\frac{2}{3}$.

In Exercises 49–54, perform the indicated subtraction.

49. $-\dfrac{3}{5} - \left(-\dfrac{1}{2}\right)$

50. $-\dfrac{3}{4} - \left(-\dfrac{2}{3}\right)$

51. $-5 - \left(-\dfrac{1}{3}\right)$

52. $-8 - \left(-\dfrac{3}{4}\right)$

53. $\dfrac{7}{16} - \left(-\dfrac{5}{8}\right)$

54. $\dfrac{5}{7} - \dfrac{9}{14}$

In Exercises 55–58, use your calculator to perform the indicated operations.

55. $-22 - 14 - (-13) - 4 + (-7)$

56. $-35 - (-14) + (-23) - (-13) + 5 - (-8)$

57. $13.84 - (-26.19) - (-27.5) - 25 - (-31.7)$

58. $-17.2 - (-82) - 45.23 + (-23.75) - (-54.1)$

In Exercises 59–62, determine the unknown number.

59. What number must be subtracted from 3 to obtain -2?

60. The difference between what number and -6 is 4?

61. If decreasing a number by 7 results in -1, what is the number?

62. If 9 less than a number is -2, what is the number?

In Exercises 63–82, perform the indicated operations.

63. $-7 - (-2)$ **64.** $-12 - (-12)$

65. $7 - (-36)$ **66.** $21 - (-35)$

67. $|-8| - |-3|$ **68.** $|-10| - (-6)$

69. $-\dfrac{2}{3} - \dfrac{4}{5}$ **70.** $-\dfrac{3}{4} - 4$

71. $-6 - 10 + (-5)$ **72.** $12 - 8 - (-2)$

73. $-16 - 16$ **74.** $-22 - 37$

75. $0 - (-32)$

76. $-72 - (-72)$

77. $23.89 - (-35.87)$

78. $-974.3 - (-678.39)$

79. $-24 - 56$

80. $35 - 47$

81. $9 - 27 + 2 - (-1)$

82. $-12 + 3 - (-8) - (-5)$

In Exercises 83–86, state the property or definition that justifies each step.

83. $-4 - (-3) + 7$

$= -4 + 3 + 7$ (a) ▨▨▨

$= -4 + (3 + 7)$ (b) ▨▨▨

$= -4 + 10$

$= 6$

84. $5 - 4 + 3 - 2$

$= 5 + (-4) + 3 + (-2)$ (a) ▨▨▨

$= -4 + (-2) + 5 + 3$ (b) ▨▨▨

$= [-4 + (-2)] + (5 + 3)$ (c) ▨▨▨

$= -6 + 8$

$= 8 + (-6)$ (d) ▨▨▨

$= 8 - 6$ (e) ▨▨▨

$= 2$

85. $-(a - b) + a$

$= (b - a) + a$ (a) ▨▨▨

$= [b + (-a)] + a$ (b) ▨▨▨

$= b + (-a + a)$ (c) ▨▨▨

$= b + 0$ (d) ▨▨▨

$= b$ (e) ▨▨▨

86. $-(-x) - (x - y)$

$= -(-x) + [-(x - y)]$ (a) ▨▨▨

$= -(-x) + (y - x)$ (b) ▨▨▨

$= x + (y - x)$ (c) ▨▨▨

$= x + [y + (-x)]$ (d) ▨▨▨

$= x + (-x + y)$ (e) ▨▨▨

$= [x + (-x)] + y$ (f) ▨▨▨

$= 0 + y$ (g) ▨▨▨

$= y$ (h) ▨▨▨

In Exercises 87–94, perform the indicated operations.

87. $7 - 4(3)$

88. $-6 - 2(5) - 1$

89. $-3 - 3^2 - 3(4)$

90. $4 - 5 \cdot 2^2 - (-3)$

91. $-2 - \sqrt{3^2 - (-16)}$

92. $|4 - 7| - \sqrt{5^2 - 4^2}$

93. $\dfrac{23 - 2^3}{0 - (-15)}$

94. $\dfrac{9 - 7 + 3^2}{8 - (-3)}$

🌐 Real-Life Applications

95. The lowest point on earth is the Dead Sea in Asia at 1290 feet below sea level. Asia also has the highest point on earth, Mount Everest, at 29,108 feet above sea level. What is the difference between the two altitudes?

96. Mauna Loa on the island of Hawaii has an elevation of 4169 feet. This peak is 32,024 feet above the ocean floor near Hawaii. How deep is the ocean floor?

97. By the end of June, the rainfall in Biloxi was 2.3 inches above the average for the year. In July and September, the rainfall was below the monthly average by 3.1 and 1.7 inches, respectively, but the August rainfall was 0.6 inches above average. By the end of September, how far above or below the yearly average was the total rainfall?

98. Four weeks prior to an election, a candidate led her opponent by 6 percentage points. During the next two weeks, her lead increased by 1 and 2 points, respectively, but in the last two weeks, she dropped 5 and then 7 points in the polls. What was her status on election day?

Data Analysis: *Elevations in the United States*

The mean elevation of the continental United States is 2500 feet. The following table shows the highest points in six selected states.

State	Highest Point	Altitude Difference (in feet)
Colorado	Mount Elbert	14,431
Florida	Iron Mountain	325
Illinois	Charles Mound	1,241
Rhode Island	Durfee Hill	805
Tennessee	Clingmans Dome	6,642
Washington	Mount Rainier	14,408

(Source: U.S. Geological Survey.)

99. For each state in the table, record the difference between the altitude of the highest point and the mean elevation of the continental United States. Use positive numbers for altitudes above the mean and negative numbers for altitudes below the mean.

100. Calculate the mean of the differences for the six entries in the table. Why is the mean difference so large?

101. How many miles high is Mount Rainier?

102. One foot is approximately 0.3 meter. Express the elevation of Clingmans Dome in meters.

Challenge

In Exercises 103 and 104, let a represent a negative number and b represent a positive number. Determine the sign of the expression.

103. (a) $a - b$ (b) $b - a$

104. (a) $a - |a|$ (b) $|a| - a$

In Exercises 105–108, insert grouping symbols so that the value of the expression is the given number.

105. $3 - 7 - 2$; -2

106. $2 - 5 - 7 + 9$; -19

107. $2 \cdot 3 - 2 - 5$; -3

108. $4 - 2 \cdot 5 - 2$; 8

1.7 | Multiplication

Sign Rules • Multiplying Fractions • Evaluating Numerical Expressions •
Real-Life Applications

Sign Rules

From arithmetic, we know that the product of any number and 0 is 0. We call this fact
the Multiplication Property of 0.

> **Multiplication Property of 0**
>
> For any real number a, $a \cdot 0 = 0 \cdot a = 0$.

We now consider the sign rules for multiplying two nonzero numbers when at least
one factor is negative.

Exploring the Concept

Multiplying with Negative Numbers

In the following table, we see two patterns of products as the first factor decreases
and the second factor remains constant.

Column A	Column B
$3 \cdot 4 = \quad 12$	$3 \cdot (-4) = -12$
$2 \cdot 4 = \quad 8$	$2 \cdot (-4) = \quad -8$
$1 \cdot 4 = \quad 4$	$1 \cdot (-4) = \quad -4$
$0 \cdot 4 = \quad 0$	$0 \cdot (-4) = \quad 0$
$-1 \cdot 4 = -4$	$-1 \cdot (-4) = \quad 4$
$-2 \cdot 4 = -8$	$-2 \cdot (-4) = \quad 8$

Think About It

In arithmetic, multiplication can be
regarded as repeated addition. For
example, $3 \cdot 5$ means $5 + 5 + 5$ and
$3(-5)$ means $(-5) + (-5) + (-5)$.
Why is it not possible to regard
a product such as $(-3)(-5)$
in this way?

The last two results in column A are based on the observation that the products are
decreasing by 4. This pattern suggests that when the factors have unlike signs, the
product is negative. Also, if we multiply the absolute values of the factors, we
obtain the absolute value of the product.

These rules are carried over to the start of column B. The last two results in column
B follow from the fact that the products are increasing by 4. This pattern suggests
that when both factors are negative, the product is positive. The absolute value of
the product is obtained by multiplying the absolute values of the factors.

The multiplication rules suggested by these patterns can be proved to be true.

> **Multiplication Rules for Nonzero Numbers**
>
> If a and b are any nonzero real numbers, then the sign of the product ab is
>
> 1. positive if a and b have like signs.
> 2. negative if a and b have unlike signs.
>
> The absolute value of the product ab is equal to the product of the absolute values
> of a and b.

Using the Sign Rules to Perform Multiplication

Perform the indicated operation. Use your calculator in parts (d) and (e).

(a) $3(-6)$ (b) $(-8)(-4)$ (c) $(-67)(0)$

(d) $-56 \cdot 14$ (e) $-3.78(-6.3)$

Figure 1.15

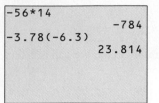

Solution

(a) $3(-6) = -18$ Unlike signs

(b) $(-8)(-4) = 32$ Like signs

(c) $(-67)(0) = 0$ Multiplication Property of 0

(d) and (e) See Fig. 1.15.

LEARNING TIP

Compare the use of parentheses in a product and in a sum. In the expression $4 + (-3)$, the plus symbol indicates addition. However, the expression $4(-3)$ indicates multiplication because no operation symbol appears between the 4 and the -3.

Note: Because we can easily determine the sign of a product in advance, it is not necessary to enter signs in your calculator when you use it to multiply. For example, in part (e) of Example 1, we know that the product will be positive, so we can simply enter 3.78 (6.3).

The sign rules for multiplication lead to the following generalization about signs in products.

Property of Signs in Products

For any real numbers a and b,

1. $a(-b) = (-a)b = -ab$ 2. $(-a)(-b) = ab$

We can expand the sign rules for multiplication to include products with more than two factors. If a product has an even number of negative factors, then the product is positive; if a product has an odd number of negative factors, then the product is negative. This fact is useful for determining the sign of a product in advance. For example, we know that $(-2)^{11}$ means 11 factors of -2. Because there is an odd number of negative factors, the result is negative. Similarly, $(-3)^6$ is positive.

EXAMPLE 2

Determining the Sign of a Product with More than Two Factors

Determine the sign of each product. Then determine the product.

(a) $-2(-1)(-3)(-2)(-1)$ (b) $(-1)(-5)(-2)(-3)$

(c) $(-2)^7$ (d) $(-5)^4$

Solution

(a) $-2(-1)(-3)(-2)(-1) = -12$ Odd number of negative factors

(b) $(-1)(-5)(-2)(-3) = 30$ Even number of negative factors

(c) $(-2)^7 = -128$ Odd number of negative factors

(d) $(-5)^4 = 625$ Even number of negative factors

Consider the following products in which one factor is -1.

$$-1(3) = -3 \qquad -1(-5) = 5 \qquad 4(-1) = -4 \qquad -17(-1) = 17$$

In each case, the result of multiplying a number by -1 is the opposite of the number. This fact is generalized in the following property.

> **Multiplication Property of -1**
>
> For any real number a, $a(-1) = -1a = -a$.

Note: The Multiplication Property of -1 entitles us to write the symbol $-x$ as $-1x$ if we want to. As we will see, this property will be useful in simplifying expressions.

Multiplying Fractions

Recall from arithmetic that we multiply fractions by multiplying the numerators and denominators.

> **Multiplication Rule for Fractions**
>
> For any real numbers a, b, c, and d, where $b \neq 0$ and $d \neq 0$,
>
> $$\frac{a}{b} \cdot \frac{c}{d} = \frac{ac}{bd}$$

Note: Common factors in any numerator and denominator can be divided out before performing the multiplication.

EXAMPLE   **3** | **Multiplying Fractions**

Determine the following products.

(a) $-\dfrac{2}{3} \cdot \dfrac{5}{7} = -\dfrac{10}{21}$

(b) $-\dfrac{4}{7}\left(-\dfrac{3}{5}\right) = \dfrac{12}{35}$

(c) $\dfrac{5}{8}\left(-\dfrac{16}{25}\right) = \dfrac{5}{8}\left(-\dfrac{8 \cdot 2}{5 \cdot 5}\right) = -\dfrac{2}{5}$ Divide out common factors before multiplying.

Evaluating Numerical Expressions

The sign rules for multiplication are used for evaluating numerical expressions that involve products.

EXAMPLE 4

Evaluating Products

Determine the following products.

(a) $-274(-19) = 5206$

(b) $(-5)(-4)(3)(2)(-1) = -120$ Odd number of negative factors

(c) $(-1)(-2)(-3)(-4)(-5)(-84)(-93)(0) = 0$ Multiplication Property of 0

(d) $-\dfrac{1}{2}\left(\dfrac{1}{3}\right)\left(\dfrac{2}{5}\right) = -\dfrac{1}{15}$ Divide out common factors before multiplying.

(e) $\left(-\dfrac{3}{2}\right)\left(-\dfrac{4}{3}\right)\left(-\dfrac{5}{4}\right)\left(-\dfrac{6}{5}\right) = 3$ Even number of negative factors

(f) $(-0.08)(126)(-0.47) = 4.7376$

The next example uses the Order of Operations for evaluating numerical expressions.

EXAMPLE 5

Evaluating Numerical Expressions

Evaluate the following expressions.

(a) $5 + (-3)(-4) = 5 + 12 = 17$ Multiply first, then add.

(b) $\sqrt{-2(3) + 15} = \sqrt{-6 + 15} = \sqrt{9} = 3$ Begin inside the grouping symbol.

(c) $\dfrac{-4 \cdot 2 + 3^2}{-5 - (-6)} = \dfrac{-4 \cdot 2 + 9}{-5 - (-6)}$ The fraction bar is a grouping symbol.

$ = \dfrac{-8 + 9}{-5 + 6}$ In the numerator, work with first the exponent, then the product, then the sum.

$ = \dfrac{1}{1} = 1$

(d) $\dfrac{-1 + \sqrt{(-1)^2 - 4(1)(-2)}}{2 \cdot 1}$

$ = \dfrac{-1 + \sqrt{1 - 4(1)(-2)}}{2 \cdot 1}$ Begin with the exponent under the square root symbol.

$ = \dfrac{-1 + \sqrt{1 - (-8)}}{2}$ Evaluate the product next and then the difference.

$ = \dfrac{-1 + \sqrt{9}}{2}$ Evaluate the numerator.

$ = \dfrac{-1 + 3}{2} = \dfrac{2}{2} = 1$

EXAMPLE 6

Translating into Products

In each part, write the described product and perform the multiplication.

(a) the product of 4 and -3

(b) The factors are -8, -5, and 2.

(c) two-thirds of -18

(d) -2 raised to the third power

Solution

(a) $4(-3) = -12$

(b) $-8(-5)(2) = 80$

(c) $\dfrac{2}{3} \cdot (-18) = -12$ Note that "of" means "times."

(d) $(-2)^3 = -8$

Real-Life Applications

Multiplication of negative numbers may be needed to describe changes in, for example, the Dow-Jones average, interest rates, or weather reports.

EXAMPLE **7**

Changes in the Temperature

When a cold front moved across Chicago, the temperature dropped 6 degrees per hour for 4 consecutive hours. If the temperature was 20° before the cold front moved through, what was the temperature at the end of the 4-hour period?

Solution

The temperature change each hour was $-6°$. Therefore, the total change for the 4-hour period was $4(-6)$ or $-24°$. The temperature after the cold front moved through was $20 + (-24) = -4°$.

Quick Reference 1.7

Sign Rules   • Multiplication Property of 0

For any real number a, $a \cdot 0 = 0 \cdot a = 0$.

• The following are the rules for multiplying two nonzero real numbers a and b.

1. The sign of the product ab is

(a) positive if a and b have like signs.

(b) negative if a and b have unlike signs.

2. The absolute value of the product ab is equal to the product of the absolute values of a and b.

• Property of Signs in Products

For any real numbers a and b,

1. $a(-b) = (-a)b = -ab$ 2. $(-a)(-b) = ab$

• If a product has an even number of negative factors, then the product is positive; if a product has an odd number of negative factors, then the product is negative.

• Multiplication Property of -1

For any real number a, $a(-1) = -1a = -a$.

Multiplying Fractions • Multiplication Rule for Fractions

For any real numbers a, b, c, and d, where $b \neq 0$ and $d \neq 0$, $\dfrac{a}{b} \cdot \dfrac{c}{d} = \dfrac{ac}{bd}$.

Evaluating Numerical Expressions • Numerical expressions are evaluated according to the Order of Operations along with the sign rules and the properties of real numbers.

Speaking the Language 1.7

1. A product is always 0 whenever one of the ▨▨▨▨ is 0.
2. A product of two nonzero numbers is positive if the numbers have ▨▨▨▨ signs, whereas the product is negative if the numbers have ▨▨▨▨ signs.
3. Multiplying a number by -1 changes the number to its ▨▨▨▨ .
4. A product is negative if it has an ▨▨▨▨ number of negative factors.

Exercises 1.7

Concepts and Skills

1. Explain how to determine the product of two numbers with like signs and two numbers with unlike signs.
2. Suppose a product consists of 17 factors. Eight factors are positive, eight factors are negative, and one factor is zero. Evaluate the product and state the property that justifies your answer.

In Exercises 3–18, determine the product.

3. $7(-4)$
4. $8(-5)$
5. $-5(2)$
6. $-7(6)$
7. $(-3)(-4)$
8. $(-6)(-10)$
9. $-6(-2)$
10. $-8(-5)$
11. $-10 \cdot 10$
12. $-9 \cdot 9$
13. $11(-4)$
14. $8(-7)$
15. $-7 \cdot 1$
16. $-1 \cdot 0$
17. $(0)(-45)$
18. $(-1)(-92)$

In Exercises 19–22, predict the sign of the product. Then use your calculator to multiply.

19. $(-2.3)(-2.7)$
20. $(-4.5)(-3.4)$
21. $2.1(-2.3)$
22. $0.5(-4.3)$

In Exercises 23–26, fill in the blank to make the equation true.

23. ▨▨▨ $\cdot 7 = -28$
24. $-10 \cdot$ ▨▨▨ $= 20$
25. $-8 \cdot$ ▨▨▨ $= 8$
26. $-4 \cdot$ ▨▨▨ $= 0$

27. Suppose a product consists of nine nonzero factors and six of the factors are negative. Explain how to determine the sign of the product.

28. Which of the following products is easier to evaluate without a calculator? Why?

(i) $(-1)(-1)(-1)(-1)(-1)(-1)(-1)(-1)$

(ii) $(1.37)(6.42)$

In Exercises 29–34, determine the product. Use your calculator to verify your results.

29. $(-2)(-3)(-4)$
30. $-5(7)(-2)$
31. $(-2)(-1)(3)(2)$
32. $6(-1)(4)(-2)$
33. $(-1)(-2)(3)(1)(-6)$
34. $(-1)(-1)(2)(-3)(-5)$

In Exercises 35–40, perform the indicated operations.

35. $(-3)^5$

36. $(-5)^3$

37. $(-2)^8$

38. $(-3)^4$

39. $(-1)^{33}$

40. $(-1)^{28}$

41. What property guarantees that the following is true: $-1(a + b) = -(a + b)$?

42. In your own words, explain how to multiply two or more fractions.

In Exercises 43–48, multiply the fractions. Use your calculator to verify your results.

43. $-\dfrac{1}{3} \cdot \dfrac{3}{5}$

44. $-\dfrac{4}{5} \cdot \left(-\dfrac{1}{2}\right)$

45. $-5 \cdot \left(-\dfrac{2}{3}\right)$

46. $-\dfrac{3}{4} \cdot 8$

47. $-\dfrac{34}{5} \cdot \left(-\dfrac{25}{4}\right)$

48. $-\dfrac{21}{4} \cdot \left(-\dfrac{16}{3}\right)$

In Exercises 49–62, perform the indicated operations.

49. $(-5.76)(-4.95)$

50. $-0.34 \cdot 25.3$

51. $-5|-4|$

52. $|-9| \cdot |-10|$

53. $5(-2)(4)$

54. $(-2)(-1)(-2)(-3)$

55. $-\dfrac{3}{7} \cdot \left(-\dfrac{4}{9}\right)$

56. $\dfrac{3}{5} \cdot \dfrac{5}{7} \cdot \left(-\dfrac{7}{9}\right)$

57. $|9| \cdot |-8|$

58. $|-7| \cdot (-6)$

59. $-4(-24)$

60. $-14 \cdot 10$

61. $(-7)(-2)(0)(2)(-3)$

62. $(-1)(-1)(-4)(-1)(-3)$

In Exercises 63–70, perform the indicated operations.

63. $-9 - (-3) \cdot 2$

64. $-6(1 - 4) - 5^2$

65. $(-4)^2 - 5(-4)$

66. $(-2)^3 - (-2)^2$

67. $\dfrac{-3 + \sqrt{(-3)^2 - 4(2)(-2)}}{2}$

68. $\dfrac{-4(3) - (-3)(-2)(-4)}{3 - (-1)}$

69. $-3(4 - 6)^3 \div (2^2 \cdot 3)$

70. $15 - 3(7 + 3) - (-2)^3$

In Exercises 71–74, write a numerical expression that models the problem and then evaluate it.

71. What is the product of 29 and 0.63?

72. If the two factors are -1.8 and 1.8, what is the product?

73. What is $-\dfrac{3}{4}$ of 32?

74. What is twice the product of 5 and -3?

In Exercises 75–78, state the property that justifies the step.

75. $-6 \cdot 2 \cdot (-1) \cdot 3$

$= 2 \cdot 3 \cdot (-6) \cdot (-1)$ (a) ▬▬▬

$= (2 \cdot 3) \cdot [(-6) \cdot (-1)]$ (b) ▬▬▬

$= 6 \cdot 6$

$= 36$

76. $12 \cdot \dfrac{1}{5} \cdot (-10) \cdot \left(-\dfrac{1}{3}\right)$

$= 12 \cdot \left(-\dfrac{1}{3}\right) \cdot (-10) \cdot \dfrac{1}{5}$ (a) ▬▬▬

$= \left[12 \cdot \left(-\dfrac{1}{3}\right)\right] \cdot \left[-10 \cdot \dfrac{1}{5}\right]$ (b) ▬▬▬

$= -4 \cdot (-2)$

$= 8$

77. $-\dfrac{1}{6}(6x) = \left(-\dfrac{1}{6} \cdot 6\right)x$ (a) ▬▬▬

$= -1x$

$= -x$ (b) ▬▬▬

78. $6\left[\dfrac{1}{2}(x + 1)\right] = \left(6 \cdot \dfrac{1}{2}\right)(x + 1)$ (a) ▬▬▬

$= 3(x + 1)$

$= 3x + 3$ (b) ▬▬▬

In Exercises 79–86, find two numbers whose sum is the first number and whose product is the second number.

79. $5, 6$

80. $-6, 8$

81. $1, -12$

82. $2, -15$

83. $-8, 12$

84. $-10, 24$

85. $-7, -8$

86. $-4, -21$

In Exercises 87–90, determine the unknown number.

87. The product of what number and −8 is 24?

88. Five times what number is −20?

89. What number multiplied by −3 is −18?

90. If the product of three factors is 0 and two of the factors are 120 and −813, what is the third factor?

Real-Life Applications

91. Newlyweds obtained a loan of $1075 for the purchase of apartment furnishings. After five monthly payments of $225.39, the loan was paid off. How much interest was paid?

92. A homeowner in Milwaukee bought a 40-pound bag of rock salt to melt ice on his driveway. If he used 20 ounces of salt each day for 3 weeks, how many pounds of salt were left? (There are 16 ounces in 1 pound.)

93. Instead of recording the actual number of strokes, a golfer used the following method to record her scores on each hole.

eagle: −2 (2 below par)

birdie: −1 (1 below par)

par: 0 (standard score)

bogey: +1 (1 above par)

double bogey: +2 (2 above par)

At the end of the round, the golfer found that her scores on the 18 holes were as follows.

eagle: 1

birdie: 3

par: 9

bogey: 2

double bogey: 3

If par for the course was 72, what was the golfer's score?

94. On a certain 50-question standardized test, students receive 1 point for correct answers. A deduction of 1 point is made for incorrect answers, but there is no deduction for answers left blank. Suppose that a student answers 30 questions correctly and leaves 6 answers blank. What is the test score?

Modeling with Real Data

95. The accompanying table shows the population (in millions) of selected countries, along with the annual per-person consumption of 8-ounce servings of Coca-Cola products in those countries.

Country	Population (millions)	Annual Per-Person Consumption (8-ounce servings)
Brazil	162	122
China	1221	4
Germany	82	201
Mexico	94	322
Russia	147	6
United States	263	343
Zimbabwe	11	60

(Source: Coca-Cola Company.)

(a) Add a column to the table with the heading *Total Annual Consumption (billions of gallons)*, and fill in this column. (There are 128 ounces in a gallon.)

(b) On the average, in which country does every person consume nearly one serving of a Coca-Cola product per day?

96. In a recent year, 12.7 billion cases of Coca-Cola products were sold, and 95% of the sales were in continents other than Africa. (Source: Coca-Cola Company.)

How many cases of Coca-Cola products were sold in Africa?

Challenge

In Exercises 97–100, let *a* and *b* represent negative numbers and let *c* represent a positive number. What is the sign of the given expression?

97. $a \cdot |b| \cdot c$

98. abc

99. $a \cdot a$

100. $a \cdot b \cdot b$

In Exercises 101 and 102, insert grouping symbols to make the equation true.

101. $-5 \cdot 6 - 8 = 10$

102. $-2^4 - 9 = 7$

1.8   Division

Definition and Rules   •   Dividing with Fractions   •   Real-Life Applications

Definition and Rules

We have seen that an indicated division can be written in several ways:

$$19 \div 7 = 19/7 = \frac{19}{7}$$

Recall that the number 19 is called the **dividend** and the number 7 is called the **divisor.** The expression and the result of the division are both called the **quotient.**

The following is one way that we can define division in terms of multiplication.

> **Definition of Division**
>
> For any real numbers a, b, and c, with $b \neq 0$,
>
> $$a \div b = c \quad \text{if} \quad c \cdot b = a.$$

Because division is defined in terms of multiplication, the rule for dividing signed numbers follows directly from the rule for multiplying signed numbers.

> **Performing Division with Real Numbers**
>
> For any real numbers a and b, $b \neq 0$, the quotient $\dfrac{a}{b}$ is
>
> 1. positive if a and b have like signs.
> 2. negative if a and b have unlike signs.
>
> The absolute value of the quotient $\dfrac{a}{b}$ equals the quotient of the absolute values of a and b.

In our definition of division, we require the divisor to be a *nonzero* number. We can see why this is so if, for example, we try to evaluate $5 \div 0$. The definition states that there is a number c such that $c \cdot 0 = 5$, but this is impossible. If we try to evaluate $0 \div 0$, then there is a number c such that $c \cdot 0 = 0$. But this is true for any number c. (We say that $0 \div 0$ is *indeterminate*.)

In short, dividing by 0 is not defined. If you attempt to divide by 0 with your calculator, you will receive an error message.

EXAMPLE **1**   **Performing Division**

Evaluate each of the following quotients, if possible. Use your calculator for part (f).

(a) $\dfrac{-20}{-4} = 5$ $5(-4) = -20$

(b) $\dfrac{36}{-9} = -4$ $-4(-9) = 36$

(c) $\dfrac{-1}{0}$ is undefined   Division by 0 is undefined.

(d) $\dfrac{-18}{3} = -6$ $-6(3) = -18$

(e) $\dfrac{0}{-12} = 0$ $0(-12) = 0$

(f) $\dfrac{-16.2}{-2.7} = 6$ $6(-2.7) = -16.2$

Because the sign rules for division are the same as the sign rules for multiplication, the following generalization can be stated for signs in quotients.

Property of Signs in Quotients

For real numbers a and b $(b \neq 0)$, each of the following statements is true.

$$-\frac{a}{b} = \frac{-a}{b} = \frac{a}{-b} \qquad\qquad \frac{-a}{-b} = \frac{a}{b} \qquad\qquad -\frac{-a}{-b} = \frac{-a}{b} = \frac{a}{-b} = -\frac{a}{b}$$

In words, if a fraction has a single opposite sign, the sign can be placed in the numerator, in the denominator, or in front of the fraction. If a fraction has two opposite signs, they can both be removed. If a fraction has three opposite signs, any two of them can be removed.

EXAMPLE 2

Opposite Signs in Quotients

In each part write the given quotient in two other ways.

(a) $-\dfrac{5}{8}$ (b) $\dfrac{t}{-3}$

Solution

(a) $-\dfrac{5}{8} = \dfrac{-5}{8} = \dfrac{5}{-8}$

(b) $\dfrac{t}{-3} = \dfrac{-t}{3} = -\dfrac{t}{3}$

EXAMPLE 3

Opposite Signs in Quotients

Rewrite each quotient with fewer opposite signs.

(a) $\dfrac{-(x+3)}{-2}$ (b) $-\dfrac{-x}{-y}$

Solution

(a) $\dfrac{-(x+3)}{-2} = \dfrac{x+3}{2}$

(b) $-\dfrac{-x}{-y} = -\dfrac{x}{y}$ or $\dfrac{-x}{y}$ or $\dfrac{x}{-y}$

Dividing with Fractions

In arithmetic, the rule for dividing by a fraction is to "invert and multiply." This is an informal way of stating the following rule.

Think About It

Survey friends and family with the following question:"What is 10 divided by one-half?" Why do so many people say 5? Show how you would explain their error.

> **Division Rule for Fractions**
>
> For any real numbers a, b, c, and d, where b, c, and d are nonzero numbers,
>
> $$\frac{a}{b} \div \frac{c}{d} = \frac{a}{b} \cdot \frac{d}{c}$$

This rule is useful when either the dividend or the divisor is a fraction. In such cases we divide by multiplying the dividend by the reciprocal of the divisor.

Note: In the following example, we approximate some results with a rounded-off decimal. The symbol $\approx$ means *is approximately equal to*.

EXAMPLE **4**

Dividing with Fractions

Determine the value of each quotient.

(a) $-\dfrac{1}{3} \div 5$ (b) $\dfrac{5}{7} \div \left(-\dfrac{3}{4}\right)$ (c) $-6 \div \left(-\dfrac{2}{3}\right)$

LEARNING TIP

When operations involve fractions, writing all numbers as fractions may be helpful. Recall that an integer can be written with a denominator of 1.

Solution

(a) $-\dfrac{1}{3} \div 5 = -\dfrac{1}{3} \cdot \dfrac{1}{5}$ Multiply by the reciprocal of 5.

$\qquad\quad = -\dfrac{1}{15}$

$\qquad\quad \approx -0.067$ Rounded to the nearest thousandth

(b) $\dfrac{5}{7} \div \left(-\dfrac{3}{4}\right) = \dfrac{5}{7} \cdot \left(-\dfrac{4}{3}\right)$ Multiply by the reciprocal of $-\dfrac{3}{4}$.

$\qquad\qquad\quad = -\dfrac{20}{21}$

$\qquad\qquad\quad \approx -0.95$ Rounded to the nearest hundredth

(c) $-6 \div \left(-\dfrac{2}{3}\right) = \dfrac{-6}{1} \cdot \left(-\dfrac{3}{2}\right) = 9$

EXAMPLE **5**

Evaluating Numerical Expressions

Evaluate the following expressions.

(a) $\dfrac{1 - 3^2}{4} = \dfrac{1 - 9}{4} = \dfrac{-8}{4} = -2$

(b) $(-8)^2 \div \dfrac{16}{-4} = 64 \div \dfrac{16}{-4} = 64 \div (-4) = -16$

(c) $\dfrac{2(-3) + \sqrt{9}}{(6 - 8)^3 - 1} = \dfrac{2(-3) + 3}{(-2)^3 - 1} = \dfrac{-6 + 3}{-8 - 1} = \dfrac{-3}{-9} = \dfrac{1}{3} \approx 0.33$

EXAMPLE 6

Translating into Quotients

In each part write the described quotient and perform the division.

(a) -21 divided by 7

(b) The dividend is 10 and the divisor is -2.

(c) the quotient of -18 and -3

Solution

(a) $\dfrac{-21}{7} = -3$

(b) $\dfrac{10}{-2} = -5$

(c) $\dfrac{-18}{-3} = 6$ The first given number is the dividend, and the second given number is the divisor.

Real-Life Applications

If a quantity has decreased over a period of time, calculating the average decrease may require dividing a negative number by a positive number.

EXAMPLE 7

Average Change in Temperature

The temperature dropped 32°F in a 4-hour period. What was the average hourly change in temperature?

Solution

The average change was $\dfrac{-32}{4}$ or -8°F per hour.

Quick Reference 1.8

Definition and Rules

- In the expression $a \div b$, a is called the **dividend,** and b is called the **divisor.** Both the expression and the value of the expression are called the **quotient.**

- For real numbers a, b, and c, with $b \neq 0$, $a \div b = c$ if $c \cdot b = a$.

- The sign rules for division are the same as the sign rules for multiplication. The absolute value of a quotient is the quotient of the absolute values of the dividend and divisor.

- Dividing 0 by any nonzero number is permitted and the result is always 0. However, division by 0 is undefined.

- Property of Signs in Quotients

  For real numbers a and b ($b \neq 0$), the following statements are true.

$$-\frac{a}{b} = \frac{-a}{b} = \frac{a}{-b} \qquad \frac{-a}{-b} = \frac{a}{b} \qquad -\frac{-a}{-b} = \frac{-a}{b} = \frac{a}{-b} = -\frac{a}{b}$$

Dividing with Fractions

• Division Rule for Fractions

For any real numbers a, b, c, and d, where b, c, and d are nonzero numbers,

$$\frac{a}{b} \div \frac{c}{d} = \frac{a}{b} \cdot \frac{d}{c}$$

• When either the dividend or the divisor is a fraction, divide by multiplying the dividend by the reciprocal of the divisor.

Speaking the Language 1.8

1. In the quotient $x \div y$, x is called the _____ and y is called the _____.
2. The sign rules for division are the same as the sign rules for _____.
3. To perform $\dfrac{a}{b} \div \dfrac{x}{y}$, we multiply $\dfrac{a}{b}$ by the _____ of $\dfrac{x}{y}$.
4. If $a \neq 0$, then $0 \div a$ is _____, whereas $a \div 0$ is _____.

Exercises 1.8

Concepts and Skills

1. For the statement $\frac{-12}{2} = -6$, what do we call -12, 2, and -6?

2. Explain how to determine the quotient of two numbers with
   (a) like signs.   (b) unlike signs.

In Exercises 3–20, divide, if possible.

3. $8 \div (-4)$

4. $-14 \div 2$

5. $\dfrac{15}{-5}$

6. $\dfrac{-24}{6}$

7. $-40 \div (-5)$

8. $-60 \div (-10)$

9. $-16 \div (-2)$

10. $-92 \div (-92)$

11. $\dfrac{-80}{-5}$

12. $\dfrac{-8}{8}$

13. $-21 \div 7$

14. $55 \div (-55)$

15. $\dfrac{-20}{5}$

16. $\dfrac{56}{-7}$

17. $-4 \div 0$

18. $\dfrac{-2}{0}$

19. $\dfrac{0}{-1}$

20. $0 \div (-35)$

In Exercises 21–28, perform the indicated operations.

21. $-6 \div 3 \div (-4)$

22. $-30 \div (-5) \div (-2)$

23. $24 \div [-6 \div (-2)]$

24. $-12 \div [-6 \div (-3)]$

25. $-16 \cdot 2 \div 8$

26. $20 \div 5 \cdot 3$

27. $18 \div (-6 \cdot 3)$

28. $4 \cdot [-8 \div (-16)]$

In Exercises 29–32, predict the sign of the quotient. Then use your calculator to perform the operation.

29. $-5.9 \div (-2.7)$

30. $42.4 \div (-13.5)$

31. $-4.7 \div 9.3$

32. $-25.7 \div (-21.2)$

33. Show two other ways to write $\dfrac{x}{-y}$ and state the property that justifies your answer.

34. Describe how a fraction with more than one opposite sign can be written with fewer opposite signs.

In Exercises 35–40, rewrite the given fraction with fewer opposite signs. Express the result with no opposite sign in the denominator.

35. $\dfrac{-5}{-8}$

36. $-\dfrac{-6}{-11}$

37. $-\dfrac{3}{-5}$

38. $-\dfrac{-4}{7}$

39. $-\dfrac{-y}{-5}$

40. $\dfrac{-c}{-(c+1)}$

41. In your own words, how do we divide when the divisor is a fraction?

42. In terms of signs, how does the opposite of a nonzero real number x differ from the reciprocal of x?

In Exercises 43–48, determine

(a) the opposite of the given number

(b) the reciprocal of the given number

43. 5   **44.** -8 **45.** $-\dfrac{3}{4}$

46. $\dfrac{2}{5}$ **47.** 0   **48.** $-\dfrac{4}{5}$

In Exercises 49–54, perform the indicated division. Use your calculator only to verify your results.

49. $\dfrac{5}{16} \div \left(-\dfrac{3}{8}\right)$ **50.** $-\dfrac{3}{4} \div \left(-\dfrac{3}{8}\right)$

51. $-5 \div \left(-\dfrac{2}{3}\right)$ **52.** $-7 \div \left(-\dfrac{3}{4}\right)$

53. $\left(-\dfrac{3}{4}\right) \div 8$ **54.** $\left(-\dfrac{7}{8}\right) \div 16$

In Exercises 55–72, perform the indicated operations.

55. $-\dfrac{2}{3} \div \dfrac{8}{27}$ **56.** $\dfrac{6}{7} \div \left(-\dfrac{3}{28}\right)$

57. $-56 \div 4 \cdot 2$ **58.** $12 \div (-2 \cdot 3)$

59. $-7 \div \left(-\dfrac{2}{5}\right)$ **60.** $-\dfrac{1}{6} \div 6$

61. $\dfrac{9}{-3}$ **62.** $\dfrac{-48}{-6}$

63. $86.47 \div (-4.47)$

64. $-95.3 \div (-57.37)$

65. $-20 \div (-10) \div (-6)$

66. $72 \div (-6 \div 3)$

67. $\dfrac{|-48|}{|-12|}$ **68.** $\dfrac{|-42|}{|7|}$

69. $27 \div (-3)$ **70.** $-27 \div (-9)$

71. $\dfrac{0}{0}$ **72.** $-11 \div 0$

In Exercises 73–78, perform the indicated operations.

73. $\dfrac{3 - (-9)}{-5 - (-1)}$ **74.** $\dfrac{4(7 - 9)}{2} - \dfrac{30}{-5}$

75. $-18 \div (-3)^2$ **76.** $|-8 \div 4| \div (2 - 2^2)$

77. $\dfrac{5(3 - 4) - 5^2}{3(9) - 2(1 + 5)}$ **78.** $\sqrt{\dfrac{20 + (-4)^2}{-5 - 2(-3)}}$

In Exercises 79–82, write a numerical expression that models the problem and then evaluate that expression.

79. What is the quotient of -16.4 and -4.1?

80. What is zero divided by negative ten?

81. What is -20 divided into -10?

82. What is 12 divided by one-half?

In Exercises 83–86, state the property or definition that justifies the step.

83. $\dfrac{3}{7} + \dfrac{8}{7} = 3 \cdot \dfrac{1}{7} + 8 \cdot \dfrac{1}{7}$ (a) ▨▨▨▨

$\qquad = \dfrac{1}{7}(3 + 8)$ (b) ▨▨▨▨

$\qquad = \dfrac{3 + 8}{7}$ (c) ▨▨▨▨

$\qquad = \dfrac{11}{7}$

84. $\dfrac{2x}{3} = 2x \cdot \dfrac{1}{3}$ (a) ▨▨▨▨

$\qquad = 2 \cdot \dfrac{1}{3} \cdot x$ (b) ▨▨▨▨

$\qquad = \left(2 \cdot \dfrac{1}{3}\right)x$ (c) ▨▨▨▨

$\qquad = \dfrac{2}{3}x$

85. For $x \neq 0$, $\dfrac{x^2}{x} = x^2 \cdot \dfrac{1}{x}$ (a) ▨▨▨▨

$\qquad = x \cdot x \cdot \dfrac{1}{x}$

$\qquad = x \cdot \left(x \cdot \dfrac{1}{x}\right)$ (b) ▨▨▨▨

$\qquad = x \cdot 1$ (c) ▨▨▨▨

$\qquad = x$ (d) ▨▨▨▨

86. For $c \neq 0$, $\dfrac{a + b}{c}$

$\qquad = (a + b) \cdot \dfrac{1}{c}$ (a) ▨▨▨▨

$\qquad = a \cdot \dfrac{1}{c} + b \cdot \dfrac{1}{c}$ (b) ▨▨▨▨

$\qquad = \dfrac{a}{c} + \dfrac{b}{c}$ (c) ▨▨▨▨

In Exercises 87–90, determine the unknown number.

87. The quotient of what number and -7 is -2?

88. Ten divided by what number is -5?

89. If the quotient is 10 and the divisor is -4, what is the dividend?

90. What is the divisor if the dividend is -8 and the quotient is 8?

🌐 Real-Life Applications

91. The total change in the stock market over an 8-day period was -39 points. What was the average daily change?

92. An instructor grades a set of 20-point quizzes by indicating the total number of incorrect answers. Here are the scores for six of the students: -5, -1, -7, -4, -5, -2.

   (a) What was the average number of incorrect answers?

   (b) What was the average percentage of correct answers?

✏️ Modeling with Real Data

93. The table shows the number of nuclear weapons in selected nations.

Country	Number of Nuclear Weapons
United States	10,400
Russia	12,000
France	450
China	400
United Kingdom	260
Israel	150

(Source: National Resources Defense Council.)

   (a) What is the total number of weapons for the six countries listed in the table?

   (b) Write fractions to model each country's portion of the total weapons in the table.

94. Refer to Exercise 93.

   (a) Write a fraction to describe the portion of the weapons in the four countries other than the United States and Russia.

   (b) Which country's number of weapons is approximately half of the total?

Data Analysis: *Women in the Military*

In 1998, the number of women who were serving in the U.S. armed forces was 196,487. The table shows the number of female officers and enlisted women in the four branches.

| | Women in U.S. Military | |
Branch	Officers	Enlisted
Army	10,367	60,787
Navy	7,777	42,261
Marines	854	8,928
Air Force	11,971	53,542

(Source: Department of Defense.)

95. How many women served in each branch of the military?

96. In each branch, how many times as many women were enlisted as were officers? (Round to the nearest tenth.)

97. Round the numbers in the table to the nearest thousand. Then write fractions to model the portion of women in each branch who were enlisted.

98. To the nearest hundredth, what portion of all women in the military are in each branch?

Challenge

In Exercises 99–102, let a represent a negative number and let b represent a positive number. Determine the sign of the expression.

99. $\dfrac{a}{b}$

100. $\dfrac{a}{-b}$

101. $\dfrac{-a}{b}$

102. $\dfrac{|a|}{-b}$

In Exercises 103 and 104, evaluate the given quotient.

103. $\dfrac{-\dfrac{5}{8}}{-\dfrac{3}{4}}$

104. $\dfrac{\dfrac{14}{3}}{\dfrac{-7}{6}}$

Chapter 1 Review Exercises

Section 1.1

1. Write the set of integers that are greater than -4 but are not natural numbers.

2. Draw a number line graph of
$$A = \left\{ \sqrt{7}, 1.\overline{3}, -3.5, -1, -\frac{12}{5} \right\}$$

3. Write the decimal name for the given rational number and state whether the decimal is terminating or repeating.

 (a) $-\dfrac{5}{8}$ (b) $\dfrac{1}{6}$

4. Write two inequalities to describe the fact that a number n is at least -1.

5. Assume that w is a whole number and k is an integer. Write the set described by the given inequality.

 (a) $w \le 3$ (b) $k < 3$

6. Insert $<$, $>$, or $=$ to make each statement true.

 (a) $-\dfrac{2}{3}$ _____ $-\dfrac{3}{4}$ (b) $\dfrac{7}{8}$ _____ 0.875

 (c) -173 _____ $\dfrac{1}{173}$

7. Indicate whether each of the following statements is true or false.

 (a) Every irrational number is also a real number.

 (b) $0.47 \ne 0.\overline{47}$

 (c) Every rational number can be written as a terminating decimal.

 (d) The number 5 is a natural number, a whole number, an integer, a rational number, and a real number.

8. Describe the decimal names of the irrational numbers.

Section 1.2

9. Name the operation associated with each of the following words.

 (a) factor (b) dividend

 (c) difference (d) addend

10. Translate the given expression into words.

 (a) -7 (b) $4 - 7$ (c) $-a$

11. Write the product $b \cdot b \cdot b \cdot b$ in exponential form.

12. Write $(-3)^3$ as a product.

13. What is $\sqrt{2116}$?

14. Evaluate $-|-9|$.

15. What is the opposite of the opposite of -10?

16. Evaluate the given numerical expression.

 (a) $18 + 5 \cdot 3$

 (b) $(6 - 2)^2 - (4 - 1)$

 (c) $\sqrt{5^2 - 4(1)(6)}$

17. Insert $<$, $>$, or $=$ to make each statement true.

 (a) $|-2| + |5|$ _____ $2 + 5$

 (b) $\dfrac{4}{|-2|}$ _____ $-\dfrac{4}{2}$

 (c) $|0|$ _____ $|-1|$

18. Under what condition is $|x| = -x$?

Section 1.3

In Exercises 19–23, name the property illustrated.

19. $3x \cdot \dfrac{1}{3x} = 1, x \ne 0$

20. $2(6x) = (2 \cdot 6)x$

21. $x + (3 + y) = x + (y + 3)$

22. $1 \cdot (a + 4) = a + 4$

23. $-5c + 5c = 0$

24. Name the property that justifies each step.

$$-1x + 2\left(\frac{1}{2}x\right) = -1x + \left(2 \cdot \frac{1}{2}\right)x \quad \text{(a)} \underline{\quad\quad}$$
$$= -1x + 1x \quad \text{(b)} \underline{\quad\quad}$$
$$= 0 \quad \text{(c)} \underline{\quad\quad}$$

25. Use the Distributive Property to multiply $6(x + 2)$.

26. Use the Distributive Property to factor $3a + 12$.

27. What name do we give to the sum of a number and its additive inverse?

28. Use the properties of real numbers to simplify each expression. Assume that n is a natural number.

 (a) $(4 - 3)^n$ (b) $(4 - 4)^n$

 (c) $\dfrac{n + 1}{n + 1}$ (d) $\dfrac{n - n}{n}$

Section 1.4

In Exercises 29–33, determine the sum.

29. $-8 + (-11)$

30. $-25 + 9$

31. $17.4 + (-8.1)$

32. $-|-3| + |-1|$

33. $-23 + (-1) + 23$

34. What properties allow us to reorder addends and to group positive and negative addends together?

35. What is the sum if the addends are -16, 18, and -3?

36. The sum of three numbers is -6. If two of the addends are represented by $-c$ and $-(-c)$, what is the third addend?

37. Evaluate $(-9 + 7)^2 + (-11 + 4)$.

38. On June 1 a person's checking account balance was $463.71. On June 2 he wrote a check for $600, and on June 3 he deposited $300. If he was charged $20 for his overdrawn account, what was his account balance on June 4?

Section 1.5

39. Which of the following rational numbers is not equivalent to $\frac{14}{21}$?

(i) $\dfrac{20}{30}$ (ii) $\dfrac{7}{10}$ (iii) $\dfrac{6}{9}$

40. Prime factor 36.

41. Rename $-\frac{5}{8}$ so that the denominator is 40.

42. Simplify $-\frac{20}{35}$.

43. What is the LCD for $\frac{1}{8}$ and $-\frac{5}{12}$?

44. Add $-\frac{6}{11}$ and $\frac{3}{11}$.

45. Find the sum of $-\frac{2}{9}$ and $\frac{4}{15}$.

46. Evaluate $\frac{1}{8} + (-5) + 5\frac{1}{2}$.

47. In order for $\dfrac{n}{6}$ to be reduced, what must be true about n?

48. A sheet of aluminum is rolled to reduce its thickness by $\frac{11}{80}$ of an inch. If the new thickness is $\frac{1}{16}$ of an inch, what was the original thickness?

Section 1.6

49. What changes must be made to $-5 - (-7)$ in order to perform the subtraction?

In Exercises 50–53, determine the difference.

50. $-3 - 9$

51. $18 - (-17)$

52. $4 - 11$

53. $-12 - (-3)$

54. Rewrite $-(x - 8)$ without grouping symbols.

55. If the subtrahend of an expression is -9 and the minuend is -2, what is the difference?

56. Find the difference of $-\frac{8}{9}$ and $-\frac{1}{3}$.

57. Evaluate $\dfrac{16 - (-8)}{|2 - 20| - (7 - 13)}$.

58. If a number line is drawn through the Missouri cities of St. Louis, Columbia, and Kansas City, with Columbia at the origin, the coordinates of St. Louis and Kansas City are $+113$ and -144, respectively. What is the distance between these two cities?

Section 1.7

In Exercises 59–63, multiply.

59. $7(-4)$

60. $(-8.1)(-3.75)$

61. $(-2)(-1)(-7)(-6)$

62. $-|-2| \cdot |-5|$

63. $-\dfrac{14}{3} \cdot \dfrac{9}{7}$

64. Evaluate each of the following expressions.

(a) $(-4)^3$ (b) $(-1)^{38}$

65. If the factors of a product are $(-1)^5$ and $(-4)^2$, what is twice the product?

66. Find two numbers whose sum is -12 and whose product is 35.

67. Evaluate $\sqrt{(-6)^2 - 4(1)(-16)}$.

68. A weight-loss clinic recorded the one-week gains and losses for 20 people.

Number of People	Weight Gain (+) or Loss (−)
2	−7
1	−5
6	−3
3	−2
2	−1
2	0
3	+1
1	+4

What was the total net gain or loss for the 20 people?

Section 1.8

In Exercises 69–73, divide.

69. $\dfrac{-28}{4}$

70. $-16 \div (-2)$

71. $48 \div (-4) \div (-3)$

72. $\dfrac{-671.58}{9.1}$

73. $\dfrac{3}{16} \div \left(-\dfrac{9}{4}\right)$

74. Compare $\dfrac{0}{-8}$ with $\dfrac{-8}{0}$.

75. If the dividend is -6 and the divisor is $-\frac{2}{3}$, what is the quotient?

76. Rewrite each fraction with fewer opposite signs. Express the result with no opposite sign in the denominator.

(a) $-\dfrac{3}{-7}$

(b) $-\dfrac{-x}{-4}$

(c) $\dfrac{-(-a)}{-b}$

77. Evaluate $\dfrac{(-3)^2 - |2 - (-12)|}{4(-1) - 1}$.

78. In Exercise 68, what was the average weight gain or loss for the 20 people?

Chapter 1 Test

1. Indicate whether each statement is true or false. If the statement is false, explain why.

(a) All rational and irrational numbers are real numbers, but no rational number is also an irrational number.

(b) A repeating decimal is an example of a nonterminating decimal.

(c) The fraction $\pi/3$ is an example of a rational number.

(d) No number between -5 and -2 is a whole number.

2. Draw a number line graph of $\left\{2, -\sqrt{3}, 0.\overline{72}, -\frac{9}{4}, 3.4\right\}$.

3. Write two inequalities to describe the fact that a number n is at most 7.

4. A numerical expression consists of 4 factors of 2. Write the expression in exponential form and evaluate it.

5. Evaluate each of the following.

(a) $\sqrt{492.84}$

(b) $-|-4|$

(c) $-(-4)$

6. State the word used for the given number.

(a) 7 in the quotient $-21 \div 7$

(b) -3 in the sum $-3 + 11$

(c) -9 in the product $(-4)(-9)$

7. Name the property that is illustrated.

(a) $(x + 4) + 7 = x + (4 + 7)$

(b) $-x = -1x$

(c) $-5y + 5y = 0$

(d) $x(3 + 5) = (3 + 5)x$

8. In each part, use the Distributive Property.

(a) Factor $4x + 12$.   (b) Multiply $3(a + 2)$.

9. What is the product of any nonzero number and its reciprocal? How do you know?

In Questions 10–15, perform the indicated operations.

10. $-7(-3)$

11. $-23 + 22$

12. $-17 - (-10)$

13. $-\dfrac{-30}{-6}$

14. $-\dfrac{5}{6} + \dfrac{11}{20}$

15. $-\dfrac{12}{5} \div \dfrac{3}{20}$

16. Evaluate $\dfrac{(-2)^3 - (-6)}{|-8 - 2| + (-9)}$.

17. Simplify $\dfrac{27}{6}$.

18. Rewrite $-(-3 - k)$ without grouping symbols.

19. Find two numbers whose sum is -5 and whose product is -24.

20. In the 95th Congress (1977–1979), there were 292 Democrats in the House of Representatives. The following table shows the changes in the number of Democrats in succeeding sessions.

Congress	Years	Net Change in the Number of Democrats
96th	1979–81	-15
97th	1981–83	-35
98th	1983–85	$+27$
99th	1985–87	-16
100th	1987–89	$+5$

How many Democrats were in the House of Representatives during the 100th Congress?

(Source: Neilson Media Research.)

Overall, the television ratings for the Democratic and Republican national conventions have declined since 1968. The data in the bar graph can be modeled by ***algebraic expressions.*** *Such expressions can be used to write* ***equations*** *and* ***inequalities*** *that we can use to compare ratings and to describe trends. (For more on this real-data problem, see Exercises 89–92 at the end of Section 2.7.)*

Algebra Basics, Equations, and Inequalities

Chapter Snapshot

This chapter opens with a discussion of algebraic expressions and how to evaluate them algebraically and graphically. We then discuss equations whose solutions we estimate graphically. By applying the properties of equations, we learn how to obtain exact solutions algebraically. We conclude with graphing and algebraic methods for solving inequalities.

Warm-Up Skills

The following exercises review concepts and skills that you will need in Chapter 2.

In Exercises 1 and 2, use the Distributive Property to perform the indicated operation.

1. Multiply $3(x - 5)$.
2. Factor $4x + 4y$.
3. Use an associative property to multiply $-4(3y)$.

In Exercises 4 and 5, evaluate the given expression.

4. $2(-3)^2 - 4(-3)(2) - 5^2$
5. $7 - 2(5 - 1)$
6. Graph the set $\left\{-4, \frac{5}{2}, 0, \sqrt{25}\right\}$ on a number line.

In Exercises 7–9, insert $<$, $>$, or $=$ to make the statement true.

7. $2(-2) + 5 \rule{1cm}{0.4pt} 1 - (-2)$
8. $\frac{1}{3}(-12) - 16 \rule{1cm}{0.4pt} 2(-12) + 4$
9. $2(-1) + 1 \rule{1cm}{0.4pt} \dfrac{3(-1) - 1}{2}$

10. Explain why $x - 10$ can be written as $-10 + x$.

In Exercises 11 and 12, use the given property to complete the statement.

11. Multiplication Property of -1: $-a = \rule{1cm}{0.4pt} \cdot a$
12. Associative Property of Multiplication:

$$6\left[\frac{5}{6}(y + 3)\right] = \rule{1cm}{0.4pt} (y + 3)$$

| 2.1 | **Algebraic Expressions** |

Evaluating Algebraic Expressions • Terms and Coefficients • Equivalent Expressions • Combining Like Terms • Removing Grouping Symbols • Simplifying Algebraic Expressions

Evaluating Algebraic Expressions

In Chapter 1 we described a numerical expression as any combination of numbers and operations. A numerical expression may also contain grouping symbols to indicate the order of operations.

Similarly, an **algebraic expression** is any combination of numbers, variables, grouping symbols, and operations. Each of the following is an algebraic expression.

$$x^2 + 3x - 4 \qquad \frac{a - 4}{5} \qquad 3y + 2(y - 4) \qquad 2x - 3y$$

A numerical expression has a specific value that can be obtained by performing the indicated operations. An algebraic expression does not have a specific value until the variables in the expression have been replaced with numbers.

Evaluating Algebraic Expressions

1. Replace every occurrence of each variable with the given value for that variable.
2. Evaluate the resulting numerical expression.

EXAMPLE **1** **Evaluating Algebraic Expressions with One Variable**

Evaluate the given algebraic expression for the given value of the variable.

(a) $2x - 7$; $x = 4$ (b) $t^2 - 3t + 1$; $t = 5$

(c) $-x^2$; $x = -3$

LEARNING TIP

An important part of evaluating an expression is applying the Order of Operations correctly.

Solution

(a) $2x - 7 = 2(4) - 7$ Replace x with 4.

$\qquad\qquad = 8 - 7$ Evaluate the numerical expression.

$\qquad\qquad = 1$

(b) $t^2 - 3t + 1 = 5^2 - 3(5) + 1$ Replace t with 5.

$\qquad\qquad\qquad = 25 - 15 + 1$ Evaluate the numerical expression.

$\qquad\qquad\qquad = 11$

(c) $-x^2 = -1 \cdot x^2$ Multiplication Property of -1

$\qquad\quad = -1 \cdot (-3)^2$ Replace x with -3.

$\qquad\quad = -1 \cdot 9$ Square -3; then multiply.

$\qquad\quad = -9$

EXAMPLE 2

Evaluating Algebraic Expressions with More Than One Variable

Evaluate the given algebraic expression for the given values of the variables.

(a) $2x - 3(y - 4)$; $x = -1, y = 2$

(b) $\dfrac{a + 3b}{2a - b}$; $a = -2, b = 3$

(c) $(3x + yz)(2y - z^2)$; $x = 2, y = 0, z = 3$

Solution

(a) $2x - 3(y - 4) = 2(-1) - 3(2 - 4)$ Replace x with -1 and y with 2.

$\qquad\qquad\qquad = 2(-1) - 3(-2)$ Parentheses first.

$\qquad\qquad\qquad = -2 + 6$ Then multiply.

$\qquad\qquad\qquad = 4$ Then add.

(b) $\dfrac{a + 3b}{2a - b} = \dfrac{-2 + 3(3)}{2(-2) - 3} = \dfrac{-2 + 9}{-4 - 3} = \dfrac{7}{-7} = -1$ Substitute $a = -2, b = 3$.

(c) $(3x + yz)(2y - z^2)$

$\qquad = [3(2) + (0)(3)][2(0) - 3^2]$ Substitute $x = 2, y = 0, z = 3$.

$\qquad = [6 + 0][0 - 9]$ Evaluate inside the brackets.

$\qquad = 6(-9) = -54$

We can also use a calculator to evaluate an expression. For now we will limit our discussion to expressions with one variable, and we will consider only one of several possible methods.

⚿ Keys to the Calculator

Store

We assign a value to a variable by *storing* the value in the variable.

Evaluate

You can use your calculator to evaluate an expression for a given (stored) value of the variable. To let $x = 3$ and evaluate $x^2 - 5$, press

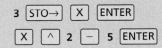

3 [STO→] [X] [ENTER]

[X] [^] 2 [−] 5 [ENTER]

```
3→X
                 3
X^2-5
                 4
```

EXAMPLE 3

Evaluating an Expression on the Home Screen

Evaluate $3x^2 + 5x - 6$ for $x = -4$.

Solution

Store -4 in x and then enter the expression on the home screen. As shown in Figure 2.1, when $x = -4$, the value of the expression is 22.

Figure 2.1

```
-4→X
                         -4
3X²+5X-6
                         22
```

Terms and Coefficients

The addends of an expression are called **terms.** A **constant term** is a term with no variable.

Note: We can think of the terms of an expression as the parts of the expression that are separated by plus or minus signs that are not inside grouping symbols.

Because the expression $6x^2 - 8x + 7$ can be written as $6x^2 + (-8x) + 7$, the terms are $6x^2$, $-8x$, and 7. The constant term is 7. Because $x + y$ can be written as $x + y + 0$, the constant term is 0.

Recall that in an expression such as $5xy$, the factors are 5, x, and y. The numerical factor in a term is called the **numerical coefficient** or simply the **coefficient.** Thus the coefficient of $5xy$ is 5. If the coefficient is not specifically indicated, it is understood to be 1 or -1.

Term	*Coefficient*	
$y = 1y$	1	Multiplicative Identity Property
$-ab = -1ab$	-1	Multiplication Property of -1

EXAMPLE **4**

Terms and Coefficients

For each expression, identify the terms and coefficients.

Expression	*Terms*	*Coefficients*	
(a) $2ab - b^2 + 8$	$2ab$	2	
	$-b^2$	-1	
	8	8	
(b) $y + 3x$	y	1	
	$3x$	3	
(c) $\dfrac{x}{3} + 2(x - 1)$	$\dfrac{x}{3}$	$\dfrac{1}{3}$	$\dfrac{x}{3} = \dfrac{1x}{3} = \dfrac{1}{3}x$
	$2(x - 1)$	2	

Note: In part (c) of Example 4, the minus sign in $2(x - 1)$ is inside grouping symbols and thus is not considered when we count the terms. Therefore, $2(x - 1)$ is a single term.

Equivalent Expressions

Two or more expressions may look different but have the same value for any permissible replacement of the variable. The following table shows the values of two different expressions, $3x + 5x$ and $8x$, for selected values of x.

x	-4	-1	0	2	7	100
$3x + 5x$	-32	-8	0	16	56	800
$8x$	-32	-8	0	16	56	800

Observe that the two expressions have the same values for the indicated values of x. In fact, these expressions have the same value for *any* replacement of x. We call such expressions **equivalent expressions.**

Definition of Equivalent Expressions

Two expressions are **equivalent expressions** if they have the same value for any permissible replacement(s) of the variable(s).

Note: Two expressions may be equivalent even though some replacements for the variable are not permitted. For example, the expressions $\dfrac{x}{x}$ and 1 are equivalent because they have the same value for any replacement of x except 0.

The remainder of this section explains how to **simplify** an expression. As we will see, this means writing an equivalent expression that contains no grouping symbols and that has no terms that can be combined.

Combining Like Terms

Two terms are **like terms** if both are constant terms or if both have the same variable factors with the same exponents.

Like terms	*Unlike terms*
$-2y, 9y$	$5x, 3$
$x^2, 5x^2$	$3x, 3x^2$
$-3a^2b, -2a^2b$	ab^2, a^2b

When an expression contains like terms, we can simplify the expression by *combining like terms*. To simplify the expression $4n + 3n$, we combine like terms by using the Distributive Property.

$$4n + 3n = (4 + 3)n \qquad \text{Distributive Property}$$
$$= 7n$$

Similarly, we can combine like terms in the difference $5x^2 - 6x^2$.

$$5x^2 - 6x^2 = (5 - 6)x^2 \qquad \text{Distributive Property}$$
$$= -1x^2$$
$$= -x^2 \qquad \text{Multiplication Property of } -1$$

With practice you should be able to perform the second step mentally and simply write the result. Note that the terms and the result are all like terms.

> **Combining Like Terms**
> 1. Identify the terms that are like terms.
> 2. Add or subtract the coefficients as indicated.
> 3. Retain the common variables.

EXAMPLE 5 | **Combining Like Terms**

Combine like terms.

(a) $-4x + x = -4x + 1x = (-4 + 1)x = -3x$

(b) $6y^2 - (-3y^2) = 6y^2 + 3y^2 = (6 + 3)y^2 = 9y^2$

(c) $3a - 5a - 2a = (3 - 5 - 2)a = -4a$

We can use the Commutative and Associative Properties of Addition to group like terms together.

$$7x + 5 - 3x + 2 = (7x - 3x) + (5 + 2) = 4x + 7$$

Again, you should be able to reorder and regroup mentally in order to omit the second step.

EXAMPLE 6 | **Combining Like Terms**

Think About It

Why can the terms of $7(x - 1) + 3(x - 1)$ be combined without removing the parentheses, but the terms of $7(x - 1) + 3(x + 1)$ cannot be combined in this way?

Combine like terms.

(a) $5m + 6n - 7n + m + 3 = 6m - n + 3$

(b) $3x - 7y - 5 + 7y - 3x + 4 = -1$

(c) $a^2 + 3a - 4a + 3a^2 = 4a^2 - a$

(d) $3xy^2 - 2x^2y - xy^2 + 7x^2y = 2xy^2 + 5x^2y$

Removing Grouping Symbols

We also simplify an expression by removing grouping symbols. Example 7 shows how the Associative Property of Multiplication can be used to remove grouping symbols in products.

EXAMPLE 7 | **Removing Grouping Symbols in Products**

Remove the grouping symbols.

(a) $-5(2x) = (-5 \cdot 2)x = -10x$

(b) $-\dfrac{2}{3}(-12y^2) = \left(\dfrac{-2}{3} \cdot \dfrac{-12}{1}\right)y^2 = 8y^2$

Sometimes we can use the Distributive Property to remove grouping symbols.

EXAMPLE **8**   **Removing Grouping Symbols with the Distributive Property**

Remove the grouping symbols.

(a) $5(2x + 3) = 5(2x) + 5(3) = 10x + 15$

(b) $-4(5x - 6) = -4(5x) - (-4)(6) = -20x - (-24) = -20x + 24$

(c) $4(2a + 5b - 3) = 4(2a) + 4(5b) - 4(3)$
$$= 8a + 20b - 12$$

The Distributive Property also applies when an expression inside grouping symbols is preceded by an opposite sign or by a minus sign.

$-(a + b - c) = -1(a + b - c)$	Multiplication Property of -1
$= -1a + (-1)b - (-1)c$	Distributive Property
$= -a + (-b) - (-c)$	Multiplication Property of -1
$= -a - b + c$	Definition of subtraction

An opposite sign or a minus sign preceding the expression in grouping symbols has the effect of changing the signs of each term inside the group.

EXAMPLE **9**   **Expressions Preceded by Opposite or Minus Signs**

Remove the grouping symbols.

(a) $-(6y - 1) = -6y + 1$

(b) $a - (-5 + 4m - 2n) = a + 5 - 4m + 2n$

The following summarizes the procedure for using the Distributive Property to remove grouping symbols.

Removing Grouping Symbols with the Distributive Property

If an expression has the form $a(b + c)$, remove grouping symbols by using the Distributive Property to multiply.

1. For expressions of the form $(b + c)$, the number a is understood to be 1, and the result is $b + c$. In other words, the grouping symbols can simply be removed.

2. For expressions of the form $-(b + c)$, the number a is understood to be -1, and the result is $-b + (-c)$. In other words, the grouping symbols can be removed by changing the sign of each term in the group.

Simplifying Algebraic Expressions

The following summarizes the procedure for simplifying an algebraic expression.

Simplifying an Algebraic Expression

1. Remove grouping symbols.

2. Combine like terms.

EXAMPLE 10

Simplifying Algebraic Expressions

Simplify the expression.

(a) $2(3 - 2a) + 3(a - 2)$

$\quad = 6 - 4a + 3a - 6$ Distributive Property

$\quad = -a$ Combine like terms.

LEARNING TIP

To avoid errors as you simplify an algebraic expression, pay close attention to signs and write the complete expression at each step.

(b) $-3(x + 3y - 4) + 2(x - 5)$

$\quad = -3x - 9y + 12 + 2x - 10$ Distributive Property

$\quad = -x - 9y + 2$ Combine like terms.

(c) $(5x + 4) - (2x - 3)$

$\quad = 5x + 4 - 2x + 3$ Change the signs in $(2x - 3)$.

$\quad = 3x + 7$ Combine like terms.

(d) $-(2x^2 + x - 6) - 2(x - 2)$

$\quad = -2x^2 - x + 6 - 2x + 4$ Distributive Property

$\quad = -2x^2 - 3x + 10$ Combine like terms.

Sometimes an expression contains *nested* grouping symbols—that is, grouping symbols inside of other grouping symbols. Recall from the Order of Operations that we begin with the innermost grouping symbols.

EXAMPLE 11

Simplifying Expressions with Nested Grouping Symbols

Simplify the expression.

$2 - 5[6 - (5 + 2x)] = 2 - 5[6 - 5 - 2x]$ Remove parentheses.

$\qquad\qquad\qquad\quad = 2 - 5[1 - 2x]$ Simplify inside the brackets.

$\qquad\qquad\qquad\quad = 2 - [5 - 10x]$ Distributive Property

$\qquad\qquad\qquad\quad = 2 - 5 + 10x$ Remove brackets.

$\qquad\qquad\qquad\quad = 10x - 3$ Combine like terms.

Quick Reference 2.1

Evaluating Algebraic Expressions

- An **algebraic expression** is any combination of numbers, variables, grouping symbols, and operations.
- To evaluate an algebraic expression,
  1. replace every occurrence of each variable with the given value for that variable and
  2. evaluate the resulting numerical expression.

Terms and Coefficients

- The addends of an expression are called **terms.** A **constant term** is a term with no variable.
- The numerical factor in a term is called the **coefficient.**

Equivalent Expressions

- Two expressions are **equivalent expressions** if they have the same value for any permissible replacement(s) for the variable(s).

- We **simplify** an expression by writing an equivalent expression with no grouping symbols and no terms that can be combined.

Combining Like Terms

- Two terms are **like terms** if both are constant terms or if both have the same variable factors with the same exponents.
- To combine like terms,
  1. identify the terms that are like terms,
  2. add or subtract the coefficients as indicated, and
  3. retain the common variables.

Removing Grouping Symbols

- If an expression has the form $a(b + c)$, remove grouping symbols by using the Distributive Property to multiply.
  1. For expressions of the form $(b + c)$, the number a is understood to be 1, and the result is $b + c$. In other words, the grouping symbols can simply be removed.
  2. For expressions of the form $-(b + c)$, the number a is understood to be -1, and the result is $-b + (-c)$. In other words, the grouping symbols can be removed by changing the sign of each term in the group.

Simplifying Algebraic Expressions

- To simplify an algebraic expression,
  1. remove grouping symbols and
  2. combine like terms.
- For expressions with nested grouping symbols, begin by removing the innermost grouping symbols.

Speaking the Language 2.1

1. In the expression $2x + 5y$, we call $5y$ a(n) ▒▒▒▒▒▒, and we call 5 the ▒▒▒▒▒▒ .

2. Because $3(x + 4)$ and $3x + 12$ have the same value for any replacement of x, the expressions are called ▒▒▒▒▒▒ expressions.

3. In the expression $4a + 7a$, we say that $4a$ and $7a$ are ▒▒▒▒▒▒ terms. The process of combining them to obtain $11a$ is called ▒▒▒▒▒▒ the expression.

4. For $-(a - 5)$, we can remove the parentheses by ▒▒▒▒▒▒ of each term inside the parentheses.

Exercises 2.1

Concepts and Skills

1. In what way is a numerical expression a special kind of algebraic expression?

2. What must be known before an algebraic expression can be evaluated?

In Exercises 3–10, evaluate each expression for the given value of the variable.

3. $2x - 7$;   3

4. $-x - 2$; -5

5. $-3(1 - x)$; -1

6. $4(x - 5)$;   2

7. $x - (1 - 3x)$; -2

8. $6 - 2(x - 2)$;   2

9. $2x^2 + 3x + 2$;   3

10. $3x^2 + 5$; -4

In Exercises 11–16, evaluate each expression for the given values of the variables.

11. $3x - y$; $x = 2, y = -4$

12. $xy + y - 3$; $x = 0, y = 3$

13. $x^2 + xy + 3y^2$; $x = -3, y = 1$

14. $2x^2 - 3xy - y^2$; $x = 0, y = 5$

15. $\dfrac{5 - y}{3 - x}$; $x = 2, y = -1$

16. $\dfrac{x - y}{x + y}$; $x = 2, y = -3$

In Exercises 17–20, evaluate each expression for $a = 3$, $b = -2$, and $c = 4$.

17. $\dfrac{a + b}{c}$

18. $\dfrac{b - c}{2a}$

19. $(a + b)^2 - 3c$

20. $(a - b)^2 - (b - c)^2$

In Exercises 21–24, use your calculator to evaluate the given expression on the home screen.

21. $x - 2(x - 1)$; $x = -3.92$

22. $x^2 - 3x + 1$; $x = 2.1$

23. $\dfrac{x + 5}{2x - 1}$; $x = -1.2$

24. $\dfrac{x + 3}{3x}$; $x = 0.2$

▼ **25.** Compare the number of terms in $2x + 1$ and in $2(x + 1)$. In $2(x + 1)$, what do we call 2 and $(x + 1)$?

▼ **26.** If $x = 2$, then x^2 and $2x$ have the same value. Does this mean that x^2 and $2x$ are equivalent expressions? Explain.

In Exercises 27–32, identify the terms and coefficients in each expression.

27. $3y + x - 8$

28. $x - y + 6z + 1$

29. $x^2 + 3x - 6$

30. $x^2 - x$

31. $7b - c + \dfrac{3a}{7}$

32. $m - 5m^2 + \dfrac{n}{3}$

In Exercises 33–36, fill in the blanks to make the equations true.

33. $3x + 8x = ($ ▨ $+$ ▨ $)x =$ ▨ x

34. $x - 5x = ($ ▨ $-$ ▨ $)x =$ ▨ x

35. $-y + 3y = ($ ▨ $+$ ▨ $)y =$ ▨ y

36. $5b - 6b = ($ ▨ $-$ ▨ $)b =$ ▨ b

In Exercises 37–40, combine the like terms.

37. $-3x - 5x$

38. $n + 3n$

39. $4t^2 - t^2$

40. $xy + 5xy$

In Exercises 41–56, combine the like terms.

41. $6x - 7x + 3$

42. $4y - 7 - 3y$

43. $9x - 12 - 3x + 5$

44. $b + a - 6a - 4b$

45. $2m - 7n - 3m + 6n$

46. $x + 5 - y - 6$

47. $10 - 2b + 3b - 9$

48. $7b - 3b - b$

49. $2a^2 + a^3 - a^2 - 5a^3$

50. $y^3 - 2y^2 - 8y^2 + 3y^3$

51. $ab + 3ac - 2ab + 5ac$

52. $3mn - 4m + 5m + 6mn$

53. $\dfrac{1}{3}a + \dfrac{1}{2}b - a + \dfrac{2}{3}b$

54. $\dfrac{x}{4} + \dfrac{y}{2} - \dfrac{x}{2} + \dfrac{y}{6}$

55. $2.1x - 4.2y + 0.3y + 1.2x$

56. $0.2m + 1.2m - 4.3n - 2.6m$

▼ **57.** What is the procedure for simplifying an algebraic expression?

▼ **58.** Describe two ways to remove the parentheses from the expression $-(a + b)$.

In Exercises 59–64, remove the grouping symbols and simplify the product.

59. $-5(3x)$

60. $9(-2y)$

61. $-\dfrac{1}{4}(4y)$

62. $\dfrac{1}{3}(6a)$

63. $-\dfrac{5}{6} \cdot \left(-\dfrac{9}{10}x\right)$

64. $\dfrac{2}{3}\left(\dfrac{3}{2}y\right)$

In Exercises 65–74, use the Distributive Property to simplify each expression.

65. $5(4 + 3b)$

66. $-2(3a - 7)$

67. $-3(5 - 2x)$

68. $5(2x - 3)$

69. $2(3x + y)$

70. $-2(x + 5)$

71. $-3(-x + y - 5)$

72. $4(2x + y - 6)$

73. $-\dfrac{3}{5}(5x + 20)$

74. $\dfrac{2}{3}(6x - 15)$

In Exercises 75–80, write the given expression without grouping symbols.

75. $-(x - 4)$

76. $-(3 + 5x)$

77. $-(3 + n)$

78. $-(2a - 3b + 7d)$

79. $-(2x + 5y - 3)$

80. $-(-4x - y + 7)$

In Exercises 81–92, simplify the expression.

81. $2(x - 3) + 4$

82. $4(x - 2) - 3x$

83. $5 - 3(x - 1)$

84. $3 - 2(1 - 2y)$

85. $-2(x - 3y) + y + 2x$

86. $2(4x - 1) + 3(1 - 2x)$

87. $(2a + 7) + (5 - a)$

88. $-(x + 3) + (4 + 2x)$

89. $2 + (2a + b) + a - b$

90. $2a - 7(1 + a) + 9$

91. $4(1 - 3x) - (3x + 4)$

92. $9 - 3(1 + 3t) + 3(3t - 2)$

In Exercises 93–96, simplify the expression.

93. $2[5 - (2x - 3)] + 3x$

94. $-[2x + (4 - 3x)] - x$

95. $4 + 5[x - 3(x + 2)]$

96. $8x - 7[6 + (-5 + 2x)]$

Real-Life Applications

97. It is 8 A.M., and a student has a final exam at 6 P.M. The exam grade is predicted by $8t + 20$, where t is the number of study hours. What is the predicted grade if the student studies from now until 4 P.M.?

98. The expression $5x - 5 + 4(1 - 2x)$ describes the temperature change (in °F) x hours after sunset. Write the expression in simplified form and determine the temperature change 4 hours after sunset.

99. When a certain heating and air conditioning contractor performs in-home service, the total cost (in dollars) is represented by $55h + 75$, where h is the number of hours on the job. If this contractor were to spend 2.8 hours at your home, what amount would you owe?

100. Because you use your cell phone only for emergencies, your billing plan is $15 per month plus 90 cents per minute of phone use. Your monthly bill can be described by $0.9m + 15$, where m is the total number of minutes that you use your phone. If you make three calls lasting 4 minutes, 7 minutes, and 6 minutes, what is your cell phone bill for that month?

Challenge

In Exercises 101 and 102, simplify.

101. $2(3n + 5) - 7 - [3(n + 1) - 2(n - 4)]$

102. $a - [b - (a - b)] - (a - b)$

In Exercises 103–106, use your calculator to evaluate the expression for the given values of the variables.

103. $\dfrac{x + y}{2x + 3y}$; $\quad x = 1.2, y = -3.1$

104. $\dfrac{xy}{3 + y}$; $\quad x = -2.1, y = -1.9$

105. $2a^2 + (c - 4)^2$; $\quad a = -5.3, c = 2.15$

106. $4 - w^2 + 2z$; $\quad w = 1.9, z = 4.25$

2.2 | The Coordinate Plane

Familiar Graphs • Graphs in Mathematics • Coordinate Systems on a Calculator • Modeling with Real Data

Familiar Graphs

Bar graphs and line graphs appear frequently in the news media.

Exploring the Concept

Reading Graphs

The bar graph in Figure 2.2 and the line graph in Figure 2.3 are visual ways of showing the relationship between a person's weight and the number of calories that are burned when a person walks at a normal pace for 30 minutes. In both graphs, weights are given along the horizontal axis, and calories are reported along the vertical axis.

Figure 2.2   **Figure 2.3**

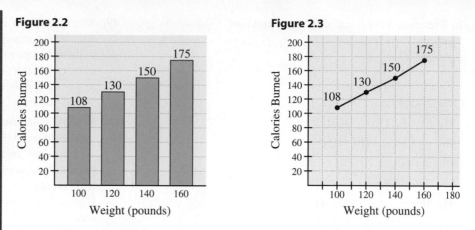

We can begin at either axis and read to the other axis. The following are examples of information that can be obtained from either graph.

1. Beginning at 120 pounds on the horizontal axis, we move up to the top of the bar (or to the line in the line graph) and then left to the vertical axis. The corresponding number is 130 calories.

2. Beginning at 175 calories on the vertical axis, we move right to the top of the bar (or to the line in the line graph) and then down to the horizontal axis. The corresponding number is 160 pounds.

We can use the data in Figures 2.2 and 2.3 to create the following table, which shows each pairing of weight and calories consumed. In the third column, we show the pairings in the form (weight, calories).

Weight	Calories Consumed	(Weight, Calories)
100	108	(100, 108)
120	130	(120, 130)
140	150	(140, 150)
160	175	(160, 175)

Note that in the pairings (weight, calories), the first number of the pair is the number from the horizontal axis, and the second number of the pair is the number from the vertical axis.

Figure 2.4 is another kind of graph, in which only the points that represent data are shown.

Figure 2.4

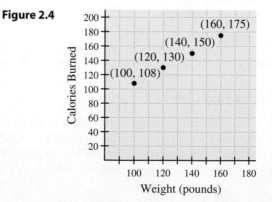

Graphs in Mathematics

Bar graphs and line graphs are used to show the relationship between one set of numbers (reported along the horizontal axis) and another set of numbers (reported along the vertical axis). In mathematics, graphs are constructed in a similar manner.

To visualize certain sets of real numbers, we use a number line. However, to visualize how the numbers x in one set are paired with the numbers y in another set (possibly the same set), we need to use two number lines.

Conventionally, we represent the x-values along a horizontal number line called the **x-axis** and the y-values along a vertical number line called the **y-axis.** These two number lines intersect at a point called the **origin.** (See Fig. 2.5.) This arrangement is called a **rectangular coordinate system,** a **Cartesian coordinate system,** or simply a **coordinate plane.**

Figure 2.5

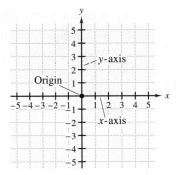

Note: Roughly speaking, you can think of a *plane* as the flat, unbounded surface (such as a piece of paper, a chalkboard, or your calculator screen) on which the axes are drawn.

In a coordinate plane, positive numbers are indicated with *tick marks* to the right of the origin on the x-axis and above the origin on the y-axis. Negative numbers are indicated with tick marks to the left of the origin on the x-axis and below the origin on the y-axis.

The number lines divide the plane into four **quadrants** that are numbered counterclockwise. (See Fig. 2.6.)

The location of each point in a coordinate plane can be described by a pair of numbers called an **ordered pair.** For example, the ordered pair $(3, -4)$ describes a point P that is 3 units to the right of the origin and 4 units below the origin. (See Fig. 2.7.)

Figure 2.6

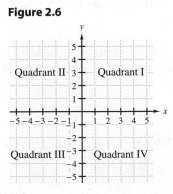

Figure 2.7

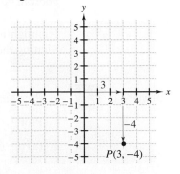

Note: It is very important to remember that the first number of an ordered pair indicates horizontal direction and distance and that the second number of the pair indicates vertical direction and distance.

LEARNING TIP

Think of the coordinate plane as a map with the *x*-axis oriented east/west and the *y*-axis oriented north/south. The first coordinate in a pair indicates the distance east or west, and the second coordinate indicates the distance north or south.

When we associate an ordered pair of numbers with a point in the coordinate plane, we call the numbers **coordinates** of the point. For point $P(3, -4)$, we call 3 the **x-coordinate** of point P and -4 the **y-coordinate** of point P. Each point in the coordinate system is associated with exactly one ordered pair of numbers, and each ordered pair of numbers is associated with exactly one point in the coordinate system.

EXAMPLE 1

Ordered Pairs and Points

Use Figure 2.8 to answer each of the following questions.

(a) What are the coordinates of point P?

(b) Which point has the coordinates $(-2, 4)$?

(c) Which point is in Quadrant IV?

(d) What are the coordinates of the point in Quadrant III?

Think About It

Do you think that the latitude and longitude system for the earth is the same as a rectangular coordinate system?

Figure 2.8

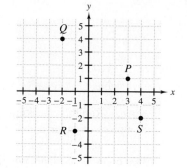

Solution

(a) $(3, 1)$

(b) Point $Q(-2, 4)$

(c) Point $S(4, -2)$

(d) $(-1, -3)$

Note: In Example 1, the two different points Q and S have the same coordinates but in reverse order. Pairs of numbers are called *ordered* pairs because the order of the coordinates is significant.

Highlighting the point corresponding to an ordered pair is called **plotting** the point. Plotting the points corresponding to the ordered pairs of a set is called **graphing** the set.

EXAMPLE 2

Plotting Points

Plot the points associated with each of the following ordered pairs.

(a) $(-6, 5)$ (b) $(-3, 0)$ (c) $(0, 4)$ (d) $(5, -3)$

Solution

Figure 2.9

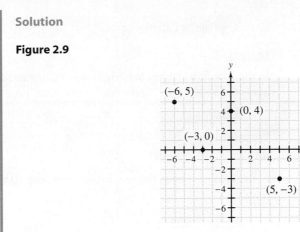

Coordinate Systems on a Calculator

We can produce a coordinate system on a graphing calculator. Moreover, we can decide what portion (*window*) of the coordinate plane we want to be displayed in the viewing rectangle, and we can set the distance (*scale*) between the tick marks on either axis.

We begin with the calculator screens that we use most often.

⊙⟊ **Keys to the Calculator**

Home

The home screen is the calculator's primary screen for basic operations, such as performing arithmetic and evaluating expressions.

To display the home screen from any other screen, press [2nd] [QUIT]. To erase the home screen, press [CLEAR].

```
Plot1  Plot2 Plot3
\Y₁=
\Y₂=
\Y₃=
\Y₄=
\Y₅=
\Y₆=
\Y₇=
```

Y Screen

For purposes to be described later, we enter algebraic expressions on the Y screen. To access this screen, press [Y=].

To erase an entry on the Y screen, move the cursor to anywhere on the entry and press [CLEAR].

Graph

Pressing the [GRAPH] key will display a coordinate system.

Your calculator has certain built-in settings that you can use to display the graph screen. You can also customize the display with your own settings. In the figures, the numbers that appear at the ends of the axes indicate the minimum and maximum values for those axes.

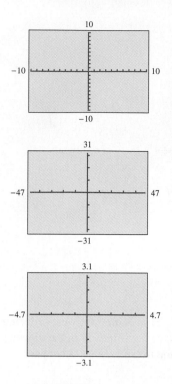

🔑 Keys to the Calculator

Standard

In the *standard* setting, the origin is at the center, and the tick marks on both axes represent 1 unit. To produce this screen, press:

ZOOM   6:ZStandard

Integer

In the *integer* setting, the tick marks represent 10 units. To produce this screen, press:

ZOOM   8:ZInteger   ENTER

Decimal

In the *decimal* setting, the tick marks represent 1 unit but the window is smaller than the window displayed with the standard setting. To produce this screen, press:

ZOOM   4:ZDecimal

Window

The window screen displays the current viewing window settings. You can choose the minimum and maximum values for each axis as well as the values that the tick marks represent.

To access the window menu, press WINDOW. Use the up and down arrow keys to enter values.

Suppose that we set the x-axis from -8 (Xmin) to 12 (Xmax), with the tick marks representing 4 units (Xscl). The y-axis is set from -6 (Ymin) to 18 (Ymax), with the tick marks representing 3 units (Yscl). When we press the GRAPH key, the coordinate system is displayed with those settings.

Note: Selecting the integer or decimal setting will not necessarily display the origin at the center of the viewing window. One method for displaying the origin at the center is first to select standard, and then select integer or decimal.

A *general cursor* is used to move around a graph screen.

🔑 Keys to the Calculator

Cursor

If you are on the graph screen, you can activate the general cursor by pressing any of the arrow keys. The arrow keys are also used to move the cursor around the screen.

The figure shows the integer setting with the cursor on the point $(-20, 15)$.

For the integer setting, x-coordinates are displayed as integers; for the decimal setting, x-coordinates are displayed in tenths.

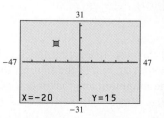

Exploring the Concept

Points in the Coordinate Plane

Figure 2.10 shows a coordinate system with the integer setting.

Figure 2.10

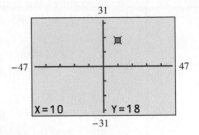

Note that the cursor is on the point (10, 18) and that both coordinates are positive. As you move your cursor around in Quadrant I, you will find that both coordinates are positive for any point in that quadrant.

Similarly, moving your cursor around in the other three quadrants will reveal the following sign patterns.

Quadrant I	*Quadrant II*	*Quadrant III*	*Quadrant IV*
$(+, +)$	$(-, +)$	$(-, -)$	$(+, -)$

By moving the cursor along the two axes, we can obtain information about the coordinates of those points. Figure 2.11 shows the cursor on the point (15, 0) on the *x*-axis.

Figure 2.11

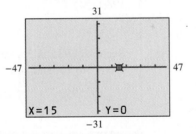

As we move the cursor back and forth, we see that every point of the *x*-axis has a *y*-coordinate of 0. Similarly, moving the cursor up and down along the *y*-axis reveals that every point of the *y*-axis has an *x*-coordinate of 0.

1. *x*-axis: All points have the form $(a, 0)$.
2. *y*-axis: All points have the form $(0, b)$.

Being familiar with the coordinate patterns for points in the coordinate plane will be very useful in the topics that follow.

Modeling with Real Data

At the beginning of this section, we saw how real data can be represented with a bar graph or a line graph. We also showed how data can be written as ordered pairs, which can then be represented by points of a graph.

Real data are often related to time. Example 3 shows a method for writing ordered pairs of such data.

EXAMPLE **3**

Ordered Pairs of Real Data

The table shows the dramatic increase in the number of cellular phone antenna sites during the period 1985–2005.

Year	Cellular Phone Antenna Sites
1985	599
1995	21,000
2005	117,920

(Source: Cellular Telecommunications Industry Association.)

We can write the data as ordered pairs as follows.

(1985, 599)   (1995, 21,000)   (2005, 117,920)

Rather than writing the actual year, a more convenient method is to let the first coordinate be the number of years since some base year. For example, if the first coordinate is the number of years *since 1985,* then 1985 corresponds to 0, 1995 corresponds to 10, and 2005 corresponds to 20. Now the ordered pairs are

(0, 599)   (10, 21,000)   (20, 117,920)

Write the ordered pairs for which the first coordinate is the number of years since 1980.

Solution

Because 1985 is 5 years after 1980, 1985 corresponds to 5. Similarly, 1995 corresponds to 15, and 2005 corresponds to 25.

(5, 599)   (15, 21,000)   (25, 117,920)

Quick Reference 2.2

Familiar Graphs
- Bar graphs and line graphs show the relationship between two sets of data. Such graphs can be read from either axis to the other axis.
- Data from one set can be paired with data from another set, and the pairings can be represented with points.

Graphs in Mathematics
- The **rectangular coordinate system** (or **Cartesian coordinate system** or **coordinate plane**) consists of a horizontal number line called the **x-axis** and a vertical number line called the **y-axis.** The axes intersect at a point called the **origin.**
- In a coordinate plane, positive numbers are indicated with *tick marks* to the right of the origin on the *x*-axis and above the origin on the *y*-axis. Negative numbers are indicated with tick marks to the left of the origin on the *x*-axis and below the origin on the *y*-axis.

- The axes divide the plane into four **quadrants** that are numbered counterclockwise.
- The location of each point in a coordinate plane can be described with a pair of numbers called an **ordered pair.** The first number of the pair is the *x*-**coordinate,** and the second number is the *y*-**coordinate.**
- Highlighting the point corresponding to an ordered pair is called **plotting** the point. Plotting the points corresponding to the ordered pairs of a set is called **graphing** the set.

Coordinate Systems on a Calculator
- When we produce a coordinate system on a calculator, we must select a *window* (the minimum and maximum *x*- and *y*-values for the axes) and a *scale* (the distance between tick marks on the axes).
- The sign patterns for the coordinates in each of the quadrants are as follows:

Quadrant I	*Quadrant II*	*Quadrant III*	*Quadrant IV*
$(+, +)$	$(-, +)$	$(-, -)$	$(+, -)$

- Every point of the *x*-axis has a *y*-coordinate of 0. Every point of the *y*-axis has an *x*-coordinate of 0.

Speaking the Language 2.2

1. In a rectangular coordinate system, the point at which the *x*-axis and the *y*-axis intersect is called the ▨▨▨▨ .
2. Each point in the coordinate plane is associated with a(n) ▨▨▨▨ pair.
3. The *x*-axis and the *y*-axis divide the coordinate plane into regions called ▨▨▨▨ .
4. Every point of the ▨▨▨▨ has an *x*-coordinate of 0.

Exercises 2.2

Concepts and Skills

1. In a coordinate system, what are the two number lines and their point of intersection called?
2. What do we know about any point of the *x*-axis? of the *y*-axis?

In Exercises 3–14, without plotting the given point, name the axis or quadrant in which the point lies.

3. $(3, 7)$

4. $(-0.3, 4.7)$

5. $\left(-6, \dfrac{-9}{4}\right)$

6. $\left(\dfrac{5}{8}, \dfrac{-2}{3}\right)$

7. $\left(-\sqrt{5}, 6\right)$

8. $(-\pi, -3)$

9. $\left(4.9, \dfrac{-12}{5}\right)$

10. $\left(\sqrt{6}, \sqrt{2}\right)$

11. $(-6.27, 0)$

12. $(0, -4.78)$

13. $\left(0, \sqrt{7}\right)$

14. $\left(\dfrac{19}{7}, 0\right)$

In Exercises 15–18, determine the coordinates of the points.

15.

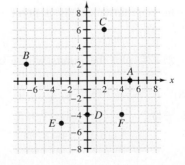

16.

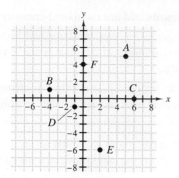

17.

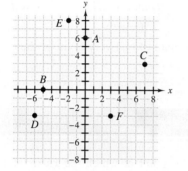

18.

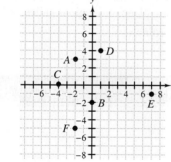

19. Why do we consider $(3, 1)$ and $(1, 3)$ to be different pairs?

20. Under what circumstances is a point not in any quadrant?

In Exercises 21–24, plot the points whose coordinates are given.

21. (a) $(1, 6)$ (b) $(-3, 1)$

    (c) $(-1, -1)$ (d) $(1, -5)$

    (e) $(-3, 0)$ (f) $(0, 5)$

22. (a) $(-6, 2)$ (b) $(-3, -2)$

    (c) $(3, -3)$ (d) $(-7, 0)$

    (e) $(0, 2)$ (f) $(2, 2)$

23. (a) $(7, -2)$ (b) $(0, 0)$

    (c) $(-1, 5)$ (d) $(6, 0)$

    (e) $(-6, -4)$ (f) $(7, 3)$

24. (a) $(0, -6)$ (b) $(9, -4)$

    (c) $(6, 1)$ (d) $(4, 0)$

    (e) $(-8, -9)$ (f) $(-2, 4)$

In Exercises 25–32, graph the given set. Predict the next ordered pair in the set.

25. $\{(-2, 0), (-1, 1), (0, 2), (1, 3), \ldots\}$

26. $\{(-1, -5), (0, -4), (1, -3), (2, -2), \ldots\}$

27. $\{(-2, 4), (0, 0), (2, -4), (4, -8), \ldots\}$

28. $\{(-4, -2), (-2, -1), (0, 0), (2, 1), \ldots\}$

29. $\{(7, 2), (7, 4), (7, 6), (7, 8), \ldots\}$

30. $\{(-3, -3), (0, -3), (3, -3), (6, -3), \ldots\}$

31. $\{(2, 1), (4, -1), (6, -3), (8, -5), \ldots\}$

32. $\{(-1, -3), (1, 1), (3, 5), (5, 9), \ldots\}$

33. Give the signs of a and b if $P(a, b)$ lies in

    (a) Quadrant IV.

    (b) Quadrant I.

    (c) Quadrant II.

    (d) Quadrant III.

34. What is true about x and y if $P(x, y)$ lies on the

    (a) x-axis to the left of the origin?

    (b) y-axis below the origin?

    (c) y-axis above the origin?

    (d) x-axis to the right of the origin?

In Exercises 35–44, information about the coordinate(s) of a point is given. In which *possible* quadrants or on which axes could the point lie?

35. The first coordinate is positive.

36. The second coordinate is 0.

37. The second coordinate is negative.

38. The first coordinate is -3.

39. The product of the coordinates is positive.

40. The first coordinate is negative.

41. The first coordinate is 0.

42. The second coordinate is positive.

43. The product of the coordinates is negative.

44. The second coordinate is 5.

45. In each part, the coordinates of a point are described. In which *possible* quadrants or on which axes could the point lie?

(a) $(a, 0)$, $a > 0$ (b) $(0, b)$, $b < 0$

(c) $(a, -a)$, $a > 0$ (d) (a, a), $a \neq 0$

46. In each part, the location of $Q(b, a)$ is given. Determine the location of $P(a, b)$.

(a) $Q(b, a)$ is in Quadrant II.

(b) $Q(b, a)$ is in Quadrant IV.

(c) $Q(b, a)$ is in Quadrant I.

(d) $Q(b, a)$ is in Quadrant III.

In Exercises 47–52, the location of point P is described. Use the integer setting and the general cursor to locate the point on your calculator's graphing screen. Write the coordinates of point P.

47. Point P is on the y-axis 13 units above the x-axis.

48. Point P is on the x-axis directly below $(-14, 7)$.

49. Point P is directly above $(9, 0)$, and the coordinates are equal.

50. Point P is directly to the left of $(0, -11)$, and the coordinates are equal.

51. Point P is directly below $(-12, 13)$, and the y-coordinate is 6.

52. Point P is directly to the right of $(-17, -20)$, and the x-coordinate is -5.

In Exercises 53–56, the coordinates of the vertices of a geometric figure are given. Draw the figure. [A **vertex** (*pl:* vertices) of a figure is a point where two sides intersect.]

53. Rectangle: $(1, 7), (1, -2), (5, 7), (5, -2)$

54. Rectangle: $(-1, 5), (6, 5), (-1, 1), (6, 1)$

55. Triangle: $(3, 1), (3, -2), (8, 1)$

56. Triangle: $(0, 0), (3, -1), (5, 3)$

In Exercises 57–60, use the integer setting and the general cursor. Start at point $A(12, -8)$ and move your cursor to point B according to the given directions. Then write the coordinates of point B.

57. Move 16 units left and then up 14 units.

58. Move 7 units right and then down 5 units.

59. Move 5 units right and then up 8 units.

60. Move 6 units left and then down 6 units.

In Exercises 61–64, calculate the perimeter and the area of the figure with the given vertices.

61. Rectangle: $(-4, 1), (-4, 6), (-1, 1), (-1, 6)$

62. Square: the endpoints of one side are $(3, 2)$ and $(6, 2)$.

63. Square: the endpoints of a diagonal are $(-3, 3)$ and $(0, 0)$.

64. Rectangle: the endpoints of a diagonal are $(-4, 0)$ and $(0, 5)$.

In Exercises 65–68, the first coordinates of five points are given, and the second coordinate of each is described.

(a) Determine the second coordinate of each point.

(b) Let x represent the first coordinate and write a general expression for the second coordinate.

65. $(-3, \_), (-1, \_), (1, \_), (3, \_), (5, \_)$; the second coordinate is 2 more than the first coordinate.

66. $(2, \_), (3, \_), (4, \_), (5, \_), (6, \_)$; the second coordinate is 5 less than the first coordinate.

67. $(-3, \_), (-1, \_), (0, \_), (1, \_), (3, \_)$; the second coordinate is the opposite of the first coordinate.

68. $(-4, \_), (-2, \_), (0, \_), (2, \_), (4, \_)$; the second coordinate is the absolute value of the first coordinate.

Real-Life Applications

69. An architectural drawing for a small rectangular manufacturing facility is laid out in a coordinate system, with each tick mark representing 1 foot. Three of the vertices of the rectangular work space are $(10, -10), (-40, -30)$, and $(10, -30)$.

(a) What are the coordinates of the fourth vertex?

(b) What is the area of the work space?

70. On a grid map, the towns of Arthur, Belton, Conrad, and Davis are the consecutive vertices of a rectangle. The coordinates (in miles) of Arthur and Conrad are $(-20, -30)$ and $(40, -10)$, respectively. A truck started in Belton, which is directly north of Arthur, and made deliveries to Arthur, Davis, and Conrad, in that order, before returning to Belton. How many miles did the truck travel?

Modeling with Real Data

71. The percentage of household income spent on food has decreased to less than half that spent in 1930. The following table shows the percentages for selected years.

Year	Percentage of Income
1930	25%
1960	17%
2000	11%

(Source: American Farm Bureau Federation.)

(a) Suppose you depict the information as ordered pairs with the year as the first coordinate and then plot the ordered pairs. In which quadrant would the points lie?

(b) Write the information as ordered pairs with the number of years since 1930 as the first coordinate.

72. Numerous medicines are undergoing clinical trials or are awaiting approval by the federal government. The following table shows the number of such drugs for selected diseases.

Disease	Number of Medicines Awaiting Approval
Cancer	354
Heart disease and stroke	104
Diseases of aging	191

(Source: New Medicines in Development, 1999.)

Suppose that the information is displayed in a coordinate system with the y-axis associated with the number of medicines.

(a) Which point would be the greatest distance above the x-axis?

(b) Suppose that a horizontal line is drawn 200 units above the x-axis. What is the significance of the points that lie below this line?

Data Analysis: *Employee Training*

Companies often offer training to both new and current employees. The following table gives a list of training areas, along with the percentages of companies that offer training in those areas.

Basic computer	68%
New technologies	61%
Customer service	57%
Organizational skills	56%
Interpersonal communication	47%
Sales/marketing	35%
Written communication	35%

(Source: Olsen Staffing Services.)

Suppose that the information in the table is displayed in a coordinate system with the y-axis associated with percentages.

73. Which points would be the same distance from the x-axis?

74. Which point would be the greatest distance above the x-axis?

75. Suppose that a horizontal line is drawn 50 units above the x-axis. What is the significance of the points that lie above the line?

76. Suppose that a horizontal line is drawn 40 units above the x-axis and that a second horizontal line is drawn 60 units above the x-axis. What is the significance of the points that lie between the lines?

Challenge

In Exercises 77–80, the coordinates of the endpoints of a line segment are given. Determine the coordinates of the midpoint of the segment.

77. $(-2, 3), (6, 3)$ **78.** $(-5, -2), (-1, -2)$

79. $(-4, 1), (-4, 3)$ **80.** $(3, -3), (3, 7)$

81. One side of a rectangle has vertices $(-2, -3)$ and $(7, -3)$. Another side is 8 units long. Point P is a third vertex in Quadrant I. Determine the coordinates of point P.

82. A square whose sides are 8 units long is drawn with its center at the origin. What are the coordinates of the four vertices?

2.3 | The Graph of an Expression

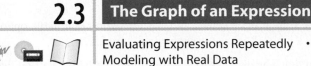

Evaluating Expressions Repeatedly   •   Evaluating Expressions with Graphs   •
Modeling with Real Data

Evaluating Expressions Repeatedly

In Section 2.1 we considered one method for using a calculator to evaluate an algebraic expression. On the home screen, we store the value of the variable and then enter the expression. The calculator reports the value of the expression.

In this section, we consider three additional methods for evaluating an expression. We begin with the calculator's table feature.

⌥ **Keys to the Calculator**

Table

For an expression that has been entered on the Y screen, a table can be produced to show values of the variable in the X column and the corresponding values of the expression in the Y column.
To access your calculator table, press

[2nd] [TABLE]

Table Set

To prepare a calculator table, you can select the initial x-value, the increments of the x-values, and the method by which table values are displayed. To access the table setup menu, press

[2nd] [TBL SET]

Use the arrow keys to select options.

```
TABLE SETUP
 TblStart=-3
 ΔTbl=2
Indpnt: Auto Ask
Depend: Auto Ask
```

1. In the figure, TblStart=−3 means that −3 is the first displayed x-value.
2. ΔTbl=2 means that the x-values will be listed in increments of 2.
3. Setting both options to Auto provides a complete table for both variables. Setting Indpnt: to Ask and Depend: to Auto allows you to enter an x-value of your choice; the corresponding value of the expression will be reported.

EXAMPLE 1

Using a Calculator Table

(a) Set both table options to Auto. Then evaluate $2x - 5$ for $x = 0$, $x = 6$, and $x = 17$.

(b) Set Indpnt: to Ask. Then evaluate $2x - 5$ for $x = 431$.

Solution

(a) First, we enter $2x - 5$ on the Y screen as Y_1. After setting both table options to Auto, we produce a table of values, as shown in Figure 2.12.

Using the up and down arrow keys to scroll through the table, we find that the value of $2x - 5$ for $x = 0, 6$, and 17, are $-5, 7$, and 29, respectively.

Figure 2.12

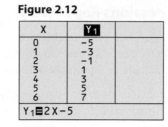

Figure 2.13

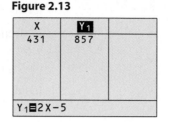

(b) After we set Indpnt: to Ask, we press the table key to obtain a blank table. We enter the x-value of 431 and, as shown in Figure 2.13, we see that the corresponding value of $2x - 5$ is 857.

Note: Setting both table options to Auto is usually better when an expression is to be evaluated repeatedly. The Ask option is usually better for single evaluations, especially when the x-value is not an integer and/or considerable scrolling would be needed to find it in the table.

Our second method for evaluating an expression also begins with entering the expression on the Y screen.

🔑 Keys to the Calculator

Y Var

You can evaluate an expression by entering it on the Y screen and storing the given value of the variable on the home screen. The value of the expression is the value of Y.

Suppose that you have entered $3x - 2$ on the Y screen as Y_1 and that you have stored 5 in X. On the home screen, the current value of $3x - 2$ can be determined simply by entering Y_1.

To retrieve the value of Y_1, use these keys

[VARS] [Y-VARS] [1:Function] [1:Y₁] [ENTER]

```
5→X
                    5
Y₁
                   13
```

EXAMPLE 2

Evaluating an Expression on the Y Screen

Evaluate $x + 4$ for $x = -6, -4, 0, 2$, and 5.

Solution

Begin by entering the expression $x + 4$ as Y_1 on the Y screen (see Fig. 2.14). In effect, we are letting $y_1 = x + 4$.

Figure 2.14

Figure 2.15

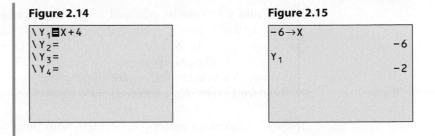

Now return to the home screen, store the first value of x, and retrieve Y_1, which holds the value of $x + 4$ (see Fig. 2.15). Repeat this process for the remaining values of x. The following table summarizes the results.

x	-6	-4	0	2	5
$x + 4$	-2	0	4	6	9

Evaluating Expressions with Graphs

We can use a graph to visualize the relationship between the values of a variable and the corresponding values of an expression containing that variable.

Exploring the Concept

The Graph of an Expression

For the expression $x + 4$ in Example 2, we can write ordered pairs in which the first coordinate corresponds to the value of the variable x and the second coordinate corresponds to the value of the expression.

x	-6	-4	0	2	5
$x + 4$	-2	0	4	6	9
$(x, x + 4)$	$(-6, -2)$	$(-4, 0)$	$(0, 4)$	$(2, 6)$	$(5, 9)$

LEARNING TIP

The graph of an expression is a picture of the infinitely many values of the variable and the corresponding values of the expression.

In Figure 2.16 we have plotted the points associated with the ordered pairs in the table.

Figure 2.16

Figure 2.17

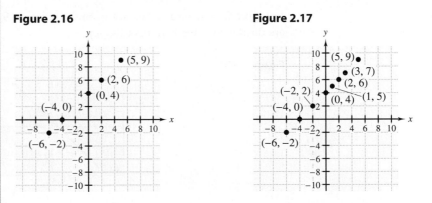

We can continue to evaluate $x + 4$ for other values of x. In Figure 2.17, we have plotted three additional points obtained by evaluating $x + 4$ for $x = -2$, 1, and 3. Note that the more points we plot, the more the graph begins to resemble a straight line.

Because $x + 4$ can be evaluated for any real number value of x, there is no end to the number of possible pairs and points. We say that there are *infinitely many* such points $(x, x + 4)$. We call the graph of all pairs $(x, x + 4)$ the *graph of the expression* $x + 4$. The process of plotting these points is called *graphing the expression*. Although it is impossible to plot all the points $(x, x + 4)$, we can use a calculator to produce a graph of many of the points.

Note: A calculator cannot plot infinitely many points, but it can rapidly plot as many points as the current window settings allow.

When we evaluated $x + 4$ repeatedly in Example 2, we entered the expression on the Y screen. We indicated that we were storing $x + 4$ as Y_1 by writing $y_1 = x + 4$. Now produce the graph of $y_1 = x + 4$ with the integer setting. (See Fig. 2.18.)

Figure 2.18

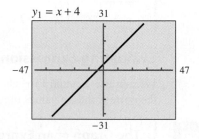

$y_1 = x + 4$

⊙⚿ Keys to the Calculator

Trace

On the graphing screen, a second kind of cursor can be activated. As this *tracing cursor* is moved to each point of a graph, the coordinates of that point are given.

To activate the tracing cursor, press TRACE. Use the left and right arrow keys to move the tracing cursor along the graph.

In Figure 2.19 the tracing cursor is on the point $(5, 9)$, and the coordinates of the point are displayed at the bottom of the screen.

Figure 2.19

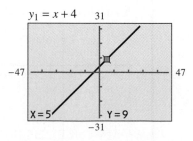

$y_1 = x + 4$

Using the tracing cursor on the graph of an expression gives us another method for evaluating the expression. When we trace to any point of the graph, the first coordinate is the value of x, and the second coordinate is the value of the expression.

As is the case with bar graphs and line graphs, we can read graphs of expressions from either axis to the other axis. Given a value of the variable, we can learn the value of the expression (reading from the x-axis to the y-axis), or given a value of the expression, we can learn the value of the variable (reading from the y-axis to the x-axis).

EXAMPLE **3**

Reading from Either Axis

Use the integer setting to produce the graph of $9 - x$. Trace the graph to complete the following table.

x	-18	-7	0		
$9 - x$				-3	-16
$(x, 9 - x)$	$(-18,\ \ )$	$(-7,\ \ )$	$(0,\ \ )$	$(\ \ , -3)$	$(\ \ , -16)$

LEARNING TIP

As you trace the graph of an expression and observe the displayed coordinates, keep in mind that the first coordinate is the value of the variable and the second coordinate is the value of the expression.

Solution

Figure 2.20 shows the graph of the expression $9 - x$. For the first entry in the table, we trace to the point whose first coordinate is -18. By reading the second coordinate of that point, we see that the corresponding value of the expression is 27.

Figure 2.20 **Figure 2.21**

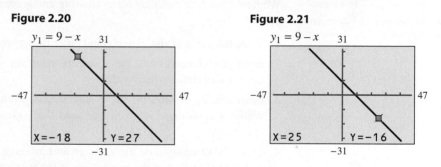

For the last entry in the table, we trace to the point whose y-coordinate is -16. (See Fig. 2.21.) By reading the first coordinate of that point, we see that the corresponding value of x is 25.

We can complete the other entries in the table in the same way.

x	-18	-7	0	12	25
$9 - x$	27	16	9	-3	-16
$(x, 9 - x)$	$(-18, 27)$	$(-7, 16)$	$(0, 9)$	$(12, -3)$	$(25, -16)$

As we see in Example 4, the graph of an expression is not always a straight line. However, we can still use the graph to evaluate the expression.

EXAMPLE **4**

A Graph That Is Not a Line

Use the integer setting to produce the graph of $y_1 = 0.05x^2$. Then use the graph and the tracing cursor for each of the following.

(a) Evaluate $0.05x^2$ for $x = -18, 9, 17$.

(b) Determine the x-value such that the value of $0.05x^2$ is 7.2.

Solution

(a) Figure 2.22 shows the graph of $0.05x^2$ with the tracing cursor on $(-18, 16.2)$. Therefore, when $x = -18$, the value of $0.05x^2$ is 16.2. With additional tracing, we find that when $x = 9$, $0.05x^2 = 4.05$ and that when $x = 17$, $0.05x^2 = 14.45$.

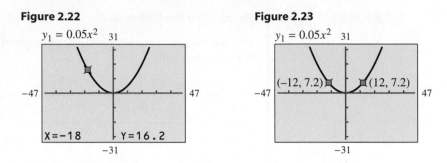

Figure 2.22

$y_1 = 0.05x^2$

Figure 2.23

$y_1 = 0.05x^2$

(b) We trace to a point whose y-coordinate is 7.2. (See Fig. 2.23.) Note that there are two such points: $(12, 7.2)$ and $(-12, 7.2)$. Therefore, $0.05x^2$ has a value of 7.2 when x is either 12 or -12.

Think About It

Suppose that you want to graph an expression whose value is the same for every x-value. Describe the graph.

We now have four methods for evaluating an algebraic expression with a graphing calculator.

1. Store the value of the variable and enter the expression on the home screen.

2. Enter the expression on the Y screen, store the value of the variable on the home screen, and retrieve the y-value.

3. Enter the expression on the Y screen and produce the graph. Trace to the point whose x-coordinate is given and read the y-coordinate, which is the value of the expression.

4. Enter the expression on the Y screen and produce a table. Scroll to the given x-value and read the value of the expression in the Y column.

Modeling with Real Data

A *model* is a representation of a thing (such as a model airplane), a concept (such as an economic theory), or data. In mathematics, we can use an algebraic expression to model certain conditions or information.

EXAMPLE 5

Modeling Real Data with an Expression

In Section 2.2, Example 3, we presented a table to show the increase in the number of cellular phone antenna sites during the period 1985–2005.

Year	Cellular Phone Antenna Sites
1985	599
1995	21,000
2005	117,920

(Source: Cellular Telecommunications Industry Association.)

The expression $7.5t^3$ can be used to model the information in the table. In this model, t represents the number of years since 1980.

(a) What values of t represent the years in the table?

(b) How many cellular phone antenna sites does the model indicate for the 3 years in the table?

(c) Use the model to estimate the number of antenna sites in the year 2000.

Solution

(a) Because t represents the number of years since 1980,

Year	Value of t
1985	5
1995	15
2005	25

(b)

t	Rounded Value of $7.5t^3$
5	$7.5(5)^3 = 7.5(125) = 938$
15	$7.5(15)^3 = 7.5(3375) = 25,313$
25	$7.5(25)^3 = 7.5(15,625) = 117,118$

(c) For the year 2000, $t = 20$.
$$7.5t^3 = 7.5(20)^3 = 7.5(8000) = 60,000$$

Note: As shown in Example 5, a mathematical model usually cannot be expected to provide results that are completely accurate. Different models can be used to describe the same data. In general, the more accurate the results, the more valid we consider the model to be.

Quick Reference 2.3

Evaluating Expressions Repeatedly

- To evaluate an expression with a calculator, we can enter the expression on the Y screen as Y_1. Then

  (a) display a table of x-values and the corresponding values of the expression, or

  (b) on the home screen, store the value of x and retrieve the value of Y_1.

Evaluating Expressions with Graphs

- We can relate the value of an expression for a given value of the variable with an ordered pair in the form (value of the variable, value of the expression).

- We call the graph of all such pairs the *graph of the expression*. The process of plotting these points is called *graphing the expression*. When we graph an expression, the x-coordinate of each point is the value of the variable, and the y-coordinate is the corresponding value of the expression.

- When we produce the graph of an expression, we can activate a *tracing cursor* that can be moved to the displayed points of the graph. The coordinates of the point that is highlighted by the tracing cursor are displayed.

- With a graphing calculator, we can read the graph of an expression in either of two ways.

  1. To evaluate an expression for a given x-value, trace the graph to the point whose x-coordinate is the given x-value. The value of the y-coordinate at that point corresponds to the value of the expression.

  2. To determine the x-value for which an expression has a given value, trace the graph to the point whose y-coordinate is the given value of the expression. Then the x-coordinate is the desired value of x.

Speaking the Language 2.3

1. If you enter an expression on the Y screen and produce a table of values, the numbers in the Y column are values of the ▨▨▨▨▨ .

2. Each point of the graph of an expression has two coordinates. The first coordinate is the value of the ▨▨▨▨▨ , and the second coordinate is the value of the ▨▨▨▨▨ .

3. The coordinates of the points of a calculator graph can be displayed by moving a(n) ▨▨▨▨▨ cursor along the graph.

4. To evaluate an expression by entering it on the home screen, you must first ▨▨▨▨▨ the value of the variable.

Exercises 2.3

Concepts and Skills

1. If an expression is to be evaluated repeatedly, what is the advantage of using the Y screen rather than the home screen?

2. To evaluate $x^2 + 1$, we can enter the expression as Y_1 on the Y screen, store the value of the variable on the home screen, and retrieve Y_1. What does this value of Y_1 represent?

In Exercises 3–14, use a calculator table to evaluate the given expression for each of the given values of x.

3. $x - 6$; $x = -1, 0, 4, 11$

4. $x + 3$; $x = -7, -2, 3, 15$

5. $-2.1x + 7$; $x = -8, 3, 6, 12$

6. $3x - 5.6$; $x = -6, -1, 5, 11$

7. $x^2 + 4x - 21$; $x = -9, -4, 3, 5$

8. $25 - x^2$; $x = -7, -2, 6, 7$

9. $2(x + 8)$; $x = -8, -1, 2, 10$

10. $-4(x - 1)$; $x = -3, 1, 9, 20$

11. $|2x + 1|$; $x = -5, -1, -0.5, 8$

12. $|9 - x|$; $x = -4, 6, 9, 14.3$

13. $\dfrac{x + 1}{2x}$; $x = -5, 1, 25, 50$

14. $\dfrac{2x - 4}{x - 2}$; $x = -12, -7, 612, 1995$

In Exercises 15–20, use the method of retrieving Y_1 to evaluate the expression in the second row of the table for the x-values given in the first row. Then write the results as ordered pairs in the third row.

15.

x	-2	1	4	7	10
$x - 3$					
$(x, x - 3)$					

16.

x	-8	-5	-2	1	4
$x + 6$					
$(x, x + 6)$					

17.

x	3	7	0	-4	-5
$2x + 5$					
$(x, 2x + 5)$					

18.

x	1	2	-1	-3	-4
$1 - 3x$					
$(x, 1 - 3x)$					

19.

x	1	4	6	-3	-5
$x^2 - 3x - 4$					
$(x, x^2 - 3x - 4)$					

20.

x	-6	-4	1	2	4
$6 - x - x^2$					
$(x, 6 - x - x^2)$					

21. Suppose that you use the Y screen to evaluate $\dfrac{x}{x - 1}$ for $x = -3, 1, 4, 9$. For one of these x-values you obtain an error message. Why?

22. Suppose that you decide to use a graph to evaluate $2x + 1$ for $x = 5$. How do you know when the tracing cursor is on the right point?

In Exercises 23–26, the given figure shows the graph of a certain expression. Use the graph to determine the value of the expression for the given value of x.

23. $x = 3$　　　　**24.** $x = -2$

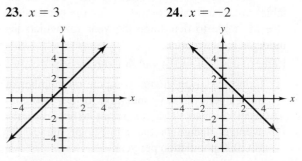

25. $x = -1$　　　　**26.** $x = 0$

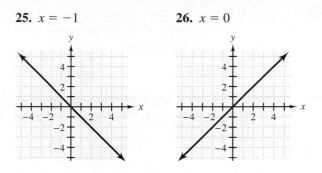

In Exercises 27–32, enter the given expression as Y_1 on the Y screen and produce the graph with the integer setting. Then trace the graph to estimate the value of the expression for each given value of x.

27. $2x - 10$;　$x = 2, 5, 7$

28. $7 + x$;　$x = -2, -5, -10$

29. $\dfrac{1}{3}x + 5$;　$x = -30, -9, 12$

30. $9 - \dfrac{3}{4}x$;　$x = -12, 4, 26$

31. $-(3x + 1)$;　$x = -6, 0, 5$

32. $2x - (x + 1)$;　$x = -11, 7, 28$

33. Suppose that $P(3, 7)$ is a point of the graph of an expression. Explain the significance of the two coordinates.

34. Suppose that you have produced the graph of $x + 3$ and you want to trace the graph to learn the value of the variable for which the expression has a value of 8. How will you know when the tracing cursor is on the right point?

In Exercises 35–38, use a graph to fill in the blanks in the table.

35.

x	2		0		-1	
$x - 4$		0		1		-6

36.

x	-1		-3		2	
$x + 3$		1		-1		7

37.

x		-8		16		28
$\dfrac{1}{4}x + 3$	-5		0		4.5	

38.

x	▬ -16 ▬ 2 ▬ 12
$6 - \dfrac{1}{2}x$	2.5 ▬ -4 ▬ -10 ▬

In Exercises 39–44, use a graph to determine the value of x such that the expression has the value given in parts (a)–(d).

39. $x + 7$

    (a) 10   (b) 16

    (c) 2   (d) -3

40. $5 - x$

    (a) 9   (b) 6

    (c) 2   (d) -3

41. $7 - 2x$

    (a) 9   (b) 3

    (c) -3 (d) -9

42. $3x + 11$

    (a) 2   (b) 8

    (c) 11   (d) 14

43. $10 - 0.5x$

    (a) 21.5   (b) 7

    (c) 0   (d) -6.5

44. $0.4x - 8$

    (a) -4 (b) -16.8

    (c) 0   (d) 2

In Exercises 45–48, use a graph to determine all values of x such that the expression has the value given in parts (a)–(d).

45. $9x - x^2$

    (a) -22 (b) -10

    (c) 8   (d) 14

46. $x^2 + 2x - 35$

    (a) 0   (b) -20

    (c) -11 (d) 13

47. $|x - 5|$

    (a) 17   (b) 8

    (c) 0   (d) 4

48. $|7 - 2x|$

    (a) 21   (b) 7

    (c) 15   (d) 27

In Exercises 49–52, from the set $\{-1, 1, -2, 2\}$ choose the value for which both expressions have the same value.

49. $2x + 1$, $4 - x$

50. $1 - 3x$, $x + 5$

51. $3x$, $8 - x$

52. $5x - 2$, $3x - 6$

🌐 Real-Life Applications

53. For a family reunion, 17 pizzas were ordered. If x people attended and each family member ate half of a pizza, the expression $17 - 0.5x$ represents the amount of pizza left at the end of the gathering. Produce the graph of this expression and trace it to estimate the amount of pizza left if 28 people attended the reunion.

54. The cost (in dollars) of a car rental is given by the expression $0.2x + 20$, where x is the number of miles driven. Produce the graph of this expression and trace it to estimate the number of miles driven when the rental cost is $28.

📐 Modeling with Real Data

55. The following table shows the number of registered nurses per 100,000 population during the period 1998–2000. With t representing the number of years since 1995, the data can be modeled by $-5.5t + 972$.

Year	Number of Registered Nurses (per 100,000 Population)
1998	956
1999	949
2000	945

(Source: American Nursing Association.)

For values of t corresponding to the years in the table, produce a table of values for the expression. Use the table to determine the year for which the model is least accurate.

56. Refer to the data and model in Exercise 55.

    (a) Using the given model, compute the number of nurses per 100,000 population for any two consecutive years. Use the result to estimate the average change in the number of nurses per 100,000 population per year.

(b) Use the data in the table to compute the average per-year decrease in the number of nurses per 100,000 population from 1998 to 2000.

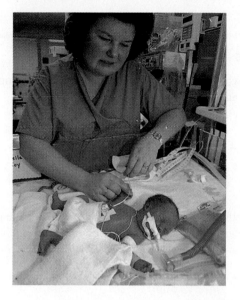

Data Analysis: *African-American Elected Officials*

The accompanying bar graph shows there has been a dramatic increase in the number of African-American elected officials since the passage of the Voting Rights Act in 1965. (Source: Joint Center for Political and Economic Studies.)

Figure for 57–60

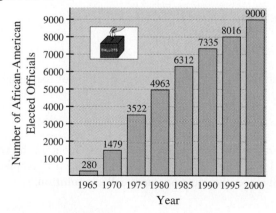

The expression $253.6x + 675.3$ can be used to model the number of elected officials. In this expression, x represents the number of years since 1965.

57. Use a calculator to evaluate the model expression for each of the years indicated in the bar graph. Summarize your results in the following table.

Year	x	Value of the Expression
1965	0	
1970	5	
1975	10	
1980	15	
1985	20	
1990	25	
1995	30	
2000	35	

58. For what year is the model expression the best approximation of the actual data in the bar graph?

59. For what year is the model expression the least accurate approximation of the actual data?

60. Assuming the model expression is valid, what is the projected number of African-American elected officials in the year 2005?

Challenge

In Exercises 61–64, suppose that you have produced the graph of the expression. Describe how the y-coordinate of each point of the graph compares to the x-coordinate.

61. $2x + 1$

62. $\dfrac{1}{2}x - 3$

63. $2(x + 1)$

64. $7 - x$

In Exercises 65 and 66, use the Y screen to evaluate the expression for the given values of the variables.

65. $c^2 + 5k$

   (a) $c = 0.5$, $k = 6.1$

   (b) $c = -0.5$, $k = 6.1$

66. $\dfrac{a + b + c}{3}$

   (a) $a = 85$, $b = 91$, $c = 73$

   (b) $a = 56$, $b = 79$, $c = 84$

2.4 Equations and Estimated Solutions

Introduction to Equations • Testing for Solutions • Estimating Solutions of Equations • Special Cases

Introduction to Equations

An **equation** is a statement that two expressions have the same value. It is important to recognize the difference between an algebraic expression and an equation.

Expression	*Equation*
$7x - 4$	$7x - 4 = -5$
$x^2 + 4x + 3$	$x^2 + 4x + 3 = 0$
$3(x - 5) + 1 + 2x$	$3(x - 5) = 1 + 2x$

The expression to the left of the equality symbol is called the *left side* of the equation, and the expression to the right is called the *right side* of the equation.

$$\underbrace{x^2 - 7x + 1}_{\text{left side}} = \underbrace{6 - 8x}_{\text{right side}}$$

An equation may be true or false. If an equation has no variables, we can determine whether the equation is true simply by evaluating both sides of the equation.

$7 - 5 = 5 - 7$ False

$3 + 2 \cdot 8 = 20 - 1$ True

Even if an equation contains a variable, the equation may be true for any permissible replacement of the variable. Such equations are called **identities.** The following equations are all identities.

$6(3x) = 18x$ Associative Property of Multiplication

$y + 3 = 3 + y$ Commutative Property of Addition

$3(x - 5) = 3x - 15$ Distributive Property

Some equations, called **contradictions,** are false for all replacements of the variable. The following are contradictions.

$0 \cdot y = 4$ $x + 1 = x + 5$

A **conditional equation** may be true or false depending on the replacement for the variable(s). For example, $5x + 1 = 11$ is true if x is replaced with 2 and false if x is replaced with 1. A value of the variable that makes the equation true is called a **solution.**

Testing for Solutions

Given an equation, we can test any number to determine whether it is a solution of the equation.

EXAMPLE **1**

Testing for Solutions

For each equation, determine whether the given replacement for the variable is a solution.

(a) $1 - 3x = -5$;   2

(b) $2x + 5 = 1 - x$; -2

Solution

(a) $1 - 3x = -5$

$1 - 3(2) = -5$ Replace x with 2.

$1 - 6 = -5$ Evaluate the left side.

$-5 = -5$ True

Therefore, 2 is a solution.

(b) $2x + 5 = 1 - x$

$2(-2) + 5 = 1 - (-2)$ Replace x with -2.

$-4 + 5 = 1 + 2$ Evaluate both sides.

$1 = 3$ False

Therefore, -2 is not a solution.

The following example illustrates one way in which a calculator can be used to test for solutions.

EXAMPLE **2**

Using a Calculator to Test for Solutions

Use a calculator to determine whether 4 is a solution of $2(x - 1) = x + 2$.

Solution

Figure 2.24 shows that when 4 is stored in **x**, the left side $2(x - 1)$ and the right side $x + 2$ both have a value of 6. Therefore, 4 is a solution.

Figure 2.24

```
4→X
                    4
2(X-1)
                    6
X+2
                    6
```

Estimating Solutions of Equations

To this point we have only tested given numbers to determine whether they were solutions of equations. A more important skill is solving an equation. To **solve** an equation means to determine each value of the variable that is a solution.

Because there are many types of equations, the methods for solving them vary. The first type of equation that we will learn to solve is a **linear equation in one variable.**

> **Definition of a Linear Equation in One Variable**
>
> A **linear equation in one variable** is an equation that can be written in the form $Ax + B = 0$, where A and B are real numbers and $A \neq 0$.

The following are examples of linear equations.

$$-3x + 14 = 0 \qquad 2x - 9 = 21 \qquad 3(x - 1) = 3 - x$$

The first equation is in the form $Ax + B = 0$. As we will see later, the other two equations can also be written in that form.

In the remainder of this section, we consider ways to estimate the solution of a linear equation by using a graph. Example 3 illustrates the process.

EXAMPLE 3

Estimating the Solution of a Linear Equation

Use the integer setting and a graph to estimate the solution of $3x - 11 = 16$.

Solution

Figure 2.25 shows the graph of the expression $3x - 11$. Because $3x - 11 = 16$, we trace the graph to the point whose y-coordinate is 16.

Figure 2.25

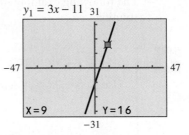

From the display we see that the value of $3x - 11$ is 16 when $x = 9$. Thus 9 is the estimated solution of the equation.

Note: We can easily verify that the estimated solution in Example 3 is the actual solution.

$$3x - 11 = 16$$
$$3(9) - 11 = 16 \qquad \text{Replace } x \text{ with the estimated solution 9.}$$
$$27 - 11 = 16 \qquad \text{Evaluate the left side.}$$
$$16 = 16 \qquad \text{True}$$

Figure 2.26

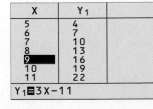

We can also estimate the solution of the equation in Example 3 by producing a table of values for the expression $3x - 11$. In Figure 2.26 we see that the value of $3x - 11$ is 16 when x is 9. Thus the solution of the equation is 9.

Another approach to estimating a solution is to graph the expressions on both sides of the equation.

Estimating Solutions by Graphing

Consider the equation $2x + 4 = x - 6$. To estimate the value of x for which the expressions are equal, we can display a table of values for each expression: $y_1 = 2x + 4$, $y_2 = x - 6$. (See Fig. 2.27.)

From the table, we see that when x is -10, both $2x + 4$ and $x - 6$ have a value of -16. Therefore, -10 is the estimated solution.

Figure 2.28 shows the graphs of the left and right sides of the equation with the tracing cursor on the point of intersection $(-10, -16)$.

Figure 2.27

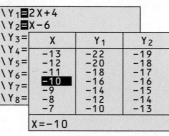

Figure 2.28

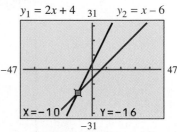

Because the point of intersection is a point that belongs to both graphs, we conclude that both expressions have a value of -16 when x is -10. Thus the point of intersection corresponds to the estimated solution of the equation.

Estimating Solutions by Graphing

1. Graph each side of the equation.

2. Trace to the point of intersection of the graphs.

3. The x-coordinate of the point of intersection is the estimated solution of the equation. (The y-coordinate is the corresponding value of each side of the equation.)

Note: Because we have already entered the left and right sides on the Y screen in order to produce the graphs, we can easily use the calculator to verify the estimated solution by storing it in the variable and evaluating the two expressions.

When we use a calculator to graph the two sides of an equation, the displayed point of intersection may not represent the exact solution. Example 4 shows how we can magnify the display to improve our estimate of the coordinates of the point of intersection.

⌨ Keys to the Calculator

Zoom In

You can magnify any portion of a graph. Move the tracing cursor to the point where you wish to magnify the graph. Then press

[ZOOM] [2:Zoom In] [ENTER]

EXAMPLE 4

Estimating a Solution by Graphing

Use the integer setting and the graphing method to estimate the solution of the equation $21 - 2x = \frac{1}{3}x + 1$.

Solution

Figure 2.29 shows the graphs of the left and right sides of the equation. The tracing cursor is on the apparent point of intersection, and the estimated solution is 9. However, it is easy to verify that 9 is not the exact solution of the equation.

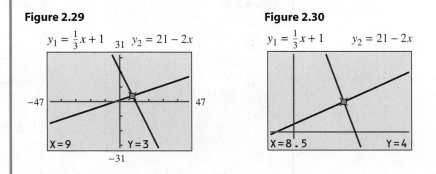

Figure 2.29

Figure 2.30

In Figure 2.30, we have magnified the display by zooming in on the point of intersection, and we have traced again to that point. Now the estimated solution is 8.5. Although this is still not the exact solution, additional zooms will provide better and better estimates. It can be shown that the exact solution is $\frac{60}{7}$, which is approximately 8.57.

Note: The zoom-and-trace procedure can be used repeatedly. We will usually stop when the estimated solution is accurate to the nearest hundredth.

Special Cases

Using a calculator for estimating solutions may lead us to suspect that an equation is an identity or a contradiction.

EXAMPLE 5

Special Cases of Equations

Estimate the solutions of the following equations.

(a) $x - 7 = x + 8$

(b) $x - 10 = -10 + x$

Solution

(a) We enter $x - 7$ as Y_1 and $x + 8$ as Y_2, and we produce the table shown in Figure 2.31. The table indicates that the values of Y_2 are always 15 greater than the values of Y_1. This suggests that $x - 7$ and $x + 8$ can never be equal, and so the equation has no solution.

Figure 2.31

X	Y₁	Y₂
-3	-10	5
-2	-9	6
-1	-8	7
0	-7	8
1	-6	9
2	-5	10
3	-4	11

Figure 2.32

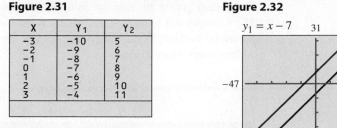

In Figure 2.32, we have produced the graphs of $x - 7$ and $x + 8$. The graphs do not appear to intersect, which suggests that the equation is a contradiction and has no solution.

(b) We use a similar approach to the equation $x - 10 = -10 + x$. Figure 2.33 is a table of values for the expressions $x - 10$ and $-10 + x$. The table indicates that the two expressions have the same values for all values of x. This suggests that every real number is a solution.

Figure 2.33

X	Y₁	Y₂
-3	-13	-13
-2	-12	-12
-1	-11	-11
0	-10	-10
1	-9	-9
2	-8	-8
3	-7	-7

Figure 2.34

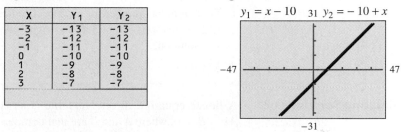

In Figure 2.34, the graphs of $x - 10$ and $-10 + x$ appear to coincide, which means that every point of the graph is a point of intersection. Therefore, we conclude that the equation is an identity and every real number is a solution.

It is often convenient to describe the solution(s) of an equation with a set called the **solution set.** For example, the solution of $x - 3 = 5$ is 8, and we write the solution set $\{8\}$. If an equation has two solutions such as -2 and 4, we would write the solution set $\{-2, 4\}$.

Because the equation in part (a) of Example 5 has no solution, the solution set has no elements. We call this set the **empty set,** and we denote it with the symbol $\emptyset$.

In part (b), every real number is a solution, so the solution set is **R,** the set of all real numbers. Except for $\emptyset$ and **R,** we will not use set notation to describe solutions.

Note: The equations in Example 5 illustrate the limitations of a graphing calculator. Are the graphs in part (a) really parallel, or do they only appear to be parallel? Do the graphs in part (b) really coincide? Eventually, we will need to find ways of being certain.

Producing graphs to estimate solutions of equations is an excellent way to visualize the problem and to understand what it means to solve an equation. However, we must be aware that graphs are merely suggestive. Until a solution is verified, it must be considered only an estimate.

Quick Reference 2.4

Introduction to Equations

- An **equation** is a statement that two expressions have the same value.
- An **identity** is an equation that is true for all permissible replacements of the variable. A **contradiction** is an equation that is false for all permissible replacements of the variable.
- A **conditional equation** is an equation that is true or false depending on the replacement for the variable. A replacement that makes an equation true is called a **solution.**

Testing for Solutions

- We can determine whether a number is a solution of an equation in either of the following ways.
  1. Substitute the number for each occurrence of the variable. If the resulting numerical expressions on the left and right sides of the equation are equal, the number is a solution.
  2. Use a calculator to store the number and evaluate the two sides of the equation.

Estimating Solutions of Equations

- A **linear equation in one variable** is an equation that can be written in the form $Ax + B = 0$, where A and B are real numbers and $A \neq 0$.
- In general, we can estimate the solution of an equation as follows.
  1. Graph each side of the equation.
  2. Trace to the point of intersection of the graphs.
  3. The x-coordinate of the point of intersection is the estimated solution of the equation.
- We can also estimate a solution by producing a table of values for the left and right sides of the equation. The x-value for which the two expressions have the same value is the solution of the equation.

Special Cases

- If the graphs of the expressions on the two sides of an equation never intersect, then there is no point of intersection, and the equation has no solution. The equation is a contradiction, and the solution set is $\emptyset$.
- If the graphs of the expressions on the two sides of an equation coincide, then every point of the two graphs is a point of intersection, and every real number is a solution. The equation is an identity, and the solution set is **R.**

Speaking the Language 2.4

1. The statement that $2x + 3$ and $5 - x$ have the same value is called a(n) ▨▨▨▨▨ . The value of x for which the statement is true is called a(n) ▨▨▨▨▨ .

2. An equation of the form $Ax + B = 0$, $A \neq 0$, is called a(n) ▨▨▨▨▨ equation in one variable.

3. The equation $x + 3 = 5$ is true if x is 2 and false otherwise. This equation is an example of a(n) ▨▨▨▨▨ equation.

4. A solution of an equation is represented by a(n) ▨▨▨▨▨ of the graphs of the two sides of the equation.

Exercises 2.4

Concepts and Skills

1. What is the difference between an expression and an equation?

2. What is the difference between a conditional equation and an identity?

In Exercises 3–10, identify expressions and equations.

3. $2x + 3$

4. $3x = 7$

5. $x - y = 2$

6. $2x + 3y - 1$

7. $x^2 + x = 2$

8. $x^2 + 3x + 2$

9. $\dfrac{x + 1}{3}$

10. $y = \dfrac{3y}{x - 1}$

11. What is a solution of an equation?

12. What is a linear equation in one variable?

In Exercises 13–20, without using a calculator, determine whether the given number is a solution of the equation. Answer with "yes" or "no."

13. $x - 5 = 9$;   14

14. $7 - x = 9$;   2

15. $5 - 3x = 11$;   2

16. $4x + 9 = 5$; -1

17. $10x - 3 = 5x + 1$;   1

18. $1 - 2x = 3x + 1$;   0

19. $x^2 + 3x + 2 = 0$; -1

20. $x(x - 5) = 6$;   1

In Exercises 21–28, use your calculator to determine whether the given number is a solution of the equation. Answer with "yes" or "no."

21. $3x - 4 = -1$;   1

22. $8 - 3x = 2$; -2

23. $x + 4 = 2 - x$; -1

24. $x - 3 = 4 - x$;   1

25. $3x + 5 = 2x + 2$; -3

26. $2(x - 1) = x$;   2

27. $x^2 + 2x = 0$;   2

28. $4 - x^2 = 0$; -2

In Exercises 29–40, use the integer setting and estimate the solution of the given equation by tracing the graph.

29. $2x - 11 = 17$

30. $5 - 3x = 23$

31. $12 - x = 0$

32. $x + 9 = 0$

33. $9 = 21 - 3x$

34. $6 = -x - 15$

35. $15 - 0.3x = 19.2$

36. $0.4x - 18 = -12.4$

37. $\dfrac{1}{4}x - 6 = -14$

38. $\dfrac{1}{2}x - 14 = 4$

39. $9 = -\dfrac{1}{5}x + 5$

40. $11 = 8 + \dfrac{1}{3}x$

41. Describe a graphing method you can use to estimate the solution of the following equation.
$$3x + 1 = 7 - 5x$$

42. Suppose that you have used a graphing method to estimate the solution of an equation and the point of intersection is $P(a, b)$. What is the significance of the coordinates a and b?

In Exercises 43–48, use the graphing method to estimate the solution of the given equation.

43. $5x + 3 = x - 2$

44. $2x - 3 = 4 + 6x$

45. $3x = 1 - 4x$

46. $-4x = 5x + 4$

47. $\frac{3}{2}x + 1 = \frac{1}{3}x - 1$

48. $7x + \frac{1}{6} = \frac{1}{2} + 6x$

In Exercises 49–66, use a table or the graphing method to estimate the solution of the given equation.

49. $2x + 27 = 6 - x$

50. $x - 36 = -x$

51. $x + 4 = 2x + 16$

52. $2x - 27 = x - 5$

53. $\frac{1}{3}x - 16 = 2x + 4$

54. $-\frac{2}{3}x + 26 = \frac{1}{2}x + 5$

55. $\frac{1}{2}x + 19 = \frac{1}{5}x + 13$

56. $-\frac{1}{4}x - 13 = -\frac{3}{2}x + 7$

57. $1.25x - 2.5 = -0.5 + x$

58. $24.5 - 2.1x = 1.3x + 7.5$

59. $x - 3 = 3(x + 5)$

60. $-5(2 + 0.3x) = 2(x - 5)$

61. $3(x - 5) - 2x = 5(3 - x) + 4x$

62. $2(x + 12) = -x$

63. $5 - 2(6 - x) = 5(x + 4)$

64. $6(x - 3) - x = 12$

65. $\frac{x - 5}{2} = \frac{x}{3}$

66. $\frac{2x - 1}{5} = \frac{x}{2}$

67. Explain how each of the following special cases can be recognized graphically and state the solution set.

(a) A contradiction　　(b) An identity

68. Without graphing, state the solution set for each of the following. Explain how you know.

(a) $x + 5 = 5 + x$

(b) $2(x - 3) = 2x - 6$

(c) $0x = 9$

In Exercises 69–86, use the graphing method to estimate the solution(s) of the given equation. In the case of an identity or contradiction, write the solution set.

69. $2x + 8 = 8 + x$

70. $x + 6 - 2x = 7 + x - 1$

71. $4 - x = 12 - x$

72. $2x + 7 = 2x - 9$

73. $x - 8 = -(8 - x)$

74. $-2(x - 6) = 12 - 2x$

75. $2x = 2(x - 3) - 1$

76. $9 - x = x + 2(1 - x)$

77. $2(2x + 3) = 2 + 4(x + 1)$

78. $2x - (x + 10) = -(10 - x)$

79. $3(6 - x) = \frac{3}{2}x - 18$

80. $5(x + 1) - 3x = \frac{1}{2}(x + 10)$

81. $2 - 0.5x = -0.25x - 4$

82. $1.5x - 16 = 22 - 0.4x$

83. $2x + 1 = 4(x - 2) - 2x$

84. $4(x - 1) = 1 + 4(x + 1)$

85. $8 - (2 + x) = 2x + 3(2 - x)$

86. $2(x + 1) + 1 = 4x - (2x - 3)$

Modeling with Real Data

87. The table shows a comparison of the average ticket prices for NFL and NBA games during the period 1991–1999. The average ticket prices can be modeled by the expressions $2.6x + 21.6$ for the NFL and $3.1x + 18.4$ for the NBA, where x represents the number of years since 1990.

Year	Average Ticket Price	
	NFL	**NBA**
1991	$25.21	$23.24
1993	28.68	27.12
1996	35.74	34.08
1999	45.93	48.73

(Source: Team Marketing Report.)

(a) Write an equation to determine the year in which an NFL ticket would cost $58. Then estimate the solution of the equation.

(b) Write an equation to determine the year in which an NBA ticket would cost $68. Then estimate the solution of the equation.

88. Refer to the model expressions in Exercise 87.

(a) Write an equation that indicates that the prices of an NFL ticket and an NBA ticket are the same.

(b) Estimate the solution of the equation. Interpret the result.

Challenge

In Exercises 89 and 90, first use the standard setting to produce the graphs of the left and right sides of the given equation. What is the apparent solution set of the equation? Then use the integer setting to produce the graphs. What do you conclude?

89. $0.4x - 5 = 6 + 0.3x$

90. $-x - 7 = -1.2x + 5$

In Exercises 91–94, use the graphing method with the integer setting to estimate the solution of the given equation.

91. $|x - 12| = 18$

92. $-9 = |x + 9|$

93. $0.1x^2 - 4x + 20 = -25$

94. $-0.1x^2 - 2x + 10 = -20$

2.5 | Properties of Equations

Equivalent Equations · Addition Property of Equations · Multiplication Property of Equations

Equivalent Equations

Consider the following three equations.

$$2x - 1 = 7 \qquad 2x = 8 \qquad x = 4$$

It is easy to verify that 4 is the solution of all three equations. In fact, we say that the equation $x = 4$ can be solved *by inspection* because the solution is obvious. These equations are examples of **equivalent equations.**

> **Definition of Equivalent Equations**
>
> **Equivalent equations** are equations that have exactly the same solutions.

One way in which equivalent equations can be produced is to exchange the left and right sides of the equation.

$$6 = x - 5$$
$$x - 5 = 6 \qquad \text{Exchanging the sides of the equation produces an equivalent equation.}$$

A second way of producing equivalent equations is to replace any expression with an equivalent expression. This is normally accomplished by simplifying an expression.

$$2(x + 3) + x = 7$$
$$2x + 6 + x = 7 \qquad \text{Remove parentheses with the Distributive Property.}$$
$$3x + 6 = 7 \qquad \text{Combine like terms.}$$

These three equations are all equivalent.

In Section 2.4 we estimated solutions of equations graphically. Solving an equation *algebraically* means using algebraic operations to produce equivalent equations until we obtain one that can be solved by inspection.

EXAMPLE **1**

Solving an Equation Algebraically

Solve $3x - 2x = 9$ algebraically.

Solution

$$3x - 2x = 9$$
$$x = 9 \qquad \text{Combine like terms.}$$

The equation $x = 9$ is equivalent to the original equation, and it can be solved by inspection. The solution is 9.

As we see in Example 1, if an equation can be written in the form $x = $ number, then the number is the solution. Thus our strategy is to write equivalent equations until the variable is alone on one side of the equation. We call this *isolating* the variable.

Sometimes more than just simplifying is needed in order to isolate the variable. In the remainder of this section, we consider two properties of equations that are essential to solving an equation algebraically.

Addition Property of Equations

Adding the same number to both sides of an equation results in an equivalent equation.

LEARNING TIP

Think of the two sides of an equation as balanced (equal). To maintain the balance, we must perform the same operations on both sides of the equation.

Addition Property of Equations

For real numbers a, b, and c, the equations $a = b$ and $a + c = b + c$ are equivalent equations.

In words, we can add the same number to both sides of an equation.

Note: Because subtraction is defined in terms of addition, we can also subtract the same number from both sides of an equation.

Exploring the Concept

Solving with the Addition Property of Equations

To solve $x - 6 = 5$, our goal is to isolate the variable x by eliminating 6 from the left side. We can accomplish this by adding 6 to both sides of the equation.

$$x - 6 = 5$$
$$x - 6 + 6 = 5 + 6 \qquad \text{Addition Property of Equations}$$
$$x + 0 = 11$$
$$x = 11$$

All of these equations are equivalent, and the last one can be solved by inspection. We can easily verify that 11 is the solution of the original equation.

$$x - 6 = 5$$
$$11 - 6 = 5 \qquad \text{Replace x with 11.}$$
$$5 = 5 \qquad \text{True}$$

Note: Although algebraic methods lead to an exact solution, the solution should be verified as a check against errors. In our remaining examples, we will leave the verification of solutions to you.

EXAMPLE 2

Solving with the Addition Property of Equations

Solve the following equations.

(a) $-8 + y = -8$ (b) $3 = x + 5$

Think About It

By adding 10 to both sides of the equation $x - 7 = 4$, we obtain the equivalent equation $x + 3 = 14$. How does the graphing method reveal that the equations are equivalent? Why do the graphs of the two sides of each equation not intersect at the same point?

Solution

(a) $-8 + y = -8$

$-8 + y + 8 = -8 + 8$ Add 8 to both sides.

$y = 0$ The variable is isolated.

The solution is 0.

(b) Although it is not necessary, you may wish to exchange the sides of an equation to place the variable on the left side.

$3 = x + 5$

$x + 5 = 3$ Exchange sides.

$x + 5 - 5 = 3 - 5$ Subtract 5 from both sides.

$x = -2$ The variable is isolated.

The solution is -2.

If a variable appears on both sides of an equation, we eliminate the variable term from one side of the equation.

EXAMPLE 3

Solving Equations When the Variable Is on Both Sides

Solve $5x = 3 + 4x$.

Solution

We subtract $4x$ from both sides of the equation so that the variable is on only one side.

$5x = 3 + 4x$

$5x - 4x = 3 + 4x - 4x$ Subtract 4x from both sides.

$1x = 3 + 0$

$x = 3$

The solution is 3.

Sometimes we begin by simplifying the expressions on both sides of the equation.

EXAMPLE 4

Equations that Require Simplifying

Solve $-4x + 5x - 6 = 7 + 3$.

Solution

$$-4x + 5x - 6 = 7 + 3$$
$$x - 6 = 10 \qquad \text{Combine like terms.}$$
$$x - 6 + 6 = 10 + 6 \qquad \text{Add 6 to both sides.}$$
$$x = 16$$

The solution is 16.

For some equations it may be necessary to use the Addition Property of Equations more than one time.

EXAMPLE 5

Repeated Use of the Addition Property of Equations

Solve $7x - 4 = 6x + 5$.

Solution

$$7x - 4 = 6x + 5$$
$$7x - 4 - 6x = 6x + 5 - 6x \qquad \text{Subtract } 6x \text{ from both sides.}$$
$$x - 4 = 5 \qquad \text{Combine like terms.}$$
$$x - 4 + 4 = 5 + 4 \qquad \text{Add 4 to both sides.}$$
$$x = 9$$

The solution is 9.

Multiplication Property of Equations

The second property that we use to solve equations algebraically is the Multiplication Property of Equations.

Multiplication Property of Equations

For real numbers a, b, and c, $c \neq 0$, the equations $a = b$ and $ac = bc$ are equivalent equations.

In words, we can multiply both sides of an equation by the same nonzero number.

To solve $4x = 12$, we can multiply both sides of the equation by $\frac{1}{4}$. Equivalently, we can divide both sides by 4.

$$4x = 12 \qquad\qquad 4x = 12$$
$$\frac{1}{4}(4x) = \frac{1}{4}(12) \qquad\qquad \frac{4x}{4} = \frac{12}{4}$$
$$\left(\frac{1}{4} \cdot 4\right)x = 3 \qquad\qquad 1x = 3$$
$$1x = 3 \qquad\qquad x = 3$$
$$x = 3$$

Note that we cannot isolate x by subtracting 4 from both sides because $4x - 4$ cannot be simplified.

EXAMPLE 6

Using the Multiplication Property of Equations

Solve each equation.

(a) $-5x = 4$

(b) $-n = 7$

Solution

(a) $-5x = 4$

$$\frac{-5x}{-5} = \frac{4}{-5}$$ Divide both sides by -5.

$$x = -\frac{4}{5}$$ The variable is isolated.

The solution is $-\frac{4}{5}$.

(b) $-n = 7$

$-1n = 7$ Multiplication Property of -1

$$\frac{-1n}{-1} = \frac{7}{-1}$$ Divide both sides by -1.

$n = -7$

The solution is -7. The same result would be obtained by multiplying both sides of $-1n = 7$ by -1.

LEARNING TIP

Use the Addition Property of Equations to eliminate a term; use the Multiplication Property of Equations to eliminate a coefficient.

Note: When we use the Addition Property of Equations, the sign of the number that we add to both sides is the *opposite* of the sign of the term we are trying to eliminate. The term is thereby eliminated because the sum of two opposites is 0. When we use the Multiplication Property of Equations, the sign of the number by which we multiply or divide both sides is the *same* as that of the coefficient we are trying to eliminate. The coefficient is thereby eliminated because we are dividing out the common factor.

When the coefficient of a variable term is an integer, it is usually easiest to divide both sides of the equation by the coefficient. When the coefficient is a fraction, multiplying both sides by the reciprocal of the fraction is an efficient way to isolate the variable. When we use this technique, we must recall the rules for multiplying fractions. The following pairs of terms are equivalent.

$$\frac{x}{5} \quad \text{and} \quad \frac{1}{5}x \qquad \frac{3}{4}x \quad \text{and} \quad \frac{3x}{4}$$

EXAMPLE 7

Eliminating Fractional Coefficients

Solve each equation.

(a) $\frac{2}{3}x = 16$ (b) $\frac{y}{2} = -3$

Solution

(a) $\dfrac{2}{3}x = 16$

$\dfrac{3}{2} \cdot \dfrac{2}{3}x = \dfrac{3}{2} \cdot 16$ Multiply both sides by the reciprocal of $\dfrac{2}{3}$.

$x = 24$ The product of the reciprocals is 1.

The solution is 24.

(b) $\dfrac{y}{2} = -3$

$\dfrac{1}{2}y = -3$ $\dfrac{y}{2} = \dfrac{1}{2}y$

$2 \cdot \left(\dfrac{1}{2}y\right) = 2(-3)$ Multiply both sides by the reciprocal of $\dfrac{1}{2}$.

$y = -6$

The solution is -6.

Sometimes we must simplify before applying the Multiplication Property of Equations.

EXAMPLE 8

Combining Like Terms

Solve $45 = 9x - 4x$.

Solution

$9x - 4x = 45$ Exchange sides (optional).

$5x = 45$ Combine like terms.

$\dfrac{5x}{5} = \dfrac{45}{5}$ Divide both sides by 5.

$x = 9$

The solution is 9.

Quick Reference 2.5

Equivalent Equations
- **Equivalent equations** are equations that have exactly the same solutions.
- Two ways in which equivalent equations can be produced are
  1. exchanging the two sides of the equation, and
  2. replacing any expression with an equivalent expression.
- To solve an equation algebraically, our goal is to isolate the variable. Then the equation in the form $x = $ number can be solved by inspection.

Addition Property of Equations
- For real numbers a, b, and c, the equations $a = b$ and $a + c = b + c$ are equivalent equations. In words, we can add (or subtract) any number to (or from) both sides of an equation, and the result will be an equivalent equation.

- To solve an equation, we use the Addition Property of Equations to eliminate constant or variable terms of our choice from one side of the equation.
- It is usually best to simplify the sides of an equation before employing the Addition Property of Equations.

Multiplication Property
of Equations

- For real numbers a, b, and c, $c \neq 0$, the equations $a = b$ and $ac = bc$ are equivalent equations. In words, we can multiply (or divide) both sides of an equation by any nonzero number, and the result will be an equivalent equation.
- When the coefficient of a variable term is an integer, it is usually easiest to divide both sides by the coefficient. When the coefficient is a fraction, multiplying both sides by the reciprocal of the fraction is an efficient way to isolate the variable.
- It is usually best to simplify the sides of an equation before employing the Multiplication Property of Equations.

Speaking the Language 2.5

1. Equivalent equations have exactly the same ▨▨▨▨▨ .
2. The goal in solving an equation algebraically is to ▨▨▨▨▨ the variable.
3. To solve $5x + 3 = 4x + 8$, we should use the ▨▨▨▨▨ Property of Equations twice.
4. To solve $\frac{2}{3}x = 7$, a good strategy is to ▨▨▨▨▨ both sides by the ▨▨▨▨▨ of $\frac{2}{3}$.

Exercises 2.5

Concepts and Skills

▼ **1.** What are equivalent equations?

▼ **2.** From the following list, select the pair of equations that are not equivalent and explain your choice.

(i) $-7 = x$ (ii) $5x - 2x = 0$
     $x = -7$ $3x = 0$
(iii) $x + 3 = 5$ (iv) $2(x + 1) = 8$
     $x + 7 = 9$ $2x + 2 = -8$

In Exercises 3–6, solve the given equation.

3. $7x - 6x = 9$ **4.** $4 = 2x - x$

5. $1.8x - 0.8x = 3.5$ **6.** $12 = \frac{2}{9}x + \frac{7}{9}x$

In Exercises 7–10, to solve the given equation, which of the two steps is better to perform first on both sides?

7. $x - 5 = 0$

(i) Add 5.   (ii) Add -5.

8. $-4 = 4 + x$

(i) Add 4.   (ii) Add -4.

9. $5x = 4x - 6$

(i) Add 6.   (ii) Add $-4x$.

10. $7y = 1 + 6y$

(i) Subtract $7y$.   (ii) Subtract $6y$.

In Exercises 11–20, use the Addition Property of Equations to solve the equation.

11. $x + 5 = 9$ **12.** $x - 2 = -3$

13. $-4 = x - 6$ **14.** $0 = 3 + x$

15. $25 + x = 39$ **16.** $-21 = -32 + x$

17. $x + 27.6 = 8.9$ **18.** $-13.9 + x = -20.7$

19. $12 = x + 12$ **20.** $x - 5 = -5$

In Exercises 21–28, use the Addition Property of Equations to solve the equation.

21. $3x = 2x - 1$ **22.** $-7x = 3 - 8x$

23. $7x = 5 + 6x$ **24.** $2x = x - 5$

25. $-4x + 2 = -3x$

26. $-x = 5 - 2x$

27. $9x - 6 = 10x$

28. $-6x - 5 = -5x$

In Exercises 29–32, use the Addition Property of Equations to solve the equation.

29. $-3a + 5a = 9 + a + 6$

30. $3 + 2y + 4 = -8y + 11y$

31. $3x + 3 - 3x + 7 = 6x - 6 - 5x$

32. $-8 + 9 = 5x - 7 - 4x$

In Exercises 33–40, use the Addition Property of Equations to solve the equation.

33. $7x - 9 = 8 + 6x$

34. $11t + 5 = 1 + 10t$

35. $1 - 3y = 4 - 4y$

36. $-7x - 5 = 6 - 8x$

37. $5x + 3 = 6x + 1$

38. $x - 5 = 2x - 5$

39. $1 - 5x = 1 - 4x$

40. $-6x - 3 = 6 - 5x$

41. Why do we *add* 3 to both sides to solve $x - 3 = 12$ but *divide* both sides by 3 to solve $3x = 12$?

42. To solve $-y = 6$, should we multiply both sides by -1 or divide both sides by -1?

In Exercises 43–46, to solve the given equation, which of the two steps is better to perform first on both sides?

43. $-4x = 20$

  (i) Add 4.   (ii) Divide by -4.

44. $-9 = -x$

  (i) Multiply by -1.   (ii) Add x.

45. $\frac{1}{5}x = 2$

  (i) Multiply by 5.   (ii) Divide by 5.

46. $\frac{x}{7} = 0$

  (i) Multiply by 7.   (ii) Divide by 7.

47. To solve $\frac{3}{4}x = 15$, which of the following is an appropriate first step? Explain.

   (i) Multiply by $\frac{4}{3}$.

   (ii) Divide by $\frac{3}{4}$.

   (iii) Multiply by 4 and then divide by 3.

48. Which of the following expressions are equivalent? Explain.

   (i) $\frac{2}{7}x$ (ii) $\frac{2x}{7}$ (iii) $\frac{2}{7x}$

In Exercises 49–56, use the Multiplication Property of Equations to solve the equation.

49. $5x = 20$

50. $-6x = 18$

51. $48 = -16x$

52. $-36 = 12x$

53. $-y = -3$

54. $-x = -4$

55. $-3x = 0$

56. $12x = 0$

In Exercises 57–62, use the Multiplication Property of Equations to solve the equation.

57. $\frac{x}{9} = -4$

58. $\frac{x}{6} = 1$

59. $\frac{-t}{4} = -5$

60. $\frac{-x}{3} = 6$

61. $\frac{3}{4}x = 9$

62. $-8 = -\frac{2}{5}y$

In Exercises 63–68, solve the given equation.

63. $2x - 7x = 0$

64. $5t - 2t = -12$

65. $y - 6y = -25$

66. $-5y - 6y = 121$

67. $8 = 3t - 4t$

68. $96 = 9n + 7n$

In Exercises 69–78, solve the given equation.

69. $9x = 8x$

70. $4x + 10 = 5x$

71. $-4x = -28$

72. $8t = 8$

73. $5 - 7x = -6x$

74. $3x = 2x - 6$

75. $-3 + x = 0$

76. $-10 = -7 + x$

77. $\frac{1}{4} = -2y$

78. $-t = -\frac{1}{2}$

In Exercises 79–84, given the first equation in each pair, state the property used to write the second equivalent equation.

79. $5(x - 1) = 12$
  $5x - 5 = 12$

80. $3 \cdot \left(\frac{1}{3}x\right) = 3 \cdot 4$
  $\left(3 \cdot \frac{1}{3}\right)x = 3 \cdot 4$

81. $\frac{4}{3}x = 20$
  $4x = 60$

82. $\frac{y}{6} = 2$
  $y = 12$

83. $7 + 8y = 7y$
  $7 + y = 0$

84. $-y = 8$
  $y = -8$

Data Analysis: *Online Sales of CDs*

The accompanying bar graph shows that the sales of CDs online increased dramatically from 1997 to 1999. (Source: Jupiter Communications.)

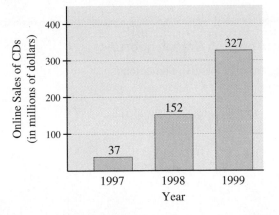

If we let x represent the number of years since 1995, an expression that models the sales (in millions of dollars) of CDs online is $145x - 263$.

85. Solve the equation $145x - 263 = 1042$.

86. What does the solution in Exercise 85 represent?

87. Write an equation for which sales are $2.2 billion.

88. Solve the equation in Exercise 87 and interpret the solution.

Challenge

In Exercises 89–92, solve for x.

89. $x + a = b$ **90.** $2x = y$

91. $5x - a = b$ **92.** $\dfrac{2}{3}x - \dfrac{1}{3}y = 0$

In Exercises 93–96, determine the value of k that makes the given number a solution of the equation.

93. $x - k = 0$; -4 **94.** $5 = kx$; 10

95. $kx + 1 = 7$; 2 **96.** $kx - 7 = -1$; 3

97. If $2x - 1 = 5$, evaluate $3x - 4$.

98. If $7x + 3x = 20$, evaluate $\dfrac{1}{2}x + 5$.

2.6 | Solving Linear Equations

The Solving Routine   •   Clearing Fractions   •   Special Cases

The Solving Routine

In Section 2.5 we gained some experience in solving linear equations in one variable. The following is a summary of the complete solving routine. The first step, clearing fractions, is an optional step that will be discussed later in this section.

> **The Solving Routine for Linear Equations in One Variable**
>
> 1. Clear fractions by multiplying both sides by the LCD of all the fractions.
> 2. Use the Distributive Property to simplify both sides.
>   (a) Remove grouping symbols.
>   (b) Combine like terms.
> 3. If a variable term appears on both sides, use the Addition Property of Equations to eliminate the variable term from one side.
> 4. Use the Addition Property of Equations to isolate the variable term.
> 5. Use the Multiplication Property of Equations to isolate the variable itself.
> 6. Solve the resulting equation by inspection and verify the solution.

LEARNING TIP

Following the order of the steps in the solving routine is usually a good approach to solving linear equations. However, the nature of an equation may make changing the order more efficient. For instance, you might remove grouping symbols before clearing fractions. With experience you will recognize such situations.

Verifying solutions can be done by substitution or with a calculator. Again, we will not show this step in the following examples.

In our first example, we apply both the Addition and Multiplication Properties of Equations to isolate the variable.

EXAMPLE 1

Solving Linear Equations: Applying Both Properties

Use the properties of equations to solve the given equations.

(a) $\quad -3x - 7 = 8$

$$-3x - 7 + 7 = 8 + 7 \qquad \text{Add 7 to isolate the variable term.}$$

$$-3x = 15$$

$$\frac{-3x}{-3} = \frac{15}{-3} \qquad \text{Divide by } -3 \text{ to isolate the variable.}$$

$$x = -5$$

The solution is -5.

(b) $\quad 10 - 4x = x$

$$10 - 4x + 4x = x + 4x \qquad \text{Add } 4x \text{ to eliminate the variable on the left.}$$

$$10 = 5x \qquad \text{Combine like terms.}$$

$$\frac{10}{5} = \frac{5x}{5} \qquad \text{Divide by 5 to isolate the variable.}$$

$$2 = x$$

The solution is 2.

The next example illustrates steps 3–6 of the solving routine.

EXAMPLE 2

Solving Linear Equations: Eliminating Variable Terms

Solve the given equation.

$$8 + 7x = 9x - 2$$

$$8 + 7x - 9x = 9x - 2 - 9x \qquad \text{Eliminate } 9x \text{ from the right side.}$$

$$8 - 2x = -2 \qquad \text{Combine like terms.}$$

$$8 - 2x - 8 = -2 - 8 \qquad \text{Subtract 8 to isolate the variable term.}$$

$$-2x = -10 \qquad \text{Divide by } -2 \text{ to isolate the variable.}$$

$$x = 5$$

The solution is 5.

In our next example, we consider equations in which terms must be combined and in which variable terms appear on both sides.

EXAMPLE **3**   **Solving Linear Equations: Combining Terms**

Solve the given equation.

$$4 + 7y + 3y - 25 = 3y - 2 + y$$

$$-21 + 10y = 4y - 2 \qquad \text{Combine like terms.}$$

$$-21 + 10y - 4y = 4y - 2 - 4y \qquad \text{Eliminate } 4y \text{ from the right side.}$$

$$-21 + 6y = -2 \qquad \text{Combine like terms.}$$

$$6y = 19 \qquad \text{Add 21 to isolate the variable term.}$$

$$y = \frac{19}{6} \qquad \text{Divide by 6 to isolate the variable.}$$

The solution is $\frac{19}{6}$.

EXAMPLE **4**   **Solving Linear Equations: Removing Grouping Symbols**

Solve the given equations.

(a) $3(x - 1) - 5x = 6$

$$3x - 3 - 5x = 6 \qquad \text{Remove parentheses with the Distributive Property.}$$

$$-2x - 3 = 6 \qquad \text{Combine like terms.}$$

$$-2x = 9 \qquad \text{Add 3 to isolate the variable term.}$$

$$x = -\frac{9}{2} \qquad \text{Divide by } -2 \text{ to isolate the variable.}$$

The solution is $-\frac{9}{2}$.

(b) $3(x + 1) + x = 10 - (x + 7)$

$$3x + 3 + x = 10 - x - 7 \qquad \text{Remove parentheses.}$$

$$4x + 3 = 3 - x \qquad \text{Combine like terms.}$$

$$5x + 3 = 3 \qquad \text{Eliminate the variable term on the right.}$$

$$5x = 0 \qquad \text{Subtract 3 to isolate the variable term.}$$

$$x = 0 \qquad \text{Divide by 5 to isolate the variable.}$$

The solution is 0.

Clearing Fractions

If an equation involves fractions, we can usually obtain a simpler equation by eliminating the fractions. The Multiplication Property of Equations allows us to multiply
both sides of the equation by any nonzero number. If we multiply both sides by the
LCD of all the fractions and then simplify, the resulting equation will have no fractions. We call this process *clearing fractions*.

Note: Because of the Distributive Property, multiplying both sides means multiplying *every term* on both sides.

EXAMPLE 5

Solving Linear Equations: Clearing Fractions

Solve the given equations.

(a) $\quad \dfrac{3}{4}y + \dfrac{5}{8} = \dfrac{3}{4}$ The LCD of the three fractions is 8.

$\quad 8 \cdot \dfrac{3}{4}y + 8 \cdot \dfrac{5}{8} = 8 \cdot \dfrac{3}{4}$ Multiply both sides (every term) by 8 to clear the fractions.

$\quad 6y + 5 = 6$

$\quad 6y = 1$ Subtract 5 to isolate the variable term.

$\quad y = \dfrac{1}{6}$ Divide by 6 to isolate the variable.

The solution is $\frac{1}{6}$.

(b) $\quad \dfrac{t}{3} + \dfrac{1}{2} = \dfrac{t}{2} + 1$ The LCD of the fractions is 6.

$\quad 6 \cdot \dfrac{t}{3} + 6 \cdot \dfrac{1}{2} = 6 \cdot \dfrac{t}{2} + 6 \cdot 1$ Multiply both sides by 6 to clear the fractions.

$\quad 2t + 3 = 3t + 6$

$\quad 2t + 3 - 2t = 3t + 6 - 2t$ Eliminate $2t$ from the left side.

$\quad 3 = t + 6$

$\quad -3 = t$ Subtract 6 to isolate the variable.

The solution is -3.

EXAMPLE 6

Simplifying and Clearing Fractions

Think About It

The solving routine for linear equations works well. However, suppose that you attempted to apply the routine to solving the equation $x^2 + x = 2$. At what point in the solving routine would you encounter difficulty? Use the graphing method to estimate the solutions of the equation.

Solve the given equation.

$$\dfrac{y}{2} - 1 = \dfrac{5}{6}(y + 3) + \dfrac{1}{2}$$

$6 \cdot \dfrac{y}{2} - 6 \cdot 1 = 6 \cdot \dfrac{5}{6}(y + 3) + 6 \cdot \dfrac{1}{2}$ Clear the fractions.

$3y - 6 = \left(6 \cdot \dfrac{5}{6}\right)(y + 3) + 3$ Associative Property of Multiplication

$3y - 6 = 5(y + 3) + 3$

$3y - 6 = 5y + 15 + 3$ Remove parentheses.

$3y - 6 = 5y + 18$

$-6 = 2y + 18$ Eliminate $3y$ from the left side.

$-24 = 2y$ Subtract 18 to isolate the variable term.

$-12 = y$ Divide by 2 to isolate the variable.

The solution is -12.

Special Cases

In Section 2.4 where we used the graphing method to estimate solutions, we encountered the equation $x - 7 = x + 8$, for which there appeared to be no solution, and the equation $x - 10 = -10 + x$, for which every real number appeared to be a solution. Solving these equations algebraically confirms that they are special cases.

$$x - 7 = x + 8 \qquad\qquad\qquad x - 10 = -10 + x$$
$$x - 7 - x = x + 8 - x \qquad\qquad x - 10 - x = -10 + x - x$$
$$-7 = 8 \quad \text{False} \qquad\qquad\qquad -10 = -10 \quad \text{True}$$

$$\text{Solution set: } \varnothing \qquad\qquad\qquad \text{Solution set: } \mathbf{R}$$

EXAMPLE **7**

Solving Linear Equations: Special Cases

Solve the given equations.

(a) $\quad 3(2x - 1) = 2(4 + 3x)$

$\qquad\quad 6x - 3 = 8 + 6x$ $\qquad$ Remove grouping symbols.

$\quad 6x - 3 - 6x = 8 + 6x - 6x$ $\qquad$ Eliminate the variable term from one side.

$\qquad\qquad\quad -3 = 8$ $\qquad$ False

Because the resulting equation is false, the original equation is a contradiction, and the solution set is $\varnothing$.

(b) $\quad 2(3x + 6) = 7x - (x - 12)$

$\qquad\quad 6x + 12 = 7x - x + 12$ $\qquad$ Remove grouping symbols.

$\qquad\quad 6x + 12 = 6x + 12$ $\qquad$ Combine like terms.

$\quad 6x + 12 - 6x = 6x + 12 - 6x$ $\qquad$ Eliminate the variable term from one side.

$\qquad\qquad\quad 12 = 12$ $\qquad$ True

Because the resulting equation is true, the original equation is an identity, and the solution set is $\mathbf{R}$.

Summary of Special Cases

If solving a linear equation leads to an equivalent equation with no variable, then

1. if the resulting equation is false, the original equation is a contradiction, and the solution set is $\varnothing$;

2. if the resulting equation is true, the original equation is an identity, and the solution set is $\mathbf{R}$.

Quick Reference 2.6

The Solving Routine • The following is a summary of the solving routine for linear equations in one variable.

1. Clear fractions by multiplying both sides by the LCD of all the fractions.

2. Use the Distributive Property to simplify both sides.

(a) Remove grouping symbols.

(b) Combine like terms.

3. If a variable term appears on both sides, use the Addition Property of Equations to eliminate the variable term from one side.

4. Use the Addition Property of Equations to isolate the variable term.

5. Use the Multiplication Property of Equations to isolate the variable itself.

6. Solve the resulting equation by inspection and verify the solution.

Clearing Fractions
- If an equation contains one or more fractions, an optional step is to clear the fractions. This is done by multiplying both sides (every term) of the equation by the LCD of all the fractions.

Special Cases
- If solving a linear equation leads to an equivalent equation with no variable, then

  1. if the resulting equation is false, the original equation is a contradiction, and the solution set is Ø;

  2. if the resulting equation is true, the original equation is an identity, and the solution set is **R**.

Speaking the Language 2.6

1. To simplify both sides of an equation by removing grouping symbols and combining like terms, we use the ▨▨▨▨ Property.

2. In the solving routine, we use the ▨▨▨▨ Property of Equations to isolate the variable term, and we use the ▨▨▨▨ Property of Equations to isolate the variable itself.

3. Multiplying both sides of an equation by the LCD of all the fractions is known as ▨▨▨▨ the fractions.

4. Solving $2(x + 1) = 2x + 2$ leads to the equivalent equation $0 = 0$. Thus the original equation is a(n) ▨▨▨▨.

Exercises 2.6

Concepts and Skills

1. To solve the equation $2x - 5 = 7$, should we begin by dividing both sides by 2 or by adding 5 to both sides? Why?

2. According to the Multiplication Property of Equations, if we multiply both sides of an equation by any *nonzero* number, the resulting equation is equivalent to the original equation. Using an example, explain why this would not be true if we multiplied both sides by 0.

In Exercises 3–12, solve the given equation.

3. $5x + 14 = 4$

4. $2 = 4x + 6$

5. $2 - 3x = 0$

6. $7 = 7 - 4x$

7. $0.2 - 0.37t = -0.91$

8. $6.7a - 17.5 = 42.8$

9. $7x = 2x - 5$

10. $4t + 6 = -6t$

11. $3x = 5x + 14$

12. $-7y = y - 12$

In Exercises 13–22, solve the given equation.

13. $4x - 7 = 6x + 3$

14. $9 + 7x = x + 3$

15. $5x + 2 = 2 - 3x$

16. $-2 - 6x = x - 2$

17. $x + 6 = 5x - 10$

18. $2x - 7 = 4 + 3x$

19. $5 - 9x = 8x + 5$

20. $7x - 2 = 2x + 13$

21. $9 - x = x + 15$

22. $7x - 4 = -4 + 5x$

In Exercises 23–28, solve the given equation.

23. $3x + 8 - 5x = 2 - x + 2x - 3$

24. $12x - x = 7 + 10x - 5x + 8$

25. $3 + 5y - 24 + 3y = y - 1 + y$

26. $10 + 3t - 3 = 2t + 3 - 7t$

27. $5 - 5x + 8 - 4 = -9 + x$

28. $3x - 8 - 5x + 2 = -12$

In Exercises 29–40, solve the given equation.

29. $3(x + 1) = 1$

30. $4(3t + 5) = 44$

31. $5(2x - 1) = 3(3x - 1)$

32. $3(x + 7) + 4(3 - x) = 0$

33. $-10 = 3x + 8 + 4(x - 1)$

34. $12 = 5x - 7 + 3(2x - 1)$

35. $4(x - 1) = 5 + 3(x - 6)$

36. $5 - (4x + 9) = 8 + 2x$

37. $6x - (5x - 3) = 0$

38. $0 = 3x - (x - 4)$

39. $2 - 2(3x - 4) = 8 - 2(5 + x)$

40. $12 - (t + 4) = 5(2 - 3t)$

41. To clear fractions from an equation, we multiply both sides by some number n. Explain how to determine n.

42. To clear the fraction from $\frac{1}{2}(x - 3) = 5$, we multiply both sides by 2. Is the result on the left side $x - 3$ or $2x - 6$? Why?

In Exercises 43–50, solve the given equation.

43. $\frac{2}{3}x - \frac{2}{3} = \frac{5}{3}$ **44.** $\frac{3}{5} - \frac{2}{5}x = 1$

45. $1 + \frac{3t}{4} = \frac{1}{4}$ **46.** $\frac{1}{2}t - \frac{1}{4} = \frac{1}{3}t$

47. $\frac{1}{2}x - 2 = 1 + \frac{2}{3}x + \frac{1}{6}$

48. $2 + \frac{5}{6}y - \frac{3}{4} = \frac{3}{4}y + 1$

49. $\frac{x + 5}{3} = -\frac{x}{2} + 5$

50. $3 - \frac{2t + 7}{3} = \frac{2t}{7}$

In Exercises 51–56, solve the given equation.

51. $\frac{3}{5}(x - 1) = 9$ **52.** $-\frac{4}{7}(3 - x) = 4$

53. $\frac{x}{3} + 2 = \frac{3}{2}(x + 3) - 6$

54. $\frac{1}{2}x + \frac{1}{5}(x - 1) = x - \frac{1}{5}$

55. $\frac{3}{7}x = \frac{5}{14}(x + 1) + \frac{1}{7}(8 - x)$

56. $\frac{1}{2}(5 - x) = \frac{3}{4}x - \frac{1}{2}(2x - 1)$

In Exercises 57–60, solve the given equation.

57. $7.35x + 5.1 = 2.6x - 0.6$

58. $4.9x + 6.31 - 2.5x = 13.75$

59. $2.7(3x - 1.5) + 6.3 = 9.1x + 6.55$

60. $1.4(2.1 - 3.2x) - 4.1x = 3x - 11.535$

In Exercises 61–70, classify the given equation as an identity, a contradiction, or a conditional equation. If the equation is a conditional equation, determine its solution; otherwise, write the solution set.

61. $x + 7 = 7 + x$

62. $-2 + y = y - 2$

63. $x = x + 3$

64. $y - 2 = y + 1$

65. $2x = 3x$

66. $x + 1 = 4$

67. $6t + 7 = 7$

68. $0 = 18t - 7t$

69. $0x = 3$

70. $-x = -1x$

In Exercises 71–82, solve the given equation. If the equation is an identity or a contradiction, write the solution set.

71. $6x - 3 - x = 6x + 5$

72. $5x + 1 - 3x = -5 + 2x + 6$

73. $2 + 3x + 5 = 6x + 7 - 6x$

74. $6 + 3x - 1 - 4x = x + 1 - 2x$

75. $\frac{3}{4}x + \frac{5}{2} = \frac{1}{2}(5 + x) + \frac{1}{4}x$

76. $\frac{1}{2}(x + 1) + \frac{x}{4} = -\frac{1}{2}$

77. $4(x - 2) + 8 = 6(x + 1) - 2x$

78. $7(1 - x) = 2 + 5(1 + x) - 7x$

79. $2(3x - 4) = 2 - 6(1 - x)$

80. $6(x + 1) + 1 = 2(3x - 1)$

81. $3x - 1 - 2(x - 4) = 2x + 7 - x$

82. $5(x + 3) - 3x = 2(x + 5) + 5$

Modeling with Real Data

83. The table shows the estimated amount that parents need to pay fully for a child's college education. The total amounts (in thousands of dollars) for a four-year college education at a public and at a private college can be modeled by $2.28x + 15.7$ and $5.88x + 40.58$, respectively, where x represents the number of years after 1990.

| Year | Cost of Education | |
	Public	Private
1997	$31,637	$81,754
2004	47,571	122,930

(Source: Merrill Lynch.)

(a) What does each of the following equations model?

    (i) $5.88x + 40.58 = 150$

    (ii) $2.28x + 15.7 = 2(31.637)$

(b) Solve each equation in part (a). Approximately what year is represented by each solution?

84. Refer to the model expressions in Exercise 83.

(a) What does the following equation model?

    $2.28x + 15.7 = 5.88x + 40.58$

(b) Solve the equation in part (a) and interpret the solution.

Data Analysis: *Olympics 100-Meter Run*

The accompanying bar graph shows the winning times for the men's and women's 100-meter run in the 1956 and 2000 Olympics. (Source: World Almanac.)

The following expressions model the winning times, where t is the number of years since 1956.

Men: $-0.014t + 10.5$

Women: $-0.017t + 11.5$

Figure for 85–88

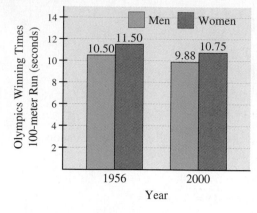

85. Use the models to estimate the winning times for men and women in the 1976 Olympics.

86. Equate the two model expressions and solve the resulting equation.

87. From the solution in Exercise 86, predict the year when the winning times for men and women will be the same. What are the projected winning times in that year? (Ignore the possibility that the year is not an Olympic year.)

88. Are the results in Exercise 87 possible? probable? Comment on the validity of the model expressions beyond 2000.

Challenge

In Exercises 89–92, solve the given equation. If the equation is an identity or a contradiction, write the solution set.

89. $x - 5\{x - 5[x - 5(x - 5)]\} = 1$

90. $3\{2x - 2[4 - (x - 7)]\} = 6$

91. $x - [2(x - 3) - (x + 1)] = -3(x - 1) + 4 + 3x$

92. $\dfrac{3x - 4}{3} - (x + 1) = -[x - (x + 5)]$

In Exercises 93 and 94, solve for x.

93. $ax + b = c, \quad a \neq 0$

94. $2(a - c) - 7x = 0$

In Exercises 95 and 96, solve the given equation.

95. $2t^2 - 15 = t(t + 5) + t^2$

96. $3x(2x + 5) = 4(x^2 + 15) + 2x^2$

2.7 | Inequalities: Graphing Methods

Inequalities and Solutions • Estimating Solutions of Linear Inequalities •
Special Cases • Double Inequalities

Inequalities and Solutions

In Chapter 1 we used inequality symbols to describe the order of real numbers. In this section, we relate algebraic expressions using inequalities.

$$x \geq 5$$ The value of x is greater than or equal to 5.

$$3x - 7 < 14$$ The value of $3x - 7$ is less than 14.

$$4 - 5x > 9 - 6x$$ The value of $4 - 5x$ is greater than the value of $9 - 6x$.

$$2x \leq x + 1$$ The value of $2x$ is less than or equal to the value of $x + 1$.

Any value of the variable that makes an inequality true is called a **solution.** The set of all solutions of an inequality is called the **solution set.**

Although there are many similarities between the ways we solve equations and inequalities, there are also some important differences. For example, the only solution of the equation $4 - 3x = 7$ is -1. But as Example 1 shows, the inequality $4 - 3x \leq 7$ has more than one solution.

EXAMPLE **1**

Testing Solutions of an Inequality

Which members of the set $S = \{-3, -1, 0, 2\}$ are solutions of $4 - 3x \leq 7$?

Solution

$x = -3$	$x = -1$	$x = 0$	$x = 2$
$4 - 3x \leq 7$	$4 - 3x \leq 7$	$4 - 3x \leq 7$	$4 - 3x \leq 7$
$4 - 3(-3) \leq 7$	$4 - 3(-1) \leq 7$	$4 - 3(0) \leq 7$	$4 - 3(2) \leq 7$
$4 + 9 \leq 7$	$4 + 3 \leq 7$	$4 - 0 \leq 7$	$4 - 6 \leq 7$
$13 \leq 7$	$7 \leq 7$	$4 \leq 7$	$-2 \leq 7$
False	True	True	True

Three members of set S are solutions: $-1, 0,$ and 2.

When we write $x \geq 5$, we mean that x represents 5 or any number larger than 5. Some of the many numbers that are solutions of this inequality are 5, 7, 6.2, $\frac{25}{3}$, and $\sqrt{50}$. Because such an inequality has infinitely many solutions, it is helpful to visualize the solutions by graphing them on a number line. We call this *graphing the solution set of the inequality* or simply *graphing the inequality*. We show the solutions of $x \geq 5$ by highlighting the interval of the number line corresponding to numbers that are 5 or larger. (See Fig. 2.35.)

Figure 2.35

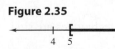

We use the symbol [to indicate that the left endpoint of the interval is included. In Example 2, we see how other symbols are used to indicate whether an interval endpoint is included.

EXAMPLE **2**   **Number Line Graphs of Inequalities**

Draw a number line graph of each inequality.

(a) $x > -3$ (b) $x < 2$

(c) $0 < x \leq 7$ (d) $-8 \leq x < -2$

LEARNING TIP

Recall that an inequality can be written in two ways. For instance, $x < 3$ and $3 > x$ describe the same values. Remember that the inequality symbol points toward the smaller value. Sketching a number line graph may help you recognize equivalent inequalities.

Solution

(a) Note that the symbol (means that -3 is not included.

Figure 2.36

(b) Note that the symbol) means that 2 is not included.

Figure 2.37

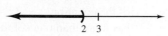

(c) The inequality $0 < x \leq 7$ means that x is between 0 and 7, including 7 but not 0. It is read, "0 is less than x, and x is less than or equal to 7." Note that the symbol] means that 7 is included.

Figure 2.38

(d) **Figure 2.39**

Estimating Solutions of Linear Inequalities

In this section and the next, we focus on one type of inequality.

> **Definition of a Linear Inequality in One Variable**
>
> A **linear inequality in one variable** can be written as $Ax + B > 0$, where $A \neq 0$. A similar definition can be stated for $<$, $\geq$, and $\leq$.

We can use a calculator to estimate solutions of an inequality. The techniques are similar to those used to estimate solutions of equations.

Exploring the Concept

Estimating Solutions of an Inequality

For the inequality $x - 2 \leq 9$, we can enter the left side, $x - 2$, as Y_1 and the right side, 9, as Y_2.

Figure 2.40 shows a calculator table of values. The table shows that $x - 2 = 9$ when $x = 11$. Because the inequality symbol is $\leq$, 11 is a solution. Also, we can scroll to see that the value of $x - 2$ is less than 9 for all x-values less than 11. We conclude that the solutions of $x - 2 \leq 9$ are 11 and all numbers less than 11.

Figure 2.40

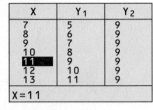

X	Y₁	Y₂
7	5	9
8	6	9
9	7	9
10	8	9
11	9	9
12	10	9
13	11	9

X = 1 1

Figure 2.41

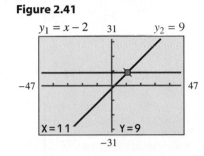

Figure 2.41 shows the graphs of the left and right sides of the inequality $x - 2 \leq 9$. The point of intersection is $(11, 9)$, which indicates that $x - 2 = 9$ when x is 11.

To the left of the point of intersection, the graph of $x - 2$ is below the graph of 9. For these points the value of the expression $x - 2$ is less than 9, so the x-coordinates are solutions of the inequality $x - 2 < 9$. Thus 11 and all x-values less than 11 are solutions of $x - 2 \leq 9$. The solutions may also be illustrated with a number line graph (see Fig. 2.42).

Rather than describe the solutions of $x - 2 \leq 9$ with a set, we will simply write $x \leq 11$, and we will follow this convention in the future.

Figure 2.42

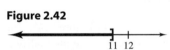

The following is a summary of the graphing method for estimating the solutions of an inequality.

Estimating Solutions of Inequalities Graphically

1. Enter the left-side expression y_1 as Y_1 and the right-side expression y_2 as Y_2.

2. Produce the graphs of y_1 and y_2.

3. Trace to the point of intersection and estimate the x-coordinate. For $\leq$ and $\geq$ inequalities, this x-coordinate is a solution.

4. Trace the graph of y_1 to determine the following:
   (a) If the inequality has the form $y_1 < y_2$ or $y_1 \leq y_2$, find where the graph of y_1 is below the graph of y_2.
   (b) If the inequality has the form $y_1 > y_2$ or $y_1 \geq y_2$, find where the graph of y_1 is above the graph of y_2.
   The x-coordinates of all such points are solutions.

5. Write an inequality that describes the solutions. (The solutions can also be given as a solution set, or they can be described with a number line graph.)

EXAMPLE **3**

Estimating Solutions of Linear Inequalities

Use the graphing method to estimate the solutions of each inequality. Draw a number line graph of the solutions.

(a) $-8 < 7 - x$ (b) $0.5x + 16 \geq -x - 8$

Solution

(a) Figure 2.43 shows the graphs of $y_1 = -8$ and $y_2 = 7 - x$. The point of intersection is $(15, -8)$, but 15 is not a solution because the inequality is in the form $y_1 < y_2$.

LEARNING TIP

Remember that the values of an expression (y-values) are associated with the vertical axis. Thus, for the inequality $A < B$, the values of expression A must be less than (or *below*) the values of expression B.

Figure 2.43

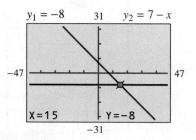

The graph of y_1 is below the graph of y_2 to the left of $(15, -8)$. Therefore, the solutions can be described by $x < 15$. (See Fig. 2.44.)

Figure 2.44

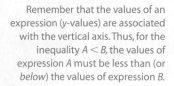

(b) The graphs of $y_1 = 0.5x + 16$ and $y_2 = -x - 8$ are shown in Figure 2.45. The point of intersection is $(-16, 8)$, and -16 is a solution because the inequality is in the form $y_1 \geq y_2$.

Think About It

Use the graphing method to estimate the solutions of the following inequalities.

(a) $|x| \leq 4$

(b) $|x| \geq -2$

(c) $|x| \leq -2$

Figure 2.45

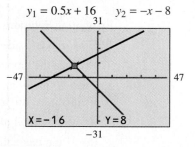

The graph of y_1 is above the graph of y_2 to the right of $(-16, 8)$, so $x \geq -16$ describes the solutions. (See Fig. 2.46.)

Figure 2.46

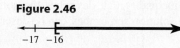

Special Cases

Earlier we saw that the solution set for an equation might be Ø, the empty set, or **R,** the set of all real numbers. There are similar special cases for inequalities.

EXAMPLE **4**

Special Cases of Inequalities

Use your calculator to estimate the solutions of each inequality.

(a) $x - 6 < x + 8$ (b) $x - 6 > x + 8$

Solution

We enter $x - 6$ as Y_1 and $x + 8$ as Y_2. Figure 2.47 shows a table of values for the two expressions, and Figure 2.48 shows their graphs. Note that the graphs appear to be parallel.

Figure 2.47

X	Y₁	Y₂
4	-2	12
5	-1	13
6	0	14
7	1	15
8	2	16
9	3	17
10	4	18

Figure 2.48

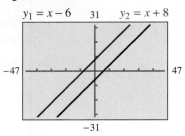

(a) In the table in Figure 2.47, we see that the value of $x - 6$ is *always* less than the value of $x + 8$. In Figure 2.48, the graph of $x - 6$ *always* appears to be below the graph of $x + 8$. We conclude that the inequality $x - 6 < x + 8$ is true for all real numbers, and the solution set is **R.**

(b) The reverse is true for $x - 6 > x + 8$. The table shows that the value of $x - 6$ is *never* greater than the value of $x + 8$, and the graph of $x - 6$ is *never* above the graph of $x + 8$. Therefore, there is no value of x for which $x - 6 > x + 8$ is true. The solution set is Ø.

Note: The conclusions in Example 4 are based on the fact that the graphs of y_1 and y_2 appear to be parallel. In the next section, we will see how to verify these conclusions algebraically.

Double Inequalities

To indicate that the value of an expression is between two quantities, we use a **double inequality.** If the value of $x + 5$ is between -9 and 15, then we write $-9 < x + 5$ *and* $x + 5 < 15$. The more compact form for this double inequality is $-9 < x + 5 < 15$.

We can easily extend our graphing methods to estimate the solutions of a double inequality.

EXAMPLE **5**

Estimating the Solutions of a Double Inequality

Use your calculator to estimate the solutions of $-9 < x + 5 < 15$.

Solution

We enter -9 as Y_1, $x + 5$ as Y_2, and 15 as Y_3.

Figure 2.49

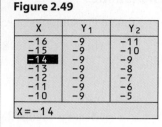

X	Y₁	Y₂
-16	-9	-11
-15	-9	-10
-14	-9	-9
-13	-9	-8
-12	-9	-7
-11	-9	-6
-10	-9	-5

X=-14

Figure 2.50

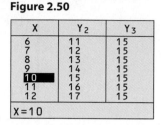

X	Y₂	Y₃
6	11	15
7	12	15
8	13	15
9	14	15
10	15	15
11	16	15
12	17	15

X=10

Figure 2.49 shows that $y_1 = y_2$ (that is, $-9 = x + 5$) when $x = -14$. Figure 2.50 shows that $y_2 = y_3$ (that is, $x + 5 = 15$) when $x = 10$. By scrolling through the table, we find that the values of $x + 5$ are *between* -9 and 15 when x is *between* -14 and 10. Therefore, the solutions are $-14 < x < 10$.

Figure 2.51 shows the graphs of $y_1 = -9$, $y_2 = x + 5$, and $y_3 = 15$. By tracing, we estimate the points of intersection to be $(-14, -9)$ and $(10, 15)$.

Figure 2.51

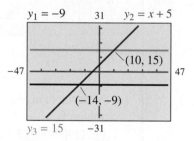

The portion of the graph of y_2 that lies between the graphs of y_1 and y_3 is the segment between the two points of intersection. Therefore, the solutions are described by the double inequality $-14 < x < 10$. (See Fig. 2.52.)

Figure 2.52

$\xleftarrow{\hspace{1em}} \underset{-15\ -14}{(\hspace{4em})} \underset{10\ 11}{\hspace{1em}} \xrightarrow{\hspace{1em}}$

Quick Reference 2.7

Inequalities and Solutions

- The order of algebraic expressions is described with an inequality.

- Any value of the variable that makes an inequality true is called a **solution.** The set of all solutions of an inequality is called the **solution set.**

- We can determine whether a number is a solution of an inequality by substituting the number for the variable and evaluating the expressions on the two sides of the inequality. If the resulting inequality is true, the number is a solution.

- Because inequalities typically have infinitely many solutions, it is helpful to visualize the solutions by graphing them on a number line. We call this *graphing the solution set of the inequality* or simply *graphing the inequality*.

- When we graph an inequality, the symbols [and] mean that the endpoints of an interval are included in the graph; the symbols (and) mean that the endpoints are not included.

Estimating Solutions of Linear Inequalities

- A **linear inequality in one variable** can be written in the form $Ax + B > 0$, where $A \neq 0$. A similar definition can be stated for $<$, $\geq$, and $\leq$.
- The following summarizes the procedure for using a calculator to estimate the solutions of an inequality.

  1. Enter the left-side expression y_1 as Y_1 and the right-side expression y_2 as Y_2.

  2. Produce the graphs of y_1 and y_2.

  3. Trace to the point of intersection and estimate the x-coordinate. For $\leq$ and $\geq$ inequalities, this x-coordinate is a solution.

  4. Trace the graph of y_1 to determine the following:

     (a) If the inequality has the form $y_1 < y_2$ or $y_1 \leq y_2$, find where the graph of y_1 is below the graph of y_2.

     (b) If the inequality has the form $y_1 > y_2$ or $y_1 \geq y_2$, find where the graph of y_1 is above the graph of y_2.

     The x-coordinates of all such points are solutions.

  5. Write an inequality that describes the solutions. (The solutions can also be given as a solution set, or they can be described with a number line graph.)

Special Cases

- Special cases arise when the graphs of the left and right sides of an inequality are parallel or coincide.
- For these special cases, the solution set is Ø or **R,** depending on the inequality symbol.

Double Inequalities

- A **double inequality** is used to indicate that the value of an expression is between two quantities.
- To estimate the solutions of a double inequality $y_1 < y_2 < y_3$ graphically, graph all three expressions and trace to estimate the points of intersection. Then observe where the graph of y_2 is between the graphs of y_1 and y_3. The solutions are the x-coordinates of the points in that interval.

Speaking the Language 2.7

1. When you use the graphing method, the x-coordinate of the point of intersection is a solution of an inequality only if the inequality symbol is ▨▨▨▨▨▨ or ▨▨▨▨▨▨ .

2. The inequality $3x + 5 < 0$ is an example of a(n) ▨▨▨▨▨▨ .

3. The graphs of $x + 3$ and $x + 7$ are parallel lines. If these expressions are the sides of an inequality, then the solution set is either ▨▨▨▨▨▨ or ▨▨▨▨▨▨ .

4. To use the graphing method to estimate the solutions of a(n) ▨▨▨▨▨▨ inequality of the form $y_1 < y_2 < y_3$, we look for the portion of the graph of y_2 that is ▨▨▨▨▨▨ the graphs of y_1 and y_3.

Exercises 2.7

Concepts and Skills

1. In your own words, explain what we mean when we write $x + 3 \le 7$.

2. What do we mean by the *solution set* of an inequality?

In Exercises 3–8, for the given inequality determine which of the given numbers is a solution.

3. $2 + x > 7$; 3, 5, 9, 11

4. $5 - x \le 3$; $-8, -1, 2, 5$

5. $4 \le 1 - 3x$; $-4, -2, -1, 0$

6. $2x - 7 > 5$; 0, 2, 6, 10

7. $-3 \le 2x - 7 < 5$; $-1, 2, 5, 6, 9$

8. $5 < 2 - 3x \le 14$; $-6, -4, -1, 1$

9. When we draw a number line graph of an inequality, what symbols do we use to indicate whether an endpoint is included in the graph?

10. If $-3 < x \le 7$, describe the possible values of x.

In Exercises 11–22, draw a number line graph of the given inequality.

11. $x < 7$
12. $x > 2$

13. $x > -3$
14. $x < -4$

15. $1 \le x$
16. $6 \ge x$

17. $x \le -5$
18. $x \ge -1$

19. $-5 < x < -1$
20. $0 \le x \le 7$

21. $1 \le x < 4$
22. $-3 < x \le 0$

In Exercises 23–30, translate the given description into an inequality.

23. n is nonnegative.
24. n is positive.

25. x is at most 3.
26. x is at least -3.

27. n is negative.
28. n is not positive.

29. x is no more than 6.
30. x is no less than 1.

In Exercises 31–36, write an inequality that corresponds to the given number line graph. Use x for your variable.

31.

32.

33.

34.

35.

36.

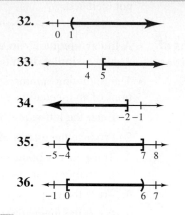

37. Suppose that y_1 and y_2 represent the expressions on the left and right sides of an inequality. If the graphs of y_1 and y_2 intersect, under what conditions does the point of intersection represent a solution?

38. Suppose that y_1 and y_2 represent the expressions on the left and right sides of an inequality and that the graphs of y_1 and y_2 are lines that intersect at $(-3, 7)$. Describe how to solve $y_2 > y_1$.

In Exercises 39–42, y_1 and y_2 represent the two sides of a linear inequality. Each figure shows the graphs of y_1 and y_2 along with their point of intersection. Use the figure to determine the solutions of the given inequality. Draw a number line graph of the solution set.

39. $y_1 > y_2$

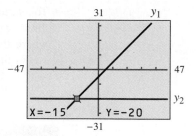

40. $y_1 \le y_2$

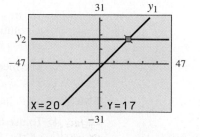

41. $y_1 \le y_2$

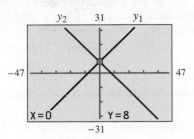

42. $y_1 > y_2$

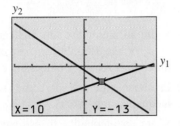

In Exercises 43–50, use the graphing method to estimate the solutions of the given inequality.

43. $x + 8 > -9$

44. $15 - 2x > -13$

45. $10 - x \le 15$

46. $\frac{1}{2}x - 9 \le 11$

47. $-13 > 2x - 7$

48. $-6 < -3x - 18$

49. $21 \ge 3x$

50. $-14 < 2x$

In Exercises 51–58, use the graphing method to estimate the solutions of the given inequality.

51. $\frac{1}{2}x > 2x$

52. $-x \ge -\frac{3}{2}x$

53. $3 - 2x > x - 3$

54. $x + 23 \le -x - 7$

55. $\frac{1}{2}x - 6 \le x - 2$

56. $3 - x \ge x - 3$

57. $7 + 0.8x < 19 - 0.2x$

58. $9 - 2.35x \ge 0.65x + 27$

59. Suppose that y_1 and y_2 represent the expressions on the left and right sides of an inequality and that the graphs of y_1 and y_2 are parallel. If the graph of y_1 is above the graph of y_2, describe the possible solution set of the inequality.

60. Suppose that y_1 and y_2 represent the expressions on the left and right sides of an inequality and that the graphs of y_1 and y_2 coincide. Describe the possible solution set of the inequality.

In Exercises 61 and 62, y_1 and y_2 represent the two sides of a linear inequality. Each figure shows the graphs of y_1 and y_2. Use the figure to estimate the solution set of the given inequality.

61. $y_1 > y_2$

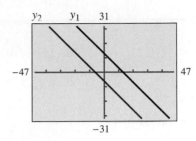

62. $y_1 \ge y_2$

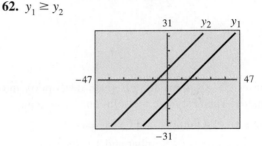

In Exercises 63 and 64, use the graphing method to compare the estimated solution sets for the inequalities in each pair.

63. (a) $x + 12 \ge -12 + x$

(b) $x + 12 \le -12 + x$

64. (a) $1.5x + 9 < -6 + 1.5x$

(b) $1.5x + 9 \ge -6 + 1.5x$

In Exercises 65–72, use the graphing method to estimate the solutions of the given double inequality. Draw a number line graph of the solution set.

65. $-14 \le x - 3 \le 20$

66. $-11 < x + 5 < 28$

67. $-18 < x + 1 \le 14$

68. $0 \le x - 1 < 19$

69. $-5 \le 4 - x < 17$

70. $-9 < 7 - x \le 9$

71. $-13 < 2x - 1 < 15$

72. $-12 \le 1.5x + 3 \le 15$

In Exercises 73–84, use the graphing method to estimate the solutions of the given inequality. For special cases, use Ø or **R** to describe the solution set.

73. $5 - x \geq -4$ **74.** $\frac{2}{3}x + 7 < 15$

75. $x - 7 > 8 + x$ **76.** $5 - x < -x - 10$

77. $-x < \frac{4}{3}x + 21$ **78.** $2x - 7 \geq x - 3$

79. $x < x + 12$ **80.** $6 + 2x > 2x - 15$

81. $x < -x$

82. $-x + 5 \leq 2x - 4$

83. $-\frac{1}{2}(x + 20) \geq -\frac{1}{2}x$

84. $\frac{1}{2}x \geq -8 + 0.5x$

In Exercises 85–88, translate the given description into an inequality. Then estimate the solutions by graphing.

85. The sum of a number and 6 is at least 3.

86. The difference of a number and 1 is less than 6.

87. The difference of 8 and a number is greater than −2.

88. The sum of 4 and a number is at most −5.

Data Analysis: *Political Convention TV Ratings*

As shown in the bar graph, the TV ratings for the Democratic and Republican conventions have declined. (Source: Neilsen Media Research.)

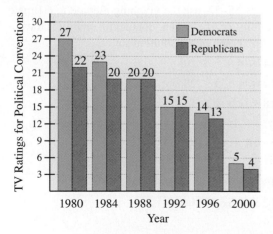

The TV ratings for the Democratic conventions can be modeled by the expression $-1.01x + 27.5$, and the model for the Republican conventions is $-0.83x + 23.9$. In both models, x represents the number of years since 1980.

89. (a) Write an inequality indicating that the TV rating for the Democratic convention was at least 20.

    (b) Estimate the solution of the inequality in part (a).

90. Repeat Exercise 89 for the Republican convention.

91. (a) Write an inequality that indicates that the ratings for the Republican convention exceeded the ratings for the Democratic convention.

    (b) Estimate and interpret the solution of the inequality in part (a).

92. (a) Write an inequality to estimate the years for which the ratings for the Democratic convention exceeded the ratings for the Republican convention by more than 1 point.

    (b) Estimate and interpret the solution of the inequality in part (a).

Challenge

In Exercises 93 and 94, use the graphing method to estimate the solution set of the given inequality.

93. $x + 1 \leq x + 1 < x + 1$

94. $2(x - 3) - x \leq x - 6 \leq -(6 - x)$

In Exercises 95 and 96, use the graphing method to compare the estimated solution sets for the inequalities in each pair.

95. (a) $2(x + 1) < 2x + 2$

    (b) $2(x + 1) \geq 2x + 2$

96. (a) $13 - x \leq -x + 13$

    (b) $13 - x > -x + 13$

In Exercises 97–100, use the graphing method to estimate the solution set of the given inequality.

97. $x + 1 \geq x + 1$

98. $x - 2 < x - 2$

99. $2(x + 5) > 10 + 2x$

100. $\frac{3}{2}x + 10 \leq 10 + 1.5x$

2.8 | Inequalities: Algebraic Methods

Addition Property of Inequalities • Multiplication Property of Inequalities •
Special Cases • Double Inequalities

Addition Property of Inequalities

In Section 2.7 we used graphing methods to estimate the solutions of inequalities. Now we turn to algebraic methods that provide exact solution sets.

The procedure for solving a linear inequality is nearly identical to the procedure for solving a linear equation. The strategy is to write an equivalent inequality in which the variable is isolated. To do this, we use properties of inequalities that are much like the previously discussed properties of equations.

> **Addition Property of Inequalities**
>
> For real numbers a, b, and c, the inequalities $a < b$ and $a + c < b + c$ are equivalent. A similar result holds for $>$, $\leq$, and $\geq$.

In words, the Addition Property of Inequalities allows us to add the same number to both sides of an inequality. We can also subtract the same number from both sides. We can use this property to isolate the variable term just as we did for equations.

EXAMPLE 1

Using the Addition Property of Inequalities

Solve each inequality and graph the solution set.

(a)
$$6 - 4x < 3 - 5x$$
$$6 - 4x + 5x < 3 - 5x + 5x \qquad \text{Add } 5x \text{ to both sides so that the variable term is on one side.}$$
$$6 + x < 3$$
$$6 + x - 6 < 3 - 6 \qquad \text{Subtract 6 from both sides to isolate the variable term.}$$
$$x < -3 \qquad \text{See Figure 2.53.}$$

(b)
$$-3 \leq y - 2$$
$$-3 + 2 \leq y - 2 + 2 \qquad \text{Add 2 to both sides to isolate the variable term.}$$
$$-1 \leq y$$
$$y \geq -1 \qquad \text{See Figure 2.54.}$$

Figure 2.53

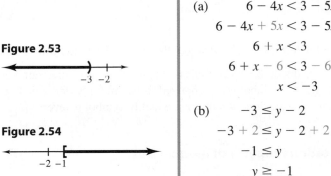

−3 −2

Figure 2.54

−2 −1

Note: As with equations, we can exchange the left and right sides of inequalities. However, we must take care to retain the direction of the inequality symbol. In part (b) of Example 1, exchanging the sides of the inequality $-1 \leq y$ also changes the inequality symbol to $\geq$.

Multiplication Property of Inequalities

We begin by investigating the effect of multiplying both sides of an inequality by a positive number and a negative number.

Exploring the Concept

Multiplying Both Sides of an Inequality

Consider the following true inequalities.

$$2 < 7 \qquad\qquad -1 > -7 \qquad\qquad -6 < 3$$

Observe that multiplying both sides of each inequality by 4 results in a true inequality.

$$4(2) \rule{1cm}{0.4pt} 4(7) \qquad 4(-1) \rule{1cm}{0.4pt} 4(-7) \qquad 4(-6) \rule{1cm}{0.4pt} 4(3)$$
$$8 < 28 \qquad\qquad -4 > -28 \qquad\qquad -24 < 12$$

However, if we multiply both sides by -2, the direction of the inequality symbol must be reversed to obtain a true inequality.

$$-2(2) \rule{1cm}{0.4pt} -2(7) \qquad -2(-1) \rule{1cm}{0.4pt} -2(-7) \qquad -2(-6) \rule{1cm}{0.4pt} -2(3)$$
$$-4 > -14 \qquad\qquad 2 < 14 \qquad\qquad 12 > -6$$

We can conclude from this experiment that multiplying both sides of an inequality by a nonzero number is permissible. However, if we multiply by a negative number, then we must reverse the direction of the inequality symbol.

Think About It

A counterexample is an example that is used to show that some proposition is not true in all instances. To show that subtraction is not commutative, $3 - 5 \neq 5 - 3$ is a counterexample. Can you find counterexamples for the following propositions?

(a) If $a < b$, then $a^2 < b^2$.

(b) If $a < b$, then $\dfrac{1}{a} < \dfrac{1}{b}$.

These results suggest the following property of inequalities.

> ### Multiplication Property of Inequalities
>
> For real numbers a, b, and c,
> 1. if $c > 0$, then the inequalities $a < b$ and $ac < bc$ are equivalent.
> 2. if $c < 0$, then the inequalities $a < b$ and $ac > bc$ are equivalent.
>
> A similar result holds for $>$, $\leq$, and $\geq$.

Because division is defined in terms of multiplication, the same rules hold for division.

In words, we can multiply or divide both sides of an inequality by a positive number, and the direction of the inequality symbol is not changed. If we multiply or divide both sides by a negative number, then the direction of the inequality symbol is reversed.

EXAMPLE **2**

Using the Multiplication Property of Inequalities

Solve each inequality and graph the solution set.

(a) $-6x \geq 24$

$$\dfrac{-6x}{-6} \leq \dfrac{24}{-6} \qquad \text{Reverse the direction of the inequality symbol.}$$

$$x \leq -4 \qquad \text{See Figure 2.55.}$$

(b) $3x < -15$

$$\dfrac{3x}{3} < \dfrac{-15}{3} \qquad \text{The direction of the inequality symbol is unchanged.}$$

$$x < -5 \qquad \text{See Figure 2.56.}$$

Figure 2.55

Figure 2.56

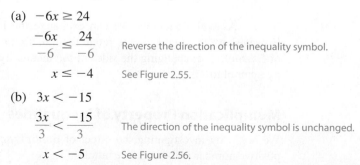

Figure 2.57

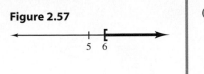

(c) $\qquad -\frac{1}{3}x \le -2$

$-3 \cdot \left(-\frac{1}{3}x\right) \ge -3 \cdot (-2)$ Reverse the direction of the inequality symbol.

$x \ge 6$ See Figure 2.57.

The algebraic procedure for solving a linear inequality is the same as that for linear equations. The only additional point to remember is to reverse the direction of the inequality symbol if you multiply or divide both sides by a negative number.

The Solving Routine for Linear Inequalities in One Variable

1. Clear fractions by multiplying both sides by the LCD of all the fractions.
2. Use the Distributive Property to simplify both sides.
   (a) Remove grouping symbols.
   (b) Combine like terms.
3. If a variable term appears on both sides, use the Addition Property of Inequalities to eliminate the variable term from one side.
4. Use the Addition Property of Inequalities to isolate the variable term.
5. Use the Multiplication Property of Inequalities to isolate the variable itself.
6. The resulting inequality describes the solutions of the original inequality. The solutions can also be given by a solution set or described by a number line graph.

Note: Again, we will use the inequality in step 6 (rather than a solution set) to describe the solutions of inequalities.

EXAMPLE **3**

Solving Linear Inequalities

Solve the given inequality.

LEARNING TIP

Write the complete inequality at each step. In most cases the last step is to divide by the coefficient of the variable. Thus you will usually not need to reverse the inequality symbol before the last step.

(a) $\qquad \dfrac{11}{6} < \dfrac{1}{2} - \dfrac{1}{3}x$

$6 \cdot \dfrac{11}{6} < 6 \cdot \dfrac{1}{2} - 6 \cdot \dfrac{1}{3}x$ Clear fractions. The LCD is 6.

$11 < 3 - 2x$

$8 < -2x$ Subtract 3 from both sides.

$\dfrac{8}{-2} > \dfrac{-2x}{-2}$ Divide both sides by -2.

$-4 > x$

Note that we reversed the inequality symbol when we divided by -2. The inequality can also be written as $x < -4$.

(b) $0.3x - 0.5 > 1.3$

$\qquad 0.3x > 1.8 \qquad$ Add 0.5 to both sides.

$\qquad \dfrac{0.3x}{0.3} > \dfrac{1.8}{0.3} \qquad$ Divide both sides by 0.3.

$\qquad x > 6$

Dividing both sides by 0.3 did not affect the inequality symbol.

(c) $-7 + 3(x - 1) \leq 2(5x + 2)$

$\qquad -7 + 3x - 3 \leq 10x + 4 \qquad$ Remove grouping symbols.

$\qquad -10 + 3x \leq 10x + 4$

$\qquad -10 - 7x \leq 4 \qquad$ Subtract 10x from both sides.

$\qquad -7x \leq 14 \qquad$ Add 10 to both sides.

$\qquad \dfrac{-7x}{-7} \geq \dfrac{14}{-7} \qquad$ Divide both sides by −7 and reverse the inequality symbol.

$\qquad x \geq -2$

Special Cases

In Section 2.7, we considered two special cases of inequalities in which the graphs of the left and right sides appeared to be parallel. We can now verify the conclusions that we drew about the solution sets.

$$x - 6 < x + 8 \qquad\qquad x - 6 > x + 8$$
$$x - 6 - x < x + 8 - x \qquad\qquad x - 6 - x > x + 8 - x$$
$$-6 < 8 \quad \text{True} \qquad\qquad -6 > 8 \quad \text{False}$$

Solution set: **R** $\qquad\qquad$ Solution set: Ø

The algebraic solving process detects these special cases when we obtain an equivalent inequality with no variable.

EXAMPLE $\boxed{4}$

Solving Inequalities: Special Cases

Solve the given inequality.

(a) $x + 2(x - 1) < 3(x + 1)$

$\qquad x + 2x - 2 < 3x + 3 \qquad$ Remove grouping symbols.

$\qquad 3x - 2 < 3x + 3 \qquad$ Combine like terms.

$\qquad -2 < 3 \qquad$ Subtract 3x from both sides.

Because $-2 < 3$ is true, the original inequality is true for all real numbers. The solution set is **R.**

(b) $3(x - 1) - x \geq 2x - 1$

$\qquad 3x - 3 - x \geq 2x - 1 \qquad$ Remove grouping symbols.

$\qquad 2x - 3 \geq 2x - 1 \qquad$ Combine like terms.

$\qquad -3 \geq -1 \qquad$ Subtract 2x from both sides.

Because $-3 \geq -1$ is false, there is no number for which the original inequality is true. The solution set is Ø.

Double Inequalities

When we solve a double inequality algebraically, we use the Addition and Multiplication Properties of Inequalities on all three parts of the inequality.

EXAMPLE **5**

Solving a Double Inequality Algebraically

Solve the given inequality.

(a) $-3 < 2x + 7 < 9$ The goal is to isolate the variable in the middle.

$-3 - 7 < 2x + 7 - 7 < 9 - 7$ Subtract 7 from all three parts to isolate the variable term.

$-10 < 2x < 2$

$\dfrac{-10}{2} < \dfrac{2x}{2} < \dfrac{2}{2}$ Divide all three parts by 2 to isolate the variable.

$-5 < x < 1$

(b) $3 \le 3 - 2x \le 12$

$3 - 3 \le 3 - 2x - 3 \le 12 - 3$ Subtract 3 from all three parts to isolate the variable term.

$0 \le -2x \le 9$

$\dfrac{0}{-2} \ge \dfrac{-2x}{-2} \ge \dfrac{9}{-2}$ Divide all three parts by -2 to isolate the variable.

$0 \ge x \ge -\dfrac{9}{2}$

Note that both inequality symbols were reversed when we divided by -2. The inequality can also be written $-\frac{9}{2} \le x \le 0$.

Quick Reference 2.8

Addition Property of Inequalities

- For real numbers a, b, and c, the inequalities $a < b$ and $a + c < b + c$ are equivalent. A similar result holds for $>$, $\le$, and $\ge$.

Multiplication Property of Inequalities

- For real numbers a, b, and c,
  1. if $c > 0$, then the inequalities $a < b$ and $ac < bc$ are equivalent.
  2. if $c < 0$, then the inequalities $a < b$ and $ac > bc$ are equivalent.

  A similar result holds for $>$, $\le$, and $\ge$.

- The routine for solving a linear inequality in one variable is nearly identical to the routine for solving linear equations.
  1. Clear fractions by multiplying both sides by the LCD of all the fractions.
  2. Use the Distributive Property to simplify both sides.
     (a) Remove grouping symbols.
     (b) Combine like terms.
  3. If a variable term appears on both sides, use the Addition Property of Inequalities to eliminate the variable term from one side.

4. Use the Addition Property of Inequalities to isolate the variable term.

5. Use the Multiplication Property of Inequalities to isolate the variable itself.

6. The resulting inequality describes the solutions of the original inequality. The solutions can also be given by a solution set or described by a number line graph.

Special Cases   • When we solve an inequality algebraically, we may obtain an equivalent inequality with no variables.

1. If the equivalent inequality is true, then the solution set of the original inequality is **R.**

2. If the equivalent inequality is false, then the solution set of the original inequality is Ø.

Double Inequalities   • To solve a double inequality algebraically, use the Addition and Multiplication Properties of Inequalities on all three parts of the inequality to isolate the variable.

Speaking the Language 2.8

1. The inequalities $x < y$ and $ax < ay$ are equivalent provided that the number a is ▨▨▨ .

2. If we multiply both sides of $3x > 5$ by -1, we will obtain an equivalent inequality only if we ▨▨▨ the inequality symbol.

3. The inequality $x - 2 \le x$ is ▨▨▨ for all values of x, and so the solution set is ▨▨▨ .

4. The goal in solving the double inequality $-3 < x + 2 \le 1$ is to isolate ▨▨▨ .

Exercises 2.8

Concepts and Skills

1. Compare the Addition Property of Inequalities and the Addition Property of Equations.

2. Compare the Multiplication Property of Inequalities and the Multiplication Property of Equations.

In Exercises 3–10, insert an inequality symbol to make the conclusion true.

3. If $-4 < 1$, then $-4(-1)$ ▨▨▨ $1(-1)$.

4. If $-2 > -7$, then $3(-2)$ ▨▨▨ $3(-7)$.

5. If $a > b$, then $3 + a$ ▨▨▨ $3 + b$.

6. If $a < b$, then $a - 6$ ▨▨▨ $b - 6$.

7. If $a < b$, then $-4a$ ▨▨▨ $-4b$.

8. If $a > b$, then $2a$ ▨▨▨ $2b$.

9. If $a > b$, then $\dfrac{a}{-5}$ ▨▨▨ $\dfrac{b}{-5}$.

10. If $a < b$, then $-\dfrac{2}{3}a$ ▨▨▨ $-\dfrac{2}{3}b$.

In Exercises 11–18, use the Addition Property of Inequalities to solve the inequality.

11. $x + 5 < 1$

12. $x - 6 \ge 2$

13. $-3 \le 2 + x$

14. $0 > -3 + x$

15. $5x - 1 \ge -1 + 4x$

16. $2 - 3x < 1 - 4x$

17. $x - 2 > 2x + 6$

18. $5 - 7x \le -6x$

In Exercises 19–28, use the Multiplication Property of Inequalities to solve the inequality.

19. $3x > 12$

20. $2x \le -8$

21. $-5x < 15$

22. $-8 \leq -4x$

23. $0 \leq -x$

24. $-y > -2$

25. $\dfrac{3}{4}x \leq 12$

26. $10 > -\dfrac{5}{6}y$

27. $\dfrac{x}{-5} \leq 1$

28. $\dfrac{-x}{3} > -8$

29. To solve $-3.2x < -8.9$, suppose that you decide to clear the decimals as follows.

$$-10(-3.2x) < -10(-8.9)$$
$$32x < 89$$

Is the resulting inequality equivalent to the original inequality? Explain.

30. Suppose that you want to solve $-\frac{3}{4}x > 9$. The following are appropriate first steps.

(a) Multiply both sides by 4.

(b) Multiply both sides by $\frac{4}{3}$.

(c) Multiply both sides by $-\frac{4}{3}$.

How would the results differ?

In Exercises 31–40, solve the inequality.

31. $5 + 3x > -4$

32. $2x - 7 \leq 5$

33. $3 - x \leq 5$

34. $-x - 2 > -1$

35. $3 - 4x \leq 19$

36. $9 - 7x > 2$

37. $5x + 5 \leq 4x - 2$

38. $2x - 3 < 3x + 1$

39. $6 - 3x > x - 2$

40. $7 - 3x < 22 - 8x$

In Exercises 41–54, solve the inequality.

41. $7 + x \leq 3(1 - x)$

42. $5(x + 1) < 2x - 1$

43. $6(2 - 3x) > 11(3 - x)$

44. $3(x + 3) \leq 2(x + 4)$

45. $3y + 2(2y + 1) > 11 + y$

46. $5(x + 1) + 2 < 3(x - 1)$

47. $6(t - 3) + 3 \leq 2(4t + 3) + 5t$

48. $2(x - 7) + 3 > 7 + 4(x - 2)$

49. $2x + \dfrac{1}{2} \geq \dfrac{1}{3}$

50. $\dfrac{7}{4}x - 1 \geq \dfrac{5}{2}$

51. $\dfrac{1}{2}t - \dfrac{3}{4} < -\dfrac{1}{3}t$

52. $\dfrac{5}{4} < 3x + \dfrac{1}{2}$

53. $-1.6t - 1.4 \geq 4.2 - 3t$

54. $0.25x - 1 \leq 0.5x + 1.25$

55. When we solve an inequality algebraically, we sometimes obtain an equivalent inequality that has no variable. How do we determine the solution set of such an inequality?

56. The Addition Property of Inequalities refers to a *single* inequality, but we still use this property to solve *double* inequalities such as $-2 < x - 5 < 7$. Why are we entitled to do that?

In Exercises 57–62, determine the solution set of the given inequality.

57. $3(x - 2) \geq 2x + x$

58. $2x < 2(x + 1)$

59. $2(3 - x) + x \leq 6 - x$

60. $4(x + 1) + x - 4 < 5(x - 1)$

61. $2x - 5 + x \leq -3(2 - x)$

62. $4(2 - x) \leq 3x + 8 - 7x$

In Exercises 63–72, solve the double inequality.

63. $4 < x + 2 < 11$

64. $-1 < x - 6 < 5$

65. $10 \leq -5x < 35$

66. $-8 < -2x < 10$

67. $0 \leq \dfrac{y}{3} \leq 2$

68. $-2 < \dfrac{y}{5} < 1$

69. $-5 < 2x - 1 < -1$

70. $-7 \leq 2 + 3x < 5$

71. $-2 \leq 4 - 3x \leq 4$

72. $1 < 1 - x \leq 7$

In Exercises 73–86, solve the inequality.

73. $-3x + 5 \geq x - 1$

74. $10x + 7 < 2x - 5$

75. $5x - 6 \geq 5x - 1$

76. $7 + 6x < 6x - 1$

77. $x + 5 \leq x + 7$

78. $3x - 2 < 3x$

79. $1 - 3(x + 2) < 7$

80. $3(x - 3) \geq 1 + 2(x - 5)$

81. $2x + 1 > \dfrac{3x - 1}{2}$

82. $\dfrac{x + 10}{4} \leq 2x - 1$

83. $2(x + 3) < 2x + 1$

84. $3x - 2 \geq 4x - (x + 7)$

85. $2.3y + 0.75 \geq 4.2$

86. $4.4 - 1.3y < 5.05$

Data Analysis: *Employed Women*

In 1940, the percentage of women who were employed was 27.4%. By 2000 the percentage had risen to 61.4%. (Source: Statistical Abstract of the United States.)

Assuming a constant increase during this period, the expression $0.57x + 27.4$ can be used to model the percentage of employed women, where x is the number of years since 1940. In Exercises 87–90, we use the model $y_1 = 0.57x + 27.4$.

87. Write and solve an inequality to predict the years for which the percentage of employed women will exceed 65%.

88. The percentage of unemployed women for any year x is 100% minus the percentage of employed women. Simplify the expression $100 - y_1$ and let y_2 represent the result.

89. Solve the inequality $y_1 < y_2$ to the nearest whole number.

90. How can the result in Exercise 89 be interpreted?

Challenge

In Exercises 91–94, solve the given inequality with the given conditions on a and b.

91. $x + a < b$

92. $ax < 7, \quad a > 0$

93. $ax + b \leq 0, \quad a < 0$

94. $\dfrac{x}{a - b} < 1, \quad a < b$

In Exercises 95 and 96, solve.

95. $x(2x - 5) > 2x^2 + 5x + 20$

96. $x(x + 1) \geq x^2$

In Exercises 97 and 98, solve the double inequality.

97. $3 + x < 2x - 4 < x + 6$

98. $x^2 - 2 \leq x(x - 1) < x^2 + 5$

Chapter 2 Review Exercises

Section 2.1

1. Evaluate the expression for the given value of the variable.

(a) $-4x + 7; \quad x = -6$

(b) $-x^2 + 1; \quad x = 1$

2. Evaluate the expression for the given values of the variables.

(a) $-2(x - y) + 3(x + y); \quad x = 4, y = -3$

(b) $\dfrac{3a + 2b}{c}; \quad a = 2, b = 5, c = 4$

3. Evaluate $7 + 2x - x^2$ for $x = -3$ on the home screen of your calculator.

4. Identify the terms and coefficients of each expression.

(a) $x^2 - 2xy + 7$ (b) $\dfrac{a}{2} + 5(a + 1)$

5. Combine like terms.

(a) $3xy + x - 5xy + 4x$

(b) $8c^2 - 6c^2 + 5c^2$

6. Simplify $\dfrac{3}{5}(-10x)$.

7. Simplify $-3(2a - b + 4)$.

8. Simplify each expression.

(a) $-(2x - 1)$ (b) $x - (4 - 2x)$

9. Simplify $-(x + 2y) - 4(3x - y + 1)$.

10. Simplify $2 - 2[x - (3 - x)]$.

Section 2.2

For Exercises 11–14, refer to the accompanying figure.

Figure for 11–14

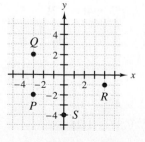

11. What are the coordinates of point R?

12. Which point is in Quadrant III?

13. Which point has the coordinates $(-3, 2)$?

14. Which point is on the y-axis?

15. Describe the set of all points in a coordinate plane whose y-coordinates are 0.

16. What are the signs of the coordinates of all points in Quadrant IV?

17. What name is given to the point $(0, 0)$?

18. If the coordinates of a point have the same signs, in what possible quadrants could the point lie?

Section 2.3

19. Use your calculator to evaluate $2x - 3$ for $x = -4$, $-1, 3$, and 7.

20. Use the graph of $0.5x + 7$ to evaluate the expression for $x = -6, 0$, and 15.

21. Use the graph of $-x + 12$ to estimate the value of x for which $-x + 12 = -6$.

22. Use a graph to estimate two values of x for which $0.02x^2 = 18$.

23. If you produce the graph of $7 - 2x$ and trace to the point whose x-coordinate is 5, what will the y-coordinate be? Why?

24. Which of the following expressions has a graph that contains the point $(-2, 5)$?
    (i) $3x + 1$ (ii) $x + 5$
    (iii) $1 - 2x$

Section 2.4

25. Which of the following is an example of a conditional equation?
    (i) $x + 3 = x + 5$
    (ii) $x - 3 = 3 - x$
    (iii) $x + 3 = 3 + x$
    (iv) $2(x - 3) = 2x - 3$

26. Which member of the set $\{-3, -1, 0, 2\}$ is a solution of $4x - 5 = 9x$?

27. Determine whether -3 is a solution of the equation $x^2 + 1 = -3x + 1$ by using your calculator to evaluate both sides of the equation.

28. Suppose that you store -5 in your calculator and then enter $2(x + 1) = 1 - x$. What result will you obtain, and what does it mean?

29. Use the graphing method to estimate whether 4 is a solution of $-2x + 15 = 7$.

30. What do we call an equation that can be written as $Ax + B = 0$, where $A \neq 0$?

31. Suppose that you have used a graph to estimate the solution of $14 - 3x = 20$, and the tracing cursor is on the point that reveals the solution. To what point have you traced?

32. Use the graphing method to estimate the solution of $-0.5x + 7 = 2x - 3$.

33. Describe the graphs of the left and right sides of an equation when the equation is
    (a) a contradiction.   (b) an identity.

34. Use the graphing method to estimate the solution(s) of the given equation.
    (a) $2(x - 1) + 8x = -(12 - 5x)$
    (b) $\frac{1}{2}(x + 18) = -6 + 0.5x$
    (c) $2x - (x - 1) = 1 + x$

Section 2.5

35. Given the equation $2(x + 1) = x$, which of the following steps would not produce an equivalent equation?
    (i) Divide the left and right sides by 4.
    (ii) Exchange the left and right sides.
    (iii) Add x to the left side and subtract x from the right side.
    (iv) Replace $2(x + 1)$ with $2x + 2$.

36. To solve $5x = 20$, we can multiply both sides by $\frac{1}{5}$. What other step would produce the same result?

37. According to the solving routine, for which of the following equations is isolating the variable term the next step in solving the equation?
    (i) $2x = 8$
    (ii) $2x + 3 = 8$
    (iii) $2x + 3 = x + 8$

38. To solve $\frac{3}{4}x = 9$, each of the following is an appropriate first step.
    (i) Multiply both sides by 4.
    (ii) Multiply both sides by $\frac{4}{3}$.
    (iii) Divide both sides by 3.
    Which of these steps would isolate the variable?

In Exercises 39–44, determine the solution of the equation.

39. $7 = 9x - 8x$

40. $-14 + x = -16$

41. $-5x + 8 = -4x$

42. $-9x + 7 + 10x = 8$

43. $-\dfrac{x}{3} = -6$ **44.** $6x - 7x = -\dfrac{1}{4}$

Section 2.6

In Exercises 45–56, solve the equation. In the case of an identity or contradiction, write the solution set.

45. $9 = 2x - 6$

46. $2x - 15 = 6 - 5x$

47. $6 + 3x - 4 - 7x - 2 = 6 - 5x - 10$

48. $3 + 2(x - 4) - x = -2$

49. $\dfrac{4}{3}x + \dfrac{14}{9} = \dfrac{x}{9} + \dfrac{1}{3}$

50. $\dfrac{2}{5}(3x - 1) = x$

51. $2(x - 3) - 3(x + 1) = 0$

52. $8 - 2[3 - 2(x + 1)] = x + 3$

53. $3x + 7 - x = 4 + 2x + 3$

54. $8 - 5x = 2x + 9 - 7x$

55. $-3(x - 4) = 4x - (7x - 13)$

56. $-4x + 1 + 3x = -(x - 1)$

Section 2.7

57. Which of the following is a correct interpretation of $x \le 5$?

   (i) x is at least 5.

  (ii) 5 is greater than x.

 (iii) $x = 5$ or $x < 5$.

▼ **58.** To estimate the solutions of an inequality graphically, we graph the two sides of the inequality. What is the next step?

▼ **59.** If the graphing method is used to estimate the solutions of an inequality, under what conditions does the point of intersection of the graphs represent a solution?

60. Which member of the set $\{-1, 2, 3, 5\}$ is a solution of the given inequality?

  (a) $3 - 4x > -5$

  (b) $4 \le x + 1 \le 5$

61. Translate the given description into an inequality.

  (a) A number is at most 2.

  (b) A number is between -1 and 3, including 3.

62. Use the graphing method to estimate the solutions of the given inequality. Draw a number line graph of the solution set.

  (a) $2 - 5x > -8$

  (b) $-x < x + 4$

63. Suppose that $8 + x$ and $x - 6$ are the two sides of a $<$ or $>$ inequality. Write the inequality so that the solution set is the following.

  (a) **R**   (b) Ø

64. Use the graphing method to write the solution set of each inequality.

  (a) $3(x + 1) < 3x - 8$

  (b) $-(1 - 2x) < 2(x + 3)$

Section 2.8

65. Which of the following inequalities is equivalent to $x < 7$?

   (i) $x - 7 > 0$

  (ii) $-7x > 0$

 (iii) $7 - x > 0$

66. Suppose that solving an inequality leads to $3 \le 3$. What is the solution set of the original inequality?

In Exercises 67–74, solve the inequality.

67. $-8 \ge 3 - x$ **68.** $3x + 11 < 2x - 4$

69. $-\dfrac{3}{4}x \le 15$ **70.** $\dfrac{1}{3} - \dfrac{1}{4}x > \dfrac{1}{6}(x - 3)$

71. $-3(x + 1) < -(x - 1)$

72. $2.7 + 1.9x \ge 4.1x + 0.5$

73. $-12 < -4x \le 16$ **74.** $-16 \le 7x - 2 < 12$

In Exercises 75 and 76, write the solution set.

75. $3(x - 1) - (x + 1) \le 2(x + 4)$

76. $2 - [x - (3 + x)] \le -(x + 8) + x$

Chapter 2 Test

1. Evaluate $\dfrac{2x}{y} - (x + y)$ for $x = 2$ and $y = -4$.

2. Combine like terms: $7a - b + 3b - 6a$.

3. Simplify $-(x - 1) + 4(2x + 3)$.

4. What is the coefficient of $\dfrac{x - 9}{2}$? Why?

5. Name the quadrant in which all points have the coordinate sign pattern $(-, +)$.

6. Describe the set of all points in the coordinate plane whose x-coordinates are 0.

7. If the graph of the expression $3x - 2$ contains a point whose x-coordinate is -4, what is the y-coordinate? Why?

8. Use the graphing method to estimate the value of x for which $6 - 2x = -4$.

9. Use a graph to estimate the value of x for which the expressions $0.9x - 4$ and $6 - 0.1x$ have the same value.

10. Use the graphing method to estimate the solution of each equation.

   (a) $-x + 19 = 3$ (b) $4x + 3 = -2 + 5x$

11. What name do we give to an equation if the graphs of the two sides

   (a) coincide?   (b) are parallel?

12. Use a graph to determine the number of x-values for which $x^2 + 1 = -4$.

In Questions 13–18, solve the equations algebraically. In the case of a contradiction or an identity, write the solution set.

13. $14x - 20 = 1 + 13x$

14. $4 - \dfrac{x}{2} = 5$

15. $\dfrac{2}{3}(x - 5) = x + 2$

16. $-2(x + 3) - (4 - x) = -x - 10$

17. $5 - 5(1 - x) = 3 + 5x$

18. $\dfrac{x}{3} + 2 = \dfrac{5x}{6} - 1$

19. Translate the following sentence into an inequality: 3 less than a number is at least -2.

20. Use the graphing method to estimate the solutions of $3 - 5x < x - 9$. Graph the solution set.

21. Suppose that the left and right sides of an inequality are $9 - x$ and $-11 - x$. What is the solution set if the inequality symbol is as follows? Explain your answers.

   (a) $\leq$ (b) $>$

In Questions 22–25, solve the inequality.

22. $-(11 - x) < 2x + 3(x - 1)$

23. $x + \dfrac{1}{4} \geq -\dfrac{4x}{5} - 1.55$

24. $-2 < -\dfrac{x}{3} \leq 5$ **25.** $-4 \leq 2(x - 7) < 0$

The accompanying pie graph shows certain ranges of household incomes and the percentages of home buyers whose incomes fell within those ranges.

A large number of application problems use percents to compare and analyze information. For example, we can easily see from the graph that more than half of the home buyers had incomes of less than $61,000. For more on this real-data application, see Exercises 75–78 in Section 3.2.

Income Range	
Less than $30,000	(1)
$30,000–$40,000	(2)
$41,000–$50,000	(3)
$51,000–$60,000	(4)
At least $61,000	(5)

16% 18% 41% 10%

4 3
5 2
1

(Source: Chicago Title and Trust.)

Modeling and Applications

Chapter Snapshot

Because ratios and proportions, percents, and formulas all play a significant role in solving applications problems, we devote the first half of this chapter to those fundamental topics. We then learn about translating information into expressions and equations, and we discuss a general problem-solving strategy. The chapter concludes with a variety of examples and exercises to help you apply your knowledge of algebra to solving real-life problems.

Warm-Up Skills

The following exercises review concepts and skills that you will need in Chapter 3.

In Exercises 1–6, write and evaluate a numerical expression that models the given description.

1. Nine more than twice -6
2. The quotient of -8 and 12
3. Three-fourths of -28
4. The difference between -3 and -10
5. Six less than 2
6. Five times the sum of 7 and 2
7. Evaluate the expression $\frac{9}{5}C + 32$ for $C = 30$.
8. Simplify the expression $4.75x + 3.25(20 - x)$.
9. Compare the procedure for solving $-3x = 15$ and the procedure for solving $-3x \le 15$.

In Exercises 10–12, solve the given equation.

10. $\dfrac{x+5}{3} = \dfrac{5}{2}$

11. $x - 0.32x = 578$

12. $\dfrac{1}{2}[x - (90 - x)] = 5$

3.1 Ratio and Proportion

Ratio • Proportion

Ratio

We use a **ratio** to compare one quantity to another. For example, a team's win–loss ratio is a comparison of the team's wins to its losses. The ratio of stocks to bonds in your investment portfolio is a comparison of the value of your stock holdings to the value of your bonds. Two-cycle engines require a certain ratio of oil to gasoline for the fuel mixture.

We express a ratio as a quotient of two quantities. For example, the **ratio of a to b** is written as $\dfrac{a}{b}$. (Sometimes ratios are written as $a{:}b$, but we will not use this notation.)

Because a ratio is a quotient, we operate with ratios as we do with all fractions. For instance, if a class consists of 3 men and 6 women, the ratio of men to women is $\dfrac{3}{6} = \dfrac{1}{2}$. Note that we usually write ratios in reduced form. In this case, we are saying that there are 3 men for every 6 women, or, equivalently, that there is 1 man for every 2 women.

EXAMPLE 1

Writing Ratios

A basketball team has a record of 24 wins and 9 losses. What is the ratio of

(a) wins to losses?

(b) wins to number of games played?

Solution

(a) $\dfrac{\text{wins}}{\text{losses}} = \dfrac{24}{9} = \dfrac{8}{3}$ The team won 8 games for every 3 losses.

(b) $\dfrac{\text{games won}}{\text{games played}} = \dfrac{24}{33} = \dfrac{8}{11}$ The team won 8 of every 11 games played.

A ratio is valid if the units are the same or if the units refer to entirely different measurements. For example, ratios may indicate miles per hour or dollars per pound. However, if the units are different but refer to the same measurement, such as length, weight, or time, then we need to convert the units so that they are the same before we can make a comparison.

EXAMPLE 2

Converting Units

(a) What is the ratio of 12 ounces to 2 pounds?

(b) What is the ratio of 4 feet to 5 yards?

Solution

(a) $\dfrac{12}{32} = \dfrac{3}{8}$ 1 pound is 16 ounces; 2 pounds is 2 · 16 or 32 ounces.

(b) $\dfrac{4}{15}$ 1 yard is 3 feet; 5 yards is 3 · 5 or 15 feet.

The *unit price* of an item is the ratio of the total price to the number of units of the item. For example, if a 5-ounce box of chocolates costs $2.10, then the unit price is the ratio $\dfrac{2.10}{5}$, or 0.42. The unit price is 42¢ per ounce.

EXAMPLE **3**

Determining the Unit Price

Suppose laundry detergent is available in the following sizes.

Size	Price
40 ounces	$3.50
4 pounds, 6 ounces	5.25
6 pounds, 4 ounces	9.10

In terms of unit price, which size is the best buy?

Solution

Convert each size to ounces and determine the unit price (price per ounce) by dividing the price by the size in ounces. Recall that 1 pound is 16 ounces.

Size	Size (ounces)	Price	Unit Price (dollars per ounce)
40 ounces	40	$3.50	0.0875
4 pounds, 6 ounces	70	5.25	0.075
6 pounds, 4 ounces	100	9.10	0.091

The 4-pound, 6-ounce size is the most economical because it has the lowest unit price, 7.5¢ per ounce.

Proportion

A **proportion** is an equation that states that two ratios are equal. If the ratio of a to b is the same as the ratio of c to d, then we write $\dfrac{a}{b} = \dfrac{c}{d}$, which we read "$a$ is to b as c is to d." As with other equations involving fractions, we can eliminate, or clear, the fractions in a proportion by multiplying both sides by the LCD.

$$bd \cdot \dfrac{a}{b} = bd \cdot \dfrac{c}{d}$$
$$ad = bc$$

The same result can be obtained by multiplying according to the following pattern.

$$\frac{a}{b} = \frac{c}{d}$$

$$ad = bc$$

This procedure is sometimes called **cross-multiplication,** and the terms ad and bc are called **cross-products.**

Property of Cross-Products

For real numbers a, b, c, d, where $b \neq 0$ and $d \neq 0$, $\dfrac{a}{b} = \dfrac{c}{d}$ if and only if $ad = bc$.

This property gives us an efficient way to solve a proportion.

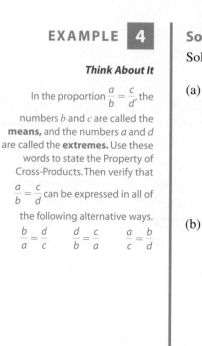

EXAMPLE 4

Think About It

In the proportion $\dfrac{a}{b} = \dfrac{c}{d}$, the numbers b and c are called the **means,** and the numbers a and d are called the **extremes.** Use these words to state the Property of Cross-Products. Then verify that $\dfrac{a}{b} = \dfrac{c}{d}$ can be expressed in all of the following alternative ways.

$$\frac{b}{a} = \frac{d}{c} \qquad \frac{d}{b} = \frac{c}{a} \qquad \frac{a}{c} = \frac{b}{d}$$

Solving Proportions by Cross-Multiplying

Solve the given proportion.

(a) $\dfrac{20}{x} = \dfrac{5}{36}$

$20 \cdot 36 = 5 \cdot x$ Cross-multiply.

$720 = 5x$

$144 = x$ Divide by 5 to isolate the variable.

The solution is 144.

(b) $\dfrac{x + 3}{x - 1} = \dfrac{5}{4}$

$4(x + 3) = 5(x - 1)$ Cross-multiply.

$4x + 12 = 5x - 5$ Distributive Property

$12 = x - 5$ Subtract 4x from both sides.

$17 = x$ Add 5 to both sides.

The solution is 17.

The conditions of many applications problems can be described with a proportion.

EXAMPLE 5

Solving Application Problems with Proportions

While reviewing your budget, you notice that groceries cost a total of $206.40 for the last 3 weeks. At that rate, how much should you budget for groceries for the next 7 weeks?

Solution

Let x = cost of groceries for 7 weeks.

$$\frac{\text{cost}}{\text{weeks}} = \frac{x}{7} = \frac{206.40}{3}$$

$3x = 7(206.40)$ ⟶ Cross-multiply.

$x = 481.60$ ⟶ Divide both sides by 3.

You should budget \$481.60 for groceries for the next 7 weeks.

EXAMPLE 6

Interpreting the Results of a Survey

A survey of 180 dentists found that they recommend Bristle toothbrushes by 5 to 4. Of the surveyed dentists, how many recommend Bristle?

Solution

If x = the number of dentists who recommend Bristle, then $180 - x$ = the number of dentists who do not recommend Bristle. Use the ratio of the number of dentists who recommend Bristle to the number who do not recommend Bristle to write a proportion.

$$\frac{\text{number who recommend}}{\text{number who do not recommend}} = \frac{5}{4} = \frac{x}{180 - x}$$

$5(180 - x) = 4x$ ⟶ Cross-multiply.

$900 - 5x = 4x$ ⟶ Distributive Property

$900 = 9x$ ⟶ Add 5x to both sides.

$100 = x$ ⟶ Divide both sides by 9.

Of the 180 dentists surveyed, 100 recommend Bristle toothbrushes.

Two triangles are called **similar triangles** if the measures of their corresponding angles are equal. A theorem from geometry states that the lengths of the corresponding sides of similar triangles are proportional.

Note: For a triangle ABC, such as the one shown in the following example, we use the symbol $\overline{AB}$ to refer to the *side* of the triangle, and we use the symbol AB to refer to the *length* of the side.

EXAMPLE 7

Similar Triangles

In Figure 3.1, the triangles are similar triangles. Determine AB and AC.

Figure 3.1

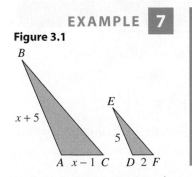

Solution

For these triangles, $\overline{AB}$ corresponds to $\overline{DE}$, and $\overline{AC}$ corresponds to $\overline{DF}$. The lengths of the corresponding sides are proportional.

$$\frac{x + 5}{5} = \frac{x - 1}{2} \qquad \frac{AB}{DE} = \frac{AC}{DF}$$

$$2(x + 5) = 5(x - 1) \qquad \text{Cross-multiply.}$$
$$2x + 10 = 5x - 5 \qquad \text{Distributive Property}$$
$$10 = 3x - 5 \qquad \text{Subtract } 2x \text{ from both sides.}$$
$$15 = 3x \qquad \text{Add 5 to both sides.}$$
$$5 = x \qquad \text{Divide both sides by 3.}$$

$AB = x + 5 = 5 + 5 = 10$ units, and $AC = x - 1 = 5 - 1 = 4$ units.

Quick Reference 3.1

Ratio
- A **ratio** is a comparison of one quantity to another. The **ratio of a to b** is written as the fraction $\dfrac{a}{b}$. We simplify and operate with ratios as we would with any other fraction.

- If the units in a ratio are different but refer to the same measurement, such as length, weight, or time, we need to convert the units so that they are the same in order to make a valid comparison.

- A commonly used ratio is a *unit price*, which is the ratio of the total price of an item to the number of units of the item.

Proportion
- A **proportion** is an equation that states that two ratios are equal.

- Property of Cross-Products: For real numbers a, b, c, and d, where $b \neq 0$ and $d \neq 0$, $\dfrac{a}{b} = \dfrac{c}{d}$ if and only if $ad = bc$. In words, we can eliminate the fractions in a proportion by **cross-multiplication,** a procedure in which ad and bc are called **cross-products.**

- Two triangles are called **similar triangles** if the measures of their corresponding angles are equal. If two triangles are similar triangles, then the lengths of their corresponding sides are proportional.

Speaking the Language 3.1

1. We express a comparison of one quantity to another by way of a(n) ▬▬▬▬ .

2. A statement that two ratios are equal is called a(n) ▬▬▬▬ .

3. The Property of ▬▬▬▬ allows us to rewrite $\dfrac{a}{b} = \dfrac{c}{d}$ as $ad = bc$.

4. If two triangles are ▬▬▬▬ triangles, then the lengths of their corresponding sides are proportional.

Exercises 3.1

Concepts and Skills

1. What is a ratio?

2. Why is a ratio of women to men valid, whereas a ratio of feet to inches is not valid?

In Exercises 3–10, write a ratio that represents a comparison of the given quantities.

3. The number of men to women in a class if there are 12 men and 8 women

4. The number of Democrats to Republicans in the Senate if there are 48 Democrats and 52 Republicans

5. The number of fiction to nonfiction books in a library if there are 780 volumes of nonfiction and 600 volumes of fiction

6. The salary to the number of hours worked if the salary is $342 for 36 hours of work

7. The amount of fertilizer required to the actual size of a yard if 25 pounds of fertilizer are required for 5000 square feet of yard

8. The height of a tree to the length of its shadow if a 76-foot tree casts a 20-foot shadow

9. The ratio of total cholesterol to HDL cholesterol is an indicator of the risk of heart disease. If a person's total cholesterol level is 173 mg/dl and the HDL level is 42 mg/dl, what is the ratio of total cholesterol to HDL?

10. The ratio of BUN to creatinine is used to detect kidney disease. What is the ratio of BUN to creatinine if the BUN level is 15 mg/dl and the creatinine level is 0.90 mg/dl?

In Exercises 11–14, write the ratio.

11. Balls to strikes for a pitcher who throws 120 pitches, of which 55 were balls

12. Yards rushing to yards passing for a football team if the team has 460 total yards and 210 yards rushing

13. The area of a circular pool to the distance around the pool if the radius is 16 feet

14. The area of a square to the perimeter of the square if the length of a side is 2 meters

In Exercises 15–18, write the ratio.

15. A small pond has a population of 500 blue gill, 350 bass, and 250 sunfish. Write the following ratios.

    (a) The number of bass to the number of all other fish

    (b) The number of blue gill to the number of bass

    (c) The number of sunfish to the number of blue gill

16. A garden center stocks three types of flowering fruit trees: 75 pear trees, 40 cherry trees, and 60 peach trees. Write the following ratios.

    (a) The number of pear trees to the number of all other trees

    (b) The number of peach trees to the number of cherry trees

    (c) The number of cherry trees to the total number of trees

17. The grade distribution in an English class is 3 A's, 5 B's, 10 C's, 3 D's, and 2 F's. Write the following ratios.

    (a) The number of A's and B's to the number of C's

    (b) The number of C's to the number of D's and F's

    (c) The number of A's, B's, and C's to the total number of grades

18. A portfolio consists of $8000 in stock, $5000 in bonds, $20,000 in real estate, and $7000 in a money market fund (cash). Write the following ratios.

    (a) The combined value of the bonds and cash to the value of the entire portfolio

    (b) The value of the stock to the value of the bonds

    (c) The value of the real estate to the value of the remainder of the portfolio

    (d) The value of the cash to the value of the stocks and bonds combined

In Exercises 19–26, what is the ratio of the two given quantities?

19. 70 meters to 50 meters

20. 12 pounds to 9 pounds

21. 3 pints to 2 quarts

22. 17 nickels to 2 dollars

23. 5 yards to 20 feet

24. 9 hours to 2 days

25. 3 dollars to 15 quarters

26. 24 ounces to 2 pounds

27. What do we mean by the unit price of an item?

28. Suppose that a retailer advertises tires for $89.85 each or four for $359. How enticing is the retailer's offer? Why?

In Exercises 29–36, determine the unit price.

29. An 8-ounce box of vermicelli costs 55 cents.

30. Five pounds of sugar costs $1.69.

31. A package of 24 paper plates costs $1.99.

32. A box of 30 washers costs $2.89.

33. The cost of 12 square yards of carpet is $223.44.

34. The price of a 10-foot piece of lumber is $6.80.

35. A bottle of 60 vitamin tablets is $7.60.

36. A campaign advertisement costing $4500 results in 1200 votes.

In Exercises 37–40, which size is the most economical?

37. Ketchup

Size	Price
(i) 14 ounces	$0.81
(ii) 28 ounces	1.59
(iii) 64 ounces	3.46

38. Nails

Size	Price
(i) 50 nails	$2.27
(ii) 80 nails	3.84
(iii) 125 nails	5.94

39. Shampoo

Size	Price
(i) 7 ounces	$2.97
(ii) 11 ounces	3.88
(iii) 15 ounces	5.69

40. Aspirin

Size	Price
(i) 24 tablets	$2.56
(ii) 60 tablets	5.98
(iii) 100 tablets	9.10

41. What is a proportion?

42. Explain why the proportion $\dfrac{3}{x} = \dfrac{5}{4}$ is equivalent to the proportion $\dfrac{x}{3} = \dfrac{4}{5}$.

In Exercises 43–56, solve the given proportion.

43. $\dfrac{x}{32} = \dfrac{7}{8}$

44. $\dfrac{20}{6} = \dfrac{r}{24}$

45. $\dfrac{16}{9} = \dfrac{4}{x}$

46. $\dfrac{6}{r} = \dfrac{9}{2}$

47. $\dfrac{x+5}{3} = \dfrac{5}{2}$

48. $\dfrac{4}{3} = \dfrac{x-3}{2}$

49. $\dfrac{2-x}{2} = \dfrac{x+3}{4}$

50. $\dfrac{2x-5}{2} = \dfrac{5x+1}{4}$

51. $\dfrac{4}{r-3} = \dfrac{5}{r-2}$

52. $\dfrac{7}{x-3} = \dfrac{2}{x+2}$

53. $\dfrac{r+3}{r-2} = \dfrac{3}{4}$

54. $\dfrac{x+6}{x+2} = \dfrac{5}{3}$

55. $\dfrac{3}{2} = \dfrac{2x+1}{x-5}$

56. $\dfrac{7}{4} = \dfrac{2x-1}{x+3}$

In Exercises 57–60, the given triangles are similar triangles. Determine the indicated length(s).

57. *PQ*

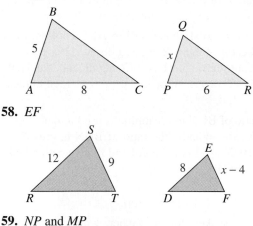

58. *EF*

59. *NP* and *MP*

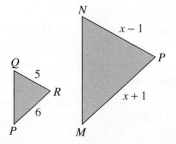

60. *BC* and *EF*

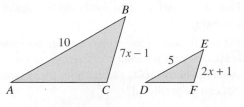

![Triangle ABC with side AB = 10, and 7x − 1 along BC; smaller triangle DEF with DE = 5 and 2x + 1 along EF]

Real-Life Applications

In Exercises 61–78, write a proportion that describes the given conditions, solve it, and answer the question. In exercises involving a sample or survey, assume that the given data accurately reflect the population at large.

61. Five tomatoes cost 95 cents. What is the cost of three tomatoes?

62. A recipe for apple muffins calls for $\frac{3}{4}$ cup of applesauce to make 18 muffins. How much applesauce is needed to make 2 dozen muffins?

63. If 4 ounces of uncooked macaroni yields $2\frac{1}{4}$ cups of cooked macaroni, how many ounces of uncooked macaroni are needed if a recipe calls for 5 cups of cooked macaroni?

64. Three small eggs can be substituted for 2 large eggs. If a recipe calls for 5 large eggs, how many small eggs should be used?

65. An 8-ounce cup of mocha cappuccino has 110 calories. How many calories are in a 6-ounce serving?

66. If 6 of 150 light bulbs are defective, how many defective bulbs are in a case of 200 bulbs?

67. If a boat travels 8 miles in 24 minutes, how far does it travel in $\frac{3}{4}$ hour?

68. It takes 6 hours to pump 1000 gallons of water into a tank. How long will it take to pump 1200 gallons into the tank?

69. The plans for a house have a scale of 1 inch to 3 feet.

(a) What is the length of a room that is 2.2 inches on the plan?

(b) How long on the plan is a room that is 21 feet?

70. A yard stick casts a 4-foot shadow. How high is a tree whose shadow is 48 feet long?

71. The force of gravity is six times greater on earth than it is on the moon. What is the weight of a 120-pound woman on the moon?

72. Forty tagged black bears are released in Smoky Mountain National Park, and then a group of 45 bears is selected from the general population. Of these 45 bears, 6 have tags. What is the total population of black bears in the park?

73. A candidate won an election by a 3-to-2 margin. How many votes did the winner receive if 1320 people voted?

74. Suppose that the owner of a house with an assessed value of $110,000 pays $1600 in property tax. How much is the property tax on a house valued at $89,000?

75. If doctors recommend aspirin over other headache remedies by a margin of 4 to 3, out of 140 doctors, how many recommend aspirin?

76. If customers buy vanilla ice cream rather than chocolate by a margin of 5 to 3, out of 320 customers, how many buy chocolate?

77. In a city with 20,000 homes, a survey of 250 homes indicated that 175 watched the Super Bowl. How many households in the city watched the Super Bowl?

78. Suppose that the exchange rate is 1 British pound for $1.50 U.S. How many British pounds can a person buy for $80 U.S.?

In Exercises 79–84, write a proportion and solve.

79. The ratio of two numbers is 3 to 5. The sum of the numbers is 168. What are the numbers?

80. The ratio of two numbers is 7 to 3. The difference of the numbers is 20. What are the numbers?

81. Making curtains for 8 windows takes 7 more yards of material than are required for 6 windows. How many yards are required for 9 windows?

82. Making 6 quarts of ice cream takes 1.5 more cups of sugar than making 4 quarts. How much sugar is used to make 8 quarts?

83. Painting 5 rooms takes 4.5 more gallons of paint than painting 2 rooms. How much paint is needed for 7 rooms?

84. Shipping a 5-pound package costs $4.35 more than shipping a 2-pound package. What is the cost to ship a 6-pound package?

Modeling with Real Data

85. A report indicated that 17 out of 20 college students had e-mail addresses. (Source: Scott Resource Group.) What is the estimated number of students who had an e-mail address at a college with 2800 students?

86. In a recent year, law enforcement agents made 20,000 arson arrests, and juveniles accounted for 13 of every 25 of these arrests. (Source: U.S. Justice Department.) How many adults were arrested for arson?

Data Analysis: *Sports Injuries*

The following table shows the rate of serious injuries per 100,000 participants in various sports.

Sport	Rate of Injury per 100,000
Football	2171
Bicycling	904
Golf	104

(Source: National Safety Council.)

Assume that the ratios given in the table are current and apply in each of the following exercises.

87. Out of how many golfers would we expect one serious injury?

88. Suppose that an average football team has 50 players. How many serious injuries should we expect the team to have?

89. The MS-150 Bike Tour is an annual, nationwide fund-raising event for multiple sclerosis research. If the event involves 1000 riders, how many serious injuries can be predicted?

90. Suppose that in a given year there were 14 serious injuries in sport X and 53 serious injuries in sport Y. Does this imply that sport Y is more dangerous? Why?

Challenge

In Exercises 91 and 92, determine the best buy for the given amount of the item. Then determine the best buy if you have a 50-cent coupon.

91. *Rice*   *Price*

16 ounces	$1.25
32 ounces	2.39
48 ounces	3.49

92. *Laundry Soap*   *Price*

42 ounces	$ 3.49
6 pounds, 2 ounces	7.99
14 pounds	15.99

93. A large subdivision consists of a total of 1000 individual homes, condominiums, and townhouses. There are equal numbers of townhouses and condominiums. If 3 of every 5 residences are individual homes, how many townhouses are there?

94. If $\frac{a}{b} = \frac{c}{d}$, show that $\frac{a}{a+b} = \frac{c}{c+d}$.

95. If x and y are nonzero numbers, show that the following proportion is not possible.

$$\frac{x+y}{x} = \frac{y-x}{y}$$

96. A car traveled 150 miles in the same time that a truck traveled 125 miles. If the rate of the car was 10 mph faster than that of the truck, what was the rate of the truck?

3.2 | Percents

Decimals, Fractions, and Percents • Percent Problems • Percent Increase and Decrease •
Real-Life Applications

Decimals, Fractions, and Percents

A commonly used ratio is the comparison of a number to 100. We call such a ratio a **percent,** which means *for every 100*. The symbol for percent is %.

If we say that 35% of a student body takes mathematics, we mean that 35 of every 100 students are enrolled in a mathematics class. Because a percent is a ratio, we can write 35% as $\frac{35}{100}$ or 0.35. That is, all percents can be expressed as fractions or decimals.

To express a percent as a fraction, we translate the phrase *for every 100* to *divided by 100*. Therefore, the initial form of the fraction has a denominator of 100. However, we often wish to simplify the fraction.

$$12\% = \frac{12}{100} = \frac{3}{25}$$

To express a percent as a decimal, we make use of the fact that dividing a number by 100 shifts the decimal point of the number two places to the left.

$$64\% = \frac{64}{100} = 0.64$$

EXAMPLE **1**

Converting Percents Into Fractions and Decimals

Convert the given percent into a simplified fraction and into a decimal.

(a) $80\% = \dfrac{80}{100} = \dfrac{4}{5}$

$80\% = 0.80$

(b) $170\% = \dfrac{170}{100} = \dfrac{17}{10}$

$170\% = 1.70$

(c) $9.8\% = \dfrac{9.8}{100} = \dfrac{9.8 \cdot 10}{100 \cdot 10} = \dfrac{98}{1000} = \dfrac{49}{500}$

$9.8\% = 0.098$

(d) $0.04\% = \dfrac{0.04}{100} = \dfrac{0.04 \cdot 100}{100 \cdot 100} = \dfrac{4}{10,000} = \dfrac{1}{2500}$

$0.04\% = 0.0004$

We can also convert a decimal into a percent.

$$0.48 = \frac{48}{100} = 48\%$$

To omit the intermediate step, we can simply move the decimal point two places to the right and affix the % symbol.

$$0.006 = 0.6\%$$

Converting a fraction into a percent is easily accomplished by simply performing the indicated division and then changing the resulting decimal into a percent.

$$\frac{11}{20} = 0.55 = 55\%$$

EXAMPLE 2

Converting Decimals and Fractions Into Percents

Convert the given decimal or fraction into a percent.

(a) $0.25 = 25\%$

(b) $\dfrac{2}{25} = 0.08 = 8\%$

(c) $0.125 = 12.5\%$

(d) $1 = 1.00 = 100\%$

(e) $1\dfrac{2}{5} = 1.40 = 140\%$

(f) $\dfrac{1}{400} = 0.0025 = 0.25\%$ or $\dfrac{1}{4}\%$

Percent Problems

Knowing that 35% of a student body takes mathematics does not reveal the actual number of mathematics students. To determine that number, we must multiply the total number of students by the percent who take mathematics. Thus, if the school has a total of 560 students, then 35% of 560 is $(0.35)(560) = 196$. Therefore, 196 students take mathematics.

Note: Recall that the word *of* means *times*. Also, to perform arithmetic with a percent, we convert the percent into a decimal or fraction. When you use a calculator, working with decimals is usually easier.

Percent problems usually involve three basic quantities.

1. $P = $ a percent.
2. $B = $ the original quantity, which follows the word *of* in a percent problem.
3. $A = $ the resulting amount that is the percent of the original quantity.

These three quantities are related as follows:

$$P \cdot B = A$$

If we know any two of these quantities, we can easily solve for the third quantity.

EXAMPLE 3

Solving Percent Problems

(a) What is 25% of 440?

(b) What percent of 225 is 36?

(c) 42% of what number is 184.8?

Solution

(a) We are given a percent $P = 25\% = 0.25$ and the original quantity $B = 440$.

$$P \cdot B = A$$

$$0.25 \cdot 440 = A \qquad \text{Replace } P \text{ with 0.25 and } B \text{ with 440.}$$

$$110 = A \qquad \text{The resulting amount}$$

Therefore, 25% of 440 is 110.

(b) We are given the original quantity $B = 225$ and the resulting amount $A = 36$.

$$P \cdot B = A$$

$$P \cdot 225 = 36 \qquad \text{Replace } B \text{ with 225 and } A \text{ with 36.}$$

$$P = \frac{36}{225} = 0.16 = 16\% \qquad \text{Divide both sides by 225.}$$

Therefore, 36 is 16% of 225.

(c) We are given a percent $P = 42\% = 0.42$ and the resulting amount $A = 184.8$.

$$P \cdot B = A$$

$$0.42 \cdot B = 184.8 \qquad \text{Replace } P \text{ with 0.42 and } A \text{ with 184.8.}$$

$$B = \frac{184.8}{0.42} \qquad \text{Divide both sides by 0.42.}$$

$$B = 440 \qquad \text{The original quantity}$$

Therefore, 42% of 440 is 184.8.

Percent Increase and Decrease

A percent is often used to indicate the relative amount of increase or decrease from one number to another. The following are examples.

Increases	*Decreases*
Percent increase in sales	Percent decrease in costs
Markup over wholesale price	Discount from retail price
Percent increase in attendance	Percent decrease in jobs

To increase (or decrease) a number by a given percent, we multiply the number by the percent and then add the result to (or subtract the result from) the original number.

EXAMPLE **4**

Increasing and Decreasing a Number by a Given Percent

(a) What is 30 increased by 150%?

(b) What is 80 decreased by 25%?

LEARNING TIP

Doubling an amount means a 100% increase; decreasing an amount by 50% results in half the original amount. Use familiar benchmarks such as these to think about the reasonableness of your answers.

Solution

(a) 150% of 30 is $(1.50)(30) = 45$.   Multiply the number by the percent.

$$30 + 45 = 75 \qquad \text{Add the result to the original number.}$$

When 30 is increased by 150%, the result is 75.

(b) 25% of 80 is (0.25)(80) = 20.   Multiply the number by the percent.

80 − 20 = 60   Subtract the result from the original number.

When 80 is decreased by 25%, the result is 60.

To calculate the percent increase or decrease from an original number to a new number, we begin by determining the positive difference between the two numbers. Then the percent increase or decrease is the ratio of this difference to the original number.

EXAMPLE 5

Calculating Percent Increases and Decreases

(a) What is the percent increase from 70 to 77?

(b) What is the percent decrease from 40 to 28?

Solution

(a) $77 - 70 = 7$ Find the difference between the two numbers.

$\dfrac{7}{70} = 0.10 = 10\%$ Find the ratio of the difference to the original number.

The percent increase from 70 to 77 is 10%.

(b) $40 - 28 = 12$ Find the difference between the two numbers.

$\dfrac{12}{40} = 0.30 = 30\%$ Find the ratio of the difference to the original number.

The percent decrease from 40 to 28 is 30%.

Real-Life Applications

Most application problems involving percents fall into one of the three previously discussed categories.

EXAMPLE 6

Sale Price of a Used Car

A demonstrator car is sold for 78% of its sticker price. What was the sticker price of the demonstrator if it is now priced at $14,430?

Solution

We know that the percent $P = 78\% = 0.78$ and that the sale price $A = 14,430$. Let $B =$ the sticker price.

$0.78 \cdot B = 14,430$ Replace P with 0.78 and A with 14,430.

$B = 18,500$ Divide both sides by 0.78.

The sticker price of the demonstrator was $18,500.

EXAMPLE 7

Seed Germination

If a gardener knows that 85% of all tomato seeds germinate, how many plants should she expect if she plants 240 seeds?

Solution

We know that the percent $P = 85\% = 0.85$ and that the original number of seeds $B = 240$. Let $A =$ the number of plants she should expect.

$$0.85 \cdot 240 = A \qquad \text{Replace } P \text{ with 0.85 and } B \text{ with 240.}$$
$$204 = A$$

The gardener should expect 204 seeds to germinate.

EXAMPLE 8

Percent of Income Spent on Groceries

A family spends $96 per week on groceries. If the family income is $480 per week, what percent of their income is spent on groceries?

Think About It

Consider two salary options: (1) $20,000 and an $1100 raise each year, and (2) $20,000 and a 5% raise each year. Which is the better option initially? Which is the better long-term option?

Solution

We know that the family income $B = 480$ and that the amount spent on groceries $A = 96$. Let $P =$ the percent of the total income spent for groceries.

$$P \cdot 480 = 96 \qquad \text{Replace } B \text{ with 480 and } A \text{ with 96.}$$

$$P = \frac{96}{480} = 0.20 = 20\%$$

Therefore, 20% of the family income is spent on groceries.

Quick Reference 3.2

Decimals, Fractions, and Percents

- A **percent** is a ratio that compares a number to 100.

- To express x percent as a fraction, we write $\frac{x}{100}$. To express a percent as a decimal, drop the % symbol and move the decimal point two places to the left.

- To express a decimal as a percent, we move the decimal point two places to the right and affix the % symbol. To express a fraction as a percent, we write the decimal name of the fraction and then convert this decimal to a percent.

Percent Problems

- Percent problems usually involve three basic quantities.
  1. $P =$ a percent.
  2. $B =$ the original quantity, which follows the word *of* in a percent problem.
  3. $A =$ an amount that is the percent of the original quantity.

  These three quantities are related by the formula $P \cdot B = A$.

Percent Increase and Decrease

- To increase (or decrease) a number by a given percent, we multiply the number by the percent and then add the result to (or subtract the result from) the original number.
- To calculate the percent increase or decrease from an original number to a new number, we begin by determining the positive difference between the two numbers. Then the percent increase or decrease is the ratio of this difference to the original number.

Speaking the Language 3.2

1. A ratio that compares a number to 100 is called a(n) ▭ .
2. The phrase *30% of 50* means 0.30 ▭ 50.
3. Expressing a percent as a decimal involves moving the decimal point ▭ places to the ▭ ; expressing a decimal as a percent involves moving the decimal point ▭ places to the ▭ .
4. To increase a number by 20%, we multiply the number by 0.20 and add the result to ▭ .

Exercises 3.2

Concepts and Skills

1. What is the meaning of the word *percent*?

2. Explain how to
   (a) convert a percent to a decimal.
   (b) convert a percent to a fraction.

In Exercises 3–10, change the given percent to (a) a decimal and (b) a simplified fraction.

3. 28%
4. 8%
5. 235%
6. 120%
7. 17.2%
8. 0.12%
9. $5\frac{3}{4}\%$
10. $\frac{1}{2}\%$

In Exercises 11–18, change the given decimal to a percent.

11. 0.375
12. 0.069
13. 0.55
14. 0.7
15. 6.9
16. 1.12
17. 0.0002
18. 0.0069

In Exercises 19–24, change the given fraction to a percent.

19. $\frac{7}{8}$
20. $\frac{8}{25}$
21. $\frac{7}{15}$
22. $\frac{5}{6}$
23. $1\frac{3}{4}$
24. $\frac{8}{5}$

In Exercises 25–42, solve the percent problem.

25. What is 12% of 658?
26. What is 30% of 84?
27. 3% of what number is 5.4?
28. 48% of what number is 36?
29. 85 is what percent of 125?
30. 27 is what percent of 180?
31. 19% of what number is 57?
32. 70% of what number is 322?
33. What number is 120% of 345?
34. What number is 250% of 74?
35. 128 is what percent of 1600?
36. What percent of 15 is 36?
37. What is $6\frac{1}{3}\%$ of 9600?

38. What is 10.75% of 24,800?

39. What percent of 340 is 425?

40. What percent of 8400 is 21?

41. 15 is $\frac{1}{4}$% of what number?

42. 279 is 124% of what number?

43. Suppose that a student attempts to determine the value of 10 increased by 12% by multiplying 10 by 0.12. The result is 1.2. Is 1.2 the answer? Explain.

44. Suppose that a student attempts to calculate the percent decrease from 30 to 20. The student's answer is $\frac{10}{20} = 0.50 = 50\%$. What error has the student made?

In Exercises 45–52, determine the described number.

45. 575 increased by 34%

46. 90 increased by 80%

47. 150 increased by 140%

48. 2000 increased by 230%

49. 230 decreased by 48%

50. 6 decreased by 20%

51. 1600 decreased by $\frac{1}{2}$%

52. 390 decreased by $\frac{1}{4}$%

In Exercises 53–56, determine the described percent increase or decrease.

53. (a) percent increase from 4 to 5

(b) percent decrease from 5 to 4

54. (a) percent increase from 30 to 75

(b) percent decrease from 75 to 30

55. (a) percent increase from 230 to 253

(b) percent decrease from 253 to 230

56. (a) percent increase from 80 to 96

(b) percent decrease from 96 to 80

🌎 Real-Life Applications

57. Suppose that 60% of a club's members are women. From the following list, identify the statement that is true and explain why the other two statements are false.

(i) The club has 100 members of whom 60 are women.

(ii) The club could have 5 members of whom 2 are men.

(iii) The club could have 40 women and 30 men.

58. The second congressional district is 30% minority, and the fifth congressional district is 20% minority. Does this mean that more minorities live in the second district than in the fifth district? Why?

59. The average attendance at a baseball team's home games rose from 28,000 to 35,000 per game. What was the percent increase in attendance?

60. In an English class, 12.5% of the students received an A. If 4 students received an A, how many students were in the class?

61. The total rainfall for April was 15% above the normal average of 14 inches. What was the total rainfall for the month?

62. If a student had answered one more question correctly, she would have received an A (90%) on the final exam. If she answered 35 questions correctly, how many questions were on the exam?

63. After decreasing its work force by 20%, a company had 2452 employees. How many did the company have before the layoffs?

64. The number of freshmen at a college fell from 1250 to 1150. What was the percent decrease?

65. A lot of light bulbs is rejected if more than $1\frac{1}{2}$% of the lot is defective. If 54 defective bulbs are found in a lot of 3650, should the lot be rejected?

66. Funding for a local recreation program increased from $140,000 to $161,000.

(a) What was the percent increase?

(b) What percent of the original funding is the current funding?

(c) What percent of the current funding was the original funding?

67. A person pays 28% federal tax and 8.5% Social Security tax. What total tax will that person pay on an income of $24,300?

68. A basketball player completed 76% of her free throws. If she made 19 shots, how many free throws did she attempt?

69. The Dow Jones Industrial Average dropped by 2% to 11,221. What was the opening average?

70. At the All-Star break, a major league baseball team has won 62% of its games. If this winning percentage continues, will the team win 100 games? (A major league season has 162 games.)

71. Tax revenue fell from $17.5 million to $16.8 million.

(a) What was the percent decrease in revenue?

(b) What percent of the original revenue is the current revenue?

72. The accompanying figure shows the layout of a yard that includes a rectangular garden, a patio, and a circular pool.

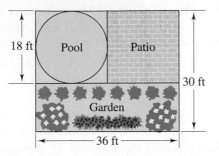

Determine the percent of the yard occupied by

(a) the garden.   (b) the patio.   (c) the pool.

(d) Why do the three results not add up to 100%?

Modeling with Real Data

73. A survey of children ages 7–12 revealed that 80% of the children in the United States dream of being rich. For six countries, the table shows the percentage of children ages 7–12 who dream of being rich.

Country	Percentage
United States	80%
Japan	64%
France	63%
Germany	61%
United Kingdom	59%
China	47%

(Source: A.B.C. Research.)

(a) In which countries in the table do more than $\frac{3}{5}$ of the children want to be rich?

(b) Suppose approximately 35 million children of ages 7–12 live in the United States. How many aspire to be rich?

74. The amount spent on hospital care is predicted to rise 36% to $569.1 billion between 2000 and 2005. (Source: Health Care Financing Administration.) How much was spent on hospital care in 2000?

Data Analysis: *Home Buyer Income*

Pie graphs are often used to show how certain categories of data compare with one another. The accompanying pie graph shows certain ranges of household incomes and the percentages of home buyers whose incomes fell within those ranges. (Source: Chicago Title and Trust.)

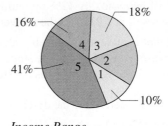

Income Range

Less than $30,000	(1)
$30,000–$40,000	(2)
$41,000–$50,000	(3)
$51,000–$60,000	(4)
At least $61,000	(5)

75. Because the pie graph shows all possible ranges of household incomes, what should be the sum of the percents for the five categories?

76. What percent of the home buyers had incomes in the $30,000–$40,000 range?

77. Suppose that a real estate analyst claims that the number of home buyers with incomes of at least $61,000 was nearly the same as the number of home buyers with incomes of at most $50,000. Would you agree with this claim? Why?

78. What percentage of home buyers had incomes between $30,000 and $60,000?

Challenge

79. If your monthly salary is $2000, which of the following is better?

(i) A 4% raise followed by a 7% raise

(ii) A 3% raise followed by an 8% raise

80. Compare the following.

(a) The percent increase from 8 to 10

(b) The percent decrease from 10 to 8

(c) The percent 8 is of 10

(d) The percent 10 is of 8

81. In real estate, a listing agent lists a house for sale. If she finds a buyer for the house, she receives 6% of the selling price as her commission. However, if another agent finds a buyer, the listing agent must split the commission with that agent. Moreover, both agents must give one-fourth of their commissions to their respective brokerage firms. If each brokerage firm received $900 for the sale of a home, what was the sale price?

82. If the price of one ounce of gold falls by 4% to $326.40, what percent increase is needed to raise the price to its original value? Express your answer to the nearest tenth of a percent.

3.3 Formulas

Evaluating with Formulas · Solving Formulas

Evaluating with Formulas

A **formula** is an equation that uses an expression to represent a specific quantity. For example, suppose that you invest or borrow a certain amount of money (*principal*). The expression Prt represents the amount of simple interest that is earned or owed. In this expression, P represents the principal, r is the simple interest rate expressed as a decimal, and t is the number of years. The formula for the interest I is $I = Prt$.

EXAMPLE 1

Calculating Simple Interest

What is the simple interest on a 6-month loan of $1700 at an interest rate of 8%?

Solution

Remember that the 8% interest rate must be expressed as a decimal, 0.08. Also, because 6 months is half of one year, the value of t is 0.5.

$$I = Prt \qquad \text{Formula for the amount of interest } I$$
$$I = 1700(0.08)(0.5) \qquad P = 1700, r = 0.08, t = 0.5$$
$$I = 68$$

The interest is $68.

Note: Although a formula is simply an equation, we can think of a formula as a set of instructions for calculating some specified quantity. In Example 1, $I = Prt$ instructs us to calculate I by finding the product of P, r, and t.

Many formulas come from geometry. Some of these formulas involve the number π, which is the ratio of the circumference of a circle to its diameter. The number π is an irrational (nonterminating, nonrepeating decimal) number approximately equal to 3.1416. Most calculators have a π key.

A + B

Alpha

Variables other than x are entered with the [Alpha] key. To enter $a + b$, press:

[Alpha] [A] [+] [Alpha] [B]

EXAMPLE **2**

Using Geometry Formulas

The circumference C of a circle with diameter d is given by the formula $C = \pi d$. The area A of a circle is given by $A = \pi r^2$, where r is the radius. What are the circumference and the area of a circle with a radius of 3 feet?

Solution

The diameter of a circle is twice its radius. Because the radius is 3 feet, the diameter is 6 feet. Figure 3.2 shows how we store the values for d and r and evaluate the formulas for the circumference and the area. The circumference is approximately 18.85 feet, and the area is approximately 28.27 square feet.

Figure 3.2

```
6→D
                    6.00
πD
                   18.85
3→R
                    3.00
πR²
                   28.27
```

Another important formula from geometry is contained in the **Pythagorean Theorem.**

The Pythagorean Theorem

A triangle is a right triangle if and only if the sum of the squares of the **legs** is equal to the square of the **hypotenuse.** Symbolically, $a^2 + b^2 = c^2$.

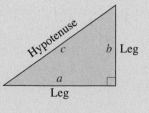

Note: The legs of a right triangle are **perpendicular,** which means that they form a right angle. The hypotenuse is the longest side of a right triangle and lies directly across from the right angle.

We can use the formula $a^2 + b^2 = c^2$ in the Pythagorean Theorem to determine whether a triangle is a right triangle.

EXAMPLE **3**

Determining Whether a Triangle Is a Right Triangle

The lengths of the sides of a triangle are given. Determine whether the triangle is a right triangle.

(a) 3, 5, 4

(b) 7, 4, 6

Think About It

The trapezoid in Figure 3.3 is called a right trapezoid because angles A and B are right angles. Which of the four dimensions would you need to know in order to determine the distance from B to D? Why?

Figure 3.3

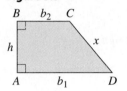

Solution

(a) $a^2 + b^2 = 3^2 + 4^2 = 9 + 16 = 25$ $a = 3, b = 4$

$c^2 = 5^2 = 25$ The longest side $c = 5$.

Therefore, $a^2 + b^2 = c^2$, and the triangle is a right triangle.

(b) $a^2 + b^2 = 4^2 + 6^2 = 16 + 36 = 52$ $a = 4, b = 6$

$c^2 = 7^2 = 49$ The longest side $c = 7$.

Therefore, $a^2 + b^2 \neq c^2$, and the triangle is not a right triangle.

Solving Formulas

A formula is usually given in the form in which it is most commonly used. For instance, the formula $d = rt$ tells us how to calculate distance d in terms of the rate r and time t. However, if we know d and t, we can solve an equation to determine r.

EXAMPLE **4**

Distance, Rate, and Time

Use the formula $d = rt$ to determine the speed (rate) of a sprinter who runs 200 meters in 24 seconds.

Solution

$d = rt$ Formula for distance

$rt = d$ Because we are solving for r, we may want to exchange sides.

$r \cdot 24 = 200$ $t = 24, d = 200$

$r \approx 8.33$ Divide both sides by 24.

The speed is approximately 8.33 meters per second.

Suppose we need to compute the rate r for several different values of d and t. Then, rather than solving an equation each time, it may be more convenient to find an explicit formula for r. The process of isolating a variable in a formula is called *solving the formula* for the variable. Because a formula is simply an equation, the procedure for solving a formula is exactly the same as the procedure for solving equations.

EXAMPLE 5

Solving a Formula

Solve the formula $d = rt$ for r. Then use the new formula to determine the rate of sprinters who run the following distances in the given times.

(a) 100 meters in 11 seconds

(b) 200 meters in 25 seconds

(c) 500 meters in 1 minute, 15 seconds

Solution

$d = rt$	The given formula for d
$rt = d$	Treat r as the variable and t and d as constants.
$\dfrac{rt}{t} = \dfrac{d}{t}$	To isolate r, we divide both sides by t.
$r = \dfrac{d}{t}$	The formula is now solved for r.

Now we can easily use a calculator to evaluate r by dividing d by t.

	d (meters)	t (seconds)	r (meters/second)
(a)	100	11	9.09
(b)	200	25	8.00
(c)	500	75	6.67

The following examples illustrate the process of solving other formulas for given variables.

EXAMPLE 6

Perimeter of a Rectangle

The perimeter P of a rectangle is given by the formula $P = 2L + 2W$, where L and W are the rectangle's length and width, respectively.

(a) Solve the formula for L.

(b) Use your new formula to determine the length of a rectangle whose perimeter is 64 inches and whose width is 10 inches.

Solution

LEARNING TIP

When you solve a formula, highlight the variable that you are trying to isolate by circling it or writing it in color.

(a)
$2L + 2W = P$	We exchange sides to place L on the left.
$2L + 2W - 2W = P - 2W$	Subtract $2W$ from both sides to isolate the variable term.
$2L = P - 2W$	
$L = \dfrac{P - 2W}{2}$	Divide both sides by 2 to isolate L.

(b) $L = \dfrac{P - 2W}{2}$ The new formula for L

$L = \dfrac{64 - 2(10)}{2}$ $P = 64, W = 10$

$L = 22$

The length is 22 inches.

EXAMPLE **7**   **Test Average**

To determine the average A of three test scores, we use the formula

$$A = \frac{a + b + c}{3}$$

where a, b, and c are the scores on the three tests.

(a) Solve this formula for c.

(b) Suppose that your scores for the first two tests were 84 and 72. Use the formula in part (a) to determine the grade you would need on the third test so that your average would be 70, so that it would be 80, and so that it would be 90.

Solution

(a) $$\frac{a + b + c}{3} = A$$ The formula for A with sides exchanged

$$3 \cdot \frac{a + b + c}{3} = 3A$$ Multiply both sides by 3 to clear the fraction.

$$a + b + c = 3A$$

$$c = 3A - a - b$$ Subtract a and b from both sides to isolate c.

(b) We can use a calculator to perform the arithmetic. For a 70 average, your third test score must be

$$c = 3(70) - 84 - 72 = 54$$

For an 80 average, your third test score must be

$$c = 3(80) - 84 - 72 = 84$$

For a 90 average, your third test score must be

$$c = 3(90) - 84 - 72 = 114$$

(Note that you cannot achieve a 90 average.)

A **trapezoid** is a four-sided figure with exactly one pair of parallel sides, as shown in Figure 3.4. The formula for the area A of a trapezoid is $A = \frac{1}{2}(b_1 + b_2)h$.

Figure 3.4

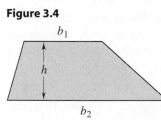

EXAMPLE **8**   **Solving the Formula for the Area of a Trapezoid**

Solve the formula for the area of a trapezoid for b_1.

Solution

$$\frac{1}{2}(b_1 + b_2)h = A$$ The formula for A with sides exchanged

$$2 \cdot \frac{1}{2}(b_1 + b_2)h = 2A$$ Multiply both sides by 2 to clear the fraction.

$$(b_1 + b_2)h = 2A$$

$$\frac{(b_1 + b_2)h}{h} = \frac{2A}{h}$$ Divide both sides by h to isolate $b_1 + b_2$.

$$b_1 + b_2 = \frac{2A}{h}$$

$$b_1 = \frac{2A}{h} - b_2$$ Subtract b_2 from both sides to isolate b_1.

As we study graphs in later chapters, we will encounter equations containing the variables x and y. Sometimes it will be necessary to solve such equations for y.

In Section 1.5, we stated the rule $\dfrac{a}{c} + \dfrac{b}{c} = \dfrac{a + b}{c}$. In reverse, the rule is $\dfrac{a + b}{c} = \dfrac{a}{c} + \dfrac{b}{c}$. In words, to divide a sum by a number, we divide each term of the sum by the number. We use this technique in Example 9.

EXAMPLE 9

Solving Equations for a Specified Variable

Solve the equation $3x - 2y = 8$ for y.

Solution

$$3x - 2y = 8 \qquad \text{The variable term is } 2y.$$
$$-3x + 3x - 2y = -3x + 8 \qquad \text{Add } -3x \text{ to both sides to isolate the variable term.}$$
$$-2y = -3x + 8$$
$$\frac{-2y}{-2} = \frac{-3x}{-2} + \frac{8}{-2} \qquad \text{Divide both sides by } -2 \text{ to isolate } y.$$
$$y = \frac{3}{2}x - 4$$

Quick Reference 3.3

Evaluating with Formulas
- A **formula** is an equation that uses an expression to represent a specific quantity.
- An important formula is contained in the **Pythagorean Theorem:** A triangle is a right triangle if and only if the sum of the squares of the **legs** is equal to the square of the **hypotenuse.** If the longest side of a triangle has length c, and the other two sides have lengths a and b, then the triangle is a right triangle if $a^2 + b^2 = c^2$.

Solving Formulas
- A formula is usually given in the form in which it is most commonly used. However, we can solve a formula for any of its variables.
- The procedure for solving a formula for a given variable is exactly the same as the procedure for solving equations.

Speaking the Language 3.3

1. The equation $A = LW$ is used to calculate the area A of a rectangle with width W and length L. This equation is called a(n) ▨▨▨ .
2. The longest side of a right triangle is called the ▨▨▨ .
3. The relationship among the sides of a right triangle is described by the ▨▨▨ .
4. The process of writing the formula $A = 2b + c$ as $c = A - 2b$ is called ▨▨▨ the formula for c.

Exercises 3.3

Concepts and Skills

1. What specific purpose does a formula serve?

2. The formula for the area A of a triangle is $A = \frac{1}{2}bh$, where b and h are the base and height of the triangle. Use this formula to write instructions for finding the area of a triangle.

In Exercises 3 and 4, the principal P, rate r, and time t are given. Determine the interest I. ($I = Prt$)

	P	r	t
3.	$2500	7.0%	1 year
4.	$3600	9.5%	4 months

In Exercises 5 and 6, determine the perimeter P and area A of a rectangle with the given length L and width W. ($P = 2L + 2W$, $A = LW$)

5. $L = 10$ feet, $W = 3$ feet
6. $L = 8.42$ km, $W = 5.25$ km

In Exercises 7 and 8, determine the circumference C and the area A of a circle with the given radius r or diameter d. ($C = 2\pi r$, $A = \pi r^2$)

7. $r = 3.2$ meters
8. $d = 2.7$ inches

In Exercises 9–18, determine the value of the variable on the left side of the given formula by evaluating the expression on the right side.

9. $F = \dfrac{9}{5}C + 32$ Celsius to Fahrenheit temperature conversion

  (a) $C = 10$
  (b) $C = 32$

10. $C = \dfrac{5}{9}(F - 32)$ Fahrenheit to Celsius temperature conversion

  (a) $F = 41$
  (b) $F = 90$

11. $d = rt$ Distance, rate, and time

  (a) $r = 55, t = 3$
  (b) $r = 16, t = 4$

12. $P = a + b + c$ Perimeter of a triangle

  (a) $a = 4, b = 7, c = 5$
  (b) $a = 9.6, b = 14.3, c = 8.0$

13. $V = LWH$ Volume of a rectangular solid

  (a) $L = 5, W = 3, H = 2$
  (b) $L = 7.5, W = 2.1, H = 1.6$

14. $P = \dfrac{KT}{V}$ Pressure, temperature, and volume of a gas

  (a) $K = 3, T = 46, V = 6$
  (b) $K = 1, T = 25, V = 10$

15. $A = \dfrac{1}{2}bh$ Area of a triangle

  (a) $b = 12.3, h = 6.1$
  (b) $b = 8, h = 12$

16. $I = Prt$ Simple interest

  (a) $P = 500, r = 0.06, t = \dfrac{2}{3}$
  (b) $P = 1270, r = 0.0325, t = 0.5$

17. $A = \pi r^2$ Area of a circle

  (a) $r = 5$
  (b) $r = 3.7$

18. $V = \pi r^2 h$ Volume of a cylinder

  (a) $r = 3, h = 5$
  (b) $r = 5.4, h = 12.1$

19. When we use the Pythagorean Theorem, what do we mean by the hypotenuse and how do we identify which side of a triangle it is?

20. Suppose that you try to construct a triangle whose sides are 5 inches, 7 inches, and 1 foot. Why is it not necessary to use the Pythagorean Theorem to determine whether the result is a right triangle?

In Exercises 21–24, the lengths of the sides of a triangle are given. Determine whether the triangle is a right triangle.

21. 6, 9, 12
22. 5, 12, 13
23. 8, 17, 15
24. 26, 21, 10

You can use your calculator in the following two ways to test for right triangles.

   (i) Enter $A^2 + B^2$ as y_1 and C^2 as y_2. Then store the values of A, B, and C, and retrieve the values of y_1 and y_2. If $y_1 = y_2$, the triangle is a right triangle.

   (ii) After you store the values of A, B, and C, enter $A^2 + B^2 = C^2$. If the returned value is 1 (true), the triangle is a right triangle.

In Exercises 25–28, use both of the preceding methods to determine whether the triangle with the given dimensions is a right triangle.

25. $a = 21, b = 28, c = 35$

26. $a = 1, b = 2.4, c = 2.6$

27. $a = 3.1, b = 4, c = 4.9$

28. $a = 43, b = 57, c = 71$

🌐 Real-Life Applications

29. A driver traveled 50 kilometers in 45 minutes. Determine the average speed (rate) in km/hr.

30. A stock car driver warmed up a car by making 10 laps around a 5-mile race track. If the total time was 35 minutes, what was the average speed (rate) in miles per hour?

31. An archer shoots an arrow upward at an initial rate of 100 feet per second. The height h (in feet) of the arrow after t seconds is given by $h = 100t - 16t^2$. Determine the height of the arrow after the following elapsed times.

   (a) 1 second   (b) 3 seconds   (c) 6.25 seconds

32. A major league pitcher throws a ball upward at an initial rate of 130 feet per second. The height h (in feet) of the ball after t seconds is given by the formula $h = 130t - 16t^2$. Determine the height of the ball after the following elapsed times.

   (a) 2 seconds   (b) 4 seconds   (c) 5 seconds

33. A basketball backboard is mounted on a 15-foot metal post that is buried in the ground 3 feet deep. A supporting cable runs from the top of the post to the ground at a point that is 5 feet from the base of the post. If the cable is 13 feet long, is the post perpendicular to the ground?

34. A wheelchair access ramp 30 feet long runs from a sidewalk to a porch that is 30 inches higher than the sidewalk. If the distance along the ground from the beginning of the ramp to the bottom of the porch is 28 feet, is the ground perfectly level?

Concepts and Skills

35. In what way(s) is the procedure for solving a formula for a given variable different from the procedure for solving an equation?

36. For the formula $W = \dfrac{a}{3} + b$, describe how you would determine the value of a if

   (a) you have one pair of values for W and b.

   (b) you have 12 pairs of values for W and b.

In Exercises 37–50, solve the given formula for the indicated variable.

37. $P = 4s;$ s

38. $P = 2L + 2W;$ W

39. $A = LW;$ L

40. $V = LWH;$ H

41. $I = Prt;$ P

42. $d = rt;$ r

43. $C = \pi d;$ d

44. $C = 2\pi r;$ r

45. $C = \dfrac{5}{9}(F - 32);$ F

46. $F = \dfrac{9}{5}C + 32;$ C

47. $A = \dfrac{1}{2}bh;$ b

48. $A = \dfrac{1}{2}(b_1 + b_2)h;$ h

49. $A = \dfrac{1}{2}(a + b);$ b

50. $P = a + b + c;$ b

In Exercises 51–62, solve for y.

51. $x + y = 8$

52. $2x + y = -7$

53. $5x - y = 7$

54. $-3x - y = -2$

55. $7x + 2y = 10$

56. $5x - 3y = 15$

57. $4y - 3x + 4 = 0$

58. $6y - x - 3 = 0$

59. $y - 3 = 2(x - 4)$

60. $y + 5 = -4(x + 2)$

61. $\dfrac{3}{4}x - \dfrac{2}{5}y = \dfrac{1}{2}$

62. $\dfrac{2}{3}x + \dfrac{1}{2}y = \dfrac{1}{6}$

🌐 Real-Life Applications

In Exercises 63 and 64, solve the given formula for the appropriate variable. Then evaluate the formula for the given data.

63. A commercial freezer must hold 1200 cubic feet of food. What must the length be for the given dimensions? (The volume of a box is $V = LWH$.)

	Width	Height
(a)	15 ft	8 ft
(b)	10 ft	10 ft
(c)	8 ft, 3 in.	9 ft, 2 in.

64. The average of three test grades a, b, and c and a final exam grade d that counts as two tests is given by $A = \dfrac{a + b + c + 2d}{5}$. Determine the grade needed on the final exam to receive the given average.

	Test 1	Test 2	Test 3	Average
(a)	85	92	72	80
(b)	95	90	0	70

In Exercises 65 and 66, the formula for an item's retail price R is $R = W + rW$, where W is the wholesale price. The formula for an item's discounted price D is given by $D = R - rR$. In these formulas r is the percent of markup or discount.

65. (a) Use the formula for R to determine the retail price for an item if the wholesale price was $78 and the percent markup was 32%.

(b) Solve this formula for r and use the result to determine the percent markup if the retail price is $448.50 and the wholesale price was $390.

66. (a) Use the formula for D to determine the discounted price for an item if the retail price was $96 and the percent of discount was 30%.

(b) Solve this formula for r and use the result to determine the percent of discount if the original retail price was $56 and the discounted price is $30.80.

Modeling with Real Data

67. The table shows the average daily high and low January temperatures for selected cities. Use the formula $F = 1.8C + 32$ to convert the temperatures from Celsius to Fahrenheit. (Round answers to the nearest whole number.)

City	High		Low	
	°C	°F	°C	°F
Bombay, India	31	(a)	17	(b)
Moscow, Russia	−6	(c)	−13	(d)
Sydney, Australia	26	(e)	18	(f)

(Source: Environmental Data Service.)

68. The table gives ocean depths in fathoms. Use the formula $F = 6f$ (F represents feet and f represents fathoms) to write the depth in feet.

Location	Depth	
	Fathoms	Feet
Mariana Trench	5973	(a)
Puerto Rico Trench	4705	(b)
Java Trench	3896	(c)

(Source: Defense Mapping Agency.)

Challenge

In Exercises 69–74, solve the formula for the indicated variable.

69. $ax + by = c$; y

70. $y = mx + b$; m

71. $m = \dfrac{y - b}{x}$; y

72. $y - a = m(x - b)$; y

73. $y - a = m(x - b)$; m

74. $\dfrac{x}{a} + \dfrac{y}{b} = 1$; y

In Exercises 75–78, solve the formula for the indicated variable.

75. $A = P + Prt$; P

76. $A = \dfrac{1}{2}ah + \dfrac{1}{2}bh$; h

77. $E = IR + Ir$; I

78. $R = \dfrac{R_1 R_2}{R_1 + R_2}$; R_1

3.4 | Translation

Expressions with One Number • Expressions with Two Numbers • Implied Conditions and Formulas

In this section we consider ways of thinking about the information in a problem and of translating that information into mathematical representations. Application problems vary in their content and complexity, so discussing every possible translation that we might encounter is not practical. The following gives us some experience with some of the more basic and common translations.

Expressions with One Number

Most application problems include at least one basic operation that must be represented symbolically. The following lists illustrate some typical word phrases and their translations into algebraic expressions.

Addition

Phrase	Expression
A number increased by 7	$x + 7$
5 more than a number	$x + 5$
A number added to -3	$-3 + x$
The sum of -4 and a number	$-4 + x$

Subtraction

Phrase	Expression
6 less than a number	$x - 6$
The difference between a number and -7	$x - (-7)$
A number decreased by 4	$x - 4$
-6 subtracted from a number	$x - (-6)$

Multiplication

Phrase	Expression
5 times a number	$5x$
Twice a number	$2x$
Half of a number	$\frac{1}{2}x$
A number multiplied by -3	$-3x$
The product of 7 and a number	$7x$
5% of a number	$0.05x$

LEARNING TIP

As you encounter other phrases or words associated with the basic operations, add those to the list. Also create a list of words that translate into equations or inequalities. For instance, *at most* translates into an inequality with the symbol $\leq$.

Division

Phrase	Expression
7 divided by a number	$\dfrac{7}{x}$
The quotient of a number and -5	$\dfrac{x}{-5}$
The ratio of a number and 8	$\dfrac{x}{8}$

In addition to translating phrases (which do not have verbs) into expressions, we usually need to translate sentences (which do have verbs) into equations. The most frequently used verb in sentence-to-equation translations is the word *is*. Other examples of words that translate into an equality symbol are *results in* and *obtain*.

When translating, we must be able to distinguish between an expression and an equation.

Phrase	*Expression*
Three times a number	$3x$

Sentence	*Equation*
Three times a number is 12.	$3x = 12$

EXAMPLE **1**

Translating Into Expressions and Equations

Translate the given phrase or sentence into an algebraic expression or equation. In each part, let x represent the unknown number.

(a) The difference between 3 and 4 times a number

(b) Half of the sum of a number and 6

(c) Five less than a number is -3.

(d) The quotient of twice a number and 3 is the number.

Solution

(a) $3 - 4x$

(b) $\dfrac{1}{2}(x + 6)$ Parentheses indicate that we find the sum first.

(c) $x - 5 = -3$ Note that *is* translates into $=$.

(d) $\dfrac{2x}{3} = x$

Expressions with Two Numbers

Application problems often involve two unknown quantities. We can assign any variable to represent one of the quantities. Then we must find a way to represent the other unknown quantity in terms of that same variable.

EXAMPLE **2**

Representing Two Unknown Quantities

In each part, a variable is assigned to one of the unknown quantities. Represent the other unknown quantity with an expression that uses the same variable.

(a) One car is traveling r miles per hour and another car is traveling 20 miles per hour faster.

(b) The length of a rectangle is 3 less than twice the width W.

(c) The main parking lot has p parking spaces, and a second parking lot has two-thirds as many spaces as the main lot.

Solution

(a) r = speed of one car; $r + 20$ = speed of the other car.

(b) W = width; $2W - 3$ = length.

(c) p = number of spaces in the main lot; $\frac{2}{3}p$ = number of spaces in the second lot.

Sometimes we know the sum of two numbers, but we don't know either of the numbers. In such cases, if x represents one of the numbers, then the other number is the sum minus x.

For geometry problems of this type, the following two definitions may be useful.

Two angles are

1. **complementary** if the sum of their measures is 90°.
2. **supplementary** if the sum of their measures is 180°.

EXAMPLE 3 **Representing Two Numbers When Their Sum Is Known**

In each part, a variable has been assigned to one of the unknown numbers. Write an expression that represents the other number.

(a) Represent the number of men in a group of 45 people if the number of women is w.

(b) A board 14 feet long is cut into two pieces. Represent the length of one piece if the length of the other piece is x.

(c) A total of $5000 is divided into two investments. If one investment is d dollars, represent the amount of the other investment.

(d) The measure of one of two complementary angles is y. Represent the measure of the other angle.

Solution

(a) $45 - w$

(b) $14 - x$

(c) $5000 - d$

(d) $90 - y$ The angles are complementary, so the sum of their measures is 90°.

Application problems sometimes involve consecutive integers, which are integers that follow in order one after the other. The following are examples of consecutive integers.

Consecutive integers:	5, 6, 7, 8	$-3, -2, -1$
Consecutive even integers:	4, 6, 8	$-2, 0, 2$
Consecutive odd integers:	9, 11, 13, 15, 17	$-1, 1, 3$

One way to represent such numbers is to assign a variable to the first integer in the list and then represent the other consecutive integers in terms of that variable.

Consecutive integers:	$n, n + 1, n + 2, \ldots$
Consecutive even integers:	$n, n + 2, n + 4, \ldots$
Consecutive odd integers:	$n, n + 2, n + 4, \ldots$

Note: The representations for consecutive even and odd integers appear the same because the difference between the numbers is 2 for both even and odd integers. However, the variable represents an *even* integer in one case and an *odd* integer in the other case.

EXAMPLE 4

Representing Consecutive Integers

Translate the given information into a simplified expression.

(a) The sum of two consecutive odd integers

(b) The difference between three times one number and twice a larger number, where the numbers are consecutive integers

Solution

(a) If n = the first odd integer, then

$n + 2$ = the next consecutive odd integer.

$n + (n + 2) = 2n + 2$

(b) If n = the first integer, then

$n + 1$ = the next consecutive integer.

$3n - 2(n + 1) = 3n - 2n - 2 = n - 2$

Implied Conditions and Formulas

Sometimes application problems do not specifically include all the information needed to translate conditions into expressions and equations. In such cases we may need to rely on our common knowledge about such things as the value of a coin. The translation may require a formula with which we are expected to be familiar.

EXAMPLE 5

Think About It

(a) Suppose you want to write part (b) of Example 5 in dollars. What expression should you write?

(b) In part (f) of Example 5, what expression would represent the volume of the solution that is not alcohol?

Translations Using Common Knowledge or Formulas

Write a simplified expression for the described quantity.

(a) The perimeter of a rectangle if the length is 4 more than the width

(b) The value in cents of a collection of 20 nickels and dimes

(c) The price per pound of pecans if 20 pounds costs d dollars

(d) The distance traveled in t hours at 52 mph

(e) The price of a television set after a 28% discount

(f) The amount of alcohol in x liters, of which 40% is alcohol

Solution

(a) If W = the width of the rectangle, then
$W + 4$ = the length of the rectangle.

The formula for the perimeter P of a rectangle is $P = 2L + 2W$, where L and W are the length and the width.
The perimeter is $2(W + 4) + 2W = 2W + 8 + 2W = 4W + 8$.

(b) If n = the number of nickels, then
$20 - n$ = the number of dimes.

Because each nickel is worth 5 cents and each dime is worth 10 cents, the total value in cents is

$$5n + 10(20 - n) = 5n + 200 - 10n = 200 - 5n$$

(c) The price per pound is found by dividing the total cost (d dollars) by the number of pounds (20 pounds).

Therefore, the price per pound is $\dfrac{d}{20}$.

(d) The formula for distance d is $d = rt$, where r is the rate (52 mph) and t is the time (t hours).

Therefore, the distance is $52 \cdot t = 52t$.

(e) If p = the original price of the television set, then
$0.28p$ = the amount of the discount.

The sale price is $p - 0.28p = 1p - 0.28p = 0.72p$.

(f) The number of liters of alcohol is 40% of the total volume x.

Therefore, the expression is $0.4x$.

Quick Reference 3.4

Expressions with One Number

- At least one of the four basic operations is usually involved in translating conditions into expressions.

- To solve application problems, we must be able to translate word phrases into expressions and sentences into equations.

- The distinguishing feature of a sentence-to-equation translation is the existence of a verb, which is represented by the equality symbol.

Expressions with Two Numbers

- When an application problem involves two unknown quantities, we can assign any variable to represent one of the quantities. Then we must find a way to represent the other unknown quantity in terms of that same variable.

- If we know the sum S of two numbers and one of those numbers is x, then the other number is $S - x$.

- The following definitions are useful for certain application problems in geometry.
Two angles are

1. **complementary** if the sum of their measures is 90°.

2. **supplementary** if the sum of their measures is 180°.

- Consecutive integers are integers that follow in order one after the other. The following are some representations involving consecutive integers.

Consecutive integers: $n, n + 1, n + 2, \ldots$
Consecutive even integers: $n, n + 2, n + 4, \ldots$
Consecutive odd integers: $n, n + 2, n + 4, \ldots$

Implied Conditions and Formulas

- Application problems do not always specifically state relationships or formulas that are assumed to be common knowledge.

Speaking the Language 3.4

1. When we translate a sentence into an equation, the verb is represented by the ▨▨▨▨ symbol.

2. If the sum of the measures of two angles is 180°, the angles are called ▨▨▨▨ ; if the sum is 90°, the angles are called ▨▨▨▨ .

3. If n represents an integer, then n and $n + 2$ represent either ▨▨▨▨ integers or ▨▨▨▨ integers.

4. If the sum of two numbers is y and one of the numbers is x, then the other number can be represented by ▨▨▨▨ .

Exercises 3.4

Concepts and Skills

▼ 1. How can you know whether to translate information into an expression or into an equation?

▼ 2. If your purchases total x dollars and you hand the clerk a five-dollar bill, what expression represents your change? Why?

In Exercises 3–8, let x represent the number and translate the given phrase into an algebraic expression.

3. The difference of a number and -7

4. The sum of -2 and a number

5. The quotient of a number and 2

6. 9% of a number

7. Two less than twice a number

8. 9 more than $\frac{2}{3}$ of a number

In Exercises 9–14, let x represent the number and translate the given sentence into an equation.

9. One-fourth of a number is 8.

10. Three more than twice a number is 9.

11. One less than 6 times a number is 11.

12. Three less than a number is $\frac{1}{4}$ the number.

13. The difference between twice a number and 1 is 5 more than the number.

14. Four times a number is 4 more than 3 times the number.

In Exercises 15–22, let x represent the number and translate the given phrase or sentence into an algebraic expression or equation.

15. Four less than 3 times a number

16. The difference between a number and 5

17. Five less than a number is -8.

18. Six more than a number is 0.

19. The difference between 2 and 3 times a number

20. The sum of 3 times a number and 2

21. The sum of 5 and twice a number is 9.

22. The difference of 10 and a number is 7.

In Exercises 23–26, a variable is assigned to one of the unknown quantities. Represent the other unknown quantity with an expression that uses the same variable.

23. The height of a person who is 6 inches shorter than a person x feet tall

24. The age a person was 3 years ago if the person is presently t years old

25. The length of a rectangle if the length is three times the width W

26. The width of a rectangle if the width is 12 less than twice the length L

In Exercises 27–32, write expressions that represent each of the described quantities.

27. Two numbers such that one number is 8 more than the other

28. Two numbers such that one number is 45% of the other

29. The number of games won by each team if one team won 7 more games than the other

30. The number of votes each candidate received if one received 20 more votes than the other

31. The lengths of two wires if the length of one is 5 units more than half the length of the other

32. The lengths of two boards if one is 3 feet shorter than 5 times the length of the other

33. Suppose that the measures of two angles are represented by x and $180 - x$. What are these angles called? Why?

34. Explain why three consecutive integers can be represented, in order, by $n - 1$, n, and $n + 1$.

In Exercises 35–38, a variable is assigned to one of the unknown quantities. Represent the other unknown quantity with an expression that uses the same variable.

35. The measure of one of two supplementary angles if the other angle has a measure of x degrees

36. The measure of one of two complementary angles if the other angle has a measure of y degrees

37. The number of miles remaining in a 130-mile trip if m miles have been driven

38. The number of dimes in a collection of 50 nickels and dimes if there are n nickels

In Exercises 39–42, write expressions to describe each of the two unknown quantities.

39. A 90-foot board is cut in two pieces.

40. The amounts of two investments totaling $6000

41. The measures of two complementary angles

42. The measures of two supplementary angles

In Exercises 43–46, write an expression for the described quantity.

43. The next consecutive integer if n is an integer

44. The next consecutive even integer if n is an even integer

45. The next two consecutive even integers if x is an odd integer

46. The next two consecutive odd integers if x is an even integer

In Exercises 47–50, translate the given information into a simplified expression.

47. The sum of three consecutive integers

48. The sum of three consecutive odd integers

49. The sum of two consecutive even integers

50. The sum of two consecutive integers

In Exercises 51–60, write an expression for the described quantity.

51. The amount of acid in x liters of which 30% is acid

52. The amount of butterfat in y ounces of milk that is 4% butterfat

53. The number of correct answers on a test with q questions if 92% of the questions were answered correctly

54. The number of bass in a pond if 35% of the fish population p are bass

55. The cost of a bicycle rental for d days at $18 per day

56. A person's salary for h hours of work at $7 per hour

57. The distance traveled in t hours at 60 mph

58. The distance traveled in 5 hours at r mph

59. The increase in the population of a county at the end of y years if the population grew steadily by 2500 people per year

60. The number of degrees that the temperature dropped after h hours if the temperature fell steadily at 2° per hour.

In Exercises 61–66, let x represent the initial quantity. Then write a simplified expression for the resulting quantity.

61. The cost of a blouse after a 25% markup

62. The price of a pair of shoes after a 35% discount

63. A salary after 34% in taxes were withheld

64. The amount in an account after a 2% service charge

65. The cost of an automobile after a 5% tax has been added

66. The price of a house plus the real estate agent's commission of 9%

In Exercises 67–82, write an expression for the described quantity.

67. The distance traveled in one day if the number of miles driven in the morning is 40 less than twice the number of miles driven in the afternoon

68. The perimeter of a rectangle if its length is twice its width

69. The perimeter of a rectangle if its width is 7 less than its length

70. The perimeter of a triangle if the lengths of two sides are consecutive integers and the length of the third side is twice that of the smaller side

71. The price of a rental car at $35 plus 30 cents per mile

72. A tuition of $160 plus $58 per semester hour

73. The total simple interest earned on $5000 if part is invested at 7% and the remainder is invested at 8%

74. The total yearly dividend income on 250 shares of stock if some shares pay $2.75 per share and the remainder pay $1.50 per share

75. The value in cents of a collection of 30 dimes and nickels

76. The value of 100 tickets if some of the tickets cost $9 each and the remaining tickets cost $7 each

77. The amount of acid in 50 liters if x liters contain 25% acid and the remainder contains 40% acid

78. The amount of insecticide in an 8-gallon mixture if g gallons contain 12% insecticide and the remainder contains 15% insecticide

79. The cost of 20 pounds of grapes and strawberries if grapes cost $1.20 per pound and strawberries cost $1.50 per pound

80. The cost of 12 gallons of gasoline if 87-octane costs $1.10 per gallon and 92-octane costs $1.30 per gallon

81. The salary for 40 hours' work if $6 per hour is paid for some of the hours and $8 per hour is paid for the remaining hours

82. The number of points scored in basketball if 18 shots were made and some of the shots were two-pointers and all others were three-pointers

Modeling with Real Data

83. Since 1990, the cost of administering federal agencies that regulate social and economic programs has increased by about $\frac{1}{3}$ billion dollars each year to $16 billion in 2000. (Source: Center for the Study of American Business.) Assume that the rate of increase continues, and let x represent the number of years after 2000.

(a) Write an expression for the cost for any year after 2000.

(b) Write an equation to model the year in which the cost will be $20 billion.

84. During a period prior to 1998, the average cost (tuition, room, and board) per year at a public college increased by about $947 each year to a total of $7628 in 1998. (Source: U.S. Department of Education.) Assume that the rate of increase continues, and let x represent the number of years after 1998.

(a) Write an expression for the cost for any year after 1998.

(b) Write an equation to model the year in which the cost will be $15,000.

Challenge

In Exercises 85–88, translate the given information into an expression.

85. The difference of the squares of x and 7

86. The sum of the cubes of 2 and y

87. The square of the sum of n and 3

88. The cube of the difference of y and 2

In Exercises 89 and 90, translate the given information into an expression with more than one variable.

89. The cost of x pounds of chocolates at $5.50 per pound and y pounds of mints at $1.25 per pound

90. The total points scored if a player makes x free throws, y two-point shots, and z three-point shots

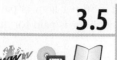

3.5 | Modeling and Problem Solving

General Strategy   •   Number Problems   •   Piece Lengths   •   Geometric Figures

General Strategy

There is such diversity among application problems that no single procedure for solving them can be stated. However, it is possible to outline a general approach that serves as a way of thinking about the problem and as a strategy for organizing information and using algebraic techniques for solving the problem. The following guidelines should be useful to you in solving most application problems that involve either equations or inequalities.

General Strategy for Solving Application Problems

1. Read the problem carefully to get a sense of what the problem is about. Make sure you understand what information is given and what the problem is asking you to determine. Try to state the problem in your own words as though you were explaining it to someone else.

2. If appropriate, draw a figure or diagram or make a chart or table to organize information.

3. Assign a variable to the unknown quantity. In defining the variable, be specific.

4. If more than one unknown quantity is involved, represent each quantity with an expression in the same variable.

5. Translate the information in the problem into an equation or an inequality.

6. Solve the equation or inequality.

7. Answer the question. (The solution in step 6 is not necessarily the answer to the question.)

8. Check the answer to see whether it meets the conditions stated in the problem.

Note: If the equation or inequality that you have written is incorrect, then checking the solution by substitution is not helpful. For this reason, check answers by verifying that they meet the conditions stated in the problem.

In this section and the next, we illustrate several types of applications. These examples should not suggest to you that every application problem can be fit neatly into some category. However, many different applications do have similar characteristics and can be solved with some of the methods we describe. In these examples, we include the step numbers to show how the general approach is used in solving the problem. (Step 2 is shown only when it is applicable.)

Admittedly, many of the problems that we consider here are not realistic applications. Their primary purpose is to assist you in developing the approaches and skills needed in solving any application problem.

Number Problems

In a sense, all application problems are number problems because the described conditions and the answer are all numerical. However, we begin with examples of problems that involve only numbers, with no reference to units.

EXAMPLE **1**

Determining an Unknown Number

(a) Nine less than twice a number is -1. What is the number?

(b) Seven more than $\frac{3}{4}$ of a number is 2 more than the number. What is the number?

Solution

① In both parts, the goal is to determine the described number.

③ (a) Let x = the number.

⑤ $\qquad 2x - 9 = -1$ $\qquad$ Nine less than twice the number is -1.

⑥ $\qquad\quad 2x = 8$ $\qquad$ Add 9 to both sides to isolate $2x$.

$\qquad\quad \dfrac{2x}{2} = \dfrac{8}{2}$ $\qquad$ Divide both sides by 2 to isolate x.

$\qquad\qquad x = 4$

⑦ $\qquad$ The number is 4.

⑧ $\qquad$ Check: Twice 4 is 8, and 9 less than 8 is -1.

③ (b) Let x = the number.

⑤ $\qquad \dfrac{3}{4}x + 7 = x + 2$ $\qquad$ Seven more than $\frac{3}{4}$ of the number is 2 more than the number.

⑥ $\qquad -\dfrac{1}{4}x + 7 = 2$ $\qquad$ Subtract x from both sides.

$\qquad\quad -\dfrac{1}{4}x = -5$ $\qquad$ Subtract 7 from both sides.

$\qquad -4 \cdot \left(-\dfrac{1}{4}x\right) = -4(-5)$ $\qquad$ Multiply both sides by -4.

$\qquad\qquad\quad x = 20$

⑦ $\qquad$ The number is 20.

⑧ $\qquad$ Check: First, $\frac{3}{4}$ of 20 is 15, and 7 more than 15 is 22. Also, 2 more than 20 is 22.

EXAMPLE **2**

Determining Two Unknown Numbers

(a) One number is six more than another. The difference of five times the smaller number and twice the larger number is 15. What are the numbers?

(b) The sum of two numbers is 36. If three times the smaller number is subtracted from the larger number, the result is at least 4. What is the greatest value of the smaller number?

Solution

① (a) The goal is to determine the two described numbers.

③   If x = the smaller number, then
④ $x + 6$ = the larger number.

⑤ $5x - 2(x + 6) = 15$ Subtract twice the larger number from 5 times the smaller number.

⑥ $5x - 2x - 12 = 15$ Distributive Property

              $3x - 12 = 15$ Combine like terms.

                  $3x = 27$ Add 12 to both sides.

                    $x = 9$ Divide both sides by 3.

⑦   The smaller number is 9. The larger number is $9 + 6$ or 15.

⑧   Check: Five times the smaller number 9 is 45. Twice the larger number 15 is 30. The difference of 45 and 30 is 15.

① (b) The goal is to determine the greatest value of the smaller number.

③   If x = the smaller number, then
④ $36 - x$ = the larger number.

⑤ $(36 - x) - 3x \geq 4$ "At least" means greater than or equal to.

⑥ $36 - x - 3x \geq 4$ Remove parentheses.

              $36 - 4x \geq 4$ Combine like terms.

                 $-4x \geq -32$ Subtract 36 from both sides.

                   $x \leq 8$ Divide both sides by -4. Reverse the inequality symbol.

⑦   The greatest value of the smaller number is 8.

⑧   Check: The larger number is $36 - 8 = 28$, and 3 times the smaller number 8 is 24. The difference of 28 and 24 is 4.

In Section 3.4, we discussed the meaning of consecutive integers and the usual ways of representing them.

EXAMPLE 3 | **Consecutive Integers**

(a) What are three consecutive odd integers whose sum is 69?

(b) For three consecutive even integers, the sum of the first two integers is equal to the third integer. What is the largest integer?

Solution

① (a) The goal is to determine the three described integers.

③   If x = the first odd integer, then
④ $x + 2$ = the second odd integer, and
      $x + 4$ = the third odd integer.

⑤ $x + (x + 2) + (x + 4) = 69$ The sum of the three integers is 69.

⑥ $x + x + 2 + x + 4 = 69$ Remove parentheses.

              $3x + 6 = 69$ Combine like terms.

                  $3x = 63$ Subtract 6 from both sides.

                    $x = 21$ Divide both sides by 3.

⑦   The integers are 21, 23, and 25.

⑧   Check: the sum of 21, 23, and 25 is 69.

① (b) The goal is to determine the largest of the three described integers.

③   If x = the first even integer, then
④ $x + 2$ = the second even integer, and
          $x + 4$ = the third even integer.

⑤ $x + (x + 2) = x + 4$ The sum of the first two integers equals the third integer.
⑥ $2x + 2 = x + 4$ Remove parentheses and combine like terms.
            $x + 2 = 4$ Subtract x from both sides.
                $x = 2$ Subtract 2 from both sides.

⑦   The largest integer is $x + 4 = 2 + 4 = 6$.

⑧   Check: The sum of the first two integers, 2 and 4, is 6, which is the largest integer.

Piece Lengths

In this type of problem, some object of a given length is divided into two or more pieces, and the relative lengths of these are described. To write the equation, observe that the sum of the lengths of the pieces is equal to the original object's total length.

EXAMPLE 4

Lengths of Pieces of Wire

A 57-foot wire is cut into three pieces. The second piece is 7 feet longer than the first piece, and the third piece is twice as long as the second piece. How long is each piece?

Solution

① The goal is to determine the lengths of the three pieces of wire.

② Figure 3.5 is useful for assigning the variable to the length of the first piece and for representing the lengths of the other two pieces.

Figure 3.5

③

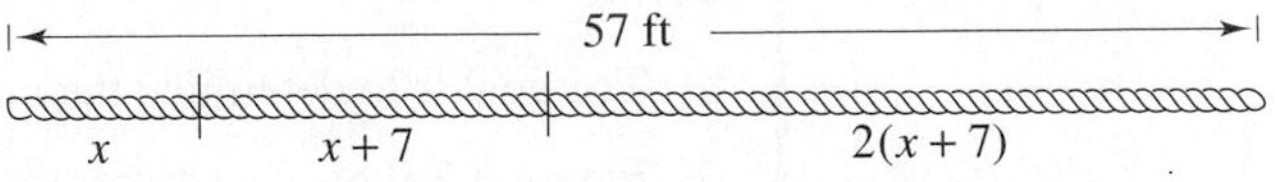

④

⑤ $x + (x + 7) + 2(x + 7) = 57$ The total length is 57 feet.
⑥ $x + x + 7 + 2x + 14 = 57$ Distributive Property
        $4x + 21 = 57$ Combine like terms.
            $4x = 36$ Subtract 21 from both sides.
              $x = 9$ Divide both sides by 4.

⑦ The lengths of the pieces are 9 feet, 16 feet, and 32 feet.

⑧ Check: The sum of 9, 16, and 32 is 57. The length of the second piece is 16 feet, which is 7 feet longer than the first piece. The length of the third piece is 32 feet, which is twice the length of the second piece.

LEARNING TIP

Writing "word equations" may assist you in writing a mathematical equation. For instance, "length of first piece + length of second piece + length of third piece = total length."

Geometric Figures

A fundamental theorem in geometry states that the sum of the measures of the angles of a triangle is 180°. This theorem is often used in application problems involving triangles.

EXAMPLE **5**

The Angles of a Triangle

The measure of the second angle of a triangle is $\frac{3}{4}$ that of the first angle, and the measure of the third angle is 40° less than the measure of the first angle. What is the measure of each angle?

Solution

① The goal is to determine the measures of the three angles of a triangle.

② Figure 3.6 shows that x represents the measure of one angle (the first). The figure also shows how the measures of the other two angles are represented.

Figure 3.6

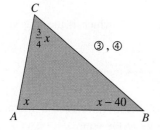

⑤ $x + \frac{3}{4}x + (x - 40) = 180$ The sum of the measures of the angles is 180°.

⑥ $\frac{11}{4}x - 40 = 180$ Remove parentheses and combine like terms.

$\frac{11}{4}x = 220$ Add 40 to both sides.

$x = 80$ Multiply both sides by $\frac{4}{11}$.

⑦ The angles are 80°, 60°, and 40°.

⑧ Check: The measure of the second angle is 60°, which is $\frac{3}{4}$ of 80°. The measure of the third angle is 40°, which is 40° less than 80°. The sum of 80°, 60°, and 40° is 180°.

Rectangle problems involve the area ($A = LW$) or the perimeter ($P = 2L + 2W$), where L and W are the length and the width of the rectangle.

EXAMPLE **6**

The Dimensions of a Rectangle

The perimeter of a rectangle is at most 60 cm. The length is 10 cm less than three times the width. What is the maximum width of the rectangle?

Solution

① The goal is to determine the maximum (greatest) width of a rectangle.

② Figure 3.7 shows the representations of the length and width of the rectangle.

Figure 3.7

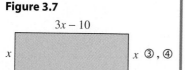

$2L + 2W = P$ Formula for the perimeter of a rectangle

⑤ $2(3x - 10) + 2x \le 60$ Substitute x for W and 3x − 10 for L. "At most" means less than or equal to.

⑥ $6x - 20 + 2x \le 60$ Distributive Property

$8x - 20 \le 60$ Combine like terms.

$8x \le 80$ Add 20 to both sides.

$x \le 10$ Divide both sides by 8.

⑦ The maximum width of the rectangle is 10 cm.

⑧ Check: Three times the width is 30, and 10 less than 30 is 20, which is the length. Therefore, the perimeter is 60.

Note: In rectangle problems, it is usually easier to assign a variable to the dimension that is not described. For instance, in Example 6, the sentence that begins "The length is . . ." means that the length is being described in terms of width, so we assign the variable to the width. If the sentence began "The width is . . . ," we would assign the variable to the length.

EXAMPLE 7

Think About It

For a triangle, the sum of the measures of the angles is 180°, and for a rectangle the sum is 360°. What is your conjecture about the sum of the measures of the angles of a pentagon (five sides)? a hexagon (six sides)? a geometric figure of n sides?

Complementary and Supplementary Angles

The sum of the measures of the complement and the supplement of an angle is 120°. What is the measure of the angle?

Solution

① The goal is to determine the measure of an angle.

③ If x = the measure of the angle, then
④ $90 - x$ = the measure of the complement, and
    $180 - x$ = the measure of the supplement.

⑤ $(90 - x) + (180 - x) = 120$ The sum of the complement and the supplement is 120.

⑥ $90 - x + 180 - x = 120$ Remove parentheses.

   $270 - 2x = 120$ Combine like terms.

   $-2x = -150$ Subtract 270 from both sides.

   $x = 75$ Divide both sides by -2.

⑦ The measure of the angle is 75°.

⑧ Check: The measure of the complementary angle is 15°, and the measure of the supplementary angle is 105°. The sum of those measures is 120°.

Exercises 3.5

Concepts and Skills

1. For a number problem, a student lets n represent the number and writes an equation to describe the given conditions. After solving the equation and obtaining 3, the student substitutes 3 for n in the equation and finds that it checks. Explain why this approach is not a sufficient way of checking the answer to an application problem.

2. How are problems involving piece lengths and problems involving the measures of the three angles of a triangle similar? What information must be given for piece length problems that is not needed for triangle problems?

Numbers

In Exercises 3–12, a description of an unknown number is given. Determine the number.

3. When a number is subtracted from 7, the result is -3.

4. Two-thirds of a number is 42.

5. Five less than 3 times a number is 4.

6. Four less than two-thirds of a number is 84.

7. Four more than 3 times a number is at most 37.

8. The difference of a number and 3 times the number is greater than 4.

9. Twice the difference of a number and 7 is 14.

10. When the sum of twice a number and 9 is divided by 3, the result is 7.

11. Five less than two-thirds of a number is 9 less than the number.

12. Five more than 3 times a number is 1 more than the number.

In Exercises 13–18, a description of two unknown numbers is given. Determine the two numbers.

13. One number is 7 more than another. The sum of the numbers is less than 25.

14. One number is 3 times another. Half the sum of the numbers is at least 8.

15. The sum of two numbers is 36. If 3 times the smaller number is subtracted from the larger number, the result is 4.

16. The difference of two numbers is 6. The difference of 5 times the smaller number and twice the larger number is 15.

17. The difference of two numbers is 5. If the smaller number is added to 4 times the larger number, the result is −10.

18. The sum of two numbers is 10. The difference of twice one number and the other number is 8.

In Exercises 19–26, determine all of the described integers.

19. The sum of two consecutive odd integers is −32.

20. The sum of two consecutive even integers is 78.

21. The sum of three consecutive even integers is 102.

22. The sum of three consecutive odd integers is −117.

23. The sum of the first two of three consecutive integers is 5 less than three times the third integer.

24. For three consecutive even integers, twice the largest integer is equal to the smallest integer.

25. For three consecutive odd integers, twice the first integer plus the sum of the other two integers is 114.

26. For three consecutive integers, the sum of the first integer, twice the second integer, and three times the third integer is −64.

Piece Lengths

In Exercises 27–30, determine the lengths of all the pieces.

27. A cable 35 feet long is cut into three pieces so that the length of the second piece is 4 times the length of the first piece, and the third piece is twice as long as the first piece.

28. A pipe 24 feet long is cut into three pieces. The first piece is twice as long as the second piece, and the third piece is 6 feet longer than the second piece.

29. A bolt of cloth 100 yards long is cut into three pieces. The second piece is 15 yards longer than the first piece, and the length of the third piece is 5 yards more than twice the length of the first piece.

30. One hundred yards of fencing is cut into four pieces to enclose a four-sided field. The north and south sides of the field have the same length. The east side is 3 times as long as the south side. The length of the west side is 5 yards less than twice the length of the north side.

Geometric Figures

31. An **isosceles triangle** is a triangle with at least two angles that have the same measure. What are the measures of the angles of an isosceles right triangle. Explain how you know.

32. The sum of the measures of the angles of a triangle is 180°. Does this mean that the angles of a triangle are supplementary? Why?

In Exercises 33–36, determine the measures of all three angles of the described triangle.

33. The measure of the second angle of a triangle is twice the measure of the first angle, and the measure of the third angle is 30° more than the measure of the second angle.

34. The measure of the second angle of a triangle is 5° less than the measure of the first angle. The third angle is 3 times as large as the second angle.

35. The first angle of an acute triangle is twice as large as the second angle. The measure of the third angle is 20° less than the measure of the first. (An **acute triangle** is a triangle in which all angles have measures less than 90°.)

36. The measure of the first angle of an obtuse triangle is 5° more than twice the measure of the second angle. The measure of the third angle is 50° more than the sum of the measures of the other two. (An **obtuse triangle** is a triangle in which one angle has a measure that is greater than 90°.)

In Exercises 37–40, determine the dimensions of the described rectangle.

37. The perimeter is 30 meters. The length is three more than twice the width.

38. The length is twice the width, and the perimeter is not less than 42 yards.

39. The width is $\frac{3}{4}$ of the length, and the perimeter does not exceed 70 feet.

40. The perimeter is 22 inches, and the width is 1 inch less than half the length.

In Exercises 41–44, determine the measure of the described angle.

41. The measure of the complement of an angle is one-fourth the measure of its supplement.

42. The supplement of an angle is 3 times as large as its complement.

43. The sum of the measures of the complement and supplement of an angle is 170°.

44. The difference of the measures of the supplement of an angle and 3 times the complement is 50°.

In Exercises 45–48, determine the measures of both the described angles.

45. Two angles are supplementary. The measure of one angle is 5 times the measure of the other angle.

46. Two angles are supplementary. The measure of one angle is 30° less than twice the measure of the other angle.

47. Two angles are complementary. The measure of one angle is 30° more than twice the measure of the other angle.

48. Two angles are complementary. Half the difference of the measures of the angles is 5°.

🌐 Real-Life Applications

49. A local referendum passed by 137 votes. If 993 votes were cast, how many voters voted against the referendum?

50. In a box of 84 cookies there are 2.5 times as many chocolate chip cookies as sugar cookies. How many of each kind of cookie are in the box?

51. On a softball team the right fielder has twice as many hits as the shortstop, and the catcher has 3 more hits than the other two players combined. If the 3 players have a total of 27 hits, how many hits does the right fielder have?

52. During a training session, an athlete swam 5 miles less than she ran and biked 3 miles less than 4 times the distance she ran. If her total distance was 34 miles, how far did she travel in each event?

53. A wallpaper border 30 feet in length is to be installed along the top of 3 walls of a rectangular kitchen. (One length of the kitchen has cabinets and does not require wallpaper.) If the length of the kitchen is 6 feet less than twice the width, what are the dimensions of the room?

54. At least 112 feet of fencing is needed to enclose three sides of a small rectangular plot of ground. (An existing wall along one length of the plot makes fencing unnecessary.) If the width of the plot is 7 feet less than the length, what is the minimum length of the plot?

55. The maximum enrollment in a class is 30 students. In a certain class, there are 6 fewer men than women. What is the greatest number of women that could be in the class?

56. A guard scored 32 points. She made twice as many two-point shots as three-point shots, and she made four free throws. How many two-point and three-point shots did she make?

57. The population of a small town is 9700 and is decreasing by 230 each year. In how many years will the population be 7860?

58. The freshman enrollment at a college is 1260 and is increasing by 120 each year. In how many years will the freshman enrollment be 1620?

59. A personal library with 64 volumes has three non-fiction books for each fiction book. How many books of each type are in the collection?

60. Food for hummingbirds must contain four times as much water as sugar. How much sugar and how much water are in 15 cups of hummingbird food?

61. One-sixth of the length of a stick is in the ground, and snow covers 9 inches of the stick. If one-third of the length of the stick is above the snow, how long is the stick?

62. A pole indicates the water level in a flash flood area. If 8 inches of the pole is above the water, $\frac{1}{9}$ of the pole is in the ground, and $\frac{2}{3}$ of the pole is in the water, how long is the pole and how deep is the water?

63. The cost of cellular phone service is either $38 per month plus 36¢ per minute or $29 per month plus 51¢ per minute. How many minutes must a person use the phone before the $29-per-month plan is more expensive?

64. One car rental company charges $42 per day plus 10¢ per mile. Another company charges $31 per day plus 21¢ per mile. How many miles must a person drive for the $42 plan to be less expensive?

65. A store stocks twice as much regular coffee as decaffeinated and three-fourths as much premium coffee as decaffeinated. If the store has 90 pounds of coffee, how many pounds of each type are in the store?

66. In a physics class of 36 students, 5 more students received B's than A's and twice as many received C's as A's. The number of students with grades below C was five less than twice the number who received A's. Determine the number of students in each grade category.

67. A warehouse membership store charges an annual fee of $49 for membership. Members pay an additional 2% on all purchases, whereas nonmembers (who pay no annual fee) pay 9% on all purchases. How much must a person purchase to make membership worthwhile?

68. A salesperson at an electronics store has the option of receiving a salary of $150 per week plus a 5% commission on all sales or a salary of $225 per week plus a 3% commission. For what weekly sales is the $150-per-week option the best?

69. The perimeter of a triangle is 7 feet. The first side is two-thirds as long as the second side, and the length of the third side is 1 foot less than the length of the second side. How long is each side?

70. A wire 23 inches long is bent to form a triangle. The second side is 5 inches longer than half the length of the first side, and the third side is 2 inches longer than half the combined lengths of the other two sides. How long is each side?

71. What was the low temperature for the day if the average temperature for the day was less than 72°F and the high was 79°F?

72. The total cost of dinner for two was at least $21.69. One person's bill was twice that of the other person's. What was the minimum amount of the smaller bill?

Data Analysis: *Internet Gambling Sites*

As the number of gambling sites on the Internet has increased, so has the revenue from online gambling. The bar graph shows the amount (in billions of dollars) of such revenue. (Source: Christiansen/Cummings Associates.)

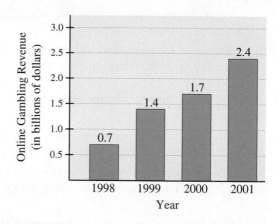

The expression $0.54t - 0.88$ models the revenue (in billions of dollars), where t represents the number of years since 1995.

73. To the nearest hundredth, what is the average annual increase in revenue?

74. Assume that the average increase determined in Exercise 73 remains valid. Let t represent the number of years since 1998 and write an expression to model the revenue for any year after 1998.

75. For which two of the years shown in the graph is the model most accurate?

76. Write and solve an equation that indicates that the revenue is $6.4 billion. Interpret the solution.

Challenge

77. The difference of two consecutive even integers is 2. What are the numbers?

78. The average of a number and its opposite is 0. What is the number?

79. If twice the largest of three consecutive odd integers is 6 more than the sum of the other two, what are the integers?

80. If twice the second of three consecutive even integers is equal to the sum of the smallest and largest, what are the integers?

3.6 | Applications

Things of Value • Percents in Business • Liquid Solutions • Distance, Rate, and Time • Past and Future Conditions

In this section we consider applications that are somewhat more realistic than those in Section 3.5. Some of the topics involve more familiar relationships that serve as the basis for writing equations.

Things of Value

Every business transaction involves the exchange of things with a monetary value. The items of value may be as simple as coins or groceries, or they may involve huge property or stock holdings or corporate mergers.

Application problems in this category are based on the fact that the total value or cost of a collection of items is found by multiplying the number of items by the value or cost of each item (the *unit value* or *unit cost*).

(Number of items) · (Unit Value or Cost) = Total Value or Cost

Note: The unit value and the total value must be expressed in the same monetary units.

EXAMPLE 1

The Value of a Collection of Coins

A change dispenser contains nickels, dimes, and quarters having a total value of $5.00. There are 2 fewer quarters than dimes and 3 times as many nickels as dimes. How many coins of each type are in the dispenser?

Solution

If $x =$ the number of dimes, then
  $x - 2 =$ the number of quarters, and
  $3x =$ the number of nickels.

A table is useful for organizing the information in the problem. Note that the headings of the table reflect the relationship presented previously. We use cents as the monetary unit.

	Number of Coins	Value of Each Coin	Total Value in Cents
Dimes	x	10	$10x$
Quarters	$x - 2$	25	$25(x - 2)$
Nickels	$3x$	5	$5(3x)$
Total			500

$10x + 25(x - 2) + 5(3x) = 500$	The total value is 500 cents.
$10x + 25x - 50 + 15x = 500$	Remove parentheses.
$50x - 50 = 500$	Combine like terms.
$50x = 550$	Add 50 to both sides.
$x = 11$	Divide both sides by 50.

There are 11 dimes, $11 - 2$ or 9 quarters, and $3 \cdot 11$ or 33 nickels.

Mixture problems typically involve two or more commodities, each with a different unit cost. Certain amounts of the commodities are combined to form a mixture. Usually, the problem is to determine the amount of each commodity to use so that the mixture will have a given unit cost.

Note: The unit cost of the mixture must always be between the unit costs of the commodities.

EXAMPLE 2

A Mixture of Pecans and Almonds

Pecans costing $4.75 per pound are to be mixed with almonds costing $3.25 per pound. How many pounds of each should be used to obtain a 20-pound mixture costing $3.70 per pound?

Solution

If x = the number of pounds of pecans in the mixture, then

$20 - x$ = the number of pounds of almonds in the mixture.

	Weight in Pounds	Cost per Pound	Total Cost in Dollars
Pecans	x	4.75	$4.75x$
Almonds	$20 - x$	3.25	$3.25(20 - x)$
Total	20	3.70	$3.70(20)$

$4.75x + 3.25(20 - x) = 3.70(20)$	The sum of the costs of the items equals the total cost.
$4.75x + 65 - 3.25x = 74$	Distributive Property
$1.50x + 65 = 74$	Combine like terms.
$1.5x = 9$	Subtract 65 from both sides.
$x = 6$	Divide both sides by 1.5.

The mixture must consist of 6 pounds of pecans and $20 - 6$ or 14 pounds of almonds.

Percents in Business

In Section 3.3, we used the formula $I = Prt$ to calculate the simple interest I where P is the principal, r is the interest rate, and t is the time in years.

EXAMPLE **3**

Dual Investments with Simple Interest

Part of a $14,000 investment earns 9% simple interest, and the remainder earns 7% simple interest. If the total interest income for 1 year is $1094, how much is invested at each rate?

Solution

If x = the amount of money invested at 9%, then
   14,000 − x = the amount of money invested at 7%.

Because the investment is for one year, $t = 1$, and the interest formula becomes $I = Pr$. Again, we use a table to organize the information in the problem.

	Rate, r	Amount, P	Interest, I
9% Investment	0.09	x	$0.09x$
7% Investment	0.07	14,000 − x	$0.07(14,000 − x)$
Total		14,000	1094

$0.09x + 0.07(14,000 − x) = 1094$	The sum of the interest incomes for the rates equals the total interest of $1094.
$0.09x + 980 − 0.07x = 1094$	Distributive Property
$0.02x + 980 = 1094$	Combine like terms.
$0.02x = 114$	Subtract 980 from both sides.
$x = 5700$	Divide both sides by 0.02.

The investments are $5700 at 9% and $14,000 − $5700, or $8300, at 7%.

EXAMPLE **4**

Commission on a Sale

A dealer agrees to sell your used car for a 15% commission. For how much must the dealer sell the car if you want to receive $5950?

Solution

If x = the dealer's price for the car, then
   $0.15x$ = the dealer's commission.

$x − 0.15x = 5950$	The amount you receive is the dealer's price minus the commission.
$1.00x − 0.15x = 5950$	Multiplicative Identity Property
$0.85x = 5950$	Combine like terms.
$x = 7000$	Divide both sides by 0.85.

The dealer must charge $7000.

Liquid Solutions

If a bottle contains 80 ounces of a liquid solution consisting of 20 ounces of acid, then $\frac{20}{80}$ or 25% of the solution is acid. The usual way of stating this fact is to say that the bottle contains a 25% acid solution.

EXAMPLE 5

Liquid Solutions

How many liters of a 60% chlorine solution should be added to 6 liters of a 20% solution to obtain a 30% solution?

Solution

Let x = the number of liters of the 60% solution to be added. The volume of chlorine in each of the solutions is the percent of chlorine times the volume of the solution. Again, we use a table to organize the information.

	Percent of Chlorine	Volume of Solution (liters)	Volume of Chlorine (liters)
20% Solution	0.20	6	0.20(6) or 1.2
60% Solution	0.60	x	0.60x
30% Solution	0.30	$x + 6$	0.30($x + 6$)

$1.2 + 0.60x = 0.30(x + 6)$	The sum of the volumes of chlorine in the 20% and 60% solutions equals the volume of chlorine in the 30% solution.
$1.2 + 0.6x = 0.3x + 1.8$	Distributive Property
$1.2 + 0.3x = 1.8$	Subtract 0.3x from both sides.
$0.3x = 0.6$	Subtract 1.2 from both sides.
$x = 2$	Divide both sides by 0.3.

Add 2 liters of the 60% solution.

Distance, Rate, and Time

The three components of motion are distance (d), speed or rate (r), and time (t). These quantities are related by the formula $d = rt$. When we work with this type of problem, we must be sure that the units are consistent.

EXAMPLE 6

Think About It

Suppose that in a distance, rate, and time problem, such as Example 6, the time is given in minutes rather than hours but the rate is in miles per hour. Cite two possible changes either of which could be made in order to write the equation.

Distance, Rate, and Time

A car and a truck leave Tulsa, Oklahoma, at the same time. The truck travels east at a constant speed of 61 mph, and the car travels west at a constant speed of 67 mph. How long will it be before they are 320 miles apart?

Solution

Let t = the number of hours until the vehicles are 320 miles apart. Note that both vehicles travel for the same amount of time.

	Rate	Time	Distance
Car	67	t	67t
Truck	61	t	61t

As Figure 3.8 indicates, the sum of the distances traveled by the two vehicles is 320 miles.

Figure 3.8

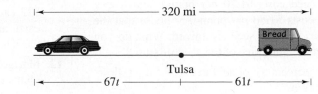

$$67t + 61t = 320 \qquad \text{The sum of the distances is 320 miles.}$$
$$128t = 320 \qquad \text{Combine like terms.}$$
$$t = 2.5 \qquad \text{Divide both sides by 128.}$$

The two vehicles are 320 miles apart after 2.5 hours.

Past and Future Conditions

In problems of this type, we are given a description of certain current conditions. Then we are told how those conditions differ at some other time, either in the past or in the future.

EXAMPLE   7

Ranch Sizes Now and in the Future
The R^2 Ranch has 3 times as many acres as the Bar H Ranch. If each rancher were to purchase an additional 300 acres, the R^2 Ranch would be only twice as large as the Bar H Ranch. How large is each ranch now?

Solution

If $x =$ the current size (in acres) of the Bar H Ranch, then
$3x =$ the current size (in acres) of the R^2 Ranch.

	Current Size	Increased Size
R^2 Ranch	$3x$	$3x + 300$
Bar H Ranch	x	$x + 300$

$$3x + 300 = 2(x + 300) \qquad \text{The increased size of } R^2 \text{ is twice the increased size of Bar H.}$$
$$3x + 300 = 2x + 600 \qquad \text{Distributive Property}$$
$$x + 300 = 600 \qquad \text{Subtract } 2x \text{ from both sides.}$$
$$x = 300 \qquad \text{Subtract 300 from both sides.}$$

At present, the Bar H Ranch has 300 acres, and the R^2 Ranch has 900 acres.

Exercises 3.6

Real-Life Applications

1. Rye grass seed costs $1.20 per pound, and fescue grass seed costs $1.80 per pound. Suppose that the two varieties of grass seed are mixed. What do you know about the unit price of the seed mixture?

2. What is the percentage of acid in each of the following bottles? Why?

   (a) Bottle A contains pure water.

   (b) Bottle B contains pure acid.

Things of Value

3. A coin dispenser containing only quarters and dimes has a total of 28 coins worth $4.15. How many coins of each type are in the dispenser?

4. If a collection of 39 coins consisting only of nickels and dimes is worth $2.70, how many coins of each type are there?

5. A collection of pennies, nickels, and dimes is worth $1.38. The number of nickels is 2 less than the number of dimes, and the number of pennies is 4 more than three times the number of dimes. How many coins of each type are there?

6. A parking meter contains only nickels and dimes. There are five more nickels than dimes, and the total value is $3.70. How many nickels and how many dimes are there?

7. A cash register contains five-, ten-, and twenty-dollar bills worth a total of $160. There are 2 fewer twenty-dollar bills than ten-dollar bills, and there are twice as many five-dollar bills as ten-dollar bills. How many bills of each kind are there?

8. A wallet contains $123 in one-, five-, and ten-dollar bills. The number of ones is 3 less than the number of tens, and the number of fives is twice the number of tens. How many bills of each kind are there?

9. A post office handled a total of 2020 pieces of first- and second-class mail, and its total receipts were $638.20. The cost per piece was 34 cents for first-class mail and 25 cents for second-class mail. How many pieces of each class of mail were processed?

10. Cashews costing $4.50 per pound and peanuts costing $2.50 per pound were mixed to obtain 12 pounds of nuts worth $40. How many pounds of each kind of nut were used?

11. A symphony orchestra sold both CD's and tapes of one of its concerts. If the number of CD's sold at $12 each was twice the number of tapes sold at $7 each, and the total receipts were $4030, how many CD's and how many tapes were sold?

12. Marigold plants cost $1.25 per tray, and salvia plants cost $1.50 per tray. If a person bought a total of 36 trays for $50, how many trays of each kind of flower were purchased?

Percents in Business

In Exercises 13–18, assume that all interest problems involve simple interest.

13. Of a total of $8500, part was invested at a 6% interest rate and part at 7%. If the total interest income for 1 year was $558, how much was invested at each rate?

14. Of a total of $15,000, part was invested at an 8% interest rate and part at 11%. The total interest income for the year was $1332. How much was invested at each rate?

15. If the amount of money invested at 9% was $900 more than the amount invested at 12%, and if the total interest income for the year was $984, how much was invested at each rate?

16. The amount of money a person borrowed at 10% was $1000 more than the amount the same person borrowed at 7.5%. If the total interest charge for the year was $922.50, how much was borrowed at each rate?

17. The total receipts (including a 5% sales tax) collected for one day at a gift shop were $2482.20. How much of the total amount represents sales, and how much represents sales tax?

18. After a 32% discount, the price of a printer was $578. What was the original price?

Liquid Solutions

19. How many milliliters of a 4% solution of medication must be added to 7 ml of a 1% solution to obtain a 3% solution of the medication?

20. How much pure water must be added to 6 liters of a 30% acid solution to obtain a 12% acid solution?

21. How many gallons of water should be added to 6 gallons of 80% insecticide to make a 15% insecticide solution?

22. How many liters of pure acid should be added to 22 liters of a 30% acid solution to obtain a 45% acid solution?

23. Suppose that two cars leave the same point at the same time and travel distances of $55t$ and $60t$ miles, respectively, where t is the number of hours of driving time. For the following situations, how would you represent the distance between the cars after t hours? Explain.

   (a) The cars travel in opposite directions.

   (b) The cars travel in the same direction.

24. Suppose that a person is 20 years old and her cousin is 10 years old.

   (a) What is the ratio of their ages now?

   (b) Determine the ratio of their ages 5 years, 10 years, 20 years, and 50 years from now.

   (c) As the two people age, what happens to the ratio of their ages?

   (d) Explain why the ratio can never reach 1.

Distance, Rate, and Time

25. Two cyclists leave the same point at the same time. One travels north 7 mph faster than the other, who travels south. After 3 hours they are 147 miles apart. How far did each cyclist travel?

26. At 8:00 A.M., two cars leave cities that are 290 miles apart and travel toward each other, one at 54 mph and the other at 62 mph. At what time do the cars meet?

27. Two people walk in opposite directions on a 15-mile loop trail. One walks 2 mph faster than the other. How fast does each person walk if they meet in $1\frac{1}{2}$ hours?

28. Two cars traveled in opposite directions. The slower car, traveling at 45 mph, left 3 hours before the faster car, which traveled at 52 mph. When they were 620 miles apart, how many hours had each car been traveling?

29. A person jogs 2 mph faster than she walks. If she has covered 4.5 miles after walking for 45 minutes and jogging for 15 minutes, what are her walking and jogging speeds?

30. Two planes to Seattle leave Bangor at the same time. One plane flies at 325 mph, and the other flies at 450 mph. After how many hours will the planes be 150 miles apart?

Past and Future Conditions

31. A brick house is twice as old as the stucco house next door. Ten years ago, the brick house was 3 times as old as the stucco house. How old is each house now?

32. A restaurant has been in business three times as long as a sandwich shop. In 6 years, the restaurant will be only twice as old as the sandwich shop. How long has each been in business?

33. The shortstop has played in four times as many consecutive games as the left fielder. If they both play 10 more consecutive games, the shortstop will have played in twice as many consecutive games as the left fielder. In how many consecutive games has each played now?

34. The balance in a savings account is twice the balance in a checking account. If $300 were transferred from checking to savings, then the savings account balance would be 5 times the checking account balance. What is the current balance in each account?

35. A farm has three times as many acres planted in corn as in wheat. Next year, when each field will be increased by 200 additional acres, the number of acres of corn will be twice the number of acres of wheat. How many acres are currently planted in corn?

36. An American flag is four times as high as a state flag. If each flag were raised 4 feet, the American flag would be only three times as high as the state flag. How high is each flag?

Miscellaneous Applications

37. One person walks through an airport at a speed of 120 feet per minute. Another person, starting at the same point and walking at the same rate, is walking on a moving sidewalk that moves at 100 feet per minute. After how many seconds will the person on the moving sidewalk be 40 feet ahead of the first person?

38. The historical society sold cookbooks for $10.50 but charged $13.00 for each book it had to mail. The number of cookbooks that were mailed was two-thirds the number of cookbooks that were not mailed. If the total receipts were $4542.50, how many cookbooks were sold? How many were mailed?

39. If the income on an 8% investment was $314 less than the income on an 11% investment, and the total amount invested was $7000, how much was invested at each rate?

40. How many quarts of pure antifreeze should be added to 6 quarts of a 40% antifreeze solution to make a 50% antifreeze solution?

41. A real estate agent receives an 8% commission on the purchase price of a home. If the seller of a home wants to net $103,500, what must the purchase price of the home be?

42. How many gallons of paint with no blue pigment must be mixed with 5 gallons of paint with 4% blue pigment to obtain paint with 2.5% blue pigment?

43. An investment at 15% yielded $192 more interest income than an investment at 18%. If the total amount invested was $3700, how much was invested at each rate?

44. One grade of crude oil priced at $18 per barrel is mixed with a higher grade of oil priced at $21 per barrel to produce 60 barrels priced at $20 per barrel. How many barrels of each grade were used?

45. One printer prints 2 pages per minute more than another printer. If together they print 70 pages in 7 minutes, at what rate does each printer print?

46. A sociology class has four times as many men as women. If 7 more men and 7 more women were added to the class, the number of men would be 3 times the number of women. How many men and how many women are in the class now?

47. A car dealer plans to advertise that all cars will be sold at a 25% discount off the sticker price. If the dealer wants to receive $6900 for a particular car, what should the sticker price be?

48. A community theater sold reserved seat tickets for $8 and general admission tickets for $4.50. The sale of reserved seats exceeded the sale of general admission seats by 80%, and the total receipts were $2740.50. How many tickets of each type were sold?

49. The faculty parking lot has one-third as many spaces as one student parking lot. If a construction project were to reduce the number of spaces in each lot by 120, the faculty lot would have one-fifth the number of parking spaces remaining in the student lot. How many parking spaces are in each lot now?

50. A person worked at two jobs for a total of 48 hours. One job paid $6.50 per hour, and the other paid $8.40 per hour. If the total pay was $342.40, how many hours did the person work at each job?

51. The interest charges on a loan at 6% were $272 more than the interest charges on a loan at 14%. If the total amount borrowed was $6200, how much was borrowed at each rate?

52. Two marathon runners began running at the same time. One ran at 7 mph, and the other ran at 7.75 mph. How long did it take the faster runner to have a 1-mile lead?

53. A person receives a 15% student discount on a season ticket to an amusement park. The student receives an additional $20 off the discounted price for purchasing the ticket on Tuesday. If the student pays $82, what is the regular price of the ticket?

54. How many ounces of pure silver must be added to 16 ounces of a 70% silver alloy to obtain a 76% silver alloy?

55. A speculator purchased land a year ago. Adjusted for 12% inflation, the price would now be $63,280. What was the original price of the land?

56. Two hikers started from opposite ends of a 29-mile trail. One hiker walked 1.25 mph slower then the other hiker, and they met after 4 hours. How fast did each hiker walk?

57. A stock portfolio valued at $6120 has 15 more shares of Coca-Cola than of AT&T. If the price of Coca-Cola is $40 per share and the price of AT&T is $52 per share, how many shares of each are in the portfolio?

58. One archeological site is currently 3 times as old as another, but 200 years ago it was 5 times as old. How old is each site now?

59. A person charged a total of $1480 on two credit cards. The monthly interest charge on the 18% card was $9 more than the interest charge on the 21% card. How much was charged on each card?

60. How many pounds of mixed nuts containing 60% peanuts by weight must be added to 5 pounds of mixed nuts containing 20% peanuts to obtain a mixture that is 50% peanuts?

61. Two cars leave Omaha at the same time and both drive west, one at 65 mph and the other at 57 mph. How long will it be before the cars are 20 miles apart?

62. The yearly dividend income from 350 shares of stock is $153.50. Part of the stock is McDonald's, paying 40¢ per share, and the remainder is Nations-Bank, paying 46¢ per share. How many shares of each does the person own?

Data Analysis: *Energy from Coal*

Estimates of energy use indicate that the reliance on coal will increase. The bar graph shows the projected amount of energy (in quadrillion BTUs) produced by coal that will be used in the United States. (Source: U. S. Energy Information Administration.)

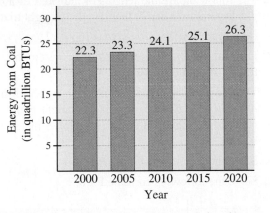

63. For the 20-year period illustrated in the graph, determine the average yearly change in the use of energy produced by coal.

64. Let t represent the number of years after 2000. Write an expression for the amount of coal energy used in any year after 2000.

65. Use the expression in Exercise 64 to predict the amount of coal energy used in 2004.

66. Write and solve an equation to predict the year in which the amount of coal energy will be 30 quadrillion BTUs.

Challenge

67. In a collection of coins, the number of dimes is 3 more than the number of pennies and 1 less than twice the number of nickels. The combined number of nickels and quarters is 1 less than the combined number of pennies and nickels. If the total value of the coins is $3.23, how many coins of each kind are in the collection?

68. A couple buying a home for a certain purchase price makes a $10,000 down payment. To obtain a mortgage, they must pay $230 for an appraisal and $50 for a credit check. In addition, they agree to pay half of the real estate agent's 6% commission, which is based on the purchase price. Finally, the couple must pay 2% interest on the mortgage loan in advance and $900 in closing costs. If the couple's total debt, including the down payment, the mortgage, and all other fees, is $137,480, what is the purchase price of the home?

69. A woman is 4 years older than her brother. Six years ago, her age was two-thirds the age her brother will be in 8 years. In how many years will her brother be 25 years old?

70. At 8:00 A.M., a cyclist left her home and traveled 3 miles at 12 mph. When she realized she had forgotten her water bottles, she returned home at 14 mph. She spent 10 minutes at home and then set out again at 12 mph. If she arrived at her destination at 10:38 A.M., what was the distance (to the nearest mile) from her home to the destination?

Chapter 3 Review Exercises

Section 3.1

1. The 101st Congress had 57 Democrats and 43 Republicans in the Senate. What was the ratio of

(a) Republicans to Democrats?

(b) Democrats to the total number of senators?

2. What is the ratio of 64 quarts to 8 gallons?

3. What is the unit cost of an item that weighs 2 pounds, 4 ounces, and costs a total of $5.40?

4. Solve the proportion $\dfrac{x+1}{8} = \dfrac{2x-1}{12}$.

5. An exit poll showed that 38 voters favored a bond issue, whereas 22 voters opposed it. If the poll is accurate, of the 2880 people who voted, how many voted in favor of the bond issue?

6. Suppose that $\triangle ABC$ is similar to $\triangle PQR$. If $AB = 5$, $AC = 7$, and $PQ = 2$, what is PR?

7. Suppose that the ratio of fiction to nonfiction books in a library is 3 to 2. How many nonfiction books are there in a collection of 10,000 volumes?

8. If it costs $0.50 more to buy 5 pounds of an item than it costs to buy 3 pounds, what is the cost of 9 pounds?

9. A 2-ounce serving of potato chips has 540 mg of sodium. How many milligrams of sodium are in a 5-ounce serving?

10. Write $3x = 5y$ as a proportion with x as the numerator of the first ratio. (There are two answers.)

11. Suppose that $\triangle ABC$ is similar to $\triangle PQR$. Which of the following proportions is valid?

(i) $\dfrac{AB}{AC} = \dfrac{PQ}{PR}$ (ii) $\dfrac{BC}{QR} = \dfrac{AB}{PQ}$

(iii) $\dfrac{QP}{RP} = \dfrac{BA}{CA}$

12. Suppose that the exchange rate is 1 British pound for $1.50 U.S. What is the cost in dollars of an item that sells for 62 pounds?

Section 3.2

13. Convert the given percent into a simplified fraction and into a decimal.

(a) 47%   (b) 110%

(c) 0.02%

14. Convert the given decimal or fraction into a percent.

(a) 0.6   (b) $\dfrac{9}{40}$

(c) 5

15. What is 32% of 80?

16. What percent of 68 is 17?

17. 98% of what number is 147?

18. What is 54 increased by 20%?

19. What is the percent decrease from 30 to 24?

20. A bakery increased production of chocolate chip cookies by 18% to 59 dozen cookies. How many dozen cookies did the bakery produce before the increase?

21. The average temperature in January was 12% below the normal average of 45°F. What was the average temperature for the month?

22. Suppose that 23% of the population of a community contracted the flu. To know the actual number of flu victims, what additional information is needed and how would it be used?

23. Show that the percent increase from 16 to 20 is not the same as the percent decrease from 20 to 16.

24. Show how to convert $\frac{11}{25}$ into a percent without writing the fraction as a decimal.

Section 3.3

25. Use the simple interest formula to compute the interest on $240 at 18% for 6 months.

26. To the nearest tenth of an inch, what is the circumference of a circle whose radius is 9 inches?

27. The following are the lengths of the sides of a triangle. Determine whether the triangle is a right triangle.

(a) 24, 51, 45   (b) 47, 36, 27

28. Determine the average speed of an automobile that travels 210 miles in 4 hours.

29. The area A of a trapezoid is given by the formula $A = \frac{1}{2}h(b_1 + b_2)$, where h is the height and b_1 and b_2 are the lengths of the bases. Solve the formula for b_2.

30. Solve the formula $A = P + Prt$ for r.

31. Solve the following formula for b.

$$Q = \frac{a + b + c}{3}$$

32. Solve the equation $2x + 3y = 6$ for y.

33. Solve the following equation for y.

$$\frac{x}{2} + \frac{y}{6} = \frac{1}{3}$$

34. A 26-foot ladder rests against a wall. The top of the ladder is 24 feet from the ground, and the bottom of the ladder is 10 feet from the base of the wall. Explain how you can tell whether the wall is perpendicular to the ground.

35. The formula for converting Fahrenheit temperatures F to Celsius temperatures C is $C = \frac{5}{9}(F - 32)$. Solve the formula for F and then determine the Fahrenheit equivalent of 15°C.

36. If the circumference of a circle is 18π feet, what is the area of the circle? (Express your answer in terms of π.)

Section 3.4

37. Translate the given information into an expression or equation. Use x as your variable.

(a) Five less than twice a number

(b) Twice a number is 4 more than the number.

38. In each part of this exercise, a variable is assigned to one of the unknown quantities. Represent the other unknown quantity with an expression that uses the same variable.

(a) The sum of two numbers, one of which is *n*, is 18.

(b) The width of a rectangle is 3 less than half its length *L*.

39. If *c* feet are cut off a 10-foot board, what is the length of the remaining piece?

40. If the measure of an angle is *m*, what is the measure of its supplement?

41. If the largest of three consecutive integers is *x*, what expressions represent the other two integers?

42. Write an expression for the described quantity.

(a) The value of *n* nickels in cents

(b) The distance traveled in 2 hours at *r* mph

(c) The amount of acid in *x* quarts of a 12% acid solution

43. Write a simplified expression for the described quantity.

(a) The perimeter of a rectangle if its width is 3 less than its length

(b) The cost of a refrigerator after a 30% discount

44. Assign a variable to the unknown quantity and translate the information into an expression.

(a) The cost of a rental car at $32 plus 30 cents per mile

(b) The total distance traveled in 8 hours by a car that averaged 40 mph for part of the trip and 50 mph for the remainder of the trip

Section 3.5

In Exercises 45–52, solve the application problem.

45. Seven less than two-thirds of a number is −3. What is the number?

46. The sum of two numbers is −1. If twice the smaller number is added to 3 times the larger number, the result is 2. What are the numbers?

47. The sum of the first and last of three consecutive integers is the second integer. What is the largest integer?

48. A 100-foot ribbon is cut into three pieces. The second piece is 3 times as long as the first piece. The third piece is 4 feet longer than the sum of the lengths of the other two pieces. How long is each piece?

49. The measure of one angle of a triangle is one-fifth the measure of the second angle. The measure of the third angle is half the sum of the measures of the other two angles. What is the measure of the largest angle?

50. The perimeter of a rectangle is 38 inches, and its length is 2 inches less than twice its width. What is the width of the rectangle?

51. The measure of the supplement of an angle is 10° less than 3 times the measure of the complement of the angle. What is the measure of the angle?

52. A rental car bill for 1 week was $215.20. The rental costs $120 per week plus 17 cents per mile. How many miles was the car driven?

Section 3.6

In Exercises 53–60, solve the application problem.

53. The change drawer in a cash register contains only nickels, dimes, and quarters. There are twice as many nickels as dimes, and there are 3 fewer dimes than quarters. If the value of all the coins is $6.15, how many quarters are there?

54. Tickets to a concert cost $22 at the box office, but a $2 service charge was assessed on each ticket purchased through TicketKing. The total receipts, including the service charges, were $54,400. If twice as many tickets were sold at the box office as were sold through TicketKing, how much money was taken in at the box office?

55. A person invested $12,000, partly in a mutual fund paying 7% and partly in a money market fund paying 4%. If the total simple interest income was $750, how much was invested in the mutual fund?

56. If you want to net $850 on the sale of an antique chair, at what price must the chair be sold by an auctioneer who charges a 15% sales commission?

57. A bottle contains 1 quart of a 20% alcohol solution. How much water must be added to the bottle to dilute the liquid to a 15% alcohol solution?

58. Two people 10 miles apart begin to travel toward each other. One is walking at 5 mph and the other is riding a bicycle at 15 mph. How long will it take for the two people to meet?

59. On a certain hospital floor, there are 3 times as many nurse's aides as registered nurses. If an additional registered nurse were assigned to the floor, there would be twice as many aides as nurses. How many aides are there?

60. All goods in a store have been marked up 20% and are subject to a 5% tax on the retail price. What was the wholesale price of an item that costs $25.20, including tax?

Chapter 3 Test

1. What is the unit cost of an item weighing 1 pound, 6 ounces, if the cost of the item is 99 cents?

2. At a certain college, the ratio of tenured faculty to nontenured faculty is 3 to 2. If there are 400 faculty members, how many are nontenured?

3. The two triangles in the figure are similar triangles. Determine AB and DE.

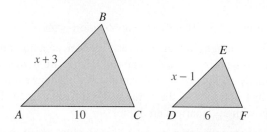

4. Only 70% of the people invited to a wedding were able to attend. If 56 people attended the wedding, how many were invited?

5. A company decreased its work force by 8% to 184 employees. How many people lost their jobs?

6. In a community of 1000 voters, 63% voted for candidate A and 38% voted for candidate B. How many people voted twice?

7. Suppose that you have an 8% mortgage and the balance due is $71,000. How much of your next monthly payment will go toward interest?

8. The area A of a triangle is given by $A = \frac{1}{2}bh$, where b is the length of the base and h is the height.
   (a) Solve the formula for b.
   (b) Determine the length of the base of a triangle with an area of 56 square inches and a height of 8 inches.

9. The width W of a rectangle is given by the formula $W = \dfrac{P}{2} - L$, where P is the perimeter of the rectangle and L is the length. Write a formula for repeated calculations of perimeters given widths and lengths.

10. Solve the equation $5x - 4y = 20$ for y.

11. The lengths of the sides of a triangle are 2 feet, 4.25 feet, and 3.75 feet. Explain how to determine whether the triangle is a right triangle.

12. If a class of 23 students consists of w women, write an expression that represents the number of men.

13. If the measure of an angle is y, what do we call the angle whose measure is $90 - y$?

14. If the second of three consecutive odd integers is n, what expression represents the sum of the three integers? Why?

15. The sum of two numbers is 0. One number is 18 less than one-fifth of the other number. What is the positive number?

16. A rope 60 feet long is cut into three pieces. The second piece is 4 times as long as the first piece, and the third piece is 12 feet shorter than the second piece. How long is each piece?

17. The sum of the measures of the complement and the supplement of an angle is 156°. What is the measure of the angle?

18. One kind of tea worth $3.40 per pound is blended with another kind of tea worth $6.10 per pound. To the nearest pound, how many pounds of each kind of tea should be mixed to obtain a 50-pound blend worth $4.91 per pound?

19. At 10:00 A.M., a truck leaves a rest stop in Akron, Ohio, and heads east at 50 mph. At 11:00 A.M., a car leaves the same rest stop and heads west at 55 mph. At what time will the car and the truck be 260 miles apart?

20. A bottle contains 200 cubic centimeters of a 5% acid solution. To the nearest tenth of a cubic centimeter, how much pure acid should be added to the bottle to strengthen the solution to 15% acid?

Cumulative Test, Chapters 1–3

1. Compare the decimal names of rational numbers with the decimal names of irrational numbers.

2. Insert $<$, $>$, or $=$ to make the statement true.

    (a) $-(-x)$ ▬▬ x

    (b) $-|-5|$ ▬▬ 5

    (c) $5 + 2(-3)$ ▬▬ $(5 + 2)(-3)$

3. What property justifies the statement $3\left(\frac{1}{3}x\right) = \left(3 \cdot \frac{1}{3}\right)x$?

4. If the addends are -4, 6, -7, and 3, what is the sum?

5. Determine the difference of 6 and $-\frac{2}{3}$.

6. Evaluate $-8 \div (-4) \cdot (-3)$.

7. Evaluate $-3x + x^3 - 5$ for $x = -2$ on the home screen of your calculator.

8. Simplify $-(x - 3a) - 2(x - a - 1)$.

9. What are the signs of the coordinates of all points in Quadrant IV?

10. If you produce the graph of the expression $1 - 5x$ and trace to the point whose x-coordinate is 3, what is the y-coordinate? Why?

11. What do we call the following equations?

    (a) $2(x + 1) = 2x + 2$ (b) $2(x + 1) = 2x + 1$

12. Solve $-3x + 2 - (x - 4) = 2x + 8$.

13. Use the graphing method to estimate the solutions of $3 - x < 5$. Draw a number line graph of the solution set.

14. A county commissioner won her election by a margin of 4 to 3. If 56,000 people voted, how many voted for her opponent?

15. A company's annual gross sales of \$521,520 represent a 6% increase over the previous year's sales. What were the gross sales last year?

16. Solve the equation for y.

    $$\frac{x}{2} - 5y = 20$$

17. Write a simplified expression for the perimeter of a rectangle whose width is 4 feet less than half its length L.

18. Tickets to the county fair cost \$8 for adults and \$4 for children. Four hundred senior citizens were admitted with a 10% discount off the adult ticket price. If \$14,080 was collected from the total of 2200 people who went to the fair, how many children's tickets were sold?

19. The measure of one unknown angle of a right triangle is 6° more than twice the measure of another angle. What are the measures of the three angles of the triangle?

20. A person begins walking along a path at a rate of 4 mph. Ten minutes later, a jogger begins at the same point and runs along the same path at 7 mph. To the nearest minute, how long does it take the jogger to catch up to the walker?

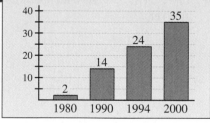

AVERAGE NUMBER OF MICROCOMPUTERS IN AUTOMOBILES

(Source: Motorola, Miller Communications.)

In recent years the design of automobiles has included an increasing amount of electronic technology. The accompanying bar graph shows the average number of microcomputers in automobiles in selected years.

The data in the graph can be modeled by a **linear equation in two variables.** The graph of this kind of equation is a line with certain properties such as **slope** and **intercepts.** What are the implications of such data for mechanics and engineers in the auto industry? (For more on this real data problem, see Exercises 95–98 in Section 4.4.)

Properties of Lines

Chapter Snapshot

In this chapter we introduce linear equations in two variables and examine in detail the properties of their graphs. In particular, we learn about slope and its role in graphing lines, writing equations of lines, and determining whether lines are parallel or perpendicular. We conclude with a discussion of linear inequalities and their graphs.

Warm-Up Skills

The following exercises review concepts and skills that you will need in Chapter 4.

1. Use a graph of the following set to predict the next ordered pair in the set.
   $\{(-2, -2), (0, -1), (2, 0), (4, 1), (6, 2), \ldots\}$

2. Suppose that (a, b) is a point of an axis. What can you conclude about the coordinates?

In Exercises 3–5, solve the given equation for y.

3. $\dfrac{x}{2} - \dfrac{y}{4} = 3$ 4. $\dfrac{y - 3}{x + 1} = 2$ 5. $3x + 2y = 12$

6. Use the given values of x and y to evaluate $\dfrac{3}{2}x - y - 1$.

   (a) $x = -4, y = -7$ (b) $x = 4, y = 5$

7. Evaluate $\dfrac{y_2 - y_1}{x_2 - x_1}$ for the given values of the variables.

   (a) $x_1 = 2, y_1 = -4, x_2 = 1, y_2 = -4$
   (b) $x_1 = 6, y_1 = -2, x_2 = 1, y_2 = -3$

8. Suppose that you start at point $A(-1, 3)$ and move right 2 units and down 5 units to point B. What are the coordinates point B?

9. If a line is perpendicular to the y-axis, then it is _____ to the x-axis. If a line is parallel to the y-axis, then it is _____ to the x-axis.

10. The product of what number and $\frac{2}{3}$ is -1?

11. Suppose that you multiply both sides of an inequality by a number. If the number is positive, then the direction of the inequality symbol is _____ . If the number is negative, then the direction of the inequality symbol is _____ .

12. Solve $3 - 2y < 11$.

4.1 Linear Equations in Two Variables

Solutions of Linear Equations · Graphs of Linear Equations

Solutions of Linear Equations

In Chapter 3 we learned how to solve linear equations in one variable. When a problem involves two unknown quantities, it is often easier to assign a different variable to each unknown quantity and then describe the conditions of the problem with an equation in those two variables.

EXAMPLE 1

Using One and Two Variables to Describe Conditions

The sum of two numbers is 9. If possible, describe this information with an equation in (a) one variable and (b) two variables.

Solution

(a) If we let x represent one number, then $9 - x$ represents the other number. However, no more information is given, so we cannot write an equation in one variable.

(b) If we let x represent one number and y represent the other number, we can write the equation $x + y = 9$.

In Example 1, the equation $x + y = 9$ is an example of a special kind of equation called a **linear equation in two variables.**

Definition of a Linear Equation in Two Variables

A **linear equation in two variables** is an equation that can be written in the *standard form $Ax + By = C$*, where A, B, and C are real numbers, and A and B are not both zero.

The following are other examples of linear equations in two variables and their standard forms.

Equation	*Standard Form*	
$5x + 7y = -3$	$5x + 7y = -3$	$A = 5, B = 7, C = -3$
$y = 3x - 8$	$-3x + y = -8$	Subtract $3x$ from both sides.
$x - 4 = 0$	$x + 0y = 4$	The coefficient of y is 0.
$y = -6$	$0x + y = -6$	The coefficient of x is 0.

The following equations are not linear equations in two variables.

$y = x^2$ The exponent on each variable must be no greater than 1.

$y = \dfrac{5}{x} - 14$ There can be no variable in the denominator of a fraction.

$xy = 7$ No term can contain both variables.

For linear equations in one variable, a solution is any replacement for the variable that makes the equation true. If an equation has two variables, we must find two replacements, one for each variable. For instance, suppose we replace x with 3 and y with 2 in the equation $2x + y = 8$.

$$2x + y = 8$$
$$2(3) + 2 = 8 \qquad \text{Replace } x \text{ with 3 and } y \text{ with 2.}$$
$$6 + 2 = 8$$
$$8 = 8 \qquad \text{True}$$

Because the resulting equation is true, we say that this pair of replacements is a **solution** of the equation. We write the solution as the *ordered pair* (3, 2), where the first coordinate is the value of x, and the second coordinate is the value of y. A solution of an equation is said to *satisfy* the equation.

Note: We call (3, 2) an *ordered* pair because the order of the coordinates is significant. For example, observe that (2, 3) is not a solution of the previous equation.

Definition of Solution of a Linear Equation in Two Variables

A **solution** of a linear equation in two variables is an ordered pair of numbers (x, y) that satisfies the equation.

EXAMPLE 2

Testing Solutions of a Linear Equation in Two Variables

Select the ordered pairs from set A that are solutions of the equation $2x - y = 6$.

$$A = \{(4, 2), (-1, -8), (0, 6)\}$$

Solution

For each pair, replace x with the first coordinate and y with the second coordinate.

(4, 2)	*(−1, −8)*	*(0, 6)*
$2x - y = 6$	$2x - y = 6$	$2x - y = 6$
$2(4) - 2 = 6$	$2(-1) - (-8) = 6$	$2(0) - 6 = 6$
$8 - 2 = 6$	$-2 + 8 = 6$	$0 - 6 = 6$
$6 = 6$ True	$6 = 6$ True	$-6 = 6$ False

The pairs (4, 2) and (−1, −8) are solutions.

We have already seen two significant differences between the solution sets of linear equations in one and two variables.

1. A solution of a linear equation in one variable is a number, whereas a solution of a linear equation in two variables is a *pair* of numbers.
2. A linear equation in one variable has exactly one solution. Example 2 shows that a linear equation in two variables has more than one solution.

Graphs of Linear Equations

Testing whether a given ordered pair is a solution of an equation requires only simple substitution. *Determining* solutions requires other techniques.

Figure 4.1

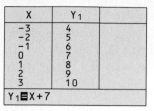

X	Y$_1$	
−3	4	
−2	5	
−1	6	
0	7	
1	8	
2	9	
3	10	

Y$_1$◨X + 7

The Graph of an Equation

Consider the equation $y - x = 7$. We can determine solutions of this equation by selecting a value for either variable and calculating the corresponding value of the other variable. However, a more common practice begins with solving the equation for y: $y = x + 7$. Then we can select x-values and easily calculate the corresponding y-values. The form $y = x + 7$ also allows us to use a calculator for determining solutions. Figure 4.1 is a table of values for the expression $x + 7$.

As we learned in Chapter 2, the entries in Figure 4.1 can be used to create ordered pairs of the form $(x, x + 7)$. However, $x + 7$ and y have the same value. Thus the pair $(x, x + 7)$, which we used to graph the expression $x + 7$, is the same as the pair (x, y), which is a solution of the equation $y = x + 7$.

By scrolling through the table, we can see many other ordered pairs that are solutions of $y = x + 7$. Each solution can be represented by a point in the coordinate plane. Figure 4.2 shows just a few of the points that represent solutions.

LEARNING TIP

When you graph an equation, the more points you plot, the easier it is to recognize a pattern.

Figure 4.2

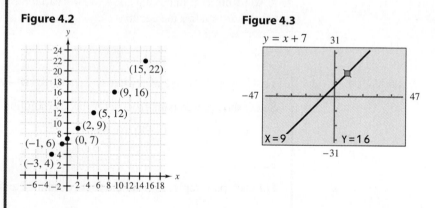

Figure 4.3

Because there are infinitely many replacements for x, the equation $y = x + 7$ has infinitely many solutions. The calculator graph in Figure 4.3 shows that the solutions of $y = x + 7$ can be represented by points of a line. The cursor highlights one solution, $(9, 16)$.

Observe that the graph of the expression $x + 7$ is the same as the **graph of the equation** $y = x + 7$.

Think About It

Produce the graphs of the following equations. Judging by the graphs, which do you think are linear equations? Does your conclusion agree with the definition of a linear equation in two variables?

$$y = x^2 \quad y = \frac{1}{x} \quad y = x \quad y = \sqrt{x}$$

The Graph of an Equation

The **graph of an equation** in two variables is the set of all points in the coordinate plane that correspond to solutions of the equation. The process of determining solutions and drawing the graph is called **graphing the equation.**

We can state the following generalizations about the graph of a linear equation in two variables.

1. The graph is a straight line.
2. Every solution of the equation is represented by a point of the line.
3. Every point of the line represents a solution of the equation.

We can produce the graph of a linear equation in two variables by determining solutions algebraically, plotting points, and drawing the line.

EXAMPLE **3**

Determining Points and Sketching the Graph

Complete the table so that the pairs (x, y) are solutions of the linear equation $y = -2x + 5$. Then sketch the graph of the equation.

x	y	(x, y)
	9	
0		
	-7	
2		

Solution

For y = 9	*For x = 0*	*For y = -7*	*For x = 2*
$9 = -2x + 5$	$y = -2(0) + 5$	$-7 = -2x + 5$	$y = -2(2) + 5$
$4 = -2x$	$y = 0 + 5$	$-12 = -2x$	$y = -4 + 5$
$-2 = x$	$y = 5$	$6 = x$	$y = 1$

The completed table is as follows.

x	y	(x, y)
-2	9	$(-2, 9)$
0	5	$(0, 5)$
6	-7	$(6, -7)$
2	1	$(2, 1)$

Figure 4.4

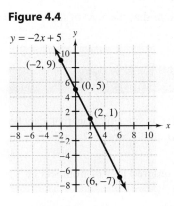

$y = -2x + 5$

Figure 4.4 shows the plotted points and the graph of the equation.

Note: Only two points are needed to determine the exact location of a line. In Example 3 we did not need all four points to sketch the graph. However, we recommend that you plot at least three points as a check against errors.

To use a calculator to produce a graph, the equation must be solved for y in order to enter it on the Y screen. Then we can trace the graph to estimate solutions of the equation.

EXAMPLE **4**

Determining Solutions from the Graph

Use the integer setting to produce the graph of $x + y = 10$ on your calculator. Then determine the missing coordinates in the following ordered pairs so that they are solutions of $x + y = 10$.

$(-4, \rule{0.5cm}{0.15cm}), (\rule{0.5cm}{0.15cm}, -7), (4, \rule{0.5cm}{0.15cm}), (\rule{0.5cm}{0.15cm}, 20)$

Solution

$$x + y = 10 \qquad \text{Solve the equation for } y.$$
$$y = -x + 10 \qquad \text{Subtract } x \text{ from both sides.}$$

Enter the expression $-x + 10$ as Y_1, and use the integer setting to produce the graph. Figure 4.5 shows the graph with the tracing cursor on $(-4, 14)$. Trace the graph to determine the other missing values.

Figure 4.5

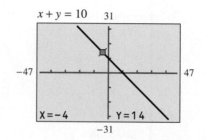

The pairs are $(-4, 14)$, $(17, -7)$, $(4, 6)$, $(-10, 20)$. Note that the sum of the coordinates is always 10.

Quick Reference 4.1

Solutions of Linear Equations

- A **linear equation in two variables** in two variables is an equation that can be written in the *standard form* $Ax + By = C$, where A, B, and C are real numbers, and A and B are not both zero.

- A **solution** of a linear equation in two variables is an ordered pair of numbers (x, y) that satisfies the equation.

Graphs of Linear Equations

- A linear equation in two variables has infinitely many solutions. Each solution can be represented by a point in the coordinate plane.

- The **graph of an equation** in two variables is the set of all points in the coordinate plane that correspond to solutions of the equation. The process of determining solutions and drawing the graph is called **graphing the equation.**

- If a linear equation is in the form $y =$ expression, then graphing the expression is equivalent to graphing the solutions of the equation.

- To sketch the graph of a linear equation, determine several solutions, plot the corresponding points, and draw a line through the points.

- To produce the graph of a linear equation on your calculator, solve the equation for y and enter the resulting expression. Solutions of the equation can be estimated by tracing the graph.

Speaking the Language 4.1

1. The equation $2x + 3y = 7$ is an example of a(n) ▒▒▒▒▒▒ equation in two ▒▒▒▒▒▒.
2. Every solution of $x - y = 2$ is in the form of a(n) ▒▒▒▒▒▒.
3. The graph of a linear equation in two variables is a(n) ▒▒▒▒▒▒.
4. The graph of the equation $y = 2x + 1$ is the same as the graph of the expression ▒▒▒▒▒▒.

Exercises 4.1

Concepts and Skills

1. The standard form of a linear equation in two variables is $Ax + By = C$. The definition states that A or B, but not both, can be 0. Why?

2. For an equation to be a linear equation in two variables, the exponent on each variable must be no greater than 1. Does this mean that $x^1y^1 = 3$ is a linear equation in two variables? Why?

In Exercises 3–10, write an equation that models the given information.

3. The total number of men and women in a class is 24.

4. There are 4 more men than women in a group of people.

5. Two angles are supplementary.

6. Two angles are complementary.

7. A 12-foot board is cut into two pieces.

8. A 30-foot rope is cut into three pieces. One piece is 8 feet long.

9. Brazilian and Turkish coffee are mixed into a 12-pound blend.

10. The retail price of an item is $12 more than the wholesale price.

In Exercises 11–18, use the word *yes* or the word *no* to indicate whether the given equation is a linear equation in two variables.

11. $x + 3y = -3$

12. $x + y^2 = 1$

13. $\dfrac{2}{x} = y + 1$

14. $x = y - 5$

15. $y = 3$

16. $3x - 5 = 0$

17. $xy = 12$

18. $x^3 + y = 0$

19. If $x = 1$ and $y = 3$, then $x + 2y = 7$ is a true equation. Does it matter whether we write the solution as $(1, 3)$ or $(3, 1)$? Why?

20. What do we mean when we say that an ordered pair *satisfies* a linear equation in two variables?

In Exercises 21–26, identify each ordered pair that is a solution of the given equation.

21. $y = -2x - 5$
    $(5, -15), (-1, -7), (0, -5), (-6, 7)$

22. $x + 2y = -2$
    $\left(-7, \frac{5}{2}\right), (-2, 0), (0, -2), (2, -2)$

23. $3 - y = 2x$
    $(1, -1), (6, -9), (2, -1), (-4, 11)$

24. $-2x + 5y = -5$
    $(-5, -3), \left(4, \frac{3}{5}\right), (0, -1), (-1, 0)$

25. $3y - 4x = 6$
    $(0, 2), (3, 6), (2, 0), \left(1, \frac{10}{3}\right)$

26. $\dfrac{3}{2}x - y - 1 = 0$
    $(-4, -7), (0, 1), (6, 8), (-2, -4)$

27. Describe the procedure for producing the graph of $3x - y = 4$ on your calculator.

28. Suppose that you have produced the graph of an equation on your calculator and traced the graph to the point $(-3, 4)$. What is the meaning of this point?

In Exercises 29–34, complete the table so that the pairs (x, y) are solutions of the given equation. Then sketch the graph of the equation.

29. $y = -x$

x	4		-3
y		2	

30. $y = 2x - 3$

x	0		-2
y		-1	

31. $y = 4x + 1$

x	1		-1
y		1	

32. $y = 3 - 2x$

x	-1		3
y		-1	

33. $x = 3 - 2y$

x			-1
y	0	-3	

34. $x - 3y = 5$

x		2	-7
y	1		

In Exercises 35–42, determine three solutions of the equation and sketch the graph.

35. $y = -x + 4$

36. $y = 2x - 5$

37. $y = x$

38. $y = \frac{1}{3}x$

39. $y = -3x + 2$

40. $y = -x$

41. $y = \frac{4x}{3} - 4$

42. $y = 6 - \frac{1}{2}x$

In Exercises 43–50, write the given equation in the form required for producing the graph on a calculator.

43. $2x + y = 3$

44. $x - 4y = 0$

45. $2 = 6x - 3y$

46. $-x + 4y = 12$

47. $5y + 2x - 10 = 0$

48. $6 = 3y - x + 12$

49. $\frac{x}{2} - \frac{y}{4} = 3$

50. $\frac{x}{5} + \frac{2y}{3} = -1$

In Exercises 51–58, determine three solutions of the equation and sketch the graph. Then produce the graph of the equation on your calculator and compare your sketch with the calculator graph.

51. $y + x = 0$

52. $y - 2x = 3$

53. $2x - y - 3 = 0$

54. $x + 3y = 12$

55. $4x - 3y = 15$

56. $x + 2y + 10 = 0$

57. $7x = 28 - 4y$

58. $2x = 3y$

In Exercises 59–64, produce the graph of the given equation on your calculator. Then, by tracing the graph, fill in the blanks so that the ordered pairs are solutions of the equation.

59. $y = -x + 7$

(a) $(4, \rule{1cm}{0.4pt})$

(b) $(\rule{1cm}{0.4pt}, 7)$

(c) $(-2, \rule{1cm}{0.4pt})$

(d) $(\rule{1cm}{0.4pt}, 0)$

60. $y = \frac{1}{2}x$

(a) $(6, \rule{1cm}{0.4pt})$

(b) $(\rule{1cm}{0.4pt}, -4)$

(c) $(-3, \rule{1cm}{0.4pt})$

(d) $\left(\rule{1cm}{0.4pt}, \frac{7}{2}\right)$

61. $x + y = 8$

(a) $(\rule{1cm}{0.4pt}, 15)$

(b) $(-2, \rule{1cm}{0.4pt})$

(c) $(\rule{1cm}{0.4pt}, 8)$

(d) $(8, \rule{1cm}{0.4pt})$

62. $x - y = -3$

(a) $(9, \rule{1cm}{0.4pt})$

(b) $(-5, \rule{1cm}{0.4pt})$

(c) $(\rule{1cm}{0.4pt}, -10)$

(d) $(\rule{1cm}{0.4pt}, 0)$

63. $3x - y = -2$

(a) $(-3, \rule{1cm}{0.4pt})$

(b) $(5, \rule{1cm}{0.4pt})$

(c) $(\rule{1cm}{0.4pt}, -10)$

(d) $(\rule{1cm}{0.4pt}, 8)$

64. $2x + 3y = 1$

(a) $(\rule{1cm}{0.4pt}, -9)$

(b) $(\rule{1cm}{0.4pt}, 5)$

(c) $(-13, \rule{1cm}{0.4pt})$

(d) $(8, \rule{1cm}{0.4pt})$

In Exercises 65 and 66, which equation is graphed in the figure?

65. (i) $x - 2y = 4$

(ii) $2y - x = 4$

(iii) $2y - x = -4$

66. (i) $3x + y = 6$

(ii) $3x - y = 6$

(iii) $3x + y = -6$

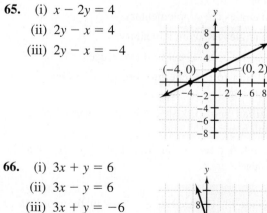

In Exercises 67 and 68, identify the equation graphed in the figure. Then determine the values of a and b.

67. (i) $x - 2y = 16$

(ii) $3x + y = 6$

(iii) $2x + y = 9$

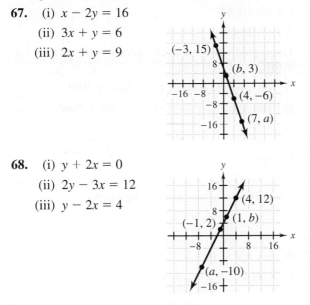

68. (i) $y + 2x = 0$

(ii) $2y - 3x = 12$

(iii) $y - 2x = 4$

In Exercises 69–76, write two ordered pairs whose coordinates satisfy the given conditions. Then translate the conditions into an equation. Finally, produce the graph of the equation and trace it to verify that your two ordered pairs are represented by points of the graph.

69. The y-coordinate is 3 more than the x-coordinate.

70. The y-coordinate is 1 less than the x-coordinate.

71. The y-coordinate is twice the x-coordinate.

72. The y-coordinate is half the x-coordinate.

73. The sum of the coordinates is 7.

74. The difference of the x- and y-coordinates is -2.

75. The coordinates are opposites.

76. The coordinates are equal.

Real-Life Applications

In Exercises 77–80, write an equation to describe the given information. Then use the equation to answer the question.

77. A taxi fare y is $5 plus $1.20 per mile x. What is the cost of a 9-mile trip?

78. In a hardware store, the difference in the number x of hammers and the number y of screwdrivers is 14. If there are 26 hammers, how many screwdrivers are in stock?

79. The number y of women attending a gathering is 3 more than twice the number x of men. If there are 25 women present, how many men are attending?

80. At an animal shelter, the ratio of the number x of dogs to the number y of cats is 3 to 2. If there are 18 cats, how many dogs are there?

Modeling with Real Data

81. The table shows the percentage of women in the United States House of Representatives. The linear equation $y = \frac{1}{2}x - 2$ models the data, where y represents the percentage of women in the House, and x represents the number of years since 1970.

Year	Number of Female Members	Percentage of Total Membership
1981	21	4.8
1983	22	5.1
1985	23	5.3
1987	23	5.3
1989	29	6.7
1991	28	6.4
1993	47	10.8
1995	47	10.8
1997	51	11.7
1999	58	13.3

(Source: National Women's Political Caucus.)

(a) Use a graph of the model equation to complete the table.

x	11	(ii)	23	(iv)
y	(i)	7.5	(iii)	11.5

(b) For which of the years represented by entries in the table in part (a) is the model most accurate?

82. Use the model in Exercise 81 to estimate

(a) the year in which $\frac{1}{6}$ of the members of the House will be women.

(b) the percentage of women in the House in 2005.

83. The average income tax refund increased from $916 in 1990 to $1668 in 1999. The data in the table can be modeled by the linear equation $y = 87x + 832$, where x is the number of years since 1990, and y represents the average refund (in dollars).

Year	Average Refund
1990	916
1993	1013
1996	1244
1998	1573
1999	1668

(Source: Internal Revenue Service.)

Verify that each ordered pair is a solution of the equation. Then interpret each solution.

(a) (2, 1006)

(b) (7, 1441)

84. Use a graph of the model equation in Exercise 83 to estimate

(a) the average income tax refund in 2001.

(b) the year in which the average refund will be $2050.

Challenge

In Exercises 85–90, determine the value of k that makes the ordered pair a solution of the equation.

85. $y = kx - 5$; $(1, -2)$

86. $y = k - 2x$; $(-3, -1)$

87. $3y - x = k$; $(k - 1, k + 1)$

88. $kx + 2y = 1$; $(-1, k - 1)$

89. $\frac{1}{3}x - \frac{2}{5}y = -2$; $(6, k)$

90. $2.6x - 1.8y = 15.8$; $(4, k)$

4.2 Intercepts and Special Cases

Intercepts • Special Cases

Intercepts

In Section 4.1, we found that we can produce a picture of the solutions of an equation. We call this picture a graph of the equation, and for linear equations in two variables, the graph is a straight line.

Of special interest are the points where a graph crosses the axes.

Definition of Intercepts

A point where a graph intersects the x-axis is called an **x-intercept,** and a point where a graph intersects the y-axis is called a **y-intercept.**

If we trace the graph of $y = 2x + 18$, we see that the x-intercept is $(-9, 0)$ and the y-intercept is $(0, 18)$. See Figures 4.6(a) and 4.6(b).

Figure 4.6

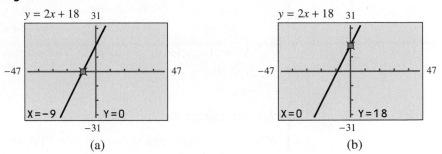

(a)   (b)

In general, because any point on the x-axis has a y-coordinate of 0, the second coordinate of the x-intercept is always 0. Likewise, because any point on the y-axis has an x-coordinate of 0, the first coordinate of the y-intercept is always 0. These facts give us an algebraic method for determining intercepts exactly.

Determining Intercepts Algebraically

For any equation in two variables, the intercepts of the graph of the equation can be determined as follows.

1. To determine the y-intercept, replace x with 0 and solve for y.
2. To determine the x-intercept, replace y with 0 and solve for x.

EXAMPLE 1

Graphing and Algebraic Methods for Determining Intercepts

(a) Use the graphing method to estimate the intercepts of the graph of $y = 2x - 15$.

(b) Determine the intercepts algebraically.

Solution

(a) Figure 4.7 shows the graph of $y = 2x - 15$. The tracing cursor is on the point $(0, -15)$, which is the estimated y-intercept.

When we trace to estimate the x-intercept, we discover that no displayed point has a y-coordinate of 0. The cursor jumps over the x-axis from $(7, -1)$ to $(8, 1)$. If we zoom in on either of those points, we obtain a magnified look at where the line crosses the x-axis. Figure 4.8 shows the result of zooming in. We estimate the x-intercept to be $(7.5, 0)$.

Figure 4.7   **Figure 4.8**

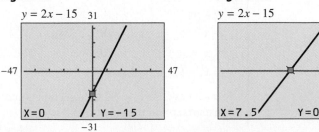

(b) *y-Intercept*   *x-Intercept*

$y = 2x - 15$ $y = 2x - 15$

$y = 2(0) - 15$ Replace x with 0 $0 = 2x - 15$ Replace y with 0
                  and solve for y.   and solve for x.
$y = 0 - 15$ $15 = 2x$

$y = -15$ $7.5 = x$

As we predicted graphically, the *y*-intercept is $(0, -15)$, and the *x*-intercept is $(7.5, 0)$.

EXAMPLE 2

Determining Intercepts Algebraically

Determine the intercepts of $5x + 2y = -20$ algebraically.

Solution

LEARNING TIP

Keep reminding yourself that a *y*-intercept is a point of the *y*-axis, where the first coordinate is always 0. Similarly, an *x*-intercept is a point of the *x*-axis, where the second coordinate is always 0.

x-Intercept   *y-Intercept*

$5x + 2y = -20$ $5x + 2y = -20$

$5x + 2(0) = -20$ Replace y with 0 $5(0) + 2y = -20$ Replace x with 0
                    and solve for x.   and solve for y.
$5x = -20$ $2y = -20$

$x = -4$ $y = -10$

The *x*-intercept is $(-4, 0)$, and the *y*-intercept is $(0, -10)$.

Exploring the Concept

Determining the y-Intercept from the Equation

Figure 4.9 shows the graphs of the following equations.

$y = x + 15$ $y = x + 6$ $y = x$ $y = x - 12$

Figure 4.9

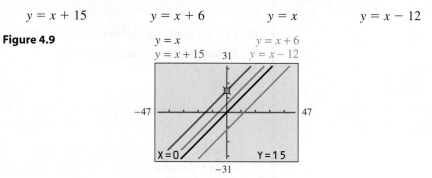

The tracing cursor is on the *y*-intercept of the graph of $y = x + 15$. The *y*-intercepts of the other graphs can also be estimated by tracing.

Equation	y-Intercept	Constant Term
$y = x + 15$	$(0, 15)$	15
$y = x + 6$	$(0, 6)$	6
$y = x + 0$	$(0, 0)$	0
$y = x - 12$	$(0, -12)$	-12

In each case, observe that the second coordinate of the *y*-intercept is the same as the constant term in the equation.

You may have observed that whenever we solve a linear equation in two variables for y, we obtain an equation in the form $y = ax + b$, where a and b are real numbers. The term ax is the *variable term*, and the term b is the *constant term*.

We can determine the y-intercept of the graph of any linear equation $y = ax + b$.

$$y = ax + b$$

$$y = a(0) + b \qquad \text{Replace } x \text{ with 0.}$$

$$y = b$$

Thus the y-intercept is $(0, b)$. This verifies our observation that the second coordinate of the y-intercept is the same as the constant term of the equation.

Determining the y-Intercept from the Equation

If a linear equation in two variables is given, we can determine the y-intercept of the graph as follows.

1. Write the equation in the form $y = ax + b$.
2. The y-intercept of the graph is $(0, b)$.

EXAMPLE **3**

Think About It

Suppose that a linear equation is written in the form $\dfrac{x}{a} + \dfrac{y}{b} = 1$, where a and b are not zero. What are the intercepts? Write the equation $2x + 3y = 12$ in a form that makes the intercepts obvious.

Determining the y-Intercept from the Equation

Determine the y-intercept of the graph of each equation.

(a) $y = 4x - 3$ \qquad (b) $3x + 2y = 10$ \qquad (c) $4x - y = 7$

Solution

(a) $y = 4x - 3$ \qquad\qquad The equation is in the form $y = ax + b$ with $b = -3$.

   The y-intercept is $(0, -3)$.

(b) Write the equation in the form $y = ax + b$.

$$3x + 2y = 10$$
$$2y = -3x + 10 \qquad \text{Subtract } 3x \text{ from both sides.}$$
$$y = -\frac{3}{2}x + 5 \qquad \text{Divide both sides (each term) by 2.}$$

   The y-intercept is $(0, 5)$. \qquad $b = 5$

(c) $4x - y = 7$
$$-1y = -4x + 7 \qquad \text{Subtract } 4x \text{ from both sides.}$$
$$y = 4x - 7 \qquad \text{Divide both sides by } -1.$$

   The y-intercept is $(0, -7)$. \qquad $b = -7$

Special Cases

For the linear equation in two variables $Ax + By = C$, A or B (but not both) can be 0. Thus one of the variables could be missing.

Equation	*Standard Form*	
$y = 7$	$0x + y = 7$	$A = 0, B = 1$
$x = -6$	$x + 0y = -6$	$A = 1, B = 0$

We refer to equations with one variable missing as special cases.

First we consider equations in the form $y = $ constant.

Exploring the Concept

Graphs of Equations in the Form $y = $ Constant

The equation $y = 17$ means that every point of the graph has a y-coordinate of 17. Figure 4.10 shows the graphs of $y = 17$, $y = -8$, and $y = -22$.

Figure 4.10

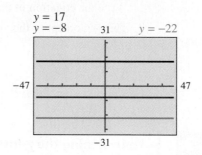

Observe that each line is a horizontal line; that is, each line is parallel to the x-axis. Tracing each graph reveals that the y-coordinate of every point is a constant that indicates how far above or below the x-axis the line is drawn. Finally, we observe that the y-intercepts are $(0, 17)$, $(0, -8)$, and $(0, -22)$, as indicated by the constant term in each equation. None of the graphs has an x-intercept.

Now we consider equations in the form $x = $ constant.

Exploring the Concept

Graphs of Equations in the Form $x = $ Constant

Just as an equation such as $y = 5$ indicates that the y-coordinate is always 5, an equation such as $x = 7$ indicates that the x-coordinate of each point of the graph is 7. The following are a few solutions of the equation $x = 7$.

$$(7, -3), (7, 0), (7, 4), (7, 9)$$

Similarly, the following are a few solutions of the equation $x = -6$.

$$(-6, -1), (-6, 3), (-6, 7), (-6, 12)$$

Figure 4.11

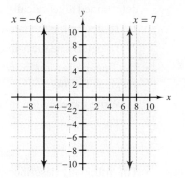

In Figure 4.11, we see that each line is a vertical line; that is, each line is parallel to the y-axis. The constant indicates how far to the left or right of the y-axis the line is drawn. The x-intercepts are $(7, 0)$ and $(-6, 0)$. Neither line has a y-intercept.

Note: The graph of $y = $ constant (horizontal line) can be produced on a calculator by entering the constant on the Y screen. However, the graph of $x = $ constant (vertical line) cannot be produced in this way because the equation cannot be solved for y.

The graph of $y = 0$ is the x-axis, so every point is an x-intercept. The graph of $x = 0$ is the y-axis, so every point is a y-intercept. In the following summary, we exclude the special cases in which the constant is 0.

Summary of Special Cases

Equation	Graph	Intercepts
$y = b$	horizontal	y-intercept: $(0, b)$
		x-intercept: If $b \neq 0$, none
$x = c$	vertical	x-intercept: $(c, 0)$
		y-intercept: If $c \neq 0$, none

Quick Reference 4.2

Intercepts
- A point where a graph intersects the x-axis is called an **x-intercept,** and a point where a graph intersects the y-axis is called a **y-intercept.**
- We can estimate intercepts graphically as follows.
  1. Trace the graph to the point whose y-coordinate is 0 to estimate the x-coordinate of the x-intercept.
  2. Trace the graph to the point whose x-coordinate is 0 to estimate the y-coordinate of the y-intercept.

- For any equation in two variables, the intercepts of the graph of the equation can be determined algebraically as follows.
    1. To determine the y-intercept, replace x with 0 and solve for y.
    2. To determine the x-intercept, replace y with 0 and solve for x.
- If a linear equation in two variables is given, we can determine the y-intercept of the graph as follows.
    1. Write the equation in the form $y = ax + b$.
    2. The y-intercept of the graph is $(0, b)$.

Special Cases
- The following describes the special cases of linear equations in the form $y = b$ and $x = c$.

Equation	Graph	Intercepts
$y = b$	horizontal	y-intercept: $(0, b)$
		x-intercept: If $b \neq 0$, none
$x = c$	vertical	x-intercept: $(c, 0)$
		y-intercept: If $c \neq 0$, none

Speaking the Language 4.2

1. If the graph of an equation intersects the x-axis, then the point of intersection is called a(n) ▨▨▨▨▨ .

2. To trace a graph to its y-intercept, we move the cursor to the point whose ▨▨▨▨▨ is 0.

3. The ▨▨▨▨▨ coordinate of the y-intercept of the graph of $y = ax + b$ is the ▨▨▨▨▨ term b.

4. A graph of a linear equation with no y-intercept is a(n) ▨▨▨▨ line, whereas the graph of an equation of the form $y = c$ is a(n) ▨▨▨▨ line.

Exercises 4.2

Concepts and Skills

1. What do we mean by the x- and y-intercepts of a graph?

2. Describe how to
    (a) estimate intercepts graphically.
    (b) determine intercepts algebraically.

In Exercises 3–10, use the graphing method to estimate the intercepts of the graph of the given equation.

3. $y = 3x + 21$

4. $y = 14 - 2x$

5. $y = -\dfrac{3}{2}x + 15$

6. $y = -\dfrac{3}{7}x + 12$

7. $x = -2y - 18$

8. $x = 2y + 22$

9. $5x + 2y = 40$

10. $4x - 3y = 48$

11. Given a linear equation in two variables, describe two algebraic ways in which the y-intercept of the graph can be determined.

12. How is it possible that the graph of a linear equation in two variables could have just one intercept?

In Exercises 13–20, determine the intercepts algebraically.

13. $3x = 5y$

14. $2x = y + 4$

15. $\dfrac{x}{2} - \dfrac{y}{4} = 1$

16. $\dfrac{1}{3}x - \dfrac{1}{2}y = 3$

17. $3x - 7y - 21 = 0$

18. $y - 4x = 7$

19. $8x + 3y = 12$

20. $5x - 4y = 10$

In Exercises 21–30, write the given equation in the form $y = ax + b$. Then determine the y-intercept.

21. $2y + 3x = 18$

22. $2x - 7y = 21$

23. $x - 3y - 18 = 0$

24. $0 = 3y - x - 15$

25. $3y - 11 = 4$

26. $y - x = 0$

27. $y - 4 = 2(x + 3)$

28. $y + 4 = -\dfrac{1}{2}(x - 6)$

29. $\dfrac{x}{4} + \dfrac{y}{3} = 1$

30. $\dfrac{3}{2}x - \dfrac{1}{4}y = 1$

In Exercises 31–34, select the equation(s) whose graph has the given intercept.

31. The y-intercept is $(0, -2)$.

   (i) $3x - y = 2$ (ii) $3y + 2x = -4$

   (iii) $4x = 3(y + 2)$

32. The y-intercept is $(0, 1)$.

   (i) $2x = 5(1 - y)$ (ii) $x + y + 1 = 0$

   (iii) $y - x = 1$

33. The x-intercept is $(5, 0)$.

   (i) $-(x - 5) = 2y$

   (ii) $3y - x = -5$

   (iii) $x + y + 5 = 0$

34. The x-intercept is $(-3, 0)$.

   (i) $x - y + 3 = 0$

   (ii) $x + y + 3 = 0$

   (iii) $-x + y - 3 = 0$

In Exercises 35–38, select the equation whose y-intercept is not the same as the y-intercept for the other two equations.

35. (i) $3x - y = -5$ (ii) $x + y = 5$

   (iii) $2x - y = 5$

36. (i) $y = 2(x + 1)$ (ii) $y - x = -2$

   (iii) $4x + y + 2 = 0$

37. (i) $y = 3(x - 2)$ (ii) $y = 2(x + 3)$

   (iii) $x - y = 6$

38. (i) $y - 2x = 0$ (ii) $7x - y = 0$

   (iii) $x = y - 1$

39. If b is a nonzero constant, describe the graph of $y = b$, and determine the intercepts, if any.

40. If c is a nonzero constant, describe the graph of $x = c$, and determine the intercepts, if any.

In Exercises 41–44, complete the table so that the pairs (x, y) are solutions of the given equation. Then sketch the graph of the equation.

41. $y = -2$

x	4	0	-7
y			

42. $x = -5$

x			
y	2	-1	5

43. $x + 3 = 5$

x			
y	4	0	-7

44. $y - 4 = 0$

x	2	-1	5
y			

In Exercises 45–52, determine three solutions of the equation and sketch the graph. If possible, produce the graph of the equation on your calculator and compare your sketch with the calculator graph.

45. $y = -4$

46. $y = 1$

47. $x = 3$

48. $x = -5$

49. $x + 5 = -2$

50. $y + 3 = 0$

51. $6 + y = 4$

52. $7 = x - 1$

In Exercises 53–60, without graphing, determine whether the graph of the equation is horizontal, vertical, or neither.

53. $3x - 12 = 0$

54. $3y = 0$

55. $y = 5x$

56. $x = 2y$

57. $-3y = 6$

58. $x + 4 = 1$

59. $y = 7 - 0x$

60. $2x - y = 0$

In Exercises 61–64, write two ordered pairs whose coordinates satisfy the given conditions. Then model the conditions with an equation. If possible, produce the graph of the equation and trace it to verify that your two ordered pairs are represented by points of the graph.

61. The x-coordinate is always 7.

62. The y-coordinate is always 2.

63. The difference between twice the y-coordinate and 6 is always 8.

64. Three more than the x-coordinate is always 12.

In Exercises 65–68, write an equation of the given graph.

65.

66.

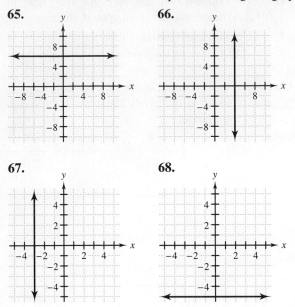

67.

68.

In Exercises 69–72, the graph of a linear equation in two variables is described. Write an equation.

69. A line contains $(4, -2)$ and is parallel to the y-axis.

70. A line contains $(-1, -5)$ and is parallel to the x-axis.

71. A line contains $(-3, 7)$ and is perpendicular to the y-axis.

72. A line contains $(2, 6)$ and is perpendicular to the x-axis.

Real-Life Applications

In Exercises 73–76, use the indicated assignments of variables and write an equation in the form $y = ax + b$. Then use the integer setting to produce the graph of the equation.

73. A window washer has 22 windows to wash. Let x represent the number of windows that have been washed and y represent the number of windows that have not yet been washed.

(a) Trace the graph to estimate the intercepts.

(b) How can the intercepts be interpreted for this problem?

74. A boy is 3 years younger than his sister. Let x represent the boy's age and y represent his sister's age.

(a) Trace the graph to estimate the intercepts.

(b) For this problem only one of the intercepts is meaningful. Why? Interpret the meaningful intercept.

75. A tool rental store rents a high-powered drill for $5 plus $1.25 per hour. Let x represent the number of hours for which the drill was rented and y represent the rental charge.

(a) Trace the graph to estimate the cost of renting the drill for 6 hours.

(b) Trace the graph to estimate the y-intercept and then interpret it.

76. A person is reading a 20-page report at a rate of 1 page per minute. Let x represent the number of minutes the person has been reading and y represent the number of pages that remain to be read.

(a) Trace the graph to estimate the intercepts.

(b) How can the intercepts be interpreted for this problem?

Data Analysis: *Pager Subscribers*

As the capability of pagers has grown to include receiving e-mail, news, and other information from the Internet, their use has increased. The bar graph shows the number (in millions) of pager subscribers. (Source: Strategis Group for Personal Communication Assoc.)

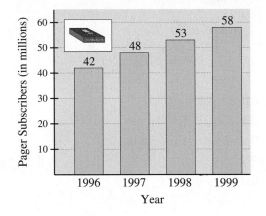

77. The data can be modeled by the linear equation $y = 5.3x + 37$, where x is the number of years after 1995. Without graphing or substitution, determine the y-intercept of the graph of the model equation. What does the point represent?

78. Determine the x-intercept for the model in Exercise 77. What does the point represent?

79. Another model for the data is the linear equation $y = 5.3x + 10.5$, where x is the number of years after 1990. Without graphing or substitution, determine the y-intercept of the graph of the model equation. What does the point represent?

80. Which point of the graph of the model in Exercise 79 represents the same information as the *y*-intercept that you determined in Exercise 77?

Challenge

In Exercises 81–84, give an example of an equation whose graph meets the given conditions.

81. The graph does not contain any points in Quadrant I or Quadrant IV.

82. The graph does not contain any points in Quadrant I or Quadrant II.

83. The *y*-intercept of the graph is $(0, -4)$.

84. The *x*-intercept of the graph is $(7, 0)$, and there is no *y*-intercept.

In Exercises 85 and 86, determine the intercepts of the graph.

85. $3y - x - 6 = 2 - y + x$

86. $2(x - y) + y = 3(4 - x + y)$

4.3 | Slope of a Line

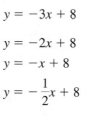

Rise and Run • The Slope Formula • Special Cases

Rise and Run

In Section 4.2 we found that when a linear equation is solved for *y*, the result is in the form $y = ax + b$. Moreover, we can look at the constant term *b* to learn that the *y*-intercept of the graph is $(0, b)$. Now we investigate the effect of the number *a* on the graph.

Exploring the Concept

The Effect of *a* on the Graph of $y = ax + b$

Figure 4.12 shows the graphs of the following equations in group 1, and Figure 4.13 shows the graphs of the equations in group 2.

Group 1

$$y = \frac{1}{2}x + 8$$

$$y = x + 8$$

$$y = 2x + 8$$

$$y = 3x + 8$$

Group 2

$$y = -3x + 8$$

$$y = -2x + 8$$

$$y = -x + 8$$

$$y = -\frac{1}{2}x + 8$$

Figure 4.12

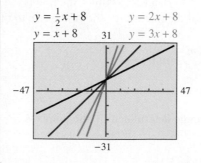

Figure 4.13

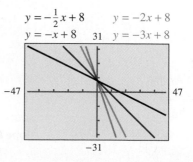

The constant term of every equation is 8, and so all of their graphs have the same y-intercept, $(0, 8)$.

The lines in group 1 rise from left to right, and the coefficient of x is positive in each case. The lines in group 2 fall from left to right, and the coefficient of x is negative in each case. Note that in both groups the lines become steeper as the absolute value of the coefficient increases.

These results suggest the following general statements regarding the graph of $y = ax + b$.

1. The sign of a determines whether the line rises or falls from left to right.

2. The absolute value of a determines the steepness of the line.

The word *steepness* is too vague for our purposes. To be more precise, we need a numerical measure of a line's steepness. We call this numerical measure the **slope** of the line.

Consider the lines L_1 and L_2 in Figure 4.14. Intuitively, we say that L_2 is steeper than L_1. Now suppose that we wished to move from point A to point B. One path is from A to P (a change called the **rise**) and then from P to B (a change called the **run**). For L_1 the rise is considerably less than the run, whereas for L_2 the rise is considerably greater than the run. This observation suggests that the slope of a line can be determined by comparing the rise of the line to the run. We can use a ratio to make such comparisons.

Figure 4.14

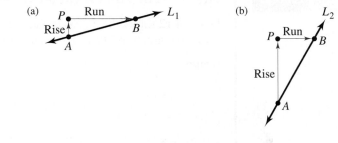

Definition of the Slope of a Line

Given a line and two points A and B of the line, the **rise** of the line is the vertical change from A to B, and the **run** of the line is the horizontal change from A to B. The **slope** m of the line is the ratio of the rise to the run.

$$\text{Slope} = m = \frac{\text{rise}}{\text{run}}$$

Note: The triangle determined by the points A, P, and B in Figure 4.14 is called the **slope triangle**.

The definition of slope does not indicate what two points to use for determining the rise and the run. Example 1 clarifies this matter.

EXAMPLE 1

Figure 4.15

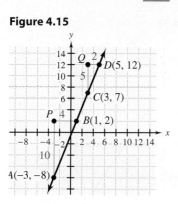

Using Points of a Line to Determine the Slope

Referring to Figure 4.15, determine the slope of the line by using

(a) points A and B.

(b) points C and D.

Solution

(a) Using the slope triangle APB, we see that the rise (AP) is 10 units and that the run (PB) is 4 units.

$$m = \frac{\text{rise}}{\text{run}} = \frac{10}{4} = \frac{5}{2}$$

(b) Using the slope triangle CQD, we see that the rise (CQ) is 5 units and that the run (QD) is 2 units.

$$m = \frac{\text{rise}}{\text{run}} = \frac{5}{2}$$

The slope is the same in parts (a) and (b) of Example 1. This suggests that the slope of a line is the same regardless of the points that we use to determine it. The conjecture can be proved. Any two points of a line can be used to determine the slope of that line.

As Example 2 shows, we can choose either point as the starting point.

EXAMPLE 2

Determining the Slope from a Graph

Determine the slope of the given line in Figure 4.16.

Figure 4.16

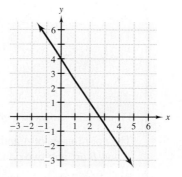

Solution

Choose two points of the line and construct a slope triangle. Figure 4.17 shows the points $A(2, 1)$ and $B(4, -2)$, along with the slope triangle APB. We can move down from A to P (rise $= -3$) and then right from P to B (run $= +2$). Then the slope is

$$\frac{-3}{2} = -\frac{3}{2}.$$

Figure 4.17   **Figure 4.18**

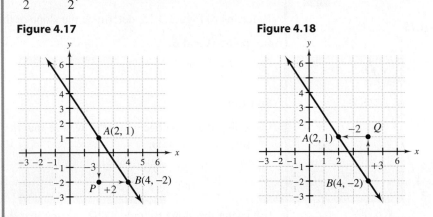

In Figure 4.18, we use the same two points A and B, along with the slope triangle AQB. We can move up from B to Q (rise $= +3$) and then move left from Q to A (run $= -2$). Then the slope is $\dfrac{3}{-2} = -\dfrac{3}{2}$. No matter which point we use as the starting point, the slope is the same.

The Slope Formula

In Example 2, we determined the rise and the run by counting units along the slope triangles APB and AQB. We used positive signs for *up* and *right* and negative signs for *down* and *left*.

Exploring the Concept

A Formula for Slope

Observe that the rise of the line in Example 2 can also be determined by simply subtracting the y-coordinates of A and B. Likewise, the run can be determined by subtracting the x-coordinates of points A and B. The following shows two ways in which we can use this method to calculate the slope of the line that contains $A(2, 1)$ and $B(4, -2)$.

Rise	*Run*	*Slope*	
$1 - (-2) = 3$	$2 - 4 = -2$	$\dfrac{3}{-2} = -\dfrac{3}{2}$	Subtract the coordinates of B from those of A.
$-2 - 1 = -3$	$4 - 2 = 2$	$\dfrac{-3}{2} = -\dfrac{3}{2}$	Subtract the coordinates of A from those of B.

Figure 4.19 shows two general points A and B of a line, along with the slope triangle.

The coordinates of A and B are written in general form with *subscripts* to indicate that A and B are any two points of the line. We read x_1 as "x sub 1" and y_1 as "y sub 1."

The rise can be found by $y_1 - y_2$, and the run can be found by $x_1 - x_2$.

Figure 4.19

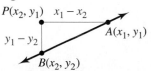

Because slope is the ratio of rise to run, we can write the following formula for the slope of a line.

Think About It

(a) Suppose a line has points only in Quadrants I and II. What is the slope of the line?
(b) What are the possible slopes of a line with no point in Quadrant I?

The Slope Formula

The slope m of a line containing the points (x_1, y_1) and (x_2, y_2) is given by

$$m = \frac{\text{rise}}{\text{run}} = \frac{y_1 - y_2}{x_1 - x_2} = \frac{y_2 - y_1}{x_2 - x_1}, \quad x_1 \neq x_2$$

Note: The choice of which point is (x_1, y_1) and which point is (x_2, y_2) is arbitrary. However, we must be consistent when using the Slope Formula. For example, if we use $y_1 - y_2$ to calculate the rise, then we must use $x_1 - x_2$ to calculate the run.

Because (x_1, y_1) and (x_2, y_2) are the coordinates of any two points of a line, the Slope Formula allows us to calculate the slope of the line without reference to a slope triangle.

EXAMPLE   3

Determining Slope with the Slope Formula

Determine the slope of the line containing the given points.

(a) $(-3, 1)$ and $(5, -1)$
(b) $(-1, -2)$ and $(-4, -8)$

Solution

(a) We choose $(-3, 1)$ for (x_1, y_1) and $(5, -1)$ for (x_2, y_2).

$$m = \frac{y_2 - y_1}{x_2 - x_1} = \frac{-1 - 1}{5 - (-3)} = \frac{-1 - 1}{5 + 3} = \frac{-2}{8} = -\frac{1}{4}$$

If we subtract the coordinates in the reverse order, we obtain the same slope.

$$m = \frac{y_1 - y_2}{x_1 - x_2} = \frac{1 - (-1)}{-3 - 5} = \frac{1 + 1}{-3 - 5} = \frac{2}{-8} = -\frac{1}{4}$$

(b) We choose $(-1, -2)$ for (x_1, y_1) and $(-4, -8)$ for (x_2, y_2).

$$m = \frac{y_2 - y_1}{x_2 - x_1} = \frac{-8 - (-2)}{-4 - (-1)} = \frac{-8 + 2}{-4 + 1} = \frac{-6}{-3} = 2$$

EXAMPLE   4

Determining the Slope from an Equation of a Line

What is the slope of the line determined by $y = x - 6$?

Solution

Produce the graph of $y = x - 6$ and trace it to determine two points of the line. Two such points are $(8, 2)$ and $(3, -3)$. Now use the Slope Formula.

$$m = \frac{y_2 - y_1}{x_2 - x_1} = \frac{-3 - 2}{3 - 8} = \frac{-5}{-5} = 1$$

Special Cases

In Section 4.2, we considered the special cases in which one variable is missing from the linear equation. The forms of these equations are $y = $ constant (horizontal line) and $x = $ constant (vertical line). We now consider the slope of such lines.

EXAMPLE **5**

Horizontal and Vertical Lines

Determine the slope of the following lines.

(a) The equation of the line is $y = -5$.

(b) The line contains the points $(4, 0)$ and $(4, 5)$.

Solution

(a) Figure 4.20 shows the graph of $y = -5$. Two points of this horizontal line are $(1, -5)$ and $(4, -5)$. We calculate the slope with the Slope Formula.

$$m = \frac{y_2 - y_1}{x_2 - x_1} = \frac{-5 - (-5)}{4 - 1} = \frac{-5 + 5}{4 - 1} = \frac{0}{3} = 0$$

Because the y-coordinates of a horizontal line are the same, the rise is always 0. Therefore, the slope of any horizontal line is 0.

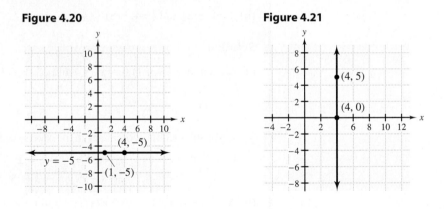

Figure 4.20

Figure 4.21

(b) Figure 4.21 shows the vertical line containing $(4, 0)$ and $(4, 5)$. If we attempt to use the Slope Formula, we obtain 0 in the denominator.

$$m = \frac{y_2 - y_1}{x_2 - x_1} = \frac{5 - 0}{4 - 4} = \frac{5}{0}$$

Because division by 0 is not defined, the Slope Formula does not apply. The slope of a vertical line is not defined. (Recall that the condition $x_1 \neq x_2$ is contained in the statement of the Slope Formula.)

Note: *Zero* and *undefined* do not mean the same thing. A horizontal line has a slope, and it is 0. The slope of a vertical line is undefined.

We can now add to our summary of special cases.

Summary of Special Cases

Equation	Graph	Intercepts	Slope
$y = b$	horizontal	y-intercept: $(0, b)$	0
		x-intercept: If $b \neq 0$, none	
$x = c$	vertical	x-intercept: $(c, 0)$	undefined
		y-intercept: If $c \neq 0$, none	

Quick Reference 4.3

Rise and Run

- For the graph of $y = ax + b$,
  1. if a is positive, the line rises from left to right.
  2. if a is negative, the line falls from left to right.

- Given a line and two points A and B of the line, the **rise** of the line is the vertical change from A to B, and the **run** of the line is the horizontal change from A to B. The **slope** m of the line is the ratio of the rise to the run.

$$\text{Slope} = m = \frac{\text{rise}}{\text{run}}$$

- Any two points of a line can be used to determine the slope of the line. Also, it does not matter which point we choose as the starting point.

- A **slope triangle** is useful in visualizing rise and run.

The Slope Formula

- The slope m of a line containing the points (x_1, y_1) and (x_2, y_2) is given by

$$m = \frac{\text{rise}}{\text{run}} = \frac{y_1 - y_2}{x_1 - x_2} = \frac{y_2 - y_1}{x_2 - x_1}, \quad x_1 \neq x_2$$

Special Cases

- The following is a summary of the special cases.

Equation	Graph	Intercepts	Slope
$y = b$	horizontal	y-intercept: $(0, b)$	0
		x-intercept: If $b \neq 0$, none	
$x = c$	vertical	x-intercept: $(c, 0)$	undefined
		y-intercept: If $c \neq 0$, none	

Speaking the Language 4.3

1. The graph of $y = ax + b$ ▨▨▨▨▨ from left to right if a is negative.
2. If P and Q are points of a line, then the horizontal change from P to Q is called the ▨▨▨▨▨, and the vertical change is called the ▨▨▨▨▨.

3. The ratio of rise to run is called the _____ .

4. The slope of a vertical line is _____ , whereas the slope of a horizontal line is _____ .

Exercises 4.3

Concepts and Skills

1. Consider a very steep, but not vertical, cliff. From the bottom to the top of the cliff, how would you compare the rise with the run?

2. Describe a line that has a

  (a) positive slope.

  (b) negative slope.

In Exercises 3–10, refer to the graph to determine the slope of the line.

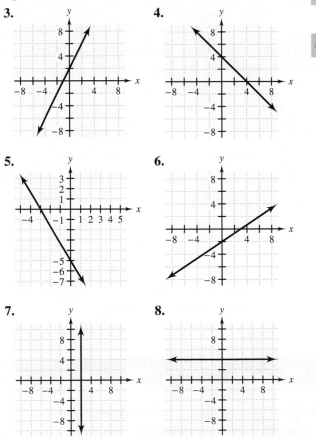

3.

4.

5.

6.

7.

8.

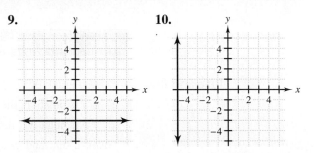

9.

10.

11. Suppose that you know the coordinates of four points of a line. Which points can be used to calculate the slope of the line with the Slope Formula?

12. Suppose that you use $x_2 - x_1$ to represent the run from $A(x_1, y_1)$ to $B(x_2, y_2)$. What expression must you use to represent the rise? Why?

In Exercises 13–36, determine the slope of the line that contains the given points.

13. $(1, 3), (4, 9)$ **14.** $(-2, 1), (2, 13)$

15. $(3, 3), (5, 5)$ **16.** $(0, 0), (6, -6)$

17. $(-2, -4), (0, -6)$ **18.** $(-5, 4), (1, 0)$

19. $(-1, 5), (3, 5)$ **20.** $(2, -4), (-1, -4)$

21. $(-6, 0), (-6, 2)$ **22.** $(1, -7), (1, 3)$

23. $(-5, -3), (7, -1)$

24. $(-3, 5), (6, -7)$

25. $(-8, -2), (-4, -6)$

26. $(-7, -4), (-3, -1)$

27. $(6, -2), (1, -3)$ **28.** $(-2, 3), (-6, 4)$

29. $(2, -5), (17, -11)$ **30.** $(0, 7), (-9, 0)$

31. $(-3, 7), (-7, -3)$ **32.** $(-5, 8), (10, 2)$

33. $(5.4, -7.1), (3.8, 6)$

34. $(0, -0.45), (-9.43, -4.62)$

35. $\left(\frac{5}{6}, \frac{5}{9}\right), \left(\frac{1}{6}, \frac{1}{9}\right)$ **36.** $\left(-\frac{3}{8}, \frac{1}{2}\right), \left(\frac{1}{8}, -\frac{5}{2}\right)$

37. Suppose that the slope of a line containing (x_1, y_1) and (x_2, y_2) is $\frac{3}{5}$. Does this mean that $y_1 - y_2 = 3$ and $x_1 - x_2 = 5$? Explain.

38. We know that the graph of $2x + 3y = 6$ is a line. Describe the steps you would take to determine the slope of the line.

In Exercises 39–46, determine two solutions of the equation and use the points to determine the slope of the line.

39. $x + 2y = 0$

40. $x - y = 7$

41. $2x - 5 = 7$

42. $4 - 3y = 16$

43. $x - 3y - 9 = 0$

44. $3x = 4 - 2y$

45. $y = \dfrac{3x}{4} + 1$

46. $y = 5 - \dfrac{x}{2}$

47. What is the difference between a zero slope and an undefined slope?

48. State the slope of the x-axis and of the y-axis. Explain your answers.

In Exercises 49–58, determine the slope of the line.

49. $y = 4$

50. $x = 2$

51. $x = -7$

52. $y = -1$

53. $x - 4 = 9$

54. $y - 6 = -4$

55. $y + 6 = -1$

56. $x + 5 = 5$

57. $2y - 8 = 0$

58. $\dfrac{1}{2}x + 5 = 7$

In Exercises 59–64, a line is described. Determine whether the slope of the line is positive, negative, 0, or undefined.

59. The line is parallel to the y-axis.

60. The line is parallel to the x-axis.

61. The line is perpendicular to the y-axis.

62. The line is perpendicular to the x-axis.

63. The line contains a point in Quadrant II and a point in Quadrant IV.

64. The line contains a point in Quadrant I and a point in Quadrant III.

In Exercises 65–70, determine the unknown coordinate such that a line with the given slope contains both points.

65. $(0, 3), (1, y)$;   slope $= 2$

66. $(-1, y), (3, -4)$;   slope $= -\frac{3}{2}$

67. $(x, 5), (4, 3)$;   slope $= -\frac{1}{2}$

68. $(2, -1), (x, 1)$;   slope $= 1$

69. $(-2, 3), (5, y)$;   slope $= 0$

70. $(-3, 7), (x, 2)$;   slope is undefined

Modeling with Real Data

71. Home health care has become an alternative to extended hospitalization. The graph shows the growth in the number of home health care patients. (Source: Health Care Financing Administration.)

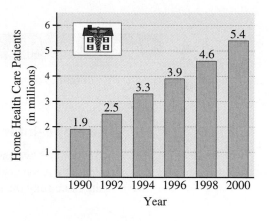

With x representing the number of years since 1990 and y representing the number (in millions) of home health care patients, the equation $20y - 7x = 36$ models the data in the graph.

(a) Determine two solutions of the model equation and use the points to calculate the slope of the line.

(b) Use the data points associated with 1990 and 2000 to calculate the slope.

72. Another model for the data in Exercise 71 is the equation $100y - 35x = 12$, where x represents the number of years since 1985 and y represents the number (in millions) of home health care patients.

(a) Determine the solutions of the equation that correspond to 1995 and 2000.

(b) Use the solutions in part (a) to calculate the slope of the line.

Challenge

In Exercises 73 and 74, determine the slope of the line containing the first two points. Then determine the unknown coordinate such that the third point is also a point of the line.

73. $(-1, -3), (1, -1), (5, y)$

74. $(-3, 2), (1, -6), (x, 6)$

In Exercises 75 and 76, determine k.

75. The slope of the line containing $(1, 3k)$ and $(-1, k)$ is $\frac{1}{2}$.

76. The slope of the line containing $(2k, -1)$ and $(-k, 5)$ is -2.

In Exercises 77 and 78, characterize the possible slope of the described line.

77. The line contains the point $(-2, 3)$, and no part of the graph is in Quadrant III.

78. The line contains the point $(-1, -4)$, and no part of the graph is in Quadrant II.

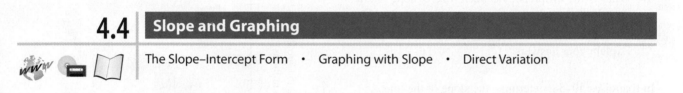

4.4 Slope and Graphing

The Slope–Intercept Form • Graphing with Slope • Direct Variation

The Slope–Intercept Form

In Section 4.3, we saw that the slope of the graph of the linear equation $y = ax + b$ is affected by a, the coefficient of x. Now that we know the Slope Formula, we can investigate more fully the role of a.

Exploring the Concept

Determining Slope from the Equation

In the following table, we consider the slopes of four lines whose equations are in the form $y = ax + b$. Each line has the same y-intercept, $(0, 5)$. A second point of each line is shown in the second column.

Equation	Second Point	Slope	a
$y = 1x + 5$	$(-5, 0)$	$\dfrac{0 - 5}{-5 - 0} = \dfrac{-5}{-5} = 1$	1
$y = 3x + 5$	$(1, 8)$	$\dfrac{8 - 5}{1 - 0} = \dfrac{3}{1} = 3$	3
$y = -2x + 5$	$(3, -1)$	$\dfrac{-1 - 5}{3 - 0} = \dfrac{-6}{3} = -2$	-2
$y = -\dfrac{2}{3}x + 5$	$(6, 1)$	$\dfrac{1 - 5}{6 - 0} = \dfrac{-4}{6} = -\dfrac{2}{3}$	$-\dfrac{2}{3}$

In the third column, we use $(0, 5)$ and the second point to calculate the slope of each line. Finally, we compare the slope to the value of a, which is the coefficient of x. In each case the slope of the line is the same as a.

These results suggest that when a linear equation is written in the form $y = ax + b$, the number a is the slope of the line. Thus the more meaningful form of the equation is $y = mx + b$, where m is the slope of the line. This form is extremely useful because it immediately reveals the slope and the y-intercept of the line.

> **The Slope–Intercept Form of a Linear Equation**
>
> The form $y = mx + b$ is called the **slope–intercept form** of a linear equation in two variables. The slope is m, and the y-intercept is $(0, b)$.

EXAMPLE **1**

Determining the Slope from the Equation

Write each equation in the slope–intercept form. Then determine the slope and the y-intercept of the graph.

(a) $y = x$

(b) $2x - 3y = 9$

Solution

(a) The equation can be written as $y = 1x + 0$, so it is already in the slope–intercept form. The slope is 1, and the y-intercept is $(0, 0)$.

(b) $2x - 3y = 9$ Solve for y.

$\qquad -3y = -2x + 9$ Subtract $2x$ from both sides.

$\qquad y = \dfrac{2}{3}x - 3$ Divide both sides (every term) by -3.

The slope is $\dfrac{2}{3}$, and the y-intercept is $(0, -3)$.

Graphing with Slope

In Section 4.3, we began with a line and used two points of the line to draw a slope triangle. We can reverse this process by using a starting point and a slope triangle to draw the line.

EXAMPLE **2**

Sketching a Line Given a Point and the Slope

Sketch a line that contains $P(2, 4)$ and has a slope of $m = -\dfrac{1}{2}$.

Solution

First, plot $P(2, 4)$. The slope is $-\dfrac{1}{2}$ or $\dfrac{-1}{2}$, so the rise is -1, and the run is 2.

From $P(2, 4)$ move down 1 unit and then right 2 units to arrive at a second point of the line, $(4, 3)$. We can continue the same pattern to locate other points of the line, but only two points are needed to draw the line. (See Fig. 4.22.)

Figure 4.22   **Figure 4.23**

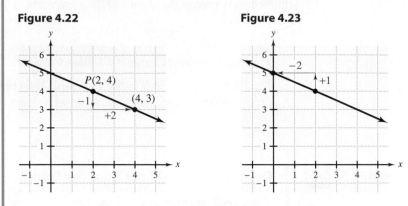

We could also write the slope as $\dfrac{1}{-2}$ and think of the rise as 1 and the run as -2.

From $P(2, 4)$ move up 1 unit and then left 2 units to the point $(0, 5)$. The resulting line, shown in Figure 4.23, is the same as the line in Figure 4.22.

Note: When using a point and the slope to sketch a line, always start the slope triangle at the *given point*, not at the origin.

We can use the procedure shown in Example 2 to sketch a graph when an equation is given. The point and slope needed to sketch the graph can be obtained from the equation written in slope–intercept form.

EXAMPLE **3**

Sketching a Line Given the Equation

Sketch the graph of each equation.

(a) $y = 2x - 5$ (b) $3x + 4y = 8$

Solution

(a) The y-intercept is $(0, -5)$, which is the starting point to plot. The slope is 2, or $\frac{2}{1}$, which means that the rise is 2 and the run is 1. From the y-intercept, locate another point of the line by moving up 2 units and right 1 unit. (See Fig. 4.24.)

Figure 4.24   **Figure 4.25**

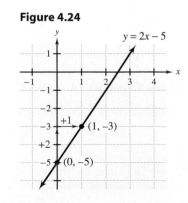
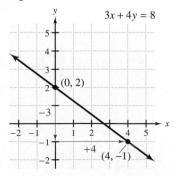

(b) Write the equation in slope–intercept form.

$$3x + 4y = 8$$

$$4y = -3x + 8 \qquad \text{Subtract } 3x \text{ from both sides.}$$

$$y = -\frac{3}{4}x + 2 \qquad \text{Divide both sides by 4.}$$

Plot the y-intercept $(0, 2)$. The slope is $\frac{-3}{4}$, so the rise is -3, and the run is 4.

From the point $(0, 2)$, locate a second point by moving down 3 units and right 4 units, as shown in Figure 4.25.

Think About It

Create a table of values with x in increments of 2 for the equation

$y = \frac{3}{2}x + 5$. Compare the change in x-values and the corresponding change in y-values with the slope. Repeat the experiment for

$y = -\frac{1}{3}x + 1$ with x in increments of 3. What conclusion can you draw?

Sketching a Line with the Slope and y-Intercept

1. Write the equation in the slope–intercept form $y = mx + b$.

2. Plot the y-intercept $(0, b)$.

3. Write the slope m as $\dfrac{\text{rise}}{\text{run}}$.

4. From the y-intercept, move up or down as indicated by the rise and left or right as indicated by the run. Plot the destination point.

5. Sketch the line through the two points.

Direct Variation

Suppose that a person earns \$8 per hour. That person's income y is given by $y = 8x$, where x is the number of hours worked. As x increases, so does y. We say that y **varies directly** with x.

Definition of Direct Variation

The value of y **varies directly** with the value of x if there is a constant k such that $y = kx$. The constant k is called the **constant of variation.**

Sometimes we say that the ratio of y to x is constant, and if $x \neq 0$, we write $\dfrac{y}{x} = k$. In this form, y is **proportional** to x, and the constant k is called the **constant of proportionality.**

The graph of the direct variation $y = kx$ is a line with a slope of k, the constant of variation, and a y-intercept of $(0, 0)$. In direct variation problems, we often begin by determining k.

EXAMPLE 4   **Cost Varies Directly with Quantity Purchased**

If 2 pounds of grapes cost \$2.50, what is the cost of 5 pounds?

Solution

The cost y varies directly with the number x of pounds purchased.

$y = kx$	Direct variation
$2.50 = k \cdot 2$	Two pounds cost $2.50.
$1.25 = k$	Divide both sides by 2 to solve for k.

The constant of variation is 1.25. Note that this is the unit price of a pound of grapes.

$y = 1.25x$	
$y = 1.25(5)$	Determine y when $x = 5$.
$y = 6.25$	

The cost of 5 pounds of grapes is $6.25.

Quick Reference 4.4

The Slope–Intercept Form

• The form $y = mx + b$ is called the **slope–intercept form** of a linear equation in two variables. The slope is m and the y-intercept is $(0, b)$.

Graphing with Slope

• To sketch a line from a given equation, follow these steps.

1. Write the equation in the slope–intercept form $y = mx + b$.

2. Plot the y-intercept $(0, b)$.

3. Write the slope m as $\dfrac{\text{rise}}{\text{run}}$.

4. From the y-intercept, move up or down as indicated by the rise and left or right as indicated by the run. Plot the destination point.

5. Sketch the line through the two points.

Direct Variation

• The value of y **varies directly** with the value of x if there is a constant k such that $y = kx$. The constant k is called the **constant of variation.**

• In the form $\dfrac{y}{x} = k$, $x \neq 0$, y is **proportional** to x, and the constant k is called the **constant of proportionality.**

• The graph of the direct variation $y = kx$ is a line with a slope of k, the constant of variation, and a y-intercept of $(0, 0)$.

Speaking the Language 4.4

1. When we solve $2x + 3y = 6$ for y, the resulting form of the equation is called the _____ form.

2. From the equation $y = -3x + 4$, we can see that the ________ is -3.
3. To sketch the graph of a linear equation, if we know only one point, then we also need to know the ________.
4. If x and y are related by $y = 3x$, then we say that y ________ with x.

Exercises 4.4

Concepts and Skills

1. When a linear equation is written in the slope–intercept form, what information is revealed?
2. Suppose that the graph of a linear equation in the slope–intercept form falls from left to right. What do you know about the coefficient of x?

In Exercises 3–16, write the equation in the slope–intercept form. Determine the slope and y-intercept of the graph.

3. $3x + y = 4$
4. $y + x - 4 = 0$
5. $y - 4x = 0$
6. $3x + y = 0$
7. $2x - y = 4$
8. $7 - y = x$
9. $2y + 10 = 0$
10. $y - 6 = 2$
11. $2x + 3y - 12 = 0$
12. $3x - 5y = 10$
13. $x - \frac{2}{3}y = 2$
14. $2x + \frac{5}{2}y - 10 = 0$
15. $\frac{1}{2}x + \frac{2}{3}y = \frac{5}{3}$
16. $\frac{3}{4}y + x = \frac{5}{2}$

In Exercises 17–20, identify which equations have graphs with the given slope.

17. $m = -\frac{3}{2}$
    (i) $2y + 3x = 0$ (ii) $3x = 10 - 2y$
    (iii) $3(y + 1) = 2x$
18. $m = 2$
    (i) $y - 2x = 1$ (ii) $2y - x = 12$
    (iii) $2x = y + 5$
19. $m = 0$
    (i) $y - 3x = 2$ (ii) $y - 3 = 2$
    (iii) $y = 3$
20. $m = -1.5$
    (i) $3x + 2y = 1$ (ii) $3x - 2y = -1$
    (iii) $6x = -4y$

21. If the slope of a line is -2, describe two ways to draw a slope triangle from a given point of the line.
22. Under what conditions would it be impossible to use a slope triangle to sketch the graph of a linear equation?

In Exercises 23–30, sketch the line with the given y-intercept and slope.

23. $(0, 0)$, $m = -2$
24. $(0, 0)$, $m = \frac{3}{2}$
25. $(0, 4)$, $m = -\frac{5}{2}$
26. $(0, -7)$, $m = 3$
27. $(0, -2)$, $m = 1$
28. $(0, 5)$, $m = \frac{7}{3}$
29. $(0, -5)$, $m = \frac{6}{5}$
30. $(0, 8)$, $m = -4$

In Exercises 31–40, sketch a line that contains the given point and has the given slope.

31. $(-5, 0)$, $m = 1$
32. $(4, 0)$, $m = -1$
33. $(2, -7)$, $m = 0$
34. $(-1, 3)$, $m = 0$
35. $(-5, -1)$, m undefined
36. $(8, -4)$, m undefined
37. $(2, 1)$, $m = -\frac{5}{4}$
38. $(-3, 2)$, $m = \frac{2}{5}$
39. $(-4, -1)$, $m = 3$
40. $(1, -3)$, $m = -2$

In Exercises 41–46, sketch the pair of lines.

41. Both lines have a slope of 2. One contains the point $(1, -1)$, the other the point $(5, 2)$.
42. Both lines have a slope of $-\frac{2}{3}$. One contains the point $(-6, 4)$, the other the point $(-2, -1)$.
43. Both lines contain the point $(3, 5)$. One has a slope of -2, the other a slope of $-\frac{1}{2}$.

44. Both lines contain the point $(4, -3)$. One has a slope of 3, the other a slope of $\frac{1}{3}$.

45. Both lines contain the point $(0, 4)$. One has a slope of $\frac{2}{3}$, the other a slope of $-\frac{3}{2}$.

46. Both lines contain the point $(-3, 0)$. One has a slope of $\frac{1}{2}$, the other a slope of -2.

In Exercises 47–58, sketch the graph of the given equation. Then produce the graph with your calculator and compare.

47. $y = -\frac{3}{2}x + 5$

48. $y = \frac{4}{3}x - 2$

49. $y = 3x + 4$

50. $y = -2x - 3$

51. $y = -4$

52. $y = 6$

53. $x = 5$

54. $x = -3$

55. $y = -x$

56. $y = x$

57. $y = \frac{x}{2}$

58. $y = -\frac{3x}{5}$

In Exercises 59–72, sketch the graph of the equation.

59. $3x - 4y = 12$

60. $2x - y = 8$

61. $y + 4 = 1$

62. $3 - y = 4$

63. $2x = 3y$

64. $x - 2y = 0$

65. $1 - 2x = 7$

66. $x + 4 = 9$

67. $x - 3y - 9 = 0$

68. $10 = 2y - 3x$

69. $x + 1 = 2y - 9$

70. $x + 5 = y - 6$

71. $\frac{x}{3} + \frac{y}{4} = 1$

72. $2 = x + \frac{y}{3}$

In Exercises 73–76, sketch the described line and state its slope.

73. Perpendicular to the y-axis; contains the point $(4, 2)$

74. Perpendicular to the x-axis; contains the point $(-3, 6)$

75. Parallel to the y-axis; contains the point $(-5, -7)$

76. Parallel to the x-axis; contains the point $(3, -8)$

77. Describe the graph of a direct variation $y = kx$.

78. In the direct variation $y = kx$, what must be true about k if y decreases as x increases? Explain.

In Exercises 79–84, determine whether the given equation represents a direct variation. Assume that x and y are positive numbers.

79. $y = 7x$

80. $xy = 3$

81. $\frac{y}{2} = x$

82. $\frac{y}{x} = 2$

83. $y = 5$

84. $y = 5x$

In Exercises 85–88, assume that y varies directly with x.

85. If y is 8 when x is 4, what is y when x is 6?

86. If y is -10 when x is -20, what is y when x is -12?

87. If y is -2 when x is -3, what is y when x is -18?

88. If y is 10 when x is 6, what is y when x is 24?

🌐 Real-Life Applications

89. If 500 bricks weigh 1175 pounds, what do 1200 bricks weigh?

90. If the length of a rectangle remains constant, then the area A of the rectangle varies directly with its width W. If the area of a rectangle is 54 square feet when the width is 9 feet, what will be the area if the length is the same but the width increases 6 feet?

91. If a recipe for 8 muffins calls for $\frac{1}{2}$ cup of sugar, how much sugar is needed to prepare 12 muffins?

92. If a person on an hourly wage earns $268.80 for a 40-hour work week, how much will the person earn in a 50-hour work week?

📏 Modeling with Real Data

93. The table shows that higher salaries are often associated with a long work week. The linear equation $12x - 25y = 100$ models the data in the table. In the model, x represents income (in thousands of dollars) and y represents the percentage of adults who work 51 hours or more per week.

Income	Percent
< $20,000	5
$20,000–39,999	12
$40,000–59,999	19
$60,000–74,999	22
≥ $75,000	32

(Source: Interep Research.)

(a) Write the model equation in the slope–intercept form.

(b) What is the slope? In what way is the value of the slope consistent with the trend indicated by the data?

94. Refer to the model equation in Exercise 93.

(a) What is the change in percentage for each $5,000 increase in salary?

(b) Use the graph of the model equation to estimate the percentage of those with income of $100,000 who work 51 hours or more.

Data Analysis: *Automobile Technology*

In recent years, the design of automobiles has included an increasing amount of electronic technology. The accompanying bar graph shows the average number of microcomputers in automobiles for selected years. (Source: Motorola, Miller Communications.)

Figure for 95–98

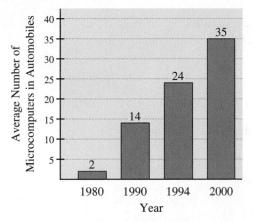

If x is the number of years since 1980 and y is the number of microcomputers in automobiles, then x and y can be related by $166x - 100y = -54$.

95. Write the model equation in slope–intercept form and determine the slope of the graph. Is this an example of direct variation?

96. What does the slope represent?

97. Use the integer setting to produce the graph of the equation.

98. Trace the graph to the point whose x-coordinate is 5 and read the corresponding y-coordinate. What is your interpretation of this point?

Challenge

99. If a linear equation is written in the standard form $Ax + By = C$, $B \neq 0$, show that the slope of the graph is $-\dfrac{A}{B}$.

100. If y varies directly with x and x varies directly with z, show that y varies directly with z.

In Exercises 101–106, what must be true about A, B, and C for the graph of $Ax + By = C$ to satisfy the given description.

101. The line has only one intercept.

102. The line is vertical.

103. The line is horizontal.

104. The line has a positive slope.

105. The line has a negative slope.

106. The line is a graph of a direct variation.

4.5 | Parallel and Perpendicular Lines

Parallel Lines • Perpendicular Lines • Applications

Parallel Lines

In geometry, two lines in the same plane that do not intersect are called **parallel** lines.

Exploring the Concept

Parallel Lines and Slopes

Figure 4.26 shows the graphs of the following equations.

$$y = x - 12 \qquad y = x - 3 \qquad y = x + 7 \qquad y = x + 15$$

Figure 4.26

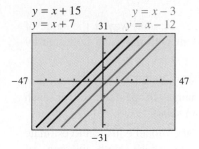

The lines are distinct (different) and appear to be parallel. The y-intercepts of the four lines are different, but all four lines have the same slope. This suggests that lines with the same slope but different y-intercepts are parallel.

Note: If two lines have the same slope and the same y-intercept, then the lines are not distinct, and so they are not parallel.

Parallel Lines and Slopes

Two distinct, nonvertical lines are **parallel** if and only if they have the same slope. Also, any two distinct vertical lines are parallel.

EXAMPLE 1

Determining Whether Lines Are Parallel

In each part, determine whether the lines are parallel.

(a) $y = 2x + 4$ and $y - 2x = 0$

(b) $y = 3 - \dfrac{1}{2}x$ and $x + 2y = 6$

(c) $4y - 3x = 4$ and $3y + 6 = 4x$

(d) $x = 3$ and $x = 8$

Solution

(a)

Equation	$y = mx + b$ Form	y-Intercept	Slope
$y = 2x + 4$	$y = 2x + 4$	$(0, 4)$	2
$y - 2x = 0$	$y = 2x + 0$	$(0, 0)$	2

From the table we see that the lines are distinct because they have different y-intercepts. Because the slopes are the same, the lines are parallel.

(b)

Equation	$y = mx + b$ Form	y-Intercept	Slope
$y = 3 - \dfrac{1}{2}x$	$y = -\dfrac{1}{2}x + 3$	$(0, 3)$	$-\dfrac{1}{2}$
$x + 2y = 6$	$y = -\dfrac{1}{2}x + 3$	$(0, 3)$	$-\dfrac{1}{2}$

Because both the y-intercept and the slope are the same for both lines, the lines are not distinct. Therefore, the lines are not parallel.

(c)

Equation	$y = mx + b$ Form	y-Intercept	Slope
$4y - 3x = 4$	$y = \dfrac{3}{4}x + 1$	$(0, 1)$	$\dfrac{3}{4}$
$3y + 6 = 4x$	$y = \dfrac{4}{3}x - 2$	$(0, -2)$	$\dfrac{4}{3}$

Because their slopes are different, the lines are not parallel.

(d) The lines are distinct vertical lines, so they are parallel.

Perpendicular Lines

Two lines are **perpendicular** if they intersect to form a right angle.

Perpendicular Lines and Slopes

Consider the following pairs of equations.

(i) $y = \dfrac{1}{3}x + 12$ (ii) $y = -\dfrac{5}{2}x + 6$

   $y = -3x - 16$ $y = \dfrac{2}{5}x - 8$

Figures 4.27 and 4.28 show the graphs of the equations.

Figure 4.27

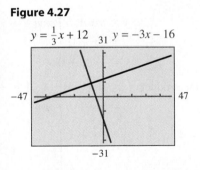

Figure 4.28

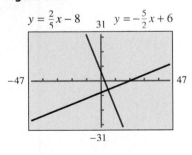

For each pair, the graphs appear to be perpendicular. If we compare the slopes, we note that in each case their product is -1.

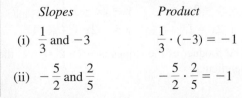

	Slopes	Product
(i)	$\dfrac{1}{3}$ and -3	$\dfrac{1}{3} \cdot (-3) = -1$
(ii)	$-\dfrac{5}{2}$ and $\dfrac{2}{5}$	$-\dfrac{5}{2} \cdot \dfrac{2}{5} = -1$

Our observation is true for any two nonvertical perpendicular lines.

> **Perpendicular Lines and Slopes**
>
> If neither of two lines is vertical, then the lines are **perpendicular** if and only if the product of their slopes is -1. Also, a vertical line and a horizontal line are perpendicular.

LEARNING TIP

If you know the slope of a non-horizontal line, then to determine the slope of a perpendicular line, simply take the opposite of the reciprocal of the slope. For instance, if the slope of a line is $\frac{2}{5}$, then the slope of a perpendicular line is $-\frac{5}{2}$.

Note: If m_1 and m_2 represent the slopes of two perpendicular lines, then the preceding rule states that $m_1 m_2 = -1$. This implies that $m_1 = -\dfrac{1}{m_2}$, if $m_2 \neq 0$, and that $m_2 = -\dfrac{1}{m_1}$, if $m_1 \neq 0$.

EXAMPLE 2

Determining Whether Lines Are Perpendicular

In each part, graph the given pair of equations in the same coordinate plane. Do the lines appear to be perpendicular? Determine algebraically whether the lines are perpendicular.

(a) $y = x$ and $y = -x$

(b) $4y - 3x = 4$ and $y = -\dfrac{4}{3}x + 5$

(c) $y + 1 = 2x$ and $y + 2x = 1$

Solution

(a) *Equation* $y = mx + b$ *Form*   *Slope*

$y = x$ $y = 1x + 0$ 1

$y = -x$ $y = -1x + 0$ -1

Because the product of the slopes is $-1 \cdot 1 = -1$, the two lines are perpendicular.

(b) *Equation* $y = mx + b$ *Form*   *Slope*

$4y - 3x = 4$ $y = \dfrac{3}{4}x + 1$ $\dfrac{3}{4}$

$y = -\dfrac{4}{3}x + 5$ $y = -\dfrac{4}{3}x + 5$ $-\dfrac{4}{3}$

Because the product of the slopes is $-\dfrac{4}{3} \cdot \dfrac{3}{4} = -1$, the two lines are perpendicular.

(c) *Equation* $y = mx + b$ *Form*   *Slope*

$y + 1 = 2x$ $y = 2x - 1$ 2

$y + 2x = 1$ $y = -2x + 1$ -2

The product of the slopes is $-2 \cdot 2 = -4$; therefore, the two lines are not perpendicular.

EXAMPLE 3

Classifying Lines as Parallel or Perpendicular

For each pair of equations, determine whether the lines are parallel, perpendicular, or neither.

(a) $y = -2x + 9$

  $6x + 3y + 15 = 0$

(b) $y = 3x - 5$

  $x + 3y = 12$

(c) $y - 2x = 5$

  $y + 2x = -3$

(d) $x = 5$

  $y = -3$

Think About It

Can every linear equation $Ax + By = C$ be written in slope–intercept form? (*Hint:* Consider $B \neq 0$ and $B = 0$.)

Solution

(a) The slope of the first line is -2.

  Solving the second equation for y, we obtain $y = -2x - 5$. Thus the slope of the second line is also -2. Because the lines are distinct (having different y-intercepts) and the slopes are the same, the lines are parallel.

(b) The slope of the first line is 3.

  Solving the second equation for y, we obtain $y = -\dfrac{1}{3}x + 4$. Thus the slope of the second line is $-\frac{1}{3}$. Because the product of the slopes is $-\frac{1}{3} \cdot 3 = -1$, the lines are perpendicular.

(c) Solve each equation for y.

  $$y = 2x + 5$$
  $$y = -2x - 3$$

  The slopes of the lines are 2 and -2. The lines are neither parallel nor perpendicular.

(d) The graph of $x = 5$ is a vertical line, and the graph of $y = -3$ is a horizontal line. Therefore, the lines are perpendicular.

Applications

Knowledge of the relationships between slopes for parallel and perpendicular lines can be useful in working with certain geometric figures.

Roughly speaking, a **quadrilateral** is a four-sided figure in a plane. A **parallelogram** is a quadrilateral whose opposite sides are parallel. A **rectangle** is a parallelogram with one angle that is a right angle. A **vertex** (plural: vertices) of a geometric figure is a point at which two sides intersect.

EXAMPLE 4

Analyzing Quadrilaterals

The four consecutive vertices of a quadrilateral are $A(-2, 0)$, $B(0, 5)$, $C(10, 1)$, and $D(8, -4)$.

(a) Show that the quadrilateral is a parallelogram.

(b) Show that the parallelogram is a rectangle.

Figure 4.29

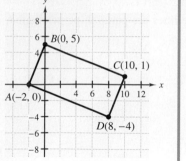

Solution

Figure 4.29 shows the quadrilateral.

(a) *Side*   *Slope*

$\overline{AB}$ $m = \dfrac{5 - 0}{0 - (-2)} = \dfrac{5}{2}$

$\overline{BC}$ $m = \dfrac{1 - 5}{10 - 0} = -\dfrac{2}{5}$

$\overline{CD}$ $m = \dfrac{-4 - 1}{8 - 10} = \dfrac{5}{2}$

$\overline{AD}$ $m = \dfrac{-4 - 0}{8 - (-2)} = -\dfrac{2}{5}$

Because they have equal slopes, the opposite sides of the quadrilateral are parallel. Therefore, the quadrilateral is a parallelogram.

(b) The product of the slopes of any two adjacent sides of the quadrilateral is -1. This means that all of the adjacent sides are perpendicular and form right angles. Therefore, the parallelogram is a rectangle.

Quick Reference 4.5

Parallel Lines

- Two lines in the same plane that do not intersect are called **parallel** lines.
- Two distinct, nonvertical lines are **parallel** if and only if they have the same slope. Also, any two distinct vertical lines are parallel.

Perpendicular Lines

- Two lines are **perpendicular** if they intersect to form a right angle.
- If neither of two lines is vertical, then the lines are **perpendicular** if and only if the product of their slopes is -1. Also, a vertical line and a horizontal line are perpendicular.

Speaking the Language 4.5

1. Lines with the same slope and different y-intercepts are ▨▨▨▨▨ .
2. Lines with the same slope and the same y-intercepts ▨▨▨▨▨ .
3. If the product of the slopes of two lines is -1, then the lines are ▨▨▨▨▨ .
4. A four-sided figure in a plane is called a(n) ▨▨▨▨▨ . If the opposites sides are parallel, the figure is called a(n) ▨▨▨▨▨ . If one pair of opposite sides are parallel, the figure is called a(n) ▨▨▨▨▨ .

Exercises 4.5

Concepts and Skills

In this exercise set, subscripts are used to identify lines and their slopes. For example, L_1 refers to line 1 and m_1 refers to the slope of line 1.

 1. If you know the equations of two lines, describe an easy way to determine whether the lines are parallel.

2. Suppose that two lines in a coordinate plane have the same slope. What additional information do we need before we can conclude that the lines are parallel?

In Exercises 3–10, the slopes of two distinct lines are given. Are the lines parallel, perpendicular, or neither?

3. $m_1 = \frac{4}{2},\quad m_2 = 2$

4. $m_1 = -1.5,\quad m_2 = -\frac{3}{2}$

5. $m_1 = -3,\quad m_2 = \frac{1}{3}$

6. $m_1 = \frac{2}{5},\quad m_2 = -\frac{5}{2}$

7. $m_1 = \frac{2}{3},\quad m_2 = \frac{3}{2}$

8. $m_1 = -3,\quad m_2 = 3$

9. $m_1 = 1,\quad m_2 = -1$

10. $m_1 = \frac{4}{5},\quad m_2 = 0.8$

In Exercises 11–18, the equations of two lines are given. Are the lines parallel, perpendicular, or neither?

11. $y = \frac{1}{4}x - 3$

$y = \frac{1}{4}x + 6$

12. $y = \frac{2}{3}x + 2$

$y = -\frac{2}{3}x - 7$

13. $y = \frac{4}{5}x - 7$

$y = -\frac{5}{4}x + 2$

14. $y = x$

$y = 6 + x$

15. $y = 3x + 7$

$3x + y = 2$

16. $y = 8 - 2x$

$2y - x - 3 = 0$

17. $2y - 3x = 4$

$2x + 3y = 6$

18. $5y - 3x - 5 = 0$

$3x = 5(y + 2)$

In Exercises 19–24, you are given two points of each of two lines. Are the lines perpendicular, parallel, or neither?

19. L_1: $(-4, 3),\quad (-1, 10)$
  L_2: $(1, -6),\quad (4, 1)$

20. L_1: $(5, 8),\quad (7, 4)$
  L_2: $(-1, 0),\quad (2, 6)$

21. L_1: $(2, 7),\quad (-3, 7)$
  L_2: $(-6, 0),\quad (-6, 5)$

22. L_1: $(7, 4),\quad (7, 9)$
  L_2: $(-3, -6),\quad (-3, 2)$

23. L_1: $(-8, 3),\quad (-4, 15)$
  L_2: $(-2, -5),\quad (7, -2)$

24. L_1: $(-1, -3),\quad (5, 0)$
  L_2: $(1, 2),\quad (4, -4)$

25. A horizontal line and a vertical line are perpendicular, but the relation $m_1 m_2 = -1$ is not valid. Why?

26. What can you conclude about lines L_1 and L_2 if $m_1 = -\dfrac{1}{m_2}$, where $m_2 \neq 0$? Why?

In Exercises 27–38, the equation of a line is given. What is the slope of a line that is

(a) parallel to the given line?

(b) perpendicular to the given line?

27. $y = x - 4$

28. $y = 1 - 2x$

29. $y = -\frac{1}{4}x + 3$

30. $y = \frac{x}{3} - 1$

31. $y = 6$

32. $y - -3$

33. $4y - 6x = 1$

34. $2y + x = 5$

35. $x + 5 = 0$

36. $3x = -6$

37. $3x - 5y - 10 = 0$

38. $8x = 9y + 3$

In Exercises 39–48, the equation of L_1 and two points of L_2 are given. Are the lines parallel, perpendicular, or neither?

39. $y = 3x + 2$; $(-1, 3), (1, 9)$

40. $y = \frac{1}{2}x - 4$; $(1, 12), (5, 4)$

41. $y = -2x + 1$; $(-2, 1), (-1, 3)$

42. $3x + 2y = 7$; $(-1, 4), (1, 1)$

43. $5x - y = -2$; $(-4, 1), (1, 0)$

44. $x + 4y = 0$; $(3, 0), (0, -2)$

45. $x - 3 = 5$; $(-2, 7), (4, 7)$

46. $1 - y = 2$; $(3, 4), (-5, 4)$

47. $2y + 1 = 3$; $(-1, 1), (4, 1)$

48. $3(2 - x) = 9$; $(-1, 0), (-1, -8)$

In Exercises 49–52, what can you conclude about m_1?

49. L_1 is parallel to the y-axis.

50. L_1 is perpendicular to the y-axis.

51. L_1 is perpendicular to the x-axis.

52. L_1 is parallel to the x-axis.

In Exercises 53–56, determine the value of k such that the given conditions are met.

53. The line whose equation is $2x + 3 = 7$ is perpendicular to the line whose equation is $y = kx$.

54. The line whose equation is $4 - y = -3$ is perpendicular to the line containing $(2, -7)$ and $(k, 5)$.

55. The line whose equation is $y = kx + 5$ is parallel to the line whose equation is $2x - y = -1$.

56. The line whose equation is $y = kx - 2$ is perpendicular to the line whose equation is $x - 2y = -6$.

Geometric Models

57. Three consecutive vertices of a rectangle are $A(2, 0)$, $B(7, 0)$, and $C(7, 4)$. What must be the coordinates of the fourth vertex?

58. Three consecutive vertices of a parallelogram are $A(0, 0)$, $B(2, 5)$, and $C(7, 5)$. What are the coordinates of the fourth vertex?

59. The consecutive vertices of a square are $A(-3, -2)$, $B(-3, 6)$, $C(5, 6)$, and $D(5, -2)$. Show that the diagonals of the square are perpendicular.

60. The consecutive vertices of a quadrilateral are $A(0, 0)$, $B(0, 5)$, $C(4, 7)$, and $D(4, 2)$. Show that the quadrilateral is a parallelogram.

61. A **trapezoid** is a quadrilateral with exactly one pair of parallel sides. The consecutive vertices of a quadrilateral are $A(11, 2)$, $B(-5, -4)$, $C(-2, 4)$, and $D(6, 7)$. Show that the quadrilateral is a trapezoid.

62. Show that the triangle whose vertices are $A(3, 1)$, $B(5, -15)$, and $C(-3, -3)$ is a right triangle.

Data Analysis: *U.S. Postal Service*

The following table compares the number of pieces of mail (in billions) handled by the U.S. Postal Service in 1995 and 1998. Also shown is the number of employees (in thousands) for those years.

Year	Pieces of Mail (billions)	Number of Employees (thousands)
1995	181	875
1998	197	905

(Source: U.S. Postal Service.)

Let t represent number of years since 1995, p the number of pieces of mail (in billions), and n the number of employees (in thousands). Then the following two equations can be used to model that data.

$$p = 5.7t + 181$$
$$n = 10t + 875$$

63. What is the interpretation of the coefficient of t in the model for the number of pieces of mail?

64. What is the interpretation of the coefficient of t in the model for the number of employees?

65. Calculate the average number of pieces of mail handled per employee in 1995 and 1998.

66. Use the results in Exercise 65 to determine the average yearly increase in the number of pieces of mail that each employee handled. Suppose that you used a linear function to model the number of pieces of mail that an employee handled each year. What would be the slope of the line?

Challenge

In Exercises 67–70, a line is described. Determine its slope and sketch the line.

67. The line contains $(2, -1)$ and is parallel to the line whose equation is $3x + y = 6$.

68. The line contains $(-1, 3)$ and is perpendicular to the line whose equation is $x - 2y = 8$.

69. The line has the same y-intercept as the line whose equation is $x + y - 3 = 0$, and it is perpendicular to the line whose equation is $2x - 3y = 0$.

70. The line has the same y-intercept as the line whose equation is $y = x - 5$, and it is parallel to the line whose equation is $y = 3x + 2$.

71. In the definition of parallel lines, why is it necessary to require that the lines be in the same plane?

72. Suppose that L_1, L_2, and L_3 are three distinct lines in the same plane. If L_1 and L_2 are both perpendicular to L_3, show that L_1 and L_2 are parallel.

4.6 | Equations of Lines

 Slope–Intercept Form • Point–Slope Form • Special Cases • Analyzing Conditions
• Real-Life Applications

In previous sections of this chapter, we began with a given linear equation, and by writing the equation in the form $y = mx + b$, we easily determined the slope and y-intercept of the line and produced the graph.

In this section we begin with given information about the graph, such as the slope, the intercepts, or other points of the line. If we are given enough information so that we can sketch the graph, then we can use the information to write an equation of the line.

Slope–Intercept Form

The slope–intercept form $y = mx + b$ provides a model for writing the equation of a line.

EXAMPLE 1

Writing an Equation of a Line Given the Slope and y-Intercept

Write an equation of the line whose y-intercept is $(0, -4)$ and whose slope is 2.

Solution

The y-intercept is $(0, -4)$, so $b = -4$. Also, the slope is 2, so $m = 2$. Use the slope–intercept form to write the equation.

$$y = mx + b$$
$$y = 2x - 4 \qquad \text{Replace } m \text{ with 2 and } b \text{ with } -4.$$

EXAMPLE 2

Writing an Equation of a Line Given the Slope and a Point

Determine an equation of the line that contains $P(2, 5)$ and whose slope is $-\frac{1}{2}$.

Solution

Because $m = -\frac{1}{2}$, we can begin by writing $y = -\frac{1}{2}x + b$. The pair $(2, 5)$ is a solution to the equation.

$$y = -\frac{1}{2}x + b$$
$$5 = -\frac{1}{2} \cdot 2 + b \qquad \text{Replace } x \text{ with 2 and } y \text{ with 5.}$$
$$5 = -1 + b \qquad \text{Solve for } b.$$
$$b = 6$$

Now complete the equation by replacing b with 6.

$$y = -\frac{1}{2}x + 6$$

Point–Slope Form

Another approach to Example 2 is to use a proportion.

Figure 4.30

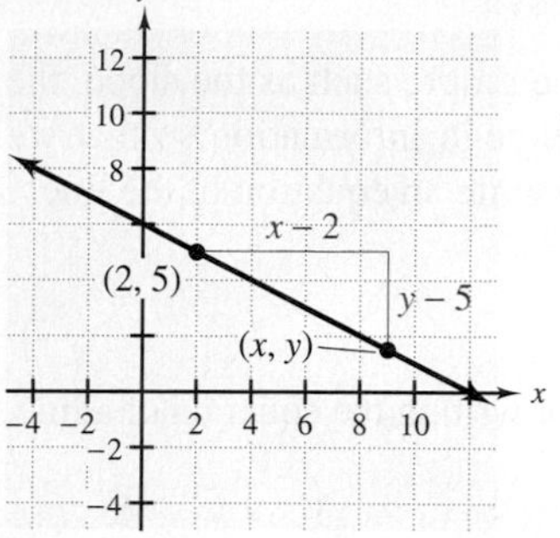

The Point–Slope Form

Knowing the slope and a point, we can sketch the graph. Figure 4.30 shows the graph of the line with slope $-\frac{1}{2}$ and containing the point $(2, 5)$. Let (x, y) be any other point of the line.

From the slope triangle, we see that the slope is the ratio $\dfrac{y - 5}{x - 2}$. Because we know that the slope is $-\frac{1}{2}$, we can write the following equation.

$$\frac{y - 5}{x - 2} = \frac{-1}{2}$$

Note that the equation is in the form of a proportion. To eliminate the fractions, we can cross-multiply.

$$2(y - 5) = -1(x - 2)$$

If we write this equation in the form $y = mx + b$, we can verify that the equation is the same as that obtained in Example 2.

$2y - 10 = -x + 2$ Distributive Property

$2y = -x + 12$ Add 10 to both sides.

$y = -\dfrac{1}{2}x + 6$ Divide both sides by 2.

Figure 4.31

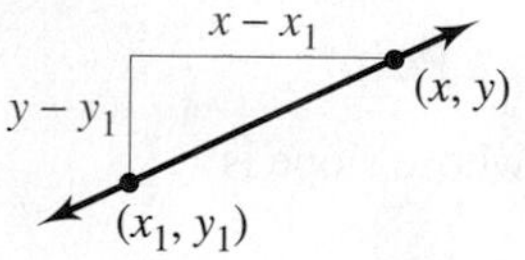

The approach of using a proportion can be generalized. Figure 4.31 shows a line that contains a given point (x_1, y_1) and any other point (x, y). From the slope triangle we see that $m = \dfrac{y - y_1}{x - x_1}$. Cross-multiply to obtain the **point–slope form** of a linear equation.

$$y - y_1 = m(x - x_1)$$

The Point–Slope Form of a Linear Equation

The **point–slope form** of a linear equation is

$$y - y_1 = m(x - x_1)$$

where m is the slope of the line and (x_1, y_1) is a fixed point of the line.

The point–slope form is particularly useful for writing an equation of a line when the slope and one point are given.

EXAMPLE 3   **Writing an Equation of a Line Given the Slope and One Point**

Use the point–slope form to write an equation of a line that has a slope of $-\frac{3}{2}$ and that contains the point $(-4, -3)$.

Solution

$$y - y_1 = m(x - x_1)$$

$$y - (-3) = -\frac{3}{2}[x - (-4)] \qquad \text{Use } m = -\frac{3}{2}, x_1 = -4 \text{ and } y_1 = -3.$$

$$y + 3 = -\frac{3}{2}(x + 4) \qquad \text{Definition of subtraction}$$

$$y + 3 = -\frac{3}{2}x - 6 \qquad \text{Distributive Property}$$

$$y = -\frac{3}{2}x - 9 \qquad \text{Subtract 3 from both sides.}$$

If we know two points of a line, we can easily sketch the graph. Therefore, we can determine the equation of the line.

EXAMPLE 4

Think About It

If three points are not collinear, how many line segments containing two of the points can you draw, and how many corresponding linear equations can you write to model the data approximately?

Writing an Equation of a Line Given Two Points

Write an equation of the line that contains the points $(-2, 5)$ and $(2, -7)$.

Solution

Use the Slope Formula to determine the value of m.

$$m = \frac{y_2 - y_1}{x_2 - x_1} = \frac{-7 - 5}{2 - (-2)} = \frac{-12}{4} = -3$$

Because both of the given points represent solutions, we can use either ordered pair in the point–slope form.

$$y - y_1 = m(x - x_1)$$

$$y - 5 = -3[x - (-2)] \qquad \text{Use } x_1 = -2, y_1 = 5, m = -3.$$

$$y - 5 = -3(x + 2)$$

$$y - 5 = -3x - 6 \qquad \text{Distributive Property}$$

$$y = -3x - 1 \qquad \text{Slope–intercept form}$$

Verify that using $(2, -7)$ results in the same equation.

Special Cases

We can use either model to write the equation of a horizontal line. However, because $m = 0$, it is easier to use the form $y = b$. For vertical lines, m is undefined, so neither model applies. Instead, we use the form $x = c$.

EXAMPLE 5

Special Cases

Write an equation of a line that

(a) is horizontal and contains $(3, -5)$.

(b) contains $(4, 2)$ and $(4, -6)$.

Solution

(a) Because the line is horizontal, we use the model $y = b$. The y-coordinates of all points of the line are the same, and the y-coordinate of the given point is -5. (See Fig. 4.32.) Therefore, an equation of the line is $y = -5$.

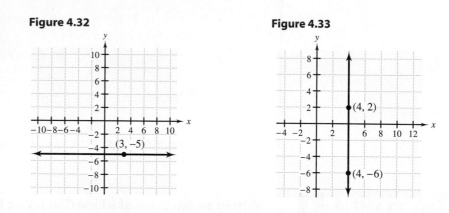

Figure 4.32

Figure 4.33

(b) The first coordinates of both given pairs are the same. This means that the line is vertical, so we use the model $x = c$. The x-coordinate of the given points and all other points of the line is 4. (See Fig. 4.33.) Therefore, an equation of the line is $x = 4$.

Analyzing Conditions

Sometimes one line is described in reference to one or more other lines. If enough information is furnished about the given line(s), we can deduce enough features of the required line to write its equation.

EXAMPLE 6

Figure 4.34

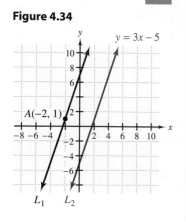

Analyzing Conditions: Parallel Lines

A line L_1 contains the point $A(-2, 1)$ and is parallel to another line L_2 whose equation is $y = 3x - 5$. Write an equation of L_1.

Solution

In Figure 4.34, L_1 is the required line, and L_2 is the given line. The equation of L_2 is $y = 3x - 5$, so $m_2 = 3$. Because the lines are parallel, m_1 must also be 3. Further, $A(-2, 1)$ represents a solution of the required equation. We use the slope–intercept form to write the equation.

$$y = mx + b$$
$$1 = 3(-2) + b \qquad \text{$m = 3, x = -2$, and $y = 1$}$$
$$1 = -6 + b \qquad \text{Solve for b.}$$
$$7 = b$$

The equation of L_1 is $y = 3x + 7$.

EXAMPLE **7**

Analyzing Conditions: Perpendicular Lines

Write an equation of the line L_1 that contains the point $A(1, -5)$ and is perpendicular to the line L_2 that contains the points $B(-1, 1)$ and $C(7, -3)$. (See Fig. 4.35.)

Figure 4.35

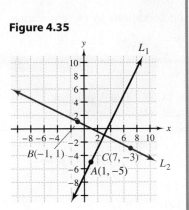

Solution

Use the Slope Formula to determine m_2.

$$m_2 = \frac{y_2 - y_1}{x_2 - x_1} = \frac{-3 - 1}{7 - (-1)} = \frac{-3 - 1}{7 + 1} = \frac{-4}{8} = -\frac{1}{2}$$

Because L_1 is perpendicular to L_2,

$$m_1 = -\frac{1}{m_2} = \frac{-1}{-\dfrac{1}{2}} = 2$$

Now we use the point–slope form with point $A(1, -5)$.

$$
\begin{aligned}
y - y_1 &= m(x - x_1) & \\
y - (-5) &= 2(x - 1) & m = 2, x_1 = 1, y_1 = -5 \\
y + 5 &= 2x - 2 & \text{Distributive Property} \\
y &= 2x - 7 & \text{Subtract 5 from both sides.}
\end{aligned}
$$

The equation of L_1 is $y = 2x - 7$.

EXAMPLE **8**

Analyzing Conditions: Special Cases

Write an equation of the line L_1 that contains $A(-1, 4)$ and is perpendicular to the line L_2 that contains $B(2, 5)$ and $C(2, -3)$. (See Fig. 4.36.)

Figure 4.36

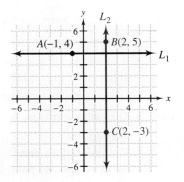

Solution

Because the first coordinates of points B and C are the same, L_2 is a vertical line. Because L_1 is perpendicular to a vertical line, L_1 is a horizontal line, and its equation form is $y = b$. The second coordinate of point A and of every other point of L_1 is 4. Therefore, the equation of L_1 is $y = 4$.

Real-Life Applications

Sometimes we need to write an equation to model the conditions in an application problem.

EXAMPLE 9 | **Modeling with a Linear Equation**

In 1997 ($x = 0$), the population of a small town was 14,000. The population is projected to be 41,000 by 2006 ($x = 9$). Assuming a linear relationship, write an equation that models the population growth during this period.

Solution

Year (x)	Population (y)	(x, y)	
1997 ($x = 0$)	14,000	(0, 14,000)	The y-intercept
2006 ($x = 9$)	41,000	(9, 41,000)	

We use the (x, y) pairs in the table to calculate m.

$$m = \frac{41{,}000 - 14{,}000}{9 - 0} = \frac{27{,}000}{9} = 3000$$

Knowing the slope and the y-intercept, we can easily write the equation.

$$y = mx + b$$
$$y = 3000x + 14{,}000 \qquad m = 3000, b = 14{,}000$$

The population y for any year x is modeled by the equation $y = 3000x + 14{,}000$.

Quick Reference 4.6

Slope–Intercept Form
- The **slope–intercept form** $y = mx + b$ is a model for writing a linear equation.
- If the slope and the y-intercept of a line are given, the model can be used directly.
- If the slope and some point A other than the y-intercept are given, replace m with the given slope and replace x and y with the coordinates of A in order to solve for b.

Point–Slope Form
- The **point–slope form** of a linear equation is $y - y_1 = m(x - x_1)$, where m is the slope of the line and (x_1, y_1) is a fixed point of the line.
- The point–slope form is particularly useful for writing an equation of a line when the slope and one point are given.
- If two points of a line are given, use the Slope Formula to calculate m. Then use m along with either of the given points in the point–slope form.

Special Cases
- The model for the equation of a horizontal line is $y = b$, where the constant is the second coordinate of any point of the line.
- The model for the equation of a vertical line is $x = c$, where the constant is the first coordinate of any point of the line.

Analyzing Conditions   • If a line is described in reference to another given line, we use what we know about the given line to determine the slope (if any) and one point of the required line. Then we use one of the equation forms to write the equation of the required line.

Speaking the Language 4.6

1. The form $y - y_1 = m(x - x_1)$ is called the [blank] form of a linear equation.

2. If you know the slope and y-intercept of a line, the easiest equation form for writing an equation of the line is the [blank] form.

3. Except for special cases, you must know the [blank] in order to write an equation of a line.

4. The form $x = c$ is used to write an equation of a [blank] line, whereas the form $y = b$ is used to write an equation of a [blank] line.

Exercises 4.6

Concepts and Skills

1. What information is needed to use the slope–intercept form to write an equation of a line?

2. If you are given the coordinates of a point of a line, how can you tell whether that point is the y-intercept?

In Exercises 3–8, write an equation of the line with the given slope and y-intercept.

3. $m = -1, \quad (0, 6)$

4. $m = 3, \quad (0, -5)$

5. $m = \dfrac{5}{2}, \quad (0, -2)$

6. $m = -\dfrac{2}{3}x, \quad (0, 4)$

7. $m = 2, \quad \left(0, \dfrac{3}{4}\right)$

8. $m = -4, \quad \left(0, \dfrac{1}{2}\right)$

In Exercises 9–14, write an equation of the line shown in each graph.

9.   **10.**

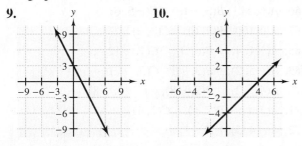

11.   **12.**

13.   **14.**

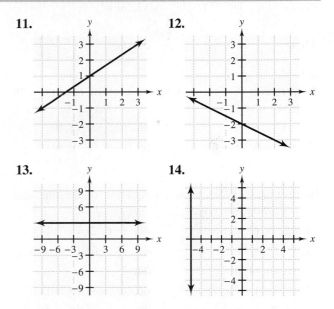

In Exercises 15–24, use the slope–intercept form to write an equation of the line that contains the given point and has the given slope. Write the equation in slope–intercept form.

15. $(-4, 0), \quad m = -3$

16. $(2, 0), \quad m = 5$

17. $(3, -2), \quad m = 1$

18. $(-1, -4), \quad m = -2$

19. $(6, 2)$, $m = -\dfrac{2}{3}$ **20.** $(12, -8)$, $m = -\dfrac{5}{4}$

21. $(-3, -7)$, $m = 0$ **22.** $(-1, 4)$, $m = 0$

23. $\left(-\dfrac{3}{2}, 3\right)$, $m = 2$ **24.** $\left(\dfrac{4}{3}, -2\right)$, $m = -6$

25. What information is needed to use the point–slope form to write an equation of a line?

26. If two points of a line are given, what step is necessary before an equation of the line can be written?

In Exercises 27–34, use the point–slope form to write an equation of the line that contains the given point and has the given slope. Write the equation in the standard form $Ax + By = C$.

27. $m = 2$, $(-1, -3)$

28. $m = -1$, $(4, 1)$

29. $m = -\dfrac{5}{2}$, $(3, -4)$ **30.** $m = \dfrac{7}{3}$, $(-1, 2)$

31. $m = 3$, $\left(\dfrac{1}{2}, \dfrac{2}{3}\right)$ **32.** $m = -2$, $\left(\dfrac{1}{5}, -\dfrac{3}{2}\right)$

33. $m = \dfrac{1}{4}$, $\left(2, \dfrac{3}{5}\right)$ **34.** $m = \dfrac{5}{6}$ $\left(\dfrac{2}{5}, \dfrac{5}{2}\right)$

In Exercises 35–38, write an equation of the line with the given intercepts.

35. $(0, 4), (2, 0)$ **36.** $(0, -3), (1, 0)$

37. $(-2, 0), (0, -5)$ **38.** $(-6, 0), (0, 6)$

In Exercises 39–50, write an equation of the line that contains the given points.

39. $(7, 1), (3, 9)$ **40.** $(-6, 2), (-2, 14)$

41. $(-2, 5), (4, 5)$ **42.** $(3, -6), (-4, -6)$

43. $(0, 6), (4, 0)$ **44.** $(-1, -2), (-6, -8)$

45. $(-8, 2), (-8, 7)$ **46.** $(9, -3), (9, -1)$

47. $(1, 2), (5, 8)$ **48.** $(-7, 5), (5, -4)$

49. $(-5, 0), (0, 7)$ **50.** $(-2, -3), (-5, -4)$

51. Explain why neither the slope–intercept form nor the point–slope form can be used to write an equation of a vertical line.

52. Describe the equation of a line that is

    (a) parallel to the y-axis.

    (b) perpendicular to the y-axis.

In Exercises 53–56, write an equation of the line that contains the given point and satisfies the additional condition.

53. $(4, 7)$;   horizontal

54. $(-6, 4)$;   vertical

55. $(4, 8)$;   no y-intercept

56. $(-3, -2)$;   parallel to the x-axis

57. When a line is described in reference to another given line, what information must we be able to deduce in order to write an equation of the required line?

58. Suppose that a line is perpendicular to a given line. Can we use either the slope–intercept form or the point–slope form to write an equation of the given line? Explain.

In Exercises 59–64, write an equation of a line that contains the given point and is

    (a) parallel to the line whose equation is given.

    (b) perpendicular to the line whose equation is given.

59. $(4, -4)$; $y = x$

60. $(-2, 1)$; $y = -2x + 5$

61. $(1, 3)$; $y = \dfrac{3}{4}x + 6$

62. $(3, -4)$; $y = -\dfrac{1}{2}x - 6$

63. $(4, -5)$; $5x - 3y = 7$

64. $(-1, 3)$; $x + 3y - 14 = 0$

In Exercises 65–72, write an equation of the line that contains point A and is

    (a) perpendicular to the line that contains points B and C.

    (b) parallel to the line that contains points B and C.

65. $A(3, 2)$; $B(-1, 1)$, $C(1, 5)$

66. $A(6, 5)$; $B(3, -6)$, $C(-5, 6)$

67. $A(0, 0)$; $B(2, 6)$, $C(3, 3)$

68. $A(-1, 4)$; $B(6, 3)$, $C(1, -12)$

69. $A(-5, 2)$; $B(3, 7)$, $C(3, -2)$

70. $A(4, -2)$; $B(2, -8)$, $C(-3, -8)$

71. $A(3, 1)$; $B(2, 4)$, $C(1, 4)$

72. $A(-4, 5)$; $B(-1, 0)$, $C(-1, 7)$

In Exercises 73–86, write an equation of the described line.

73. The line is parallel to the line $y = -3x + 8$ and contains the point $(-2, 1)$.

74. The line contains the point $(4, 6)$ and is parallel to the line $y = x$.

75. The line is parallel to the line $y = x - 5$ and has the same y-intercept as the line $y = 7 - 2x$.

76. The line contains the point $(-2, -4)$ and has the same y-intercept as the line $y = 2x - 7$.

77. The line contains the point $(0, 5)$ and is parallel to the line $y = \frac{1}{2}x - 6$.

78. The line contains the point $(0, -8)$ and is parallel to the line $y = 7$.

79. The line contains the point $(-3, 7)$ and is perpendicular to the line $y = -4$.

80. The line contains the point $(6, 1)$ and is perpendicular to the line $y = -3x - 5$.

81. The line contains the point $(1, -4)$ and is parallel to the line $x = 7$.

82. The line contains the point $(5, 3)$ and is perpendicular to the line $x = -2$.

83. The line is perpendicular to $y = -\frac{1}{2}x - 5$ and contains the point $(0, 7)$.

84. The line contains the point $(3, 0)$ and is perpendicular to the line $y = x + 4$.

85. The line contains the point $(-4, 3)$ and has the same y-intercept as the line $y = x$.

86. The line is perpendicular to $2x - y = 5$ and has the x-intercept $(-3, 0)$.

Real-Life Applications

87. On July 1, a person placed $500 in a savings account. On the first day of each subsequent month, she added $120 to the account. Write an equation that describes the total amount that was deposited at the end of any month.

88. A person borrowed $1500 from his father and repaid the loan with no interest by making monthly payments of $130. Write an equation that describes the balance of the loan following any payment.

89. A forest had 8000 trees on the day that logging began. Seven months later, 4859 trees remained. Write an equation that describes the number of trees remaining at the end of any month.

90. The cash value of a person's life insurance policy at age 30 was $800. At age 44, the cash value was $3600. Write an equation that describes the cash value of the policy for any age.

Modeling with Real Data

91. The average income (before taxes) for pediatricians changed from $126,000 in 1994 to $142,000 in 1997. (Source: American Medical Association.)

(a) Write the data as ordered pairs with the number of years since 1994 as the first coordinate and the income (in thousands of dollars) as the second coordinate.

(b) What are the slope and y-intercept of a line containing the two points?

(c) Write a linear equation to model the data.

92. The cost of a ticket to the Super Bowl increased from $100 in 1988 to $325 in 2000. (Source: USA Today.)

(a) Let x represent the number of years after 1980. Write ordered pairs to model the data.

(b) Use the pairs in part (a) to write a linear equation to model the data.

(c) According to the model, by how much will the price of a ticket change from 2000 to 2005?

Data Analysis: *Internet Music Downloading*

In one month in 2000, more than 4 million people downloaded music from the Internet. These downloads are used to promote CD sales. Online music sales rose from $37 million in 1997 to $327 million in 1999. (Source: Jupiter Communications.)

93. Suppose that x represents the number of years since 1997. Write the coordinates of two points that describe the data. Determine the slope and y-intercept of the line containing the two points.

94. Using the slope–intercept form, write the equation of the line containing the two points in Exercise 93.

95. Now suppose that x represents the number of years since 1990. Write the coordinates of two points that describe the data, and use the points to determine the slope of the line that contains those points. Compare this slope to the slope that you obtained in Exercise 93.

96. Using the slope–intercept form, write the equation of the line containing the two points in Exercise 95.

Challenge

In Exercises 97–100, determine k.

97. The line $kx - 2y = 10$ is parallel to the line $y = 7 - 5x$.

98. The line $3y + kx - 9 = 0$ is perpendicular to the line $y = -3x - 2$.

99. The line $x + ky + 3 = 0$ has the same y-intercept as the line $y = x + 1$.

100. The line $kx + 4y = 2$ has the same x-intercept as the line $x - 2y - 4 = 0$.

101. Show that when the slope–intercept form is used to write an equation of a horizontal line, the model reduces to $y = b$.

102. Show that when the point–slope form is used to write an equation of a horizontal line, the model reduces to $y = b$.

4.7 | Graphs of Linear Inequalities

Linear Inequalities   •   Graphs of Solution Sets

Linear Inequalities

The definition of a linear inequality in two variables is similar to the definition of a linear equation in two variables.

Definition of Linear Inequality in Two Variables

A **linear inequality in two variables** is an inequality that can be written in the *standard form* $Ax + By < C$, where A, B, and C are real numbers, and A and B are not both zero. A similar definition holds for $>$, $\leq$, and $\geq$.

The following are some examples of linear inequalities, along with their standard forms.

Linear Inequality	*Standard Form*
$2x - y > 6$	$2x - 1y > 6$
$y \leq x + 4$	$-1x + 1y \leq 4$
$x < 5$	$1x + 0y < 5$
$y + 1 \geq -3$	$0x + 1y \geq -4$

Note: An inequality such as $y \leq x + 4$ means that $y < x + 4$ or $y = x + 4$.

> **Definitions of a Solution and a Solution Set**
>
> A **solution** of a linear inequality in two variables is a pair of numbers (x, y) that makes the inequality true. The **solution set** is the set of all such solutions.

We verify that an ordered pair is a solution of an inequality the same way we verify a solution of an equation.

EXAMPLE **1**

Verifying Solutions of an Inequality

Determine whether $(4, 7)$, $(6, 2)$, and $(-3, -15)$ are solutions of the inequality $2x - y > 6$.

Solution

In each case, replace x with the first coordinate and y with the second coordinate.

(4, 7)	*(6, 2)*	*(−3, −15)*
$2x - y > 6$	$2x - y > 6$	$2x - y > 6$
$2 \cdot 4 - 7 > 6$	$2 \cdot 6 - 2 > 6$	$2(-3) - (-15) > 6$
$8 - 7 > 6$	$12 - 2 > 6$	$-6 + 15 > 6$
$1 > 6$ False	$10 > 6$ True	$9 > 6$ True

Only $(6, 2)$ and $(-3, -15)$ are solutions.

Graphs of Solution Sets

Just as the graph of an equation allows us to visualize the solution set of the equation, the graph of an inequality is a picture of all the solutions of the inequality.

Exploring the Concept

Graphing a Linear Inequality

Consider the inequality $y \geq x - 8$. Figure 4.37 shows the graph of the associated equation $y = x - 8$.

Figure 4.37

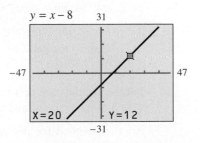

We see that $P(20, 12)$ represents a solution of $y = x - 8$, and so $(20, 12)$ is a solution of $y \geq x - 8$. In fact, every point of the line represents a solution of the inequality.

Figure 4.38 shows the general cursor at $A(20, 24)$, which is directly above point P. The y-coordinate of point A is greater than the y-coordinate of point P. Therefore, at point A, $y > x - 8$. Thus point A represents a solution of $y \geq x - 8$.

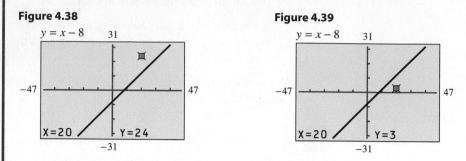

Figure 4.38

Figure 4.39

Figure 4.39 shows the general cursor at $B(20, 3)$, which is directly below point P. At point B, the y-coordinate is less than the y-coordinate of point P, so $y < x - 8$. Thus point B does not represent a solution of $y \geq x - 8$.

This experiment suggests that all points of the line and all points above the line represent solutions of $y \geq x - 8$. Points below the line do not represent solutions.

The line $y = x - 8$ is called a **boundary line.** If the points of the boundary line represent solutions, as is the case for $\leq$ and $\geq$ inequalities, then we draw a solid boundary line. For $<$ and $>$ inequalities, points of the boundary line do not represent solutions, and we draw a dashed boundary line.

Boundary lines divide the coordinate plane into two regions called **half-planes.** The graph of a linear inequality consists of one of the half-planes, which we identify with shading, and possibly the boundary line.

When a linear inequality is in the form $y < mx + b$, all points below the boundary line represent solutions. Similarly, when a linear inequality is in the form $y > mx + b$, all points above the boundary line represent solutions. Thus a method for graphing a linear inequality is as follows.

> **Graphing a Linear Inequality**
>
> 1. Solve the inequality for y.
>
> 2. Draw a solid boundary line for $\leq$ and $\geq$ inequalities; draw a dashed boundary line for $<$ and $>$ inequalities.
>
> 3. For $<$ or $\leq$ inequalities, shade the half-plane below the boundary line. For $>$ or $\geq$ inequalities, shade the half-plane above the boundary line.

For the special case of a vertical boundary line, we shade the half-plane on the left for $<$ and $\leq$ inequalities, and we shade the half-plane on the right for $>$ and $\geq$ inequalities.

Another method for deciding which half-plane to shade is to select a **test point** on either side of the boundary line and determine whether it represents a solution, as we did in Example 1. The following summarizes the test point procedure for graphing a linear inequality.

Graphing a Linear Inequality

1. Draw the boundary line as previously described.

2. Select any test point on either side of the boundary line and determine whether the point represents a solution. If the coordinates of the test point satisfy the inequality, shade the half-plane that contains the point. Otherwise, shade the other half-plane.

EXAMPLE 2

Graphing Linear Inequalities

Graph the solution set of $x + 2y < 10$.

Solution

$$x + 2y < 10 \qquad \text{Solve the inequality for } y.$$
$$2y < -x + 10 \qquad \text{Subtract } x \text{ from both sides.}$$
$$y < -\frac{1}{2}x + 5 \qquad \text{Divide both sides by 2.}$$

Draw the boundary line $y = -\frac{1}{2}x + 5$. The boundary line is dashed because the inequality does not include the equality symbol. Because the inequality is in the form $y < mx + b$, we shade the region below the boundary line.

Figure 4.40

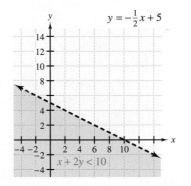

The point (3, 7) lies above the boundary line. We can test this point to determine whether it represents a solution.

$$x + 2y < 10 \qquad \text{The original inequality}$$
$$3 + 2 \cdot 7 < 10 \qquad \text{Replace } x \text{ with 3 and } y \text{ with 7.}$$
$$3 + 14 < 10$$
$$17 < 10 \qquad \text{False}$$

Because (3, 7) does not satisfy the inequality, the half-plane below the line should be shaded. This result confirms the graph in Figure 4.40.

EXAMPLE **3**

Graphing Linear Inequalities

Sketch the graph of each inequality.

(a) $3x - 2y < 8$ (b) $x \geq 3$

Solution

(a) First, solve the inequality for y.

$$3x - 2y < 8$$

$$-2y < -3x + 8 \qquad \text{Subtract } 3x \text{ from both sides.}$$

$$\frac{-2y}{-2} > \frac{-3x}{-2} + \frac{8}{-2} \qquad \text{Divide both sides by } -2 \text{ and reverse the direction of the inequality symbol.}$$

$$y > \frac{3}{2}x - 4$$

The inequality does not include the equality symbol, so we draw a dashed boundary line. Because the inequality symbol is $>$, shade above the line. (See Fig. 4.41.)

LEARNING TIP

To help you remember "above" and "below" the line, think of the boundary line as a roof. Then "above the line" means on top of the roof, whereas "below the line" means under the roof.

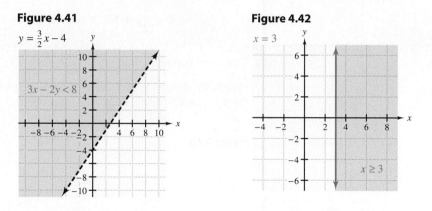

Figure 4.41

Figure 4.42

(b) The solid boundary line is the vertical line $x = 3$. The symbol $\geq$ indicates that we should shade the half-plane to the right of the line. (See Fig. 4.42.)

Sometimes inequalities are used to model the conditions of a problem.

EXAMPLE **4**

Problem Solving with Linear Inequalities

The sum of a number and twice another number is at most 12. Assign a variable to each unknown number, write an inequality, and graph the solution set.

Solution

Let x and y represent the two unknown numbers. The inequality is $x + 2y \leq 12$.

$$x + 2y \leq 12 \qquad \text{Solve for } y.$$

$$2y \leq -x + 12 \qquad \text{Subtract } x \text{ from both sides.}$$

$$y \leq -\frac{1}{2}x + 6 \qquad \text{Divide both sides by 2.}$$

Think About It

Consider the nonlinear inequality $y \geq x^2$. Produce the graph of the associated equation $y = x^2$ with your calculator. Then, on the basis of the procedure for graphing a linear inequality, describe the graph of the inequality.

Draw the solid boundary line $y = -\frac{1}{2}x + 6$ and shade the half-plane below the line. (See Fig. 4.43.)

Figure 4.43

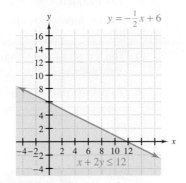

$$y = -\frac{1}{2}x + 6$$

$$x + 2y \leq 12$$

Note that all points of the boundary line and all points below it represent solutions of the inequality.

Quick Reference 4.7

Linear Inequalities

- A **linear inequality in two variables** is an inequality that can be written in the *standard form* $Ax + By < C$, where A, B, and C are real numbers, and A and B are not both zero. A similar definition holds for $>$, $\leq$, and $\geq$.

- A **solution** of a linear inequality in two variables is a pair of numbers (x, y) that makes the inequality true. The **solution set** is the set of all such solutions.

- We can determine whether a given ordered pair is a solution of a linear inequality by substituting the coordinates. To be a solution, the pair must satisfy the inequality.

Graphs of Solution Sets

- Associated with a linear inequality is a linear equation whose graph is the **boundary line** for the graph of the inequality.

- The graph of a linear inequality is a **half-plane.**

- A method for graphing a linear inequality is as follows.

  1. Solve the inequality for y.

  2. Draw a solid boundary line for $\leq$ and $\geq$ inequalities or a dashed boundary line for $<$ and $>$ inequalities.

  3. For $<$ or $\leq$ inequalities, shade the half-plane below the boundary line. For $>$ or $\geq$ inequalities, shade the half-plane above the boundary line.

- For the special case of a vertical boundary line, we shade the half-plane on the left for $<$ and $\leq$ inequalities, and we shade the half-plane on the right for $>$ and $\geq$ inequalities.

- To graph a linear inequality with the test point procedure, use the following steps.

  1. Draw the boundary line as previously described.

  2. Select any test point on either side of the boundary line and determine whether the point represents a solution. If the coordinates of the test point satisfy the inequality, shade the half-plane that contains the point. Otherwise, shade the other half-plane.

Speaking the Language 4.7

1. The inequality $5x - 7y > 2$ is an example of a(n) ▓▓▓▓▓▓▓▓ .

2. For an inequality, the graph of the associated equation is called the ▓▓▓▓▓▓▓ line.

3. The graph of the solution set of a linear inequality in two variables is called a(n) ▓▓▓▓▓▓▓▓ .

4. To decide on the region to shade, we can select a point and determine whether its coordinates satisfy the linear inequality. Such a point is called a(n) ▓▓▓▓▓▓ point.

Exercises 4.7

Concepts and Skills

1. What do we mean by a *solution* of a linear inequality in two variables?

2. What must be true for an ordered pair to be a solution of $y \geq x + 3$?

In Exercises 3–8, identify the ordered pairs that are solutions of the given inequality.

3. $y < -2x + 5$
$(-2, 3), (6, -8)$
$(-7, 19), (4, 6)$

4. $y \leq 3x - 1$
$(0, -2), (3, 8)$
$(3, 2), (-4, 10)$

5. $y \geq \dfrac{x}{2} - 5$
$(4, -3), (8, 2)$
$(-3, -7), (-6, 0)$

6. $y > -\dfrac{2}{3}x + 4$
$(-2, 6), (6, 1)$
$(-10, 9), (3, 2)$

7. $5x > 2y + 10$
$(4, 3), (0, -2)$
$(1, -4), (4, 5)$

8. $x + 3y - 3 \leq 0$
$(2, -2), (-2, 2)$
$(-3, 2), (0, 0)$

9. Suppose that you are graphing the solution set of $2x - y < 3$. A classmate advises you to shade below the boundary line because the inequality has a $<$ symbol. What is your opinion of this advice?

10. Suppose that you are graphing a linear inequality and that you have already drawn the boundary line. Describe two ways of determining which half-plane to shade.

In Exercises 11–14, match the given inequality to its graph.

11. $y \leq 2$

12. $y \geq 2x$

13. $x + 3 > 0$

14. $x + 3y < 0$

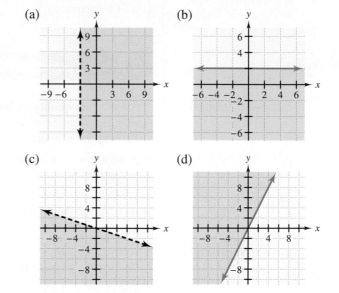

(a) (b) (c) (d)

In Exercises 15–18, the figure shows the boundary line for the graph of the given inequality. Complete the graph by shading the correct half-plane.

15. $y \leq x + 4$

16. $y > -\dfrac{1}{2}x + 3$

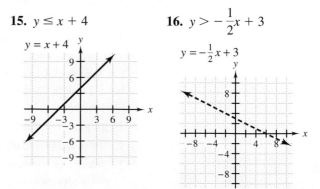

17. $y - 3 > 0$ **18.** $x + 2 \leq 0$

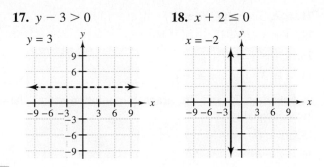

19. In graphing a linear inequality, when do we use a solid line and when do we use a dashed line for the boundary line?

20. Suppose that you are graphing the solution set of $y < 2x + 3$. Why is $(0, 3)$ not a valid test point?

In Exercises 21–32, graph the inequality.

21. $y < 2x$

22. $y \geq -x$

23. $y < x - 4$

24. $y > x + 1$

25. $y \geq -2x - 7$

26. $y \leq -3x + 1$

27. $y \leq -\dfrac{1}{2}x + 4$

28. $y > \dfrac{5}{2}x - 5$

29. $x < -4$

30. $y > 1$

31. $y \leq 2$

32. $x \geq -3$

In Exercises 33–50, graph the inequality.

33. $x - y \leq 0$

34. $y + 2x < 0$

35. $x > 2y$

36. $x \leq -3y$

37. $x + 3y < 9$

38. $3x - 4y \geq 16$

39. $x \geq 6 - 2y$

40. $x < 2y + 2$

41. $x + y > -3$

42. $5 \geq x - y$

43. $-2 < 2y - 3x$

44. $-x - y \geq -4$

45. $\dfrac{x}{3} - y < \dfrac{1}{4}$

46. $\dfrac{x}{6} - \dfrac{y}{2} > \dfrac{1}{4}$

47. $2x + 1 > y - 2$

48. $x - y \geq 2x + 3$

49. $3(x + y) \leq 2(x - y)$ **50.** $2(x - 6) < 3(y - 4)$

In Exercises 51–60, translate the information into a linear inequality.

51. One number y is at least 2 more than another number x.

52. One number y is no more than 1 less than another number x.

53. The sum of two numbers is less than 10.

54. The difference of two numbers is more than 2.

55. The difference of one number y and twice another number x is positive.

56. The sum of three times a number x and another number y is negative.

57. One number y is at most half of another number x.

58. One-third of a number y is no less than another number x.

59. The difference of one number y and 3 less than another number x is nonnegative.

60. The sum of a number y and 2 more than half a number x is nonpositive.

61. To graph the solution set of $x < 5$, we might use a number line or a coordinate plane. What do we need to know before we can draw the graph?

62. To graph $x > 3$, why do we shade the half-plane to the right of the vertical boundary line?

In Exercises 63–66, sketch the graph of the given inequality

(a) on a number line.

(b) in a coordinate plane.

63. $x > 2$ **64.** $x \leq 4$

65. $x \geq -5$ **66.** $x < -3$

Data Analysis: *Children's Age and Weight*

The table shows the average weights for children of various ages.

Age	Average Weight (pounds)	
	Boy	Girl
1	21	20
3	31	30
5	39	38
7	51	49
9	63	62
11	71	77
13	92	98

(Source: *Physicians Handbook.*)

The following equations model the data for all ages between 1 and 14. In these equations, w and a represent weight and age, respectively.

Boy: $w = 6.1a + 10.6$

Girl: $w = 6.5a + 8$

67. Write inequalities to model the age and weight combinations for a child whose weight is above the average.

68. Describe the graph of each inequality in Exercise 67.

69. Suppose that a child is considered obese if the child's weight is at least 10% above the average weight. Write inequalities to model the age and weight combinations for a child who is obese.

70. Describe the graph of each inequality in Exercise 69.

Challenge

In Exercises 71–74, graph the solution set of the inequality. (*Hint*: The graph is the intersection of the graphs of the individual inequalities.)

71. $x \geq -3$ and $x \leq 4$

72. $1 \leq x \leq 3$

73. $-1 \leq y \leq 5$

74. $x \geq 3$ and $y \geq -1$

In Exercises 75 and 76, graph the solution set of the given inequality.

75. $xy \geq 0$

76. $xy < 0$

77. Use inequalities to describe the points in

    (a) Quadrant I.

    (b) Quadrant II.

    (c) Quadrant III.

    (d) Quadrant IV.

Chapter 4 Review Exercises

Section 4.1

1. Which of the following is not a linear equation in two variables? Why?

    (i) $y = x$ (ii) $y = \dfrac{1}{x}$

    (iii) $xy = 1$ (iv) $y - x = 1$

2. Write an equation that describes the given information. Let x and y represent the two unknown numbers.

    (a) In a certain triangle, the measure of one angle is 45°.

    (b) There are 120 vehicles in a parking lot. Sixteen of the vehicles are trucks, and the rest are vans and cars.

3. Identify each ordered pair that is a solution of the equation $3x - 2y = 12$.

    (i) $(2, -3)$ (ii) $(-10, -21)$

    (iii) $(0, 6)$ (iv) $(6, 3)$

4. Determine three solutions of the equation $y = -2x + 5$ and sketch the graph.

5. Complete the table so that each pair is a solution of $x = y + 4$. Then sketch the graph.

x	0		7
y		-1	

6. Write the equation $\dfrac{1}{2}x = -3 - y$ in the form required for producing its graph on a calculator.

7. Produce the graph of $y = 2x + 1$ on your calculator. Then trace the graph to determine a and b such that $A(a, 7)$ and $B(-4, b)$ are points of the graph.

8. Suppose that the points of a line are such that the y-coordinate is always 1 less than twice the x-coordinate.

    (a) Write two ordered pairs that satisfy the given condition.

    (b) Write an equation that describes the condition.

    (c) Produce the graph of your equation and trace it to verify that the points in part (a) belong to the graph.

9. Which of the following equations have the same graph?

    (i) $x = -2(y + 5)$ (ii) $y = -0.5x + 5$

    (iii) $x + 2y = 10$

10. Suppose that every solution of a certain equation has the form (a, a). Write the equation in standard form.

Section 4.2

11. Use the graphing method to estimate the intercepts of the graph of $x = 2y - 18$.

12. Determine algebraically the intercepts of the graph of $\frac{3}{2}x - 5y = 15$.

13. Write the equation $2x - 5y - 20 = 0$ in the form $y = ax + b$ and identify the y-intercept of the graph.

14. Select the equation whose y-intercept is not the same as that of the other two equations.
    (i) $2x - 3y = -15$
    (ii) $4x + y = 5$
    (iii) $y - 4x = \ 5$

15. Describe the graph of an equation if the first coordinates of all points of the graph are the same.

16. If $(-2, 5)$ and $(3, y)$ are points of a horizontal line, what is the value of y?

17. Determine whether the graph of the equation is horizontal, vertical, or neither.
    (a) $x - 7 = 0$
    (b) $x - 7y = 0$
    (c) $7y = 0$

18. Write an equation of the line that contains $(-3, -2)$ and is
    (a) parallel to the x-axis.
    (b) perpendicular to the x-axis.

19. The cost y of a sofa is \$50 more than twice the cost x of a chair. Write an equation that describes this information and determine the intercepts of its graph.

20. If the x-intercept of a line is the origin, what is the y-intercept?

Section 4.3

21. Which of the following expressions is used in the Slope Formula?
    (i) $\dfrac{x_1 - x_2}{y_1 - y_2}$
    (ii) $\dfrac{y_1 - x_1}{y_2 - x_2}$
    (iii) $\dfrac{y_2 - y_1}{x_2 - x_1}$

22. If a line falls from left to right, what is the sign of its slope?

23. Suppose that A, B, C, D, and E are all points of a line and that their coordinates are known. Which two points can be used to calculate the slope of the line?

24. Determine the slope of a line that contains $A(-2, 5)$ and $B(7, -1)$.

25. Determine two solutions of $2x + 3y = 9$ and use them to determine the slope of its graph.

26. Tell what you know about the slope of the graph of the given equation.
    (a) $2x + 1 = -2$
    (b) $27 - y = 0$

27. Determine the unknown coordinate such that a line with the given slope contains both points.
    (a) $(-8, -2)$, $(7, y)$; slope is 0
    (b) $(0, 9)$, $(x, -3)$; slope is undefined

28. State whether the described line has a slope that is positive, negative, 0, or undefined.
    (a) The line contains $(-4, 7)$ and $(3, -1)$.
    (b) The line is perpendicular to the x-axis.
    (c) The line contains no points in Quadrant III or Quadrant IV.
    (d) The line contains the origin and a point in Quadrant I.

29. If a line contains the points $A(-4, -7)$ and $B(1, 6)$, how long are the legs of the corresponding slope triangle?

30. If the slope of line M is $\frac{1}{2}$ and the slope of line N is -5, which line is steeper?

Section 4.4

31. Write $4x - 5y = 5$ in the slope–intercept form and determine the slope and y-intercept of the graph.

32. Which equation does not have the same slope as the other two?
    (i) $3y = x + 27$
    (ii) $x - 3y = 15$
    (iii) $-x = 3y - 12$

33. Sketch a line for which $m = \frac{2}{3}$ and the y-intercept is $(0, -4)$.

34. Sketch a line that contains $(-4, 3)$ and has a slope of $-\frac{3}{5}$.

35. Sketch the graph of $\dfrac{x}{5} + y = -1$.

36. Sketch the line that is parallel to the x-axis and contains $(-9, -4)$. What is the slope of the line?

37. Describe a line with a rise of -4 and a run of 0.

38. Which of the following is an example of a direct variation?
    (i) $xy = 1$
    (ii) $\dfrac{y}{x} = 1$
    (iii) $x = 1$

39. If y varies directly with x, and y is 28 when x is 4, what is y when x is 16?

40. If a person reads 6 pages in 4 minutes, what is her reading rate? How long will she take to read a 32-page report?

Section 4.5

41. How can L_1 be parallel to L_2 but not have the same slope?

42. In each part, the slopes of two lines are given. Are the lines parallel, perpendicular, or neither?

(a) $m_1 = \frac{2}{3}$, $m_2 = \frac{3}{2}$

(b) $m_1 = -5$, $m_2 = 0.2$

(c) $m_1 = 1.25$, $m_2 = \frac{5}{4}$

43. If the equation of L_1 is $y = 1 - 3x$ and the equation of L_2 is $x + \dfrac{y}{3} = 2$, what relationship, if any, exists between L_1 and L_2?

44. If line L_1 contains $(5, 3)$ and $(7, 2)$ and line L_2 contains $(-8, 0)$ and $(-7, 2)$, what relationship, if any, exists between L_1 and L_2?

45. Suppose that $3x + 4y = 0$ is the equation of L_1. What is the slope of L_2 if L_2 is

(a) parallel to L_1?

(b) perpendicular to L_1?

46. Suppose that the slope of L_1 is undefined. What must be the slope of L_2 for the two lines to be perpendicular?

47. The equation of L_1 is $3x + 2y = 5$, and the equation of L_2 is $y = kx + 1$. If L_1 and L_2 are parallel, what is the value of k?

48. Suppose that every point of L_1 has coordinates of the form (a, a). If L_2 is parallel to L_1, what is the slope of L_2?

49. Suppose that the slopes of two sides of a triangle are m_1 and m_2. If $m_2 = -\dfrac{1}{m_1}$, what can you conclude about the triangle?

50. The consecutive vertices of a quadrilateral are $A(-2, 2)$, $B(-2, 7)$, $C(5, 9)$, and $D(5, 4)$. Show that the quadrilateral is a parallelogram.

Section 4.6

51. Write an equation of the line with a slope of $\frac{3}{5}$ and a y-intercept of $(0, -2)$.

52. Write an equation of the line whose graph is shown in the figure.

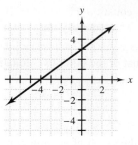

53. Use the slope–intercept form to write an equation of the line that contains $(3, -2)$ and has a slope of $-\frac{1}{2}$.

54. A line contains the points $(-3, 2)$ and $(5, 7)$. Write the standard form of the equation of the line beginning with the

(a) slope–intercept form.   (b) point–slope form.

55. Why can neither the slope–intercept nor the point–slope form be used to write an equation of the line containing $(-4, -1)$ and $(-4, 0)$?

56. Write an equation of the line that contains $(5, -4)$ and is parallel to the line $y = -\dfrac{1}{3}x + 11$.

57. Write an equation of the line that contains $(-5, 4)$ and is perpendicular to the line that contains $(0, 1)$ and $(1, 0)$.

58. Write an equation of the line that is parallel to the line whose equation is $y = -3x + 5$ and that has the same y-intercept as the line $2x - 3y = 12$.

59. Write an equation of the line that contains $(3, 7)$ and is perpendicular to the line $y = \dfrac{1}{3}x - 7$.

60. After one day, a person had read 20 pages of a 570-page novel. At the end of the twenty-second day, 130 pages remained to be read. Assuming that the person read the same number of pages each day, write an equation that describes the number of pages read at the end of any day.

Section 4.7

61. Determine whether each ordered pair is a solution of the inequality $y > 3x - 5$.

(a) $(1, -2)$ (b) $(3, 5)$

(c) $(-2, -10)$ (d) $(0, 0)$

62. What do we call the region in a plane that is above or below a boundary line?

63. Suppose that you are graphing the solution set of $x - y < 0$. Will you shade above or below the boundary line?

64. Which of the following have graphs with a solid boundary line?

   (i) $y \le x$ (ii) $y > x$

   (iii) $y \ge x$

65. After drawing a boundary line, what point can you use as a test point to determine the region to shade?

66. Graph the inequality $y \le -0.5x + 3$.

67. Graph the inequality $2x > -4(y - 1)$.

68. Model the following information with a linear inequality and graph the solution set.

   The difference of one number y and twice another number x is nonnegative.

69. Graph the inequalities in a plane.

   (a) $x < -2$ (b) $y \ge 3$

70. A collection of n nickels and d dimes is worth no more than $9.85. Write an inequality that describes this information.

Chapter 4 Test

1. Identify each ordered pair that is a solution of the equation $x + 4y = 8$.

   (i) $(4, 1)$ (ii) $(0, -2)$ (iii) $(8, 0)$ (iv) $(1, 4)$

2. Determine three solutions of $y = \dfrac{1}{2}x - 3$ and sketch its graph.

3. One of the following equations has a graph that is different from that of the other two. Why?

   (i) $y = \dfrac{2}{3}x + 1$ (ii) $2x + 3y = 3$

   (iii) $3y - 2x = 3$

4. Determine algebraically the intercepts of the graph of $\dfrac{1}{3}x - \dfrac{1}{2}y = 1$.

5. Write the equation $4y + 2x + 10 = 0$ in the form $y = mx + b$ and identify the y-intercept of the graph.

6. Write the equation of the line that contains $(8, -7)$ and is

   (a) parallel to the x-axis.

   (b) perpendicular to the x-axis.

7. Determine the slope of the line that contains $A(-3, -8)$ and $B(-5, 0)$.

8. Determine two solutions of $5x + 2y + 20 = 0$ and use them to determine the slope of its graph.

9. State whether the described line has a slope that is positive, negative, 0, or undefined.

   (a) The line contains no points in Quadrant II or Quadrant III.

   (b) The line contains $(-7, 9)$, and the x-intercept is $(4, 0)$.

   (c) The intercepts are $(-5, 0)$ and $(0, 5)$.

   (d) The line is parallel to the x-axis.

10. Write $3x = 7(y + 2)$ in the slope–intercept form and determine the slope and the y-intercept of the graph.

11. Explain how to sketch the line that contains the point $(-2, -5)$ and has a slope of $\dfrac{2}{5}$.

12. If y varies directly with x, and y is -15 when x is 3, what is y when x is -3?

13. Suppose that the equation of L_1 is $5y + x = 5$. What is the slope of L_2 if L_2 is

   (a) parallel to L_1?   (b) perpendicular to L_1?

14. If L_1 contains $(-2, 7)$ and $(8, 8)$ and L_2 contains $(-6, -3)$ and $(4, -2)$, what relationship, if any, exists between L_1 and L_2?

15. The vertices of a triangle are $A(-5, -5)$, $B(3, 3)$, and $C(7, -1)$. Show that the triangle is a right triangle.

16. Use the slope–intercept form to write an equation of the line that contains $(-5, 2)$ and has a slope of $\dfrac{3}{5}$.

17. A line contains $(3, 4)$ and $(-3, -4)$. Write the equation of the line in standard form.

18. Write an equation of the line that is perpendicular to the line $y = 2x - 7$ and that has the same y-intercept as the line $-2y = -x + 8$.

19. Graph the inequality $y > 0.2x - 5$.

20. Graph the inequalities in a plane.

   (a) $y < 4$ (b) $x \ge -5$

21. A group of x hunters and y hunting dogs has a total of at least 64 legs. Write an inequality that describes this information.

The accompanying graph shows a comparison between the percentages of online shoppers who are men and those who are women.

*The data in each category can be modeled by a linear equation, and the two equations together form a **system of equations.** By solving the system, we can project the year in which the percentages for the two categories might be the same. (For more on this real data application, see Exercises 73–76 at the end of Section 5.2.)*

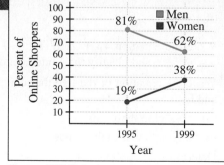

(Source: USA Today.)

Chapter Snapshot

Most of this chapter is devoted to the topic of systems of two linear equations in two variables. We study both graphing and algebraic methods for solving such systems and use these skills in solving a wide variety of application problems. In the final section we learn to graph the solutions of systems of linear inequalities.

Warm-Up Skills

The following exercises review concepts and skills that you will need in Chapter 5.

In Exercises 1–3, use the graphing method to estimate the solution(s) of the given equation. In the case of an identity or contradiction, write the solution set.

1. $\frac{3}{4}x - 16 = 17 - 2x$

2. $x + 5 = 3(x - 1) - 2x$

3. $3 - 2(1 + x) = 1 - 2x$

In Exercises 4–6, the equations of two lines are given. Are the lines parallel, perpendicular, or neither?

4. $y - 2x = 0$
   $x + y = 15$

5. $2x - 3y = -21$
   $4x - 6y = 30$

6. $2y - 3x = 14$
   $6x - 4y = -28$

7. Fill in the blanks so that the ordered pairs are solutions of the equation $3x - y = 8$.

   (a) $(3, \underline{\quad})$

   (b) $(\underline{\quad}, -2)$

8. Simplify the expression $5\left(\frac{3}{4}y + 1\right) - 2y$.

9. Compare the graphs of $y \le x + 3$ and $y < x + 3$.

In Exercises 10 and 11, write an equation that models the given information.

10. The total value of a collection of nickels and dimes is $3.45.

11. On a 306-mile trip the average speed was 48 mph for the first segment and was 62 mph for the other portion.

12. Solve the equation $4(2y - 5) - 5y = -14$.

5.1 | The Graphing Method

Definitions · Estimating Solutions · Special Cases · Applications

Definitions

In Chapter 4 we considered applications involving two unknown quantities. We described the conditions of such problems with linear equations in two variables and used the graphs of the equations to visualize solutions.

The information in a problem often leads to more than one equation. To solve such a problem, we need to find a solution that satisfies each equation.

Exploring the Concept

Systems of Equations

Suppose that we want to determine two numbers x and y whose sum is 10. The equation $x + y = 10$ models the requirement. Figure 5.1 shows the graph of $y = -x + 10$ with the cursor on $(-6, 16)$, one of the infinitely many solutions of the equation. Tracing the graph, we find that there are infinitely many pairs of numbers whose sum is 10.

Figure 5.1   **Figure 5.2**

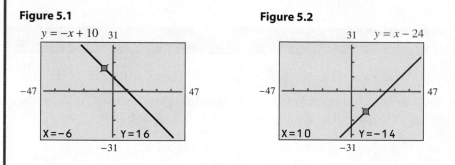

Suppose that we want to determine two numbers x and y whose difference is 24. This requirement is modeled by $x - y = 24$. The graph of $y = x - 24$ is shown in Figure 5.2 with the tracing cursor on $(10, -14)$. Again, the equation has infinitely many solutions, so there are infinitely many pairs of numbers whose difference is 24.

Now suppose that we want to determine two numbers x and y whose sum is 10 *and* whose difference is 24. The conditions are modeled by $x + y = 10$ *and* $x - y = 24$. Figure 5.3 shows the graphs of both of these equations in the same coordinate system.

Figure 5.3

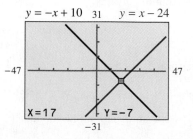

The point of intersection is $(17, -7)$. Because this point belongs to both lines, it represents a solution of both equations. Thus 17 and -7 are two numbers whose

sum is 10 *and* whose difference is 24. The equations used to model the conditions of this problem are joined by the word *and*. We say that the equations are considered *simultaneously*, which means that both equations must be satisfied. Usually, we do not explicitly include the word *and* when we write the equations.

$$x + y = 10$$
$$x - y = 24$$

Definition of a System of Equations and Its Solution

Two or more equations considered simultaneously form a **system of equations.**

If two linear equations in the same two variables are considered simultaneously, they form a **system of two linear equations in two variables.**

A **solution** of a system of two equations in two variables is a pair of numbers (x, y) that satisfies both equations.

Note: When referring to a system of two linear equations in two variables, we may say *system of equations*, or simply *system*, when the kind of equations in the system is clear from the context.

EXAMPLE **1**

Testing Solutions

Determine whether each pair is a solution of the given system of equations.

$$2x - 5y = 9$$
$$3x + y = 5$$

(a) $(7, 1)$ (b) $(2, -1)$

Solution

(a) *First Equation*   *Second Equation*

$$2x - 5y = 9 \qquad\qquad\qquad\qquad\qquad\qquad 3x + y = 5$$
$$2(7) - 5(1) = 9 \qquad \text{Replace } x \text{ with 7 and } y \text{ with 1.} \qquad 3(7) + (1) = 5$$
$$14 - 5 = 9 \qquad\qquad\qquad\qquad\qquad\qquad 21 + 1 = 5$$
$$9 = 9 \quad \text{True} \qquad\qquad\qquad\qquad\qquad 22 = 5 \quad \text{False}$$

Although $(7, 1)$ satisfies the first equation, it does not satisfy the second equation. Therefore, the pair $(7, 1)$ is not a solution of the system.

(b) *First Equation*   *Second Equation*

$$2x - 5y = 9 \qquad\qquad\qquad\qquad\qquad\qquad 3x + y = 5$$
$$2(2) - 5(-1) = 9 \qquad \text{Replace } x \text{ with 2 and } y \text{ with } -1. \qquad 3(2) + (-1) = 5$$
$$4 + 5 = 9 \qquad\qquad\qquad\qquad\qquad\qquad 6 - 1 = 5$$
$$9 = 9 \quad \text{True} \qquad\qquad\qquad\qquad\qquad 5 = 5 \quad \text{True}$$

Because the pair satisfies *both* equations, $(2, -1)$ is a solution of the system.

Estimating Solutions

At the beginning of this section, we used graphs to estimate the solution of the system $x + y = 10$ and $x - y = 24$. This graphing method can be summarized as follows.

Estimating Solutions of Systems by Graphing

1. Produce the graph of each equation in the same coordinate system.
2. Trace to estimate the point of intersection of the two graphs.
3. This point represents the estimated solution of the system of equations.

EXAMPLE 2

Estimating Solutions by Graphing

Solve the following system of equations.

$$y = 2x$$
$$y + x = 15$$

Solution

Solve the second equation for y and produce the graph of both equations in the same coordinate plane. (See Fig. 5.4.)

The point of intersection appears to be $(5, 10)$, so $(5, 10)$ is the estimated solution of the system.

We verify the solution as in Example 1.

Figure 5.4

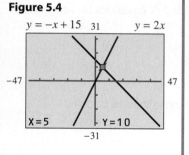

$y = -x + 15$ 31 $y = 2x$

-47 ——————— 47

X=5   Y=10

-31

$y = 2x$	$y + x = 15$	
$10 = 2(5)$	$10 + 5 = 15$	Replace x with 5 and y with 10.
$10 = 10$	$15 = 15$	Both equations are satisfied.

In the remaining examples verification is left to the reader.

EXAMPLE 3

Estimating Solutions by Graphing

Solve the system.

$$5x - 4y = 8$$
$$x + 2y = -32$$

Solution

Solve each equation for y, produce their graphs, and trace to the point of intersection. (See Fig. 5.5.)

Figure 5.5

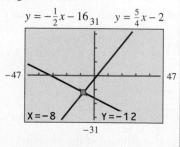

$y = -\frac{1}{2}x - 16$ 31 $y = \frac{5}{4}x - 2$

-47 ——————— 47

X=-8   Y=-12

-31

$$5x - 4y = 8 \qquad\qquad x + 2y = -32$$
$$-4y = -5x + 8 \qquad\qquad 2y = -x - 32$$
$$y = \frac{5}{4}x - 2 \qquad\qquad y = -\frac{1}{2}x - 16$$

The point of intersection $(-8, -12)$ represents the solution of the system.

Special Cases

For a system of linear equations to have a unique solution, the graphs of the equations must intersect at exactly one point, as in the previous examples. When a system has a solution, we say that the system is **consistent.** When the graphs of the equations are different, we say that the equations are **independent.**

Two other possibilities may arise when we graph systems of linear equations.

EXAMPLE 4

Special Cases

Solve the following systems of equations.

(a) $2x - 3y = -21$
     $4x - 6y = 30$

(b) $-3x + 2y = 14$
     $6x - 4y = -28$

Solution

In each part, we begin by writing the equations of the system in the slope–intercept form and producing their graphs.

(a) With the equations in the form $y = mx + b$, the system is as follows.

$$y = \frac{2}{3}x + 7 \qquad \text{Solve } 2x - 3y = -21 \text{ for } y.$$

$$y = \frac{2}{3}x - 5 \qquad \text{Solve } 4x - 6y = 30 \text{ for } y.$$

Figure 5.6

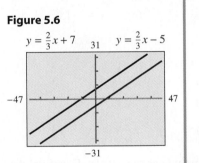

$y = \frac{2}{3}x + 7 \qquad y = \frac{2}{3}x - 5$

Figure 5.6 shows the graphs of the equations. The graphs appear to be parallel. This observation is easily confirmed by examining the equations in the slope–intercept form. We see that the slopes are the same and the y-intercepts are different. Because the lines have no point of intersection, the system has no solution.

(b) With the equations in the form $y = mx + b$, the system is as follows.

$$y = \frac{3}{2}x + 7 \qquad \text{Solve } -3x + 2y = 14 \text{ for } y.$$

$$y = \frac{3}{2}x + 7 \qquad \text{Solve } 6x - 4y = -28 \text{ for } y.$$

The slope–intercept forms of the equations reveal that the slopes and the y-intercepts of the two lines are the same, so the lines coincide. Because every point of either line is a point of intersection, the system has infinitely many solutions.

In part (a) of Example 4, the graphs of the equations are different, so the equations are independent. However, because the system has no solution, we say that the system is **inconsistent.**

In part (b) of Example 4, the graphs of the equations are the same, so we say that the equations are **dependent.** Furthermore, because the system has at least one solution (in this case infinitely many solutions), the system is consistent. The three possible outcomes when we use graphing to estimate the solution of a system of two linear equations are summarized in Figure 5.7.

Figure 5.7

(a)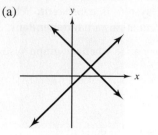

(b)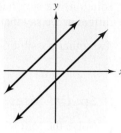

(c)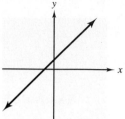

Solution is unique.
Equations are independent.
System is consistent.
The lines intersect.

No solution.
Equations are independent.
System is inconsistent.
The lines are parallel.

Infinitely many solutions.
Equations are dependent.
System is consistent.
The lines coincide.

EXAMPLE 5 Describing Systems with Slope and y-Intercept

Without graphing, describe the given systems of equations and their solutions.

(a) $3y - 2x = 3$
$2x = 3y - 3$

(b) $x = y - 4$
$2y + x = -4$

(c) $2x + y = 1$
$6x + 3y = -15$

Solution

(a) Write each equation in the slope–intercept form.

$$3y - 2x = 3 \qquad\qquad 2x = 3y - 3$$
$$3y = 2x + 3 \qquad\qquad 3y = 2x + 3$$
$$y = \frac{2}{3}x + 1 \qquad\qquad y = \frac{2}{3}x + 1$$

Think About It

Consider the system $y = ax + 4$ and $y = 2x + b$. What must be true about a and b if the graphs (a) coincide? (b) are parallel?

The lines have the same slope, $\frac{2}{3}$, and the same y-intercept, $(0, 1)$. Therefore, the lines coincide, and the equations are dependent. Each point of the line is a solution of the system, so the system is consistent. The system has infinitely many solutions.

(b) Write each equation in the slope–intercept form.

$$x = y - 4 \qquad\qquad 2y + x = -4$$
$$y = x + 4 \qquad\qquad 2y = -x - 4$$
$$y = -\frac{1}{2}x - 2$$

The slopes and y-intercepts of the two lines are different. Therefore, the lines intersect at exactly one point. The equations are independent, and the system is consistent and has exactly one solution.

(c) $2x + y = 1 \qquad\qquad 6x + 3y = -15$
$y = -2x + 1 \qquad\qquad 3y = -6x - 15$
$\qquad\qquad\qquad\qquad y = -2x - 5$

The slopes are the same, but the y-intercepts are different. Therefore, the lines are parallel, and the system has no solution. This system is inconsistent, and the equations are independent.

Applications

Using only one variable, we represent the measures of complementary angles with x and $90 - x$. Example 6 illustrates the convenience of using two variables.

EXAMPLE 6

Complementary Angles

Two angles are complementary. The measure of one angle is 6° more than one-third the measure of the other angle. What is the measure of each angle?

Solution

Let x = the measure of one angle and y = the measure of the other angle.

$$x + y = 90 \qquad \text{The sum of the measures of complementary angles is 90°.}$$

$$y = \frac{1}{3}x + 6 \qquad \text{Add 6 to } \tfrac{1}{3} \text{ of the measure of one of the angles.}$$

Figure 5.8

$y = -x + 90 \qquad y = \frac{1}{3}x + 6$

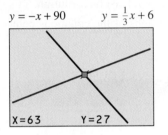

X = 63   Y = 27

Write each equation in slope–intercept form. Then graph the equations and trace to estimate the solution. (See Fig. 5.8.)

$$y = -x + 90 \qquad \text{Solve the first equation for } y.$$

$$y = \frac{1}{3}x + 6 \qquad \text{Second equation is already solved for } y.$$

The lines appear to intersect at (63, 27). We verify this by checking the conditions of the problem.

1. Because $63 + 27 = 90$, the angles are complementary.
2. Because 6 more than one-third of 63 is $\frac{1}{3}(63) + 6 = 21 + 6 = 27$, the second condition is satisfied.

The measures of the angles are 63° and 27°.

Quick Reference 5.1

Definitions
- Two or more equations considered simultaneously form a **system of equations.**
- If two linear equations in the same two variables are considered simultaneously, they form a **system of two linear equations in two variables.**
- A **solution** of a system of two equations in two variables is a pair of numbers (x, y) that satisfies both equations.
- We can verify that a given ordered pair is a solution of a system by replacing the variables with the coordinates. The pair must satisfy each equation of the system.

Estimating Solutions
- If a system of equations has a solution, it can be estimated with the following graphing method.
  1. Produce the graph of each equation in the same coordinate system.
  2. Trace to estimate the point of intersection of the two graphs.
  3. This point represents the estimated solution of the system of equations.

Special Cases
- If a system of equations has at least one solution, then the system is **consistent.** A system with no solution is **inconsistent.**
- If the graphs of the equations of a system are identical, then the equations are **dependent.** If the graphs are different, then the equations are **independent.**
- The graphs of the equations of an inconsistent system are parallel, and the system has no solution. If the equations of a system are dependent, then the graphs of the equations coincide, and the system has infinitely many solutions.

Speaking the Language 5.1

1. Two or more equations joined by the word *and* form a(n) _____ of equations.
2. A solution of a system of equations is represented by a(n) _____ of the graphs of the equations.
3. A(n) _____ system of equations is a system that has at least one solution.
4. If the equations of a system are _____, then the graphs of the equations coincide. The graphs of an inconsistent system are _____ .

Exercises 5.1

Concepts and Skills

Unless otherwise indicated, all references to systems of equations mean systems of two linear equations in two variables.

1. What do we mean by a solution of a system of equations?

2. Suppose that you have determined that a given ordered pair does not satisfy the first equation of a system of equations. Why can you conclude that the pair is not a solution of the system?

In Exercises 3–6, identify each ordered pair that is a solution of the system of equations.

3. $2x - 3y = 11$
   $x + 2y = -12$
   $(-2, -5), (0, -6), (2.5, -2)$

4. $10x + 9y - 1 = 0$
   $5x - 3y = 3$
   $(1.2, 1), \left(\frac{2}{5}, -\frac{1}{3}\right), (1, -1)$

5. $5y - 2x = 5$
   $y = 0.4x + 1$
   $(2, 1), (5, 3), (-5, -1)$

6. $2x - 6 = 0$
   $y + 2 = -3(x - 1)$
   $(3, 1), \left(-\frac{2}{3}, 3\right), (3, -8)$

In Exercises 7–12, determine the value of a, b, or c such that the given ordered pair is a solution of the system of equations.

7. $ax - 3y = 21$ $(1, -6)$
   $6x + by = 36$

8. $x + by = 1$ $(-2, 1)$
   $ax + 2y = 2$

9. $x + y = c$ $(2, 1)$
   $ax + y = -5$

10. $ax - 3y = 9$ $(-3, -2)$
    $3x - y = c$

11. $2x + by = -5$ $(-1, 3)$
    $x + 3y = c$

12. $x - y = c$ $(-4, -1)$
    $ax - 5y = 3$

13. Can a system of linear equations have exactly two solutions? Explain.

14. If a system of equations has a unique solution, what do we call the system? What do we call the equations of the system?

In Exercises 15–22, solve by graphing and verifying.

15. $y = 2x$
$y - x = 0$

16. $y + 10 = x$
$y + x = 10$

17. $y + 9 = 0$
$y = -2x + 15$

18. $y = 12$
$2y = x + 4$

19. $y = -\dfrac{4}{3}x + 23$
$3y = 2x + 15$

20. $y = x - 3$
$y + x + 17 = 0$

21. $y = \dfrac{3}{4}x - 16$
$y + 2x = 17$

22. $y = 0.25x + 13$
$y = x + 25$

In Exercises 23–26, solve by graphing and verifying.

23. $4y - 3x = -20$
$x = 4$

24. $3y - 2x = 12$
$x + 3 = 0$

25. $x + 8 = 0$
$y + 3 = 0$

26. $x = 4$
$y - 6 = 0$

27. For each row in the following list, describe the graph of the system of equations. If a particular combination is not possible, so indicate.

System	Equations	
(a) Consistent	Independent	intersect
(b) Consistent	Dependent	overlap
(c) Inconsistent	Independent	parallel
(d) Inconsistent	Dependent	not pos

28. In each part, describe the graph of a system of equations that has
(a) no solution. in con dep
(b) infinitely many solutions. con dep
(c) a unique solution. ind const

In Exercises 29–36, determine the number of solutions of the given system by writing the equations in the form $y = mx + b$.

29. $y = 3x - 2$
$y = -2x + 3$

30. $y = 2x + 3$
$y = \dfrac{1}{2}x + 4$

31. $3y - x = 15$
$3y - 10 = x + 5$

32. $2x + y = 3$
$12 - 4y = 8x$

33. $y - x = 7$
$2y + 6 = 2x$

34. $4y - 3x = 28$
$8 = 3x - 4y$

35. $2y + 8 = x$
$2x = y + 4$

36. $3x + 3y = 15$
$x = y - 5$

In Exercises 37–42, determine whether the system of equations is consistent or inconsistent by writing the equations in the form $y = mx + b$, if possible.

37. $2x + y = 4$
$2x + 3 = -y$

38. $y - 3 = 0.4x$
$5y = 2x - 25$

39. $3y - x = 6$
$6y - 12 = 2x$

40. $16x + 12y = -48$
$3y + 12 = -4x$

41. $x = 4$
$x = 4y$

42. $y + 9 = 2x$
$y = 9 - 2x$

In Exercises 43–48, determine whether the equations are dependent or independent by writing them in the form $y = mx + b$.

43. $y - 2x = 6$
$y = 2x + 6$

44. $5x + y = 7$
$y = 7 - 5x$

45. $y + 2x = 1$
$y = 5 - 2x$

46. $y + 3 = x$
$y = x + 2$

47. $y = 3x$
$y - 3 = 0$

48. $y = x - 4$
$y = 4 - x$

In Exercises 49–60, use the graphing method to estimate the solutions of the given system of equations.

49. $2x - y = 6$
$5y - 2x = 10$

50. $20 = 3x - 4y$
$2y + 16 = 3x$

51. $5x = 8y + 56$
$4y - 2.5x = 40$

52. $3x + 3y = 24$
$2x + 2y = 12$

53. $x + y = -2$
$5y - 2x = 25$

54. $7x - 4y = -4$
$x - 2y = 8$

55. $4x = 24 - 3y$
$6y + 8x = 48$

56. $2y + 22 = x$
$44 = 2x - 4y$

57. $y - 10 = 0$
$3x - 2y = 12$

58. $2y + 18 = 0$
$2x + 3y = -3$

59. $x - 5 = 6$
$2y - 2x = 10$

60. $x + 8 = 2$
$x - 3y = -36$

In Exercises 61–72, use the graphing method to estimate the solutions of the given system of equations. Use a calculator to verify your solutions.

61. $x - 2y = 6$
$2x + y = 4$

62. $8 = y - x$
$5y + 5x = 8$

63. $3x + y = 4$
$x = 3y + 6$

64. $x + y + 5 = 0$
$4x - 6y - 11 = 0$

65. $x - 2y = 5$
$3x = 6y + 8$

66. $\frac{3}{4}x - \frac{2}{3}y = 3$
$x - \frac{8}{9}y = 8$

67. $x - 2y = 5$
$2x = 4y + 10$

68. $\frac{2}{5}x - \frac{1}{4}y = 2$
$x = \frac{5}{8}y + 5$

69. $1.50a - 2b = 1$
$2.25a + 2b = 0.25$

70. $2a + 3b = 1$
$4a + 15b = 8$

71. $2y - 7 = 0$
$\frac{2}{3}x + \frac{1}{2}y = 1$

72. $3x + 5y = 1.1$
$x - 2y = 0$

In Exercises 73–76, use the graphing method to estimate the answer to the question. Verify the answer by confirming that it satisfies the conditions of the problem.

73. The sum of two numbers is 69. One number is 9 less than twice the other. Determine the two numbers.

74. The difference of two numbers is 30. The larger number is 12 more than 3 times the other. What are the two numbers?

75. Two angles are supplementary. The measure of one angle is 12° more than the measure of the other. What is the measure of each angle?

76. Two angles are complementary. The measure of one angle is 6° less than 7 times the measure of the other. What is the measure of each angle?

Modeling with Real Data

77. The table shows the number (in millions) of people in the United States aged 7 and over who played soccer in 1992 and 1997.

| Year | Soccer Participation | |
	Male	Female
1992	7.18	3.44
1997	8.30	5.35

(Source: National Sporting Goods Association.)

(a) Write linear equations to model the data for males and females. Let x equal the number of years since 1992, and let y equal the number of participants.

(b) Estimate the solution of the system of equations.

78. What does the solution in Exercise 77 represent?

79. From 1990 to 1998, the number of consumer complaints against U.S. airlines about cancellations and delays declined from 3034 to 2277, whereas the number of complaints about ticketing and boarding problems rose from 624 to 1137. (Source: U.S. Department of Transportation, Office of Consumer Affairs.)

(a) Write linear equations to model the data for each category. Let x represent the number of years since 1990, and let y represent the number of consumer complaints.

(b) Explain how you know that the system of equations has a solution.

80. Refer to the model system in Exercise 79.

(a) Use the graphing method to estimate the solution of the system.

(b) What does the solution indicate?

Challenge

In Exercises 81–84, use the graphing method to estimate the solution of the given system of three equations in two variables.

81. $y = x - 12$
$2y + x = 0$
$2y = 3x - 32$

82. $x + y = 3$
$y = \dfrac{2}{5}x + 10$
$-3x + 2y = 31$

83. $y = 9 - x$
$3y + 3x = -21$
$2x = 18 - 2y$

84. $x - y = -7$
$x + y = 7$
$2y = x - 10$

85. Determine the vertices of a triangle formed by the graphs of $y = \dfrac{2}{3}x + 13$, $2y + 3x = 0$, and $x = 12$. Is it a right triangle?

86. Determine the vertices of a quadrilateral formed by the graphs of the equations $y = x + 6$, $y = x - 8$, $y = 7$, and $y = -4$. Is it a parallelogram?

5.2 | The Addition Method

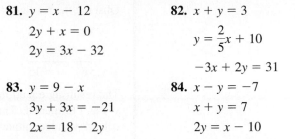

The Addition Method · Special Cases · Applications

The Addition Method

With the graphing method, we can estimate the solution of a system of equations and then verify that solution by substitution. Algebraic methods provide a more direct method for obtaining an exact solution.

In this text we will consider two algebraic methods for solving a system of equations. We begin with the addition method.

Exploring the Concept

Using the Addition Method to Solve a System

The Addition Property of Equations allows us to add the same quantity to both sides of an equation. For real numbers a, b, c, and d, if $a = b$, then $a + c = b + c$. If $c = d$, then we can extend the property.

$$a = b$$
$$\underline{c = d}$$
$$a + c = b + d$$

We call this procedure *adding the equations*. Note that because c and d have the same value, we are adding the same quantity to both sides of $a = b$. This fact is the basis for the addition method.

Consider the following system of equations.

$$3x - y = 8$$
$$2x + y = 7$$

Observe the result of adding the equations of the system.

$$\begin{array}{rl}
3x - y = & 8 \qquad \text{Note that the coefficients of } y \text{ are opposites: 1 and } -1. \\
\underline{2x + y = \; 7} & \qquad \text{Add the equations.} \\
5x \qquad = & 15 \qquad -y + y = 0
\end{array}$$

The resulting equation, $5x = 15$, contains only one variable, x, because the variable y was eliminated. This equation is a linear equation in one variable and can be solved with techniques that we already know.

$$\begin{array}{rl}
5x = 15 & \qquad \text{Divide both sides by 5.} \\
x = 3
\end{array}$$

We now know that the graphs of the two equations intersect at the point $(3, y)$. Because this pair must satisfy both equations, we can replace x with 3 in either equation and solve for y.

First Equation		*Second Equation*
$3x - y = \;\; 8$		$2x + y = 7$
$3(3) - y = \;\; 8$	Replace x with 3.	$2(3) + y = 7$
$9 - y = \;\; 8$	Now solve for y.	$6 + y = 7$
$-y = -1$		$y = 1$
$y = \;\; 1$		

The solution of the system is $(3, 1)$.

Note: Because the addition method results in the elimination of a variable, this method is sometimes called the *elimination method*.

When we use the addition method, our goal is to add the two equations and eliminate a variable. In order for us to accomplish this, the coefficients of one of the variables must be opposites. In other words, the sum of the coefficients must be 0.

If the coefficients of one variable are not opposites, we can make them opposites by applying the Multiplication Property of Equations, which allows us to multiply both sides of an equation by any nonzero number.

EXAMPLE 1 **Multiplying One Equation**

Solve the following system of equations.

$$\begin{array}{r}
x + 3y = -9 \\
4x - 2y = \;\;\; 6
\end{array}$$

Solution

Suppose that we wish to eliminate x. Adding the equations in their present form will not accomplish our goal. Thus we begin by multiplying the first equation by -4 so that the coefficients of x in the two equations will be opposites.

$$\begin{array}{rll}
-4(x + 3y) = -4\,(-9) \rightarrow -4x - 12y = & 36 & \quad \text{Multiply both sides by } -4. \\
4x - 2y = \;\;\; 6 \qquad\quad \rightarrow \underline{\;\; 4x - \;\; 2y = \;\;\; 6} & & \quad \text{Add the equations.} \\
-14y = & 42 & \quad -4x + 4x = 0 \\
y = & -3 & \quad \text{Divide both sides by } -14.
\end{array}$$

Now substitute -3 for y in either equation.

$$x + 3y = -9 \qquad \text{First equation}$$
$$x + 3(-3) = -9 \qquad \text{Replace } y \text{ with } -3.$$
$$x - 9 = -9 \qquad \text{Solve for } x.$$
$$x = \;\; 0$$

The solution of the system is $(0, -3)$.

Note: The solution of a system of equations is a *pair* of numbers. Remember to substitute the value of the first variable to obtain the value of the second variable.

To avoid multiplying both sides of an equation by a fraction, we sometimes need to multiply each equation of the system by a different number.

EXAMPLE 2

Multiplying Both Equations

Solve the following system of equations.

$$2x - 4y - \;\; 6$$
$$3x + 5y = -2$$

Solution

LEARNING TIP

To eliminate a specific variable, first determine the LCM of the coefficients of that variable. Then multiply the equations by numbers that make the resulting coefficients the LCM and its opposite.

To eliminate x, we can multiply the first equation by -3 and the second equation by 2.

$$-3(2x - 4y) = -3(6) \rightarrow -6x + 12y = -18$$
$$2(3x + 5y) = 2(-2) \rightarrow \underline{6x + 10y = \;\; -4} \qquad \text{Add the equations.}$$
$$22y = -22 \qquad {\scriptstyle -6x + 6x = 0}$$
$$y = \;\; -1$$

Substitute for y in either equation.

$$3x + 5y = -2 \qquad \text{Second equation}$$
$$3x + 5(-1) = -2 \qquad \text{Let } y = -1.$$
$$3x - 5 = -2 \qquad \text{Solve for } x.$$
$$3x = 3$$
$$x = 1$$

The solution is $(1, -1)$.

Note that it does not matter which variable we eliminate. We could have chosen to eliminate y first. To accomplish this, multiply the first equation by 5 and the second equation by 4.

Note: The Multiplication Property of Equations requires us to multiply both sides of an equation by the same number. However, it is not necessary to multiply both equations by the same number.

In order for us to use the addition method effectively, the like terms in both equations should be aligned. Usually we write the equations in standard form.

EXAMPLE  3

Using the Addition Method Twice

Solve the following system of equations.

$$2x + 1 = y$$
$$19 - 12y = 4x$$

Think About It

If the two equations of a system are in slope–intercept form, would you write them in standard form before applying the addition method? What is an alternative?

Solution

$$2x + 1 = y \quad \rightarrow 2x - \quad y = -1 \qquad \text{Write each equation in standard form.}$$
$$19 - 12y = 4x \rightarrow 4x + 12y = \quad 19$$

To eliminate y, multiply the first equation by 12.

$$24x - 12y = -12 \qquad\qquad 12(2x - y) = 12(-1)$$
$$\underline{\quad 4x + 12y = \quad 19\quad} \qquad \text{Add the equations.}$$
$$28x \qquad = \quad 7$$
$$x = \quad \frac{7}{28} = \frac{1}{4}$$

Normally we would replace x with $\frac{1}{4}$ and solve for y. To avoid fractions, an alternative is to return to the original system (in standard form) and eliminate x.

$$-4x + \quad 2y = \quad 2 \qquad\qquad \text{Multiply both sides of the first equation by } -2.$$
$$\underline{\quad 4x + 12y = 19\quad} \qquad \text{Add the equations.}$$
$$14y = 21$$
$$y = \frac{21}{14} = \frac{3}{2}$$

The solution is $\left(\frac{1}{4}, \frac{3}{2}\right)$.

If the coefficients in a system are fractions, it is usually convenient to clear the fractions before applying the addition method.

EXAMPLE 4

Clearing Fractions

Solve the following system of equations.

$$\frac{1}{2}x - \frac{1}{3}y = -1$$
$$2x + \frac{1}{2}y = \quad 7$$

Solution

Clear the fractions in each equation by multiplying by the LCD.

$$6\left(\frac{1}{2}x - \frac{1}{3}y\right) = 6(-1) \rightarrow 3x - 2y = -6$$
$$2\left(2x + \frac{1}{2}y\right) = 2(7) \quad \rightarrow 4x + \quad y = \quad 14$$

$$3x - 2y = -6$$
$$\underline{8x + 2y = 28} \quad \text{To eliminate } y, \text{ multiply both sides of the second equation by 2.}$$
$$11x \quad = 22 \quad \text{Add the equations.}$$
$$x = 2$$

$$2x + \frac{1}{2}y = 7 \quad \text{Second equation}$$

$$2(2) + \frac{1}{2}y = 7 \quad \text{Replace } x \text{ with 2.}$$

$$4 + \frac{1}{2}y = 7$$

$$\frac{1}{2}y = 3 \quad \text{Subtract 4 from both sides.}$$

$$y = 6 \quad \text{Multiply by 2.}$$

The solution is $(2, 6)$.

The following summarizes the addition method for systems that have unique solutions.

Using the Addition Method to Solve a System of Equations

1. Write each equation in the standard form $Ax + By = C$.
2. If necessary, multiply one or both equations by a nonzero number to obtain opposite coefficients of one of the variables.
3. Add the equations to eliminate that variable. The result is an equation in one variable.
4. Solve the resulting equation.
5. Substitute this value into either original equation and solve for the other variable.
6. Check the solution.

Special Cases

In Section 5.1, we observed that a system of equations does not always have a unique solution.

EXAMPLE 5

Special Cases

Use the addition method to solve each of the following systems. Then use the slope–intercept forms of the equations to interpret the results.

(a) $2x - 6y = 8$
$\quad\;\; x - 3y = 4$

(b) $4x - 2y = 10$
$\quad\;\; y - 1 = 2x$

Solution

(a) $\quad 2x - 6y = 8$
$\quad \underline{-2x + 6y = -8} \quad \text{Multiply the second equation by } -2.$
$\qquad\quad\; 0 = 0 \quad \text{True}$

Note that both variables were eliminated and the resulting equation is true. In the slope–intercept form, both equations of the system are $y = \frac{1}{3}x - \frac{4}{3}$. Thus the graphs coincide, the equations are dependent, and the system has infinitely many solutions. Note, however, that not every ordered pair is a solution. Only the infinitely many points of the line represent solutions.

(b) Write the equations in standard form.

$$4x - 2y = 10 \rightarrow \quad 4x - 2y = 10$$
$$y - 1 = 2x \quad \rightarrow -2x + \ y = \ \ 1$$

$$
\begin{array}{ll}
4x - 2y = 10 & \\
\underline{-4x + 2y = \ \ 2} & \text{Multiply the second equation by 2.} \\
\qquad\quad 0 = 12 & \text{False}
\end{array}
$$

Again, both variables were eliminated, but this time the resulting equation is false.

In slope–intercept form, the equations of the system are as follows.

$$y = 2x - 5$$
$$y = 2x + 1$$

Now we can see that the graphs are parallel, the system is inconsistent, and there is no solution.

Example 5 shows that applying the addition method may result in the elimination of both variables. We can state the following generalizations.

1. If the resulting equation is true (identity), the equations are dependent, and the system has infinitely many solutions.

2. If the resulting equation is false (contradiction), the system is inconsistent and has no solution.

Applications

In Example 6, we apply the addition method to solving an application problem involving a system of equations.

EXAMPLE 6

Ratios

One number is 8 more than another. The ratio of the smaller number increased by 10 to the larger number decreased by 6 is $\frac{5}{3}$. What are the two numbers?

Solution

Let $x =$ the smaller number and $y =$ the larger number. Write the first equation in standard form.

$$
\begin{array}{ll}
y = x + 8 & \text{The larger number is 8 more than the smaller number.} \\
-x + y = 8 & \text{Standard form}
\end{array}
$$

Next describe the revised numbers.

$x + 10 =$ smaller number increased by 10 and

$y - 6 =$ larger number decreased by 6.

$$\frac{x + 10}{y - 6} = \frac{5}{3}$$ Write the ratio of the revised numbers.

$3(x + 10) = 5(y - 6)$ Cross-multiply.

$3x + 30 = 5y - 30$ Distributive Property

$3x - 5y = -60$ Write the equation in standard form.

The two equations form the following system.

$-x + y = \quad 8$ First equation

$3x - 5y = -60$ Second equation

We use the addition method to solve the system.

$-3x + 3y = \quad 24$ Multiply the first equation by 3.

$\underline{\quad 3x - 5y = -60}$ Add the equations.

$\qquad -2y = -36$ Solve for y.

$\qquad y = \quad 18$

$y = x + 8$ Use the first equation to determine x.

$18 = x + 8$ Replace y with 18.

$x = 10$

The numbers are 10 and 18.

Quick Reference 5.2

The Addition Method

- The addition method (sometimes called the elimination method) is an algebraic technique for solving a system of equations. By *adding the equations* of the system, we try to eliminate one of the variables.

- For a system of equations with a unique solution, the addition method is as follows.

  1. Write each equation in the standard form $Ax + By = C$.

  2. If necessary, multiply one or both equations to obtain opposite coefficients of one of the variables.

  3. Add the equations to eliminate that variable. The result is an equation in one variable.

  4. Solve the resulting equation.

  5. Substitute this value into either original equation and solve for the other variable.

  6. Check the solution.

- Sometimes it is convenient to clear any fractions from the equations before applying the addition method.

- If the solution for the first variable is a fraction, it may be convenient to return to the original system and apply the addition method again to solve for the other variable.

Special Cases • Applying the addition method may result in the elimination of both variables.

1. If the resulting equation is true (identity), the equations are dependent, and the system has infinitely many solutions.

2. If the resulting equation is false (contradiction), the system is inconsistent and has no solution.

Speaking the Language 5.2

1. Adding equations of a system to eliminate a variable is called the ▨▨▨▨ or ▨▨▨▨ method.

2. In order for us to eliminate a variable by adding equations, the coefficients of the variable must be ▨▨▨▨.

3. If adding equations results in a false equation with no variable, the original system is ▨▨▨▨.

4. When we solve a system of two linear equations in two variables, there are three possible outcomes: a ▨▨▨▨ solution, ▨▨▨▨ solution, or ▨▨▨▨ solutions.

Exercises 5.2

Concepts and Skills

1. Under what circumstances will adding equations result in the elimination of a variable?

2. What is the best way to write the equations of a system when
   (a) we estimate the solution by graphing?
   (b) we use the addition method to solve the system?

In Exercises 3–8, estimate the solution by graphing. Then use the addition method to solve the system.

3. $y = x - 7$
   $x + y = 3$

4. $x + y = -12$
   $y = 2x$

5. $2x + y = -15$
   $y = x + 6$

6. $y = -2x + 8$
   $-2x + 3y = 0$

7. $y = 5 - x$
   $2x + 2y = -16$

8. $6x - 3y = 21$
   $y = 2x - 7$

In Exercises 9–18, solve with the addition method.

9. $2x + y = 2$
   $x - y = -11$

10. $3x + y = -6$
    $4x - y = 6$

11. $x + 5y = 13$
    $-x + 6y = 9$

12. $-3x + y = 5$
    $3x - 2y = -7$

13. $x - 6y = -11$
    $3x + 6y = 7$

14. $2x + 3y = -5$
    $-2x - y = 7$

15. $4x - 9y = -7$
    $-4x + 5y = -5$

16. $7x - 8y = -38$
    $-4x + 8y = 20$

17. $\dfrac{1}{5}x - \dfrac{3}{4}y = 4$
    $-\dfrac{2}{5}x + \dfrac{3}{4}y = -2$

18. $-\dfrac{4}{3}x + \dfrac{5}{8}y = -7$
    $\dfrac{4}{3}x - \dfrac{3}{8}y = 1$

19. Why is the addition method sometimes called the elimination method?

20. Consider the following system of equations.

$$4x + 3y = 15$$
$$2x - 5y = 1$$

Describe what to do in order to

(a) eliminate x.

(b) eliminate y.

In Exercises 21–32, use the addition method to solve the system of equations.

21. $5x + y = 12$
$3x + y = 2$

22. $x + 4y = 7$
$x - y = 7$

23. $3x + 7y = 8$
$6x - y = 1$

24. $4x + 3y = 3$
$7x + 9y = 24$

25. $x - 3y = 5$
$2x - 6y = -7$

26. $12x - 15y = 4$
$4x - 5y = 6$

27. $9x - 2y = 7$
$4x + 8y = 12$

28. $2x - 7y = 5$
$x + 14y = 20$

29. $4x - y = -3$
$-8x + 2y = 6$

30. $6x - y = 3$
$12x - 2y = 6$

31. $15x - 8y = 8$
$30x + 7y = -7$

32. $7x + 32y = -11$
$-12x - 16y = -20$

In Exercises 33–44, use the addition method to solve the system of equations.

33. $3x - 4y = 11$
$2x + 3y = -4$

34. $3x + 7y = 14$
$5x - 4y = -8$

35. $5x - 2y = 1$
$2x - 3y = 7$

36. $4x + 7y = 4$
$3x + 2y = 3$

37. $3x - 5y = 6$
$\dfrac{1}{2}x - \dfrac{5}{6}y = 1$

38. $2x + y = 6$
$3x + \dfrac{3}{2}y = 9$

39. $3x - 2y = 13$
$2x + 3y = 0$

40. $4x - 3y = -1$
$5x - 4y = -3$

41. $2x + 5y = -4$
$x + \dfrac{5}{2}y = 2$

42. $3x + y = 6$
$4x + \dfrac{4}{3}y = 12$

43. $6x + 5y = 14$
$4x + 7y = 2$

44. $8x + 3y = -2$
$7x - 4y = -15$

45. When solving a system of equations, we can use the Multiplication Property of Equations to adjust the coefficients in the equations. What is another way in which this property might be used?

46. Suppose that when we use the addition method to solve a system of equations, both variables are eliminated. Describe the solution set if the resulting equation is

(a) $0 = 1$.   (b) $0 = 0$.

In Exercises 47–66, solve the given system of equations.

47. $2x = 3y - 5$
$3x = 2y$

48. $3x + 12 = 5y$
$10 - 2y = 4x$

49. $x + y + 2 = 0$
$4 - 3y = 3x$

50. $2x - 8 = y$
$6x = 10 + 3y$

51. $3x = 4 + 2y$
$y = x + 3$

52. $y = 3x + 10$
$3x = 4 - y$

53. $2x = y + 1$
$5y - 1 = 2x$

54. $y + 3x = 1$
$y = 2 - 5x$

55. $10 - y - 3x = 0$
$0 = 5 - 6x - 2y$

56. $2x = 7 + 4y$
$12y + 21 = 6x$

57. $8x = 3y + 11$
$-3 = 7y + 4x$

58. $y = -4x + 11$
$2x - 3 = -3y$

59. $3x = 5 - 5y$
$4x = 5y - 5$

60. $x = 3x + 3y$
$2x + 2y = 4 + y$

61. $\dfrac{1}{2}x + \dfrac{1}{5}y = 0$
$\dfrac{1}{5}x + \dfrac{1}{3}y = 0$

62. $\dfrac{2}{3}x + \dfrac{1}{2}y = \dfrac{5}{2}$
$\dfrac{2}{5}x - y = \dfrac{1}{5}$

63. $2x - \dfrac{1}{3}y = 2$
$5x + \dfrac{1}{6}y = 11$

64. $-\dfrac{1}{2}x + 7y = 18$
$\dfrac{1}{4}x - 3y = -8$

65. $\dfrac{4}{7}x + \dfrac{1}{2}y = -1$
$\dfrac{4}{5}x - 5y = 10$

66. $\dfrac{4}{5}x + \dfrac{3}{5}y = 1$
$\dfrac{3}{8}x - \dfrac{1}{4}y = 1$

In Exercises 67–70, write a system of equations and use the addition method to solve the system.

67. A number is 1 more than half a larger number. The ratio of the smaller number decreased by 3 to the larger number decreased by 2 is $\frac{2}{5}$. What are the numbers?

68. The difference of a number and 3 times a smaller number is 11. The ratio of the larger number decreased by 2 to the smaller number increased by 4 is $\frac{3}{2}$. What are the numbers?

69. Two angles are complementary. The difference between 3 times the measure of the smaller angle and the measure of the larger angle is 6°. What is the measure of each angle?

70. Two angles are supplementary. The sum of half the measure of the larger angle and the measure of the smaller angle is 125°. What is the measure of each angle?

✎ Modeling with Real Data

71. The following table shows the findings of a survey to determine the percentage of men and the percentage of women who access the Internet at least once per month.

	Percentage	
Year	**Men**	**Women**
1998	25.3	19.6
1999	34.8	30.4

(Source: Media Mark Research Inc.)

(a) Let x represent the number of years since 1998, and let y represent the percentage. Write a system of equations to model the data for men and women.

(b) Solve the system of equations in part (a).

72. Repeat Exercise 71 with x representing the number of years since 1990 and y representing the percentage.

Data Analysis: *Online Shoppers*

The following graph shows a comparison between the percentages of online shoppers who are men and those who are women. (Source: *USA Today.*)

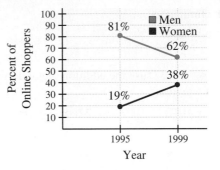

If we let x represent the number of years since 1990, the percentages y of online shoppers can be modeled by the following equations.

Men: $y = -4.75x + 104.75$

Women: $y = 4.75x - 4.75$

73. What do the slopes of the graphs of these two equations indicate?

74. How do you know that this system of equations has a unique solution?

75. Use the addition method to solve the system of equations.

76. What does the solution that you obtained in Exercise 75 represent?

Challenge

In Exercises 77 and 78, write the solution of the system in terms of a and b.

77. $x + y = a$
$x - y = b$

78. $y - ax = b, \quad a \neq 1$
$y - x = -b$

In Exercises 79 and 80, determine k such that the system has infinitely many solutions.

79. $x - 2y = k$
$4y - 2x = 10$

80. $kx - 3y = 21$
$x - y = 7$

In Exercises 81 and 82, determine k such that the system has no solution.

81. $kx + 14y = 2$
$2x - 7y = 3$

82. $3x + 4y = 0$
$-12x + ky = 4$

5.3 | The Substitution Method

The Substitution Method   •   Special Cases   •   Applications

The Substitution Method

In this section we consider a second algebraic technique for solving systems of equations. Although the addition method can always be used, the substitution method offers a good alternative in some cases.

The substitution method is based on the fact that we obtain an equivalent equation when we replace a variable with an expression that is equal to that variable. The substitution method is most useful when it is easy to solve for one variable in an equation.

EXAMPLE 1

Solving with the Substitution Method

Solve the following system of equations.

$$3x + 2y = 11$$
$$y = x - 2$$

Solution

Because $y = x - 2$, we can replace y in the first equation with the expression $x - 2$.

$3x + 2y = 11$	First equation
$3x + 2(x - 2) = 11$	Substitute $x - 2$ for y.
$3x + 2x - 4 = 11$	The resulting equation has only one variable.
$5x - 4 = 11$	
$5x = 15$	
$x = 3$	

Now use the second equation to determine y.

$y = x - 2$	Second equation
$y = 3 - 2$	Replace x with 3.
$y = 1$	

Verify that $(3, 1)$ is the solution of the system.

The following summarizes the substitution method.

LEARNING TIP

The method that you choose to solve a system influences the form in which you write the equations. To use the graphing method, write the equations in slope–intercept form. For the addition method, write the equations in standard form. For the substitution method, solve one equation for one of the variables.

The Substitution Method

1. Solve one equation for one of the variables.
2. Substitute the expression for that variable in the other equation. The resulting equation will have only one variable.
3. Solve the resulting equation.
4. Substitute this value into either original equation and solve for the other variable.
5. Check the solution.

Note that we can begin by solving either equation for either variable. Choose the equation for which a variable is easily isolated. In particular, if the coefficient of a variable in one equation is 1 or -1, that variable is often a convenient choice.

EXAMPLE **2**

Solving with the Substitution Method

Solve the following system of equations.

$$2y - x = 5$$
$$4x - 5y = -14$$

Solution

Because the coefficient of x is -1 in the first equation, it is easy to isolate x. Therefore, we solve the first equation for x.

$2y - x = 5$	First equation
$-x = -2y + 5$	Subtract $2y$ from both sides.
$x = 2y - 5$	Divide both sides by -1.

Now substitute $2y - 5$ for x in the second equation.

$4x - 5y = -14$	Second equation
$4(2y - 5) - 5y = -14$	Replace x with $2y - 5$. The equation now involves only y.
$8y - 20 - 5y = -14$	Distributive Property
$3y - 20 = -14$	
$3y = 6$	
$y = 2$	

LEARNING TIP

The solution of a system of equations is a pair of numbers. After you solve the equation in one variable, remember to sustitute back to determine the value of the other variable.

Use the first equation in the form $x = 2y - 5$ to determine x.

$x = 2y - 5$	
$x = 2 \cdot 2 - 5$	Replace y with 2.
$x = 4 - 5$	
$x = -1$	

The solution of the system is $(-1, 2)$.

Sometimes, as in the next example, there is no obvious choice of a variable to isolate. Here we use the substitution method for illustration, although the addition method is a better alternative.

EXAMPLE **3**

Solving with the Substitution Method

Use the substitution method to solve the system of equations.

$$5x - 2y = -2$$
$$3y - 4x = -4$$

Think About It

We said that the addition method is a better alternative for solving the system in Example 3. Why is it better?

Solution

We arbitrarily choose to solve for x in the second equation.

$$3y - 4x = -4 \qquad \text{Second equation}$$

$$-4x = -3y - 4 \qquad \text{Subtract } 3y \text{ from both sides.}$$

$$x = \frac{3}{4}y + 1 \qquad \text{Divide both sides by } -4.$$

Now replace x in the first equation with $\frac{3}{4}y + 1$.

$$5x - 2y = -2 \qquad \text{First equation}$$

$$5\left(\frac{3}{4}y + 1\right) - 2y = -2 \qquad \text{Replace } x \text{ with } \frac{3}{4}y + 1.$$

$$\frac{15}{4}y + 5 - 2y = -2 \qquad \text{Distributive Property}$$

$$\frac{7}{4}y + 5 = -2 \qquad \frac{15}{4}y - 2y = \frac{15}{4}y - \frac{8}{4}y = \frac{7}{4}y$$

$$\frac{7}{4}y = -7 \qquad \text{Multiply by } \frac{4}{7}.$$

$$y = -4$$

Finally, determine the value of x.

$$x = \frac{3}{4}y + 1$$

$$x = \frac{3}{4}(-4) + 1 \qquad \text{Replace } y \text{ with } -4.$$

$$x = -3 + 1$$

$$x = -2$$

The solution is $(-2, -4)$.

Special Cases

In Section 5.2, we found that special cases arise in which both variables are eliminated when we use the addition method. In the next example, we illustrate how to recognize these special cases when we use the substitution method.

EXAMPLE **4**

Special Cases

Use the substitution method to solve each system of equations.

(a) $2y + 6x = -24$
    $3x + y = 6$

(b) $6x - 2y = 10$
    $y + 5 = 3x$

Solution

(a) Solve the second equation for y.

$$3x + y = 6 \qquad \text{Second equation}$$
$$y = -3x + 6 \qquad \text{Subtract } 3x \text{ from both sides.}$$

Substitute $-3x + 6$ for y in the first equation.

$$2y + 6x = -24 \qquad \text{First equation}$$
$$2(-3x + 6) + 6x = -24 \qquad \text{Replace } y \text{ with } -3x + 6.$$
$$-6x + 12 + 6x = -24 \qquad \text{Distributive Property}$$
$$12 = -24$$

The resulting equation is false (contradiction). As we saw when we used the addition method, this means that the system has no solution.

(b) Solve the second equation for y.

$$y + 5 = 3x \qquad \text{Second equation}$$
$$y = 3x - 5 \qquad \text{Subtract 5 from both sides.}$$

Substitute $3x - 5$ for y in the first equation.

$$6x - 2y = 10 \qquad \text{First equation}$$
$$6x - 2(3x - 5) = 10 \qquad \text{Replace } y \text{ with } 3x - 5.$$
$$6x - 6x + 10 = 10$$
$$10 = 10$$

The resulting equation is true (identity). This means that the system has infinitely many solutions.

In general, when we use the substitution method to solve a system of equations, if the resulting equation is

1. a contradiction, the system has no solution.
2. an identity, the system has infinitely many solutions.

Applications

In Example 5, we illustrate how to use the substitution method to solve an application problem involving a system of equations.

EXAMPLE 5

Dimensions of a Rectangle

The length of a rectangle is 1 less than twice its width. If the perimeter is 46 feet, what are the dimensions?

Solution

Let $x = $ width and $y = $ length.

$$y = 2x - 1 \qquad \text{Subtract 1 from 2 times the width.}$$
$$2x + 2y = 46 \qquad \text{The formula for perimeter } P \text{ is } P = 2L + 2W.$$

We use the substitution method to solve the system.

$$2x + 2y = 46 \qquad \text{Second equation}$$
$$2x + 2(2x - 1) = 46 \qquad \text{Replace } y \text{ with } 2x - 1 \text{ from the first equation.}$$
$$2x + 4x - 2 = 46 \qquad \text{Distributive Property}$$
$$6x - 2 = 46 \qquad \text{Combine like terms.}$$
$$6x = 48 \qquad \text{Add 2 to both sides.}$$
$$x = 8 \qquad \text{Divide both sides by 6.}$$

$$y = 2x - 1 \qquad \text{Use the first equation to determine } y.$$
$$y = 2(8) - 1 \qquad \text{Replace } x \text{ with 8.}$$
$$y = 16 - 1$$
$$y = 15$$

The length is 15 feet, and the width is 8 feet.

Quick Reference 5.3

The Substitution Method
- The substitution method is a second algebraic technique for solving systems of equations. It is most useful when one equation is solved (or can be easily solved) for one of the variables.
- The substitution method is as follows.
  1. Solve one equation for one of the variables.
  2. Substitute the expression for that variable in the other equation. The resulting equation will have only one variable.
  3. Solve the resulting equation.
  4. Substitute this value into either original equation and solve for the other variable.
  5. Check the solution.

Special Cases
- When we use the substitution method to solve a system of equations, if the resulting equation is
  1. a contradiction, the system has no solution.
  2. an identity, the system has infinitely many solutions.

Speaking the Language 5.3

1. When one equation is already solved for one of the variables, the ▬▬▬▬ method is useful for solving a system.
2. If one equation of a system of equations is $3y - x = 7$, then solving the equation for ▬▬▬▬ would be a good choice when using the substitution method to solve the system.

3. Using the substitution method could lead to an equation with no variable. If that equation is true, then the equations of the system are ▨▨▨▨ .

4. No algebraic method is needed to solve the system $y = 2x - 7$ and $y = 2x + 5$ because the graphs of the equations are ▨▨▨▨ .

Exercises 5.3

Concepts and Skills

1. Consider the following system of equations.
$$5x + 3y = 11$$
$$2x + y = 5$$
Compare the first step in solving the system with
(a) the addition method.
(b) the substitution method.

2. Under what circumstances is the substitution method a good choice for solving a system of equations?

In Exercises 3–12, use the substitution method to solve the system of equations.

3. $8x + y = 9$
$y = 3 - 2x$

4. $y = 6x + 31$
$-2x - y = 1$

5. $x = 3y - 6$
$x - y = 8$

6. $3x + y = 11$
$x = 2y - 1$

7. $x = 5$
$3x - 4y = 3$

8. $y = -3$
$4x - 3y = 1$

9. $y = -2x - 7$
$3x + 2y = -10$

10. $7x + 5y = 0$
$y = 4 - 3x$

11. $\frac{1}{2}x - 4y = 6$
$x = 4 + 6y$

12. $x = 3y - 12$
$\frac{1}{3}x + y = 8$

In Exercises 13–24, use the substitution method to solve the system of equations.

13. $2x + y = 4$
$2x - y = 0$

14. $x + 2y = 11$
$2x - y = 12$

15. $4x - 5y = -20$
$x + 2y = 8$

16. $2y - x = 7$
$2x + 3y = 14$

17. $y + 5 = 0$
$2x + 3y = 1$

18. $6 - x = 0$
$3x + y = 3$

19. $2x - 6y = 10$
$7 = 2x - 5y$

20. $3x + 2y = 10$
$2y + 2 = 5x$

21. $5x - 4y = 4$
$2x - 3 = 3y$

22. $8x - 5y = 4$
$2x - 7y = -22$

23. $\frac{1}{6}x + \frac{1}{4}y = \frac{1}{3}$
$\frac{1}{3}x - \frac{1}{2}y = -\frac{4}{3}$

24. $\frac{1}{3}x - \frac{1}{6}y = -1$
$\frac{3}{2}x + \frac{1}{2}y = \frac{1}{2}$

25. Suppose that you are using the substitution method to solve a system of equations, and after you make a substitution, the resulting equation is $-2x = -2x$. Describe the graphs of the two equations and state the solution set.

26. Suppose that you are using the substitution method to solve a system of equations, and after you make a substitution, the resulting equation is $7 = 3$. Describe the graphs of the two equations and state the solution set.

In Exercises 27–40, use the substitution method to solve the system of equations.

27. $y = 2x$
$x - 2y = -9$

28. $x = y - 1$
$3y - x = 1$

29. $3x - 12y = 0$
$x - 4y = 2$

30. $3x - y = 5$
$y - 3x = 5$

31. $x - 2y = 3$
$6y - 3x = -9$

32. $3x - y = -2$
$-6x + 2y = 4$

33. $x + 4y = 2$
$x + 3y = -1$

34. $4x + 3y = 18$
$2x - 4y = -2$

35. $2y = 6x$
$3x - y = 0$

36. $2x - 4y = 3$
$8y - 4x = -6$

37. $2x + 3y = 0$
$x - 5y = 0$

38. $2x + 3y = 6$
$2x - 4y = -8$

39. $x + 5y = 2$
$3x = 5 - 15y$

40. $x - 3y = -1$
$6y - 2x = 3$

In Exercises 41–44, write a system of equations and solve the system algebraically.

41. The difference of two numbers is 16. The sum of the larger number and 3 times the smaller number is -24. What are the two numbers?

42. One number is 4 less than another. The difference of twice the smaller number and 5 times the larger number is -11. What are the numbers?

43. The width of a rectangle is 1 foot less than half its length. If the perimeter is 64 feet, what are the dimensions?

44. The length of a rectangle is 4 inches more than its width. If the perimeter is 48 inches, what are the dimensions?

Data Analysis: *Research and Development Funding*

From 1990 to 1999, federal funding for research and development increased from \$68 billion to \$75 billion. The following table shows the spending (in billions of dollars) in three categories.

Year	Energy	Environment and Natural Resources	General Science
1990	2.7	1.4	2.4
1999	1.5	2.0	4.6

(Source: Media Mark Research Inc.)

45. For each category, write a linear equation to model the data. (Let x represent the number of years since 1990, and let y represent the amount spent in billions of dollars.)

46. Judging by the slopes of the lines, for which category is funding increasing the most?

47. Consider the model equations for energy and environment as a system of equations. Solve the system. Round the results to the nearest integer and interpret the solution.

48. Consider the model equations for environment and general science as a system of equations. Solve the system. Round the results to the nearest integer and interpret solution the system.

Challenge

49. Determine k_1 such that the system in (a) has no solution, and determine k_2 such that the system in (b) has infinitely many solutions.

(a) $k_1 y + 12 = x$ (b) $3y - 2x = k_2$

$3y - x = 21$ $y = \dfrac{2}{3}x + 1$

In Exercises 50 and 51, use the substitution method to solve the system of three linear equations in three variables.

50. $x + 2y - z = -6$ **51.** $x - 2 = 5$
$y + z = 4$ $x + 2y = 13$
$3z = 9$ $2x - y + z = 9$

52. Determine the vertices of a quadrilateral bounded by the graphs of the equations $y = x - 5$, $y + x = 3$, $x + y + 2 = 0$, and $y - x = 7$. What specific kind of quadrilateral is this figure?

53. Determine the vertices of a triangle bounded by the graphs of $x + 3 = 0$, $y = 6$, and $6y - 7x = 15$. An **isosceles triangle** is a triangle in which two of the sides are of equal length. Is this triangle an isosceles triangle?

5.4 | Applications

Mixture Problems • Motion Problems • Percents • Past and Future Conditions

In the first three sections of this chapter, we considered simple application problems involving numbers and geometric figures. To solve such problems, we used two variables to represent the unknown quantities and wrote a system of two linear equations to describe the conditions of the problem. Then we used the addition method or the substitution method to solve the system.

In this section, we consider other kinds of application problems. Some of the problems are similar to those in previous chapters. The purpose of reconsidering them here is to learn how to solve such problems with systems of equations.

Mixture Problems

In this category, items of different values form a collection or mixture that has a total value. Typically, we are asked to find the quantity of each item in the mixture.

EXAMPLE **1** | **A Collection of Coins**

A vending machine contains only dimes and quarters. The total value of the 30 coins in the machine is $4.80. How many coins of each kind are there?

Solution

Let x = the number of dimes and y = the number of quarters. A table is helpful in organizing the information.

	Number of Coins	Unit Value (cents)	Total Value (cents)
Dimes	x	10	$10x$
Quarters	y	25	$25y$
Total	30		480

Write the system of equations.

$$x + y = 30 \qquad \text{The total number of coins is 30.}$$
$$10x + 25y = 480 \qquad \text{The total value is 480 cents.}$$

Use the addition method to solve the system of equations.

$$-10x - 10y = -300 \qquad \text{Multiply the first equation by } -10.$$
$$\underline{10x + 25y = 480} \qquad \text{Add the equations.}$$
$$15y = 180$$
$$y = 12$$

$$x + y = 30 \qquad \text{Use the first equation to determine } x.$$
$$x + 12 = 30 \qquad \text{Replace } y \text{ with 12.}$$
$$x = 18$$

There are 18 dimes and 12 quarters.

EXAMPLE **2** | **A Blend of Teas**

Cinnamon tea worth $4 per pound is mixed with apple tea worth $2.50 per pound. How many pounds of each tea should be used to obtain 18 pounds of a mixture worth $3 per pound?

Solution

Let x = the number of pounds of cinnamon tea and y = the number of pounds of apple tea. In the table, the total value of each tea and of the blend is the weight times the unit price.

	Weight (pounds)	Unit Price (dollars/pound)	Total Value (dollars)
Cinnamon Tea	x	4.00	$4.00x$
Apple Tea	y	2.50	$2.50y$
Blend	18	3.00	54.00

Now write the system.

$$x + \quad y = 18 \qquad \text{The total weight is 18 pounds.}$$
$$4.00x + 2.50y = 54.00 \qquad \text{The total value is \$54.00.}$$

We use the addition method to solve the system.

$$-4.00x - 4.00y = -72 \qquad \text{Multiply the first equation by } -4.00.$$
$$\underline{4.00x + 2.50y = \quad 54} \qquad \text{Add the equations.}$$
$$-1.50y = -18$$
$$y = \quad 12$$

$$x + \ y = 18 \qquad \text{Use the first equation to determine } x.$$
$$x + 12 = 18 \qquad \text{Replace } y \text{ with 12.}$$
$$x = \ 6$$

The blend consists of 6 pounds of cinnamon tea and 12 pounds of apple tea.

Motion Problems

Distance d, rate r, and time t are related by the formula $d = rt$. Systems of equations are particularly useful when motion problems involve two travelers or when the speed of wind or water affects the rate of travel.

EXAMPLE **3**

Motion Affected by Water Current

A barge makes a 240-mile trip up river in 15 hours and returns in 8 hours. What are the speeds of the barge and of the water current?

Solution

Let $x =$ the speed of the barge in still water and $y =$ the speed of the current. In the table, note that the current reduces the upstream rate and increases the downstream rate.

	Rate (mph)	Time (hours)	Distance (miles)
Upstream	$x - y$	15	240
Downstream	$x + y$	8	240

LEARNING TIP

To help understand the effect of current, think of walking on a moving sidewalk. Walking with the sidewalk, you move faster; walking against the sidewalk, you move slower.

Use the formula $d = rt$ to write the system of equations.

$$15(x - y) = 240$$
$$8(x + y) = 240$$

Simplify each equation and use the addition method to solve the system of equations.

$$
\begin{array}{ll}
x - y = 16 & \text{Divide both sides of the first equation by 15.} \\
\underline{x + y = 30} & \text{Divide both sides of the second equation by 8.} \\
2x \quad\;\; = 46 & \text{Add the equations.} \\
x = 23 &
\end{array}
$$

$$
\begin{array}{ll}
x + y = 30 & \text{Use the second equation to determine } y. \\
23 + y = 30 & \text{Replace } x \text{ with 23.} \\
y = 7 &
\end{array}
$$

The barge travels at 23 mph in still water, and the speed of the current is 7 mph.

EXAMPLE **4**

Two Travelers in the Same Direction

A car leaves Denver and travels north at 65 mph. Thirty minutes later, a truck leaves the same location and travels along the same route at 60 mph. How long will each vehicle have traveled when the car is 50 miles ahead of the truck?

Solution

Let x = number of hours that the car travels and y = the number of hours that the truck travels.

	Rate (mph)	Time (hours)	Distance ($d = rt$) (miles)
Car	65	x	$65x$
Truck	60	y	$60y$

Think About It

Suppose that in Example 4, both the car and the truck traveled at 60 mph. Write and solve the system of equations and explain the solution.

$$
\begin{array}{ll}
x - \quad y = 0.5 & \text{The difference in time is 30 minutes, or 0.5 hour.} \\
65x - 60y = 50 & \text{The difference in distance is 50 miles.}
\end{array}
$$

Use the addition method to solve the system of equations.

$$
\begin{array}{ll}
-60x + 60y = -30 & \text{Multiply the first equation by } -60. \\
\underline{65x - 60y = \quad 50} & \text{Add the equations.} \\
5x \quad\quad\;\; = \quad 20 & \\
x = \quad 4 &
\end{array}
$$

$$
\begin{array}{ll}
x - y = 0.5 & \text{Use the first equation to determine } y. \\
4 - y = 0.5 & \text{Replace } x \text{ with 4.} \\
-y = -3.5 & \\
y = \quad 3.5 &
\end{array}
$$

The vehicles will be 50 miles apart 4 hours after the car leaves and 3.5 hours after the truck leaves.

Percents

Applications involving percents include problems such as dual interest rates and liquid solutions.

EXAMPLE 5

Dual Interest Rates

To repair flood damages to an antique store, the owner borrowed $9000, part at 9% simple interest and part at 11% simple interest. If the total interest for one year was $844, how much did the owner borrow at each rate?

Solution

Let x = the amount borrowed at 9% and y = the amount borrowed at 11%. The simple interest paid is the interest rate (expressed as a decimal) times the amount of money borrowed.

	Amount Borrowed	Interest Rate	Interest Paid
9% Loan	x	0.09	$0.09x$
11% Loan	y	0.11	$0.11y$
Total	9000		844

Write the system of equations.

$$x + \quad y = 9000 \qquad \text{The total amount borrowed is \$9000.}$$
$$0.09x + 0.11y = \quad 844 \qquad \text{The total interest due is \$844.}$$

Use the addition method to solve the system of equations.

$$
\begin{aligned}
-0.09x - 0.09y &= -810 \qquad \text{Multiply the first equation by } -0.09.\\
\underline{0.09x + 0.11y} &= \underline{\quad 844} \qquad \text{Add the equations.}\\
0.02y &= \quad 34\\
y &= 1700
\end{aligned}
$$

$$
\begin{aligned}
x + \quad y &= 9000 \qquad \text{Use the first equation to determine } x.\\
x + 1700 &= 9000 \qquad \text{Replace } y \text{ with 1700.}\\
x &= 7300
\end{aligned}
$$

The owner borrowed $7300 at 9% and $1700 at 11%.

Recall that the amount of substance in a liquid solution is expressed as a percentage concentration.

$$(\text{Volume of solution}) \cdot (\text{Concentration of substance}) = \text{Volume of substance}$$

EXAMPLE 6

Liquid Solutions

How many gallons each of a 45% insecticide solution and a 30% solution should be mixed to obtain 24 gallons of a 35% solution?

Solution

Let x = the required number of gallons of the 45% solution and y = the required number of gallons of the 30% solution.

	Volume of Solution (gallons)	Concentration of Insecticide	Volume of Insecticide (gallons)
45% Solution	x	0.45	$0.45x$
30% Solution	y	0.30	$0.30y$
35% Solution	24	0.35	8.4

Write the system of equations.

$$x + y = 24 \qquad \text{The sum of the volumes of solution is 24 gallons.}$$
$$0.45x + 0.30y = 8.4 \qquad \text{The sum of the volumes of insecticide is 8.4 gallons.}$$

Use the addition method to solve the system of equations.

$$
\begin{array}{ll}
-0.45x - 0.45y = -10.8 & \text{Multiply the first equation by } -0.45. \\
\underline{0.45x + 0.30y = 8.4} & \text{Add the equations.} \\
-0.15y = -2.4 & \\
y = 16 &
\end{array}
$$

$$
\begin{array}{ll}
x + y = 24 & \text{Use the first equation to determine } x. \\
x + 16 = 24 & \text{Replace } y \text{ with 16.} \\
x = 8 &
\end{array}
$$

Use 8 gallons of the 45% solution and 16 gallons of the 30% solution.

Past and Future Conditions

This type of application problem typically involves two things or people and presents information about certain current conditions and information about those same conditions at some other time in the past or future.

EXAMPLE 7

Ages of Artifacts

Two artifacts, a knife and a piece of pottery, were discovered in 1995. The knife was twice as old as the pottery. In 1920 the knife was three times as old as the pottery. When was each piece made?

Solution

Let x = the age of the pottery in 1995 and y = the age of the knife in 1995.

$$
\begin{array}{ll}
y = 2x & \text{In 1995 the knife's age was twice the age of the pottery.} \\
2x - y = 0 & \text{Standard form}
\end{array}
$$

In 1920 the age of each piece was 75 years less than the age in 1995. Thus we let $x - 75 =$ the age of the pottery in 1920, and $y - 75 =$ the age of the knife in 1920.

$$y - 75 = 3(x - 75)$$ In 1920 the knife's age was three times the pottery's age.

$$y - 75 = 3x - 225$$ Distributive Property

$$-3x + y = -150$$ Standard form

Use the addition method to solve the system of equations.

$$
\begin{array}{rl}
2x - y = 0 & \text{Equation describes ages in 1995.}\\
-3x + y = -150 & \text{Equation describes ages in 1920.}\\
\hline
-x = -150 & \text{Add the equations.}\\
x = 150 &
\end{array}
$$

$$y = 2x$$ Use the first equation to determine y.

$$y = 2(150)$$ Replace x with 150.

$$y = 300$$

In 1995 the knife was 300 years old, and the pottery was 150 years old. The knife was made in 1695 and the pottery in 1845.

EXAMPLE 8

Relative Sizes of Two Parks

A city has two parks. Olde Mill Park is 4 times as large as Dupree Park. If the size of each park were increased by 60 acres, Olde Mill Park would be only 3 times as large as Dupree Park. How large is each park now?

Solution

Let $x =$ the number of acres in Dupree Park and $y =$ the number of acres in Olde Mill Park.

$$y = 4x$$ Currently, Olde Mill is 4 times the size of Dupree.

$$4x - y = 0$$ Standard form

If each park is increased by 60 acres, then $x + 60 =$ the size of Dupree Park, and $y + 60 =$ the size of Olde Mill Park.

$$y + 60 = 3(x + 60)$$ Olde Mill would be 3 times the size of Dupree.

$$y + 60 = 3x + 180$$ Distributive Property

$$-3x + y = 120$$ Standard form

Use the addition method to solve the system of equations.

$$
\begin{array}{rl}
4x - y = 0 & \text{Equation describing current size}\\
-3x + y = 120 & \text{Equation describing increased size}\\
\hline
x = 120 & \text{Add the equations.}
\end{array}
$$

$$y = 4x$$ Use the first equation to determine y.

$$y = 4(120)$$ Replace x with 120.

$$y = 480$$

Currently, Dupree Park has 120 acres, and Olde Mill Park has 480 acres.

Exercises 5.4

Real-Life Applications

1. What is the relationship among a number of items, the unit values of the items, and the total value of the items?

2. What is the effect of the water current on the rate of a boat that travels upstream and downstream?

Mixtures

3. A collection of nickels and quarters contains 43 coins and is worth $4.95. How many coins of each kind are in the collection?

4. A jar contains dimes and nickels worth $3.45. The number of dimes is 6 less than 4 times the number of nickels. How many coins of each kind are in the jar?

5. A cashier has $275 in $5 bills and $20 bills. The number of $5 bills is 1 less than 3 times the number of $20 bills. How many bills of each kind does the cashier have?

6. A wallet contains $57 in $1 and $5 bills. The number of ones is 1 more than twice the number of fives. How many bills of each kind does the wallet contain?

7. Basil at $6.20 per ounce and oregano at $3.70 per ounce are mixed to obtain 20 ounces of a blend worth $5.20 per ounce. How many ounces of each herb are used?

8. Gladioli are worth $6 per dozen, and dahlias are worth $20 per dozen. A garden shop plans to sell a mixture of 10 dozen of these plants for $11.60 per dozen. How many dozens of each type of plant should be included in the mixture?

Motion

9. An airplane flew 800 miles in 4 hours with the wind. The return trip against the wind required 5 hours. What were the speed of the wind and the speed of the plane?

10. A boat traveled 210 miles in 10 hours against the current. The boat then returned with the current in 6 hours. What were the speeds of the boat and of the current?

11. A canoeist paddled downstream 14 miles in 2 hours. Later, the canoeist took the same amount of time to paddle 6 miles upstream. What were the speeds of the canoe and the current?

12. An airplane completed a 3150-mile trip against the jet stream in 9 hours. The return trip required 7 hours. What were the speed of the plane and the wind speed?

13. A 306-mile automobile trip required 5.5 hours. During part of the trip, the car traveled on a freeway at an average speed of 62 mph. For the remainder of the trip, the car traveled on secondary roads at an average speed of 48 mph. How far did the car travel on each type of road?

14. An athlete running at 6 mph and later cycling at 24 mph completed a 61-mile course in 3.75 hours. How far did she cycle and how far did she run?

15. Suppose that a solution is 20% acid. What additional information do you need in order to know the actual amount of acid in the solution? Once that information is known, how do you calculate the amount of acid in the solution?

16. Suppose that an automobile is presently x years old. What is your interpretation of the expressions $x - 2$ and $x + 3$?

Percents

All interest applications involve simple interest.

17. A savings account pays 4% per year, and a certificate of deposit pays 5.5% per year. How much is in each account if a total investment of $8250 earns $414 in interest for 1 year?

18. To pay college expenses, a student borrowed $1450, part from the college student loan fund at 3% and part from the bank at 7%. If the interest for the year was $67.10, how much did the student borrow from each source?

19. A discount store earns profits from retail sales of 20% on clothing items and 15% on other items. If the total sales for one day were $9950 and the total profits for the day were $1677.50, what were the sales in each category?

20. A real estate agent receives a 9.5% commission on total sales of properties sold for $100,000 or less and an 8% commission on total sales of properties sold for over $100,000. If the agent's total sales for the year were $521,000 and his total commission was $43,465, what were the agent's total sales in each category of property?

21. How many liters each of a 75% chlorine solution and a 45% chlorine solution should be used to obtain 15 liters of a 65% chlorine solution?

22. A hospital employee needs to mix 70% disinfectant solution with 45% solution to obtain 7.5 gallons of a 60% solution. How many gallons of each concentration are needed?

Past and Future Conditions

23. A car is twice as old as a truck. Two years ago, the car was four times as old as the truck. How old is each vehicle now?

24. One historic site is three times as old as another. In 25 years, it will be only twice as old. How old is each site now?

25. This year a farmer planted four times as many acres in wheat as in corn. Next year, if he increases the size of each field by 20 acres, he will have three times as many acres in wheat as in corn. How many acres of each grain did he have this year?

26. The number of vacationers per week who visited Lake Helen this year is five times the number who visited Clear Lake. If each lake has 40 more visitors per week next year, the number at Lake Helen will be three times the number at Clear Lake. How many vacationers per week visited each lake this year?

27. In a certain residential area, the number of single-family homes is three times the number of apartment units. If the number of apartment units decreases by 50 and the number of single-family homes increases by the same amount, there will be five times as many homes as apartments. How many dwellings of each kind are there now?

28. In a certain town, the number of acres of undeveloped land is four times the number of acres of commercial property. If 100 acres of undeveloped land becomes commercial, the number of acres of undeveloped land will be twice the number of acres of commercial property. How many acres of each type of property are there now?

Miscellaneous Applications

29. An auto dealer has three times as many cars as trucks. If there are 48 vehicles altogether, how many of each kind does the dealer have?

30. The number of people who attended a matinee was 20 less than the number at the evening performance. If a total of 300 people attended the two performances, how many attended each?

31. A fruit company sells special 15-pound gift boxes containing pears and apples for $18.75. If apples are $1.10 per pound and pears are $1.40 per pound, how many pounds of each fruit are included in the box?

32. Deluxe mixed nuts sell for $4 per pound. The mixture contains cashews that cost $4.50 per pound and pecans that cost $3.25 per pound. How many pounds of each type of nut does 12 pounds of mixed nuts contain?

33. A trucker drove 87 miles. He spent 45 minutes on icy roads and 1.5 hours on wet roads. His speed on icy roads was 4 mph less than half his speed on wet roads. What was the speed under each condition?

34. A person drove from her home to the Atlanta airport in 1 hour and then boarded a plane for a 1.5-hour flight to Cleveland. The plane traveled 320 mph faster than the car, and the total distance from the person's home to Cleveland was 600 miles. What were the person's driving speed and the plane's speed?

35. The numerator of a fraction is 4 less than the denominator. If the numerator is increased by 2 and the denominator is increased by 3, the resulting fraction equals $\frac{1}{2}$. What is the original fraction?

36. The difference of the numerator and denominator of a fraction is 4. If the numerator is decreased by 1 and the denominator is doubled, the resulting fraction equals $\frac{4}{5}$. What is the original fraction?

37. A city park worker has 30 linear feet of lumber to build a rectangular sandbox for the children's play area. If the length should be twice the width, what will the dimensions of the sandbox be?

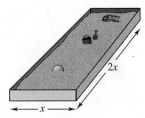

38. It takes 2300 feet of fencing to enclose a rectangular pasture. The length of the pasture is 200 feet less than twice its width. What are the dimensions of the pasture?

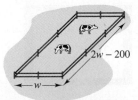

39. A recycling drive received $50.56 for a collection of 842 cans and bottles. If the cans were worth 5¢ each and the bottles 8¢ each, how many of each were collected?

40. The day's box office receipts for a movie were $1431. Tickets were $4.50 for the matinee and $6 for the evening show. If a total of 258 people attended the two shows, how many attended each show?

41. A prescription calls for 60 ml of a solution containing 30% medication. How many milliliters of a 25% solution and a 40% solution should be combined?

42. A bricklayer needs to make a cleaning solution by mixing 60% acid solution with 35% acid solution to obtain 20 liters of 50% solution. How many liters of each solution should be used?

43. The total number of points scored in a football game was 45. The losing team scored 3 points more than half the number of points that the winning team scored. What was each team's score?

44. A community has 660 housing units (apartment units and single-family homes). The number of homes is 180 more than the number of apartment units. How many dwellings of each type are there?

45. A basketball team made 50% of the free throws it attempted. If the team is successful on the next 20 consecutive free throws, its success rate will increase to 75%. Currently, how many free throws has the team attempted, and how many attempts were successful?

46. A gardener planted one-fourth as many marigolds as zinnias. If the gardener were to add 30 plants of each type, the number of marigolds would be one-half the number of zinnias. How many of each plant did the gardener plant originally?

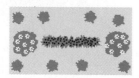

47. A strip of wood 40 inches long is cut into four pieces and glued together to make a rectangular picture frame. The frame's length is 4 inches more than its width. What are the dimensions of the frame?

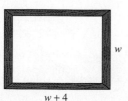

$w + 4$

48. A triangular lot is enclosed by 180 feet of fencing. One side of the lot is 70 feet long. The second side is 40 feet shorter than the third side. What are the dimensions of the lot?

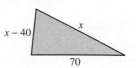

49. One biker left Durango and traveled north. Another biker left Silverton, which is 50 miles north of Durango, and cycled south along the same road. The southbound biker averaged 5 mph more than the northbound biker, and they met in 2 hours. What was the speed of each biker?

50. A car and a motorcycle left a rest stop on route I-40 at the same time. The car traveled east, and the motorcycle traveled west. If the car's average speed was 18 mph less than that of the motorcycle, what was the speed of each if they were 305 miles apart after $2\frac{1}{2}$ hours?

51. It costs $1176 to buy two kinds of carpet to cover 72 square yards of floor space in a home. If the carpet for the upstairs area costs $18 per square yard and the carpet for the basement recreation area costs $14 per square yard, how many square yards of carpet will be used on each level?

52. An investor purchased three times as many shares of Duke Power at $37 per share as Wrigley at $51 per share. If the total investment was $9720, how many shares of each company's stock were purchased?

53. In a small town, the number of Republicans is twice the number of Democrats. Two years ago each party had 800 fewer members, and the number of Republicans was three times the number of Democrats. What is the current number of Democrats and Republicans?

54. For a reception, a caterer prepared three times as many cheese straws as cucumber sandwiches. After 30 minutes, 50 of each item had been eaten, and the number of cheese straws remaining was 5 times the number of cucumber sandwiches. How many of each item did the caterer prepare?

55. An alloy of 45% copper was mixed with an alloy of 20% copper to obtain 30 pounds of alloy containing 25% copper. How many pounds of each alloy were used?

56. A nature store has two types of wild bird food. One contains 35% sunflower seed, and the other contains 28% sunflower seed. How much of each type is needed to obtain 35 pounds of bird food consisting of 30% sunflower seed?

57. A bond issue passed by a 3-to-2 margin. If 860 people voted, how many voted for the bond issue and how many voted against it?

58. In a can of mixed nuts, the ratio of peanuts to all other nuts is 5 to 3. In a 16-ounce can, how many ounces of peanuts and how many ounces of other nuts are there?

59. At Taco Supreme, 5 tacos and 2 colas cost $5.50. The bill for 3 tacos and 1 cola is $3.15. What is the price of each item?

60. The cost of 5 gallons of paint and 8 rolls of wallpaper is $253. Seven gallons of paint and 5 rolls of wallpaper cost $224. What is the price of each item?

Modeling with Real Data

61. As farming has become a more competitive business, the number of small farms has declined, whereas the number of large farms has remained about the same.

	Number of Farms (thousands)		
	1987	**1992**	**1997**
Under 10 acres	183	166	154
1000–1999 acres	102	102	101

(Source: U.S. Census Bureau.)

(a) Use the data for 1987 and 1997 to write linear equations that model the data for each farm size. (Let x represent the number of years since 1987, and let y represent the number of farms.)

(b) Solve the system of equations in part (a) to determine the year in which the numbers of farms in both categories will be the same.

62. Using the data for 1992 and 1997, repeat parts (a) and (b) of Exercise 61. (Let x represent the number of years since 1987.)

Data Analysis: *Baseball Payrolls*

The table shows the payrolls for the New York Yankees, Detroit Tigers, and Boston Red Sox for the years 1994 and 1999.

	Payroll (millions of dollars)		
	New York Yankees	**Detroit Tigers**	**Boston Red Sox**
1994	47.5	41.1	36.3
1999	94.4	49.4	80.4

(Source: Major League Player Relations Committee.)

63. Use the given data to write a linear equation to model the data for each team. (Let x represent the number of years since 1994, and let y represent the payroll in millions of dollars.)

64. Consider the models for New York and Detroit as a system of equations. Solve the system and interpret the solution.

65. Consider the models for Boston and Detroit as a system of equations. Solve the system and interpret the solution.

66. Consider the models for New York and Boston as a system of equations. How do you know that the system has no solution for years beyond 1994?

Challenge

67. A rancher on horseback traveling at 8 mph rides the fence lines of a rectangular ranch in 3 hours. If the width of the ranch is 2 miles less than its length, what are the dimensions of the ranch?

68. The perimeter of an isosceles triangle is 21 inches. The shortest side is 3 inches less than either of the two equal sides. What are the triangle's dimensions?

69. A cat saw a dog 20 feet to its right and started to run left toward a tree. At the same instant, the dog saw the cat and began running toward it. The dog's running speed (in feet per second) is $\frac{4}{5}$ that of the cat. In 2 seconds, the cat had reached the tree, and the dog was 26 feet from the tree. How fast did each animal run?

70. Tiles 1 foot square are used to cover a floor that is 10 feet wide and 12 feet long. The layout grid requires four times as many solid-color tiles as patterned tiles. How many tiles of each type are needed?

5.5 | Systems of Linear Inequalities

Verifying Solutions • Graphing Solution Sets

Verifying Solutions

In Section 4.7, we defined a linear inequality in two variables. In many problems involving inequalities, we need two or more inequalities to describe fully the conditions of the problem.

> **Definition of a System of Inequalities**
>
> Two or more inequalities considered simultaneously form a **system of inequalities**. A **solution** of a system of inequalities is an ordered pair of numbers that satisfies each inequality in the system.

EXAMPLE 1

Verifying Solutions

In each part, determine whether the ordered pair is a solution of the following system of inequalities.

$$x - 4y > 3$$
$$y \geq 2x - 6$$

(a) $(-4, -2)$ (b) $(6, -4)$

Solution

(a) Replace x with -4 and y with -2.

$x - 4y > 3$	$y \geq 2x - 6$
$-4 - 4(-2) > 3$	$-2 \geq 2(-4) - 6$
$-4 + 8 > 3$	$-2 \geq -8 - 6$
$4 > 3$ True	$-2 \geq -14$ True

Because both inequalities are true, the pair $(-4, -2)$ is a solution of the system.

(b) Replace x with 6 and y with -4.

$x - 4y > 3$	$y \geq 2x - 6$
$6 - 4(-4) > 3$	$-4 \geq 2(6) - 6$
$6 + 16 > 3$	$-4 \geq 12 - 6$
$22 > 3$ True	$-4 \geq 6$ False

Although the pair $(6, -4)$ satisfies the first inequality, it does not satisfy the second inequality. Therefore, this pair is not a solution of the system.

Graphing Solution Sets

A solution of a system of equations is represented by a point that belongs to the graph of each equation. Thus a point of intersection of the lines corresponds to a solution of the system. Similarly, a solution of a system of inequalities is represented by a point that belongs to the graph of each inequality. Because the graph of an inequality is a half-plane, the points that belong to the intersection of the half-planes correspond to solutions of the system.

<table>
<tr><td>**Exploring the Concept**</td></tr>
</table>

Graphing the Solution Set of a System of Inequalities

Consider the following system of inequalities.

$$y \leq x + 3$$
$$x + 2y > -4$$

To graph this system, we begin by graphing each inequality. To graph $y \leq x + 3$, graph the boundary line $y = x + 3$. Recall that we use a solid line because the inequality includes the equality symbol. The symbol $\leq$ indicates that the other solutions are represented by the half-plane below the line. (See Fig. 5.9.)

Figure 5.9

Figure 5.10

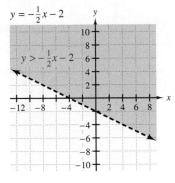

To graph the second inequality, first solve for y.

$$x + 2y > -4$$
$$2y > -x - 4$$
$$y > -\frac{1}{2}x - 2$$

Because this inequality does not include the equality symbol, we graph the boundary line $y = -\frac{1}{2}x - 2$ with a dashed line. Shade above the boundary line because y is larger than the expression on the right side. (See Fig. 5.10.)

Points that represent solutions of the system of inequalities belong to both half-planes. Therefore, we represent the solution set by shading the intersection of the two half-planes. (See Fig. 5.11.)

Figure 5.11

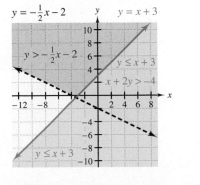

LEARNING TIP

Think of the graph of each inequality as being drawn on a sheet of clear plastic. Place one on top of the other. The points where the shaded regions overlap represent the solutions of the system.

The preceding results suggest the following procedure.

Graphing the Solution Set of a System of Inequalities

1. Graph each inequality in the same coordinate system.
2. Shade the intersection of the half-planes.

EXAMPLE **2**   **Graphing a System of Linear Inequalities**

Graph the solution set of the following system of inequalities.

$$4x + 3y < -21$$
$$x - 2y > 2$$

Figure 5.12

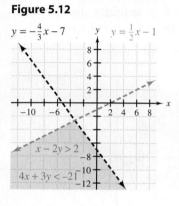

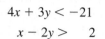

Solution

Solve each inequality for y.

First Inequality	*Second Inequality*
$4x + 3y < -21$	$x - 2y > 2$
$3y < -4x - 21$	$-2y > -x + 2$
$y < -\dfrac{4}{3}x - 7$	$y < \dfrac{1}{2}x - 1$

Note that both boundary lines are dashed lines. The solution set is represented by the intersection of the two half-planes. (See Fig. 5.12.)

EXAMPLE **3**   **Graphing a System of Linear Inequalities**

Graph the solution set of the following system of inequalities.

$$x \geq -2$$
$$y \leq 3$$

Solution

Note that the solid boundary line $x = -2$ is a vertical line and the solid boundary line $y = 3$ is a horizontal line. (See Fig. 5.13.)

Figure 5.13

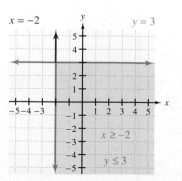

EXAMPLE **4**

System of Inequalities with No Solution

Graph the solution set of the following system of inequalities.

$$y \leq -x - 6$$
$$y \geq 4 - x$$

Solution

The graph of each inequality is shown in Figure 5.14. Observe that there are no points common to the two half-planes. The system of inequalities has no solution.

Figure 5.14

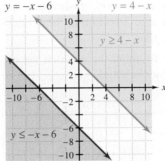

Think About It

Suppose that a movie producer is casting a role for a man who is between the ages of 40 and 50, inclusive, who weighs no more than 170 pounds. Letting a and w represent age and weight, write a system of inequalities that describes all of the requirements.

Quick Reference 5.5

Verifying Solutions
- Two or more inequalities considered simultaneously form a **system of inequalities**. A **solution** of a system of inequalities is an ordered pair of numbers that satisfies each inequality in the system.
- To determine whether a given ordered pair is a solution of a system of inequalities, test the pair in each inequality. To be a solution, the pair must satisfy each inequality.

Graphing Solution Sets
- A solution of a system of inequalities is represented by a point that belongs to the graph of each inequality. Because the graph of an inequality is a half-plane, the points that belong to the intersection of the half-planes correspond to solutions.
- To graph the solution set of a system of inequalities, use the following procedure.
  1. Graph each inequality in the same coordinate system.
  2. Shade the intersection of the half-planes.

Speaking the Language 5.5

1. Two or more inequalities connected with the word *and* form a(n) ▒▒▒▒▒ .
2. Any ordered pair that satisfies each inequality of a system of inequalities is called a(n) ▒▒▒▒▒ .

3. The solution set of a system of inequalities is represented by the ▓▓▓▓▓▓ of the graphs of the inequalities.

4. Suppose that d represents distance in a system of inequalities. Then an implied condition is $d \geq 0$ because distance is ▓▓▓▓▓▓.

Exercises 5.5

Concepts and Skills

1. Suppose that the graph of a system of inequalities has one solid boundary line and one dashed boundary line. Does the point of intersection of these lines represent a solution? Explain.

2. Suppose that you want to graph the solution set of the following system of inequalities.

$$3x - 2y \leq 6$$
$$2x + 3y \geq -4$$

What is a good first step?

In Exercises 3–6, identify each ordered pair that is a solution of the system of inequalities.

3. $y < 2x - 5$ (i) $(3, -5)$
   $2x + 3y < 6$ (ii) $(0, 0)$

4. $x - y \geq 1$ (i) $(2, 2)$
   $3x > y + 2$ (ii) $(2, -2)$

5. $2x - y \leq 3$ (i) $(5, 7)$
   $x + 2y \geq 19$ (ii) $(5, 8)$

6. $y > x - 4$ (i) $(1, 5)$
   $y < x + 3$ (ii) $(-5, 1)$

In Exercises 7–20, graph the solution set of the system of inequalities.

7. $y \geq -3x + 5$
   $y < x$

8. $y < x - 1$
   $y \geq -x + 3$

9. $y \geq -2x - 1$
   $y \geq \frac{1}{2}x - 6$

10. $y < \frac{1}{2}x$
    $y < x - 6$

11. $y > x$
    $y < x + 4$

12. $y < 2x + 5$
    $2x - y < 7$

13. $y < x - 5$
    $x + 2y \leq -4$

14. $y > 2x - 3$
    $x \leq 3y - 1$

15. $x - y > 2$
    $x + y < 8$

16. $x + y \geq -2$
    $2x + y \leq -5$

17. $y - x \leq 6$
    $2x + 3y \geq -12$

18. $x - 4y < 24$
    $3x + 2y > 2$

19. $x - y > 5$
    $x + 2y \leq -4$

20. $y - x + 7 \geq 0$
    $x - 3y + 3 \geq 0$

In Exercises 21–26, graph the solution set of the system of inequalities.

21. $y \leq 8$
    $2y > x$

22. $2y - 3x \leq 18$
    $x + 4 > 0$

23. $x > 2$
    $y > -3$

24. $x \leq 1$
    $y < 6$

25. $x \geq 0$
    $y \leq 2x + 1$

26. $y > x - 4$
    $y \leq 0$

27. Suppose that a system of two linear inequalities is such that the boundary lines do not coincide. If the system has no solution, describe the graphs of the inequalities.

28. Consider the graph of a system of two linear inequalities, each solved for y. If the solid boundary lines are parallel, and the solution set is represented by shading between the boundary lines, what do you know about the two inequality symbols?

In Exercises 29–34, graph the solution set of the system of inequalities.

29. $y > 4 - x$
    $x + y < -4$

30. $y - 2x \geq 0$
    $y + 5 \leq 2x$

31. $x \leq 2y + 8$
    $2y - x \geq 2$

32. $4y + 3x - 24 \geq 0$
    $0 \geq 4 - 3x - 4y$

33. $y > \frac{3}{2}x - 7$
    $y < \frac{3}{2}x + 2$

34. $y \leq -\frac{1}{3}x + 5$
    $y \geq -\frac{1}{3}x$

In Exercises 35–38, graph the solution set of the system of inequalities.

(In many application problems, the unknown quantities x and y are nonnegative numbers. Thus $x \geq 0$ and $y \geq 0$ are two additional conditions in the problem, and these inequalities must be included in the system of inequalities. These inequalities limit the graph of the solution set to Quadrant I and to points of the axes that border this quadrant.)

35. $x + y < 5$
$\quad x \geq 0$
$\quad y > 0$

36. $x + 2y \geq 6$
$\quad x \geq 0$
$\quad y \geq 0$

37. $y \geq x$
$\quad x \geq 0$
$\quad y \geq 0$

38. $y < 2x - 3$
$\quad x \geq 0$
$\quad y \geq 0$

In Exercises 39 and 40, write the system of three linear inequalities illustrated in the given graph.

39.   **40.**

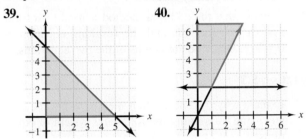

In Exercises 41 and 42, graph the solution set of the system of inequalities.

(Recall that the inequality $a \leq x \leq b$ means that $x \geq a$ and $x \leq b$. Thus each system in the following two exercises consists of four inequalities.)

41. $0 \leq x \leq 4$
$\quad 0 \leq y \leq 3$

42. $-2 \leq x \leq 3$
$\quad -4 \leq y \leq 1$

Challenge

In Exercises 43–46, describe the solution set of the system of inequalities.

43. $y \geq x - 5$
$\quad y \leq x - 5$

44. $y \geq \dfrac{1}{2}x + 3$
$\quad y \leq \dfrac{1}{2}x + 3$

45. $y < 2x + 1$
$\quad y > 2x + 1$

46. $y < \dfrac{3}{4}x - 2$
$\quad y > \dfrac{3}{4}x - 2$

In Exercises 47–50, graph the solution set of the system of inequalities.

47. $2x + y \geq 8$
$\quad y \leq 8$
$\quad x \leq 4$

48. $y \leq 2x + 3$
$\quad y \geq 3$
$\quad x \leq 4$

49. $2x - y \leq 7$
$\quad x + y \geq 2$
$\quad x \geq 0$
$\quad y \geq 0$

50. $x - y \geq 5$
$\quad x + y \leq 0$
$\quad x \geq 0$
$\quad y \geq 0$

Chapter 5 Review Exercises

Section 5.1

1. What do we mean by a *solution* of a system of equations in two variables?

2. Describe three ways in which two lines can be oriented with respect to each other in a coordinate plane.

3. Identify each ordered pair that is a solution of the following system of equations.
$$y = 7x - 4$$
$$5x - 2y = -1$$
(i) $(-1, 3)$ (ii) $(1, 3)$ (iii) $(1, -3)$

4. Determine a and b such that $(-2, 3)$ is a solution of the following system of equations.

$$3x + by = 6$$
$$ax - 2y = -4$$

5. Use the graphing method to estimate the solution of the following system of equations.

$$7x - 3y = 15$$
$$y = 2x - 5$$

6. Solve the following system of equations.

$$5y - 2x = 4$$
$$x - 3 = 0$$

7. Use the slope–intercept form of each equation of the system of equations to determine the number of solutions of the system.

(a) $2y - x = -10$ (b) $2x - 3y = -3$

$\dfrac{1}{2}x - y = -3$ $2y - 3x = -2$

(c) $4x = 5y - 15$

$y - 3 = \dfrac{4}{5}x$

8. For each part of Exercise 7, describe the system of equations and describe the equations of the system.

9. Use the graphing method to estimate the solution of the following system of equations.

$$\frac{1}{2}x + \frac{1}{3}y = -1$$
$$y = x - 23$$

10. If the following system of three linear equations has a unique solution, what is the value of k?

$$kx + 8y = 0$$
$$x + 2 = 0$$
$$y - 1 = 0$$

Section 5.2

In Exercises 11–15, use the addition method to solve the system of equations.

11. $2x + 3y = -11$
$\ 5x - 3y = 25$

12. $-4x + 11y = 7$
$\ \ 2x - 33y = -31$

13. $y = -6x + 19$
$\ 3x = 7 - y$

14. $\dfrac{4}{3}x + \dfrac{5}{3}y = 1$

$\dfrac{1}{10}x - \dfrac{4}{5}y = 1$

15. $5x + 3y = 11$
$\ -2x + 7y = -29$

In Exercises 16 and 17, use the addition method to determine the solution set.

16. $5x - 6y = 24$

$y + 4 = \dfrac{5}{6}x$

17. $4x - 7y = -7$

$\dfrac{4}{7}x - 1 = y$

18. If the following system of equations has a unique solution, use the addition method to show that $a \ne 3$.

$$ax + 3y = 5$$
$$x + y = 1$$

19. Add the equations in the following system.

$$x = 1$$
$$y = 2$$

Does the resulting equation indicate that the system has infinitely many solutions?

20. Suppose that the addition method is used to solve a system of equations and the resulting equation is $0 = -1$. Describe the graphs of the two equations.

Section 5.3

21. Explain why the substitution method is not the best technique for solving the following system of equations.

$$3x - 5y = 10$$
$$2x + 4y = 7$$

22. Describe two ways to solve the following system of equations with the substitution method.

$$2x + y = 5$$
$$x + 2y = -4$$

In Exercises 23–28, use the substitution method to solve the system of equations. Identify inconsistent systems and systems whose equations are dependent.

23. $y = -x + 1$
$y = 2x + 19$

24. $x + 5y = -19$
$2x - 3y = 14$

25. $\dfrac{2}{3}x - 4y = 10$

$x = 3$

26. $-x + 3y = -2$

$y = \dfrac{1}{3}x + 5$

27. $7y = 6x + 21$

$y - 3 = \dfrac{6}{7}x$

28. $\dfrac{3}{7}x + \dfrac{4}{11}y = \dfrac{61}{77}$

$\dfrac{4}{11}y = \dfrac{28}{77}x$

29. Consider the following system of equations.

$$2x + 3y = 7$$
$$7 + y = 4x$$

Does substituting $2x + 3y$ for 7 in the second equation lead to a solution of the system?

30. Why is the substitution method not appropriate for solving the following system of equations?

$$x + 3 = 5$$
$$y - 4 = 1$$

Section 5.4

31. The difference of two numbers is 25. The sum of the larger number and twice the smaller number is -2. What are the numbers?

32. One number is 3 less than another number. If both numbers are doubled, their sum is -10. What are the numbers?

33. The measure of the larger of two complementary angles is 2° more than 7 times the measure of the smaller angle. What are the measures of the angles?

34. The length of a rectangular desk is 1 foot less than twice the width. The perimeter of the desk is 30 feet. What are the dimensions of the desk?

35. A baseball card collector has a set of old cards worth $12 each and a set of newer cards worth $3 each. The entire collection is worth $300. If the collector has 25 more newer cards than old cards, how many cards are in each set?

36. In one day a ship's bakery used flour costing $0.22 per pound and eggs worth $0.90 per dozen. The total cost of these two items was $984. If the number of dozens of eggs used was two-thirds the number of pounds of flour used, how many pounds of flour were used?

37. A commuter lives 53 miles from her downtown office, and her trip home takes 1 hour and 10 minutes. She begins the trip on a bus that averages 26 mph and then drives the rest of the way at an average speed of 60 mph. How long does each part of the trip take?

38. A boat traveled 45 miles downstream in 3 hours and returned upstream in 5 hours. What were the speeds of the boat and the current?

39. A real estate agent earns a 6% commission on the sale of a home and a 10% commission on the sale of

vacant land. Last year the agent earned $66,000 in commissions on total sales of $1 million. How much did the agent earn on the sale of homes?

40. Some 75% alcohol solution is to be mixed with a solution that is $\frac{1}{3}$ alcohol to produce 20 pints of a 50% alcohol solution. How many pints of each concentration should be used?

41. In an old library, there were 1.5 times as many fiction books as nonfiction books. In the new, expanded library, 2000 books of each type were added to the collection, and the number of nonfiction books is now $\frac{3}{4}$ the number of fiction books. How many books of each type are in the new library?

42. The perimeter of a rectangular town square is 240 yards. If the length were decreased by 10 yards and the width were increased by 10 yards, the town square would truly be a square. What are the actual dimensions of the town square?

Section 5.5

43. Give two advantages of solving each inequality of a system of linear inequalities for y.

44. Suppose that there are x boys and y girls in a kindergarten class, and there are fewer than 30 students in the class. Explain why three inequalities are needed to describe these conditions, and state what the inequalities are.

45. Identify each ordered pair that is a solution of the following system of inequalities.

$$y < -2x - 5 \qquad \text{(i)} \ (-4, 3)$$
$$x - y \le 3 \qquad \text{(ii)} \ (-5, -1)$$

In Exercises 46–49, graph the solution set of the system of inequalities.

46. $y \ge x + 3$
$x + y \le 5$

47. $y < 4$
$2x - y \le 3$

48. $y < \frac{1}{2}x + 5$
$x - 2y < 6$

49. $2x - 3y > -12$
$x \le 0$
$y \le 0$

50. Consider the triangle with vertices $A(0, 0)$, $B(0, 4)$, and $C(4, 0)$. Write a system of inequalities whose solution set is represented by the triangle and its interior.

Chapter 5 Test

1. Determine a and b such that $(2, -3)$ is a solution of the following system of equations.

$$ax + y = -9$$
$$x - by = 8$$

2. Solve the following system of equations by graphing the equations and verifying the solution.

$$y = 7 + x$$
$$2x + 2y = 2$$

3. Describe the graphs of the equations of a system for each set of conditions.

   (a) The system is consistent, and the equations are dependent.

   (b) The system is inconsistent, and the equations are independent.

   (c) The system is consistent, and the equations are independent.

4. Without graphing or solving algebraically, show that this system of equations has a unique solution.

$$2x + y = 1$$
$$y = 2x + 1$$

In Questions 5–8, use the addition method to solve the system of equations.

5. $3x - 2y = -4$
$3y - 2x = -9$

6. $\dfrac{1}{2}x - \dfrac{1}{4}y = 1$

   $y = 7x - 4$

7. $3y - 4x = -15$

   $y = \dfrac{4}{3}x + 2$

8. $\dfrac{1}{2}x - y = 8$

   $x = 2(y + 8)$

9. When we use the substitution method to solve a system of equations, what is the first step?

10. Use the substitution method to solve the following system of equations.

$$-5x + y = 14$$
$$3x - 2y = -14$$

11. Use the substitution method to solve the following system of equations. Describe the solution set depending on the value of a.

$$2ax - ay = 2$$
$$y = 2x + 1$$

12. Use the substitution method to solve the following system of equations.

$$4 - x = 2$$
$$3x + y = -1$$

13. The sum of two numbers is 17. Half the larger number is 10 more than the smaller number. What are the numbers?

14. The length of a rectangular office is 7 feet more than its width. If the perimeter is 74 feet, what are the dimensions of the office?

15. When the clerk at a checkout counter added a roll of 50 dimes to the cash register drawer, the total value of the nickels, dimes, and quarters was $8.70. If there were 2 fewer quarters than nickels, how many nickels were in the drawer?

16. A plane flying against the wind requires 2 hours for a flight of 740 miles. Flying with the wind, the plane can travel 1290 miles in 3 hours. What is the speed of the plane?

17. To buy property costing $25,000, a person borrowed the down payment from her father at a 6% simple interest rate and the rest from a bank at a 10% simple interest rate. She paid off the loan in 1 year, and the total interest paid was $2260. What was the down payment?

18. A person's home is 4 years older than his neighbor's home. In 6 years, his home will be $1\frac{1}{2}$ times as old as his neighbor's home. How old is his neighbor's home now?

In Questions 19 and 20, graph the solution set of the system of inequalities.

19. $x + y \geq 4$
$x - y < -1$

20. $2x - 3y \geq -12$
$x \leq 0$
$y \geq 0$

Cumulative Test, Chapters 4–5

1. Complete the table so that each pair is a solution of $x - y = 7$.

x		6	
y	-2		3

2. Produce the graph of $y = \dfrac{1}{2}x - 5$ on your calculator.

   Then trace the graph to determine a and b so that $A(a, -3)$ and $B(-4, b)$ are points of the graph.

3. Determine algebraically the intercepts of the graph of $4x - 3y = -12$.

4. Write an equation of the line that contains $P(2, -5)$ and is

   (a) parallel to the y-axis.

   (b) perpendicular to the y-axis.

5. Determine the slope of a line that contains $P(-3, 5)$ and $Q(7, -4)$.

6. State whether the described line has a slope that is positive, negative, 0, or undefined.

   (a) The line contains $(2, 5)$ and $(-2, 5)$.

   (b) The line is perpendicular to the x-axis.

   (c) The line contains $(-3, 9)$ and $(3, 8)$.

7. Use the slope–intercept form of $5x - 3y = 9$ to determine the slope and y-intercept of the graph.

8. Sketch the line that has a slope of $-\frac{3}{4}$ and contains $P(-7, 5)$.

9. Which of the following equations has a graph that is not parallel to the other two?

   (i) $3x - 4y = 20$ (ii) $5y - 3x = 25$

   (iii) $y = \dfrac{3}{4}x + 4$

10. The equation of L_1 is $2x + 5y = 20$, and the equation of L_2 is $y = kx - 3$. If L_1 is perpendicular to L_2, what is the value of k?

11. Write the standard form of the equation of the line containing $A(-6, 2)$ and $B(5, 11)$.

12. Write the slope–intercept form of the equation of L_1 that contains $(2, 3)$ and is parallel to L_2, whose equation is $4x + y = -3$.

13. When we are graphing a linear inequality in two variables, what must be true before the inequality symbol can be used to determine the half-plane to be shaded?

14. Graph the inequality $4(y - 2) < -3x$.

15. Without solving or graphing, determine the number of solutions of the following system of equations.

   $$3x - 2y = -8$$
   $$y = 1.5x - 3$$

16. Use the graphing method to estimate the solution of the following system of equations.

   $$-x + y = -5$$
   $$2x - 3y = 22$$

In Questions 17 and 18, use the addition method to solve the system of equations.

17. $5x + 2y = -17$
    $3x - 5y = -4$

18. $\dfrac{1}{2}x - \dfrac{1}{3}y = -1$
    $\dfrac{1}{4}x + \dfrac{1}{5}y = 5$

19. Use the substitution method to solve the following system of equations.

   $$3y + x = 23$$
   $$4x - 2y = -34$$

20. The measure of one angle is $2°$ more than the measure of a second angle. If the angles are supplementary, what is the measure of the smaller angle?

21. A brother and sister live 11.5 miles from their school. They walk at 3 mph from their home to the school bus stop and then travel on the bus at an average speed of 20 mph. If the trip from home to school takes 1 hour, how far is it from their home to the bus stop?

22. An 18% medicinal solution is mixed with a 6% solution to obtain 12 ounces of a 13% solution. How many ounces of the 6% solution should be used?

23. Sketch the solution set of the following system of inequalities.

   $$2x + y \le 4$$
   $$5x > 2y - 10$$

24. Describe the solution set of the following system of inequalities.

   $$x + 2y > 10$$
   $$y < -\dfrac{1}{2}x - 3$$

In 1988, 1992, 1996, and 2000, the number of volunteers who worked at the Olympic Games increased.

Both the number of participating countries and the number of volunteers can be modeled with **polynomials.** The number of volunteers per country can be determined by dividing the polynomials, and the trend can then be observed. (For more on this real data application, see Exercises 87–90 at the end of Section 6.6.)

Year	Olympic City	Number of Countries	Number of Volunteers
1988	Seoul	159	27,000
1992	Barcelona	169	30,000
1996	Atlanta	197	35,000
2000	Sydney	198	45,000

(Source: International Olympic Committee.)

Chapter 6

Exponents and Polynomials

Chapter Snapshot

This chapter opens with the rules for operating with exponents. We then turn to polynomials and the basic operations that can be performed with them. We conclude with the introduction of negative exponents and a discussion of scientific notation.

Warm-Up Skills

The following exercises review concepts and skills that you will need in Chapter 6.

In Exercises 1–3, evaluate the given numerical expression.

1. (a) 2^5 (b) $(-5)^3$

2. (a) $\dfrac{4^3}{4^3}$ (b) $\dfrac{3^2}{3^5}$

3. (a) $5 \cdot 10^3$ (b) $\dfrac{1 + 7^2}{10^2}$

In Exercises 4–6, evaluate the expression for the given value(s) of the variable(s).

4. $x^2 - 5x - 2$; $x = 3$

5. $3 - 5t$; $t = -2$

6. $2x + 3y^2 - 4xy$; $x = 2, y = -1$

In Exercises 7 and 8, write the given expression without grouping symbols.

7. $-(-x^2 + 3x - 1)$ 8. $3(6 - 5x)$

In Exercises 9–12, simplify the expression.

9. $-2(x - 2y + 1) + 3(x + 1)$

10. $(x + 1) - 4(x + 2) + 3(4 - x)$

11. $(x^2 - 2x + 6) + (3x^2 + 2x - 5)$

12. $(7x - 5) - (4x - 3)$

6.1 | Properties of Exponents

Product Rule for Exponents • Power to a Power Rule • Product to a Power Rule •
Quotient to a Power Rule • Quotient Rule for Exponents • The Zero Exponent

Recall that in Chapter 1 we defined a positive integer exponent as a way to indicate repeated multiplication of a value by itself. The exponent indicates the number of times the base is used as a factor.

$$4 \cdot 4 \cdot 4 = 4^3 \qquad x^2 = x \cdot x \qquad b^1 = b$$

Identifying the base is important when we evaluate an exponential expression.

$$2^5 = 2 \cdot 2 \cdot 2 \cdot 2 \cdot 2 = 32 \qquad \text{The base is 2.}$$
$$(-7)^2 = (-7)(-7) = 49 \qquad \text{The base is } -7.$$
$$-7^2 = -1 \cdot 7^2 = -1 \cdot 49 = -49 \qquad \text{The base is 7.}$$

In this section we introduce several properties of exponents.

Product Rule for Exponents

To multiply exponential expressions, we can use the definition of an exponent.

$$b^3 \cdot b^4 = \underbrace{(b \cdot b \cdot b)}_{b^3}\underbrace{(b \cdot b \cdot b \cdot b)}_{b^4}$$
$$= b \cdot b \cdot b \cdot b \cdot b \cdot b \cdot b$$
$$= b^7 \qquad \text{There are 7 factors of } b.$$

This result suggests that when the bases are the same, we can multiply exponential expressions by adding the exponents.

The Product Rule for Exponents

For any real number b and any positive integers m and n,

$$b^m \cdot b^n = b^{m+n}$$

In words, to multiply exponential expressions with like bases, retain the base and add the exponents.

Note: In order for us to use the Product Rule for Exponents, the expression must be a product, and the bases must be the same. The rule does not apply to $5^3 + 5^2$ because this expression is not a product, nor does it apply to $3^2 \cdot 5^2$ because the bases are not the same. Also, we add the exponents, but the base of the result remains the same.

EXAMPLE 1

LEARNING TIP

Remember that $a = a^1$. You may find that writing the understood exponent 1 is useful as you apply the properties of exponents.

Using the Product Rule for Exponents

Multiply the given exponential expressions.

(a) $y^3 \cdot y^5 = y^{3+5} = y^8$

(b) $7^2 \cdot 7^4 = 7^{2+4} = 7^6$

(c) $a \cdot a^3 \cdot a^2 = a^{1+3+2} = a^6$

Sometimes we use the Product Rule for Exponents along with other properties of real numbers. Recall that the associative and commutative properties of multiplication allow us to reorder and regroup factors.

EXAMPLE 2 · **Using the Product Rule for Exponents**

Multiply the given exponential expressions.

(a) $-5x(4x^2) = -5 \cdot 4 \cdot x \cdot x^2$ Reorder and regroup the factors.

   $= -20x^3$ Product Rule for Exponents

(b) $(2x^4)(3x^7) = 2 \cdot 3 \cdot x^4 \cdot x^7 = 6x^{4+7} = 6x^{11}$

(c) $(x^4y)(x^2y^3) = x^4 \cdot x^2 \cdot y \cdot y^3 = x^6y^4$

Power to a Power Rule

The expression $(b^2)^3$ is an exponential quantity b^2 raised to the power 3. We call such an expression a *power to a power*. We can use the Product Rule for Exponents to simplify or evaluate a power to a power.

$$(2^5)^3 = 2^5 \cdot 2^5 \cdot 2^5 = 2^{15}$$
$$(b^2)^4 = b^2 \cdot b^2 \cdot b^2 \cdot b^2 = b^8$$

These results suggest that we can raise a power to a power by multiplying the exponents.

> **The Power to a Power Rule**
>
> For any real number b and any positive integers m and n,
>
> $(b^m)^n = b^{mn}$
>
> In words, to raise an exponential expression to a power, retain the base and multiply the exponents.

EXAMPLE 3 | **Using the Power to a Power Rule**

Raise the exponential expression to the given power.

(a) $(6^2)^5 = 6^{2 \cdot 5} = 6^{10}$ (b) $(a^4)^3 = a^{4 \cdot 3} = a^{12}$

EXAMPLE 4 | **Combining Properties**

Think About It

Explain why a^{mn} can be written as $(a^m)^n$ or as $(a^n)^m$.

Perform the indicated operations.

$(y^3)^4(y^2)^5 = y^{12} \cdot y^{10}$ Power to a Power Rule used twice

   $= y^{12+10}$ Product Rule for Exponents

   $= y^{22}$

Product to a Power Rule

The following illustrates how to simplify exponential expressions in which the base is a product.

$$(xy)^3 = (xy)(xy)(xy) \qquad \text{Definition of an exponent}$$
$$= (x \cdot x \cdot x)(y \cdot y \cdot y) \qquad \text{Reorder and regroup factors.}$$
$$= x^3y^3$$

We can omit the intermediate steps in the procedure by observing that each factor of the product is raised to the power 3. This is the basis for the Product to a Power Rule.

The Product to a Power Rule

For any real numbers a and b and any positive integer n,

$$(ab)^n = a^nb^n$$

In words, to raise a product to a power, raise each factor to that power.

EXAMPLE **5**

Using the Product to a Power Rule

Raise the product to the given power.

(a) $(xy)^5 = x^5y^5$ 　　　　(b) $(3n)^2 = 3^2n^2 = 9n^2$

EXAMPLE **6**

Combining Properties

Perform the indicated operations.

(a) $(2x^4)^3 = 2^3(x^4)^3$ 　　　　Product to a Power Rule
$$= 2^3x^{4 \cdot 3} \qquad \text{Power to a Power Rule}$$
$$= 8x^{12}$$

(b) $(x^2y^5)^4 = (x^2)^4 \cdot (y^5)^4$ 　　　　Product to a Power Rule
$$= x^8y^{20} \qquad \text{Power to a Power Rule}$$

Quotient to a Power Rule

In the following, we consider an exponential expression in which the base is a quotient.

$$\left(\frac{2}{5}\right)^3 = \left(\frac{2}{5}\right)\left(\frac{2}{5}\right)\left(\frac{2}{5}\right) = \frac{2 \cdot 2 \cdot 2}{5 \cdot 5 \cdot 5} = \frac{2^3}{5^3}$$

We can write the result directly by observing that both the numerator and the denominator are raised to the power.

The Quotient to a Power Rule

For any real numbers a and b, $b \neq 0$, and any positive integer n,

$$\left(\frac{a}{b}\right)^n = \frac{a^n}{b^n}$$

In words, to raise a quotient to a power, raise both the numerator and the denominator to that power.

EXAMPLE 7

Using the Quotient to a Power Rule

Raise the quotient to the given power.

(a) $\left(\dfrac{a}{b}\right)^5 = \dfrac{a^5}{b^5}$

(b) $\left(\dfrac{x}{-2}\right)^4 = \dfrac{x^4}{(-2)^4} = \dfrac{x^4}{16}$

EXAMPLE 8

Combining Properties

Perform the indicated operations.

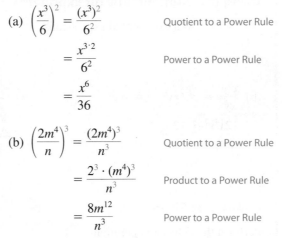

(a) $\left(\dfrac{x^3}{6}\right)^2 = \dfrac{(x^3)^2}{6^2}$ Quotient to a Power Rule

$= \dfrac{x^{3\cdot2}}{6^2}$ Power to a Power Rule

$= \dfrac{x^6}{36}$

(b) $\left(\dfrac{2m^4}{n}\right)^3 = \dfrac{(2m^4)^3}{n^3}$ Quotient to a Power Rule

$= \dfrac{2^3 \cdot (m^4)^3}{n^3}$ Product to a Power Rule

$= \dfrac{8m^{12}}{n^3}$ Power to a Power Rule

Quotient Rule for Exponents

To divide exponential expressions, we can use the Product Rule for Exponents and then divide out the common factors.

$$\frac{2^5}{2^3} = \frac{2^3 \cdot 2^2}{2^3} = 2^2$$

$$\frac{b^{11}}{b^4} = \frac{b^4 \cdot b^7}{b^4} = b^7$$

These results suggest that when the bases are the same, we can divide exponential expressions by subtracting the exponents.

The Quotient Rule for Exponents

For any real number b, $b \neq 0$, and for positive integers m and n, where $m > n$,

$$\frac{b^m}{b^n} = b^{m-n}$$

In words, to divide exponential expressions with the same base, retain the base and subtract the exponents.

Note: The Quotient Rule for Exponents applies only if the bases are the same. Also, the expression $\frac{a^5}{a^3}$ is an example of a *quotient of exponential expressions,* whereas the expression $\left(\frac{a}{b}\right)^3$ is an example of a *quotient to a power.* Knowing the difference is essential to applying the correct rule when simplifying.

As stated, the Quotient Rule for Exponents refers only to quotients in which the exponent in the numerator is greater than the exponent in the denominator. Later we will consider the cases $m = n$ and $m < n$.

EXAMPLE 9

Using the Quotient Rule for Exponents

Divide the given exponential expressions.

(a) $\dfrac{y^7}{y^4} = y^{7-4} = y^3$

(b) $\dfrac{x^4 y^6}{x y^3} = x^{4-1} y^{6-3} = x^3 y^3$

(c) $\dfrac{12x^4 y^7}{3x^3 y^3} = \dfrac{12}{3} \cdot \dfrac{x^4}{x^3} \cdot \dfrac{y^7}{y^3} = 4 \cdot x^{4-3} \cdot y^{7-3} = 4xy^4$

EXAMPLE 10

Combining Properties

Perform the indicated operations.

LEARNING TIP

We often have several ways to approach exponent problems. For example, in part (a) we could use the Quotient to a Power Rule.

$\left(\dfrac{a^4}{a}\right)^2 = \dfrac{(a^4)^2}{(a)^2}$

$= \dfrac{a^8}{a^2}$

$= a^{8-2}$

$= a^6$

(a) $\left(\dfrac{a^4}{a}\right)^2 = (a^{4-1})^2$ Quotient Rule for Exponents

$= (a^3)^2$

$= a^6$ Power to a Power Rule

(b) $\dfrac{(n^2)^3}{n^4 \cdot n} = \dfrac{n^6}{n^4 \cdot n}$ Power to a Power Rule

$= \dfrac{n^6}{n^5}$ Product Rule for Exponents

$= n$ Quotient Rule for Exponents

The Zero Exponent

Suppose that we try to extend the Quotient Rule for Exponents as follows.

$$\frac{4^3}{4^3} = 4^{3-3} = 4^0$$

So far, our only experience with exponents has involved natural number exponents. If we intend to use the Quotient Rule for Exponents in this way, it is necessary to assign a meaning to the exponent 0.

Because any nonzero number divided by itself equals 1, we know that the value of $\dfrac{4^3}{4^3}$ is 1. If 4^0 and 1 both represent the value of $\dfrac{4^3}{4^3}$, then we conclude that $4^0 = 1$. This reasoning is the basis for our definition of the zero exponent.

> **Definition of the Zero Exponent**
>
> For any real number b, $b \neq 0$, $b^0 = 1$.

As before, when we evaluate an expression with a zero exponent, it is important to identify the base of the exponent.

EXAMPLE **11**   **Evaluating Expressions with a Zero Exponent**

Evaluate the given exponential expression.

(a) $32^0 = 1$ The base is 32.

(b) $(-6)^0 = 1$ The base is -6.

(c) $-6^0 = -1 \cdot 6^0 = -1 \cdot 1 = -1$ The base is 6.

(d) $3x^0 = 3 \cdot 1 = 3$ The base is x.

(e) $(3x)^0 = 1$ The base is $3x$.

The Quotient Rule for Exponents can now be applied to expressions of the form $\dfrac{b^m}{b^n}$, where $m > n$ or $m = n$. At the end of this chapter, we will consider the case $m < n$.

Quick Reference 6.1

Product Rule for Exponents

- For any real number b and any positive integers m and n,
  $$b^m \cdot b^n = b^{m+n}$$

  In words, to multiply exponential expressions with like bases, retain the base and add the exponents.

- The associative and commutative properties of multiplication allow us to reorder and regroup factors in order to apply the Product Rule for Exponents.

Power to a Power Rule

- For any real number b and any positive integers m and n,
  $$(b^m)^n = b^{mn}$$

  In words, to raise an exponential expression to a power, retain the base and multiply the exponents.

Product to a Power Rule

- For any real numbers a and b and any positive integer n,
  $$(ab)^n = a^n b^n$$

  In words, to raise a product to a power, raise each factor to that power.

Quotient to a Power Rule • For any real numbers a and b, $b \neq 0$, and any positive integer n,

$$\left(\frac{a}{b}\right)^n = \frac{a^n}{b^n}$$

In words, to raise a quotient to a power, raise both the numerator and the denominator to that power.

Quotient Rule for Exponents • For any real number b, $b \neq 0$, and for positive integers m and n, where $m > n$,

$$\frac{b^m}{b^n} = b^{m-n}$$

In words, to divide exponential expressions with the same base, retain the base and subtract the exponents.

The Zero Exponent • For any real number b, $b \neq 0$, $b^0 = 1$.

• By defining the zero exponent, we can extend the Quotient Rule for Exponents to include expressions of the form $\frac{b^m}{b^n}$, where $m > n$ or $m = n$.

Speaking the Language 6.1

1. To multiply exponential expressions with like bases, we retain the ▨▨▨ and ▨▨▨ the exponents.

2. To raise a quotient to a power, we raise the ▨▨▨ and the ▨▨▨ to that power.

3. In order for us to apply the Quotient Rule for Exponents, the ▨▨▨ must be the same.

4. Any nonzero quantity with a ▨▨▨ exponent has a value of 1.

Exercises 6.1

Concepts and Skills

1. Explain why $x^2 \cdot x^3$ can be simplified but $x^2 + x^3$ cannot.

2. Compare $x^2 \cdot x^3$ with $(x^2)^3$.

In Exercises 3–6, evaluate each expression.

3. $-4^2, (-4)^2, 4^2$

4. $(-2)^2, 2^2, -2^2$

5. $7^2, -7^2, (-7)^2$

6. $(-5)^2, 5^2, -5^2$

In Exercises 7–14, use the Product Rule for Exponents to simplify the expression.

7. $7^3 \cdot 7$

8. $(-2)^2 \cdot (-2)^5$

9. $y^5 \cdot y^3$

10. $x \cdot x^4$

11. $x^4 \cdot x^5 \cdot x$

12. $a^2 \cdot a \cdot a^8$

13. $(3x)^5 \cdot (3x)^9$

14. $(-7y)^4 \cdot (-7y)^6$

In Exercises 15–24, simplify.

15. $-4x^3(x^6)$

16. $(5y^4)y^{10}$

17. $-4x^5(-2x^7)$

18. $-6x(5x^4)$

19. $(2x^2y^5)(-5x^3y^2)$

20. $(-8r^7t^3)(-2rt^5)$

21. $(3x^3)(-2x^4)(5x)$

22. $-4x^2(3x^2)(-x^4)$

23. $(xy^2)(x^3y)(x^4y^2)$

24. $(-2a^2b)(-a^3b^2)(a^5b^3)$

In Exercises 25–30, (a) add the exponential expressions, if possible, and (b) multiply the exponential expressions.

25. $-5x^4, 7x^4$

26. $-4y^3, y^3$

27. $8y^5, 8y^4$

28. $6t^5, -7t^5$

29. $n, 3n, -6n$

30. x, x^2, x^5

In Exercises 31–36, use the Power to a Power Rule to simplify the expression.

31. $(5^2)^4$

32. $(2^6)^3$

33. $(a^3)^5$

34. $(b^8)^3$

35. $(-x^4)^7$

36. $(-y^3)^6$

In Exercises 37–44, use the Product to a Power Rule or the Quotient to a Power Rule to simplify the expression.

37. $(xy)^4$

38. $(ab)^3$

39. $(4y)^3$

40. $(6x)^2$

41. $\left(\dfrac{a}{b}\right)^5$

42. $\left(\dfrac{y}{x}\right)^8$

43. $\left(\dfrac{-2}{x}\right)^5$

44. $\left(\dfrac{y}{7}\right)^2$

In Exercises 45–60, simplify.

45. $(3x^4)^2$

46. $(5a^5)^3$

47. $(-2y^3)^5$

48. $(-4y^4)^3$

49. $\left(\dfrac{a^4}{b^3}\right)^5$

50. $\left(\dfrac{x^6}{y^8}\right)^4$

51. $\left(\dfrac{3n}{5m}\right)^3$

52. $\left(\dfrac{-2a}{bc}\right)^5$

53. $\left(\dfrac{-8x^3}{y}\right)^2$

54. $\left(\dfrac{a^2}{2b^3}\right)^3$

55. $(n^2)^4(n^3)^2$

56. $(r^2)^3(r^3)^2$

57. $3x^2(x^2)^4$

58. $-5t(t^4)^2$

59. $(2x^2)^3(-3x^5)^2$

60. $(5y^3)^2(2y^4)^3$

In Exercises 61–66, determine whether the statement is true or false.

61. $4^2 \cdot 4^4 = 16^6$

62. $3^4 \cdot 3^2 = 3^8$

63. $(t^4)^3 = (t^6)^2$

64. $-5^2 = -25$

65. $(y^2)^3 = y^8$

66. $(3 + 4)^2 = 3^2 + 4^2$

67. Describe two ways to simplify $\left(\dfrac{x^5}{x^3}\right)^2$.

68. Describe two ways to simplify $\dfrac{x^2}{x^0}$.

In Exercises 69–76, use the Quotient Rule for Exponents to simplify the expression.

69. $\dfrac{x^8}{x^5}$

70. $\dfrac{a^{12}}{a^6}$

71. $\dfrac{t^5}{t}$

72. $\dfrac{a^4}{a^3}$

73. $\dfrac{x^6 y^3}{x^2 y^2}$

74. $\dfrac{a^4 b^6}{ab^2}$

75. $\dfrac{12c^5 d^2}{9c^3 d}$

76. $\dfrac{-10a^5 b^5}{15ab^2}$

In Exercises 77–84, simplify.

77. $\left(\dfrac{a^4}{a^2}\right)^6$

78. $\left(\dfrac{7^8}{7^7}\right)^2$

79. $\dfrac{(a^2)^3}{a^5}$

80. $\dfrac{(x^4)^5}{x^7 \cdot x^2}$

81. $\dfrac{(2y^2)^3}{y^3}$

82. $\dfrac{(xy^4)^3}{x^2 y^8}$

83. $\left(\dfrac{t^2}{3t}\right)^3$

84. $\left(\dfrac{2x^5}{x^2}\right)^2$

In Exercises 85–90, evaluate.

85. 5^0

86. $(-7)^0$

87. $(-6a)^0$

88. $-(4x)^0$

89. $3x^0$

90. $-5y^0$

In Exercises 91–96, simplify.

91. $\dfrac{x^0 y^3}{x^2 y}$

92. $\dfrac{2y}{(2y)^0}$

93. $\dfrac{6x^4}{2x(3x^0)}$

94. $(3x^0)(2x^2)$

95. $(t^2)^0(3t)^2$

96. $-3a(-5ab)^0$

Geometric Models

In Exercises 97–100, write an exponential expression for the area or volume of the specified geometric figure.

97. The area of a rectangle whose sides are $3d$ and $2d^2$

98. The area of a square whose side is $3x$

99. The volume of a cube whose side is $2x$

100. The volume of a rectangular box whose length is $3x$ if its end is a square whose side is $5x$.

Modeling with Real Data

101. As the number of Medicare patients has increased, so have Medicare payments. The table shows the number (in millions) of patients and Medicare payments (in billions of dollars).

Year	Patients (millions)	Medicare Payments (billions of dollars)
1990	1.8	3.4
1995	3.2	14.0
1997	3.5	17.0

(Source: Health Care Financing Administration.)

The data can be modeled by the following exponential expressions, where x represents the number of years since 1985.

Number of patients (in millions): $\dfrac{3}{10}x$

Amount of payments
(in billions of dollars): $\dfrac{1}{2}\left(\dfrac{x}{2}\right)^2$

(a) Use the properties of exponents to write the expression for payments without parentheses.

(b) Show that the following expressions are equivalent to the given models.

Number of patients (in millions): $\dfrac{(3x)^2}{30x}$

Amount of payments
(in billions of dollars): $\dfrac{(2x)^2}{32}$

102. Refer to the models in Exercise 101.

(a) Write a simplified expression for the average payment per patient.

(b) Estimate the payment per patient in 2003.

Challenge

In Exercises 103–106, determine the value of n that makes the statement true.

103. $(x^n)^2 = x^{10}$

104. $x^n \cdot x^4 = x^7$

105. $\dfrac{x^7}{x^n} = x^4$

106. $\left(\dfrac{3}{t^2}\right)^n = \dfrac{27}{t^6}$

In Exercises 107–110, simplify.

107. $3^n \cdot 3^{n+1}$

108. $\dfrac{x^{3n}}{x^{n-1}}$

109. $(y^{2n})^n$

110. $\dfrac{(a^{2n-1})^3}{a^{3n+2}}$

6.2 Introduction to Polynomials

Terminology   •   Evaluating Polynomials   •   Real-Life Applications

Terminology

In Chapter 2 we discussed algebraic expressions. In this section we will learn about a particular kind of algebraic expression called a polynomial.

A **monomial** is a number or a product of a number and one or more variables with whole number exponents. Each of the following expressions is an example of a monomial.

$$7 \qquad y^3 \qquad -5x^2 \qquad 8t \qquad \frac{1}{2}ab \qquad 3x^2y^4$$

The following expressions are not monomials.

$\dfrac{4}{x}$ A variable cannot appear in the denominator of a fraction.

$\sqrt{x}$ A variable cannot appear under a square root symbol.

A **polynomial** is a monomial or a sum of two or more monomials. The algebraic expression $3x^3 - 5x^2 + x - 6$ is an example of a polynomial. Because this expression can be written as $3x^3 + (-5x^2) + x + (-6)$, it is the sum of the monomials $3x^3$, $-5x^2$, x, and -6. Previously we referred to the parts of an expression separated by plus signs as *terms*, so monomials are the terms of a polynomial. Also, recall that the numbers 3, -5, 1, and -6 are called *numerical coefficients* (or simply *coefficients*), and we refer to -6 as the *constant term*. The following expressions are not polynomials.

$$\frac{3}{x} - 3x + 1 \qquad \frac{3}{x} \text{ is not a monomial.}$$

$$\frac{x+2}{x-7} \qquad \text{The expression is a quotient, not a sum.}$$

LEARNING TIP

"Mono" (as in *monotone*) means one, "bi" (as in *bicycle*) means two, and "tri" (as in *triangle*) means three. "Poly" means many, but a polynomial may have just one term.

When a polynomial has only one term, we usually just call the expression a monomial. A polynomial with two nonzero terms is called a **binomial,** and a polynomial with three nonzero terms is called a **trinomial.** Polynomials with more than three terms have no special names.

EXAMPLE 1

Describing a Polynomial

Give the name of each polynomial and identify the terms and coefficients.

Polynomial	Name	Terms	Coefficients
m^2n^8	monomial	m^2n^8	1
-12	monomial	-12	-12
$x^2 - 25$	binomial	$x^2, -25$	$1, -25$
$x^3 + \dfrac{yz}{3}$	binomial	$x^3, \dfrac{yz}{3}$	$1, \dfrac{1}{3}$
$2x^2 + 3x - 6$	trinomial	$2x^2, 3x, -6$	$2, 3, -6$
$a^2 + b^2 - a + b$	polynomial	$a^2, b^2, -a, b$	$1, 1, -1, 1$

The **degree of a monomial** in one variable is the value of the exponent of the variable. The degree of a monomial with more than one variable is the sum of the exponents of the variables in that monomial. For example, the degree of $3x^5$ is 5, and the degree of x^3y^5 is $3 + 5$, or 8. The constant term 4 can be written as $4x^0$; therefore, the degree of the term is 0. In general, every nonzero constant term has degree 0.

Note: The monomial 0 can be written $0x^n$, where n is any whole number. Therefore, the degree of 0 is not defined.

EXAMPLE 2

Determining the Degree of a Monomial

Determine the degree of the given monomial.

Monomial	Degree	
(a) $2a^4$	4	The exponent is 4.
(b) -9	0	$-9 = -9x^0$. The exponent of the variable is 0.
(c) $5y$	1	$5y = 5y^1$. The exponent is 1.
(d) $2xy^6$	7	$2xy^6 = 2x^1y^6$. The sum of the exponents is $1 + 6$, or 7.

The **degree of a polynomial** is the degree of the nonzero term with the highest degree.

EXAMPLE 3

Determining the Degree of a Polynomial

Determine the degree of the given polynomial.

Polynomial	Degree	
(a) $x^4 + 4x^2 - 3$	4	The largest exponent is 4.
(b) $8 - 2x$	1	$-2x = -2x^1$. The largest exponent is 1.
(c) $7x^2y^2 - 5xy^3 + x^4y^2$	6	The term x^4y^2 has degree $4 + 2$, or 6.

The polynomial in one variable $x^4 + 4x^2 - 3$ is written in **descending order.** This means that the terms are arranged in order of descending (or decreasing) degrees. In some cases it is convenient to write a polynomial in one variable in **ascending order,** which means that the terms of the polynomial are in order of increasing degrees.

EXAMPLE 4

Think About It

We might write a polynomial with more than one variable in descending or ascending order by specifying the variable. Write the polynomial $x^3y^2 - 6x^5y^3 + xy - y^4$ in order of

(a) descending powers of x.

(b) ascending powers of y.

Writing a Polynomial in Descending and Ascending Order

Write the given polynomial in descending and ascending order.

Polynomial	Descending Order	Ascending Order
(a) $6 - 5x^2 + 2x$	$-5x^2 + 2x + 6$	$6 + 2x - 5x^2$
(b) $x^4 + x^5 - 3 + x$	$x^5 + x^4 + x - 3$	$-3 + x + x^4 + x^5$

The polynomial in part (b) of Example 4 contains no term with x^2 or x^3. We therefore say that it has *missing terms.* Sometimes it is convenient to rewrite the polynomial with coefficients of zero for the missing terms. For example, we could rewrite the polynomial in part (b) as $x^5 + x^4 + 0x^3 + 0x^2 + x - 3$.

Evaluating Polynomials

A special notation called **function notation** can be used to represent an algebraic expression and to represent its value for a given value of the variable. For example, to represent a polynomial whose variable is x, such as $x^2 + 3x - 5$, we might name the polynomial P and write $P(x) = x^2 + 3x - 5$.

The notation $P(x)$ is read "P of x" and indicates that the name of the polynomial is P and the variable is x. Just as we can use any variable when we write a polynomial, we can use letters other than P to name a polynomial. The following are other examples of polynomials written with function notation.

$$Q(x) = x^3 + 3x + 1$$
$$R(t) = 5t - 7t^2$$
$$P(w) = w^4 + 3w^3 + 2w - 1$$

Note: Be sure to read $P(x)$ as "P of x." Here the parentheses do *not* mean "P times x."

Function notation is especially convenient when we want to indicate the value of a polynomial for a particular value of the variable. If $P(x) = 5x^2 - 8x + 3$, then $P(2)$ represents the value of that expression when x is replaced by 2.

$$P(x) = 5x^2 - 8x + 3$$
$$P(2) = 5(2)^2 - 8(2) + 3 \quad \text{Replace } x \text{ with 2.}$$
$$= 20 - 16 + 3$$
$$= 7$$

EXAMPLE 5

Evaluating a Polynomial

Let $P(t) = -t^2 + 5t - 2$. Evaluate $P(-3)$.

Solution

$$P(t) = -t^2 + 5t - 2$$
$$P(-3) = -(-3)^2 + 5(-3) - 2 \quad \text{Replace } t \text{ with } -3.$$
$$= -9 - 15 - 2$$
$$= -26$$

EXAMPLE 6

Evaluating Polynomials

Let $p(x) = 3x^2 + x + 1$ and $g(t) = 1 - 4t$. Evaluate each of the following.

(a) $p(2)$

(b) $g(0)$

(c) $p(-1) - g(5)$

Solution

(a) $p(x) = 3x^2 + x + 1$

$\quad\ p(2) = 3(2)^2 + (2) + 1 \quad \text{Replace } x \text{ with 2.}$

$\quad\quad\quad = 12 + 2 + 1$

$\quad\quad\quad = 15$

(b) $g(t) = 1 - 4t$

$\quad\ g(0) = 1 - 4(0) \quad\quad\quad \text{Replace } t \text{ with 0.}$

$\quad\quad\quad = 1$

(c) $p(-1) = 3(-1)^2 + (-1) + 1 = 3 - 1 + 1 = 3$

$\quad\ g(5) = 1 - 4(5) = 1 - 20 = -19$

$\quad$ Therefore, $p(-1) - g(5) = 3 - (-19) = 22.$

Recall that we can store the value of the variable in a calculator and then evaluate the expression by entering it on the home screen or on the Y screen.

Y₁(2)

15

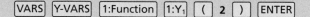

Keys to the Calculator

Evaluate Y

Using function notation, you can evaluate an expression without explicitly storing the value of the variable.

In Example 6, $p(x) = 3x^2 + x + 1$. Suppose that the expression has been entered on the Y screen as Y_1. To evaluate the expression for $x = 2$, use these keystrokes on the home screen:

[VARS] [Y-VARS] [1:Function] [1:Y₁] [( ] **2** [ )] [ENTER]

EXAMPLE **7**

Using a Calculator to Evaluate a Polynomial

Let $Q(x) = x^3 - x^2 + 4x - 1$ and $R(x) = 4x^2 - x$.

(a) Use both the home screen and the Y screen to evaluate $Q(2)$.

(b) Use the calculator's function notation feature to determine $R(-1)$ and $R(3)$.

Solution

(a) On the home screen, we store 2 in X and then enter $Q(x)$. (See Fig. 6.1.) On the Y screen, we enter $Q(x)$ as Y_1. Then we store 2 in X on the home screen and retrieve Y_1. (See Fig. 6.2.)

Figure 6.1

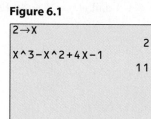

```
2→X
                    2
X^3−X^2+4X−1
                   11
```

Figure 6.2

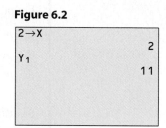

```
2→X
                    2
Y₁
                   11
```

(b) Enter the polynomial $4x^2 - x$ as Y_1 on the Y screen. Then enter $Y_1(-1)$ and $Y_1(3)$ on the home screen. (See Fig. 6.3.)

Figure 6.3

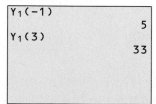

```
Y₁(−1)
                    5
Y₁(3)
                   33
```

The values are $R(-1) = 5$ and $R(3) = 33$.

We can evaluate a polynomial in more than one variable in the same way that we evaluate any other algebraic expression.

EXAMPLE **8**

Evaluating a Polynomial in More Than One Variable

Evaluate $xy^2 + 3y - 6x$ for $x = 0$ and $y = -1$.

Solution

$$xy^2 + 3y - 6x = 0(-1)^2 + 3(-1) - 6(0) \qquad \text{Replace } x \text{ with } 0 \text{ and } y \text{ with } -1.$$
$$= 0 - 3 - 0 = -3$$

Real-Life Applications

An application problem involving a polynomial sometimes requires us to evaluate the polynomial.

EXAMPLE **9**

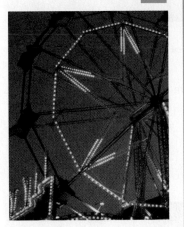

Attendance at an Amusement Park

The average daily attendance at an amusement park depends on the predicted high temperature of the day. The estimated attendance $A(t)$ can be described by a polynomial where the variable t represents the temperature (°F).

$$A(t) = -8t^2 + 1300t - 46{,}000$$

Use any calculator method to estimate the attendance for the following predicted high temperatures.

(a) 60°F   (b) 80°F   (c) 105°F

Solution

(a) $A(60) = 3200$

At 60°F the estimated attendance is 3200.

(b) $A(80) = 6800$

At 80°F the estimated attendance is 6800.

(c) $A(105) = 2300$

At 105°F the estimated attendance is 2300.

Quick Reference 6.2

Terminology
- A **monomial** is a number or a product of a number and one or more variables with whole number exponents.
- A **polynomial** is a monomial or a sum of monomials. A polynomial with two nonzero terms is a **binomial,** and a polynomial with three nonzero terms is a **trinomial.**
- The **degree of a monomial** in one variable is the value of the exponent of the variable. The degree of a monomial with more than one variable is the sum of the exponents of the variables. The degree of a constant term is 0. The monomial 0 has no defined degree.
- The **degree of a polynomial** is the degree of the nonzero term with the highest degree.

- A polynomial in one variable is in **descending order** if the terms are arranged in decreasing degrees. A polynomial in one variable is in **ascending order** if the terms are arranged in increasing degrees.

Evaluating Polynomials

- A special notation called **function notation** can be used to represent an algebraic expression and to represent its value for a given value of the variable. For a polynomial, the notation $P(x)$ is read "P of x" and indicates that the name of the polynomial is P and the variable is x.

- Function notation is especially convenient when we want to indicate the value of a polynomial for a particular value of the variable.

- Polynomials are evaluated in the same way that any algebraic expression is evaluated. Several calculator methods are available for evaluating polynomials.

Speaking the Language 6.2

1. The expressions 6, $6x^2$, and $6x^2y^5$ are all examples of ▢▢▢▢ .
2. The ▢▢▢▢ of $4x^3$ and of $-xy^2$ is 3.
3. For $P(x) = x^2 - 5x - 3$, we say that $P(x)$ is in ▢▢▢▢ notation, and the expression on the right is called a(n) ▢▢▢▢ .
4. We say that $2x^3 - 5x + 11$ is in ▢▢▢▢ order, and the polynomial has the special name ▢▢▢▢ .

Exercises 6.2

Concepts and Skills

1. Explain why the number 1 is both a monomial and a polynomial.

2. If a polynomial in one variable is written in descending order, where do you look to determine the degree of the polynomial? Why?

In Exercises 3–10, indicate with "Yes" or "No" whether the given expression is a polynomial.

3. $\dfrac{2}{x^2} + 5x + 1$ 4. $\dfrac{3}{y} - 2$

5. $y^3 + 2y^2 - y$ 6. $\dfrac{2}{3}x^2 - \dfrac{1}{2}x + 5$

7. $3xy^2 + 6y$ 8. $x^2 - 5y^3$

9. $\dfrac{x}{y} + 3y - 4$ 10. $\dfrac{6}{y} + 4xy$

In Exercises 11–18, name each polynomial as a monomial, binomial, or trinomial. If the polynomial does not have a special name, write "polynomial." Then list the terms and coefficients of each polynomial.

11. $-y - 7$ 12. $3x - 5y$

13. $-a^2b$ 14. $6x$

15. $6 - t - t^2$ 16. $y^2 + 4xy - x^2$

17. $2x^2y + 5yz - 7z + 1$

18. $5z^2 - 6z^8 + 3z - 7 + z^5$

In Exercises 19–26, identify the coefficient and the degree of each monomial.

19. $-y$ 20. $4x$

21. x^5 22. $-8y^4$

23. $5ab$ 24. $-3xy^5$

25. $2xy^2z^3$ 26. a^4bc^2

In Exercises 27–34, determine the degree of the polynomial.

27. $5 - 3x$

28. $4x^2 + 3x$

29. $8 - y^3$

30. $3y^2 - y^4 + 5y$

31. $3x^4y^2 - x^2y^3 + 2xy$

32. $2x^2 + x^2y^2 - y^3$

33. $x^2 + xyz + 3yz$

34. $x^2y + yz^4 + x^3z^3$

In Exercises 35–42, write the polynomial in
   (a) descending order.
   (b) ascending order.

35. $16 - y^2$

36. $5 + 2x - 3x^2$

37. $4 - x^2 + 3x$

38. $2y - 4 + 5y^2$

39. $t^2 - 3t^3 + 2 - t$

40. $4x - 3x^2 + 2 - 5x^3$

41. $-3x^5 + x - 6x^2 + 8x^4$

42. $6n^2 - 7n + 5 - 10n^6$

43. If we represent a polynomial with $G(x)$, what is the meaning of $G(3)$?

44. If $P(x) = -x^2$, why is $P(-1) = -1$ rather than 1?

In Exercises 45–50, evaluate the polynomial for the given values of the variable.

45. $5 + 2x$; $x = 0, -3$

46. $3x - 1$; $x = 2, -1$

47. $x^2 + 3x + 2$; $x = 1, -4$

48. $2x^2 - x - 3$; $x = 0, -2$

49. $x^4 - 3x^3 + 2x^2 + x - 5$; $x = 1, -1$

50. $x^5 + 3x^3 - 2x^2 + x$; $x = 0, 1$

In Exercises 51–58, a polynomial is represented with function notation. Fill in the blank to show how the polynomial is being evaluated.

51. $f(x) = 5x + 4$; $f(3) = 5(\quad) + 4$

52. $h(x) = 7x - 6$; $h(1) = 7(\quad) - 6$

53. $R(x) = 4 - x^2$; $R(2) = 4 - (\quad)^2$

54. $P(x) = 25 - x^2$; $P(0) = 25 - (\quad)^2$

55. $f(x) = 2x^3 - 4x + 5$;
   $f(\quad) = 2(2)^3 - 4(2) + 5$

56. $g(x) = 4x^3 + 3x^2 - 1$;
   $g(\quad) = 4(1)^3 + 3(1)^2 - 1$

57. $h(x) = -x^2 + 2x + 5$;
   $h(-2) = -(\quad)^2 + 2(\quad) + 5$

58. $g(x) = -x^4 - x^2$;
   $g(-1) = -(\quad)^4 - (\quad)^2$

In Exercises 59–64, evaluate the given polynomial as indicated.

59. $P(x) = 6 - x$
   $P(3), P(-2)$

60. $P(x) = -3x + 5$
   $P(-1), P(2)$

61. $F(x) = 2x + 9$
   $F(-3), F(4)$

62. $P(x) = 5 - 6x$
   $P(5), P(-4)$

63. $P(x) = -x^2 + x + 3$
   $P(-2), P(3)$

64. $G(x) = 3x^2 - x + 7$
   $G(0), G(1)$

In Exercises 65–68, evaluate the given polynomials as indicated.

65. $P(x) = -x + 6$ and $F(x) = 2x + 6$
   (a) $F(-4)$ (b) $P(-3)$
   (c) $F(1) - P(-2)$ (d) $P(5) + F(0)$

66. $F(x) = 5 + 3x$ and $G(x) = 7 - 2x$
   (a) $F(-2)$ (b) $G(4)$
   (c) $G(0) - F(1)$ (d) $F(-1) + G(4)$

67. $P(t) = \dfrac{1}{3}t^2$ and $Q(t) = t^3$
   (a) $P(6)$ (b) $Q(-3)$
   (c) $P(0) + Q(1)$ (d) $P(3) - Q(-1)$

68. $Q(t) = t^3 - 2t^2 + t$ and $R(t) = t - 2t^4$
   (a) $Q(-2)$ (b) $R(2)$
   (c) $Q(1) + R(-1)$ (d) $Q(0) - R(-2)$

In Exercises 69–74, use your calculator to evaluate the given polynomial as indicated. Experiment with the following techniques.
   (i)   Home screen method
   (ii) Y screen method
   (iii) Function notation method

69. $Q(x) = 9 - x^2$
   $Q(-3), Q(0)$

70. $P(x) = x^2 + 2x$
   $P(-2), P(1)$

71. $P(x) = x^3 + 2x^2 - x$
   $P(-1), P(1)$

72. $F(x) = -2x^3 - x + 5$
   $F(1), F(0)$

73. $P(x) = \frac{1}{3}x - \frac{x^2}{2}$

   $P(6), P(1)$

74. $R(x) = 6x^2$

   $R\left(\frac{1}{2}\right), R\left(\frac{1}{3}\right)$

In Exercises 75–82, evaluate the polynomial for the given values of the variables.

75. $xy + y - 3;$ $x = 0, y = 3$

76. $x - xy - y;$ $x = 7, y = 7$

77. $x^2 - y^2;$ $x = -5, y = -6$

78. $3x + 2y^2 - 4;$ $x = -1, y = -3$

79. $x^2 + 3xy + y^2;$ $x = 2, y = -3$

80. $2y^2 - x^2 + 2x;$ $x = -3, y = 0$

81. $xyz + xy - z;$ $x = 2, y = -4, z = 5$

82. $x^2 - y^2 + z^2;$ $x = -2, y = -4, z = -3$

🌐 Real-Life Applications

83. A terrarium in the shape of a cube has sides x inches long. The depth of the soil in the terrarium is y inches. Write a polynomial that describes the volume of space remaining in the terrarium.

84. Write a polynomial that describes the profit on the sale of x souvenir mugs if the total receipts are given by $0.5x^2$ and the total cost of the mugs is given by $2x$.

85. The number B of bacteria present in a culture dish 24 hours after treatment with x mg of a drug is given by $B(x) = -2x^2 - 10x + 1500$. If the bacteria are treated with 8 mg of the drug, how many bacteria remain after 24 hours?

86. A person's average weekly expenditure G for groceries is related to the age x of the person by the polynomial $G(x) = 3.25x - 20 - 0.0375x^2$. Find the average weekly grocery bill for a person who is

   (a) 20 years old.

   (b) 40 years old.

   (c) 60 years old.

✏ Modeling with Real Data

87. The graph shows credit card spending (in billions of dollars) from Thanksgiving to Christmas for various years. (Source: National Credit Card Counseling Services.)

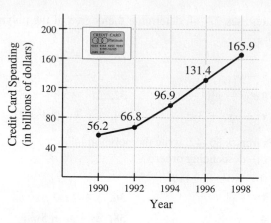

With t representing the number of years after 1990, the following polynomial models the credit card spending (in billions of dollars) between Thanksgiving and Christmas.

   $S(t) = 0.9t^2 + 6.7t + 54.1$

   (a) What is the degree of the polynomial?

   (b) Evaluate and interpret $S(12)$.

88. Refer to the model in Exercise 87.

   (a) Write the polynomial is ascending order.

   (b) Use the model to estimate the spending in 2005.

Data Analysis: *Seat Belt Use*

The bar graph shows the results of a survey to determine seat belt use by drivers in various age groups. (Source: Response Insurance.)

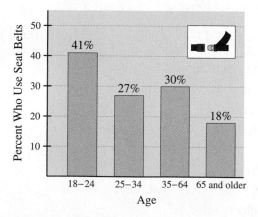

The results can be modeled by either of the following two polynomials, each written in function notation, where x is the drivers age.

   $P_1(x) = -0.34x + 44.1$

   $P_2(x) = -0.001475x^3 + 0.206x^2 - 9.14x + 155.8$

89. Use your calculator and $P_1(x)$ to estimate the percentages of drivers ages 21, 30, 55, and 70 who use seat belts. Round answers to the nearest whole number.

90. Use your calculator and $P_2(x)$ to estimate the percentages of drivers ages 21, 30, 55, and 70 who use seat belts. Round answers to the nearest whole number.

91. Compare your results from Exercises 89 and 90 to the actual data shown in the bar graph. Which polynomial appears to be a better model of the actual data?

92. Produce a table of values beginning with $x = 75$ for the function that you selected in Exercise 91. Does the model appear to be valid for older age groups?

Challenge

In Exercises 93–96, evaluate the polynomial by replacing the variable with the given expression. Simplify the result.

93. $F(x) = 4x + 5;$ $F(t + 2)$

94. $Q(t) = t^2 + 2t;$ $Q(2y)$

95. $xy^2 + 3x;$ $x = 2a, y = 3a$

96. $3xy - 2x^2 + y^2;$ $x = -4t, y = -5t$

97. Write $x^n - 3x^{3n} + 5x^{3n-1}$ in order of descending powers of x. Assume that n is a positive integer.

98. If the degree of $x^4y^n + x^2y^3$ is 7, what is n?

6.3   Addition and Subtraction

Adding Polynomials   •   Subtracting Polynomials   •   Equations and Applications

Adding Polynomials

Recall that two terms are like terms if both are constant terms or if both have the same variable factors with the same exponents. To add monomials, we use the Distributive Property to combine like terms.

$$3y + y^2 + 2y^2 - 5y = (1 + 2)y^2 + (3 - 5)y = 3y^2 - 2y$$
$$a^2b + 3ab^2 - 2ab^2 - 4a^2b = (1 - 4)a^2b + (3 - 2)ab^2 = -3a^2b + ab^2$$

To add polynomials, remove grouping symbols and combine like terms.

EXAMPLE **1**   **Adding Polynomials**

Add $(x^2 + 3x - 7) + (2x^2 - 3x + 6)$.

Solution

$$(x^2 + 3x - 7) + (2x^2 - 3x + 6)$$
$$= x^2 + 3x - 7 + 2x^2 - 3x + 6 \qquad \text{Remove the parentheses.}$$
$$= (x^2 + 2x^2) + (3x - 3x) + (-7 + 6)$$
$$= 3x^2 - 1 \qquad \text{Combine like terms.}$$

EXAMPLE **2**   **Column Addition**

Add $(x^3 + 2x^2 - 4) + (x^2 - 3x + 2) + (2x^3 + x^2 + 5x)$.

Solution

Adding polynomials in columns has the advantage of grouping like terms.

$$
\begin{array}{l}
x^3 + 2x^2 - 4 \\
 x^2 - 3x + 2 \\
\underline{2x^3 + x^2 + 5x } \\
3x^3 + 4x^2 + 2x - 2
\end{array}
$$

Note the space for the missing x-term.

Combine like terms.

EXAMPLE **3**

Think About It

Suppose that you enter $Y_1 = x + 3$, $Y_2 = 1 - 2x$, and $Y_3 = 4 - x$ in your calculator and display the table of values. How are the numbers in the Y_3 column related to the numbers in the Y_1 and Y_2 columns? Why?

Adding Polynomials

Find the sum of $(a^3 - ab + 3b)$ and $(a^3 - b^2 + 4ab)$.

Solution

$$
\begin{aligned}
&(a^3 - ab + 3b) + (a^3 - b^2 + 4ab) &&\text{Write as a sum.} \\
&= a^3 - ab + 3b + a^3 - b^2 + 4ab &&\text{Remove the parentheses.} \\
&= 2a^3 + 3ab + 3b - b^2 &&\text{Combine like terms.}
\end{aligned}
$$

Subtracting Polynomials

To subtract polynomials, we remove grouping symbols and combine like terms just as we did in addition. However, remember to change the sign of each term of the polynomial that is being subtracted.

EXAMPLE **4**

Subtracting Polynomials

Subtract $(5x^3 + 2x - 4) - (x^3 - x + 1)$.

Solution

$$
\begin{aligned}
&(5x^3 + 2x - 4) - (x^3 - x + 1) \\
&= 5x^3 + 2x - 4 - x^3 + x - 1 &&\text{Note the sign changes.} \\
&= 4x^3 + 3x - 5 &&\text{Combine like terms.}
\end{aligned}
$$

EXAMPLE **5**

Column Subtraction

Subtract $(x^3 + x^2 - x) - (x^3 - 3x - 1)$.

LEARNING TIP

When you use column subtraction, write the sign changes in a different color or circle them. Better yet, rewrite the problem with the sign changes.

Solution

Sometimes it is convenient to write the subtraction problem in columns. As in addition, we align like terms and leave spaces for missing terms.

$$
\begin{array}{l}
x^3 + x^2 - x \\
\underline{-(x^3 - 3x - 1)}
\end{array}
\rightarrow
\begin{array}{l}
x^3 + x^2 - x \\
\underline{-x^3 + 3x + 1} \\
 x^2 + 2x + 1
\end{array}
$$

The constant term is missing.

Note the sign changes.

EXAMPLE 6

Subtracting Polynomials

Subtract $(y^4 + y^2 - 6)$ from $(2y^4 - y^2 - 6)$.

Solution

$$(2y^4 - y^2 - 6) - (y^4 + y^2 - 6) \qquad \text{Write as a difference.}$$
$$= 2y^4 - y^2 - 6 - y^4 - y^2 + 6 \qquad \text{Change the signs.}$$
$$= y^4 - 2y^2 \qquad \text{Combine like terms.}$$

Addition and subtraction both may appear in the same expression.

EXAMPLE 7

Combining Operations

Perform the indicated operations.

$$(x^2 - x + 5) - (x^2 + 2x - 4) + (3x - 6)$$

Solution

$$(x^2 - x + 5) - (x^2 + 2x - 4) + (3x - 6)$$
$$= x^2 - x + 5 - x^2 - 2x + 4 + 3x - 6 \qquad \text{Remove parentheses.}$$
$$= 3 \qquad \text{Combine like terms.}$$

Equations and Applications

Equations sometimes involve the sum or difference of polynomials.

EXAMPLE 8

Solving an Equation with Polynomials

Solve $(2x^2 + 2x - 6) + (2 + 2x - x^2) = x^2 + 16$.

Solution

$$(2x^2 + 2x - 6) + (2 + 2x - x^2) = x^2 + 16 \qquad \text{Add the polynomials.}$$
$$2x^2 + 2x - 6 + 2 + 2x - x^2 = x^2 + 16 \qquad \text{Remove parentheses.}$$
$$x^2 + 4x - 4 = x^2 + 16 \qquad \text{Combine like terms.}$$
$$x^2 + 4x - 4 - x^2 = x^2 + 16 - x^2 \qquad \text{Subtract } x^2 \text{ from both sides.}$$
$$4x - 4 = 16$$
$$4x = 20$$
$$x = 5$$

EXAMPLE 9

Revenue, Cost, and Profit as Polynomials

The Donut Hole finds that the revenue from the sale of x cases of donuts is given by $R(x) = x^3 + 2x - 5$ and that the cost of producing x cases of donuts is given by $C(x) = x^3 - 3x + 10$. How many cases of donuts must the store sell to realize a profit of $30?

Solution

To determine the profit, subtract the cost from the revenue.

$$(x^3 + 2x - 5) - (x^3 - 3x + 10) = 30 \qquad \text{Profit is \$30.}$$
$$x^3 + 2x - 5 - x^3 + 3x - 10 = 30 \qquad \text{Remove parentheses.}$$
$$5x - 15 = 30 \qquad \text{Combine like terms.}$$
$$5x = 45$$
$$x = 9$$

The store must sell 9 cases of donuts to realize a profit of \$30.

Quick Reference 6.3

Adding Polynomials
- To add monomials, we use the Distributive Property to combine like terms.
- To add polynomials, remove grouping symbols and combine like terms.
- Adding polynomials in columns has the advantage of grouping like terms. Leave spaces for missing terms.

Subtracting Polynomials
- To subtract polynomials, remove grouping symbols and combine like terms. However, when you change to addition, remember that a minus symbol in front of parentheses changes the sign of each term of the polynomial that is being subtracted.

Equations and Applications
- An early step in equation solving is to simplify both sides of the equation. For equations involving polynomials, we may need to add or subtract the polynomials by removing grouping symbols and combining like terms.

Speaking the Language 6.3

1. Adding monomials is the same as _____ like terms with the _____ Property.
2. When you add polynomials in columns, leaving spaces for _____ terms helps to align like terms.
3. When performing polynomial subtraction, you must _____ of each term of the polynomial that is being subtracted.
4. A first step in adding or subtracting polynomials is to remove _____.

Exercises 6.3

Concepts and Skills

1. Prior to this section, we have used the word *simplify* to mean removing grouping symbols and combining like terms. Explain why adding or subtracting polynomials does not require any new knowledge.

2. What is an advantage of adding or subtracting polynomials in columns?

In Exercises 3–12, add the polynomials.

3. $(3x + 7) + (5x - 8)$ 4. $(2x - 6) + (-3x - 1)$

5. $(x^2 + 3x - 7) + (-x + 2)$

6. $(6x + 3) + (x^2 - 5x - 3)$

7. $(4x^2 + 3x + 5) + (2x^2 - x - 2)$

8. $(-2x^2 - x + 4) + (4x^2 + x - 7)$

9. $(2y^3 + 5y + 1) + (y^2 - 7y)$

10. $(x^2 - x^3 + 3x) + (x^2 + 5x + 2)$

11. $(x^2 - 5xy + y^2) + (3x^2 + 4xy - 2y^2)$

12. $(x^2 - xy - 3y^2) + (x^2 + xy + 2y^2)$

In Exercises 13–16, add the polynomials.

13. $4x - 7$
    $3x + 8$

14. $-2x - 5$
    $x - 1$

15. $x^2 + 4x - 6$
    $3x + 6$

16. $2x^2 - x - 1$
    $-x - 1$

In Exercises 17–22, use column addition to add the polynomials.

17. $-w^2 - 4w,$ $w^2 + 3w + 2$

18. $3y^2 + y - 4,$ $3y + 4$

19. $2x - x^2 + 4,$ $5 - x^2$

20. $x - 2x^2,$ $4 - x^2 + 3x$

21. $4x^3 + x - 8,$ $x^2 - 6x$

22. $3x^2 + 7x - 2,$ $x^2 - 3x^3 + 2 - 6x$

In Exercises 23–28, add the polynomials.

23. $2x + 3,$ $4 - 5x,$ $x^2 - 3x - 1$

24. $x^2 + 3x - 2,$ $2x + 3,$ $x^2 - 5$

25. $1 - 3w^2,$ $3w^2 + 2w - 1,$ $6 - 2w$

26. $y + 7,$ $3y - 4y^2,$ $5 - y + y^2$

27. $3x^2 + 5,$ $x^3 + 2x - 1,$ $4 - 3x^2 - 4x^3$

28. $x^3 - 3x,$ $x^2 - 4,$ $6x - x^2 - 2x^3$

In Exercises 29–32, write the resulting polynomial.

29. The sum of $6y + 2$ and $y^2 - 2$

30. The sum of $x^2 + 3x - 1$ and $7 - 8x$

31. Add $6b - b^2$ to $2b + 3$.

32. Add $3 + a - 2a^2$ to $a^2 + 4a + 4$.

In Exercises 33–44, subtract the polynomials.

33. $(6x + 1) - (3x + 2)$

34. $(a - 3) - (3 - a)$

35. $(2x^2 + 6 - 7x) - (5x^2 + x - 4)$

36. $(6x^2 + 2x - 1) - (x - 2x^2 - 3)$

37. $(5y^2 + y + 3) - (y^2 - 3)$

38. $(3t + 4) - (-5t^2 - 2t - 1)$

39. $(x^2 + 16) - (x^3 - 8x)$

40. $(2x^3 + x^2) - (3x^2 + 4)$

41. $(y^2 - 3xy - 2x^2) - (y^2 + xy + x^2)$

42. $(x^2 - y^2 + 3x) - (y^2 - x^2 + 4x)$

43. $(xy^2 + x^2y + 3) - (xy^2 - x^2y - 2x)$

44. $(xy^2 + 3xy - 4) - (x^2 - xy + 5)$

In Exercises 45–48, subtract the polynomials.

45. $6 - 5x$
    $5 + 7x$

46. $-7x - 3$
    $4x - 8$

47. $x^2 - 3x - 1$
    $2x - 3$

48. $6y^2 + y - 1$
    $-y + 5$

In Exercises 49–54, use the column method to subtract the second polynomial from the first polynomial.

49. $-t^2 - 9t,$ $t^2 - 2t + 6$

50. $3y^2 + y - 4,$ $3y + 4$

51. $6x + 7 - 2x^2,$ $1 - 4x^2$

52. $4w - 2w^2,$ $7w - 3 - 5w^2$

53. $6x^3 + 2x^2 - 5,$ $2x^2 + 4$

54. $x^2 - 8x^3 - 2x,$ $x^2 - 4x - 8x^3 + 7$

In Exercises 55–58, write the resulting polynomial.

55. Subtract $3x - 5$ from $6x + 2$.

56. Subtract $x + 4$ from $2x + 3$.

57. Determine the difference of $3x^2 - 4x + 8$ and $x^2 - 7x - 5$.

58. Determine the difference of $9 - x^2$ and $6 + 9x - 8x^2$.

In Exercises 59–64, perform the indicated operations.

59. $6x^2 - (4 - 5x) + (8x^2 + 15)$

60. $3y^2 - (5y + y^2) - (3y - 4)$

61. $(w^3 + 2w - 1) - (w^2 + w - 2) - (3w^2 - w^3)$

62. $(x^3 + 2) + (x^2 - x - 4) - (3x^2 - x^3)$

63. $(3x^2 - 5x - 1) + (4x^2 + 11) - (2x^2 - 4x - 7)$

64. $(a^2 - 4) - (a^2 - a - 4) + (a^2 + a - 1)$

In Exercises 65–68, write the expression and simplify.

65. Subtract $t^2 - 3t + 7$ from the sum of $2t - 5$ and $5t^2 + 4$.

66. Add $3y^2 - 4$ to the difference of $y^3 + 2y + 5$ and $y^2 + 3y - 4$.

67. Subtract the sum of $z^3 + 2z^2 - 5$ and $4z^2 + 3z + 2$ from $2z^3 + 5z + 6$.

68. Subtract the difference of $5y + 2$ and $y - 7$ from $4y$.

69. The equation $x^2 + 5x - 7 = 6 - 4x + x^2$ does not appear to be a linear equation in one variable. Nevertheless, our methods for solving a linear equation in one variable can be applied to this equation. Why?

70. A classmate has solved an equation with the following steps, but the solution is not the same as that given in the answer key. Why?

$$x^2 - (x^2 + 6x) = 18$$
$$6x = 18$$
$$x = 3$$

In Exercises 71–76, solve the equation.

71. $x^2 + 2x + 5 = x^2 + x + 2$

72. $4 + 3x + x^2 = x^2 + 2x + 15$

73. $(x^3 + x - 4) + (2 - x^3) = -2$

74. $(x^3 + 2x - 7) + (x - 8) = x^3$

75. $(3x + x^2) - (x^2 + x - 1) = 7 - x$

76. $(x^2 - 4x) - (3 - 4x + x^2) = 2 - x$

77. What must be added to $2x - 1$ to obtain 0?

78. What must be added to $4x + 5$ to obtain $x + 9$?

79. What must be subtracted from $4 - x$ to obtain 3?

80. What must be subtracted from $3x - 2$ to obtain $4x + 1$?

In Exercises 81 and 82, determine a, b, and c.

81. $(ax^2 + bx - 4) - (3x^2 + 2x + c) = -2x^2 - 5$

82. $(y^3 + by + 2) - (ay^3 - 2y + c) = 2y^3 + y$

Geometric Models

83. Write a polynomial to describe the perimeter of the rectangle shown in the figure.

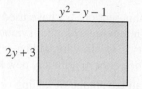

84. Write a polynomial to describe the perimeter of the triangle shown in the figure.

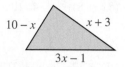

85. If the measures of two angles of a triangle are given by $x^3 + 3x$ and $x^3 - x - 50$, write a polynomial that represents the measure of the third angle.

86. Given the two right triangles in the figure, if the perimeter of the larger triangle is 18 inches more than the perimeter of the smaller triangle, what is the length of each hypotenuse?

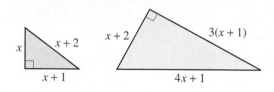

Real-Life Applications

87. The revenue from the sale of x units of a breakfast cereal is $R(x) = 2x^3 + 3x - 1$, and the cost is given by $C(x) = x^3 + x^2 + 4$. Write a polynomial to describe the profit.

88. With x representing the number (in hundreds) of visitors to a zoo, a vendor finds that the profit on sales of snow cones is given by the polynomial $S(x) = x^2 - 7x + 15$ and the profit on sales of ice cream is given by $C(x) = x^2 + 3x + 6$. How many zoo visitors are there if the profit on ice cream exceeds the profit on snow cones by $81?

89. When an object is thrown into the air, its height after t seconds is given by $-16t^2 + vt$, where v is the initial velocity. Suppose that one object is thrown upward at an initial velocity of 80 feet per second and, at the same instant, another object is thrown upward at an initial velocity of 60 feet per second. After how many seconds will the heights of the objects differ by 40 feet?

90. A company has x employees. Last year, the cost of production was x^3, the cost of health care was x^2, and the payroll was $4x + 81$. This year, health care costs increased to x^3, the cost of production decreased to x^2, and the payroll was $7x - 18$. If this year's total costs were the same as last year's, how many people does the company employ?

Modeling with Real Data

91. From 1970 to 1998, the number of households with an unmarried female as head of the household rose from 5.5 million to 12.7 million, and the number of households with an unmarried male as head of the household increased from 1.2 million to 3.9 million. (Source: Census Bureau.) Let x represent the number of years since 1970.

(a) Write a first-degree polynomial $F(x)$ to model the number of households with an unmarried female as head.

(b) Write a first-degree polynomial $M(x)$ to model the number of households with an unmarried male as head.

92. Refer to Exercise 91.

(a) What does $F(x) - M(x)$ represent?

(b) What does $F(x) + M(x)$ represent?

Challenge

In Exercises 93–96, simplify the given expression.

93. $(8x + 3) - [(2x - 1) + (x + 3)]$

94. $8 - [(2y + 7) - (4y - 5)]$

95. $7y^2 + [(5y^2 + 3) - (y^2 - 6)]$

96. $[(x^2 + 3x) - (2x - 1)] + (8x^2 - 2x)$

In Exercises 97 and 98, let $P(x) = 1 - 3x$. Write and simplify the given expression.

97. $P(2a) - P(a)$

98. $P(2 + h) - P(2)$

6.4 | Multiplication

Products of Monomials and Polynomials · Products of Polynomials · The Order of Operations · Applications

Products of Monomials and Polynomials

The most basic example of a product of polynomials is a product of two monomials. Example 1 reviews the use of properties of exponents that we discussed in Section 6.1.

EXAMPLE 1   **Multiplying Monomials**

Multiply the given monomials.

(a) $2x(3x^2) = (2 \cdot 3) \cdot (x \cdot x^2) = 6x^3$ Product Rule for Exponents

(b) $5x^3(-3x^2) = -15x^5$

(c) $(-2a^2b^6)(-a^4b) = 2a^6b^7$

To multiply a polynomial by a monomial, we use the Distributive Property to write the product as a sum of products of monomials.

EXAMPLE 2 | **Multiplying a Polynomial by a Monomial**

Multiply the polynomial by the monomial.

(a) $x^2(3x + 2) = x^2 \cdot 3x + x^2 \cdot 2 = 3x^3 + 2x^2$ Distributive Property

(b) $-3x(x^2 - 3x + 2)$

$= (-3x)(x^2) - (-3x)(3x) + (-3x)(2)$ Distributive Property

$= -3x^3 + 9x^2 - 6x$

(c) $2x^2(x^3 - 5x^2 - x + 2) = 2x^5 - 10x^4 - 2x^3 + 4x^2$

Note: As shown in part (c) of Example 2, you should eventually be able to distribute the monomial mentally and write the result directly.

Products of Polynomials

Multiplying two polynomials that both have more than one term requires repeated use of the Distributive Property.

EXAMPLE 3 | **Multiplying Two Binomials**

Multiply the given binomials.

(a) $(x + 6)(x - 8)$

$= x(x - 8) + 6(x - 8)$ Distributive Property

$= x \cdot x - x \cdot 8 + 6 \cdot x - 6 \cdot 8$ Distributive Property

$= x^2 - 8x + 6x - 48$

$= x^2 - 2x - 48$ Combine like terms.

(b) $(2x - 5)(4x - 3)$

$= 2x(4x - 3) - 5(4x - 3)$ Distributive Property

$= 8x^2 - 6x - 20x + 15$ Distributive Property

$= 8x^2 - 26x + 15$ Simplify.

In part (b) of Example 3 we saved a step by distributing the monomials mentally. In Section 6.5 we will learn some special patterns for multiplying binomials that will allow us to write the product of two binomials without any intermediate steps at all.

EXAMPLE 4 | **Multiplying a Trinomial by a Binomial**

Think About It

If you multiply a binomial by a trinomial, the result (before combining like terms) has six terms. Suppose one polynomial has m terms and the other has n terms. Before combining like terms, how many terms are in the result?

Multiply the trinomial by the binomial.

$(x + 2)(x^2 + 3x - 4)$

$= x(x^2 + 3x - 4) + 2(x^2 + 3x - 4)$

$= x \cdot x^2 + x \cdot 3x - 4 \cdot x + 2 \cdot x^2 + 2 \cdot 3x - 2 \cdot 4$

$= x^3 + 3x^2 - 4x + 2x^2 + 6x - 8$

$= x^3 + 5x^2 + 2x - 8$ Combine like terms.

Note again that the Distributive Property was needed twice in Example 4. However, steps can be saved by observing that each term of $x^2 + 3x - 4$ was multiplied by each term of $x + 2$. In general, we can write the product of two polynomials simply by multiplying each term of one polynomial by each term of the other polynomial.

EXAMPLE 5

Column Multiplication

Multiply $(2x + 3)(x^3 + 2x^2 - x - 5)$.

Solution

Arranging the problem in columns is particularly useful when one or both of the polynomials have many terms. This facilitates combining like terms.

$$
\begin{array}{r}
x^3 + 2x^2 - \ \ x - \ \ 5 \\
2x + \ \ 3 \\
\hline
3x^3 + 6x^2 - \ \ 3x - 15 \\
2x^4 + 4x^3 - 2x^2 - 10x \\
\hline
2x^4 + 7x^3 + 4x^2 - 13x - 15
\end{array}
$$

$x^3 + 2x^2 - x - 5$ multiplied by 3

$x^3 + 2x^2 - x - 5$ multiplied by $2x$

The sum of the two products

The Order of Operations

The normal Order of Operations is applicable to operations involving polynomials. When a product includes more than two factors, we can begin by multiplying any two of the factors.

EXAMPLE 6

Multiplying More Than Two Factors

Determine the product.

(a) $x(x + 3)^2$

(b) $(x - 2)(x + 1)(x - 3)$

Solution

(a) $x(x + 3)^2 = x(x + 3)(x + 3)$ $(x + 3)^2$ means $(x + 3)(x + 3)$.

$= x(x^2 + 3x + 3x + 9)$ Multiply the binomials.

$= x(x^2 + 6x + 9)$ Combine like terms.

$= x^3 + 6x^2 + 9x$ Distributive Property

(b) $(x - 2)(x + 1)(x - 3)$

$= (x - 2)(x^2 - 3x + x - 3)$ Multiply the last two binomials.

$= (x - 2)(x^2 - 2x - 3)$ Combine like terms.

$= x^3 - 2x^2 - 3x - 2x^2 + 4x + 6$ Distributive Property

$= x^3 - 4x^2 + x + 6$ Combine like terms.

Note: In part (b) of Example 6, we chose to multiply the last two binomials first. The result would have been the same if we had chosen to begin with the first two binomials.

EXAMPLE 7

Combining Operations

Perform the indicated operations: $3x^2 - (x + 2)(x - 5)$.

Solution

$$3x^2 - (x + 2)(x - 5)$$
$$= 3x^2 - (x^2 - 5x + 2x - 10) \quad \text{Multiply the binomials first.}$$
$$= 3x^2 - (x^2 - 3x - 10) \quad \text{Combine like terms.}$$
$$= 3x^2 - x^2 + 3x + 10 \quad \text{Remove the parentheses.}$$
$$= 2x^2 + 3x + 10 \quad \text{Combine like terms.}$$

Applications

EXAMPLE 8

Figure 6.4

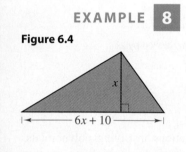

6x + 10

Describing the Area of a Triangle

Write a simplified expression for the area of the triangle in Figure 6.4.

Solution

$$A = \frac{1}{2}bh = \frac{1}{2}x(6x + 10) = 3x^2 + 5x$$

EXAMPLE 9

Consecutive Integers

The product of two consecutive odd integers is 14 more than the square of the first. What are the integers?

Solution

Let $x = $ the first odd integer and $x + 2 = $ the second odd integer.

$$x(x + 2) = x^2 + 14$$
$$x^2 + 2x = x^2 + 14 \quad \text{Multiply the polynomials.}$$
$$2x = 14 \quad \text{Subtract } x^2.$$
$$x = 7 \quad \text{Divide by 2.}$$

The first odd integer is 7, and the second odd integer is $7 + 2 = 9$.

Quick Reference 6.4

Products of Monomials and Polynomials

- To multiply monomials, use the properties of real numbers to reorder and regroup factors, and the properties of exponents discussed in Section 6.1.

- To multiply a polynomial by a monomial, we use the Distributive Property to write the product as a sum of products of monomials.

Products of Polynomials

- Multiplying two polynomials that both have more than one term requires repeated use of the Distributive Property.

- In general, we can write the product of two polynomials simply by multiplying each term of one polynomial by each term of the other polynomial.
- When one or both polynomials of a product have many terms, column multiplication has the advantage of aligning like terms.

The Order of Operations
- The Order of Operations is applicable to operations involving polynomials.

Speaking the Language 6.4

1. We use the ░░░░░░░ Property to perform all polynomial multiplication.

2. To multiply two polynomials, we can multiply each term of one polynomial by ░░░░░░░ .

3. An advantage of column multiplication is that ░░░░░░░ can be aligned in columns.

4. When three or more polynomials are to be multiplied, the ░░░░░░░ Property of Multiplication allows us to choose the two polynomials to multiply first.

Exercises 6.4

Concepts and Skills

1. When we multiply $(x + 1)(x + 2)$, the simplified product has three terms. Does the product of two binomials always have three terms? Explain.

2. Multiplying $(x + 2x + 3x)(2x + 3x + 4x)$ results in a product with nine terms, and then we combine like terms. Describe an easier way to simplify this expression.

In Exercises 3–8, multiply the monomials.

3. $-5x(3x)$

4. $-4(3x^2)$

5. $2y^2(-5y)$

6. $(-2x^4)(-4x^5)$

7. $3x^2y(xy^4)$

8. $2x^5y^2(3x^2y^3)$

In Exercises 9–20, multiply.

9. $3x^2(4x - 7)$

10. $-8y(4 - 3y)$

11. $-5x(3x^2 + x - 2)$

12. $2x(x^2 - 5x + 4)$

13. $-4x^2(x^4 - 3x^2 + 2x)$

14. $x^2(5x^3 - 2x + 9)$

15. $6x(3x^2 + 4x - 5)$

16. $-x^3(x^3 + 3x^2 - 7x)$

17. $xy^2(x^2 + 3xy + y^2)$

18. $ab(a^2b + ab^2 - 1)$

19. $x^3(xy^2 + y + 2x)$

20. $-y(x^2 - y + xy)$

In Exercises 21–30, use the Distributive Property to multiply the binomials.

21. $(x + 7)(x - 3)$

22. $(x - 2)(x - 5)$

23. $(y + 6)(y + 1)$

24. $(w - 4)(w - 2)$

25. $(2x + 1)(x - 3)$

26. $(3 - 2x)(1 + x)$

27. $(3a - 5)(2a + 1)$

28. $(4t + 1)(3t - 2)$

29. $(x + 4)^2$

30. $(3 - t)^2$

In Exercises 31–40, use the Distributive Property to multiply.

31. $(x + 2)(x^2 + 2x - 5)$

32. $(x - 3)(x^2 - x + 3)$

33. $(x - 3)(x^2 + 3x + 9)$

34. $(x + 2)(x^2 - 2x + 4)$

35. $(y - 4)(3y^2 - y + 2)$

36. $(w + 2)(2w^2 + 4w - 5)$

37. $(2x - 1)(x^2 - x - 3)$

38. $(3t + 2)(t^2 + 3t - 4)$

39. $(x^2 + x - 4)(x^2 + 2x - 1)$

40. $(x^2 - 3x - 1)(x^2 + 3x + 1)$

In Exercises 41–44, multiply.

41. $3t - 7$
$\underline{ t + 5}$

42. $6y + 7$
$\underline{ 2y - 1}$

43. $x^2 - 4x + 7$
$\underline{ x - 2}$

44. $3x^2 + x + 4$
$\underline{ 2x - 3}$

45. According to the Order of Operations, to simplify $(x - 5)(x + 3)^2$ we square $x + 3$ and then multiply the result by $x - 5$. Would we obtain the same result if we multiply $x - 5$ and $x + 3$ and then multiply the result by $x + 3$? Why?

46. When we simplify $x^2 - (x + 1)^2$, is the second step $x^2 - x^2 + 2x + 1$ or should it be $x^2 - x^2 - 2x - 1$? Why?

In Exercises 47–58, multiply.

47. $2x(x + 3)(x - 6)$

48. $-t(2t + 3)(t - 9)$

49. $(x + 1)(x - 2)(x + 3)$

50. $(2x - 3)(x + 2)(x - 1)$

51. $(4x + 3)(x - 1)(2x + 1)$

52. $(x + 1)(2x + 3)(3x - 4)$

53. $2y(y + 1)^2$

54. $-3t(1 - 2t)^2$

55. $-3x^3(x - 1)(x + 2)$ **56.** $5w^2(w + 5)(w - 2)$

57. $(x + 1)^3$ **58.** $(3 - t)^3$

In Exercises 59–66, perform the indicated operations.

59. $3x + (x + 1)(x - 4)$

60. $(x - 3)(x - 1) - 5$

61. $7x - (x + 2)(x - 3)$

62. $12 - (2x + 3)(x + 4)$

63. $(x + 2)(x - 4) + (x^2 + 2x)$

64. $(2x - 5) + (x + 1)(x - 4)$

65. $(7x - 2) - (x + 6)(x - 7)$

66. $(3x + 8) - (x + 3)(x + 4)$

In Exercises 67–72, write the expression and simplify.

67. Subtract the product of x and $x - 3$ from $2x + 5$.

68. Add $x^2 - 7$ to the product of $-2x$ and $x - 1$.

69. Add the trinomial $2x^2 + 5x - 6$ to the product of $2x - 1$ and $x + 2$.

70. Subtract the product of the binomials $x - 5$ and $x + 2$ from $x^2 + x - 4$.

71. Multiply the product of $x + 3$ and $2x - 1$ by x^2.

72. Multiply $-3y$ times the product of $1 + y$ and $2 - y$.

In Exercises 73–76, solve each equation.

73. $x(x + 3) = x^2 + 15$

74. $x(2x - 7) = 12 - 5x + 2x^2$

75. $(x + 2)(x - 3) = x(x - 2)$

76. $1 - 2x(2 - x) = (x + 1)(2x + 1)$

Geometric Models

In Exercises 77–80, write a simplified polynomial that represents the described quantity.

77. The area of a rectangle with sides represented by $2x + 1$ and $x^2 + 3x - 1$

78. The area of a triangle with base $2x - 1$ and height $6x$

79. The area of a walk around a rectangular swimming pool if the walk is 6 feet wide and the length of the pool is 7 feet more than its width x

80. The total volume of the box in the figure

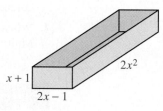

Real-Life Applications

81. The length of a poster is 3 inches more than its width. If each side were increased by 5 inches, the area would be increased by 160 square inches. What are the dimensions of the poster?

82. A fully carpeted rectangular room whose length is twice its width requires 68 square feet more carpeting than it would if it had an uncarpeted 1-foot border. What are the dimensions of the room?

83. A patio adjoining a house has a border of flowers 4 feet wide on three of its sides. (See the accompanying figure.) The width of the patio is 4 feet less than its length, and the area of the flower bed is 184 square feet. What are the dimensions of the patio?

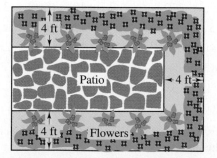

84. A quilt is 2 feet longer than it is wide. It takes an additional 28 square feet of material to make a 1-foot-wide border for the quilt. What are the dimensions of the quilt?

85. The original city limits of Glenwood formed a rectangle whose length was 1 mile more than twice its width. When the city increased its limits by 2 miles in both length and width, its area increased by 12 square miles. What were the original dimensions of the city?

86. A triangular sail has a base 2 feet longer than its height. If the boat owner installs a new sail whose base and height are both 3 feet longer than the original sail, the area is increased by 22.5 square feet. What are the dimensions of the original sail?

✏ **Modeling with Real Data**

87. The table shows the amount spent per person (age 12 and older) on home video games for selected years. Also shown is the population, age 12 and older, for those years.

Year	Amount	Population Age 12 and Older (in millions)
1996	$12.45	209
1998	17.19	214
2000	18.71	219
2002	20.04	222

(Source: Veronis, Suhler & Associates.)

If we let x represent the number of years since 1995, the following polynomials can be used to model the data.

Expenditures: $E(x) = 1.2x + 12.3$

Population: $P(x) = 2.2x + 207$

(a) What does the product of the polynomials represent?

(b) Write the polynomial $T(x)$ that is the product of the polynomials $E(x)$ and $P(x)$.

88. Refer to the models in Exercise 87.

(a) Evaluate and interpret $E(10)$, $P(10)$, and $T(10)$.

(b) What is the relationship among $E(10)$, $P(10)$, and $T(10)$?

Data Analysis: *Cable TV Subscribers*

The table shows the number (in millions) of cable TV subscribers in the United States for selected years from 1990 to 1998. Also shown are the average monthly rates for those years.

Year	Number of Subscribers (in millions)	Average Monthly Rate (in dollars)
1990	50.5	$16.78
1995	60.9	23.07
1997	64.2	26.48
1998	65.4	27.43

(Source: Cable TV Financial Data Book.)

The number n of subscribers (in millions) can be modeled by $n(x) = 1.9x + 50.8$, and the average monthly rate r can be modeled by $r(x) = 1.4x + 16.7$. For both functions, x is the number of years since 1990.

89. Evaluate $n(7) \cdot r(7)$.

90. If $p(x) = n(x) \cdot r(x)$, write an algebraic expression for $p(x)$. (Use your calculator to perform the arithmetic.) Then evaluate $p(7)$.

91. What is your interpretation of the values in Exercises 89 and 90?

92. Use the models to estimate the total monthly cable TV revenue in 2002.

Challenge

In Exercises 93 and 94, multiply.

93. $x^2(3x^{2n} + 2x^n - 5)$

94. $(x + 3)(x^{2n} - 7x^n - 6)$

In Exercises 95 and 96, multiply.

95. $(x^2 + 5x - 1)(x^3 - 2x^2 + 4x - 5)$

96. $(x^2 + 2x - 1)^3$

In Exercises 97 and 98, determine a and b.

97. $3x^3(ax^2 + bx - 4) = -6x^5 + 3x^4 - 12x^3$

98. $(ax + 2)(3x + b) = 12x^2 + 18x + 6$

6.5 | Special Products

The Product of Two Binomials (FOIL) • The Product of a Sum and Difference of the Same Two Terms • The Square of a Binomial • Applications

The Product of Two Binomials (FOIL)

Products of binomials occur so frequently in algebra that it is worth considering special patterns that can aid in carrying out the multiplications efficiently.

Exploring the Concept

Multiplying Two Binomials

Consider the product of $x + 3$ and $x + 7$, and observe the pattern that is followed.

$$(x + 3)(x + 7) = x(x + 7) + 3(x + 7)$$
$$= x \cdot x + 7x + 3x + 3 \cdot 7$$
$$= x^2 + 10x + 21$$

The first term of the result is the product of the *first terms* (x and x) of the binomials.

The next two terms are the products of the *outer terms* (x and 7) and the *inner terms* (3 and x).

The last term of the result is the product of the *last terms* (3 and 7).

We can remember the pattern by using the letters of the word FOIL.

For $(x + 3)(x + 7)$:

F stands for **F**irst terms: $(x + 3)(x + 7)$ $x \cdot x$

O stands for **O**uter terms: $(x + 3)(x + 7)$ $7 \cdot x$

I stands for **I**nner terms: $(x + 3)(x + 7)$ $3 \cdot x$

L stands for **L**ast terms: $(x + 3)(x + 7)$ $3 \cdot 7$

EXAMPLE **1**

Using the FOIL Method to Multiply Two Binomials

Use the FOIL method to determine the product $(3x - 5)(2x + 1)$.

Solution

First terms

Last terms   F   O   I   L

$$(3x - 5)(2x + 1) = 6x^2 + 3x - 10x - 5 = 6x^2 - 7x - 5$$

Inner terms

Outer terms

Note: The FOIL method is neither different from nor better than the use of the Distributive Property. However, as we will see in Chapter 7, knowing where the terms of the result come from is helpful.

EXAMPLE 2

Using the FOIL Method to Multiply Binomials

Use the FOIL method to multiply the given binomials.

$$\overset{\text{F} \qquad \text{O} \qquad \text{I} \qquad \text{L}}{}$$
(a) $(x - 4)(x - 2) = x^2 - 2x - 4x + 8 = x^2 - 6x + 8$

$$\overset{\text{F} \qquad \text{O} \qquad \text{I} \qquad \text{L}}{}$$
(b) $(2x + 3)(3x - 4) = 6x^2 - 8x + 9x - 12 = 6x^2 + x - 12$

(c) $(x - 3y)(x - 2y) = x^2 - 2xy - 3xy + 6y^2 = x^2 - 5xy + 6y^2$

(d) $(2a + b)(a - b) = 2a^2 - 2ab + ab - b^2 = 2a^2 - ab - b^2$

The Product of a Sum and Difference of the Same Two Terms

Sometimes, in a product of two binomials, the terms of the binomials are the same, but one binomial is a sum and the other binomial is a difference. This leads to an interesting result.

Exploring the Concept

A Special Product

Consider the following products.

$$(x + 4)(x - 4) = x^2 - 4x + 4x - 16 = x^2 - 16$$

$$(7 + 2y)(7 - 2y) = 49 - 14y + 14y - 4y^2 = 49 - 4y^2$$

$$(3b + 2)(3b - 2) = 9b^2 - 6b + 6b - 4 = 9b^2 - 4$$

In each case one factor is the sum of the two terms, and the other factor is the difference of the same two terms. The result is the difference of the squares of the two terms.

The pattern applies to any product in the form $(A + B)(A - B)$.

The Product of the Sum and the Difference of the Same Two Terms

$$(A + B)(A - B) = A^2 - B^2$$

In words, the product of the sum and the difference of the same two terms is the square of the first term minus the square of the second term. The result is called a *difference of two squares*.

EXAMPLE 3

The Product of the Sum and the Difference of the Same Two Terms

Use the pattern $(A + B)(A - B) = A^2 - B^2$ to multiply.

(a) $(x + 5)(x - 5) = x^2 - 5^2 = x^2 - 25$

(b) $(3 + y)(3 - y) = 3^2 - y^2 = 9 - y^2$

(c) $(2x - y)(2x + y) = (2x)^2 - y^2 = 4x^2 - y^2$

(d) $\left(x + \dfrac{2}{3}\right)\left(x - \dfrac{2}{3}\right) = x^2 - \left(\dfrac{2}{3}\right)^2 = x^2 - \dfrac{4}{9}$

(e) $(t^3 + 1)(t^3 - 1) = (t^3)^2 - 1^2 = t^6 - 1$

The Square of a Binomial

Another product that occurs frequently is the square of a binomial, such as $(y + 3)^2$. The square of a binomial can be written as the product of binomials:

$$(y + 3)^2 = (y + 3)(y + 3)$$

Then our methods for multiplying binomials apply.

The square of a binomial is written $(A + B)^2$ or $(A - B)^2$. We multiply to obtain the special pattern for squaring a binomial.

$$(A + B)^2 = (A + B)(A + B) = A^2 + AB + AB + B^2 = A^2 + 2AB + B^2$$

$$(A - B)^2 = (A - B)(A - B) = A^2 - AB - AB + B^2 = A^2 - 2AB + B^2$$

Recognizing the square of a binomial as a special product allows us to write the result with no intermediate steps.

Think About It

Write a special product pattern for $(A + B)^3$ and use it to determine $(x + 2)^3$.

> **The Square of a Binomial**
>
> $$(A + B)^2 = A^2 + 2AB + B^2$$
> $$(A - B)^2 = A^2 - 2AB + B^2$$
>
> In words, the square of a binomial is the square of the first term plus (or minus) twice the product of the two terms plus the square of the last term.

Note: The Power of a Product Rule states that $(AB)^2 = A^2B^2$. However, there is no similar rule for the power of a sum or difference: $(A + B)^2$ is *not* $A^2 + B^2$. This error can be avoided by remembering that $(A + B)^2 = (A + B)(A + B)$.

EXAMPLE 4

LEARNING TIP

For products of the form $(A + B)(A - B)$, the result is always a difference. For the square of a binomial, the sign of the middle term is plus for $(A + B)^2$ and minus for $(A - B)^2$.

Squaring a Binomial

Use the special patterns to square the given binomials.

(a) $(x + 8)^2 = x^2 + 2(x \cdot 8) + 8^2 = x^2 + 16x + 64$

(b) $(y - 5)^2 = y^2 - 2(y \cdot 5) + 5^2 = y^2 - 10y + 25$

(c) $(4a - 1)^2 = (4a)^2 - 2(4a \cdot 1) + 1^2 = 16a^2 - 8a + 1$

(d) $(2m + 3n)^2 = (2m)^2 + 2(2m \cdot 3n) + (3n)^2 = 4m^2 + 12mn + 9n^2$

(e) $\left(6x + \dfrac{1}{3}\right)^2 = (6x)^2 + 2\left(6x \cdot \dfrac{1}{3}\right) + \left(\dfrac{1}{3}\right)^2 = 36x^2 + 4x + \dfrac{1}{9}$

Applications

Special products sometimes arise in the course of solving an application problem.

EXAMPLE 5

Describing the Volume of a Box

Write a simplified polynomial describing the volume of a box with sides $2y - 5$ and $2y + 5$ and height $2y$.

Solution

$$V = LWH$$ The formula for the volume of a rectangular solid

$$V = 2y(2y - 5)(2y + 5)$$ The factors can be written in any order.

$$= 2y(4y^2 - 25)$$ $(A + B)(A - B) = A^2 - B^2$

$$= 8y^3 - 50y$$ Distributive Property

EXAMPLE 6

The Dimensions of a Parking Lot

A square parking lot is surrounded by a walk 1 meter wide. If the area of the walk is 100 square meters, what are the dimensions of the parking lot?

Solution

Let $x =$ the length of a side of the parking lot. Then $x + 2$ represents the length of the square including the walk. (See Fig. 6.5.)

The area of the walk is the difference of the areas of the larger square and the smaller square.

$$(x + 2)^2 - x^2 = 100$$ The area of the walk is 100 square meters.

$$x^2 + 4x + 4 - x^2 = 100$$ $(A + B)^2 = A^2 + 2AB + B^2$

$$4x + 4 = 100$$

$$4x = 96$$

$$x = 24$$

The length of a side of the parking lot is 24 meters.

Figure 6.5

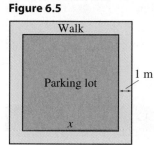

Quick Reference 6.5

The Product of Two Binomials (FOIL)

- When we use the Distributive Property to multiply two binomials, the following pattern holds for the result.

  (a) The first term of the result is the product of the *first terms* of the binomials.

  (b) The next two terms of the result are the products of the *outer terms* and the *inner terms*.

  (c) The last term of the result is the product of the *last terms*.

- The acronym FOIL can be used to recall this pattern.

The Product of a Sum and Difference of the Same Two Terms

- When the terms of two binomials are the same, but one binomial is a sum and the other binomial is a difference, the pattern of the product of the binomials is as follows.

  $$(A + B)(A - B) = A^2 - B^2$$

- In words, the product of the sum and the difference of the same two terms is the square of the first term minus the square of the second term. The result is called a *difference of two squares*.

The Square of a Binomial

- The result of squaring a binomial is given by one of the following patterns.

$$(A + B)^2 = A^2 + 2AB + B^2$$
$$(A - B)^2 = A^2 - 2AB + B^2$$

- In words, the square of a binomial is the square of the first term plus (or minus) twice the product of the two terms plus the square of the last term.

Speaking the Language 6.5

1. In the acronym FOIL, the O and the I refer to the products of the ▭ and the ▭ terms, respectively.

2. The result of multiplying binomials of the form $A + B$ and $A - B$ is described as a(n) ▭ .

3. The result of squaring $A + B$ is the ▭ of A plus ▭ the product of A and B plus the ▭ of B.

4. The special product patterns described in this section are simply shortcuts for using the ▭ Property.

Exercises 6.5

Concepts and Skills

1. Does the FOIL method apply to the following expressions? Explain.

   (i) $2x + 3(x - 1)$

   (ii) $(2x + 3)(x - 1)$

2. Which of the following products can be determined by using the Distributive Property and which can be determined with the FOIL method?

   (i) $(x + 3)(x^2 + x + 1)$

   (ii) $(x + 3)(x^2 + 1)$

   (iii) $(x + 3)(x + 1)$

In Exercises 3–26, use the FOIL method to determine the product.

3. $(x + 5)(x + 2)$

4. $(x - 8)(x - 1)$

5. $(y - 7)(y + 3)$

6. $(x - 2)(x + 5)$

7. $(8 + x)(2 - x)$

8. $(10 + a)(4 - a)$

9. $(3x + 2)(2x - 3)$

10. $(2x + 5)(x + 4)$

11. $(4x - 1)(x - 2)$

12. $(3x + 4)(x - 1)$

13. $(x + 5)(2x - 7)$

14. $(5x + 1)(x + 1)$

15. $(2a - 5)(4a - 3)$

16. $(3y + 1)(2y - 1)$

17. $(x + y)(x - 3y)$

18. $(2x + y)(x + 4y)$

19. $(5x - 4y)(4x + 5y)$

20. $(r - 2t)(3r - t)$

21. $(x^2 + 5)(x^2 - 3)$

22. $(y^2 - 1)(y^2 - 4)$

23. $(a^2 + 4)(a^2 + 2)$

24. $(7 - x^2)(1 - x^2)$

25. $(ab + 3)(ab + 2)$

26. $(8 - 3xy)(5 + xy)$

27. Compare the sum of the inner and outer terms for each of the following.

   (i) $(x + 2)^2$ (ii) $(x + 2)(x - 2)$

28. One of the following relations is false. Identify it and write the correct answer.

   (i) $(xy)^2 = x^2y^2$ (ii) $(x + y)^2 = x^2 + y^2$

In Exercises 29–42, determine the product.

29. $(x + 7)(x - 7)$

30. $(x + 6)(x - 6)$

31. $(3 + y)(3 - y)$

32. $(2 + a)(2 - a)$

33. $(3x - 5)(3x + 5)$ **34.** $(2y + 7)(2y - 7)$

35. $(6 + 5y)(6 - 5y)$ **36.** $(8 - 3b)(8 + 3b)$

37. $(3x - 7y)(3x + 7y)$ **38.** $(x + 4y)(x - 4y)$

39. $(2a + b)(2a - b)$ **40.** $(a - 5b)(a + 5b)$

41. $(x^2 + 3)(x^2 - 3)$ **42.** $(xy + 1)(xy - 1)$

In Exercises 43–54, square the binomial.

43. $(x + 2)^2$ **44.** $(x - 5)^2$

45. $(9 - x)^2$ **46.** $(7 - x)^2$

47. $(2x + 1)^2$ **48.** $(3y - 2)^2$

49. $(4 - 7x)^2$ **50.** $(2x - y)^2$

51. $(a - 3b)^2$ **52.** $(10x + y)^2$

53. $(x^3 + 2)^2$ **54.** $(x - 3y^2)^2$

In Exercises 55–70, use the applicable special product pattern to determine the product.

55. $(y - 3)(y - 2)$

56. $(y - 5)(y + 3)$

57. $(c - 5)(c + 5)$ **58.** $(n - 4)(n + 4)$

59. $(3 + b)^2$ **60.** $(10 + y)^2$

61. $(4 - 7b)(4 + 7b)$ **62.** $(3y + 2)(3y - 2)$

63. $(2y + 3)(3y + 5)$ **64.** $(5a - 2)(3a + 4)$

65. $(3x + 5)^2$ **66.** $(6x - 1)^2$

67. $(a - 6b)(b + 2a)$ **68.** $(a + b)(b - a)$

69. $(y + 6x)(y - 6x)$ **70.** $(11y + 2x)(11y - 2x)$

In Exercises 71–76, perform the indicated operations.

71. $(x^2 + 3x) + (7 + x)(7 - x)$

72. $(x + 5)(x - 5) - (2x - 15)$

73. $x^2 - (6 - x)^2$

74. $(y - 6)^2 - (y + 6)^2$

75. $y(3y + 2)^2$

76. $-2a(4 - a)(4 + a)$

In Exercises 77–82, write the described expression and then simplify it.

77. Multiply $x^2 + 4$ by the product of $x + 2$ and $x - 2$.

78. Multiply $-3y$ by the square of $2y - 1$.

79. Add the square of $x - 5$ to $1 + 2x - x^2$.

80. Subtract the square of $2 + y$ from $3 + y + y^2$.

81. Subtract the product of $2x - 5$ and $2x + 5$ from $6x^2 - 3x$.

82. Add $4x - 1$ to the product of $1 + 3x$ and $1 - 3x$.

In Exercises 83–88, solve each equation.

83. $(x + 1)(x - 2) = (x + 3)(x + 2)$

84. $(3x + 1)(x + 1) = (x - 1)(3x - 1)$

85. $(x + 1)(x - 1) = x(x + 2)$

86. $2x(8x + 3) = (4x + 1)(4x - 1)$

87. $(2x + 3)(2x - 3) = (4x + 3)(x - 1)$

88. $(x + 1)^2 - (x - 1)^2 = 12$

89. The sum of the squares of two consecutive integers is 7 less than twice the square of the second integer. What are the integers?

90. The difference of the squares of two consecutive odd integers is 32. What are the integers?

91. The square of a number is the same as the square of 1 more than the number. What is the number?

92. The square of a number is the same as the square of 3 less than the number. What is the number?

Geometric Models

In Exercises 93–96, write a simplified polynomial for the described area or volume.

93. The area of a rectangle with dimensions $3x + 7$ and $3x - 7$

94. The area of a triangle with base $x + 6$ and height $x - 6$

95. The area of a square picture frame that is 2 inches wide and that holds a picture x inches square

96. The volume of a box formed by cutting out the corners of a 15-inch metal square and folding up the sides.

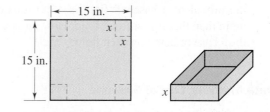

Real-Life Applications

97. The length of a campaign sign is 6 feet more than the width. If the width were increased by 3 feet, the area of the sign would be increased by 33 square feet. What are the dimensions of the sign?

98. A city has a square historic district. If an additional square, 3 blocks less on each side than the original district, is declared a historic district, the total area of the two districts will be 21 blocks less than twice the original area. How many blocks is a side of the original district?

99. A square room has an Oriental rug and an uncarpeted area of hardwood floor bordering it that is 2 feet wide on each side. If the area of the exposed hardwood border is 60 square feet, what is the size of the carpet?

100. A circular water treatment pond is surrounded by a service walkway that is 3 feet wide. If the area of the walkway is 219π square feet, what is the radius of the pond?

101. A parking lot in the shape of a trapezoid has the relative dimensions shown in the figure. If the distance between Ash Street and the back of the lot were increased by 2 feet, the area of the lot would be increased by 122 square feet. What is the frontage of the lot along Ash Street?

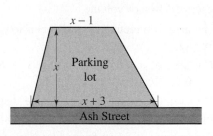

102. The length of the base path for a baseball field is 30 feet more than for a softball field. The area of the infield of a baseball field is 4500 square feet more than that of a softball field. How far is it from third base to home on each field?

Data Analysis: *Cost of Gasoline*

Since 1980 the cost of gasoline per mile traveled (adjusted for inflation) has decreased, but the part of the cost that is due to taxes has increased. The accompanying bar graph shows these trends. (Source: American Petroleum Institute.)

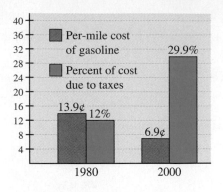

If we let x represent the number of years since 1975, the cost of gasoline per mile traveled can be modeled by the expression $-0.35x + 15.65$, and the part of the cost that is due to taxes can be modeled by $0.009x + 0.075$.

103. Multiply the two binomials and call the result $p(x)$. Use your calculator to perform the necessary arithmetic, and round all coefficients to the nearest thousandth.

104. How can $p(x)$ be interpreted?

105. Use the actual data to determine the per-mile tax in 2000. How would you use $p(x)$ to approximate this number?

106. In what year does the model predict that $\frac{1}{3}$ of the cost of gasoline will be for taxes?

Challenge

In Exercises 107–110, use the special product patterns to determine the product.

107. $(x^2 + 9)(x + 3)(x - 3)$

108. $[(x - 3) + y]^2$

109. $(2x + 1)^2 - (2x + 1)(2x - 1)$

110. $[(x + 2) + y][(x + 2) - y]$

In Exercises 111 and 112, evaluate the given polynomial as indicated.

111. $P(x) = x^2 + 3x - 1;$ $P(t + 2)$

112. $Q(x) = x^3;$ $Q(a - 1)$

In Exercises 113 and 114, determine the missing term to make the polynomial the square of a binomial.

113. $x^2 + 10x +$ ▨   **114.** $x^2 - 14x +$ ▨

6.6 | Division

Quotients of a Polynomial and a Monomial • Quotients of Polynomials

Quotients of a Polynomial and a Monomial

The most basic example of a quotient of polynomials is the quotient of two monomials. We use properties of exponents as well as the properties of real numbers to simplify the quotient.

EXAMPLE 1

Dividing Monomials

Simplify the given quotients.

(a) $\dfrac{12x^7}{6x^3} = \dfrac{12}{6} \cdot x^{7-3} = 2x^4$

(b) $\dfrac{10y^6}{15y^5} = \dfrac{10}{15} \cdot y^{6-5} = \dfrac{2}{3}y$

(c) $\dfrac{4x^5y^4}{20x^2y^4} = \dfrac{4}{20} \cdot x^{5-2} \cdot y^{4-4} = \dfrac{1}{5} \cdot x^3 \cdot 1 = \dfrac{1}{5}x^3$

Think About It

Which two quotients are the same? Why is the other not the same?

(i) $6y \div 2y$ (ii) $\dfrac{6y}{2y}$ (iii) $(6y) \div (2y)$

Recall the following rule for adding fractions.

For real numbers a, b, and c, $c \neq 0$, $\dfrac{a}{c} + \dfrac{b}{c} = \dfrac{a+b}{c}$

To determine the quotient of a polynomial and a monomial, we use this same rule but in reverse order.

> **Dividing a Polynomial by a Monomial**
>
> For monomials a, b, and c, $c \neq 0$,
>
> $$\dfrac{a+b}{c} = \dfrac{a}{c} + \dfrac{b}{c}$$
>
> In words, write the quotient as a sum of quotients of monomials.

EXAMPLE 2

Dividing a Polynomial by a Monomial

Perform the indicated division.

(a) $\dfrac{6y+9}{3} = \dfrac{6y}{3} + \dfrac{9}{3} = 2y + 3$

(b) $\dfrac{7y^4 + 21y^3}{7y^2} = \dfrac{7y^4}{7y^2} + \dfrac{21y^3}{7y^2} = y^2 + 3y$

(c) $\dfrac{15x^4 + 6x^3 + 3x}{3x} = \dfrac{15x^4}{3x} + \dfrac{6x^3}{3x} + \dfrac{3x}{3x} = 5x^3 + 2x^2 + 1$

EXAMPLE 3

Dividing a Polynomial by a Monomial

Perform the indicated division.

(a) Divide $18x^7 + 12x^5 + 3x^3$ by $6x^4$.

(b) $(-24a^7 + 4a^5 - 6a^4 + 6a^3) \div (-6a^3)$

(c) Divide $10m^4n^2 - 5m^3n^4 + 15m^2n^2$ by $5m^2n^2$.

Solution

(a) $\dfrac{18x^7 + 12x^5 + 3x^3}{6x^4} = \dfrac{18x^7}{6x^4} + \dfrac{12x^5}{6x^4} + \dfrac{3x^3}{6x^4}$

$\qquad\qquad = 3x^3 + 2x + \dfrac{1}{2x}$ $\qquad$ $\dfrac{3x^3}{6x^4} = \dfrac{3x^3}{2x \cdot 3x^3} = \dfrac{1}{2x}$

(b) $\dfrac{-24a^7 + 4a^5 - 6a^4 + 6a^3}{-6a^3} = \dfrac{-24a^7}{-6a^3} + \dfrac{4a^5}{-6a^3} + \dfrac{-6a^4}{-6a^3} + \dfrac{6a^3}{-6a^3}$

$\qquad\qquad\qquad\qquad = 4a^4 - \dfrac{2}{3}a^2 + a - 1$

(c) $\dfrac{10m^4n^2 - 5m^3n^4 + 15m^2n^2}{5m^2n^2} = \dfrac{10m^4n^2}{5m^2n^2} + \dfrac{-5m^3n^4}{5m^2n^2} + \dfrac{15m^2n^2}{5m^2n^2}$

$\qquad\qquad\qquad\qquad = 2m^2 - mn^2 + 3$

Quotients of Polynomials

Division of polynomials is similar to long division in arithmetic.

$$\begin{array}{r} 8 \quad \leftarrow \text{quotient} \\ \text{divisor} \rightarrow 23\overline{)189} \quad \leftarrow \text{dividend} \\ \underline{184} \\ 5 \quad \leftarrow \text{remainder} \end{array}$$

The result can be written in fraction form.

$$\frac{\text{dividend}}{\text{divisor}} = \text{quotient} + \frac{\text{remainder}}{\text{divisor}}$$

$$\frac{189}{23} = 8 + \frac{5}{23}$$

We check the result with the following relation.

$$\begin{array}{ccccc} (\text{divisor}) \cdot & (\text{quotient}) & + & \text{remainder} & = & \text{dividend} \\ \downarrow & \downarrow & & \downarrow & & \downarrow \\ 23 \cdot & 8 & + & 5 & = & 189 \end{array}$$

EXAMPLE 4

Dividing a Polynomial by a Binomial

Determine the following quotient: $(x^2 + 7x + 12) \div (x + 3)$.

Solution

First, rewrite $(x^2 + 7x + 12) \div (x + 3)$ in the long-division format.

$$x + 3\overline{)x^2 + 7x + 12}$$

Divide x^2, the first term in the dividend, by x, the first term in the divisor: $\dfrac{x^2}{x} = x$. Place the result above the term in the dividend containing x.

$$\begin{array}{r} x \\ x + 3 \overline{\smash{)}x^2 + 7x + 12} \end{array}$$

Now multiply the divisor $x + 3$ by x and write the result under the dividend. Align like terms.

$$\begin{array}{r} x \\ x + 3 \overline{\smash{)}x^2 + 7x + 12} \\ \underline{x^2 + 3x} \end{array} \qquad x(x + 3) = x^2 + 3x$$

Subtract the result from the dividend.

$$\begin{array}{r} x \\ x + 3 \overline{\smash{)}x^2 + 7x + 12} \\ \underline{x^2 + 3x} \\ 4x \end{array}$$

Bring down the 12 from the dividend.

$$\begin{array}{r} x \\ x + 3 \overline{\smash{)}x^2 + 7x + 12} \\ \underline{x^2 + 3x} \\ 4x + 12 \end{array}$$

Now repeat the steps. Divide the $4x$ in $4x + 12$ by x, the first term of the divisor: $\dfrac{4x}{x} = 4$. Place the 4 above the constant term of the dividend.

$$\begin{array}{r} x + 4 \\ x + 3 \overline{\smash{)}x^2 + 7x + 12} \\ \underline{x^2 + 3x} \\ 4x + 12 \end{array}$$

Multiply $x + 3$ by 4. Place the result under the like terms.

$$\begin{array}{r} x + 4 \\ x + 3 \overline{\smash{)}x^2 + 7x + 12} \\ \underline{x^2 + 3x} \\ 4x + 12 \\ \underline{4x + 12} \end{array}$$

Subtract.

$$\begin{array}{r} x + 4 \\ x + 3 \overline{\smash{)}x^2 + 7x + 12} \\ \underline{x^2 + 3x} \\ 4x + 12 \\ \underline{4x + 12} \\ 0 \end{array}$$ Quotient

Remainder

The quotient is $x + 4$, and the remainder is 0.

We can write the result in the following form.

$$\frac{x^2 + 7x + 12}{x + 3} = x + 4$$

We check the result with the following relation.

$$(\text{divisor}) \cdot (\text{quotient}) + \text{remainder} = \text{dividend}$$
$$(x + 3) \cdot (x + 4) + \quad 0 \quad = x^2 + 7x + 12 + 0$$
$$= x^2 + 7x + 12$$

EXAMPLE 5

Dividing a Polynomial by a Binomial

Determine the following quotient.

$$(x^2 - 2x - 5) \div (x - 4)$$

Solution

$$x - 4 \overline{) x^2 - 2x - 5}$$ 　 Write the quotient in long-division format.

$$\begin{array}{r} x \phantom{{}- 2x - 5} \\ x - 4 \overline{) x^2 - 2x - 5} \\ \underline{x^2 - 4x} \phantom{{}- 5} \\ 2x - 5 \end{array}$$

Divide x^2 by x: $x^2/x = x$.

Multiply $x(x - 4)$ and place the result under the dividend.

Subtract and bring down -5.

$$\begin{array}{r} x + 2 \\ x - 4 \overline{) x^2 - 2x - 5} \\ \underline{x^2 - 4x} \phantom{{}- 5} \\ 2x - 5 \\ \underline{2x - 8} \\ 3 \end{array}$$

Divide $2x$ by x: $2x/x = 2$

Multiply $2(x - 4)$.

Subtract.

The quotient is $x + 2$, and the remainder is 3.

$$\frac{x^2 - 2x - 5}{x - 4} = x + 2 + \frac{3}{x - 4}$$

Check the result.

$$(x - 4)(x + 2) + 3 = x^2 + 2x - 4x - 8 + 3 = x^2 - 2x - 5$$

Note: We stop the long-division process when the remainder is 0 or when the degree of the remainder is less than the degree of the divisor.

If either the dividend or the divisor has missing terms, it is helpful to write those terms with a zero coefficient.

EXAMPLE 6

Missing Terms

Determine each quotient.

(a) $(x^3 - 8) \div (x - 2)$

(b) $(4x^4 - 9x^2 + 1) \div (2x - 1)$

Solution

(a) $(x^3 - 8) \div (x - 2)$

The dividend has no x^2 or x term. Because we need each term for long division, we insert the terms $0x^2$ and $0x$ and write the problem in the long-division format.

$$x - 2 \overline{\smash{)}\, x^3 + 0x^2 + 0x - 8}$$

LEARNING TIP

Including missing terms helps you to align like terms. At the subtraction step, consider writing and circling the sign changes.

$$\begin{array}{r} x^2 \\ x - 2 \overline{\smash{)}\, x^3 + 0x^2 + 0x - 8} \\ \underline{x^3 - 2x^2} \\ 2x^2 + 0x \end{array}$$

Divide x^3 by x: $x^3/x = x^2$.

Multiply: $x^2(x - 2) = x^3 - 2x^2$.

Subtract and bring down $0x$.

$$\begin{array}{r} x^2 + 2x \\ x - 2 \overline{\smash{)}\, x^3 + 0x^2 + 0x - 8} \\ \underline{x^3 - 2x^2} \\ 2x^2 + 0x \\ \underline{2x^2 - 4x} \\ 4x - 8 \end{array}$$

Divide $2x^2$ by x: $2x^2/x = 2x$.

Multiply: $2x(x - 2) = 2x^2 - 4x$.

Subtract and bring down -8.

$$\begin{array}{r} x^2 + 2x + 4 \\ x - 2 \overline{\smash{)}\, x^3 + 0x^2 + 0x - 8} \\ \underline{x^3 - 2x^2} \\ 2x^2 + 0x \\ \underline{2x^2 - 4x} \\ 4x - 8 \\ \underline{4x - 8} \\ 0 \end{array}$$

Divide $4x$ by x: $4x/x = 4$.

Multiply: $4(x - 2) = 4x - 8$.

Subtract.

The quotient is $x^2 + 2x + 4$, and the remainder is 0.

$$\frac{x^3 - 8}{x - 2} = x^2 + 2x + 4$$

We verify the result as follows.

$$(x - 2)(x^2 + 2x + 4) + 0 = x^3 + 2x^2 + 4x - 2x^2 - 4x - 8$$
$$= x^3 - 8$$

(b) $(4x^4 - 9x^2 + 1) \div (2x - 1)$

Insert $0x^3$ and $0x$ for the missing terms in the dividend.

LEARNING TIP

Consider writing each division off to the side. For example, you might find that writing $\frac{4x^4}{2x} = 2x^3$ is easier than trying to divide mentally.

$$\begin{array}{r} 2x^3 + x^2 - 4x - 2 \\ 2x - 1 \overline{\smash{)}\, 4x^4 + 0x^3 - 9x^2 + 0x + 1} \\ \underline{4x^4 - 2x^3} \\ 2x^3 - 9x^2 \\ \underline{2x^3 - x^2} \\ -8x^2 + 0x \\ \underline{-8x^2 + 4x} \\ -4x + 1 \\ \underline{-4x + 2} \\ -1 \end{array}$$

The quotient is $2x^3 + x^2 - 4x - 2$, and the remainder is -1.

$$\frac{4x^4 - 9x^2 + 1}{2x - 1} = 2x^3 + x^2 - 4x - 2 + \frac{-1}{2x - 1}$$

Verify the result.

$$(2x - 1)(2x^3 + x^2 - 4x - 2) - 1$$
$$= 4x^4 + 2x^3 - 8x^2 - 4x - 2x^3 - x^2 + 4x + 2 - 1$$
$$= 4x^4 - 9x^2 + 1$$

Quick Reference 6.6

Quotients of a Polynomial and a Monomial

- To divide a monomial by a monomial, use the properties of exponents and the properties of real numbers to simplify the quotient.
- To divide a polynomial by a monomial, use the following rule.

  For monomials a, b, and c, $c \neq 0$, $\dfrac{a + b}{c} = \dfrac{a}{c} + \dfrac{b}{c}$.

  In words, write the quotient as a sum of quotients of monomials.

Quotients of Polynomials

- To divide a polynomial by a polynomial, write the problem in the long-division format and perform operations comparable to long division in arithmetic.
- If either the dividend or the divisor has missing terms, write the terms with a zero coefficient.
- We stop the long-division process when the remainder is 0 or when the degree of the remainder is less than the degree of the divisor.
- The result can be checked with the following relation.

  (divisor)(quotient) + remainder = dividend

Speaking the Language 6.6

1. Dividing $5x^4$ by $2x^2$ involves ▨▨▨▨ the exponents.
2. To divide a polynomial by a monomial, we divide ▨▨▨▨ of the polynomial by the ▨▨▨▨ .
3. When we divide a polynomial by a polynomial in long-division format, we stop dividing when the ▨▨▨▨ is 0 or when the degree of the remainder is less than the degree of the ▨▨▨▨ .
4. The result of a long division can be checked with the following relation:

   (▨▨▨▨) (quotient) + ▨▨▨▨ = ▨▨▨▨

Exercises 6.6

Concepts and Skills

1. If $A \div B = C$, what names do we give to the expressions A, B, and C?

2. Describe how to divide a polynomial of two or more terms by a monomial.

In Exercises 3–6, perform the indicated division.

3. $\dfrac{14x^4}{21x^3}$ 4. $\dfrac{18y^4}{22y}$

5. $\dfrac{3x^6y^3}{12x^6y^2}$ 6. $\dfrac{10a^4b^7}{14a^2b^3}$

In Exercises 7–22, determine the quotient.

7. $\dfrac{10x - 16}{2}$ 8. $\dfrac{10x^2 - 5x - 20}{5}$

9. $\dfrac{2y^6 + 6y^4 - 2y^2}{2y^2}$ 10. $\dfrac{a^7 + 3a^5 - a^3}{a^2}$

11. $(3x^4 - 9x^3 + 6x^2) \div (6x^2)$

12. $(9y^6 + 12y^4) \div (9y^4)$

13. $\dfrac{6a^2b^3 - 8a^3b^2}{2ab}$ 14. $\dfrac{5x^3y^2 + 15x^2y^3 + 5xy^4}{5xy^2}$

15. $(2x^4 + 12x^3 - 4x^2) \div (4x^2)$

16. $(6x^2 - 9x^3 - 12x^4) \div (-6x^2)$

17. $\dfrac{6 - 4x}{2x}$ 18. $\dfrac{9 - 6x + x^2}{3x}$

19. $(6x^3 + 6x^2 - 2x) \div (6x)$

20. $(8x^5 - 16x^3 + 6x^2) \div (8x^2)$

21. $(6x^3 - 2x^2 + 2x) \div (2x^2)$

22. $(3x^2 + 9x - 12) \div (3x)$

23. The following quotient is in the long-division format. Describe the first two steps.

$$x - 4 \overline{)\, x^3 + 3x^2 - 2x + 4}$$

24. Write an equation with no fractions that describes the relationship among the dividend, the divisor, the quotient, and the remainder.

In Exercises 25–38, determine the quotient and the remainder.

25. $(x^2 - 4x - 21) \div (x - 7)$

26. $(x^2 + 7x + 10) \div (x + 5)$

27. $(12 - x - x^2) \div (x + 4)$

28. $(12 + 4x - x^2) \div (x - 6)$

29. $\dfrac{2x^2 + 13x + 20}{x + 4}$ 30. $\dfrac{4x^2 - 9x + 2}{x - 2}$

31. $(x^2 - 2x - 5) \div (x - 5)$

32. $(x^2 + 6x + 1) \div (x + 4)$

33. $(2x^2 - 9x - 15) \div (x - 7)$

34. $(2x^2 + 3x - 20) \div (x + 6)$

35. $\dfrac{3x^2 + x - 12}{x + 2}$ 36. $\dfrac{3x^2 - 13x + 9}{x - 4}$

37. $(2x^2 + 5x + 1) \div (x + 3)$

38. $(3x^2 - 4x - 5) \div (x - 2)$

In Exercises 39–48, determine the quotient and the remainder.

39. $\dfrac{6x^2 - 7x - 3}{3x + 1}$ 40. $\dfrac{15x^2 + 14x - 8}{5x - 2}$

41. $(7 + 19x - 6x^2) \div (2x - 7)$

42. $(2 + 3x - 9x^2) \div (3x - 2)$

43. $(3x^2 + x - 2) \div (3x - 5)$

44. $(6x^2 + x - 6) \div (2x + 1)$

45. $(4x^2 - 5x - 9) \div (4x + 3)$

46. $(8x^2 - 14x + 1) \div (2x - 3)$

47. $\dfrac{6x^2 + 13x + 2}{3x + 5}$ 48. $\dfrac{15x^2 + 4x - 5}{5x - 2}$

49. Suppose that you need to divide $x^3 - 5$ by $x + 1$. When you write the problem in the long-division format, how would you write the dividend?

50. Suppose that P and Q are polynomials. If you are asked to determine the quotient of Q and P, which is the dividend and which is the divisor?

In Exercises 51–56, determine the quotient and the remainder.

51. $(4x^2 - 25) \div (2x - 5)$

52. $(16 - 49x^2) \div (7x + 4)$

53. $(x^2 - 5) \div (x - 5)$ 54. $(x^2 - 9) \div (x + 2)$

55. $\dfrac{6x^2 + 5x}{2x - 1}$ 56. $\dfrac{8x^2 - 10x}{4x + 1}$

In Exercises 57–68, determine the quotient and the remainder.

57. $(x^3 + 4x^2 + 2x - 1) \div (x + 1)$

58. $(x^3 - 3x^2 - 9x - 2) \div (x - 5)$

59. $(2x^3 - 11x^2 - 8x + 12) \div (x - 6)$

60. $(2x^3 + 7x^2 - 5x - 4) \div (x + 4)$

61. $(2x^3 - 7x^2 - x - 4) \div (x - 4)$

62. $(6x^3 - 3x^2 - 8x + 4) \div (2x - 1)$

63. $(6x^3 - x^2 - 4x - 1) \div (3x + 1)$

64. $(2x^3 - 3x^2 - 2x - 10) \div (2x + 5)$

65. $\dfrac{x^4 - 1}{x - 1}$ **66.** $\dfrac{x^3 + 27}{x + 3}$

67. $\dfrac{x^4 + 20x - 10}{x + 3}$ **68.** $\dfrac{6x^4 + x^3 + 5x + 4}{2x - 1}$

In Exercises 69–72, write the expression and simplify.

69. Divide $3x^2 + x - 4$ by $x - 1$.

70. Divide $2x^2 + 3x - 35$ by $x + 5$.

71. The quotient of $x^2 - 4x - 11$ and $x - 7$

72. The quotient of $x^2 + 12x + 16$ and $x + 8$

In Exercises 73–76, simplify.

73. $[(2x + 1)(x + 2)] \div (x + 3)$

74. $[(3x + 2)(x - 3)] \div (x - 2)$

75. $[(x^3 + 2x^2 + x) - (5x^2 + x - 5)] \div (x - 2)$

76. $(x^3 + 2x + 15) \div [(2x + 1) + (2 - x)]$

In Exercises 77–80, write an expression and simplify.

77. Divide $3x^3 + x^2 - 1$ by the sum of $2x - 2$ and $1 - x$.

78. Divide $2x^3 + 10$ by the difference of $2x$ and $x - 1$.

79. Divide the difference of x^4 and $x^3 + 28x + 40$ by $x + 4$.

80. Divide the sum of $3x^4 - 8x^2$ and $8 - 2x$ by $x - 2$.

81. The area of a rectangle is $3t^2 - 5t - 2$, and the width is $t - 2$. What is the length?

82. The area of a parallelogram is $6x^2 - x - 2$. If the base is $2x + 1$, what is the height?

83. Suppose that the value of $2x + 3$ shares of stock is $2x^3 + 3x^2 + 8x + 12$ dollars. What is the price per share?

84. It takes $2t + 1$ hours to travel $6t^2 - 7t - 5$ miles. What is the average speed?

85. The product of two integers is $n^2 + 3n + 2$. If $n + 1$ is one integer, show that the integers are consecutive integers.

86. Suppose that the product of two even integers is $n^2 - 4n + 3$. If $n - 1$ is one integer, show that the integers are consecutive even integers.

Data Analysis: *Summer Olympics*

The accompanying table shows the number of countries participating in the Summer Olympics from 1988 to 2000. The table also shows the number of volunteers working at these games.

Year	Olympic City	Number of Countries	Number of Volunteers
1988	Seoul	159	27,000
1992	Barcelona	169	30,000
1996	Atlanta	197	35,000
2000	Sydney	198	45,000

(Source: International Olympic Committee.)

If we let x represent the number of years since 1988, then the number of countries participating can be approximated by $C(x) = 4x + 160$. The number of volunteers can be modeled by the function $N(x) = 108x^2 + 164x + 27{,}150$.

87. Use long division to determine the quotient of $N(x)$ and $C(x)$. Call the quotient $Q(x)$ and the remainder R.

88. What is your interpretation of $Q(x) + \dfrac{R}{C(x)}$?

89. Evaluate and interpret $Q(12) + \dfrac{R}{C(12)}$.

90. According to the actual data for the 2000 Olympics, what was the number of volunteers per country? Does the result in Exercise 89 compare favorably with the actual data?

Challenge

In Exercises 91 and 92, determine the quotient and the remainder.

91. $\dfrac{x^2 + 3}{x^2 - 1}$ **92.** $(x^2 + 3x - 4) \div (x^2 + 2x)$

In Exercises 93 and 94, determine the described polynomial.

93. When the polynomial is divided by $x + 3$, the quotient is $2x - 5$ and the remainder is 0.

94. When $x^2 + 3x - 4$ is divided by the polynomial, the quotient is $x + 7$ and the remainder is 24.

In Exercises 95 and 96, determine k such that the remainder is 0.

95. $(x^2 + 3x + k) \div (x + 5)$

96. $(6x^2 - 5x + k) \div (2x + 1)$

97. The volume of a box is $2x^3 + 8x^2 + 6x$. Two of its sides are $2x$ and $x + 1$. What is the third side?

6.7 | **Negative Exponents**

Definition and Evaluation   •   Simplifying Expressions   •   Properties of Exponents

Definition and Evaluation

Previously we defined a positive integer exponent and a zero exponent. In this section we extend the definition of exponent to include all integers.

Exploring the Concept

Integer Exponents

To simplify the expression $\dfrac{x^2}{x^5}$, we can apply the Product Rule for Exponents.

$$\frac{x^2}{x^5} = \frac{x^2}{x^2 \cdot x^3} = \frac{1}{x^3}$$

An alternative approach is to apply the Quotient Rule for Exponents.

$$\frac{x^2}{x^5} = x^{2-5} = x^{-3}$$

Because $\dfrac{1}{x^3}$ and x^{-3} are both results of simplifying $\dfrac{x^2}{x^5}$, the two expressions must be equivalent.

This observation suggests the following definition of an integer exponent.

Definition of an Integer Exponent

For any nonzero real number b and any integer n, $b^{-n} = \dfrac{1}{b^n}$.

In words, the number b^{-n} is the reciprocal of b^n.

According to the definition of an integer exponent, all of the following are true.

$$y^{-5} = \frac{1}{y^5} \qquad a^{-1} = \frac{1}{a} \qquad 6^{-7} = \frac{1}{6^7} \qquad (-8)^{-5} = \frac{1}{(-8)^5} \qquad \left(\frac{3}{4}\right)^{-5} = \left(\frac{4}{3}\right)^5$$

Because n is any integer, we can write $b^{-n} = \dfrac{1}{b^n}$ or $b^n = \dfrac{1}{b^{-n}}$.

$$6^3 = \frac{1}{6^{-3}} \qquad \frac{1}{y^{-4}} = y^4 \qquad \frac{1}{(-7)^{-2}} = (-7)^2$$

Note: Informally, we say that whenever a *factor* is moved from the numerator to the denominator or from the denominator to the numerator, the sign of the exponent changes. Also, observe that only the sign of the exponent changes, not the sign of the base.

EXAMPLE **1**

Evaluating Expressions with Negative Exponents

Use the definition of a negative exponent to evaluate each expression.

(a) $2^{-3} = \dfrac{1}{2^3} = \dfrac{1}{8}$

(b) $(-4)^{-2} = \dfrac{1}{(-4)^2} = \dfrac{1}{16}$ The sign of the base does not change.

(c) $\left(\dfrac{1}{3}\right)^{-2} = 3^2 = 9$ The reciprocal of $\dfrac{1}{3}$ is 3.

(d) $\left(\dfrac{3}{4}\right)^{-3} = \left(\dfrac{4}{3}\right)^3 = \dfrac{64}{27}$ The reciprocal of $\dfrac{3}{4}$ is $\dfrac{4}{3}$.

(e) $3^{-1} + 6^{-1} = \dfrac{1}{3} + \dfrac{1}{6} = \dfrac{2}{6} + \dfrac{1}{6} = \dfrac{3}{6} = \dfrac{1}{2}$

LEARNING TIP

The rule $\left(\dfrac{a}{b}\right)^{-n} = \left(\dfrac{b}{a}\right)^n$ is very handy. Note how applying it to part (d) of Example 1 makes this problem easy.

Simplifying Expressions

Usually, when an exponential expression is to be simplified, we are expected to write the result with no negative exponents.

EXAMPLE **2**

Writing Expressions with Positive Exponents

Use the definition of a negative exponent to write each expression with a positive exponent.

(a) $y^{-6} = \dfrac{1}{y^6}$

(b) $(3n)^{-1} = \dfrac{1}{3n}$ The base is $3n$.

(c) $3n^{-1} = 3 \cdot n^{-1} = 3 \cdot \dfrac{1}{n} = \dfrac{3}{n}$ The base is n.

(d) $\dfrac{1}{x^{-6}} = x^6$

(e) $\dfrac{5}{a^{-3}} = 5 \cdot \dfrac{1}{a^{-3}} = 5a^3$

Sometimes the easiest way to simplify an expression with negative exponents is to move the factors with negative exponents from the numerator to the denominator or from the denominator to the numerator. As we have seen, this changes the sign of the exponent.

EXAMPLE **3**

Writing Expressions with Positive Exponents

Rewrite each expression with positive exponents.

(a) $\dfrac{3^{-2}}{5^{-4}} = \dfrac{5^4}{3^2}$

(b) $\dfrac{a^{-4}}{b^{-3}} = \dfrac{b^3}{a^4}$

(c) $\dfrac{2x^{-3}}{3y^{-2}} = \dfrac{2y^2}{3x^3}$ Only the exponents of x and y are negative. The exponent of 2 and 3 is 1.

(d) $\dfrac{x^4 y^{-2}}{6z^{-3}} = \dfrac{x^4 z^3}{6y^2}$ Only y and z have negative exponents.

(e) $-4x^{-2}y^3 = \dfrac{-4x^{-2}y^3}{1} = \dfrac{-4y^3}{x^2}$ Only x has a negative exponent.

Properties of Exponents

Assuming that the expressions are defined, all the previously stated properties of exponents are valid for any integer exponent. In particular, the Quotient Rule for Exponents can now be stated for any integer exponents.

> ### Summary of Properties of Exponents
>
> For real numbers a and b and integers m and n for which the expression is defined, the following are properties of exponents.
>
> $b^m \cdot b^n = b^{m+n}$ Product Rule for Exponents
>
> $(b^m)^n = b^{mn}$ Power to a Power Rule
>
> $(ab)^n = a^n b^n$ Product to a Power Rule
>
> $\left(\dfrac{a}{b}\right)^n = \dfrac{a^n}{b^n}$ Quotient to a Power Rule
>
> $\dfrac{b^m}{b^n} = b^{m-n}$ Quotient Rule for Exponents

The following examples illustrate the use of these properties.

EXAMPLE 4

Using the Product Rule for Exponents to Simplify Expressions

Multiply the given exponential expressions.

(a) $x^3 \cdot x^{-7} = x^{3+(-7)} = x^{-4} = \dfrac{1}{x^4}$

(b) $(-6)^{-4}(-6)^5 = (-6)^{-4+5} = (-6)^1 = -6$

(c) $2x^{-3}(5x^{-2}) = (2 \cdot 5)(x^{-3} \cdot x^{-2}) = 10x^{-5} = \dfrac{10}{x^5}$

EXAMPLE 5

Using the Power to a Power Rule to Simplify Expressions

Simplify the given exponential expressions.

(a) $(y^{-4})^{-3} = y^{(-4)(-3)} = y^{12}$

(b) $(2^{-3})^2 = 2^{-3 \cdot 2} = 2^{-6} = \dfrac{1}{2^6} = \dfrac{1}{64}$

EXAMPLE 6

Simplifying Quotients

Simplify the given quotients.

Quotient Rule for Exponents *Alternative Method*

(a) $\dfrac{x^2}{x^7} = x^{2-7} = x^{-5} = \dfrac{1}{x^5}$ $\dfrac{x^2}{x^7} = \dfrac{x^2}{x^2 \cdot x^5} = \dfrac{1}{x^5}$

(b) $\dfrac{x^{-3}}{x^{-6}} = x^{-3-(-6)} = x^3$ $\dfrac{x^{-3}}{x^{-6}} = \dfrac{x^6}{x^3} = x^3$

(c) $\dfrac{t^{-3}}{t^2} = t^{-3-2} = t^{-5} = \dfrac{1}{t^5}$ $\dfrac{t^{-3}}{t^2} = \dfrac{1}{t^3 \cdot t^2} = \dfrac{1}{t^5}$

(d) $\dfrac{a}{a^{-5}} = a^{1-(-5)} = a^6$ $\dfrac{a}{a^{-5}} = a \cdot a^5 = a^6$

EXAMPLE 7

Combining Properties

Perform the indicated operations.

Think About It

Write the expressions $a^{-2} + b^{-2}$, $(a^2 + b^2)^{-1}$, and $(a + b)^{-2}$ with positive exponents. Select values for a and b to show that the expressions are not equivalent.

(a) $(x^{-2}y^3)^{-2} = x^{(-2)(-2)}y^{3(-2)} = x^4y^{-6} = \dfrac{x^4}{y^6}$

(b) $\left(\dfrac{a^{-2}}{b^4}\right)^{-3} = \dfrac{a^{(-2)(-3)}}{b^{4(-3)}} = \dfrac{a^6}{b^{-12}} = a^6b^{12}$

(c) $\dfrac{8ab^{-2}}{12a^{-3}b^{-5}} = \dfrac{8 \cdot a \cdot a^3 \cdot b^5}{12 \cdot b^2} = \dfrac{2a^4b^3}{3}$

(d) $\dfrac{-4x^{-1}}{(x^2)^{-3}} = \dfrac{-4x^{-1}}{x^{-6}} = \dfrac{-4x^6}{x} = -4x^5$

Quick Reference 6.7

Definition and Evaluation

- For any nonzero real number b and any integer n, $b^{-n} = \dfrac{1}{b^n}$. In words, the number b^{-n} is the reciprocal of b^n.

- To evaluate an exponential expression in which the base is a constant and the exponent is negative, write the reciprocal of the base and change the sign of the exponent.

Simplifying Expressions

- To simplify an expression with negative exponents, move the factors that have negative exponents from the numerator to the denominator or from the denominator to the numerator and change the sign of the exponent.

Properties of Exponents

- Having defined a negative exponent, we can apply the Quotient Rule for Exponents to $\dfrac{b^m}{b^n}$ for any integers m and n.

- All other previously stated properties of exponents continue to be valid for negative exponents.

Speaking the Language 6.7

1. If $b \neq 0$, then b^{-n} is the _____ of b^n.

2. In the expression $\dfrac{x^{-2}y}{5}$, the _____ x^{-2} can be moved to the denominator by changing the exponent to 2. However, in the expression $\dfrac{x^{-2} + y}{5}$, the _____ x^{-2} cannot be moved to the denominator in the same way.

3. If we multiply b^m and b^n, where m and n are negative, then the product will have a _____ exponent.

4. If we raise b^m to the nth power, where m and n are negative, then the result will have a _____ exponent.

Exercises 6.7

Concepts and Skills

1. In order for the Quotient Rule for Exponents to hold for all integer exponents, why was it necessary to define negative exponents?

2. For the expression $\dfrac{x^{-2} + 1}{y}$, why is it incorrect to move x^{-2} so that the denominator is $x^2 y$?

In Exercises 3–22, evaluate the expression.

3. 6^{-2}

4. 5^{-3}

5. $(-2)^{-3}$

6. $(-4)^{-2}$

7. $\left(\dfrac{4}{3}\right)^{-2}$

8. $\left(\dfrac{5}{2}\right)^{-3}$

9. $3^{-1} + 5^{-1}$

10. $2^{-1} + 3^{-1}$

11. $(2 + 3)^{-1}$

12. $(4 + 1)^{-2}$

13. $8 \cdot 4^{-1}$

14. $-6 \cdot 3^{-2}$

15. -3^{-2}

16. -2^{-4}

17. $(3^{-2})^{-2}$

18. $(2^{-3})^{-1}$

19. $3^{-2} \cdot 3$

20. $5^7 \cdot 5^{-6}$

21. $\dfrac{4^{-3}}{6^{-2}}$

22. $\dfrac{5^{-2}}{10^{-1}}$

In Exercises 23–30, rewrite the given expressions with positive exponents. Assume all expressions are defined.

23. x^{-1}

24. y^{-3}

25. $\left(\dfrac{3}{10}\right)^{-1}$

26. $\left(\dfrac{1}{5}\right)^{-10}$

27. $(-12)^{-9}$

28. $(-6)^{-8}$

29. $\dfrac{1}{x^{-5}}$

30. $\dfrac{1}{z^{-1}}$

In Exercises 31–36, rewrite the given expressions with negative exponents. Assume that all expressions are defined.

31. $\dfrac{1}{y^5}$

32. $\dfrac{1}{a^3}$

33. a^6

34. t^2

35. $\dfrac{-7}{x^4}$

36. $\dfrac{3}{y^{10}}$

In Exercises 37–54, rewrite the given expressions with positive exponents. Assume that all expressions are defined.

37. $5x^{-4}$

38. $-3y^{-1}$

39. $\dfrac{1}{3x^{-5}}$

40. $\dfrac{5}{2t^{-4}}$

41. $(7t)^{-1}$

42. $(-5c)^{-1}$

43. $-2x^{-2}y$

44. $4x^3 y^{-4}$

45. $7a^{-1}b^{-6}$

46. $-8s^{-6}t^{-9}$

47. $\left(\dfrac{3}{x}\right)^{-1}$

48. $\left(\dfrac{t}{-2}\right)^{-1}$

49. $\dfrac{x^{-2}}{y^3}$

50. $\dfrac{x^{-2}}{y^{-1}}$

51. $\dfrac{a^{-3}}{b^2}$

52. $\dfrac{m^5}{n^{-9}}$

53. $\dfrac{4a^{-3}}{5b^{-1}}$

54. $\dfrac{-2x^{-2}}{3y^{-4}}$

55. One approach to writing $\left(\dfrac{x^{-2}}{y^{-3}}\right)^{-1}$ with positive exponents is as follows.

$$\left(\frac{x^{-2}}{y^{-3}}\right)^{-1} = \left(\frac{y^3}{x^2}\right)^{-1} \qquad \text{Definition of a negative exponent}$$

$$= \frac{y^{-3}}{x^{-2}} \qquad \text{Quotient to a Power Rule}$$

$$= \frac{x^2}{y^3} \qquad \text{Definition of a negative exponent}$$

Describe an easier approach.

56. To write $-5x^{-5}$ with a positive exponent, do we write $\dfrac{-5}{x^5}$ or $\dfrac{1}{5x^5}$? Why?

In Exercises 57–64, simplify the expression and use only positive exponents in the answer. Assume that all expressions are defined.

57. $a^{-5} \cdot a^{-2}$

58. $x^{10} \cdot x^{-7}$

59. $y^3 \cdot y^{-12}$

60. $y^{-4} \cdot y^{-1}$

61. $4y(-2y^{-5})$

62. $(-2x^{-6})(3x^4)$

63. $7a^{-7}(5a^9)$

64. $(5y^{-4})(4y^5)$

In Exercises 65–68, simplify the expression and use only positive exponents in the answer. Assume that all expressions are defined.

65. $(y^{-4})^{-2}$

66. $(x^{-2})^3$

67. $(7^6)^{-1}$

68. $(n^{-1})^{-1}$

In Exercises 69–76, simplify the expression and use only positive exponents in the answer. Assume that all expressions are defined.

69. $\dfrac{x^4}{x^{-2}}$

70. $\dfrac{y^{-5}}{y^{-2}}$

71. $\dfrac{a^{-5}}{a^4}$

72. $\dfrac{b^3}{b^7}$

73. $\dfrac{5x^{-7}}{10x^{-2}}$

74. $\dfrac{12x}{4x^{-4}}$

75. $\dfrac{10x^{-2}}{15x^3}$

76. $\dfrac{24y^3}{18y^{-1}}$

In Exercises 77–92, simplify the expression and use only positive exponents in the answer. Assume that all expressions are defined.

77. $\left(\dfrac{x^{-3}}{x^2}\right)^{-2}$

78. $\left(\dfrac{t^{-2}}{t^{-3}}\right)^{-1}$

79. $(3a^{-2})^{-1}$

80. $(-5x^5)^{-2}$

81. $(x^3y^{-2})^{-2}$

82. $(a^{-5}b^4)^{-4}$

83. $(a^{-3}b)^{-2}(a^4b^{-3})^{-1}$

84. $(x^{-2}y^2)^3(x^{-2}y)^{-2}$

85. $\left(\dfrac{a^{-4}}{b^2}\right)^{-3}$

86. $\left(\dfrac{-2}{t^{-2}}\right)^{-3}$

87. $\left(\dfrac{x^2y^{-3}}{z^{-2}}\right)^{-1}$

88. $\left(\dfrac{-5x^{-2}}{y^{-4}}\right)^2$

89. $\dfrac{2x^{-4}}{(3x^5)^{-2}}$

90. $\dfrac{(5x)^6}{x^6}$

91. $\dfrac{x^{-5}y^2}{x^{-3}y^{-1}}$

92. $\dfrac{ab^{-3}}{a^{-2}b^{-5}}$

In Exercises 93 and 94, determine the expression.

93. The product of the expression and y^{-6} is y^{-2}.

94. The quotient of the expression and a^{-2} is -3.

In Exercises 95 and 96, determine m and n. Assume that all expressions are defined.

95. $\dfrac{x^n y^{-2}}{x^{-1}y^m} = \dfrac{x^4}{y}$

96. $(a^{-2}b^n)^m = \dfrac{a^4}{b^6}$

Modeling with Real Data

97. Cheaper digital networks and increased competition have lowered the cost of wireless phone use. The table shows the cost (in cents per minute) from 1998 to 2003.

Year	Cost (in cents per minute)
1998	33
1999	28
2000	25
2001	23
2002	22
2003	20

(Source: The Strategis Group.)

With x representing the number of years since 1990, the expression $260x^{-1}$ models the cost.

(a) Write the expression with a positive exponent.

(b) Evaluate the model expression to estimate the cost in 2005.

98. Refer to the model in Exercise 97.

(a) Show that $\dfrac{260x^{-3}}{x^{-2}}$ is an equivalent model.

(b) Evaluate the model in part (a) to estimate the cost in 2005.

Data Analysis: *Online Internet Access*

The accompanying bar graph shows the dramatic increase in the average number of hours of personal online Internet access each year. (Source: Veronis, Suhler & Associates.)

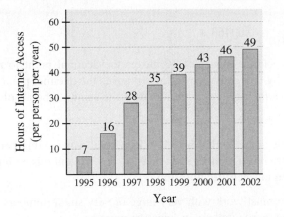

Year

If x is the number of years since 1990, then the following expression can be used to model the number of hours shown in the table.

$$H_1(x) = -x^3(105x^{-3} - 33.1x^{-2} + 2.5x^{-1} - 0.068)$$

99. Use your calculator to determine $H_1(10)$. Compare your result to the corresponding data shown in the bar graph.

100. Can H_1 be used to approximate the number of hours of Internet access in 1990? Why?

101. Use the Distributive Property and the Product Rule for Exponents to simplify the model expression. Call the result $H_2(x)$.

102. Unlike $H_1(x)$, $H_2(x)$ can be used to approximate the number of hours of Internet access in 1990. Why? Is the approximation reasonable?

Challenge

In Exercises 103–106, simplify.

103. $\dfrac{32x^3y^{-3}}{12y^2}$

104. $\dfrac{18x^{-1}y^8}{12x^{-2}y^{-2}}$

105. $\dfrac{24x^3y^{-2}}{(2x^{-1}y^{-3})^3}$

106. $\left(\dfrac{x^{-3}y^4}{x^{-6}y^{-2}}\right)^{-3}$

In Exercises 107 and 108, evaluate.

107. $\dfrac{4^{-1}}{3 + 2^{-1}}$

108. $(1 - 2^{-1})^{-1}$

6.8 | Scientific Notation

Powers of 10 • Scientific Notation to Decimal Form • Decimal Form to Scientific Notation • Calculations with Scientific Notation • Real-Life Applications

Powers of 10

The design of the decimal number system makes multiplying a number by a positive or negative power of 10 easy to do.

Exploring the Concept

Multiplying by Powers of 10

Observe the position of the decimal point in the following products. We begin with positive powers of 10.

$$4.63 \cdot 10^1 = 4.63 \cdot 10 = \qquad 46.3$$
$$4.63 \cdot 10^2 = 4.63 \cdot 10 \cdot 10 = \qquad 463.0$$
$$4.63 \cdot 10^3 = 4.63 \cdot 10 \cdot 10 \cdot 10 = 4630.0$$

Now consider these negative powers of 10.

$$4.63 \cdot 10^{-1} = \frac{4.63}{10} = 0.463$$

$$4.63 \cdot 10^{-2} = \frac{4.63}{10 \cdot 10} = 0.0463$$

$$4.63 \cdot 10^{-3} = \frac{4.63}{10 \cdot 10 \cdot 10} = 0.00463$$

Multiplying a number by a positive power of 10 moves the decimal point to the right. Multiplying a number by a negative power of 10 moves the decimal point to the left. The number of places the decimal point moves is the same as the absolute value of the exponent.

Each of the preceding products is in the form $4.63 \cdot 10^{p}$. In general, when a number is written in the form $n \cdot 10^{p}$, where n is a number such that $1 \leq n < 10$ and p is an integer, we say the number is written in **scientific notation.**

In many scientific problems, we must work with very large or very small numbers. It is often more convenient to write these numbers in scientific notation.

Note: Sometimes the multiplication symbol $\times$ is used rather than the multiplication dot. Thus 2.9×10^{6} is another acceptable way to write a number in scientific notation.

Scientific Notation to Decimal Form

To convert a number from scientific notation $n \cdot 10^{p}$ to decimal form, we multiply n by the indicated power of 10. We can do this mentally by simply moving the decimal point to the right or left, depending on the exponent on 10.

EXAMPLE 1   **Converting a Number from Scientific Notation to Decimal Form**

In each part, the given number is written in scientific notation. Write the number in decimal form.

(a) $7.5 \cdot 10^{3}$ (b) $1.93 \cdot 10^{-5}$

(c) $7.249 \cdot 10^{7}$ (d) $4.6 \cdot 10^{-2}$

Solution

(a) 7.500.   Move the decimal point 3 places to the right.

   3 places

   $7.5 \cdot 10^{3} = 7500$

(b) 0.00001.93   Move the decimal point 5 places to the left.

   5 places

   $1.93 \cdot 10^{-5} = 0.0000193$

(c) $7.249 \cdot 10^{7} = 72{,}490{,}000$ Move the decimal point 7 places to the right.

(d) $4.6 \cdot 10^{-2} = 0.046$ Move the decimal point 2 places to the left.

Decimal Form to Scientific Notation

Because $n \cdot 10^{p}$ is the scientific notation for a number, writing a number in scientific notation involves determining n and p. The following summarizes the steps for doing so.

Writing a Number in Scientific Notation

1. Place the decimal point after the first nonzero digit. The result is n.

2. Count the number of places from the decimal point to its original position. That number is the absolute value of p.

3. If the decimal point must be moved to the right to return to the original position, the exponent is positive. If the decimal point must be moved to the left to return to the original position, the exponent is negative.

EXAMPLE 2

Writing a Large Number in Scientific Notation

Write each number in scientific notation.

(a) 27,800   (b) 8,562,000,000

Solution

(a) Place the decimal point after the first nonzero digit, which is 2.

$$n = 2.7800$$

$$2.7800.$$ To return to the original location, move the decimal point 4 places to the right: $p = 4$.

$$27,800 = 2.78 \cdot 10^4$$

(b) Place the decimal point after the 8.

$$n = 8.562$$

$$8.562000000.$$ To return to the original location, move the decimal point 9 places to the right: $p = 9$.

$$8,562,000,000 = 8.562 \cdot 10^9$$

EXAMPLE 3

Writing a Small Number in Scientific Notation

Write each number in scientific notation.

(a) 0.067   (b) 0.00000000589

Solution

(a) Place the decimal point after the first nonzero digit, which is 6.

$$n = 6.7$$

$$.06.7$$ To return to the original location, move the decimal point 2 places to the left: $p = -2$.

$$0.067 = 6.7 \cdot 10^{-2}$$

LEARNING TIP

Observe that when a large number is written in scientific notation, the exponent on 10 is positive. For a small number, the exponent is negative.

(b) Place the decimal point after the 5.

$$n = 5.89$$

$$.000000005.89$$ To return to the original location, move the decimal point 9 places to the left: $p = -9$.

$$0.00000000589 = 5.89 \cdot 10^{-9}$$

Calculations with Scientific Notation

🔑 Keys to the Calculator

Scientific

Your calculator will automatically display very large and very small results in scientific notation. However, you can set the calculator so that *all* results are in scientific notation. On the first line of the MODE screen, move your cursor to Sci and press [ENTER].

```
357
              3.57E2
23.9E3
              23900
```

Line 1 of the figure shows that 357 is displayed as 3.57E2, which means $3.57 \cdot 10^2$.

You can also use the [EE] key to enter numbers in powers of ten. In line 3 of the figure, we have entered $23.9 \cdot 10^3$, with the following keystrokes:

23.9 [2nd] [EE] **3** [ENTER]

To display the result, we chose Normal on the MODE screen, but you can choose Sci to display the result in scientific notation.

EXAMPLE **4**

Scientific Notation on a Calculator

Use your calculator to perform the following operations.

(a) 125^{12} (b) $0.026 \div 30{,}000$ (c) $256{,}000 \cdot 89{,}700$

Solution

Figure 6.6 shows a typical screen display of the results. In this case, E means the exponent of 10. For answers in the form $n \cdot 10^p$, we often round off n. For example, we might write 1.455191523E25 as $1.46 \cdot 10^{25}$.

Figure 6.6

```
125^12
         1.455191523E25
0.026/30000
         8.666666667E-7
256000*89700
         2.29632E10
```

EXAMPLE **5**

Calculations with Numbers in Scientific Notation

Perform the indicated operations.

(a) $(6 \cdot 10^4)(5 \cdot 10^3) = (6 \cdot 5) \cdot (10^4 \cdot 10^3)$

                   $= 30 \cdot 10^7$ *Not yet in scientific notation*

                   $= 3.0 \cdot 10 \cdot 10^7$ *Write 30 in scientific notation.*

                   $= 3.0 \cdot 10^8$

(b) $\dfrac{8 \cdot 10^5}{4 \cdot 10^{-3}} = \dfrac{8}{4} \cdot \dfrac{10^5}{10^{-3}} = 2 \cdot 10^8$

(c) $(2 \cdot 10^4)^3 = 2^3 \cdot (10^4)^3 = 8 \cdot 10^{12}$

EXAMPLE 6

Calculations with Numbers in Scientific Notation

Use a calculator to perform each operation.

(a) $(4.9 \cdot 10^5)(3.84 \cdot 10^8)$ (b) $(7.08 \cdot 10^5)^4$

(c) $\dfrac{6.92 \cdot 10^{-9}}{3.5 \cdot 10^{-5}}$

Solution

Figure 6.7 shows a typical screen display of the results.

Figure 6.7

```
(4.9E5)(3.84E8)
              1.8816E14
(7.08E5)^4
     2.512655977E23
(6.92E-9)/(3.5E-5)
       1.977142857E-4
```

Real-Life Applications

Many real-life applications involve very large or very small numbers. Scientific notation is useful in such problems when approximations are acceptable.

EXAMPLE 7

Coffee Consumption

A report indicates that people in the United States consume an average of 26 gallons of coffee per person annually. Use scientific notation to write the total amount of coffee consumed if the population of the United States is 280 million people.

Solution

Use a calculator to multiply 26 by 280 million. People in the United States consume $7.28 \cdot 10^9$ gallons of coffee each year.

EXAMPLE **8**

Land Area of Liechtenstein

The land area of the earth is approximately 58,433,000 square miles. The country of Liechtenstein covers 61 square miles. What fractional part of the earth's land area is Liechtenstein?

Think About It

Computer users refer to kilobytes, megabytes, and gigabytes. What powers of 10 are associated with the prefixes in these words?

Solution

Use a calculator to divide 61 (the area of Liechtenstein) by 58,433,000 (the approximate land area of the earth). The land area of Liechtenstein is approximately $1.044 \cdot 10^{-6}$ or $\dfrac{1.044}{10^6}$ of the land area of the earth.

Quick Reference 6.8

Powers of 10
- Multiplying a number by 10^p moves the decimal point p places. If p is positive, the decimal point is moved to the right; if p is negative, the decimal point is moved to the left.

- When a number is written in the form $n \cdot 10^p$, where $1 \leq n < 10$ and p is an integer, we say the number is written in **scientific notation.**

Scientific Notation to Decimal Form
- To convert a number from scientific notation $n \cdot 10^p$ to decimal form, we multiply n by the indicated power of 10. We can do this mentally by simply moving the decimal point to the right or left, depending on the exponent on 10.

Decimal Form to Scientific Notation
- To write a number in scientific notation $n \cdot 10^p$, follow these steps.
  1. Place the decimal point after the first nonzero digit. The result is n.
  2. Count the number of places from the decimal point to its original position. That number is the absolute value of p.
  3. If the decimal point must be moved to the right to return to the original position, the exponent is positive. If the decimal point must be moved to the left to return to the original position, the exponent is negative.

Speaking the Language 6.8

1. A number that is written in the form $n \cdot 10^p$, where $1 \leq n < 10$ and p is an integer, is said to be in ▨▨▨▨▨▨ .

2. If a number is multiplied by 10^p, where $p > 0$, the decimal point is moved p places to the ▨▨▨▨▨▨ .

3. To write a number in scientific notation, we begin by placing the decimal point after the ▨▨▨▨▨▨ .

4. When a very small number is written in scientific notation, the exponent on 10 is ▨▨▨▨▨▨ .

Exercises 6.8

Concepts and Skills

1. If we repeatedly multiply a number by 10, what is the effect on the decimal point of the number?

2. Explain why $10 \cdot 10^4$ is not in scientific notation.

In Exercises 3–18, the given number is in scientific notation. Write the number in decimal form.

3. $2.9 \cdot 10^3$

4. $5.37 \cdot 10^2$

5. $4 \cdot 10^{-2}$

6. $6 \cdot 10^{-5}$

7. $6.39 \cdot 10^5$

8. $1.25 \cdot 10^7$

9. $6.7 \cdot 10^{-3}$

10. $9.34 \cdot 10^{-2}$

11. $7 \cdot 10^4$

12. $3 \cdot 10^8$

13. $7.56 \cdot 10^{-6}$

14. $4.04 \cdot 10^{-8}$

15. $4.285 \cdot 10^9$

16. $7.3276 \cdot 10^{10}$

17. $2.5534 \cdot 10^{-10}$

18. $9.369 \cdot 10^{-12}$

In Exercises 19–26, write the given number in scientific notation.

19. 5,600

20. 250

21. 307,000

22. 46,000

23. 52,000,000

24. 437,000,000

25. 2,000,000,000

26. 30,000,000,000

In Exercises 27–34, write the given number in scientific notation.

27. 0.03

28. 0.0025

29. 0.0000474

30. 0.000926

31. 0.00000008

32. 0.0000064

33. 0.34569

34. 0.06106

In Exercises 35–42, use your calculator to perform the indicated operation. Write the result in scientific notation $n \cdot 10^p$ with n rounded to the nearest tenth.

35. 46^{12}

36. 15^9

37. 8^{-10}

38. 25^{-8}

39. $17 \div 42,900$

40. $9.6 \div 105,000$

41. $12,237,000 \cdot 2500$

42. $259,000 \cdot 652,432$

In Exercises 43–48, the given number is written as the product of a number and a power of 10, but it is not in scientific notation. Write the number in scientific notation.

43. $470 \cdot 10^3$

44. $6470 \cdot 10^2$

45. $0.029 \cdot 10^{-2}$

46. $0.76 \cdot 10^{-5}$

47. $0.07 \cdot 10^5$

48. $0.32 \cdot 10^8$

In Exercises 49–52, write the given expression as a single number in decimal form.

49. $5 \cdot 10^2 + 2 \cdot 10^1 + 4 \cdot 10^0$

50. $8 \cdot 10^3 + 3 \cdot 10^1 + 2 \cdot 10^0$

51. $4 \cdot 10^0 + 7 \cdot 10^{-1} + 3 \cdot 10^{-2}$

52. $9 \cdot 10^0 + 5 \cdot 10^{-1} + 6 \cdot 10^{-2}$

In Exercises 53–64, perform the indicated operations without a calculator. Write the results in scientific notation.

53. $(3 \cdot 10^5)(2 \cdot 10^{-2})$

54. $(4 \cdot 10^{12})(2 \cdot 10^3)$

55. $(4 \cdot 10^4)(6 \cdot 10^8)$

56. $(7 \cdot 10^{-2})(5 \cdot 10^{-7})$

57. $\dfrac{6 \cdot 10^9}{2 \cdot 10^5}$

58. $\dfrac{9 \cdot 10^{-3}}{3 \cdot 10^7}$

59. $\dfrac{3 \cdot 10^{\;3}}{6 \cdot 10^8}$

60. $\dfrac{2 \cdot 10^{12}}{8 \cdot 10^7}$

61. $(3 \cdot 10^5)^3$

62. $(7 \cdot 10^{-5})^2$

63. $(5 \cdot 10^{-4})^3$

64. $(2 \cdot 10^6)^4$

In Exercises 65–70, use your calculator to perform the indicated operations. Write the results in scientific notation $n \cdot 10^p$ with n rounded to the nearest tenth.

65. $(4.67 \cdot 10^3)(2.3 \cdot 10^5)$

66. $(7.35 \cdot 10^{-9})(5.89 \cdot 10^3)$

67. $\dfrac{3.67 \cdot 10^9}{2.8 \cdot 10^2}$

68. $\dfrac{6.9 \cdot 10^{15}}{2.78 \cdot 10^{-4}}$

69. $(6.89 \cdot 10^6)^7$

70. $(5.34 \cdot 10^5)^{-2}$

Real-Life Applications

In Exercises 71–82, write the given number in scientific notation.

71. In 1998 the United States Postal Service handled 197,943,000,000 pieces of mail.

72. The total area of Georgia is 37,702,400 acres.

73. A picosecond is 0.000000000001 second.

74. A common bacterium measures 0.0000394 inch.

75. The salary of the President of the United States is $0.0007 per person living in the country.

76. Canada has 0.00517 of the world's total population.

77. The distance from the sun to Pluto is 3,675,000,000 miles.

78. The population of India in the year 2100 is predicted to be 1,631,800,000.

79. One pair of adult fleas could produce 100 quintillion, or 100,000,000,000,000,000,000, descendants in 6 months.

80. There were 274,000,000 books sold in college book stores in the United States in 1997.

81. The portion of the world's water supply that is in freshwater lakes and rivers is 0.000093 of the total amount of water on the earth.

82. In 2000 one share of Ford Motor Company stock represented 0.00000000088 of the company.

In Exercises 83–90, solve the problem and write each result in scientific notation.

83. A company exports automobiles to Russia. If it takes 28 rubles to buy 1 dollar, how many rubles must an importer have to buy $500,000 worth of the company's automobiles?

84. A report indicates that a total of 65.4 million cable TV subscribers pay an average of $27.43 per month. What is the total amount they pay for cable TV in 1 year?

85. What fractional part of the U.S. population plays NBA basketball? (Assume that there are 29 teams with 11 players per team and that the population of the United States is 280 million.)

86. The area of Rhode Island is 1055 square miles, and the area of Alaska is 570,833 square miles. What fraction of Alaska's area is Rhode Island's area?

87. Placed side by side, about 4000 typical cells of the human body would span 1 inch. How wide is a typical cell?

88. The diameter of an atom is between 0.1 and 0.5 nanometer. A nanometer is approximately 1/25,000,000 inch. What is the diameter in inches of an atom whose diameter is 0.2 nanometer?

89. Light travels at 186,000 miles per second. How far does light travel in 1 hour?

90. A report indicates that approximately 3.2 million gallons of ice cream are consumed per day in the United States. How many gallons are consumed in a year?

Modeling with Real Data

91. The state of Illinois produced 1,491,000,000 bushels of corn in 1999. (Source: National Corn Growers Association.) Farmers received an average of $2.25 per bushel. What was the cash value of the crop?

92. The amount of coins and currency in circulation in the United States is $1404.49 per person. (Source: U.S. Treasury.) If the population of the United States is 280 million, what is the total amount of money in circulation?

93. In 1999 the Coca-Cola Company reported that the average annual per-capita consumption of carbonated Coca-Cola products in Australia was 289 eight-ounce servings. (Source: Coca-Cola Company.) If the total population of Australia was 18,800,000, what was the total number of ounces consumed that year?

94. Of the approximately 800,000 varieties of insects, 13 varieties are considered to be endangered species. Also, of the approximately 4000 varieties of mammals, 37 are considered to be endangered species. (Source: U.S. Fish and Wildlife Service.) In each category, what fractional part of the total number of varieties is endangered?

Data Analysis: *Super Bowl Advertising*

For the Super Bowl, the cost of a 30-second television ad increased from $42,000 in 1967 to $2.2 million in 2000. The table gives the price of a 30-second ad for selected years.

Year	Price of a 30-Second Ad
1967	$ 42,000
1985	525,000
1995	1,000,000
1997	1,200,000
1999	1,600,000
2000	2,200,000

(Source: USA Today.)

95. Write the cost in 1985 and 2000 in scientific notation.

96. What was the cost per second for an ad in 1985 and 2000?

97. Suppose that during a 4-hour Super Bowl telecast in 2000, the network averaged 10 minutes of advertising per hour. What was the total revenue?

98. Using the assumptions in Exercise 97, what was the increase in the amount spent on advertising from 1967 to 2000?

Challenge

99. A land snail can travel at about 0.5 inch per minute. How many years would it take the snail to travel around the world? (Assume that the earth's circumference is 25,000 miles.)

100. The Gateway Arch in St. Louis is 630 feet high. If the thickness of paper is 0.00425 inches, how many sheets of paper would be needed to make a stack that would reach the top of the Arch?

Chapter 6 Review Exercises

Section 6.1

1. In each part, identify the base and evaluate the expression.

(a) $(-4)^2$ (b) -4^2

In Exercises 2–10, use properties of exponents to simplify the given expression.

2. $x \cdot x^2 \cdot x^5$

3. $(y^3)^4$

4. $\left(\dfrac{-2x^2}{a}\right)^3$

5. $\dfrac{9x^6}{6x^2}$

6. $5x^0$

7. $y^2(y^3)^2$

8. $(xy^2)^5(2x^3y^4)^3$

9. $\dfrac{(2x^2)^3}{(4x)(3x^4)}$

10. $\left(\dfrac{2a^4b^3}{8ab}\right)^3$

Section 6.2

11. Identify the expressions that are polynomials.

(i) $-x^2y$ (ii) $-\dfrac{y}{x^2}$

(iii) $-x^2 + y$

12. List the terms and coefficients of the polynomial $3x^2 - x + 7$.

13. Explain how to determine the degree of each of the following monomials.

(a) $4x^3$ (b) xy^4 (c) 9

14. Determine the degree of $2x^2y - 3x^3y^2 + 6y^4$.

15. Write $4x^3 + 6 - x^2 + 2x^5$ in descending order.

16. Evaluate $2y - y^2$ for $y = -4$ and $y = 3$.

17. For $P(x) = 6x + 5 + 3x^2$, determine $P(-2)$.

18. Suppose that $Q(x) = 4x^5 - 5x^2 + 7x - 8$. Use your calculator to determine $Q(3)$.

19. Evaluate $a^2b - 3abc - 2b^2c$ given $a = 1$, $b = -1$, and $c = 2$.

20. An aquarium full of water has the dimensions shown in the accompanying figure. If water is pumped out until the depth is reduced by 4 inches, write a simplified polynomial that represents the volume of water remaining in the aquarium.

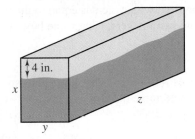

Section 6.3

21. Add.

$$(t^4 + 2t^2 + t - 4) + (2t^4 + t^3 - 4t + 3)$$

22. Use column addition to add $3 + 2x - x^2$, $7 - 6x$, and $4x^3 - x - 4$.

23. Find the sum of $3xy + 9x^2y$ and $15xy^2 - 12xy$.

24. Subtract $(5n + 2) - (3n - 2)$.

25. Use column subtraction to subtract the polynomial $x^2 - 2x - 4$ from $3x + 2$.

26. Determine the difference of $x^2 + 5x + 2$ and $5x - x^2$.

27. Perform the indicated operations.

$$(5x^2 - 6x + 7) - (x^2 + 2x - 5) + (4 + 3x - 2x^2)$$

28. Solve the equation.

$$(4 - x^2 + 7x) - (1 + 2x - x^2) = -4x$$

29. The length of a rope is given as $13x + 18$. The rope is cut into three pieces, two of which have lengths $x^2 - 2x + 8$ and $6x + 2$. Write a polynomial that represents the length of the third piece.

30. The width of a rectangular coffee table is the same as the width of a square end table. The length of the coffee table is 18 inches more than the width. If the difference of the areas of the coffee table and the end table is 540 square inches, how long is the coffee table?

Section 6.4

In Exercises 31–36, multiply.

31. $-3x^2y^3(2xy^2)$

32. $5x(2x^3 + 6x - 3)$

33. $8x - 3$
    $5x + 2$

34. $(x + 2)(x^2 - 5x + 4)$

35. $3x(x + 1)(x + 2)$

36. $(2a + 1)(a - 1)(a + 2)$

37. Simplify $(3x^2 + x - 2) - (x + 4)(x - 3)$.

38. Add $7x - 5$ to the product of $3x + 1$ and $x - 2$.

39. Solve $3x - x(x + 1) = (2 + x)(1 - x)$.

40. Consider three consecutive integers. If the product of the first two integers is equal to the product of the second two integers, what is the largest integer?

Section 6.5

In Exercises 41–46, use the applicable special product pattern to determine the product.

41. $(x + 3)(x - 1)$

42. $(x - 6)^2$

43. $(4 + a)(4 - a)$

44. $(3 + 2xy)(1 - xy)$

45. $(2a + 3b)^2$

46. $(x^2y - 3)(x^2y + 3)$

47. Simplify $(1 + 3x)(1 - 3x) - (1 - 2x)^2$.

48. Write the difference of the squares of $x + 3$ and $x - 3$ as a simplified expression.

49. Solve $(x + 1)^2 - (x + 1)(x - 1) = x$.

50. A square corner lot has a sidewalk on two sides, as shown in the accompanying figure. The sidewalk is 3 feet wide, and the total area of the sidewalk is 471 square feet. Including the sidewalk, how wide is the lot?

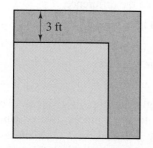

Section 6.6

51. When we use the long-division format to divide a polynomial by a polynomial, at what point in the process do we stop dividing?

In Exercises 52–55, divide.

52. $\dfrac{18x^6y^3}{15x^4y^2}$

53. $(21a^5 - 14a^4 + 28a^2) \div 7a^2$

54. $\dfrac{40x^2 + x - 6}{5x + 2}$

55. $\dfrac{x^3 - 4x}{x - 2}$

In Exercises 56–58, determine the quotient and the remainder.

56. $(3x^2 + 2x + 1) \div (x + 1)$

57. $(x^3 - 4x + 2) \div (x - 2)$

58. $[(x - 4)(x + 2)] \div (x - 1)$

59. Divide the sum of $x^3 + 3x^2 + 8$ and $2 - x^2$ by $x + 3$.

60. The area of the trapezoid in the accompanying figure is $x^3 + 4x^2 + 4x$. Determine b_1.

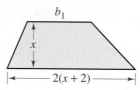

Section 6.7

In Exercises 61 and 62, evaluate the expression.

61. $(-3)^{-2}$

62. $\left(\dfrac{2}{3}\right)^{-3}$

In Exercises 63–70, simplify and write the expression with positive exponents. Assume that all expressions are defined.

63. $-5x^{-2}$

64. $\dfrac{1}{z^{-3}}$

65. $\dfrac{x^{-2}}{y^{-3}}$

66. $-6x^{-4}(-2x^3)$

67. $(-2x^{-2})^{-3}$

68. $\dfrac{x}{x^6(x^{-2})^3}$

69. $\dfrac{x^{-2}y}{x^4y^5}$

70. $\dfrac{12a^4b^{-3}}{2a^{-1}b^0}$

71. If n is an integer and $n < 0$, how can b^{-n} be written with a positive exponent?

Section 6.8

In Exercises 72 and 73, write the given number in decimal form.

72. $6.32 \cdot 10^5$ **73.** $2.8 \cdot 10^{-4}$

74. Write 93,000,000 in scientific notation.

75. Write 0.000067 in scientific notation.

76. Write $395 \cdot 10^{-7}$ in scientific notation.

In Exercises 77–81, use your calculator to perform the required operations. Write the result in scientific notation $n \cdot 10^p$ with n rounded to the nearest tenth. (In

Exercises 80 and 81, assume that the population of the United States is 250,000,000.)

77. $0.0053 \div 28{,}000$

78. 513^{12}

79. $(2.63 \cdot 10^5)^4$

80. What percent of the population do you represent?

81. To pay off the national debt in 1994, every man, woman, and child would have had to pay about $16,000 to the federal government. Approximately what was the national debt in that year?

Chapter 6 Test

In Questions 1–3, simplify the expression.

1. $(4x^3y)(-2x^2y^5)$ **2.** $(2xy^2)^5$

3. $\left(\dfrac{20x^3y^2}{4x^2y}\right)^2$

4. Write $8x + x^3 - 7 + 2x^2$ in descending order. Then determine the coefficients and state the degree of the polynomial.

5. If $P(x) = 6 + 2x - x^2$, what is $P(-2)$?

6. Evaluate $a^2b - ab^2 - ab$ for $a = -2$ and $b = -1$.

7. Add $(3x^3 + x^2 - 2x - 4) + (x^3 + 3x - 1)$.

8. Subtract $2x^2 - 5x - 1$ from $x^3 + 6x - 8$.

9. Solve $(x^2 + 3x - 8) - (2 + x^2) = x$.

10. The acronym FOIL refers to a method for multiplying certain polynomials. For what kind of polynomials is the FOIL method applicable?

In Questions 11–13, multiply.

11. $3x^2y^3(-2x^3y^6)$

12. $-2ab(a^2 - 3ab + b^2)$

13. $(2x + 1)(x - 1)(x + 4)$

In Questions 14–16, use the applicable special product pattern to determine the product.

14. $(2x - 3)(x + 5)$

15. $(2x - 5)^2$ **16.** $(pq - 7)(pq + 7)$

In Questions 17–19, determine the quotient and the remainder.

17. $(8x^7 - 12x^5 + 20x^4) \div 4x^2$

18. $(2x^3 - 5x^2 + x - 4) \div (2x + 1)$

19. $(x^4 + x^2 + 1) \div (x - 1)$

20. Evaluate -3^{-2}.

In Questions 21 and 22, simplify the expression and write the results with no negative exponents. Assume that all expressions are defined.

21. $(a^{-3}b^2)^{-4}$ **22.** $\dfrac{y^3(y^{-4})^{-5}}{(y^{-2})^{-10}}$

23. Suppose that a number is written in the scientific notation form $n \cdot 10^p$. What must be true about n and p?

24. In each part, write the given number in scientific notation.

(a) 0.00059   (b) 684,000

In Questions 25 and 26, write the result in scientific notation $n \cdot 10^p$ with n rounded to the nearest tenth.

25. $13 \div 87{,}261$ **26.** $(7.4 \cdot 10^{-3})^2$

During the period 1993–1999, attendance at movie theaters and the average price of admission both increased. The attendance and the ticket price data can be modeled with expressions whose product is a polynomial that represents the movie industry's total receipts. Each expression of the product is a **factor,** *and the polynomial is said to be in factored form. With attendance and ticket prices both increasing, we can anticipate that the total receipts are also increasing. (For more on this real data problem, see Exercises 99–102 in Section 7.6.)*

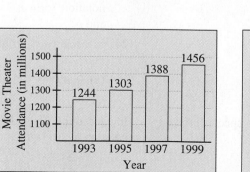

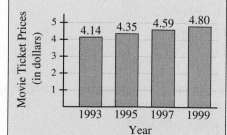

(Source: Motion Picture Association of America.)

Factoring

Chapter Snapshot

This chapter deals with the essential algebraic skill of factoring. We consider common factors, the grouping method, and special factoring patterns, along with methods for factoring trinomials. We learn one important use of factoring in solving equations and conclude with a variety of applications.

Warm-Up Skills

The following exercises review concepts and skills that you will need in Chapter 7.

In Exercises 1 and 2, use the Distributive Property to multiply.

1. $-2y(y^2 - y + 4)$
2. $ab^2(a^3b + a^2 - 1)$

In Exercises 3–8, multiply.

3. $(7y + 5)(7y - 5)$
4. $5(3 + 2a)(3 - 2a)$
5. $(4x + y)^2$
6. $(x + 7)(x - 2)$
7. $(a - 2b)(3a + b)$
8. $(x - 2)(x^2 + 2x + 4)$

In Exercises 9 and 10, determine two numbers whose product is the first number and whose sum is the second number.

9. (a) $10, -7$ (b) $8, 6$
10. (a) $-12, -4$ (b) $-15, 2$

In Exercises 11 and 12, solve the given equation.

11. $2x - 7 = 0$
12. $-3x = 0$

7.1 | Common Factors and Grouping

Greatest Common Factor   •   Common Factors in Polynomials   •   The Grouping Method

Greatest Common Factor

Recall that in the product $3 \cdot 7$ the numbers 3 and 7 are called factors. Writing a number such as 21 as the product $3 \cdot 7$ is called **factoring** the number.

We can often factor a number in more than one way.

$$18 = 9 \cdot 2 \qquad 18 = 3 \cdot 6 \qquad 18 = 2 \cdot 3 \cdot 3$$

To factor a number completely means to write the number as a product of prime numbers. For example, to factor 18 completely, we write $18 = 2 \cdot 3 \cdot 3$ or $18 = 2 \cdot 3^2$.

EXAMPLE **1** | **Prime Factorizations of Numbers**

Factor each number completely.

(a) $66 = 6 \cdot 11 = 2 \cdot 3 \cdot 11$

(b) $420 = \quad 42 \quad \cdot \quad 10$
$\qquad = \quad 6 \ \cdot \ 7 \cdot 2 \cdot 5$
$\qquad = 2 \cdot 3 \cdot 7 \cdot 2 \cdot 5$
$\qquad = 2^2 \cdot 3 \cdot 5 \cdot 7$

The **greatest common factor (GCF)** of two numbers is the largest integer that is a factor of both numbers. For instance, the GCF of 30 and 24 is 6. One way to determine the GCF of two numbers is to begin by factoring the number, as illustrated in Example 2.

EXAMPLE **2** | **Determining the GCF of Numbers**

In each part, what is the GCF of the given numbers?

(a) 15, 20, 40 (b) 84, 56 (c) 28, 45

Solution

(a) $15 = 5 \cdot 3$ Factor each number completely.
$20 = 5 \cdot 2 \cdot 2$
$40 = 5 \cdot 2 \cdot 2 \cdot 2$

Because the only common factor of all the numbers is 5, the GCF is 5.

(b) $84 = 2 \cdot 2 \cdot 3 \cdot 7 = 28 \cdot 3$
$56 = 2 \cdot 2 \cdot 2 \cdot 7 = 28 \cdot 2$

The product $2 \cdot 2 \cdot 7$ or 28 is a factor of each number, so the GCF is 28.

(c) $28 = 2 \cdot 2 \cdot 7$
$45 = 3 \cdot 3 \cdot 5$

The numbers have no common factors other than 1, so the GCF is 1.

We can determine the GCF of two or more monomials in a similar way.

$$6y^4 = 2 \cdot 3 \cdot y^4$$

$$9y^6 = 3 \cdot 3 \cdot y^4 \cdot y^2$$

Because 3 and y^4 are the greatest factors that are common to both monomials, the GCF is $3y^4$. Note that for the variable factors, we can simply look for the smallest exponent.

Determining the GCF of Two or More Monomials

To determine the GCF of two or more monomials, perform the following steps.

1. Determine the GCF of the numerical coefficients.

2. For each variable that is common to the monomials, determine the smallest exponent and write the variable with that exponent.

3. The GCF of the monomials is the product of the results of steps 1 and 2.

EXAMPLE **3**

Determining the GCF of Monomials

In each part, determine the GCF of the given monomials. Then write each monomial as a product with the GCF as one factor.

(a) $12x^2, 8x^3$ (b) $28x^2y^4, 21xy^2, 49xy^5$

Solution

(a) The GCF of 12 and 8 is 4.

The variable x is common to both monomials, and the smallest exponent is 2, so we write x^2.

The GCF is $4x^2$.

$$12x^2 = 4x^2 \cdot 3; \qquad 8x^3 = 4x^2 \cdot 2x$$

(b) The GCF of 28, 21, and 49 is 7.

The variable x is common to all three monomials, and the smallest exponent is 1, so we write x.

The variable y is common to all three monomials, and the smallest exponent is 2, so we write y^2.

The GCF is $7xy^2$.

$$28x^2y^4 = 7xy^2 \cdot 4xy^2; \qquad 21xy^2 = 7xy^2 \cdot 3; \qquad 49xy^5 = 7xy^2 \cdot 7y^3$$

Common Factors in Polynomials

Factoring a polynomial means writing the polynomial as a product of two or more polynomials. The first step in factoring a polynomial is to determine whether the terms have a common factor other than 1. To multiply a polynomial by a monomial, we use the Distributive Property. To factor a polynomial, we reverse the process. For example, to factor $12x^2 + 18x$, we begin by determining that the GCF of $12x^2$ and $18x$ is $6x$.

$$12x^2 + 18x = 6x \cdot 2x + 6x \cdot 3 \qquad \text{Factor each term with the GCF as one factor.}$$

$$= 6x(2x + 3) \qquad \text{Apply the Distributive Property.}$$

If the polynomial to be factored has integer coefficients, then we try to factor so that the resulting expression has integer coefficients. If this is not possible, we consider the expression to be a **prime polynomial.**

A polynomial is **factored completely** if each factor other than a monomial factor is prime. For example, although $12x^2 + 18x$ can be written $2x(6x + 9)$, we usually want to factor completely and write $6x(2x + 3)$.

Factoring Out a GCF from a Polynomial

1. Determine the GCF of the terms.
2. Factor each term with the GCF as one factor.
3. Use the Distributive Property to factor the polynomial with the GCF as one of the factors.
4. Verify the result by multiplying the factors. The product should be the original polynomial.

Note: Although we do not show the verification step in each example, you should always verify your result.

EXAMPLE **4**

Factoring Out a GCF from a Polynomial

Factor the given polynomial completely.

(a) $30x - 18$ (b) $8x^2 + 12x$

(c) $6a + 5b$ (d) $14a^5 + 28a^4 - 21a^3$

Solution

LEARNING TIP

After you factor out a GCF, the second factor should have the same number of terms as the original polynomial.

(a) $30x - 18 = 6 \cdot 5x - 6 \cdot 3 \qquad \text{The GCF is 6.}$

$\qquad\qquad = 6(5x - 3) \qquad \text{Distributive Property}$

To check the factorization, multiply $6(5x - 3)$. The result should be $30x - 18$.

(b) $8x^2 + 12x = 4x \cdot 2x + 4x \cdot 3 \qquad \text{The GCF is } 4x.$

$\qquad\qquad = 4x(2x + 3) \qquad \text{Distributive Property}$

(c) Because the only common factor of $6a$ and $5b$ is 1, $6a + 5b$ is prime.

(d) $14a^5 + 28a^4 - 21a^3 = 7a^3 \cdot 2a^2 + 7a^3 \cdot 4a - 7a^3 \cdot 3$

$\qquad\qquad\qquad\qquad = 7a^3(2a^2 + 4a - 3)$

EXAMPLE **5**

Factoring Out a GCF from a Polynomial

Factor the given polynomial completely.

(a) $15mn + 20n$ (b) $4a^2b^4 - 9a^3b^5$

Solution

(a) $15mn + 20n = 5n \cdot 3m + 5n \cdot 4 = 5n(3m + 4)$

(b) $4a^2b^4 - 9a^3b^5 = a^2b^4 \cdot 4 - a^2b^4 \cdot 9ab = a^2b^4(4 - 9ab)$

When a term of a polynomial is the GCF, it is helpful to write the term with a factor of 1.

EXAMPLE 6

Factoring When One Term Is the GCF

Factor the given polynomial completely.

(a) $9y + 9$ (b) $a^2b - ab$

Solution

(a) $9y + 9 = 9 \cdot y + 9 \cdot 1 = 9(y + 1)$

(b) $a^2b - ab = ab \cdot a - ab \cdot 1 = ab(a - 1)$

If the leading coefficient is negative, it is sometimes helpful to factor out a GCF with a negative coefficient.

EXAMPLE 7

Factoring Out a GCF with a Negative Coefficient

Factor completely.

(a) $-5x^2 - 10x$ (b) $-3x^3 + 3x^2 - 6x$

Solution

(a) $-5x^2 - 10x = -5x \cdot x + (-5x) \cdot 2$ Consider the GCF to be $-5x$.

$\qquad = -5x(x + 2)$

(b) $-3x^3 + 3x^2 - 6x = -3x \cdot x^2 + (-3x) \cdot (-x) + (-3x) \cdot 2$

$\qquad = -3x(x^2 - x + 2)$

Example 8 illustrates expressions in which the common factor is a binomial.

EXAMPLE 8

Factoring Out a Common Binomial Factor

Factor completely.

(a) $2x(x - 7) + 3(x - 7)$ (b) $y^2(y - 4) - 2(y - 4)$

Solution

(a) $2x(x - 7) + 3(x - 7) = (x - 7)(2x + 3)$ GCF is $x - 7$.

(b) $y^2(y - 4) - 2(y - 4) = (y - 4)(y^2 - 2)$ GCF is $y - 4$.

The Grouping Method

Occasionally it is possible to factor a polynomial by first factoring groups of terms. This method is called *factoring by grouping* or the *grouping method*.

EXAMPLE 9

Factoring by Grouping

Factor completely: $ax + ay + 3x + 3y$.

Solution

Note that there is no factor common to all terms. However, the first two terms have a common factor a, and the last two terms have a common factor 3. Group those terms and factor out the common factor in each group.

$$ax + ay + 3x + 3y$$
$$= (ax + ay) + (3x + 3y) \qquad \text{Group the terms.}$$
$$= a(x + y) + 3(x + y) \qquad \text{Factor out } a \text{ in the first group and 3 in the second group.}$$
$$= (x + y)(a + 3) \qquad \text{Factor out } (x + y).$$

Note the second step. The expression $a(x + y) + 3(x + y)$ is not in factored form. It is the sum of two terms. However, the factor $x + y$ is common to both terms, and we factor it out to complete the factorization.

Think About It

Show two groupings for which the grouping method can be used to factor the expression $x^3 + y^2 + x^2y + xy$.

The Grouping Method

1. Group the terms so that the first two terms have a common factor and the last two terms have a common factor.
2. Factor out the GCF from each group.
3. If the resulting two terms have a common binomial factor, factor it out.

Note: This summary describes the method of grouping *by pairs*. In some cases, other grouping arrangements may be needed. Also, though the grouping method is a good technique to try, there is no guarantee that it will always work. For example, although $ax + ay + bx - by$ can be grouped and written as $a(x + y) + b(x - y)$, there is no common binomial factor with which to complete the factorization.

EXAMPLE 10

Factoring by Grouping

Use the grouping method to factor the given expressions.

(a) $2ax + 2bx + a + b = 2x(a + b) + 1(a + b) = (a + b)(2x + 1)$

(b) $x^2 - x - 4x + 4 = x(x - 1) - 4(x - 1) = (x - 1)(x - 4)$

Note that in the second group it was necessary to factor out -4 rather than 4 in order to obtain a common binomial factor.

(c) $t^3 - 2t^2 + 3t - 6 = t^2(t - 2) + 3(t - 2) = (t - 2)(t^2 + 3)$

(d) $x^2 + xy - 2xy^2 - 2y^3 = x(x + y) - 2y^2(x + y) = (x + y)(x - 2y^2)$

Note: As Example 10 illustrates, special care must be taken when the second group is preceded by a minus sign. After factoring out the common factor from both groups, it is a good idea to verify by multiplication before going on. This will help to avoid sign errors.

Quick Reference 7.1

Greatest Common Factor

- Writing a number as a product of two or more factors is called **factoring** the number.
- To factor a number completely means to write the number as a product of prime numbers.
- The **greatest common factor (GCF)** of two numbers is the largest integer that is a factor of both numbers.
- To determine the GCF of two or more monomials, perform the following steps.
  1. Determine the GCF of the numerical coefficients.
  2. For each variable that is common to the monomials, determine the smallest exponent and write the variable with that exponent.
  3. The GCF of the monomials is the product of the results of steps 1 and 2.

Common Factors in Polynomials

- **Factoring a polynomial** means writing the polynomial as a product of polynomials.
- If the polynomial to be factored has integer coefficients, then we try to factor so that the resulting expression has integer coefficients. If this is not possible, then we consider the expression a **prime polynomial.**
- A polynomial is **factored completely** if each factor other than a monomial factor is prime.
- To factor out a GCF from a polynomial, perform the following steps.
  1. Determine the GCF of the terms.
  2. Factor each term with the GCF as one factor.
  3. Use the Distributive Property to factor the polynomial with the GCF as one of the factors.
  4. Verify the result by multiplying the factors. The product should be the original polynomial.
- If the leading coefficient is negative, it is sometimes helpful to factor out a GCF with a negative coefficient.

The Grouping Method

- The grouping method is a factoring technique that can sometimes be used on expressions with four or more terms.
- To factor by grouping, perform the following steps.
  1. Group the terms so that the first two terms have a common factor and the last two terms have a common factor.
  2. Factor out the GCF from each group.
  3. If the resulting two terms have a common binomial factor, factor it out.
- When the second group of the expression is preceded by a minus sign, we factor out a GCF with a negative coefficient. Care must be taken to avoid sign errors.

Speaking the Language 7.1

1. Writing $2x + 6$ as $2(x + 3)$ is called ▨▨▨▨▨▨ the expression.
2. Because 6 is the largest integer that is a factor of both 12 and 18, we call 6 the ▨▨▨▨▨▨.
3. Factoring a polynomial means writing the polynomial as a product of ▨▨▨▨▨▨.
4. A possible method for factoring a polynomial of four terms is called the ▨▨▨▨▨▨.

Exercises 7.1

Concepts and Skills

1. What does it mean to factor a number completely?

2. Suppose that three monomials with a coefficient of 1 have one variable in common. Describe a simple way of determining the GCF.

In Exercises 3–8, determine the GCF for each set of numbers.

3. 6, 15
4. 70, 60
5. 11, 18
6. 15, 28
7. 15, 12, 21
8. 6, 12, 15

In Exercises 9–14, determine the GCF of the given monomials.

9. $9y, 12y^2$
10. $5x^3, 12x^2$
11. x^2y^3, xy^5
12. $2a^4b^2, 4a^2b^3$
13. $6x^3, 3x^2, 12x$
14. $7y^4, 14y^3, 7y^3$

15. Consider the polynomial $18x - 27$.
    (a) Which expressions in the following list represent correct factorizations of the polynomial?
    (b) Which expressions represent complete factorizations of the polynomial?
        (i) $3(6x - 9)$ (ii) $9(2x - 3)$
        (iii) $-9(3 - 2x)$ (iv) $-3(-6x + 9)$

16. The Distributive Property is used to multiply two polynomials. What property do we use to factor a polynomial?

In Exercises 17–22, fill in the blank to complete the indicated factorization.

17. $56x - 32 = $ ▨▨▨ $(7x - 4)$
18. $25x^2 - 15y = $ ▨▨▨ $(5x^2 - 3y)$
19. $12x + 4 = 4($ ▨▨▨▨▨ $)$
20. $14y + 7 = 7($ ▨▨▨▨▨ $)$
21. $5x^2 + 10x + 15 = $ ▨▨▨ $(x^2 + 2x + 3)$
22. $30 - 15a + 9a^2 = $ ▨▨▨ $(10 - 5a + 3a^2)$

In Exercises 23–32, factor out the GCF.

23. $7x + 21$
24. $10 - 5y$
25. $32x + 24y - 56$
26. $42 - 21a - 35b$
27. $18a^2 - 24a$
28. $15b^2 + 12b$
29. $8x^2 - 5$
30. $11 - 15y$
31. $3x^3 - 9x^2 + 3x$
32. $18x^3 - 9x^2 - 36x$

In Exercises 33–38, factor out the GCF.

33. $10x^3y - 5x^2y^3$
34. $3x^2y + 6xy^2$
35. $24a^4b^4 + 16a^3b^4$
36. $40m^3n^2 - 70mn^5$
37. $x^2y + xy^2 - 5xy$
38. $10x^2 - 5x^2y - 5xy$

In Exercises 39–46, factor completely.

39. $24y + 8$
40. $9a - 9$
41. $21a + 35b - 7$
42. $12x - 20y + 4$
43. $12x^2 + 4x$
44. $15y^3 + 6y^2 - 3y$
45. $a^3b^4 - a^2b^4 + a^2b^3$
46. $3a^2b^2 + 12a^2 + 3a$

In Exercises 47–50, fill in the blank to complete the indicated factorization.

47. $-24y - 18z = -6 ($ ▨▨▨▨▨ $)$
48. $32y + 20z = -4($ ▨▨▨▨▨ $)$
49. $-6xy + 3y = -3y($ ▨▨▨▨▨ $)$
50. $-8x^2 + 6x = -2x ($ ▨▨▨▨▨ $)$

In Exercises 51–56, factor out the GCF and its opposite.

51. $3x - 15$
52. $-21x + 7$
53. $12 - 4x$
54. $-5 - 10y$

55. $-2x^2 + 2x + 10$

56. $-10x^2 - 5x + 20$

57. Why is $x(a + b) + y(a + b)$ not considered to be factored?

58. Why is the grouping method not applicable in attempting to factor $ax + bx - ay + by$?

In Exercises 59–66, factor out the GCF.

59. $x(x - 8) + 5(x - 8)$

60. $x(2x + 1) - 4(2x + 1)$

61. $3y(y + 3) - 2(y + 3)$

62. $x(5x - 6) + 8(5x - 6)$

63. $x(x - 4) + (x - 4)$

64. $2t(3 - 2t) + (3 - 2t)$

65. $x^2(x - 3) + 2(x - 3)$

66. $2x^2(3x - 1) - 3(3x - 1)$

In Exercises 67–82, factor by grouping.

67. $ax + 5a + 2x + 10$ **68.** $7y - 14 + by - 2b$

69. $2xy - 2x + 3y - 3$ **70.** $2xy + x + 10y + 5$

71. $xy + 3y - x - 3$ **72.** $2ay - a - 2y + 1$

73. $3x + y - 3ax + ay$ **74.** $3x - 6y - xz - 2yz$

75. $t^3 + t^2 + 3t + 3$ **76.** $x^3 + 4x^2 - 2x - 8$

77. $2x^3 + 2x^2 - x - 1$ **78.** $3x^3 - 4x^2 + 15x - 20$

79. $xz + yz + x + y$ **80.** $xz - 3x + 2yz - 6y$

81. $ax + a - x - 1$ **82.** $yz + y + z + 1$

In Exercises 83–88, use the grouping method to factor the expression. Note that in some cases you may need to factor out a GCF first, or you may need to rearrange the terms before grouping.

83. $xz - 6y - 3x + 2yz$

84. $2ax + 6b - 4a - 3bx$

85. $3a^2 + 3ab - 6a - 6b$

86. $2x^2 + 4xy + 2x + 4y$

87. $3t^4 - 6t^3 + 15t^2 - 30t$

88. $y^5 + 3y^4 + y^3 + 3y^2$

In Exercises 89–96, write NF if the expression is not factored, F if the expression is factored but not completely factored, or CF if the expression is completely factored.

89. $9x - 2(x - 4)$ **90.** $x + 5(x - 4)$

91. $2y(y - 4) + 7(y - 4)$ **92.** $(y + 1) + (y - 3)$

93. $(t + 9)(t - 6)$ **94.** $5y(x + 7y)$

95. $4(4x - 6)$ **96.** $3(3x^2 + 5x)$

Real-Life Applications

97. A window is divided into square and rectangular glass panes. The widths of the rectangular panes and the sides of the square panes are x inches. The rectangular panes are 15 inches long. Write an expression for the total area of one square pane and one rectangular pane. Then write the expression in factored form.

98. Suppose that the distance of an object from the ground is given by $24t - 16t^2$, where t represents time. Write this expression in factored form.

99. Suppose that the volume of a certain box is given by $3x^3 + 6x^2$. Factor the expression. If the length and width of the box are represented by $3x$ and x, what expression represents the height?

100. Suppose that the revenue from selling x chain saws is given by $4x^2 + 6x$. Factor the expression.

Data Analysis: *Computer Use in Jobs*

The accompanying bar graph shows that the percentage of people who use computers in their jobs has risen steadily in recent years. (Source: Human Resources Group.) If x represents the number of years since 1984, then $C(x) = 2x + 24$ models the percent of people who used computers in their jobs during this time period.

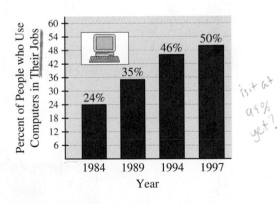

101. Factor the model expression.

102. Produce the graph of $2x + 24$ and trace it to determine the annual increase in $C(x)$.

103. How can we interpret the common factor in Exercise 101?

104. In what year does the model predict the percent to reach 70%?

Challenge

In Exercises 105 and 106, factor completely. (*Hint*: Consider grouping the terms in groups of three.)

105. $x^2 - xy + xz + 3x - 3y + 3z$

106. $xy + 2y^2 - 3yz + 2x + 4y - 6z$

In Exercises 107 and 108, fill in the blank.

107. $\dfrac{2}{3}x - \dfrac{5}{3} = \dfrac{1}{3}(\rule{2cm}{0.4pt})$

108. $\dfrac{3}{5}x^2 + \dfrac{2}{5}x + \dfrac{4}{5} = \dfrac{1}{5}(\rule{2cm}{0.4pt})$

In Exercises 109 and 110, use the method suggested in Exercises 107 and 108 to factor the expression.

109. $\dfrac{3}{4}x - \dfrac{5}{4}$

110. $\dfrac{1}{2}x^2 + \dfrac{5}{2}x - 2$

7.2 | Special Factoring

Difference of Two Squares • Perfect Square Trinomial • Sum and Difference of Two Cubes

In Chapter 6 we saw that products of certain binomials result in distinguishable patterns. We referred to these as special products. Recognizing that a polynomial is the result of a special product makes it easy to factor the polynomial.

Difference of Two Squares

We know that the product of the sum and difference of the same two terms always results in a difference of two squares.

$$(A + B)(A - B) = A^2 - B^2$$

$$(x + 5)(x - 5) = x^2 - 5^2 = x^2 - 25$$

If we begin with a binomial in the form of a difference of two squares, we can reverse the process to factor the binomial.

$$A^2 - B^2 = (A + B)(A - B)$$

$$x^2 - 9 = x^2 - 3^2 = (x + 3)(x - 3)$$

Factoring a Difference of Two Squares

We recognize a difference of two squares as two perfect squares separated by a minus sign.

$$A^2 - B^2 = (A + B)(A - B)$$

Note: The expressions $x^2 + 9$ and $x^2 - 7$ are not differences of two squares. The first expression is a sum, not a difference. In the second expression, 7 is not a perfect square. Neither expression can be factored.

EXAMPLE **1**

Factoring a Difference of Two Squares

Factor completely.

(a) $16 - y^2$

(b) $36x^2 - 25$

(c) $a^2b^2 - 9$

Solution

$$A^2 - B^2 = (A + B)(A - B)$$
$$\downarrow \quad \downarrow \quad \downarrow \quad \downarrow \downarrow \quad \downarrow$$

(a) $16 - y^2 = 4^2 - y^2 = (4 + y)(4 - y)$

$$A^2 - B^2 = (A + B)(A - B)$$
$$\downarrow \quad \downarrow \quad \downarrow \quad \downarrow \downarrow \quad \downarrow$$

(b) $36x^2 - 25 = (6x)^2 - 5^2 = (6x + 5)(6x - 5)$

(c) $a^2b^2 - 9 = (ab)^2 - 3^2 = (ab + 3)(ab - 3)$

Note: The Commutative Property of Multiplication allows us to write factors in any order. For instance, in Example 1(c), we can write the factorization as $(ab + 3)(ab - 3)$ or $(ab - 3)(ab + 3)$.

When we factor an expression, we should look for a common factor before we use other methods.

EXAMPLE **2**

Combining Methods

Factor completely.

(a) $3x^2 - 75$ (b) $9 - 81y^2$ (c) $7a^3 - 28a$

(d) $49x^3y - xy^3$ (e) $2x^3 - 16x$

Solution

(a) $3x^2 - 75 = 3(x^2 - 25)$ The GCF is 3.

$\qquad\qquad\quad\; = 3(x + 5)(x - 5)$ $x^2 - 25$ is a difference of two squares.

(b) $9 - 81y^2 = 9(1 - 9y^2) = 9(1 + 3y)(1 - 3y)$

(c) $7a^3 - 28a = 7a(a^2 - 4) = 7a(a + 2)(a - 2)$

(d) $49x^3y - xy^3 = xy(49x^2 - y^2) = xy(7x + y)(7x - y)$

(e) $2x^3 - 16x = 2x(x^2 - 8)$

Note that the factor $x^2 - 8$ cannot be factored further. The polynomial has been factored completely.

When a difference of two squares is factored, the factors still may be factorable.

EXAMPLE **3**

Factoring a Difference of Two Squares

Factor completely: $x^4 - 16$.

Solution

$$x^4 - 16 = (x^2)^2 - 4^2 \qquad \text{Write as a difference of two squares.}$$
$$= (x^2 + 4)(x^2 - 4) \qquad \text{Factor the difference of two squares.}$$
$$= (x^2 + 4)(x^2 - 2^2) \qquad \text{The second factor is a difference of two squares.}$$
$$= (x^2 + 4)(x + 2)(x - 2) \qquad \text{Factor the second factor.}$$

Perfect Square Trinomial

Another special product is the square of a binomial.

$$(A + B)^2 = A^2 + 2 \cdot A \cdot B + B^2$$

$$(x + 5)^2 = x^2 + 2 \cdot x \cdot 5 + 5^2 = x^2 + 10x + 25$$

We call the result of squaring a binomial a **perfect square trinomial.** For example, $x^2 + 10x + 25$ is a perfect square trinomial because it is the square of $(x + 5)$. Recognizing a perfect square trinomial allows us to use the results of the special product to factor.

Factoring a Perfect Square Trinomial

In a perfect square trinomial, the first and last terms are perfect squares, and the middle term is twice the product of the terms being squared.

$$A^2 + 2AB + B^2 = (A + B)^2$$
$$A^2 - 2AB + B^2 = (A - B)^2$$

Note: The expressions $x^2 + 2x + 3$ and $x^2 + 3x + 9$ are *not* perfect square trinomials. In the first expression, the last term is not a perfect square. In the second expression, the terms being squared are x and 3, but the middle term is not twice the product of x and 3.

EXAMPLE 4

Factoring a Perfect Square Trinomial

Factor completely.

(a) $y^2 + 16y + 64$ \qquad (b) $4x^2 - 28x + 49$
(c) $25x^2 + 60xy + 36y^2$

Solution

LEARNING TIP

If you suspect that a trinomial is a perfect square, write the two terms of the binomial and then check the middle term.

(a) $y^2 + 16y + 64 = y^2 + 2 \cdot y \cdot 8 + 8^2 = (y + 8)^2$

(b) $4x^2 - 28x + 49 = (2x)^2 - 2 \cdot (2x) \cdot 7 + 7^2 = (2x - 7)^2$
(c) $25x^2 + 60xy + 36y^2 = (5x)^2 + 2(5x)(6y) + (6y)^2 = (5x + 6y)^2$

Again, remember to look for a common factor before you attempt to apply a special pattern. Also, as Example 5 shows, a perfect square trinomial is not necessarily a second-degree expression.

EXAMPLE **5**

Think About It

Consider the sum of two squares $x^4 + 4y^4$. Add and subtract $4x^2y^2$ to factor the expression.

Combining Methods

Factor completely.

(a) $5x^3 - 30x^2 + 45x$ (b) $t^4 + 20t^2 + 100$

Solution

(a) $5x^3 - 30x^2 + 45x = 5x(x^2 - 6x + 9)$ The GCF is $5x$.

$\qquad\qquad\qquad\qquad\quad = 5x(x - 3)^2$

(b) $t^4 + 20t^2 + 100 = (t^2)^2 + 2 \cdot t^2 \cdot 10 + 10^2 = (t^2 + 10)^2$

Sum and Difference of Two Cubes

We can sometimes arrive at a factoring pattern by multiplying polynomials. For instance, we use the Distributive Property to determine the following products.

$$(A + B)(A^2 - AB + B^2)$$

$$= A(A^2 - AB + B^2) + B(A^2 - AB + B^2)$$

$$= A^3 - A^2B + AB^2 + A^2B - AB^2 + B^3$$

$$= A^3 + B^3$$

$$(A - B)(A^2 + AB + B^2)$$

$$= A(A^2 + AB + B^2) - B(A^2 + AB + B^2)$$

$$= A^3 + A^2B + AB^2 - A^2B - AB^2 - B^3$$

$$= A^3 - B^3$$

The two terms in each result are perfect cubes. The first result is a sum of two perfect cubes, and the second result is a difference of two perfect cubes. Reversing the steps in each case gives us a factoring pattern for the sum or difference of two cubes.

Factoring a Sum or Difference of Two Cubes

We recognize a sum or difference of two cubes as two perfect cubes separated by a plus or minus sign.

$$A^3 + B^3 = (A + B)(A^2 - AB + B^2)$$
$$A^3 - B^3 = (A - B)(A^2 + AB + B^2)$$

EXAMPLE **6**

Factoring a Sum of Two Cubes

Factor completely.

(a) $x^3 + 27$ (b) $3x^5 + 3x^2$

Solution

$$A^3 + B^3 = (A + B)(A^2 - A \cdot B + B^2)$$

(a) $x^3 + 27 = x^3 + 3^3 = (x + 3)(x^2 - x \cdot 3 + 3^2)$
$$= (x + 3)(x^2 - 3x + 9)$$

(b) $3x^5 + 3x^2 = 3x^2(x^3 + 1)$ The GCF is $3x^2$.
$$= 3x^2(x + 1)(x^2 - x + 1)$$ $x^3 + 1$ is a sum of two cubes.

EXAMPLE 7

Factoring a Difference of Two Cubes

Factor completely.

(a) $125a^3 - 64$ (b) $a^6 - 8b^3$

Solution

$$A^3 - B^3 = (A - B)(A^2 + A \cdot B + B^2)$$

(a) $125a^3 - 64 = (5a)^3 - 4^3 = (5a - 4)[(5a)^2 + (5a)(4) + 4^2]$
$$= (5a - 4)[25a^2 + 20a + 16]$$

(b) $a^6 - 8b^3 = (a^2)^3 - (2b)^3$
$$= (a^2 - 2b)[(a^2)^2 + (a^2)(2b) + (2b)^2]$$
$$= (a^2 - 2b)[a^4 + 2a^2b + 4b^2]$$

Quick Reference 7.2

Difference of Two Squares

- We recognize a difference of two squares as two perfect squares separated by a minus sign.
- To factor a difference of two squares, use the following pattern.
  $$A^2 - B^2 = (A + B)(A - B)$$
- When we factor an expression, we should look for a common factor before we use other methods.

Perfect Square Trinomial

- We call the result of squaring a binomial a **perfect square trinomial.**
- In a perfect square trinomial, the first and last terms are squares, and the middle term is twice the product of the terms being squared.
- To factor a perfect square trinomial, use one of the following patterns.
  $$A^2 + 2AB + B^2 = (A + B)^2$$
  $$A^2 - 2AB + B^2 = (A - B)^2$$

Sum and Difference of Two Cubes

- We recognize a sum or difference of two cubes as two perfect cubes separated by a plus or minus sign.
- To factor a sum or difference of two cubes, use the applicable pattern.
  $$A^3 + B^3 = (A + B)(A^2 - AB + B^2)$$
  $$A^3 - B^3 = (A - B)(A^2 + AB + B^2)$$

Speaking the Language 7.2

1. When we factor an expression of the form $A^2 - B^2$, one factor is the ▬▬▬▬ of A and B, and the other factor is the ▬▬▬▬ of A and B.

2. For a perfect square trinomial, the first and last terms are ▬▬▬▬ and the middle term is ▬▬▬▬ the product of the terms being squared.

3. The expression $A^3 + B^3$ is called a(n) ▬▬▬▬, and one of its factors is the binomial ▬▬▬▬.

4. When factoring an expression, we should look for a(n) ▬▬▬▬ before we apply a special factoring pattern.

Exercises 7.2

Concepts and Skills

1. Only one of the following binomials is a difference of two squares. Explain why the other two are not.
   (i) $x^2 - 10$
   (ii) $x^2 - 100$
   (iii) $x^2 + 100$

2. Is the following a complete factorization of the binomial $4x^2 - 36$? Why?
$$4x^2 - 36 = (2x)^2 - 6^2$$
$$= (2x + 6)(2x - 6)$$

In Exercises 3–16, factor the difference of two squares.

3. $x^2 - 25$
4. $y^2 - 49$
5. $100 - a^2$
6. $4x^2 - 9$
7. $64 - 49w^2$
8. $81 - 16a^2$
9. $a^2 - 4b^2$
10. $25 - 49a^2b^2$
11. $-x^2 + 1$
12. $-16 + 25y^2$
13. $x^2 - \dfrac{1}{4}$
14. $y^2 - \dfrac{4}{9}$
15. $x^4 - 100$
16. $25 - 9x^4$

In Exercises 17–24, factor completely.

17. $2x^2 - 8$
18. $125 - 5x^2$
19. $4x^2 - 16$
20. $16 - 36x^2$
21. $100y^5 - y^7$
22. $81t^9 - t^{11}$
23. $x^3 - xy^2$
24. $a^3b - 4ab^3$

In Exercises 25–28, factor completely.

25. $x^4 - 81$
26. $1 - y^4$
27. $x^8 - 1$
28. $y^8 - 16$

29. How can you tell at a glance that the expression $x^2 + 6x - 9$ is not a perfect square trinomial?

30. If we factor $x^2 - 2x + 1$, we obtain $(x - 1)^2$. If we rearrange the terms and write $1 - 2x + x^2$, the factored form is $(1 - x)^2$. Explain why both results are correct.

In Exercises 31–38, factor the perfect square trinomial.

31. $x^2 + 6x + 9$
32. $y^2 - 10y + 25$
33. $1 - 4y + 4y^2$
34. $49a^2 + 28a + 4$
35. $x^2 + 8xy + 16y^2$
36. $36a^2 - 60ab + 25b^2$
37. $49x^2 + 9y^2 - 42xy$
38. $16w^2 + 49z^2 - 56wz$

In Exercises 39–46, factor completely.

39. $y^4 + 6y^2 + 9$
40. $4x^4 - 20x^2 + 25$
41. $x^6 - 4x^3 + 4$
42. $9w^8 + 6w^4 + 1$
43. $5x^2 - 30x + 45$
44. $-2w^2 - 20w - 50$
45. $-49x - 28x^2 - 4x^3$
46. $3x - 6x^2 + 3x^3$

47. Will the factoring pattern for a sum of two cubes be used in factoring $x^4 + x$? Explain.

48. Name two factoring patterns that could be used to factor $x^6 - y^6$ and explain why both patterns apply.

In Exercises 49–54, factor the sum of two cubes.

49. $x^3 + 7^3$
50. $10^3 + y^3$
51. $z^3 + 1$
52. $1 + 8a^3$
53. $5 + 5y^3$
54. $81a + 3a^4$

In Exercises 55–60, factor the difference of two cubes.

55. $11^3 - t^3$ **56.** $y^3 - 6^3$

57. $64 - y^3$ **58.** $x^3 - 8$

59. $x^3 - 8y^3$ **60.** $8 - 27x^3y^3$

In Exercises 61–70, factor the sum or difference of two cubes.

61. $125 + x^3$ **62.** $b^3 - 27$

63. $64y^3 + z^3$ **64.** $a^3b^3 + 125$

65. $8a^3 - b^3$ **66.** $t^3 - 64z^3$

67. $9x^3 - 72$ **68.** $16t^3 + 16$

69. $27x^6 + 27x^9$ **70.** $8a^6 - 64a^3b^3$

In Exercises 71–92, factor completely.

71. $x^2 + 81 - 18x$ **72.** $8x + x^2 + 16$

73. $\dfrac{9}{25}t^2 - 1$ **74.** $16 - \dfrac{x^2}{49}$

75. $2m^4n + 250mn^4$ **76.** $2z^5 - 128z^2$

77. $4a^4 + 8a^3 + 4a^2$

78. $x^3y - 2x^2y^2 + xy^3$

79. $x^2 + 16$

80. $5 - t^2$

81. $-2 + 72y^2$

82. $-75 + 3z^2$

83. $100x^2 - 60xy + 9y^2$

84. $4a^2 + 36ab + 81b^2$

85. $3y^3 - 48x^2y$

86. $20x^4 - 125x^2y^2$

87. $8a^3 - 27$ **88.** $125y^3 + 64$

89. $49x^2 - 81y^6$ **90.** $x^{10} - 4y^{16}$

91. $x^6 - 27y^3$ **92.** $b^3 + 8a^9$

🌐 Real-Life Applications

93. A cardboard square x inches wide is made into a box by cutting out 3-inch corners. What is the total area of the remaining material after the corners are cut out? Write the polynomial in factored form.

94. A town park has a gazebo whose floor is a rectangle. The area of the floor is given by the trinomial $4x^2 + 28x + 49$. Write the expression in factored form. If the lengths of the sides of the gazebo are represented by binomials with integer coefficients, what do we know about the gazebo?

95. A marbletop table is placed in a cubic moving container that is $3x$ feet long on every side. The container is then placed in a cubic shipping crate whose sides are all 5 feet long. Write an expression in factored form for the volume of foam it takes to fill the unused space in the crate.

96. A washer with outer radius r fits precisely on a bolt whose diameter is 10 mm. Write an expression in factored form for the area of the washer.

Data Analysis: *Starting Salaries*

The table shows some trends in starting annual salaries in certain major fields.

	Starting Salary	
	1996	**2000**
Accounting	$30,420	$35,620
Business	28,548	36,036
Computer Science	36,960	46,656

(Source: Educaid.)

The respective salary data can be modeled by the following polynomials, where x is the number of years since 1996.

Accounting:	$A(x) = 52(25x + 585)$
Business:	$B(x) = 1872x + 28{,}548$
Computer Science:	$C(x) = 12(202x + 3080)$

97. The expression for $A(x)$ is written in factored form. What does the variable factor represent?

98. The expression for $C(x)$ is written in factored form. What does the variable factor represent?

99. Write the expression for $B(x)$ as a product so that one factor represents the weekly salary. Then write it so that one factor represents the monthly salary.

100. Is either of your results in Exercise 99 factored completely?

Challenge

In Exercises 101 and 102, factor completely. (In each part, more than one factoring method is needed.)

101. $x^4 - 8x^2 + 16$ **102.** $x^2 - 4 + xy - 2y$

In Exercises 103–106, factor completely.

103. $(x + 3)^2 - y^2$

104. $x^2 - (y - 2)^2$

105. $(x - y)^2 + 6(x - y) + 9$

106. $x^2(a + 3) - y^2(a + 3)$

7.3   Factoring Trinomials of the Form $x^2 + bx + c$

Quadratic Trinomials • Factoring Patterns • Extending the Methods

Quadratic Trinomials

Recall that when we multiply binomials such as $(x + 4)(x - 2)$, the result is a trinomial.

$$(x + 4)(x - 2) = x^2 + 2x - 8$$

We call this trinomial a **quadratic polynomial.** In general, a quadratic polynomial in one variable can be written in the form $ax^2 + bx + c$, where $a \neq 0$. The term ax^2 is called the **quadratic term,** the term bx is called the **linear term,** and the term c is called the **constant term.** The coefficient a of the quadratic term is called the **leading coefficient.**

Note: The definition of a quadratic polynomial prohibits the leading coefficient from being 0; that is, there must be a quadratic term. However, b or c can be 0, so there need not be a linear term or a constant term. Thus all of the following are quadratic polynomials.

$$2x^2 + 5x - 7 \qquad x^2 - 9 \qquad x^2 + 3x \qquad 4x^2$$

We can factor a quadratic polynomial with no constant term simply by factoring out the GCF.

$$3x^2 + 5x = x(3x + 5)$$

We also know how to factor a quadratic polynomial with no linear term provided that the polynomial is a difference of two squares.

$$4x^2 - 25 = (2x + 5)(2x - 5)$$

In this section, we will concentrate on quadratic trinomials that have all three terms and a leading coefficient a of 1.

Factoring Patterns

From our previous experience, we know that multiplying the binomials $x + m$ and $x + n$ leads to a quadratic trinomial. Therefore, when we factor the quadratic trinomial $x^2 + bx + c$, the factors should be of the form $x + m$ and $x + n$.

EXAMPLE **1** **Factoring $x^2 + bx + c, c > 0$**

Factor the given trinomial.

(a) $x^2 + 5x + 4$ (b) $x^2 - 8x + 12$

Solution

(a) Determine m and n so that $(x + m)(x + n) = x^2 + 5x + 4$.

Recalling the FOIL method for multiplying $(x + m)(x + n)$, we know that the product $m \cdot n = 4$ and $mx + nx = (m + n)x = 5x$; that is, $m + n = 5$. Thus we need to find numbers m and n such that their product is 4 and their sum is 5.

First, we list all pairs of integers whose product is 4. Then we determine the sum of each pair of numbers.

Product = 4	*Sum*
$2 \cdot 2 = 4$	$2 + 2 = 4$
$1 \cdot 4 = 4$	$1 + 4 = 5$
$-2 \cdot (-2) = 4$	$-2 + (-2) = -4$
$-1 \cdot (-4) = 4$	$-1 + (-4) = -5$

From the table, we see that $m = 1$ and $n = 4$ satisfy the requirement. Thus

$$x^2 + 5x + 4 = (x + 1)(x + 4)$$

(Because $m + n$ is positive, it would not have been necessary to include the last two entries in the table.)

(b) Determine m and n so that $(x + m)(x + n) = x^2 - 8x + 12$.

Because $m \cdot n = 12$ and $m + n = -8$, we list all integers whose product is 12 and then determine their corresponding sums.

Product = 12	*Sum*
$1 \cdot 12 = 12$	$1 + 12 = 13$
$2 \cdot 6 = 12$	$2 + 6 = 8$
$3 \cdot 4 = 12$	$3 + 4 = 7$
$-1 \cdot (-12) = 12$	$-1 + (-12) = -13$
$-2 \cdot (-6) = 12$	$-2 + (-6) = -8$
$-3 \cdot (-4) = 12$	$-3 + (-4) = -7$

Only $m = -2$ and $n = -6$ satisfy the requirement. Thus

$$x^2 - 8x + 12 = (x - 2)(x - 6)$$

(Because $m + n$ is negative, it would not have been necessary to include the first three entries in the table.)

As in Example 1, to factor $x^2 + bx + c$, where c is positive, we seek numbers m and n such that $(x + m)(x + n) = x^2 + bx + c$. This means that $m \cdot n = c$ and $m + n = b$. If b is positive, then m and n are both positive; if b is negative, then m and n are both negative.

As always, it is important to check the results by multiplying. For instance, in part (a) of Example 1, we verify that the product of the binomials is the given trinomial. Checking the middle term is particularly important.

$$(x + 1)(x + 4) = x^2 + 5x + 4$$

EXAMPLE **2**

Factoring $x^2 + bx + c$, $c > 0$

Factor each trinomial.

(a) $x^2 - 7x + 10$

(b) $x^2 + 8x + 6$

Solution

(a) $x^2 - 7x + 10$

Choose two numbers whose product is $+10$ and whose sum is -7. Note that both numbers must be negative.

Product = 10	Sum	
$-1, -10$	-11	
$-2, -5$	-7	Correct pair

Thus $x^2 - 7x + 10 = (x - 2)(x - 5)$.

Check: $(x - 2)(x - 5) = x^2 - 5x - 2x + 10 = x^2 - 7x + 10$

(b) $x^2 + 8x + 6$

Choose two numbers whose product is $+6$ and whose sum is $+8$. Both numbers must be positive.

Product = 6	Sum
$1, 6$	7
$2, 3$	5

Because no pair of numbers meets the requirements, the polynomial is prime.

Note: As shown in part (b) of Example 2, not every quadratic trinomial can be factored.

EXAMPLE **3**

Factoring $x^2 + bx + c$, $c < 0$

Factor the given trinomial.

(a) $x^2 + 2x - 15$

(b) $x^2 - 4x - 12$

Solution

(a) We need to find numbers m and n such that their product is -15 and their sum is 2.

$Product = -15$	Sum
$1 \cdot (-15) = -15$	$1 + (-15) = -14$
$-1 \cdot 15 = -15$	$-1 + 15 = 14$
$3 \cdot (-5) = -15$	$3 + (-5) = -2$
$-3 \cdot 5 = -15$	$-3 + 5 = 2$

From the table, we see that $m = -3$ and $n = 5$ satisfy the requirement. Thus

$$x^2 + 2x - 15 = (x - 3)(x + 5)$$

(Because $m + n$ is positive, the number with the larger absolute value must be positive.)

(b) We know that $m \cdot n = -12$ and $m + n = -4$.

$Product = -12$	Sum
$1 \cdot (-12) = -12$	$1 + (-12) = -11$
$2 \cdot (-6) = -12$	$2 + (-6) = -4$
$3 \cdot (-4) = -12$	$3 + (-4) = -1$
$-1 \cdot 12 = -12$	$-1 + 12 = 11$
$-2 \cdot 6 = -12$	$-2 + 6 = 4$
$-3 \cdot 4 = -12$	$-3 + 4 = 1$

The numbers $m = 2$ and $n = -6$ satisfy the requirement. Thus

$$x^2 - 4x - 12 = (x + 2)(x - 6)$$

(Because $m + n$ is negative, we could have listed only pairs of numbers in which the number with the larger absolute value is negative.)

From Example 3, we can conclude that to factor $x^2 + bx + c$, where c is negative, we seek numbers m and n such that $(x + m)(x + n) = x^2 + bx + c$. Because $c < 0$, the numbers m and n must have opposite signs. If b is positive, consider only those pairs for which the number with the larger absolute value is positive. If b is negative, the number with the larger absolute value must be negative.

EXAMPLE **4**

Factoring $x^2 + bx + c$, $c < 0$

Factor.

(a) $x^2 + x - 12$

(b) $y^2 - 10y - 24$

Solution

(a) $x^2 + x - 12$

Because the product of the numbers must be -12, the numbers m and n must have opposite signs. Because the sum of the numbers must be positive, we consider only the pairs in which the positive number has the larger absolute value.

Product = -12	*Sum*
$-1, 12$	11
$-2, 6$	4
$-3, 4$	1

The requirements are met by $m = -3$ and $n = 4$.

$$x^2 + x - 12 = (x - 3)(x + 4)$$

Check: $(x - 3)(x + 4) = x^2 + 4x - 3x - 12 = x^2 + x - 12$

(b) $y^2 - 10y - 24$

The numbers m and n must have opposite signs. We need consider only those pairs in which the negative number has the larger absolute value.

Product = -24	*Sum*
$1, -24$	-23
$2, -12$	-10
$3, -8$	-5
$4, -6$	-2

The numbers $m = 2$ and $n = -12$ meet the requirements.

$$y^2 - 10y - 24 = (y + 2)(y - 12)$$

Check: $(y + 2)(y - 12) = y^2 - 12y + 2y - 24 = y^2 - 10y - 24$

LEARNING TIP

The order in which factors are written is not important. Both $(x + 4)(x - 3)$ and $(x - 3)(x + 4)$ are correct factorizations of $x^2 + x - 12$.

Think About It

One indication that a quadratic trinomial $x^2 + bx + c$ may be prime is that b is large relative to c (for example, $x^2 + 7x + 1$). Explain why this is a reasonable conjecture.

Extending the Methods

The methods developed for factoring quadratic trinomials can be extended to other kinds of expressions. In the examples that follow, we leave verifications to you.

EXAMPLE **5**

Factoring Other Kinds of Expressions

Factor.

(a) $6w - 40 + w^2$ (b) $x^2 - 12xy + 32y^2$

(c) $t^4 + 8t^2 + 15$

Solution

(a) First write the polynomial in descending order.

$$6w - 40 + w^2 = w^2 + 6w - 40$$

Because $mn = -40$ and $m + n = 6$, we choose 10 and -4.

$$6w - 40 + w^2 = w^2 + 6w - 40 = (w + 10)(w - 4)$$

(b) $x^2 - 12xy + 32y^2$

The factorization takes the form $(x + my)(x + ny)$, but the method is the same. Because $mn = 32$ and $m + n = -12$, we choose -4 and -8.

$$x^2 - 12xy + 32y^2 = (x - 4y)(x - 8y)$$

(c) $t^4 + 8t^2 + 15$

Because the factorization takes the form $(t^2 + m)(t^2 + n)$, where $mn = 15$ and $m + n = 8$, we choose 5 and 3.

$$t^4 + 8t^2 + 15 = (t^2 + 5)(t^2 + 3)$$

EXAMPLE 6

Factoring Expressions with a Common Factor

Factor completely.

(a) $-x^2 + 3x - 2$ (b) $5x^3 + 25x^2 + 30x$

Solution

(a) $-x^2 + 3x - 2$

Factor out -1 and then factor the trinomial.

$$\begin{aligned} -x^2 + 3x - 2 &= -1(x^2 - 3x + 2) \\ &= -1(x - 2)(x - 1) \qquad \text{Choose } -2 \text{ and } -1 \text{ because the product is 2} \\ &= -(x - 2)(x - 1) \qquad \text{and the sum is } -3. \end{aligned}$$

(b) $5x^3 + 25x^2 + 30x$

Factor out the GCF $5x$ and then factor the trinomial.

$$\begin{aligned} 5x^3 + 25x^2 + 30x &= 5x(x^2 + 5x + 6) \\ &= 5x(x + 3)(x + 2) \qquad \text{Choose 3 and 2 because the product is 6} \\ &\qquad\qquad\qquad\qquad\qquad\quad \text{and the sum is 5.} \end{aligned}$$

Quick Reference 7.3

Quadratic Trinomials

- A **quadratic polynomial** in one variable is a polynomial that can be written in the form $ax^2 + bx + c$, where $a \neq 0$.

- For a quadratic polynomial, the term ax^2 is called the **quadratic term,** the term bx is called the **linear term,** and the term c is called the **constant term.** The coefficient a of the quadratic term is called the **leading coefficient.**

Factoring Patterns

- To factor $x^2 + bx + c$, we seek numbers m and n such that

$$(x + m)(x + n) = x^2 + bx + c$$

This means that $m \cdot n = c$ and $m + n = b$.

- If c is positive, then m and n have the same sign.

  1. If b is positive, m and n are both positive.

  2. If b is negative, m and n are both negative.

- If c is negative, then m and n have opposite signs.

  1. If b is positive, the number m or n with the larger absolute value is positive.

  2. If b is negative, the number m or n with the larger absolute value is negative.

Extending the Methods
- The factoring methods discussed in this section can be extended to other kinds of expressions.
- In this section, we saw examples of factoring expressions
  (a) whose terms needed to be rearranged into descending order.
  (b) that contained more than one variable.
  (c) whose degree was greater than 2.
  (d) that contained a common factor.
- In all cases, be sure to verify a factorization by multiplying the factors. The result should be the original expression.

Speaking the Language 7.3

1. A polynomial that has the form $ax^2 + bx + c$, $a \neq 0$, is called a(n) _____ polynomial.
2. When we factor $x^2 + bx + c = (x + m)(x + n)$, if c is positive, then m and n have _____ signs, whereas if c is negative, then m and n have _____ signs.
3. In $ax^2 + bx + c$, bx is called the _____ and a is called the _____.
4. We can verify a factorization by _____ the factors.

Exercises 7.3

Concepts and Skills

1. Suppose that we factor the trinomial $x^2 + bx + c$ as $(x + m)(x + n)$. If b and c are positive, what do we know about m and n?

2. Suppose that we factor the trinomial $x^2 + bx + c$ as $(x + m)(x + n)$. If b is negative and c is positive, what do we know about m and n?

In Exercises 3–8, fill in the blanks to complete the indicated factorization.

3. $x^2 + 4x + 3 = (x \quad 3)(x \quad 1)$
4. $x^2 - 3x + 2 = (x \quad 1)(x \quad 2)$
5. $y^2 + 2y - 3 = (y + \quad)(y - \quad)$
6. $y^2 - y - 2 = (y + \quad)(y - \quad)$
7. $a^2 + 5a + 6 = (a \quad)(a \quad)$
8. $a^2 - 4a - 5 = (a \quad)(a \quad)$

In Exercises 9–16, factor the trinomial.

9. $x^2 + 5x + 6$
10. $x^2 + 6x + 8$
11. $x^2 - 8x + 12$
12. $x^2 - 3x + 2$
13. $y^2 - 9y + 20$
14. $t^2 - 11t + 18$
15. $x^2 + 7x + 12$
16. $t^2 + 7t + 10$

17. Suppose that we factor the trinomial $x^2 + bx + c$ as $(x + m)(x + n)$. If b and c are negative, what do we know about m and n?

18. Suppose that we factor the trinomial $x^2 + bx + c$ as $(x + m)(x + n)$. If b is positive and c is negative, what do we know about m and n?

In Exercises 19–26, factor the trinomial.

19. $x^2 + x - 2$
20. $a^2 + a - 12$
21. $y^2 - 3y - 10$
22. $x^2 - 6x - 7$
23. $y^2 + 5y - 14$
24. $t^2 + 6t - 27$
25. $x^2 - 8x - 9$
26. $y^2 - 5y - 36$

In Exercises 27–32, rewrite the expression in descending order and factor completely.

27. $9x + 8 + x^2$
28. $7 + y^2 - 8y$
29. $-45 + z^2 - 4z$
30. $x - 42 + x^2$
31. $9x + 14 + x^2$
32. $z^2 + 10 - 7z$

33. If we wish to factor $6 - 5x - x^2$, what is the disadvantage of rewriting the expression in descending order?

34. Which of the following forms will the factorization $x^4 + 6x^2 + 8$ take? Why?

(i) $(x^3 + m)(x + n)$

(ii) $(x^2 + m)(x^2 + n)$

(iii) $(x^4 + m)(1 + n)$

In Exercises 35–44, factor completely.

35. $x^2 + 5xy + 6y^2$ **36.** $x^2 - 9xy + 14y^2$

37. $x^2 + 5xy - 14y^2$ **38.** $y^2 - 2yz - 15z^2$

39. $2 - 3xy + x^2y^2$ **40.** $a^2b^2 - 5ab - 24$

41. $x^4 + 9x^2 + 14$ **42.** $y^4 - 2y^2 - 48$

43. $x^6 - 7x^3 + 10$ **44.** $x^8 + 4x^4 - 12$

In Exercises 45–50, factor by first factoring out -1.

45. $-x^2 - 4x + 5$ **46.** $-x^2 + 4x - 3$

47. $-t^2 + 4t + 32$ **48.** $-y^2 + 2y + 48$

49. $-x^2 + 12x - 35$ **50.** $-x^2 - 8x + 33$

In Exercises 51–58, factor completely.

51. $5x^2 + 35x + 60$ **52.** $3x^2 - 9x + 6$

53. $4y^9 + 8y^8 + 4y^7$ **54.** $4t^5 + 8t^4 - 60t^3$

55. $x^4 + 3x^3 + x^2$ **56.** $3x^3 - 12x^2 + 6x$

57. $-3x^4 + 15x^3 + 18x^2$ **58.** $y^5 + 11y^4 + 28y^3$

In Exercises 59–94, factor completely.

59. $x^2 - 18x + 80$ **60.** $w^2 - 18w + 77$

61. $28 + 11x + x^2$ **62.** $27 + 12x + x^2$

63. $w^2 + 6w - 40$ **64.** $a^2 + 3a - 4$

65. $c^2 - 2c - 35$ **66.** $a^2 - a - 12$

67. $x^2 - x + 6$ **68.** $y^2 - 2y + 3$

69. $-2x^2 + 2x + 84$ **70.** $5a^2 + 35a + 50$

71. $x^4 - 8x^3 - 48x^2$ **72.** $2x^3 + 30x^2 + 100x$

73. $x^2 - 13xy + 12y^2$ **74.** $y^2z^2 - 6yz + 5$

75. $x^3 - 5x^2 + 4x$ **76.** $21y^2 - 4y^3 - y^4$

77. $-3x^2 + 12x - 36$ **78.** $4x^2 - 36x + 88$

79. $4 - 5x + x^2$ **80.** $18 - 9x + x^2$

81. $34 - 15w - w^2$ **82.** $30 - 7z - z^2$

83. $x^2 + 3xy - 4y^2$ **84.** $a^2b^2 + ab - 72$

85. $x^7 + 2x^6 - 80x^5$ **86.** $2x^4 + 12x^3 - 110x^2$

87. $x^3y + 7x^2y - 10xy$

88. $a^3b^2 - 13a^2b^2 - 40ab^2$

89. $a^2 + 9a - 36$ **90.** $x^2 + 12x - 28$

91. $a^2 - ax - 12x^2$ **92.** $y^2z^2 - 3yz - 4$

93. $x^2 - 2x + 8$ **94.** $x^2 + 4x + 21$

Real-Life Applications

95. The revenue from selling x ceiling fans is given by $2x^2 + 5x + 8$, and the cost is $x^2 - 4x$. Write an expression in factored form for the profit from selling x ceiling fans.

96. Suppose that the area of a rectangular coffee table is represented by $n^2 + 4n + 3$, where n is an integer. Write the expression in factored form. If the factors represent the dimensions of the coffee table, how would you describe the dimensions?

97. A person standing on a bridge over a river throws a rock into the air. The rock's height h above the river at any time t is given by $h(t) = -16t^2 + 48t + 64$. Write the expression in factored form with a negative numerical factor.

98. Suppose that x represents the amount spent on TV advertising. The forecasted revenue r from the sale of breakfast cereal is $r(x) = 3x^2 - 27x + 42$. Write the expression in factored form.

Modeling with Real Data

99. The table shows the percentage of Americans who have had health care coverage through their jobs. The percentage for 2002 is projected.

Year	Percentage
1990	77.7%
1995	73.9%
2002	70.4%

(Source: National Retail Federation.)

With t representing the number of years after 1990, the polynomial $-0.6t + 77.4$ models the percentage of those with health coverage through their jobs.

(a) Write the polynomial in factored form by factoring out 0.6.

(b) Use the original model expression and its factored form to predict the percentage of Americans who will have health coverage through their jobs in 2005.

100. Refer to the model in Exercise 99.

    (a) Write the polynomial in factored form by factoring out -0.6.

    (b) Suppose that you use the factored form in part (a) to estimate the percentage in 2005. How will the result compare to the result in part (b) of Exercise 99?

Challenge

In Exercises 101–104, factor completely.

101. $x^2 - 4x - 21 - wx - 3w$

102. $x^2 + \dfrac{1}{3}x - \dfrac{2}{9}$

103. $\dfrac{1}{3}x^2 - \dfrac{2}{3}x - 1$

104. $x^{2n} - 10x^n + 21$

In Exercises 105 and 106, determine all integer values of k such that the trinomial can be factored.

105. $x^2 + kx + 6$

106. $x^2 + kx - 8$

7.4 | Factoring Trinomials of the Form $ax^2 + bx + c$

Trial and Check • Applying the Grouping Method

In Section 7.3 we considered methods for factoring quadratic polynomials of the form $ax^2 + bx + c$, where $a = 1$. When the leading coefficient is not 1, those methods must be modified.

Trial and Check

One approach to factoring $ax^2 + bx + c$ is simply to try certain factors until the right combination is found or until we conclude that the polynomial is prime. This method, called the **trial-and-check method,** works best when the constant term and the leading coefficient have few factors.

Exploring the Concept

The Trial-and-Check Method

To factor $3x^2 + 5x + 2$, we begin by noting that $3x^2$ is the product of $3x$ and x. Thus the binomials must be in the following form.

    $(3x + \underline{\hspace{1cm}})(x + \underline{\hspace{1cm}})$

The product of the constant terms of the binomials must be 2. There are two possibilities: -1 and -2 or 1 and 2. Because all terms have positive coefficients, we consider only the positive factors. Therefore, there are only two possible factorizations.

    $(3x + 1)(x + 2)$ or $(3x + 2)(x + 1)$

We multiply each to determine the correct factorization.

    $(3x + 1)(x + 2) = 3x^2 + 7x + 2$ 7x is not the desired middle term.

    $(3x + 2)(x + 1) = 3x^2 + 5x + 2$ Correct

The correct factorization of $3x^2 + 5x + 2$ is $(3x + 2)(x + 1)$.

Note: The trial factors are selected in order to obtain the correct first and last terms of $ax^2 + bx + c$. Thus our main purpose in checking is to determine whether we obtain the correct middle term.

EXAMPLE 1

Using the Trial-and-Check Method

Factor the polynomial $6y^2 - y - 5$.

Solution

We can factor $6y^2$ as $3y \cdot 2y$ or as $6y \cdot y$. We begin by writing the following factors.

$$(3y + \text{\_\_\_})(2y + \text{\_\_\_}) \quad \text{or} \quad (6y + \text{\_\_\_})(y + \text{\_\_\_})$$

To obtain the last term -5, there are two possible pairs of factors: 1 and -5 or -1 and 5.

Now we list the possible factors and check the middle term for each combination.

Possible Factors	Middle Term
$(3y + 1)(2y - 5)$	$-15y + 2y = -13y$
$(3y - 5)(2y + 1)$	$3y - 10y = -7y$
$(3y - 1)(2y + 5)$	$15y - 2y = 13y$
$(3y + 5)(2y - 1)$	$-3y + 10y = 7y$
$(6y + 1)(y - 5)$	$-30y + y = -29y$
$(6y - 5)(y + 1)$	$6y - 5y = y$
$(6y - 1)(y + 5)$	$30y - y = 29y$
$(6y + 5)(y - 1)$	$-6y + 5y = -y$ Correct middle term

Because the last combination gives the correct middle term, the correct factorization is $6y^2 - y - 5 = (6y + 5)(y - 1)$.

With practice, you can check many possibilities mentally and arrive at the correct combination quickly and efficiently.

EXAMPLE 2

Using the Trial-and-Check Method

Factor completely.

(a) $5x^2 - 12x + 4$ (b) $2x^2 + 5xy - 7y^2$

(c) $-12a^2 + 20a + 8$

Solution

(a) $5x^2 - 12x + 4$

Factors of the first term are $5x$ and x. The middle term has a negative coefficient, so we try the negative factors of 4: -1 and -4 or -2 and -2.

Possible Factors	Middle Term
$(5x - 1)(x - 4)$	$-21x$
$(5x - 4)(x - 1)$	$-9x$
$(5x - 2)(x - 2)$	$-12x$ Correct middle term

The correct factorization is $5x^2 - 12x + 4 = (5x - 2)(x - 2)$.

Think About It

If a trinomial has no GCF other than 1, no factor of the trinomial can have a common factor. Explain why this is so.

(b) $2x^2 + 5xy - 7y^2$

Factors of $2x^2$ are $2x$ and x. Factors of $-7y^2$ and $7y$ and $-y$ or $-7y$ and y. Check the possible factors.

Possible Factors	Middle Term	
$(2x - y)(x + 7y)$	$13xy$	
$(2x + 7y)(x - y)$	$5xy$	Correct middle term

$\vdots$

The correct factorization is $2x^2 + 5xy - 7y^2 = (2x + 7y)(x - y)$.

(c) $-12a^2 + 20a + 8$

First factor out the GCF -4.

$$-12a^2 + 20a + 8 = -4(3a^2 - 5a - 2)$$

To factor $3a^2 - 5a - 2$, note that factors of $3a^2$ are $3a$ and a, and factors of -2 are 1 and -2 or -1 and 2. Check the possible factors.

LEARNING TIP

If the leading coefficient is negative, factoring out -1 or a negative common factor often makes factoring easier.

Possible Factors	Middle Term	
$(3a - 2)(a + 1)$	a	
$(3a + 1)(a - 2)$	$-5a$	Correct middle term

$\vdots$

The correct factorization is $-12a^2 + 20a + 8 = -4(3a + 1)(a - 2)$.

Applying the Grouping Method

A disadvantage of the trial-and-check method is that the number of possible combinations can sometimes be quite large. An alternative method is to rewrite the polynomial so that the grouping method can be applied. The following summarizes the procedure.

Factoring $ax^2 + bx + c$ with the Grouping Method

1. Evaluate the product ac.
2. Determine factors m and n of ac such that $m + n = b$.
3. Write bx as $mx + nx$.
4. Factor by grouping.

EXAMPLE 3

Factoring Trinomials with the Grouping Method

Factor $2x^2 - 7x - 4$.

Solution

The product of the leading coefficient ($a = 2$) and the constant ($c = -4$) is $2(-4)$ or -8. Now list the pairs of factors of -8 and their sums.

Factors of -8	Sum
1, -8	-7
-1, 8	7
2, -4	-2
-2, 4	2

Choose the pair whose sum is -7: 1 and -8. Use 1 and -8 to rewrite the middle term $-7x$ as $1x - 8x$.

$$2x^2 - 7x - 4 = 2x^2 + 1x - 8x - 4 \qquad \text{Write } -7x \text{ as } x - 8x.$$
$$= x(2x + 1) - 4(2x + 1) \qquad \text{Use the grouping method to factor.}$$
$$= (2x + 1)(x - 4)$$

EXAMPLE 4

Using the Grouping Method

Factor completely.

(a) $6x^2 + x - 15$ (b) $4t^2 + 27t - 7$ (c) $3x^2 - 5x + 3$

Solution

(a) $6x^2 + x - 15$

The product ac is $6(-15)$ or -90. List pairs of factors of -90 and determine the sum of each pair.

Factors of -90	Sums
-1, 90 and 1, -90	89 and -89
-2, 45 and 2, -45	43 and -43
-3, 30 and 3, -30	27 and -27
-5, 18 and 5, -18	13 and -13
-6, 15 and 6, -15	9 and -9
-9, 10 and 9, -10	1 and -1

Choose the pair whose sum is 1: -9 and 10.

$$6x^2 + x - 15 = 6x^2 - 9x + 10x - 15 \qquad \text{Write } x = -9x + 10x.$$
$$= 3x(2x - 3) + 5(2x - 3) \qquad \text{Factor each group.}$$
$$= (2x - 3)(3x + 5) \qquad \text{Distributive Property}$$

(b) $4t^2 + 27t - 7$

The product ac is $4(-7)$ or -28. List pairs of factors of -28 and their sums.

Factors of -28	Sums	
-1, 28 and 1, -28	27 and -27	27 is the coefficient of $27t$.
$\vdots$		We do not need to check other factors.

Use the numbers -1 and 28 to rewrite the polynomial.

$$4t^2 + 27t - 7 = 4t^2 + 28t - t - 7 \qquad \text{Write } 27t \text{ as } 28t - t.$$
$$= 4t(t + 7) - 1(t + 7)$$
$$= (t + 7)(4t - 1)$$

(c) $3x^2 - 5x + 3$

The product ac is 9.

Factors of 9	*Sums*
1, 9 and -1, -9	10 and -10
3, 3 and -3, -3	6 and -6

None of the pairs has a sum of -5. The polynomial is prime.

Quick Reference 7.4

Trial and Check
- To use the **trial-and-check method** for factoring $ax^2 + bx + c$, we list all possible combinations of factors whose products have the correct first and last terms. Then we select the combination, if any, whose product has the correct middle term.

- The trial-and-check method works best when the number of possible combinations is not large.

Applying the Grouping Method
- An alternative approach is to rewrite $ax^2 + bx + c$ in the form $ax^2 + mx + nx + c$ and then apply the grouping method, if possible.

- To use this method for factoring $ax^2 + bx + c$, use the following steps.

  1. Evaluate the product ac.

  2. Determine factors m and n of ac such that $m + n = b$.

  3. Write bx as $mx + nx$.

  4. Factor by grouping.

Speaking the Language 7.4

1. In the trial-and-check method for factoring, after we list all possible factors, the term that we check is the _____ term.

2. When we list possible factorizations of a trinomial, we do not need to list any binomial factors that have a(n) _____ .

3. When the number of possible combinations of factors is large, an alternative is to separate the _____ term into two terms so that the _____ method can be applied.

4. The trinomial $x^2 + x + 1$ is an example of a(n) _____ polynomial.

Exercises 7.4

Concepts and Skills

1. From the following list, identify the trinomial for which the trial-and-check method would be a good choice for factoring. Then explain why using the trial-and-check method to factor the other trinomial might be tedious.

   (i) $12x^2 - 11x - 36$

   (ii) $11x^2 - 26x - 37$

2. Suppose that you use the trial-and-check method to factor $2x^2 + 3x + 5$. What conclusion can you draw? Why?

In Exercises 3–8, fill in the blanks to complete the indicated factoring.

3. $3x^2 + 11x + 6 = (3x + \underline{})(x + \underline{})$

4. $3x^2 + 10x + 8 = (3x + \underline{})(x + \underline{})$

5. $2x^2 - 5x - 12 = (2x \underline{})(x \underline{})$

6. $9x^2 - 6x - 8 = (3x \underline{})(3x \underline{})$

7. $6x^2 - 23x + 15 = (x - 3)(\underline{})$

8. $4x^2 - 17x + 15 = (\underline{})(x - 3)$

In Exercises 9–20, factor the trinomial.

9. $2x^2 + 3x + 1$
10. $3x^2 - 10x + 3$
11. $12x^2 + x - 6$
12. $2x^2 + 11x - 6$
13. $12x^2 - 7x - 5$
14. $6a^2 - a - 7$
15. $8y^2 - 45y + 25$
16. $6t^2 - 23t + 15$
17. $4x^2 + 16x + 15$
18. $4x^2 + 11x + 6$
19. $11x^2 - x - 10$
20. $21y^2 + y - 2$

In Exercises 21–26, factor the trinomial.

21. $6x^2 + 11xy + 3y^2$
22. $5y^2 - 6yz + z^2$
23. $14a^2 + 15ab - 9b^2$
24. $14a^2 + 11ay - 15y^2$
25. $7x^2 - 33xz - 10z^2$
26. $6a^2 - 7ab - 3b^2$

27. Describe two factoring techniques that are needed to factor $2x^2 - 4x - 6$.

28. The following are two approaches to factoring the expression $2 + x - x^2$.

   (i) $2 + x - x^2 = -1(x^2 - x - 2)$
   $$= -1(x - 2)(x + 1)$$

   (ii) $2 + x - x^2 = (2 - x)(1 + x)$

   Comment on the two methods.

In Exercises 29–34, write the expression in descending order and then factor.

29. $7x + 5 + 2x^2$
30. $2 + 20x^2 - 13x$
31. $16x^2 + 9 - 24x$
32. $6y^2 - 5 + 13y$
33. $2 + 5y^2 - 11y$
34. $2x - 3 + 8x^2$

In Exercises 35–40,

(a) write the expression in ascending order and factor completely.

(b) write the expression in descending order and factor completely. (Begin by factoring out -1.)

35. $-2x^2 - 3x + 14$
36. $-3x^2 - 5x + 2$
37. $42 - x^2 + x$
38. $2x - x^2 + 15$
39. $5x + 36 - x^2$
40. $6x + 16 - x^2$

In Exercises 41–46, factor completely.

41. $10x^2 - 35x + 30$
42. $6x^2 + 3x - 84$
43. $-84x^2 - 7x + 42$
44. $-8t^3 + 20t^2 - 8t$
45. $10t^2 - 10t - 20$
46. $3y^2 + 39y + 36$

47. For the trinomial $ax^2 + bx + c$, suppose that ac is a prime number. Does this mean that the trinomial cannot be factored? Give an example to support your argument.

48. Because $13x = 12x + 1x$, we can write the following.
$$6x^2 + 13x + 5 = 6x^2 + 12x + 1x + 5$$
$$= 6x(x + 2) + (x + 5)$$

   The resulting expression does not have a common binomial factor, so we are unable to factor further. Should we conclude that the expression is a prime polynomial? Why?

In Exercises 49–96, factor completely.

49. $3x^2 + 5x + 2$
50. $2t^2 + 7t + 3$
51. $4x^2 - 11x + 6$
52. $12x^2 - 25x + 12$
53. $2x^2 + x - 3$
54. $3x^2 - 8x - 3$
55. $6x^2 - 5x + 6$
56. $3x^2 - 2x + 8$
57. $6x^3 + 4x^2 - 16x$
58. $24y^3 + 28y^2 - 12y$
59. $5t^2 - 16t + 11$
60. $3b^2 - 13b + 10$
61. $10b^2 - 7b - 12$
62. $16t^2 - 10t - 21$
63. $5x^2 + 8x + 3$
64. $6y^2 + 23y + 20$
65. $3x^2 + 5x - 2$
66. $6b^2 + b - 12$
67. $12x^2 - 20xy + 3y^2$
68. $49x^2 - 21xy + 2y^2$

69. $4a^2 + 4a - 15$

70. $3y^2 + y - 2$

71. $10y^2 - 3y - 7$

72. $18x^2 - 7x - 30$

73. $2w^6 - 3w^5 - 20w^4$

74. $10x^5 + 19x^4 + 6x^3$

75. $4x^2 + x - 5$

76. $3b^2 + 4b - 7$

77. $4x^2 + x + 3$

78. $2x^2 - x + 10$

79. $7z^2 - 5z - 18$

80. $8x^2 - 2x - 21$

81. $28t^2 + 13t - 6$

82. $14y^2 + y - 3$

83. $2x^2 - 17x + 21$

84. $3a^2 + 11a + 6$

85. $3a^2 + 8aw - 3w^2$

86. $2x^2 + 3xy - 2y^2$

87. $5x^2 + 21x + 18$

88. $7y^2 + 79y + 22$

89. $7x^2 + 4x - 3$

90. $120x^2 + 26x - 3$

91. $3a^2 - ab - 2b^2$

92. $3y^2 - 2yz - z^2$

93. $8x^2 - 2x - 45$

94. $10x^2 - 21x - 10$

95. $2a^2 - 19a + 24$

96. $3x^2 - 22x + 24$

Real-Life Applications

97. The revenue from selling x rolls of film is given by $6x + 5$. The revenue from processing x rolls of film is given by $2x^2 + 5x$. Write an expression in completely factored form for the total revenue from selling and processing x rolls of film.

98. A circular flower garden is surrounded by a grass border of uniform width. The radius of the circular garden is r. The radius of the garden including the border of grass is 1 less than twice the radius of the flowered area. Write an expression for the area of grass, simplify the expression, and then write the expression in completely factored form. Explain how you could have factored the expression before simplifying it.

99. At a software convention, a company displays its logo in the shape of a parallelogram whose area is given by $2x^2 + 9x + 4$. If the height of the logo is $x + 4$, what expression represents the length of the base?

100. The expression $3x^2 + 25x + 8$ is used to forecast the number of votes a candidate will receive, where x measures the level of expenditures for advertising. Write the expression in factored form.

Modeling with Real Data

101. The graph shows the percent of those 18 and over who voted in presidential elections in the years 1968, 1984, and 2000.

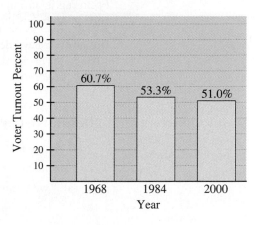

With x representing the number of years since 1960, the percent P can be modeled by the following function.

$$P(x) = 0.01x^2 - 0.78x + 66.31$$

(a) Write the trinomial with a factor of 0.01.

(b) Given that 6631 can be written only as $1 \cdot 6631$ and $19 \cdot 349$, how do you know that the expression in part (a) cannot be factored further?

(c) Use the model to predict the voter turnout in 2012.

102. Refer to the model in Exercise 101.

(a) Produce the graph of function P and trace it to estimate the presidential election year when the voter turnout was about 55%.

(b) In what future presidential year does the model predict that voter turnout will increase to 55% again?

Challenge

In Exercises 103–108, factor completely.

103. $2x^2 + \dfrac{3}{5}x - \dfrac{2}{25}$

104. $2(a + 3b)^2 - 5(a + 3b) - 3$

105. $2x^{10} - 7x^5 - 4$

106. $2x^{2n} - 3x^n - 2$

107. $36x^4 - 25x^2 + 4$

108. $7a^2(x + y) - 14a(x + y) + 7(x + y)$

7.5 | General Strategy

Factoring is an important skill. We often find it necessary to factor in the process of solving problems that require simplifying expressions or solving equations.

Because we have developed a number of factoring techniques, it is helpful to have a strategy for deciding which factoring method to use. The following summary provides an organized sequence of questions that you can ask to help you decide.

LEARNING TIP

The number of terms is a clue to determining a method of factoring.

General Strategy for Factoring an Expression

1. Is there a factor common to all terms? If so, factor out the GCF.

2. How many terms does the expression have?

   (a) Two terms:

      (i) Is the expression a difference of two squares? If so, use the special factoring form for a difference of two squares.

      $$A^2 - B^2 = (A + B)(A - B)$$

      (ii) Is the expression a sum or difference of two cubes? If so, use the special factoring form for a sum or difference of two cubes.

      $$A^3 + B^3 = (A + B)(A^2 - AB + B^2)$$
      $$A^3 - B^3 = (A - B)(A^2 + AB + B^2)$$

   (b) Three terms:

      (i) Are two of the terms perfect squares? If so, try the special factoring forms for a perfect square trinomial.

      $$A^2 + 2AB + B^2 = (A + B)^2$$
      $$A^2 - 2AB + B^2 = (A - B)^2$$

      Be sure to check that the middle term is in the form $2AB$.

      (ii) Otherwise, try the trial-and-check method or the grouping method.

   (c) Four or more terms: Try the grouping method.

Note: This summary replaces the Quick Reference for this section.

Also, keep in mind the following points.

1. Usually we want to factor completely. Always check factors to determine whether they can be factored further.

2. It may be necessary to employ more than one technique.

3. Prime polynomials cannot be factored.

In the following examples, we illustrate the use of this general strategy in deciding how to factor an expression.

EXAMPLE **1**

Think About It

Try factoring $x^3 + x + y + y^3$ by grouping the first and last pairs of terms. Does the method work? Can you find another grouping that works?

Using the General Strategy

Factor completely.

(a) $125 + 27y^3$

(b) $6x^2 - x - 2$

(c) $2x^3 - 10x^2 + 3x - 15$

(d) $5a^2 - 125$

(e) $a^3b + 3a^2b + 5ab$

Solution

(a) There are two perfect cube terms. Use the special factoring form for the sum or difference of two cubes.

$$A^3 + B^3 = (A + B)(A^2 - AB + B^2)$$
$$125 + 27y^3 = 5^3 + (3y)^3$$
$$= (5 + 3y)(5^2 - 5(3y) + (3y)^2)$$
$$= (5 + 3y)(25 - 15y + 9y^2)$$

(b) There are three terms, but the polynomial is not a perfect square trinomial. Use the trial-and-check method or grouping to factor.

$$6x^2 - x - 2 = (3x - 2)(2x + 1)$$

(c) Because there are four terms, try the grouping method.

$$2x^3 - 10x^2 + 3x - 15 = 2x^2(x - 5) + 3(x - 5)$$
$$= (x - 5)(2x^2 + 3)$$

(d) $5a^2 - 125 = 5(a^2 - 25)$ First factor out the GCF.

$\qquad\qquad\quad = 5(a + 5)(a - 5)$ $a^2 - 25$ is a difference of two squares.

(e) $a^3b + 3a^2b + 5ab = ab(a^2 + 3a + 5)$ The GCF is ab.

The last factor has three terms, but it cannot be factored.

EXAMPLE **2**

Using the General Strategy

Factor completely.

(a) $t^4 - 1$ (b) $-18y^2 + 12y - 2$

(c) $2a^3b + 9a^2b^2 + 10ab^3$ (d) $24x^2 - 41xy + 12y^2$

Solution

(a) Factor as a difference of two squares.

$$t^4 - 1 = (t^2)^2 - 1^2 = (t^2 + 1)(t^2 - 1)$$

The first factor is prime. However, the second factor is a difference of two squares.

$$t^4 - 1 = (t^2 + 1)(t + 1)(t - 1)$$

(b) $-18y^2 + 12y - 2 = -2(9y^2 - 6y + 1)$ The GCF is -2.

$\qquad\qquad\qquad = -2(3y - 1)^2$ Perfect square trinomial

(c) $2a^3b + 9a^2b^2 + 10ab^3 = ab(2a^2 + 9ab + 10b^2)$ The GCF is ab.

$\qquad\qquad\qquad = ab(2a + 5b)(a + 2b)$ Trial and check or grouping

(d) Use trial and check or grouping to factor the trinomial.

$$24x^2 - 41xy + 12y^2 = (8x - 3y)(3x - 4y)$$

Exercises 7.5

Concepts and Skills

In Exercises 1–94, factor completely.

1. $1 - x^2$

2. $36y^2 - 25$

3. $4 + 12x + 9x^2$

4. $9w^2 - 42w + 49$

5. $4y^4 + 12y^3 + 8y^2$

6. $3x^5 - 6x^4 - 9x^3$

7. $4x(2x + 5) - (2x + 5)$

8. $3y(4y - 1) - (4y - 1)$

9. $y^2 + 14y + 45$

10. $y^2 + 17y + 72$

11. $-4y^2 + 8y + 12$

12. $-7a^2 - 7a + 14$

13. $6x^2 + 11x + 5$

14. $8t^2 + 10t + 3$

15. $12x^4 + 8x^3y - 15x^2y^2$

16. $2x^3y + 5x^2y^2 - 3xy^3$

17. $56 - 15y + y^2$

18. $60 - 17t + t^2$

19. $ax + 7a + x + 7$

20. $ab - 2a + b - 2$

21. $81 + y^2$

22. $x^2 - 10$

23. $x^5y + 2x^4y^2 + x^3$

24. $2a^2x^2 + 3ax^3 - ax^2$

25. $a^2b^2 + 10ab + 25$

26. $49 - 28mn + 4m^2n^2$

27. $b^2 + b - 30$

28. $x^2 + 7x - 8$

29. $5x^2 + x - 6$

30. $4a^2 + 15a - 4$

31. $8x^2 - 9x - 14$

32. $9x^2 - 9x - 10$

33. $a(c + 2) - 2b(c + 2)$

34. $a(y - 2) + 4(y - 2)$

35. $t^2 - 5t - 24$

36. $6 - 5x - x^2$

37. $x^2 - 9x + 16$

38. $a^2 + 10a + 15$

39. $64t^{10} - 1$

40. $x^4y^6 - 36$

41. $27 - y^3$

42. $125a^3 + 8$

43. $t^3 - 5t^2 - t + 5$

44. $y^3 + 2y^2 + y + 2$

45. $-6x^4 + 9x^3 - 3x^2$

46. $-t^4 - 4t^3 + 5t^2$

47. $5x^2 + 6x - 1$

48. $10x^2 - 17x - 3$

49. $6x^3 - 9x^2 - 15x$

50. $12a^3 - 2a^2 - 2a$

51. $-2cx + c^2 - 5c$

52. $-3bx + 6b^2 + 6b$

53. $a^2c^2 + a^2d - bc^2 - bd$

54. $ax + b^2x + 3ay + 3b^2y$

55. $x^2 - 9bx + 20b^2$

56. $4x^3 + 32$

57. $2x^4 - 54x$

58. $y^2 - 12xy + 35x^2$

59. $8b^2 + 10bc + 3c^2$

60. $6b^2 + 17ab + 5a^2$

61. $3a^3 + 6a^2 - 24a$

62. $6z^3 - 6z^2 - 12z$

63. $8y^2 + 2yz - 15z^2$

64. $5a^2 + 4ax - x^2$

65. $8x^3 - 32x$

66. $27t^3 - 27t^5$

67. $x^4 + x^3 + 2x + 2$

68. $x^7 - 5x^4 + 2x^3 - 10$

69. $x^2 - 10xy - 24y^2$

70. $x^2 - 3xy - 10y^2$

71. $x^3 + 12x^2y + 36xy^2$

72. $4a^2b^2 + 4ab^3 + b^4$

73. $a^2b^2 - 8ab^3 + 24b^4$

74. $7x^4y^2 + 70x^3y^3 - 63x^2y^4$

75. $2x^4 - 10x^3 - 48x^2$

76. $-8x^5 - 4x^4 + 12x^3$

77. $8a^4 - a$

78. $8 + 8y^3$

79. $-6x^4yz^5 - 24xy^5z^4$

80. $-15a^3b^2c^2 + 5ab^2c$

81. $a^2 + 9ab - 22b^2$

82. $x^2y^2 + 4xy - 21$

83. $12x^2 - 7xz - 49z^2$

84. $20x^2 - 3bx - 9b^2$

85. $16y^4 - 1$

86. $81 - x^8$

87. $100x^2 - 180x + 81$

88. $64 + 48y + 9y^2$

89. $12y^3z^2 - y^2z^3 - yz^4$

90. $10x^5y - 3x^4y^2 - x^3y^3$

91. $2x^3 - 6x^2 + 36x$

92. $3y^3 - 27y^2 - 24y$

93. $64a^3 + b^3$

94. $y^3 - 1000z^3$

7.6 | Solving Equations by Factoring

Quadratic Equations   •   The Graphing Method   •   An Algebraic Method

Quadratic Equations

In Chapter 2 we solved *first-degree* or *linear equations* such as $2x - 3 = 5$. In this section we consider another type of equation in one variable, such as $x^2 + x - 6 = 0$. Because the largest exponent on the variable is 2, we call this equation a **second-degree** or **quadratic equation**.

> **Definition of a Quadratic Equation**
>
> A **quadratic equation** is an equation that can be written in the standard form $ax^2 + bx + c = 0$, where a, b, and c are real numbers and $a \neq 0$.

Note: The expression $ax^2 + bx + c$ in a quadratic equation is a *quadratic polynomial*, as defined in Section 7.3.

The Graphing Method

In Chapter 2 we used the graphing method to estimate the solutions of first-degree equations. We can also use the graphing method to estimate the solutions of a quadratic equation.

EXAMPLE 1

Estimating Solutions of Quadratic Equations

Produce the graph of $y = x^2 + 2x - 15$ and use the graphing method to estimate the solutions of each equation.

(a) $x^2 + 2x - 15 = 0$ (b) $x^2 + 2x - 15 = -16$

(c) $x^2 + 2x - 15 = -20$

Solution

(a) For $x^2 + 2x - 15 = 0$, we trace to the points of the graph whose y-coordinates are 0. (See Fig. 7.1.) The two x-intercepts represent the solutions: -5 and 3.

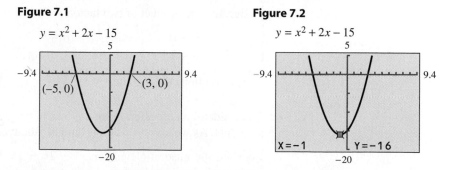

Figure 7.1

$y = x^2 + 2x - 15$

Figure 7.2

$y = x^2 + 2x - 15$

(b) For $x^2 + 2x - 15 = -16$, there is only one point whose y-coordinate is -16. (See Fig. 7.2.) The estimated solution is -1.

(c) For $x^2 + 2x - 15 = -20$, we find that there is no point of the graph with a y-coordinate of -20. Thus the equation has no solution.

Example 1 reveals two significant differences between a linear equation and a quadratic equation.

1. The graph of a linear expression is a line; the graph of a quadratic expression is called a **parabola**. (We will consider parabolas in greater detail in Chapter 10.)

2. A linear equation in one variable has exactly one solution; a quadratic equation may have zero, one, or two solutions.

An Algebraic Method

As always, the graphing method is useful for anticipating the number of solutions an equation has and for estimating what the solutions are. To determine the exact solutions of equations, however, we need algebraic methods.

From arithmetic we know that if the product of two numbers is 0, then one of the numbers must be 0. We formally state this fact as the Zero Factor Property.

The Zero Factor Property

For real numbers A and B, if $AB = 0$, then $A = 0$ or $B = 0$.

Note: As stated, the Zero Factor Property refers to the product of two numbers. The property can easily be extended to products with any number of factors.

For the Zero Factor Property to apply, two conditions must be met.

1. One side of the equation must be 0.
2. The other side of the equation must be a product.

EXAMPLE 2

Using the Zero Factor Property to Solve an Equation

Use the Zero Factor Property to solve $(x + 1)(x + 6) = 0$.

Solution

$$(x + 1)(x + 6) = 0$$

Because the product of two factors is 0, one of the factors must be zero.

$x + 1 = 0$	or $\quad x + 6 = 0$	One of the factors must be 0.
$x = -1$	or $\qquad x = -6$	Solve each first-degree equation.

The solutions are -1 and -6.

If one side of an equation is not a product, we can sometimes rewrite the expression as a product. As we know, this procedure is called *factoring the expression*.

Consider the quadratic equation $x^2 + 2x - 15 = 0$ whose solutions we estimated in Example 1. Because the left side is not a product, the Zero Factor Property does not apply. However, we can factor the left side so that the expression is a product.

$$x^2 + 2x - 15 = 0$$

$(x + 5)(x - 3) = 0$ Factor the left side. Now the Zero Factor Property applies.

$$x + 5 = 0 \qquad \text{or} \qquad x - 3 = 0 \qquad \text{Zero Factor Property}$$

$$x = -5 \qquad \text{or} \qquad x = 3 \qquad \text{Solve the two equations.}$$

Note that the two solutions agree with our estimates in Example 1.

The following summarizes the procedure for solving a quadratic equation by factoring.

Solving a Quadratic Equation by Factoring

1. Write the equation in standard form. (One side of the equation must be 0.)
2. Factor the polynomial completely.
3. Set each factor equal to zero (Zero Factor Property).
4. Solve the resulting first-degree equations.
5. Check the solutions in the original equation.

Note: In our examples, we leave step 5 to you. However, you should not infer that it is unimportant. We strongly encourage you to check all solutions.

EXAMPLE **3**

Solving Quadratic Equations by Factoring

Solve each equation.

(a) $2x^2 - x - 21 = 0$ (b) $x^2 + 8x + 16 = 0$

(c) $t^2 - 25 = 0$ (d) $3x^2 - 18x = 0$

Solution

(a) $2x^2 - x - 21 = 0$ The equation is in standard form.

      $(2x - 7)(x + 3) = 0$ Factor the trinomial.

       $2x - 7 = 0 \qquad \text{or} \qquad x + 3 = 0$ Set each factor equal to 0.

         $2x = 7 \qquad \text{or} \qquad x = -3$ Solve for x.

           $x = 3.5 \qquad \text{or} \qquad x = -3$

The two solutions are 3.5 and -3.

(b) $x^2 + 8x + 16 = 0$ The trinomial is a perfect square.

       $(x + 4)^2 = 0$ The two factors are identical.

         $x + 4 = 0$ We obtain only one first-degree equation.

           $x = -4$ Solve for x.

The only solution is -4.

LEARNING TIP

When you solve a quadratic equation by factoring, if the two factors are different, then the equation has two different solutions. If the factors are identical, then the equation has only one solution.

(c) $t^2 - 25 = 0$ Difference of two squares

    $(t + 5)(t - 5) = 0$

   $t + 5 = 0 \qquad \text{or} \qquad t - 5 = 0$ Set each factor equal to 0.

     $t = -5 \qquad \text{or} \qquad t = 5$ Solve for t.

The solutions are -5 and 5.

(d) $3x^2 - 18x = 0$

$\qquad 3x(x - 6) = 0 \qquad$ Factor out the GCF.

$\qquad 3x = 0 \qquad$ or $\qquad x - 6 = 0 \qquad$ Use the Zero Factor Property.

$\qquad x = 0 \qquad$ or $\qquad x = 6 \qquad$ Solve for x.

The solutions are 0 and 6.

EXAMPLE **4**

Solving Quadratic Equations by Factoring

Solve each equation.

(a) $x^2 + 6 = 7x$ $\qquad$ (b) $y(y + 3) = 4$

Solution

(a) $\qquad x^2 + 6 = 7x \qquad$ The equation is not in standard form.

$\qquad x^2 - 7x + 6 = 0 \qquad$ Write the equation with 0 on one side.

$\qquad (x - 6)(x - 1) = 0 \qquad$ Factor the polynomial.

$\qquad x - 6 = 0 \qquad$ or $\qquad x - 1 = 0 \qquad$ Set each factor equal to 0.

$\qquad x = 6 \qquad$ or $\qquad x = 1 \qquad$ Solve for x.

The solutions are 6 and 1.

(b) $\qquad y(y + 3) = 4 \qquad$ The equation is not in standard form.

$\qquad y^2 + 3y = 4 \qquad$ Distributive Property

$\qquad y^2 + 3y - 4 = 0 \qquad$ Write the equation with 0 on one side.

$\qquad (y + 4)(y - 1) = 0 \qquad$ Factor.

$\qquad y + 4 = 0 \qquad$ or $\qquad y - 1 = 0 \qquad$ Zero Factor Property

$\qquad y = -4 \qquad$ or $\qquad y = 1 \qquad$ Solve for y.

The solutions are -4 and 1.

The factoring method may also be applied to higher-degree equations.

EXAMPLE **5**

Think About It

Suppose that a polynomial $P(x)$ can be written as the product of n different first-degree binomials. What is the number of solutions of the equation $P(x) = 0$?

Solving Higher-Degree Equations by Factoring

Solve $x^3 - 2x^2 - 9x + 18 = 0$.

Solution

$\qquad x^3 - 2x^2 - 9x + 18 = 0$

$\qquad x^2(x - 2) - 9(x - 2) = 0 \qquad$ Use the grouping method to factor.

$\qquad (x - 2)(x^2 - 9) = 0 \qquad$ Factor out the GCF $x - 2$.

$\qquad (x - 2)(x + 3)(x - 3) = 0 \qquad$ Factor the difference of two squares.

$\qquad x - 2 = 0 \quad$ or $\quad x + 3 = 0 \qquad$ or $\quad x - 3 = 0 \qquad$ Set each factor equal to 0.

$\qquad x = 2 \quad$ or $\qquad x = -3 \quad$ or $\qquad x = 3 \qquad$ Solve for x.

The solutions are 2, -3, and 3.

To use our present algebraic method to solve $ax^2 + bx + c = 0$, we must be able to factor $ax^2 + bx + c$. For many equations, such as $x^2 + 2x + 3 = 0$, we cannot factor and so we cannot apply the Zero Factor Property. In Chapter 10 we will discuss other algebraic methods that can be used in such instances.

Quick Reference 7.6

Quadratic Equations
- A **quadratic equation** can be written in the standard form $ax^2 + bx + c = 0$, where a, b, and c are real numbers and $a \neq 0$.
- A quadratic equation is sometimes called a second-degree equation.

The Graphing Method
- The graphing method can be used to anticipate the number of solutions an equation has and to estimate what the solutions are.
- The graph of a quadratic equation is called a **parabola.**
- By exploring their graphs, we found that quadratic equations can have zero, one, or two solutions.

An Algebraic Method
- The basis for one algebraic method for solving quadratic equations (and certain other higher-degree equations) is the Zero Factor Property.

    For real numbers A and B, if $AB = 0$, then $A = 0$ or $B = 0$.

- For the Zero Factor Property to apply, two conditions must be met.
  1. One side of the equation must be 0.
  2. The other side of the equation must be a product.
- To solve a quadratic (or higher-degree) equation by factoring, follow these steps.
  1. Write the equation in standard form. (One side of the equation must be 0.)
  2. Factor the polynomial completely.
  3. Set each factor equal to zero (Zero Factor Property).
  4. Solve the resulting first-degree equations.
  5. Check the solutions in the original equation.

Speaking the Language 7.6

1. An equation of the form $ax^2 + bx + c = 0$, $a \neq 0$, is called a(n) ▢▢▢ equation.
2. The ▢▢▢ Property states that if $AB = 0$, then $A = 0$ or $B = 0$.
3. To solve a second-degree equation by factoring, we begin by writing the equation in the ▢▢▢ form.
4. In order for us to solve a second-degree equation by factoring, one side of the equation must be 0 and the other side must be a(n) ▢▢▢.

Exercises 7.6

Concepts and Skills

1. For a quadratic equation $ax^2 + bx + c = 0$, the number a cannot be 0. Why?

2. Describe two ways to use the graphing method to solve $x^2 + 3x - 6 = 12$.

3. The figure shows the graph of $y = ax^2 + bx + c$. How many solutions does the given equation appear to have?

(a) $ax^2 + bx + c = 0$

(b) $ax^2 + bx + c = -3$

(c) $ax^2 + bx + c = -4$

(d) $ax^2 + bx + c = -6$

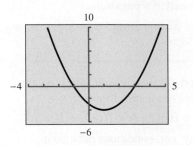

4. The figure shows the graph of $y = ax^2 + bx + c$. How many solutions does the given equation appear to have?

(a) $ax^2 + bx + c = 8$

(b) $ax^2 + bx + c = 12$

(c) $ax^2 + bx + c = -7$

(d) $ax^2 + bx + c = 9$

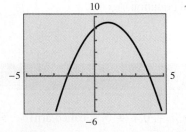

In Exercises 5–10, use the integer setting and the graphing method to estimate the solutions of the given equation.

5. $0.1x^2 - 3.5x + 25 = 0$

6. $0.05x^2 + 0.25x - 15 = 0$

7. $0.04x^2 - 0.4x + 4 = 0$

8. $-0.0625x^2 - x - 10 = 0$

9. $24 + 1.6x - 0.08x^2 = 0$

10. $-12 + 3.5x - 0.125x^2 = 0$

11. Produce the graph of $p(x) = 12 + 4x - x^2$. Then use the graph to estimate the solution(s) of each equation.

(a) $12 + 4x - x^2 = 16$

(b) $12 + 4x - x^2 = 20$

(c) $12 + 4x - x^2 = 0$

(d) $12 + 4x - x^2 = -20$

12. Produce the graph of $p(x) = x^2 - 6x - 16$. Then use the graph to estimate the solution(s) of each equation.

(a) $x^2 - 6x - 16 = -16$

(b) $x^2 - 6x - 16 = 0$

(c) $x^2 - 6x - 16 = -25$

(d) $x^2 - 6x - 16 = -26$

13. Name two conditions that must be met before the Zero Factor Property can be applied to solving an equation.

14. If we apply the Zero Factor Property to an equation of the form $(x + m)(x + n) = 0$, what is the result, and what do we do next?

In Exercises 15–28, solve the equation.

15. $(x + 5)(x - 7) = 0$ **16.** $(x + 1)(x + 4) = 0$

17. $(y - 5)(3y + 7) = 0$ **18.** $(4x - 3)(x + 8) = 0$

19. $(2x - 1)(x - 8) = 0$ **20.** $(x - 3)(3x - 5) = 0$

21. $(1 - t)(3 + t) = 0$ **22.** $(3 - 2y)(1 + 5y) = 0$

23. $5x(1 - 4x) = 0$ **24.** $-2x(x + 2) = 0$

25. $(x + 1)(x + 4)(x - 3) = 0$

26. $(x - 6)(x + 2)(x + 9) = 0$

27. $-6x(x + 9)(x - 2) = 0$

28. $x(x - 10)(x - 1) = 0$

In Exercises 29–62, solve the equation.

29. $x^2 + 6x + 8 = 0$ **30.** $x^2 - 4x - 5 = 0$

31. $x^2 + 4x - 21 = 0$ **32.** $y^2 - 6y - 40 = 0$

33. $x^2 - x = 0$ **34.** $x^2 + 3x = 0$

35. $y^2 - 64 = 0$ **36.** $t^2 - 1 = 0$

37. $x^2 + 10x + 25 = 0$ **38.** $y^2 - 14y + 49 = 0$

39. $\frac{1}{6}x^2 + x - \frac{9}{2} = 0$ **40.** $\frac{1}{4}x^2 + \frac{5}{2}x + 4 = 0$

41. $5t + t^2 - 6 = 0$ **42.** $x^2 + 6 - 7x = 0$

43. $20 + x - x^2 = 0$ **44.** $13x - 30 + x^2 = 0$

45. $5x^2 + 20x = 0$ **46.** $2x^2 - 8x = 0$

47. $4x^2 - 20x + 25 = 0$ **48.** $1 - 6x + 9x^2 = 0$

49. $49 - 4x^2 = 0$ **50.** $9x^2 - 16 = 0$

51. $6x^2 - 15x = 0$ **52.** $4x^2 + x = 0$

53. $2x^2 - 7x - 4 = 0$ **54.** $24x^2 + 7x - 5 = 0$

55. $4x^2 + 17x + 4 = 0$ **56.** $10 - x - 3x^2 = 0$

57. $2x^2 - \frac{1}{2}x - \frac{1}{4} = 0$

58. $x^2 - \frac{7}{6}x + \frac{1}{3} = 0$

59. $6x^2 = 17x - 10$

60. $11y - 2y^2 = 12$

61. $8 = 9x^2 + 14x$

62. $7x = 12 - 12x^2$

63. (a) For $2x(x - 1)$, how many factors are there?

  (b) To solve $2x(x - 1) = 0$, is it necessary to set each factor equal to 0? Why?

64. Describe the first step in solving the quadratic equation $(x + 1)(x - 3) = k$ if

  (a) $k = 0$.   (b) $k \neq 0$.

In Exercises 65–78, solve each equation.

65. $x(x + 3) - 28$ **66.** $y(y + 1) = 30$

67. $(x + 3)^2 = 16$ **68.** $(x - 5)^2 = 4$

69. $y(3 - y) = 2$ **70.** $x(6 - x) = 8$

71. $(x + 4)(x + 5) = 2$ **72.** $(x + 3)(x - 1) = 5$

73. $x(2x - 1) = 45$ **74.** $x(8x - 23) = 3$

75. $(2x + 1)(x + 2) = -1$ **76.** $(3x + 4)(x + 3) = 8$

77. $10y(y - 1) = 3 + 3y$ **78.** $12 + 4t = 3t(t - 4)$

In Exercises 79–90, solve each equation.

79. $x^3 + 3x^2 - 28x = 0$

80. $x^3 - 6x^2 + 9x = 0$

81. $8x^3 - 8x = 0$

82. $3x^3 + 24x^2 + 21x = 0$

83. $18x - 7x^2 - x^3 = 0$

84. $2x^3 - x^2 - 21x = 0$

85. $45x^3 - 80x = 0$

86. $8x^3 - 4x^2 - 60x = 0$

87. $x^3 + 2x^2 - 4x - 8 = 0$

88. $x^3 - 5x^2 - x + 5 = 0$

89. $9x^3 - 27x^2 - x + 3 = 0$

90. $16x^3 + 16x^2 - 9x - 9 = 0$

In Exercises 91–96, use the graphing method to estimate the solutions of the given equation. (You may need to make window adjustments in order to display the important features of the graph.)

91. $3.36 + x - x^2 = 0$

92. $4.2x - 1.52 - x^2 = 0$

93. $2.8x - x^2 - 1.96 = 0$

94. $x^2 + 1.4x + 0.49 = 0$

95. $x^2 - 1.4x - 1.47 = 0$

96. $x^2 + 4.7x + 4.96 = 0$

Modeling with Real Data

97. The percentage of adults who received financial assistance from their parents for education beyond high school varies by age.

Age	Percentage Who Received Assistance
18–29	27%
30–44	25%
45–59	20%
60 and over	12%

(Source: Roper Starch Worldwide.)

With x representing age, the data can be modeled by the following second-degree polynomial.

$$-0.0082(x^2 - 45x - 2800)$$

(a) Solve the following equation.

$$-0.0082(x^2 - 45x - 2800) = 0$$

(b) Which solution of the equation in part (a) is not meaningful? Why?

(c) Interpret the meaning of the other solution.

98. Refer to Exercise 97.

(a) Write an equation to determine the age for which 24.6% received assistance.

(b) Solve the equation and interpret any solution that is meaningful.

Data Analysis: *Movie Attendance and Prices*

During the period 1993–1999, attendance at movie theaters and the average price of admission both increased.

Year	Attendance (millions)	Admission Price (dollars)
1993	1244	4.14
1995	1303	4.35
1997	1388	4.59
1999	1456	4.80

(Source: Motion Picture Association of America.)

In the following models for attendance a (in millions) and average admission price p (in dollars), x is the number of years since 1990.

$$a(x) = 36x + 1132$$
$$p(x) = 0.11x + 3.80$$

99. Write a polynomial in factored form that describes the total annual receipts.

100. Produce the graph of the expression in Exercise 99. What does the shape of the graph indicate about the trend in total receipts?

101. Write an equation that describes the fact that total annual receipts reached \$10 billion. (Remember that attendance data are reported in millions.) Then use the graphing method to estimate the solution. What year does the solution represent?

102. From 1993 to 1999, did attendance or admission price increase by the greater percent?

Challenge

In Exercises 103–107, solve each equation.

103. $(2x^2 + x - 10)(3x^2 + 13x + 12) = 0$

104. $(x - 3)(x^2 + 3x - 10) = 0$

105. $(x^2 + 9)(x + 9)^2(x^2 - 9) = 0$

106. $x^4 - 81 = 0$

107. $x^4 - 10x^2 + 9 = 0$

In Exercises 108 and 109, solve for x in terms of a and b.

108. $x^2 - 2bx + ax - 2ab = 0$

109. $x^2 + 6bx + 9b^2 = 0$

7.7 | Applications

Numbers • Geometric Figures • Real-Life Applications

Sometimes the information in an application problem can be modeled with a quadratic equation. In such cases, our ability to solve the quadratic equation is a crucial part of the problem-solving process.

A general strategy for solving application problems was introduced in Section 3.5. The approach is worth reviewing because it is valid in all applications. In this section, we illustrate just a few of the wide variety of applications involving quadratic equations.

Numbers

In essence, all application problems are number problems. These examples are primarily for the purpose of practicing the problem-solving approach.

EXAMPLE **1**

Consecutive Odd Integers

The square of the smaller of two consecutive odd integers is 4 more than 3 times the other integer. Determine the integers.

Solution

If $x =$ the smaller odd integer, then

$x + 2 =$ the next consecutive odd integer.

$x^2 = 3(x + 2) + 4$	Describe the conditions of the problem.
$x^2 = 3x + 6 + 4$	Distributive Property
$x^2 = 3x + 10$	
$x^2 - 3x - 10 = 0$	Write the equation in standard form.
$(x - 5)(x + 2) = 0$	Factor.
$x - 5 = 0$ or $x + 2 = 0$	Set each factor equal to 0.
$x = 5$ or $x = -2$	Solve for x.

LEARNING TIP

Always check your solutions with the conditions of the problem. Implied conditions in the problem may lead you to exclude a solution.

Although the equation has two solutions, -2 is not a solution to the problem because it is not an odd integer. Therefore, 5 is the only solution we consider. The consecutive odd integers are 5 and 7.

EXAMPLE **2**

Numbers

Determine all numbers such that the square of the number is equal to the difference of 40 and 3 times the number.

Solution

Let $x =$ the number.

$x^2 = 40 - 3x$	Describe the conditions of the problem.
$x^2 + 3x - 40 = 0$	Write the equation in standard form.
$(x + 8)(x - 5) = 0$	Factor the trinomial.
$x + 8 = 0$ or $x - 5 = 0$	Zero Factor Property
$x = -8$ or $x = 5$	Solve for x.

Two numbers satisfy the conditions, 5 and -8.

Geometric Figures

When we work with problems involving geometric figures, a drawing of the figure can be useful for identifying the variable and labeling other important features.

EXAMPLE **3**

Determining the Dimensions of a Rectangle

The area of a rectangle is 45 square meters. The length is 1 meter less than twice the width. Determine the dimensions of the rectangle.

Figure 7.3

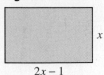

$2x - 1$

Solution

If x = the width of the rectangle, then

$2x - 1$ = the length of the rectangle. (See Fig. 7.3.)

$LW = A$	Area of a rectangle is length times width.
$x(2x - 1) = 45$	$L = 2x - 1$, $W = x$, and the area is 45.
$2x^2 - x = 45$	Distributive Property
$2x^2 - x - 45 = 0$	Write the equation in standard form.
$(2x + 9)(x - 5) = 0$	Factor the polynomial.
$2x + 9 = 0$ or $x - 5 = 0$	Solve for x.
$2x = -9$ or $x = 5$	
$x = -4.5$ or $x = 5$	

Again, the equation has two solutions, but we exclude -4.5 because length cannot be negative. The width is 5 meters, and the length is $2 \cdot 5 - 1$ or 9 meters.

Figure 7.4

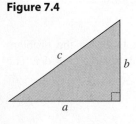

When an application involves a right triangle, the basis for the equation might be the Pythagorean Theorem. Recall that for the right triangle in Figure 7.4, the Pythagorean Theorem states that $a^2 + b^2 = c^2$, where a and b are the lengths of the legs and c is the length of the hypotenuse.

Using the Pythagorean Theorem may lead to a quadratic equation.

EXAMPLE **4**

Determining the Dimensions of a Right Triangle

One leg of a right triangle is 3, and the hypotenuse is 3 less than twice the other leg. Determine the length of each side.

Figure 7.5

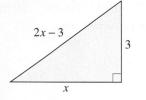

$2x - 3$

3

x

Solution

If x = the length of the other leg, then

$2x - 3$ = the length of the hypotenuse. (See Fig. 7.5.)

$a^2 + b^2 = c^2$	Pythagorean Theorem
$x^2 + 3^2 = (2x - 3)^2$	$a = x$, $b = 3$, and $c = 2x - 3$
$x^2 + 9 = 4x^2 - 12x + 9$	Square the binomial.
$3x^2 - 12x = 0$	Write the equation in standard form.
$3x(x - 4) = 0$	Factor out the common factor $3x$.
$3x = 0$ or $x - 4 = 0$	Set each factor equal to 0.
$x = 0$ or $x = 4$	Solve for x.

We exclude 0 because the length of a side cannot be 0. The legs are 3 and 4, and the hypotenuse is $2 \cdot 4 - 3$ or 5.

Real-Life Applications

Some applications involve a formula that makes writing the equation very easy.

EXAMPLE **5**

Frustration Index

A psychologist has developed an index I to measure indications of a child's frustration as the child attempts to fit blocks into a pattern. Initially, the child's frustration increases, but as the child becomes more proficient, the frustration decreases. The formula for the index is $I = -2t^2 + 20t$, where t is the number of attempts the child has made to solve the puzzle. After how many attempts is the index 42?

Solution

The graph of $I = -2t^2 + 20t$ models the child's increasing, then decreasing, frustration. In Figure 7.6(a), we estimate that the index increases to 42 at 3 attempts, and in Figure 7.6(b), we see that the index is again 42 at 7 attempts.

Figure 7.6

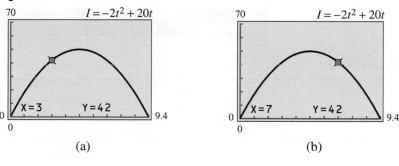

(a)   (b)

The following is an algebraic solution.

$$42 = -2t^2 + 20t$$ The index is 42.

$$2t^2 - 20t + 42 = 0$$ Write the equation in standard form.

$$2(t^2 - 10t + 21) = 0$$ The GCF is 2.

$$2(t - 3)(t - 7) = 0$$ Factor the trinomial.

$$t - 3 = 0 \quad \text{or} \quad t - 7 = 0$$ Zero Factor Property

$$t = 3 \quad \text{or} \quad t = 7$$

The frustration index is 42 after 3 attempts and after 7 attempts.

When numbers are large, a calculator can be very helpful in factoring expressions.

EXAMPLE **6**

A Sightseeing Boat's Revenue

A boat with a maximum capacity of 100 passengers offers sightseeing tours along the river and lakefront in Cleveland, Ohio. For charter groups, tickets cost $10 minus 5 cents for every passenger in the group. For how many passengers is the total revenue $180?

Solution

If x is the number of passengers, then $10 - 0.05x$ is the price of a ticket. Total revenue is given by $x(10 - 0.05x)$.

Think About It

In Example 6, the polynomial $x(10 - 0.05x)$ represents the revenue. Use a graph of this expression to estimate the maximum revenue and the number of passengers that produces the maximum revenue.

$x(10 - 0.05x) = 180$	Total revenue is \$180.
$10x - 0.05x^2 = 180$	Distributive Property
$0.05x^2 - 10x + 180 = 0$	Write the equation in standard form.
$0.05(x^2 - 200x + 3600) = 0$	Factor out 0.05 as the common factor.
$0.05(x - 20)(x - 180) = 0$	Factor the trinomial.

$$x - 20 = 0 \quad \text{or} \quad x - 180 = 0$$
$$x = 20 \quad \text{or} \quad x = 180$$

We exclude 180 as a solution because the maximum capacity of the boat is 100. The revenue is \$180 for 20 passengers.

Exercises 7.7

Concepts and Skills

Numbers

In Exercises 1–6, determine the two numbers.

1. The product of two consecutive positive odd integers is 63.

2. The product of consecutive negative integers is 56.

3. The sum of the squares of consecutive negative integers is 25.

4. The sum of the squares of consecutive positive even integers is 100.

5. One number is 11 more than another number. The product of the numbers is -30.

6. One integer is 1 more than twice another integer. The sum of their squares is 58.

In Exercises 7–16, determine the number.

7. The product of a number and 4 more than the number is 21.

8. The product of a number and 6 less than the number is 16.

9. The square of a number is equal to the difference of 24 and 5 times the number.

10. The difference between the square of a number and the number is 18 more than 6 times the number.

11. Seven more than 11 times a number is 6 times the square of the number.

12. Twice the square of a number is 10 less than 9 times the number.

13. Eight more than 10 times a number is 3 times the square of the number.

14. Twice the square of a number is 5 less than 7 times the number.

15. The sum of the square of a number and 5 times the number is the difference of 3 and the square of the number.

16. Two more than the square of a number is the difference of 5 times the number and twice its square.

Geometric Figures

In Exercises 17–24, the area A of the geometric figure is given. What are the dimensions?

17. $A = 70$

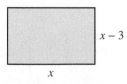

18. $A = 60$

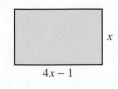

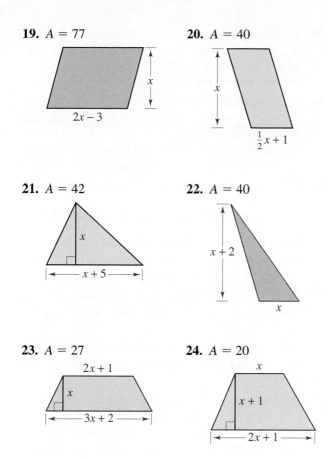

19. $A = 77$

$2x - 3$, x

20. $A = 40$

x, $\frac{1}{2}x + 1$

21. $A = 42$

x, $x + 5$

22. $A = 40$

$x + 2$, x

23. $A = 27$

$2x + 1$, x, $3x + 2$

24. $A = 20$

x, $x + 1$, $2x + 1$

In Exercises 25 and 26, the volume V of the box is given. What are the dimensions of the box?

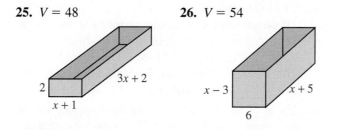

25. $V = 48$

2, $3x + 2$, $x + 1$

26. $V = 54$

$x - 3$, $x + 5$, 6

27. The area of a rectangle is 40 square feet. The length is 3 feet more than the width. What are the width and the length?

28. The width of a rectangle is 4 inches less than the length. If the area is 60 square inches, what are the width and length?

29. The base of a triangle is 1 meter less than the height. What are the base and the height if the area is 21 square meters?

30. The height of a triangle is 1 yard less than half the base. What are the base and height if the area is 12 square yards?

31. The numerical value of the area of a square is twice the perimeter of the square. What is the length of a side of the square?

32. The numerical value of the area of a circle is 6 times the circumference. What is the radius of the circle?

33. If the length of one side of a square is doubled and the other side is decreased by 4 feet, the area increases by 20 square feet. What is the length of a side of the square?

34. If the radius of a circle is increased so that it is 5 centimeters less than twice the original radius, the area is increased by 32π square centimeters. What is the original radius?

Pythagorean Theorem

In Exercises 35–38, determine the dimensions of the right triangle.

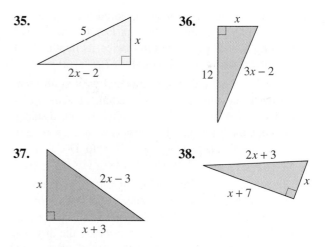

35.

5, x, $2x - 2$

36.

x, 12, $3x - 2$

37.

x, $2x - 3$, $x + 3$

38.

$2x + 3$, $x + 7$, x

In Exercises 39–44, determine the dimensions of the right triangle.

39. The hypotenuse is 25 inches, and one leg is 5 inches longer than the other leg.

40. The hypotenuse is 1 meter less than twice the length of one of the legs. The length of the other leg is 4 meters.

41. The hypotenuse is 2 yards less than 3 times the length of one of the legs. The length of the other leg is 36 feet.

42. The lengths of the sides are consecutive even integers.

43. One leg is 7 feet longer than the other leg. The hypotenuse is 8 feet longer than the shorter leg.

44. The hypotenuse is 9 centimeters longer than the shorter leg, and the longer leg is 1 centimeter shorter than twice the shorter leg.

Real-Life Applications

45. It takes 180 square feet of carpet to cover a rectangular room. If the width of the room is 3 feet less than the length, what are the dimensions of the room?

46. It takes 140 square feet of lumber to cover the floor of a rectangular platform. The length of the platform is 4 feet more than the width. What are the dimensions of the platform?

47. A 12-square-foot rectangular table is covered with a tablecloth that is 5 feet long and 4 feet wide. By how much does the tablecloth hang over on each side? (Assume the same amount of overhang on each side.)

48. A city playground manager ordered 8 cubic feet of sand to fill a sandbox. The box is 4 feet wide, and its length is 3 feet more than twice its depth. What are the dimensions of the sandbox?

49. A car left Dodge City and traveled west at 48 mph. At the same time a biker left from the same point and traveled north. After 30 minutes, the distance between the biker and the car was 6 miles more than twice the distance from the biker to Dodge City. What was the speed of the biker?

50. The length of a ladder leaning against a vertical wall is 2 feet more than the distance from the top of the ladder to the ground. The distance from the base of the ladder to the wall is 7 feet less than the distance from the top of the ladder to the ground. How long is the ladder?

51. The length of a rectangle is 1 yard more than the width. If the length of the diagonal is 5 yards, what are the dimensions of the rectangle?

52. The length of the diagonal of a rectangle is 13 inches. The width is 7 inches less than the length. What are the dimensions of the rectangle?

53. The number of ways to pair chemistry lab partners in a class of n students is given by the expression $\frac{1}{2}n(n - 1)$. If there are 300 different ways to pair lab partners, how many students are in the class?

54. In a basketball tournament with n teams, the number of possible combinations of first- and second-place finishes is given by the expression $n(n - 1)$. How many teams are in a tournament if there are 90 different combinations of first- and second-place finishes?

55. The sum of the integers 1 through n is given by $\frac{n(n + 1)}{2}$. For how many integers is the sum 55?

56. If a polygon has n sides, the number of diagonals that can be drawn is given by $\frac{1}{2}n(n - 3)$. If a polygon has 54 diagonals, how many sides does it have?

57. If $n^2 - 20n + 10$ represents the profit when n blenders are sold, how many must be sold for the profit to be \$310?

58. If $r^2 + 30r - 50$ represents the profit if r recliners are sold, how many must be sold for the profit to be \$350?

59. The number of tickets sold for a political fund-raiser is given by $n = 550 - 50p$, where p is the price of the ticket. How many tickets were sold if the revenue was \$500 and the price of each ticket was at least \$5?

60. The revenue from the sale of custom-made hiking boots is $r(x) = 3x^2 + 7$, and the cost of manufacturing the boots is $c(x) = 22x$, where x is the number of pairs made and sold. How many pairs must be made and sold to break even?

61. The expression $-2x^2 + 34x - 20$ is used to rate the effectiveness of a medication, where x is the number of milligrams in one dose of the medication. How many milligrams should be administered for an effectiveness rating of 84 if at least 5 milligrams per dose are to be administered?

62. The number of snail darters in Yellow Creek can be measured by $180 + 70x - 10x^2$, where x is an index that indicates the level of development in the area. For what level of development will the population of snail darters be 100?

63. The number of people attending a concert is estimated by $p(x) = -x^2 - 10x + 200$, where x is the price of admission. What is the price if 125 people are at the concert?

64. A company's gross sales S (in hundreds of thousands of dollars) is modeled by the polynomial $S(n) = -n^2 + 24n - 80$, where n is the number of sales representatives. If the company currently has 6 sales representatives, show that the model projects gross sales at $2.8 million. What is the minimum number of additional sales representatives needed to increase gross sales to $5.5 million?

Modeling with Real Data

65. The table shows the average per-game attendance for American League baseball teams for selected years during the period 1993–2000.

Year	Average Attendance per Game
1993	29,394
1995	25,108
1997	27,610
2000	29,600

(Source: Major League Baseball.)

With x representing the number of years since 1990, the quadratic polynomial $0.3x^2 - 3.6x + 36.2$ models the attendance (in thousands) per game.

(a) Write an equation to determine the year in which the per-game attendance is predicted to reach 36,200.

(b) Solve the equation in part (a) to determine the year after 1995 in which the average per-game attendance reached 36,200.

66. Refer to the model in Exercise 65.

(a) What do the solutions of the following equation represent?

$$0.3x^2 - 3.6x + 36.2 = 32.9$$

(b) Solve the equation in part (a). (*Hint:* Divide both sides of the equation by 0.3.)

Data Analysis: *Cost of Visits to Doctors*

The bar graph shows the increase in the amount that the average American spends each year on visits to doctors. (Source: U.S. Health Care Financing Administration.)

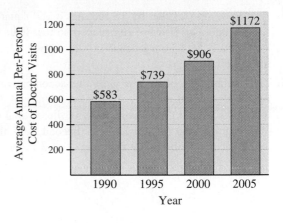

A model expression for the cost is $x^2 + 22x + 588$, where x is the number of years since 1990.

67. Write an equation expressing the fact that in some year the cost per year will be $1428.

68. Use the graphing method to estimate the solution to the equation in Exercise 67.

69. Solve the equation in Exercise 67 algebraically. (*Hint:* $840 = 20 \cdot 42$.) What year does the solution represent?

70. Use the graph to estimate the year in which the cost is expected to double from the cost in 2000.

Challenge

71. The total area of the accompanying figure is 145. Determine the dimensions of both squares.

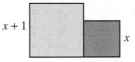

72. The area of the rectangular utility room in the figure, not including the closet, is 80 square feet. Determine the dimensions of the utility room and of the closet.

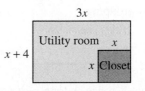

73. The area of a poster is 300 square inches. The outside dimensions of a frame around the poster are 24 inches by 19 inches. How wide is the frame?

74. A room that is 23 feet by 18 feet has a 300-square-foot rectangular Oriental rug placed so that there is a uniformly wide uncarpeted border. How wide is the border?

75. The rectangular bank lobby in the accompanying figure is carpeted on each end, and the center of the lobby is a square that is covered with tile. A total of 1000 square feet of carpet was needed, and the length of the lobby is 10 feet less than 3 times its width. What are the outside dimensions of the lobby?

Figure for 75

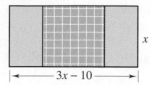

$3x - 10$

x

Chapter 7 Review Exercises

Section 7.1

1. Factor 825 completely.

2. Determine the GCF of the given numbers.
  (a) 28, 42
  (b) 45, 60, 90

3. Determine the GCF of the given monomials.
  (a) $15y^2$, $20y^3$
  (b) $6xz$, $9x^2z$, $15xz^2$

4. Factor out the GCF.
  (a) $18x^3 - 45x^2$
  (b) $16x^6 - 32x^5 + 40x^3$

5. Factor completely.
  (a) $27x^4 - 45x^3 - 9x^2$
  (b) $4x^2y - 4x$

6. For $-4x^3 + 8x^2 + 16x$, factor out the GCF and its opposite.

7. Factor $b(a^2 + 7) - (a^2 + 7)$.

8. Factor by grouping.
  (a) $at^2 - 2bt^2 + 4a - 8b$
  (b) $cx + 3x - 4c - 12$

9. Factor $x^4 + 3x^3 - 6x^2 - 18x$.

10. A box that is twice as long as it is wide is placed on a workbench (see the figure). The length of the box is the same as the width of the workbench. The length of the workbench is 1 foot less than 5 times the width of the box.

Write a completely factored expression for the uncovered area of the workbench.

Figure for 10

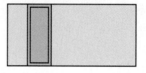

Section 7.2

11. Factor.
  (a) $16a^2 - 9$
  (b) $p^2q^2 - 4$

12. Factor.
  (a) $200 - 2x^2$
  (b) $3x^2y^2 - 27y^2$

13. Factor $b^4 - 16$ completely.

14. In $2x^2 - 12x + 18$, neither $2x^2$ nor 18 is a perfect square, but the perfect square trinomial pattern will still be used to factor the expression. Why?

15. Factor.
  (a) $x^2 - 12x + 36$
  (b) $2ax^2 + 16ax + 32a$

16. Factor.
  (a) $x^2 + 4xy + 4y^2$
  (b) $a^4 + 2a^2 + 1$

17. Factor.

   (a) $x^3 + 27$

   (b) $3 + 24a^3$

18. Factor.

   (a) $b^3 - \frac{1}{8}$

   (b) $1000 - c^3$

19. Factor $(x^2 - 1)(x^2 - 2x + 1)$ completely.

20. The number of cubic feet in a room is given by the polynomial $x^3 + 8x^2 + 16x$, where x is the height of the room. The length and the width are polynomials with integer coefficients. By factoring the expression, describe the room.

Section 7.3

21. Factor.

   (a) $c^2 + 10c + 21$

   (b) $18 + 11x + x^2$

22. Factor.

   (a) $x^2 + 9x - 10$

   (b) $z^2 - 5z - 24$

23. Factor $-y + y^2 - 30$.

24. Factor.

   (a) $a^2 - 2ab - 3b^2$

   (b) $x^2y^2 - 4xy - 12$

25. Factor $x^4 + 2x^2 - 15$.

26. Describe two ways to factor $48 - 2x - x^2$.

27. Factor.

   (a) $3y^2 - 12yz - 96z^2$

   (b) $a^2 + 7a - 10$

28. Factor completely $x^4 - 5x^2 + 4$.

29. Factor completely.

$$(x^2 - 1)(x^2 + 2x + 1)(x^2 + x - 2)$$

30. The width of a rectangle is represented by $x - 2$, and its area is represented by $x^2 + 9x - 22$. What expression represents the length of the rectangle?

Section 7.4

31. Factor.

   (a) $6x^2 + 17x + 5$

   (b) $2x^2 - 11x + 12$

32. Factor.

   (a) $10x^2 - 13x - 3$

   (b) $12x^2 - 25x + 12$

33. Factor.

   (a) $3a^2 + 13ab + 14b^2$

   (b) $2x^2 - 19xy + 9y^2$

34. Factor.

   (a) $2c^2 - 7cz - 15z^2$

   (b) $7m^2 + 12mn - 4n^2$

35. Factor by first factoring out -1.

$$-3x^2 + x + 24$$

36. Factor completely $18x^2 + 45x + 18$.

37. Factor.

   (a) $3x^2 - 4xy - y^2$

   (b) $48a^3 + 106a^2 + 56a$

38. Factor using a GCF with a negative coefficient.

$$-3x^3y + 4x^2y^2 + 4xy^3$$

39. Factor completely.

$$(2x^2 - x - 1)(4x^2 - 4x + 1)$$

40. The volume of a shipping container is given by the polynomial $2x^3 + 4x^2 - 30x$, where $2x$ is the height, and the length and width are polynomials with integer coefficients. Write expressions for the length and the width of the container.

Section 7.5

In Exercises 41–50, use the general factoring strategy to factor the expression completely.

41. $4x^2 - 16$

42. $2x^2 + 2x - 12$

43. $t^4 - 16$

44. $t^3 - 27$

45. $a^3 + a^2 + a + 1$

46. $x^4y + x^3y + x^2y$

47. $11x^2 + 19x - 6$

48. $2x^3 + 16$

49. $2x^3 + 12x^2 + 18x$

50. $3x^2 + 5x - 4$

Section 7.6

51. Use the integer setting to produce the graph of the polynomial $y = 0.1x^2 - 2x - 12$. Then use the graph to estimate the solutions of each of the following equations.

(a) $0.1x^2 - 2x - 12 = 18$

(b) $0.1x^2 - 2x - 12 = -22$

(c) $0.1x^2 - 2x - 12 = -26$

52. Solve $(x - 5)(x + 8) = 0$.

53. Solve $3x(x + 0.4)(2x - 1) = 0$.

54. The expression $x^2 + 3x - 5$ cannot be factored. Does this mean that $x^2 + 3x - 5 = 0$ has no solution? How can you tell?

In Exercises 55–60, solve the equation algebraically.

55. $x^2 - 3x - 28 = 0$

56. $-10 - x + 3x^2 = 0$

57. $x(x - 5) = 36$

58. $(2x - 1)(x + 3) = 9$

59. $x^3 - 40x = -3x^2$

60. $x^3 + x^2 = 9x + 9$

Section 7.7

In Exercises 61–70, write and solve an equation that describes the given information. Then answer the question.

61. Two brothers were born 2 years apart. The product of their ages is 63. How old is the older brother?

62. The square of a number is 45 less than 14 times the number. What is the number?

63. For three consecutive positive integers, the product of the first and third integers is 1 less than 4 times the second integer. What is the largest integer?

64. The length of a rectangular hallway is 2 feet more than 4 times the width. If the area is 42 square feet, how long is the hallway?

65. The figure shows the relative dimensions (in feet) of a ramp that was built to reach a loading dock. How long is the ramp?

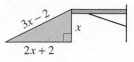

66. The sides of a picture frame are 2 inches longer than the top and bottom. When the frame was dropped, it was bent into a parallelogram. The distance between the top and bottom of the frame was reduced to 9 inches, and the area enclosed by the frame was reduced by 8 square inches. What were the original dimensions of the frame?

67. A rectangle is drawn inside a circle so that the diagonals of the rectangle intersect at the center of the circle. The radius of the circle is 8.5 inches, and the length of the rectangle is 1 inch less than twice the width. How long is the rectangle? (*Hint:* Use the fact that $288 = 8 \cdot 36$.)

68. A concession stand sells pennants at a baseball game. The number of pennants sold each day is given by $-x^2 - 12x + 160$, where x is the price (in dollars) that is charged. At what price can 27 pennants be sold?

69. The base for a stone walkway is 4 inches deep. The length of the walkway is 1 foot longer than 7 times the width. If 10 cubic feet of sand is used to construct the base, how long is the walkway?

70. A tee brace is manufactured with the relative dimensions shown in the figure. If the total area of the piece is 7 square inches, what is the length AB?

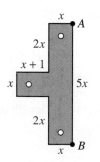

Chapter 7 Test

1. What is the GCF in the following expression?
   $$2x^5y^2 + 6x^4y^3 - 12x^5y^3$$

2. Factor $x(x - 8) - 8(x - 8)$.

3. Use the grouping method to factor $x^5 - x^4 - 2x + 2$.

In Questions 4–7, state the name given to the expression. Then factor the expression.

4. $9x^2 - 36$

5. $y^3 + 1$

6. $8a^3 - 64$

7. $4x^2 - 20x + 25$

In Questions 8 and 9, factor.

8. $x^2 + 4x - 32$

9. $a^2 - 3ab - 10b^2$

10. Factor $8x - 15 - x^2$ by first factoring out -1.

In Questions 11–14, factor.

11. $c^3 + 5c^2 - 14c$

12. $x^2 + 7x + 4$

13. $12x^2 + 8x - 15$

14. $6a^2 - 2ab - 6b^2$

15. If the area of a rectangle is $16x^2 - 66x - 27$ and the width is $2x - 9$, what is the length?

In Questions 16–19, factor completely.

16. $x^4 + 5x^2 - 6$

17. $x^3 - x^2 - 4x + 4$

18. $45x^2 + 30x + 5$

19. $a^4 - 16$

20. To solve a linear equation, we take steps to isolate the variable. Why is this procedure generally not applicable to solving quadratic equations?

In Questions 21–24, solve the given equation.

21. $6x(x - 7) = 0$

22. $x^2 - 2x - 48 = 0$

23. $x(6x - 7) = 5$

24. $x^3 - 4x^2 - x + 4 = 0$

25. Suppose that the equation $ax^2 + bx + c = 0$ has exactly one solution. What can you conclude about the expression on the left side?

In Questions 26–28, write and solve an equation that describes the given information. Then answer the question.

26. A picture 10 inches wide and 12 inches long is mounted in a frame. The total area of the picture and frame is 195 square inches. How wide is the frame?

27. A rectangular deck is 3 feet longer than it is wide. A brace is placed underneath the deck from one corner to the opposite corner along the diagonal. The brace is 6 feet longer than the width of the deck. How long is the deck?

28. It costs a lawn maintenance company $13 per hour to keep a crew of 2 people working. The revenue is given by $5x^2 - 6$, where x is the number of hours the crew works. For how many hours must the crew be on the job for the company to break even?

The bar graphs show that from 1997 to 2000 there was a substantial increase in the number of wildfires and the number of acres that were burned. The number of wildfires and the number of acres burned can both be described with polynomials. The average number of acres burned per wildfire is the quotient (called a ***rational expression***) of these polynomials. We can determine trends in the severity of wildfires by analyzing and evaluating these model expressions. (For more on this real data problem, see Exercises 51–54 at the end of Section 8.6.)

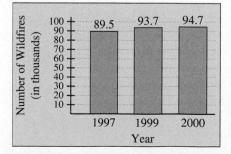

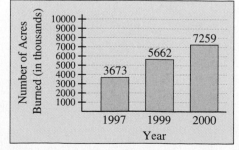

(Source: National Interagency Fire Center.)

Chapter Snapshot

In this chapter, we introduce rational expressions. We learn how to simplify them and how to perform the four basic operations with them. After a brief look at complex fractions, we study methods for solving equations with rational expressions and use this new skill to solve application problems.

Warm-Up Skills

The following exercises review concepts and skills that you will need in Chapter 8.

In Exercises 1–3, evaluate the given numerical expression.

1. (a) $\dfrac{7}{15} \cdot \dfrac{6}{11}$ (b) $\dfrac{9}{5} \div \dfrac{3}{2}$

2. (a) $\dfrac{5}{6} + \dfrac{1}{4}$ (b) $\dfrac{3}{5} - \dfrac{4}{3}$

3. $\left(2 - \dfrac{3}{2}\right) \div \left(\dfrac{1}{2} + \dfrac{4}{3}\right)$

4. Evaluate the expression $\dfrac{2x}{x - 5}$ for $x = 4, 5$, and -1.

5. Simplify $(x^2 + 6x + 7) - (x^2 + x - 3)$.

6. Describe the process for solving a linear equation and a quadratic equation.

In Exercises 7 and 8, solve the given equation.

7. $x^2 - 4x = 21$ 8. $\dfrac{t}{4} = \dfrac{t + 9}{6} - \dfrac{1}{2}$

In Exercises 9 and 10, factor the given polynomial.

9. (a) $x^2 + 6x + 9$ (b) $x^2 - 25$

10. (a) $3x^2 - 9x$ (b) $x^2 + 3x - 10$

11. Write $-w^2 + w$ by factoring out the GCF and the opposite of the GCF.

12. Simplify the given expression.

 (a) $\dfrac{18}{24}$ (b) $\dfrac{12xy^6}{15x^3y^2}$

8.1 | Introduction to Rational Expressions

Definition   •   Evaluating Rational Expressions   •   Simplifying Rational Expressions

Definition

From arithmetic we know that rational numbers can be written as the quotient $\frac{p}{q}$, where p and q are integers and $q \neq 0$. The following are examples of rational numbers.

$$\frac{5}{3} \qquad 23 \qquad \frac{0}{-4} \qquad \frac{-3}{-8}$$

The algebraic counterpart of a rational number is a **rational expression.** In fact, the two are so closely related that rational expressions are sometimes called *algebraic fractions.*

Whereas a rational number is a quotient of integers, a rational expression is a quotient of polynomials.

> **Definition of Rational Expression**
>
> A **rational expression** can be written as $\frac{P}{Q}$, where P and Q are polynomials and $Q \neq 0$.

The following are examples of rational expressions.

$$\frac{4x}{x - 5} \qquad \frac{x + 7}{x^2 + 3x - 4} \qquad x^3 + 3x - 6$$

Note: Because we can write 23 as $\frac{23}{1}$, we know that 23 is a rational number. Similarly, $x^3 + 3x - 6$ is a rational expression because we can write the polynomial as $\frac{x^3 + 3x - 6}{1}$. (Recall that 1 is a polynomial of degree 0.)

Evaluating Rational Expressions

We evaluate a rational expression in the same manner that we evaluate any algebraic expression. We replace each variable with its given value and then evaluate the resulting numerical expression.

Exploring the Concept

Evaluating Rational Expressions

Suppose that we wish to evaluate $\frac{4x}{x - 3}$ for $x = 2$ and $x = 0$. As we have seen, function notation is a useful way to indicate this.

$$r(x) = \frac{4x}{x - 3}$$

$$r(2) = \frac{4(2)}{2 - 3} = \frac{8}{-1} = -8 \qquad \text{Replace } x \text{ with 2.}$$

$$r(0) = \frac{4(0)}{0 - 3} = \frac{0}{-3} = 0 \qquad \text{Replace } x \text{ with 0.}$$

We also know that we can evaluate an expression by tracing its graph. However, the graph of a rational expression is somewhat more complicated than the graphs of linear and quadratic expressions. Figure 8.1 shows the entry of the expression $\dfrac{4x}{x - 3}$ on the Y screen and the graph of the expression.

Figure 8.1

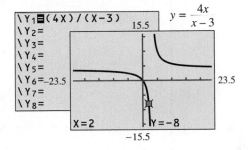

Note that the graph consists of two *branches*. The tracing cursor is on the point $(2, -8)$, which is consistent with our earlier result, $r(2) = -8$. We could also trace to see that $r(0) = 0$.

Finally, a table of values can be used to evaluate the expression.

Figure 8.2

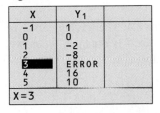

From the table in Figure 8.2, we see that $r(2) = -8$ and $r(0) = 0$, as before. However, we also observe an error message for $x = 3$. The following shows why this result occurred.

$$r(3) = \frac{4(3)}{3 - 3} = \frac{12}{0} \qquad \text{Division by 0 is undefined.}$$

Thus the rational expression $\dfrac{4x}{x - 3}$ is not defined for $x = 3$. This explains the *break* in the graph of the expression in Figure 8.1. If 3 is not a permissible replacement for x, then there can be no point of the graph with an x-coordinate of 3. In general, a rational expression is not defined for any value of the variable for which the denominator is 0.

A value that is not a permissible replacement for the variable is sometimes called a *restricted value*. When we determine restricted values, only the denominator of a rational expression needs to be checked. Any replacement is permissible in the numerator.

Determining Restricted Values

1. Set the denominator equal to 0.
2. Solve the resulting equation.
3. Any solution of the equation is a restricted value.

EXAMPLE 1

Determining Restricted Values

Determine the restricted values, if any, for each expression.

(a) $\dfrac{x+4}{x+2}$ (b) $\dfrac{x+2}{x^2-4x-21}$ (c) $\dfrac{6-x}{x^2+9}$

Solution

(a) Figure 8.3 shows the graph of the given expression with the tracing cursor on $x = -2$. Observe that no corresponding y-value is reported. This suggests that -2 is a restricted value.

Figure 8.3

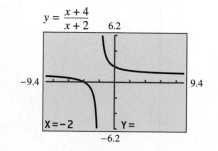

The expression is not defined if the denominator $x + 2$ has a value of 0.

$x + 2 = 0$ Set the denominator equal to 0.

$ x = -2$ Solve for x.

The restricted value is -2.

(b) The graph of the given expression is shown in Figure 8.4. The tracing cursor is on $x = -3$, but no y-value is reported. The same result is obtained when we trace to $x = 7$. Thus -3 and 7 appear to be restricted values.

Figure 8.4

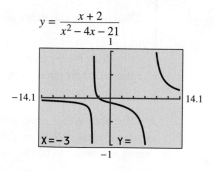

$x^2 - 4x - 21 = 0$ Set the denominator equal to 0.

$(x - 7)(x + 3) = 0$ Factor.

$x - 7 = 0$ or $x + 3 = 0$ Set each factor equal to 0.

$ x = 7$ or $ x = -3$

There are two restricted values, -3 and 7.

(c) Because the graph in Figure 8.5 has no breaks, we predict that the expression has no restricted values.

Figure 8.5

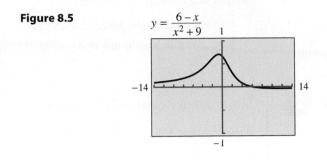

$$y = \frac{6-x}{x^2+9}$$

Because $x^2 + 9$ is never 0, there are no restricted values. The expression is defined for all values of x.

Simplifying Rational Expressions

Operations with algebraic fractions are similar to the operations that we perform with fractions in arithmetic.

A fraction is reduced or simplified if the numerator and denominator have no common factor other than 1 or -1. Recall that we use the Property of Equivalent Fractions to reduce a fraction by dividing out any factor that is common to both the numerator and denominator.

$$\frac{12}{15} = \frac{3 \cdot 4}{3 \cdot 5} = \frac{4}{5}$$

The Property of Equivalent Fractions also applies to rational expressions. A familiar example is one in which the numerator and denominator are monomials.

$$\frac{24x^4y^2}{30xy^5} = \frac{6xy^2 \cdot 4x^3}{6xy^2 \cdot 5y^3} \qquad \text{Factor.}$$

$$= \frac{4x^3}{5y^3} \qquad \text{Divide out common factors.}$$

When the numerator or the denominator of a rational expression contains more than one term, we factor in order to divide out any common factor.

$$\frac{2x-6}{5x-15} = \frac{2(x-3)}{5(x-3)} = \frac{2}{5}$$

Note: Because 3 is a restricted value, $\dfrac{2x-6}{5x-15} = \dfrac{2}{5}$ for $x \neq 3$. In our discussion of operations with rational expressions, we will usually not specifically state restricted values. It is understood that the equalities hold only for those values for which the expressions are defined.

The following summarizes the procedure for simplifying a rational expression.

> **Simplifying Rational Expressions**
>
> 1. Factor the numerator and the denominator completely.
> 2. Divide out any factor that is common to both.

EXAMPLE 2

Simplifying Rational Expressions

Simplify the given rational expressions.

(a) $\dfrac{y^2 + 6y}{3y} = \dfrac{y(y + 6)}{3y}$ Factor.

$ = \dfrac{y + 6}{3}$ Divide out the common factor.

(b) $\dfrac{x^2 - 9}{3x^2 - 9x} = \dfrac{(x + 3)(x - 3)}{3x(x - 3)}$ Factor.

$ = \dfrac{x + 3}{3x}$ Divide out the common factor.

(c) $\dfrac{x + 5}{x^2 + 3x - 10} = \dfrac{x + 5}{(x + 5)(x - 2)}$ Factor.

$ = \dfrac{1}{x - 2}$ Divide out the common factor.

(d) For $\dfrac{x^2 + 3x + 1}{x + 1}$, the numerator is prime. The expression cannot be simplified.

(e) $\dfrac{x^2y + xy^3}{xy^2} = \dfrac{xy(x + y^2)}{xy \cdot y}$ Factor.

$ = \dfrac{x + y^2}{y}$ Divide out the common factor.

Recall that we can reverse the terms in a difference by factoring out -1.

$$b - a = -1(a - b)$$

Sometimes we can use this fact to simplify a rational expression.

EXAMPLE 3

Simplifying Rational Expressions by Factoring Out -1

Simplify the given rational expressions.

(a) $\dfrac{y - 3}{3 - y} = \dfrac{1(y - 3)}{-1(y - 3)}$ Factor out -1 in the denominator.

$ = \dfrac{1}{-1}$ Divide out the common factor.

$ = -1$

(b) $\dfrac{1 - 25x^2}{5x - 1} = \dfrac{-1(25x^2 - 1)}{5x - 1}$ Factor out -1 in the numerator.

$\qquad\qquad = \dfrac{-1(5x + 1)(5x - 1)}{5x - 1}$ Factor the difference of two squares.

$\qquad\qquad = -1(5x + 1)$ or $-5x - 1$ Divide out the common factor.

(c) $\dfrac{2x^2 - 6x}{6 + x - x^2} = \dfrac{2x^2 - 6x}{-1(x^2 - x - 6)}$ Factor out -1 in the denominator.

$\qquad\qquad = \dfrac{2x(x - 3)}{-1(x - 3)(x + 2)}$ Factor the numerator and the denominator.

$\qquad\qquad = \dfrac{2x}{-1(x + 2)}$ Divide out the common factor.

Note that the result can also be written as $\dfrac{-2x}{x + 2}$ or $-\dfrac{2x}{x + 2}$.

The technique used in Example 3 can be combined with that of factoring out a common numerical factor.

EXAMPLE 4

Simplifying by Factoring Out a Negative Numerical Factor

Simplify the given rational expressions.

(a) $\dfrac{9x - 6}{8 - 12x} = \dfrac{3(3x - 2)}{-4(3x - 2)}$ Factor out -4 in the denominator.

$\qquad\qquad = -\dfrac{3}{4}$ Divide out the common factor.

(b) $\dfrac{a^2 + 2ab}{12b - 6a} = \dfrac{a(a + 2b)}{-6(a - 2b)}$ Factor out -6 in the denominator.

Because there is no common factor, this expression cannot be simplified.

(c) $\dfrac{-w^2 + w}{w^2 + w - 2} = \dfrac{-1(w^2 - w)}{w^2 + w - 2}$ Factor out -1 in the numerator.

$\qquad\qquad = \dfrac{-1w(w - 1)}{(w + 2)(w - 1)}$ Factor out w in the numerator and factor the denominator.

$\qquad\qquad = \dfrac{-w}{w + 2}$ Divide out the common factor.

Note that w in the denominator is a term, not a factor. Because we divide out only common factors, we cannot simplify the result.

With practice you will learn when to factor out a negative numerical factor. Generally speaking, if a factor in the numerator is the opposite of a factor in the denominator (remember that $a - b$ and $b - a$ are opposites), this technique is appropriate.

Quick Reference 8.1

Definition
- A **rational expression** (sometimes called an *algebraic fraction*) can be written as $\dfrac{P}{Q}$, where P and Q are polynomials and $Q \neq 0$.

Evaluating Rational Expressions
- To evaluate a rational expression, replace each variable with its given value and then evaluate the resulting numerical expression.
- All previously discussed calculator methods for evaluating expressions can be used to evaluate a rational expression.
- A rational expression is not defined for any value of the variable for which the denominator is 0. A value that is not a permissible replacement for the variable is called a *restricted value*.
- We determine restricted values as follows.
  1. Set the denominator equal to 0.
  2. Solve the resulting equation.
  3. Any solution of the equation is a restricted value.

Simplifying Rational Expressions
- To simplify a rational expression, we use the Property of Equivalent Fractions to divide out any factor that is common to both the numerator and the denominator.
  1. Factor the numerator and the denominator completely.
  2. Divide out any factor that is common to both.
- Sometimes we can create common factors in the numerator and the denominator by factoring out a negative numerical factor.

Speaking the Language 8.1

1. An expression of the form $\dfrac{P}{Q}$, where P and Q are polynomials and $Q \neq 0$, is called a(n) ▨▨▨▨ expression.

2. In the expression $\dfrac{5}{x}$, we call 0 a(n) ▨▨▨▨ value.

3. To simplify a rational expression, we divide out any ▨▨▨▨ in the numerator and the denominator.

4. A restricted value is indicated in the graph of a rational expression by a(n) ▨▨▨▨, which separates the graph into ▨▨▨▨.

Exercises 8.1

Concepts and Skills

1. Suppose that you store -2 in X and then enter $\dfrac{3x}{x+2}$ in your calculator. Explain the result.

2. Suppose that $r(x) = \dfrac{x^2 - 3}{x+1}$. What is the meaning of the symbol $r(2)$?

In Exercises 3–6, evaluate the expression for the given values.

3. $\dfrac{x}{x+4}$; $0, 4, -4$

4. $\dfrac{a+2}{2a+6}$; $-3, -2, 1$

5. $\dfrac{2t-3}{5-t}$; $-5, 5, 4$

6. $\dfrac{2y}{y^2-1}$; $0, 1, -1$

In Exercises 7–10, evaluate the expression as indicated.

7. $r(x) = \dfrac{3x}{(1-3x)^2}$; $r(1)$

8. $r(x) = \dfrac{x}{(x+2)^2}$; $r(2)$

9. $R(x) = \dfrac{x+1}{x^2-3x}$; $R(-3)$

10. $R(x) = \dfrac{x-6}{5x^2+5x}$; $R(6)$

In Exercises 11–18, determine the restricted values.

11. $\dfrac{5}{x+2}$

12. $\dfrac{-7}{x-6}$

13. $\dfrac{x^2+1}{4x-x^2}$

14. $\dfrac{a-1}{a+2a^2}$

15. $\dfrac{x}{x^2+16}$

16. $\dfrac{3-2y}{-4}$

17. $\dfrac{3x}{25-x^2}$

18. $\dfrac{6x}{x^2-3x-10}$

In Exercises 19 and 20, identify the terms in the numerator and the denominator.

19. $\dfrac{4x-3}{x-3}$

20. $\dfrac{6x+5}{x+5}$

In Exercises 21 and 22, identify the factors in the numerator and the denominator.

21. $\dfrac{4(x-4)}{3(x-4)}$

22. $\dfrac{2(x+1)}{7(x+1)}$

23. For the expression $\dfrac{x+3}{x}$, why is it incorrect to divide out x?

24. Comment on the following reasoning.

The expression $\dfrac{xy+3x}{2x}$ has no restricted values because

$$\dfrac{xy+3x}{2x} = \dfrac{x(y+3)}{2x} = \dfrac{y+3}{2}$$

and the simplified expression has no variable in the denominator.

In Exercises 25–36, simplify the rational expression.

25. $\dfrac{y^2+4y}{3y+12}$

26. $\dfrac{24-6a}{4a-a^2}$

27. $\dfrac{5x+15}{x^2-3x}$

28. $\dfrac{x+5}{x^2+4x+5}$

29. $\dfrac{4x^3+2x^2}{6x^2}$

30. $\dfrac{8x^3}{12x^3-8x^2}$

31. $\dfrac{(x+3)^2}{x^2-9}$

32. $\dfrac{a^2+a-6}{3a^2-6a}$

33. $\dfrac{3x+6}{x^2-4x-12}$

34. $\dfrac{2x^2+x-3}{2x+3}$

35. $\dfrac{w^2+2w-8}{w^2-5w+6}$

36. $\dfrac{z^2+z-6}{z^2-z-12}$

37. What do we call the expressions $2x-5$ and $5-2x$? How are these expressions related algebraically?

38. Explain why the result of dividing an expression by its opposite is -1.

In Exercises 39–46, simplify each expression.

39. $\dfrac{5-a}{a-5}$

40. $\dfrac{x-4}{4-x}$

41. $\dfrac{3y-2x}{2x-3y}$

42. $\dfrac{xy-1}{1-xy}$

43. $\dfrac{a-2b}{a+2b}$

44. $\dfrac{x+5}{x-5}$

45. $\dfrac{y-5}{10+13y-3y^2}$

46. $\dfrac{2x^2+5x-3}{1-x-2x^2}$

In Exercises 47–54, simplify the rational expression.

47. $\dfrac{12 - 4x}{x - 3}$

48. $\dfrac{14 - 2y}{y - 7}$

49. $\dfrac{9x - 3x^2}{2x - 6}$

50. $\dfrac{4x - 16}{4x - x^2}$

51. $\dfrac{2z - 2z^2}{z^2 + z - 2}$

52. $\dfrac{15a - 5a^2}{a^2 - 5a + 6}$

53. $\dfrac{2x^2 - 2}{3 - 3x}$

54. $\dfrac{2x^2 - 18}{9 - 3x}$

In Exercises 55–66, simplify.

55. $\dfrac{10a - 20b}{15a - 30b}$

56. $\dfrac{3x}{6x^2 + 15x}$

57. $\dfrac{x^2 - 2x - 8}{x^2 + 7x + 10}$

58. $\dfrac{x^2 - 5x - 6}{x^2 + x - 42}$

59. $\dfrac{9 - x^2}{x - 3}$

60. $\dfrac{-3x + 9}{2x^2 - 13x + 21}$

61. $\dfrac{x^2 - 16y^2}{x + 4y}$

62. $\dfrac{25a^2 - 4b^2}{5a - 2b}$

63. $\dfrac{y^2 - 9}{y - 3}$

64. $\dfrac{4x^2 + 4x + 1}{2x + 1}$

65. $\dfrac{10 - 15x}{3x - 2}$

66. $\dfrac{-12x + 18}{2x - 3}$

In Exercises 67 and 68, determine whether the expression is equal to 1, −1, or neither.

67. (a) $\dfrac{x + 1}{1 + x}$

(b) $\dfrac{x - 1}{x + 1}$

(c) $\dfrac{1 - x}{x - 1}$

68. (a) $\dfrac{-3y + 2}{3y - 2}$

(b) $\dfrac{3y - 2}{-1(2 - 3y)}$

(c) $\dfrac{-3y + 2}{-3y - 2}$

In Exercises 69–74, answer "yes" if the given pairs are equivalent. Otherwise, answer "no."

69. $\dfrac{x + 4}{4}$, x

70. $\dfrac{3x + 1}{3}$, $x + 1$

71. $-\dfrac{2x}{y}$, $\dfrac{-2x}{y}$, $y \neq 0$

72. $\dfrac{1 - x}{3}$, $-\dfrac{x - 1}{3}$

73. $\dfrac{x - 2}{2}$, $x - 1$

74. $\dfrac{a + 5}{a^2 + 25}$, $\dfrac{1}{a + 5}$, $a \neq -5$

Real-Life Applications

75. A city planner predicts that the number of building permits for houses will be $6x + 2x^2$ and the number of commercial building permits will be $2x$, where x is an indicator of the economy. Write a rational expression in simplified form for the ratio of housing permits to commercial building permits.

76. A poster company estimates that the revenue from selling Olympic posters will be $42x - 21$ and the number of posters sold will be $6x - 3$, where x represents the level of advertising. Write a rational expression in simplified form for the average price per poster.

77. The distance that a particle has traveled after $5x$ minutes is $3x^2 + 4x$. Write a rational expression in simplified form for the average speed of the particle.

78. The number of miles driven is $120t + 300$, and the number of gallons of fuel consumed is $6t + 15$. Write a rational expression in simplified form for the number of miles per gallon of fuel consumed.

Data Analysis: *Life Expectancy*

Since the beginning of the 20th century, the life expectancy for both men and women has increased significantly. The accompanying table compares the life expectancies in 1920, 1970, 1990, and 2000.

Life Expectancy		
Year	Men	Women
1920	53.6	54.6
1970	67.1	74.7
1990	71.6	78.4
2000	74.2	81.6

(Source: National Center for Health Statistics.)

If x is the number of years since 1900, the life expectancies for men $m(x)$ and women $w(x)$ can be modeled as follows:

$m(x) = 0.26x + 48.6$

$w(x) = 0.33x + 48.6$

79. What do the coefficients of x in the models indicate?

80. Write a rational expression $r(x)$ that compares the life expectancy of women to that of men.

81. Evaluate $r(100)$. Assuming that the models remain valid, by what percentage will the life expectancy of women exceed that of men in the year 2000?

82. Evaluate and interpret $r(0)$.

Challenge

In Exercises 83–87, simplify the rational expression.

83. $\dfrac{(x-1)^5}{(1-x)^3}$

84. $\dfrac{x^3-8}{x^2-4}$

85. $\dfrac{x^3+3x^2-4x-12}{x^2+5x+6}$

86. $\dfrac{x^2+x-6}{x^4-13x^2+36}$

87. $\dfrac{x+y-2}{(x-2)^2-y^2}$

In Exercises 88 and 89, determine the unknown numerator that makes the resulting equation true.

88. $\dfrac{\rule{1cm}{0.3cm}}{3x-6}=\dfrac{1}{x-2}$

89. $\dfrac{\rule{1cm}{0.3cm}}{x^2-3x-4}=\dfrac{1}{x+1}$

8.2 Multiplication and Division

Multiplication of Rational Expressions • Division of Rational Expressions

Multiplication of Rational Expressions

From arithmetic we know that the product of two fractions is the product of the numerators divided by the product of the denominators. Multiplication of rational expressions is performed in the same manner.

> **Multiplication of Rational Expressions**
>
> If a, b, c, and d are polynomials, where $b \neq 0$ and $d \neq 0$,
>
> $$\frac{a}{b} \cdot \frac{c}{d} = \frac{ac}{bd}$$

LEARNING TIP

If necessary, take some time to review the arithmetic of fractions. All of the arithmetic rules carry over directly to operations with rational expressions.

Note the similarity in the following two products.

$$\frac{5}{12} \cdot \frac{8}{7} = \frac{40}{84} = \frac{4 \cdot 10}{4 \cdot 21} = \frac{10}{21}$$

$$\frac{x^5}{3} \cdot \frac{6}{4x^2} = \frac{6x^5}{12x^2} = \frac{6x^2 \cdot x^3}{2 \cdot 6x^2} = \frac{x^3}{2}$$

In each of the preceding examples, we performed the multiplication and then simplified the result. However, to simplify the computation in arithmetic, recall that we usually divide out common factors before multiplying.

$$\frac{5}{12} \cdot \frac{8}{7} = \frac{5}{4 \cdot 3} \cdot \frac{4 \cdot 2}{7} = \frac{10}{21}$$

Dividing out common factors first is a convenience in arithmetic, but with rational expressions it is almost essential.

> **Procedure for Multiplying Rational Expressions**
>
> 1. Factor the numerators and denominators completely.
> 2. Divide out any factor that is common to both a numerator and a denominator.
> 3. Multiply the numerators and multiply the denominators.

Note: As in arithmetic, the common factors that can be divided out do not have to be in the same rational expression. If a factor in the numerator of one expression is the same as a factor in the denominator of another expression, the factors can be divided out.

EXAMPLE 1

Multiplying Rational Expressions

Multiply the given rational expressions.

(a) $\dfrac{5x}{4x-12} \cdot (2x^2-6x)$

$= \dfrac{5x}{4x-12} \cdot \dfrac{2x^2-6x}{1}$ Write the second expression as a fraction.

$= \dfrac{5x}{2 \cdot 2(x-3)} \cdot \dfrac{2x(x-3)}{1}$ Factor.

$= \dfrac{5x^2}{2}$ Divide out common factors.

(b) $\dfrac{x^2-4x-5}{3x^2-15x} \cdot \dfrac{6}{2x+2}$

$= \dfrac{(x-5)(x+1)}{3x(x-5)} \cdot \dfrac{3 \cdot 2}{2(x+1)}$ Factor.

$= \dfrac{1}{x}$ Divide out common factors.

(c) $\dfrac{x^2-49}{21-3x} \cdot \dfrac{12x}{x^2+9x+14}$

$= \dfrac{(x+7)(x-7)}{-3(x-7)} \cdot \dfrac{3 \cdot 4x}{(x+7)(x+2)}$ Factor.

$= \dfrac{-4x}{x+2}$ Divide out common factors.

Division of Rational Expressions

We divide rational expressions in the same way as we divide by a fraction in arithmetic: Multiply by the reciprocal of the divisor.

Think About It

Suppose that the operation $a \# b$ is defined as $\dfrac{1}{a} \cdot b$. Show that $a \# b = b \div a$.

> **Division of Rational Expressions**
>
> If a, b, c, and d are polynomials, where b, c, and d are not 0,
>
> $$\frac{a}{b} \div \frac{c}{d} = \frac{a}{b} \cdot \frac{d}{c}$$

Note the similarity in the following two quotients.

$$\frac{3}{2} \div \frac{7}{9} = \frac{3}{2} \cdot \frac{9}{7} = \frac{27}{14}$$

$$\frac{7}{x} \div \frac{x}{2y} = \frac{7}{x} \cdot \frac{2y}{x} = \frac{14y}{x^2}$$

As before, we divide out common factors, if possible, prior to multiplying.

Note: When we are performing operations with rational expressions, writing each expression as a quotient is usually helpful. For example,

$$a \div \frac{b}{c} = \frac{a}{1} \div \frac{b}{c} \quad \text{and} \quad \frac{a}{b} \div c = \frac{a}{b} \div \frac{c}{1}$$

EXAMPLE 2

Dividing Rational Expressions

Divide the given rational expressions.

(a) $\dfrac{x+5}{x} \div \dfrac{(x+5)^2}{3x}$

$= \dfrac{x+5}{x} \cdot \dfrac{3x}{(x+5)^2}$ Multiply by the reciprocal of the divisor.

$= \dfrac{x+5}{x} \cdot \dfrac{3x}{(x+5)(x+5)}$

$= \dfrac{3}{x+5}$ Divide out common factors.

(b) $\dfrac{x^2 - x - 2}{4x^2 + 8x} \div \dfrac{x^2 - 4}{4x}$

$= \dfrac{x^2 - x - 2}{4x^2 + 8x} \cdot \dfrac{4x}{x^2 - 4}$ Rewrite as a product.

$= \dfrac{(x-2)(x+1)}{4x(x+2)} \cdot \dfrac{4x}{(x+2)(x-2)}$ Factor.

$= \dfrac{x+1}{(x+2)^2}$ Divide out common factors.

(c) $\dfrac{2x^2 + x - 3}{1 + 3x - 4x^2} \div (2x + 3)$

$= \dfrac{2x^2 + x - 3}{1 + 3x - 4x^2} \div \dfrac{2x + 3}{1}$ Write the divisor as a fraction.

$= \dfrac{2x^2 + x - 3}{1 + 3x - 4x^2} \cdot \dfrac{1}{2x + 3}$ Rewrite as a product.

$= \dfrac{2x^2 + x - 3}{-1(4x^2 - 3x - 1)} \cdot \dfrac{1}{2x + 3}$ Factor out -1.

$= \dfrac{(2x+3)(x-1)}{-1(4x+1)(x-1)} \cdot \dfrac{1}{2x+3}$ Factor the trinomial.

$= \dfrac{-1}{4x + 1}$ Divide out common factors.

Quick Reference 8.2

Multiplication of Rational Expressions

- If a, b, c, and d are polynomials, where $b \neq 0$ and $d \neq 0$,

$$\frac{a}{b} \cdot \frac{c}{d} = \frac{ac}{bd}$$

- To multiply rational expressions, perform the following steps.
  1. Factor the numerators and denominators completely.
  2. Divide out any factor that is common to both a numerator and a denominator.
  3. Multiply the numerators and multiply the denominators.

Division of Rational Expressions

- If a, b, c, and d are polynomials, where b, c, and d are not 0,

$$\frac{a}{b} \div \frac{c}{d} = \frac{a}{b} \cdot \frac{d}{c}$$

- In words, to divide rational expressions, multiply by the reciprocal of the divisor.

Speaking the Language 8.2

1. We multiply rational expressions in the same way as we multiply numerical ▢▢▢▢▢ .

2. To multiply rational expressions, we factor the numerators and the denominators so that we can ▢▢▢▢▢ common factors.

3. To divide rational expressions, we multiply by the ▢▢▢▢ of the ▢▢▢▢ .

4. To multiply rational expressions, we multiply the ▢▢▢▢ and the ▢▢▢▢ .

Exercises 8.2

Concepts and Skills

1. To multiply rational expressions, we can
   (a) multiply, then divide out common factors, or
   (b) divide out common factors, then multiply.
   Explain why procedure (b) is preferred.

2. In the product $\frac{x}{5} \cdot \frac{3}{x}$, the common factor x can be divided out even though x appears in different fractions. Why?

In Exercises 3–20, multiply the rational expressions.

3. $\dfrac{3}{5x^2} \cdot \dfrac{35x^4}{12}$

4. $\dfrac{y^5}{15x^3} \cdot \dfrac{10x}{y}$

5. $\dfrac{2x + 1}{x - 5} \cdot \dfrac{15 - 3x}{2x - 1}$

6. $\dfrac{x}{4 - x} \cdot \dfrac{3x - 12}{x + 1}$

7. $(y + 3) \cdot \dfrac{y - 2}{4y + 12}$

8. $\dfrac{7x - 3}{6x - 20} \cdot (3x - 10)$

9. $\dfrac{4x - 12}{3x^2} \cdot \dfrac{6x}{2x - 6}$

10. $\dfrac{x^2 - 9}{15x^3} \cdot \dfrac{10x^4}{x^3 + 3x^2}$

11. $\dfrac{w^2 + 2w - 24}{9w^5} \cdot \dfrac{3w}{w - 4}$

12. $\dfrac{8y^3}{y^2 + 9y + 8} \cdot \dfrac{1 + y}{4y}$

13. $\dfrac{6x^2 + 18x}{x^2 - 9} \cdot \dfrac{(x - 3)^2}{3x}$

14. $\dfrac{(x + 4)^2}{10x^2 - 160} \cdot \dfrac{2}{x + 4}$

15. $\dfrac{6x - 4}{x + 1} \cdot \dfrac{1 + x}{4 - 6x}$

16. $\dfrac{3x + 15}{5 + x} \cdot \dfrac{4x - 3}{9 - 12x}$

17. $\dfrac{y^2 - 49}{8y + 56} \cdot \dfrac{y^2 - 7y}{(y - 7)^2}$

18. $\dfrac{(2w + 3)^2}{2w^2 + 3w} \cdot \dfrac{10w - 15}{4w^2 - 9}$

19. $\dfrac{3x^2 - 12x}{x^2 - x - 12} \cdot \dfrac{x + 3}{15x}$

20. $\dfrac{w^2 + 7w + 10}{8w^5} \cdot \dfrac{w^3 - 5w^2}{w^2 - 25}$

21. To divide rational expressions, what two changes do we make when we rewrite the quotient?

22. For what values of x is the following quotient not defined? Explain.

$$\dfrac{x + 2}{x - 1} \div \dfrac{x + 1}{x - 3}$$

In Exercises 23–28, determine the reciprocal of the expression.

23. x

24. $\dfrac{1}{y}$

25. $\dfrac{1}{a + 3}$

26. $z - 4$

27. $\dfrac{x + 2}{x - 1}$

28. $\dfrac{a}{a + b}$

In Exercises 29–40, divide the rational expressions.

29. $\dfrac{30a^4 b^3}{b} \div (12a^2 b^3)$

30. $\dfrac{44x^5 y^3}{xy} \div (33x^3 y)$

31. $\dfrac{12x + 6}{x^2} \div \dfrac{18x + 9}{3x}$

32. $\dfrac{x^2 - 5x}{6x^2} \div \dfrac{10x - 50}{15x}$

33. $\dfrac{x - 12}{-16} \div \dfrac{12 - x}{8}$

34. $\dfrac{2x - 1}{20} \div \dfrac{1 - 2x}{15}$

35. $\dfrac{x - 3x^2}{1 - 3x} \div (-3x^2)$

36. $\dfrac{xy - x^2 y}{x - x^2} \div (xy)$

37. $\dfrac{3(x - y)}{(2x - 5y)^2} \div \dfrac{12(y - x)}{7(2x - 5y)}$

38. $\dfrac{(a + b)^2}{24b - 12a} \div \dfrac{a + b}{4a - 8b}$

39. $\dfrac{4h}{h - 4} \div \dfrac{24h^2}{h^2 - 3h - 4}$

40. $\dfrac{5x^2 - 20x}{x^2 + 5x - 36} \div \dfrac{40x}{x + 9}$

In Exercises 41–60, perform the indicated operation.

41. $\dfrac{(y - 4)^2}{y^2 - 3y - 4} \cdot \dfrac{y^2 + 5y + 4}{y^2 - 16}$

42. $\dfrac{z^2 + 7z + 12}{z^2 + 5z + 4} \cdot \dfrac{z^2 - 7z - 8}{z^2 - 11z + 24}$

43. $\dfrac{3x^3 + 6x^2}{2x^2 + x - 6} \div (6x^2 - 15x)$

44. $\dfrac{4x^2 - 4x}{x^2 - 10x + 9} \div (4x^3 - 36x^2)$

45. $\dfrac{4d}{8d^3 - 32d} \div \dfrac{1}{d^2 + d - 6}$

46. $\dfrac{8w}{8w^3 - 20w^2} \div \dfrac{1}{2w^2 - 7w + 5}$

47. $\dfrac{z^2 - 3z - 10}{4z^2 + 8z} \cdot \dfrac{15 - 3z}{z^2 - 10z + 25}$

48. $\dfrac{n^2 - 3n - 40}{3n + 15} \cdot \dfrac{9n^2 + 72n}{n^2 - 64}$

49. $\dfrac{ab^3 + b^4}{2a^3 - a^2 b} \cdot \dfrac{ab - 2a^2}{b^5 + ab^4}$

50. $\dfrac{3x^2 - 2x^3}{6x^2 - 6x} \cdot \dfrac{2x^2 - 2x}{3x - 2x^2}$

51. $\dfrac{x^2 y^4}{x^2 - 2x - 3} \cdot \dfrac{x^2 + 3x + 2}{xy^5}$

52. $\dfrac{x^2 - 7x + 10}{x^5 y^3} \cdot \dfrac{x^8 y}{x^2 - 3x + 2}$

53. $\dfrac{a^2 + 5a + 4}{4a^3} \div \dfrac{a^2 - 16}{a^2 - 4a}$

54. $\dfrac{t^2 - 4}{t^2 - 25} \div \dfrac{6t^2 - 12t}{3t - 15}$

55. $\dfrac{3k - 6}{k^2 + 5k - 14} \div \dfrac{9k - 54}{k^2 + k - 42}$

56. $\dfrac{x^2 + 3x + 2}{x^2 - 4x - 12} \div \dfrac{2x + 2}{x^2 - 5x - 6}$

57. $\dfrac{x^2 + 5x - 24}{x^2 - 9} \div \dfrac{x^2 - 64}{x^2 + 9x + 18}$

58. $\dfrac{x^2 - x - 6}{x^2 - 4x - 12} \div \dfrac{x^2 - 5x + 6}{x^2 - 9x + 18}$

59. $\dfrac{x^2 + 2x - 8}{x^2 + 5x - 24} \cdot \dfrac{x^2 + 4x - 21}{x^2 + 5x - 14}$

60. $\dfrac{x^2 - x - 12}{x^2 + 10x + 9} \cdot \dfrac{9 + 10x + x^2}{x^2 + x - 6}$

Real-Life Applications

61. The number of hamburgers sold by a fast food chain is $\dfrac{x^2}{x+2}$ and their price is $\dfrac{4x+8}{x}$, where x is the amount of advertising for the product. Write a simplified rational expression for the amount of revenue.

62. The number of units of harvestable timber owned by a major paper company is given by $\dfrac{20x}{x+7}$, and the price per unit of timber is estimated by $\dfrac{x}{10}$. In these expressions, x represents a level of demand. Write a simplified rational expression for the value of the timber.

63. The number (in thousands) of votes a candidate receives for x dollars of campaign expenditures is $\dfrac{5x}{2x+1}$. Write a simplified rational expression for the cost per vote.

64. The number (in hundreds) of customers per month at the Bargain Barn t months after its opening is given by $\dfrac{20t}{t+5}$, and the monthly receipts (in thousands) are given by $\dfrac{300t}{t+1}$. Write a simplified rational expression for the average amount that each customer spends.

Modeling with Real Data

65. As indicated in the table, the total number of nuclear weapons worldwide is expected to decline from the levels of the 1980s. (Source: National Resources Defense Council.)

Year	Total Number of Nuclear Weapons
1980	61,472
1986	69,596
2004	22,425

With x representing the number of years since 1975, the number of weapons N (in thousands) can be modeled as follows.

$$N = 10 + \frac{590(x-3)}{(x-3)^2 + 19}$$

(a) Does the model expression have any restricted values?

(b) Evaluate the expression to estimate the number of weapons in 1982.

66. Refer to the model in Exercise 65.

(a) Evaluate the model expression to estimate the number of weapons in 1980 and in 1986.

(b) Is the model more accurate for 1980 or for 1986?

Challenge

In Exercises 67 and 68, determine the unknown numerator that makes the equation true.

67. $\dfrac{6x+9}{x-1} \cdot \dfrac{\rule{1.2cm}{0.25mm}}{2x^2+3x} = \dfrac{-3}{x}$

68. $\dfrac{12+16y}{2+y} \div \dfrac{\rule{1.2cm}{0.25mm}}{y^2+2y} = 4y$

In Exercises 69–71, perform the indicated operations.

69. $\dfrac{xy+3x-2y-6}{x^2-6x+8} \div \dfrac{y^2+5y+6}{xy-4y+2x-8}$

70. $\dfrac{2x+7}{6x} \div \left[\dfrac{1-2x}{3x} \div \dfrac{2-4x}{2x-7} \right]$

71. $\dfrac{x^3-8}{x^2-4} \cdot \dfrac{x^2+6x+8}{x^2+x-12} \div \dfrac{x^2+2x+4}{x^3-3x^2}$

72. Consider $\dfrac{x-3}{5} \cdot \dfrac{5}{x-3} = 1$. Should the graphs of $y_1 = \dfrac{x-3}{5} \cdot \dfrac{5}{x-3}$ and $y_2 = 1$ be the same? Why?

8.3 | Addition and Subtraction (Like Denominators)

Basic Rules • Adding Rational Expressions • Subtracting Rational Expressions

Basic Rules

In arithmetic we add (or subtract) fractions with the same denominator by adding (or subtracting) the numerators and retaining the common denominator. If possible, we reduce the result.

$$\frac{7}{15} + \frac{2}{15} = \frac{9}{15} = \frac{3 \cdot 3}{3 \cdot 5} = \frac{3}{5}$$

Addition and subtraction of rational expressions are performed in the same manner.

Addition and Subtraction of Rational Expressions

If a, b, and c are polynomials and $c \neq 0$, then

$$\frac{a}{c} + \frac{b}{c} = \frac{a+b}{c}$$ Addition of rational expressions

$$\frac{a}{c} - \frac{b}{c} = \frac{a-b}{c}$$ Subtraction of rational expressions

---

Compare the following.

$$\frac{3}{5} + \frac{1}{5} = \frac{4}{5} \qquad\qquad \frac{8}{15} - \frac{2}{15} = \frac{6}{15} = \frac{3 \cdot 2}{3 \cdot 5} = \frac{2}{5}$$

$$\frac{8}{x} + \frac{3}{x} = \frac{11}{x} \qquad\qquad \frac{5}{3x} - \frac{2}{3x} = \frac{3}{3x} = \frac{3 \cdot 1}{3 \cdot x} = \frac{1}{x}$$

The following summarizes the procedure for adding or subtracting rational expressions with like denominators.

Procedure for Adding or Subtracting Rational Expressions with Like Denominators

1. Add (or subtract) the numerators of each expression.
2. Retain the common denominator.
3. Simplify the result.

Adding Rational Expressions

In our first example of adding rational expressions, the resulting sums cannot be simplified.

EXAMPLE **1**

Adding Rational Expressions

Add the given rational expressions.

(a) $\dfrac{2x - 3}{x - 1} + \dfrac{2}{x - 1} = \dfrac{(2x - 3) + 2}{x - 1}$ Add numerators; retain denominator.

$= \dfrac{2x - 1}{x - 1}$ Cannot simplify the result.

(b) $\dfrac{x^2 + 3x - 2}{2x - 1} + \dfrac{2x^2 - x - 5}{2x - 1}$

$= \dfrac{(x^2 + 3x - 2) + (2x^2 - x - 5)}{2x - 1}$ Add numerators; retain denominator.

$= \dfrac{3x^2 + 2x - 7}{2x - 1}$ Cannot simplify the result.

In Example 2 the extra step of simplifying the result is required.

EXAMPLE **2**

Adding Rational Expressions and Simplifying

Add the given rational expressions.

(a) $\dfrac{2x + 1}{9x^2} + \dfrac{x - 4}{9x^2} = \dfrac{3x - 3}{9x^2}$ Add the numerators.

$= \dfrac{3(x - 1)}{3 \cdot 3x^2}$ Factor the numerator.

$= \dfrac{x - 1}{3x^2}$ Divide out the common factor 3.

(b) $\dfrac{2x^2 - 7x}{2x + 1} + \dfrac{2x - 3}{2x + 1} = \dfrac{2x^2 - 5x - 3}{2x + 1}$ Add the numerators.

$= \dfrac{(2x + 1)(x - 3)}{2x + 1}$ Factor the trinomial.

$= x - 3$ Divide out the common factor $2x + 1$.

(c) $\dfrac{x^2 + 2x}{x^2 - 16} + \dfrac{x - 4}{x^2 - 16} = \dfrac{x^2 + 3x - 4}{x^2 - 16}$ Add the numerators.

$= \dfrac{(x + 4)(x - 1)}{(x + 4)(x - 4)}$ Factor the numerator and the denominator.

$= \dfrac{x - 1}{x - 4}$ Divide out the common factor $x + 4$.

Subtracting Rational Expressions

Recall that fraction bars are grouping symbols. When we add or subtract rational expressions, each numerator must be treated as though it were enclosed in parentheses.

This is not an important consideration in addition (see Example 1) because the parentheses can simply be removed. However, in subtraction it is essential that we regard the numerators as groups. It is therefore advisable to insert parentheses as the first step. When we subtract, we change signs as we remove the parentheses and then combine the like terms.

EXAMPLE **3**

Subtracting Rational Expressions

Subtract the given rational expressions.

$$\frac{7y + 2}{5y} - \frac{2y - 1}{5y} = \frac{(7y + 2)}{5y} - \frac{(2y - 1)}{5y}$$ Insert parentheses.

$$= \frac{(7y + 2) - (2y - 1)}{5y}$$ Subtract the numerators.

$$= \frac{7y + 2 - 2y + 1}{5y}$$ Remove parentheses.
Note the sign changes.

$$= \frac{5y + 3}{5y}$$ Combine like terms.

Although $5y$ is a factor in the denominator, it is a term, not a factor, in the numerator. Thus the result cannot be simplified.

In Example 4 we insert parentheses and subtract numerators in one step. Although it is possible to insert and remove parentheses mentally, doing so invites sign errors.

EXAMPLE **4**

Subtracting Rational Expressions and Simplifying

Subtract the given rational expressions.

LEARNING TIP

Because addition and subtraction problems may seem long, you may be tempted to skip steps as you perform the operations. Writing each step completely will eliminate many common errors.

(a) $$\frac{x + 2}{1 - 2x} - \frac{3 - x}{1 - 2x} = \frac{(x + 2) - (3 - x)}{1 - 2x}$$ Insert parentheses.

$$= \frac{x + 2 - 3 + x}{1 - 2x}$$ Remove parentheses.

$$= \frac{2x - 1}{1 - 2x}$$ Combine like terms.

$$= \frac{-1(1 - 2x)}{1 - 2x}$$ Factor out -1.

$$= -1$$ Divide out the common factor $1 - 2x$.

(b) $$\frac{x^2 + 6x + 7}{x^2 + 7x + 10} - \frac{x^2 + x - 3}{x^2 + 7x + 10}$$

$$= \frac{(x^2 + 6x + 7) - (x^2 + x - 3)}{x^2 + 7x + 10}$$ Insert parentheses.

$$= \frac{x^2 + 6x + 7 - x^2 - x + 3}{x^2 + 7x + 10}$$ Remove parentheses.

$$= \frac{5x + 10}{x^2 + 7x + 10}$$ Combine like terms.

$$= \frac{5(x + 2)}{(x + 5)(x + 2)}$$ Factor the numerator and denominator.

$$= \frac{5}{x + 5}$$ Divide out the common factor $x + 2$.

Think About It

A fraction $\frac{a}{b}$ can be written as $a \cdot \frac{1}{b}$.

Use this fact along with the Distributive Property to show why the two fractions in (i) can be added in their present form but the fractions in (ii) cannot.

(i) $\frac{2}{3} + \frac{5}{3}$ (ii) $\frac{2}{3} + \frac{5}{4}$

Quick Reference 8.3

Basic Rules
- If a, b, and c are polynomials and $c \neq 0$, then

$$\frac{a}{c} + \frac{b}{c} = \frac{a+b}{c} \qquad \text{Addition of rational expressions}$$

$$\frac{a}{c} - \frac{b}{c} = \frac{a-b}{c} \qquad \text{Subtraction of rational expressions}$$

- The following summarizes the procedure for adding or subtracting rational expressions with like denominators.
  1. Add (or subtract) the numerators of each expression.
  2. Retain the common denominator.
  3. Simplify the result.

Adding Rational Expressions
- Simplifying the result of adding (or subtracting) rational expressions involves factoring the numerator and the denominator and dividing out common factors.
- The technique of factoring out a negative numerical factor can sometimes be useful in simplifying results.

Subtracting Rational Expressions
- When subtracting rational expressions, it is essential to regard each numerator as a group enclosed in parentheses. Sign errors can be avoided by inserting parentheses around numerators before performing the subtraction.

Speaking the Language 8.3

1. In order for us to add or subtract rational expressions, the expressions must have a(n) ▭ .
2. To add rational expressions with like denominators, we add the ▭ and retain the ▭ .
3. After adding or subtracting rational expressions, we should try to ▭ the result.
4. When we add or subtract rational expressions, enclosing the numerators in parentheses is a good idea because the numerators should be regarded as ▭ .

Exercises 8.3

Concepts and Skills

1. Suppose that you wish to add $\frac{5}{9} + \frac{3}{9}$. Why is reducing $\frac{3}{9}$ to $\frac{1}{3}$ not a good first step?

2. What error has been made in performing the following operation?

$$\frac{4}{x} + \frac{7}{x} = \frac{11}{2x}$$

In Exercises 3–18, add the rational expressions.

3. $\dfrac{9}{4x} + \dfrac{9}{4x}$

4. $\dfrac{7}{5y} + \dfrac{3}{5y}$

5. $\dfrac{3x+4}{x} + \dfrac{2x-9}{x}$

6. $\dfrac{3x-8y}{y} + \dfrac{2x+7y}{y}$

7. $\dfrac{3x}{1-x} + \dfrac{x-4}{1-x}$

8. $\dfrac{29-6x}{x-7} + \dfrac{x+6}{x-7}$

9. $\dfrac{x}{2x+3} + \dfrac{x-3}{2x+3}$

10. $\dfrac{x-1}{4x-1} + \dfrac{x}{4x-1}$

11. $\dfrac{y^2}{y+4} + \dfrac{4y}{y+4}$

12. $\dfrac{x^2}{x-1} + \dfrac{-x}{x-1}$

13. $\dfrac{y^2+y-5}{1-4y} + \dfrac{3y^2+2y+4}{1-4y}$

14. $\dfrac{x^2-x-8}{x+9} + \dfrac{10-x^2}{x+9}$

15. $\dfrac{x^2+10x}{x^2-4} + \dfrac{10-3x}{x^2-4}$

16. $\dfrac{x^2+3x}{x^2+x-2} + \dfrac{x^2-2}{x^2+x-2}$

17. $\dfrac{3x+7}{x^2+5x+6} + \dfrac{x+5}{x^2+5x+6}$

18. $\dfrac{2x+1}{x^2-4x-12} + \dfrac{x+5}{x^2-4x-12}$

In Exercises 19–36, subtract the rational expressions.

19. $\dfrac{5}{x} - \dfrac{4}{x}$

20. $\dfrac{7}{a} - \dfrac{2}{a}$

21. $\dfrac{t^3+8}{3t} - \dfrac{8-5t^3}{3t}$

22. $\dfrac{3z+7}{8z^4} - \dfrac{7-z}{8z^4}$

23. $\dfrac{5x+2}{x} - \dfrac{5x-2}{x}$

24. $\dfrac{2t+3}{t} - \dfrac{3-2t}{t}$

25. $\dfrac{3x}{4-3x} - \dfrac{4}{4-3x}$

26. $\dfrac{4}{5x-2} - \dfrac{10x}{5x-2}$

27. $\dfrac{x-30}{5x-9} - \dfrac{6-19x}{5x-9}$

28. $\dfrac{9x+25}{6x+7} - \dfrac{4-9x}{6x+7}$

29. $\dfrac{8-x}{x-3} - \dfrac{2+x}{x-3}$

30. $\dfrac{3x+2}{2x+1} - \dfrac{5x+3}{2x+1}$

31. $\dfrac{2x-1}{x^2-7x+6} - \dfrac{2-x}{x^2-7x+6}$

32. $\dfrac{5x+6}{x^2+10x+21} - \dfrac{3x-8}{x^2+10x+21}$

33. $\dfrac{2x^2-3x}{2x-7} - \dfrac{2x+7}{2x-7}$

34. $\dfrac{3x^2}{3x+2} - \dfrac{x+2}{3x+2}$

35. $\dfrac{3x-1}{2x^2-5x} - \dfrac{2x-1}{2x^2-5x}$

36. $\dfrac{7-15x}{6x^2-8x^3} - \dfrac{9x-11}{6x^2-8x^3}$

37. In the following subtraction problem, how might the error have been prevented?

$$\dfrac{x}{y} - \dfrac{2a+3}{y} = \dfrac{x-2a+3}{y}$$

38. In the following subtraction problem, explain how the result can be simplified.

$$\dfrac{y}{x-y} - \dfrac{x}{x-y}$$

In Exercises 39–60, perform the indicated operations.

39. $\dfrac{7x-2}{x} + \dfrac{2x-3}{x}$

40. $\dfrac{t-4}{t} + \dfrac{4}{t}$

41. $\dfrac{2x-4}{5} - \dfrac{3}{5}$

42. $\dfrac{3x}{7} - \dfrac{1-x}{7}$

43. $\dfrac{2(z+1)}{3z^2} + \dfrac{5(z-1)}{3z^2}$

44. $\dfrac{5(4t-1)}{7t^2} + \dfrac{5(3t+1)}{7t^2}$

45. $\dfrac{5x-4y}{y} - \dfrac{8x-9y}{y}$

46. $\dfrac{x-4y}{y} - \dfrac{2x-3y}{y}$

47. $\dfrac{2x-3y}{x-y} + \dfrac{2y-x}{x-y}$

48. $\dfrac{1-x}{2x-3} + \dfrac{2-x}{2x-3}$

49. $\dfrac{4x+10}{x-1} - \dfrac{7x+7}{x-1}$

50. $\dfrac{3x+5}{2-x} - \dfrac{2x+7}{2-x}$

51. $\dfrac{3t+25}{4(t+5)} + \dfrac{t-5}{4(t+5)}$

52. $\dfrac{5x-9}{3(x-2)} + \dfrac{x-3}{3(x-2)}$

53. $\dfrac{x+15}{x+7} - \dfrac{1-x}{x+7}$

54. $\dfrac{9x+1}{x-5} - \dfrac{5x+21}{x-5}$

55. $\dfrac{3(x-2)}{(x+2)(x-4)} - \dfrac{x+2}{(x+2)(x-4)}$

56. $\dfrac{3(x+5)}{(x+8)(x-1)} - \dfrac{x-1}{(x+8)(x-1)}$

57. $\dfrac{5x+4}{x^2-4} - \dfrac{x-x^2}{x^2-4}$

58. $\dfrac{7x-3}{x^2+4x-12} - \dfrac{2x+7}{x^2+4x-12}$

59. $\dfrac{x^2-4}{2x^2-x-10} + \dfrac{x^2-5x+4}{2x^2-x-10}$

60. $\dfrac{x^2-9}{x^2+6x+5} + \dfrac{6-2x}{x^2+6x+5}$

In Exercises 61–64, determine the unknown numerator that makes the statement true.

61. $\dfrac{2x-5}{x+7} + \dfrac{\rule{1cm}{0.3em}}{x+7} = \dfrac{3x+3}{x+7}$

62. $\dfrac{y+7}{5-y} - \dfrac{\rule{1cm}{0.3em}}{5-y} = \dfrac{2y-1}{5-y}$

63. $\dfrac{\rule{1cm}{0.3em}}{x^2+1} - \dfrac{2x+1}{x^2+1} = \dfrac{x^2-1}{x^2+1}$

64. $\dfrac{2x-y}{x+3y} + \dfrac{\rule{1cm}{0.3em}}{x+3y} = \dfrac{2y}{x+3y}$

In Exercises 65–68, perform the indicated operations.

65. $\dfrac{9 - 2x}{x + 3} + \dfrac{9x + 7}{x + 3} - \dfrac{x - 2}{x + 3}$

66. $\dfrac{x - 6}{x - 2} - \dfrac{2x - 2}{x - 2} + \dfrac{3x}{x - 2}$

67. $\dfrac{2z + 2}{3z} - \left[\dfrac{4z - 3}{3z} - \dfrac{5z - 6}{3z}\right]$

68. $\dfrac{x + 3}{x + 4} - \left[\dfrac{2(x - 1)}{x + 4} + \dfrac{3(2x + 1)}{x + 4}\right]$

Real-Life Applications

69. The number of votes (in thousands) that candidates C_1 and C_2 will receive in a large state election is predicted by

$$C_1(x) = \frac{x^2}{100 - x} \quad \text{and} \quad C_2(x) = \frac{100x}{100 - x}$$

where x is the number of millions of dollars spent on campaigning. (Assume that campaign spending does not exceed $100 million.)

(a) Produce the graphs of the two expressions and, simply by observation, predict the winner of the election.

(b) Trace the graphs to estimate the winning margin given that each candidate spends $8 million for campaigning.

(c) Write a simplified expression for the winning margin if each candidate spends x million dollars. Evaluate this expression for $x = 8$ and compare the result to your estimate in part (b).

70. An economist finds that the funding (in millions) for summer jobs programs from a state government and from private sources can be described by

$$G(x) = \frac{x^2}{x + 10} \quad \text{and} \quad P(x) = \frac{10x}{x + 10}$$

where x is the unemployment rate expressed as a percentage.

(a) Produce the graphs of $G(x)$ and $P(x)$. If the unemployment rate is under 10%, is the state government or the private sector the greater source of funding?

(b) By tracing both graphs, estimate the total funding when the unemployment rate is 5%.

(c) Write a simplified expression for the total funding for any unemployment rate. Then evaluate this expression for $x = 5$ and compare the result to your estimate in part (b).

71. In determining the impact on tax revenue for a community t months after acquisition of a new industry, the chamber of commerce found that the sales tax generated can be described by $\dfrac{t^2 + t - 12}{t + 3}$ and the property tax can be described by $\dfrac{2t}{t + 3}$. Write an expression to describe the amount by which the sales tax will exceed the property tax.

72. A major airline determines that the miles (in thousands) accumulated in the frequent flier program from business and personal travel can be modeled by

$$B(x) = \frac{x^4}{x^2 + 2} \quad \text{and} \quad P(x) = \frac{2x^2}{x^2 + 2}$$

where x is a measure of fare discount activity. Write a simplified expression for the total number of frequent flier miles accumulated.

Modeling with Real Data

73. The table shows the past and projected composition of the civilian labor force.

| | Civilian Labor Force (millions) | | |
Year	Men	Women	Total
1970	51.2	31.5	82.7
1980	61.5	45.5	107.0
1990	68.2	56.6	124.8
1994	70.8	60.2	131.0
2000	75.3	66.6	141.9
2005	78.7	71.8	150.5

(Source: U.S. Bureau of Labor Statistics.)

With x representing the number of years since 1970, the rational expressions represent the fraction of the labor force made up of men and of women.

Men: $\dfrac{0.77x + 52.42}{1.90x + 85.29}$

Women: $\dfrac{1.13x + 32.87}{1.90x + 85.29}$

(a) What would you expect the sum of these expressions to be.

(b) Add the expressions. Does the result agree with your prediction in part (a)?

74. Refer to the models in Exercise 73.

(a) Use first-quadrant graphs of the model expressions to predict the trends for men and women.

(b) If you wanted to trace the graphs to estimate the year when the fractions for men and women are equal, what *y*-coordinate would you seek?

Challenge

In Exercises 75–80, perform the indicated operations.

75. $\dfrac{x^2 - 1}{x^3 + 8} - \dfrac{2x - 5}{x^3 + 8}$

76. $\dfrac{a^2 - 3a}{a^2 - b^2} + \dfrac{ab - 3b}{a^2 - b^2}$

77. $\dfrac{x - 3}{x - 5} + \dfrac{x}{x + 1} \div \dfrac{x - 5}{x + 1}$

78. $(2x - 1) \cdot \dfrac{x - 4}{x^2 - 16} - \dfrac{x - 1}{x + 4}$

79. $\left[\dfrac{2x - 5}{x^2 - x} + \dfrac{x + 4}{x^2 - x} \right] \cdot \dfrac{x^2 - 1}{6x - 2}$

80. $\left[\dfrac{3x - 5}{3y} - \dfrac{2x - 7}{3y} \right] \div \dfrac{x^2 - 4}{12y^2}$

8.4 | Least Common Denominators

Renaming Fractions • Least Common Multiples • Least Common Denominators

Renaming Fractions

Fractions have infinitely many equivalent names or forms. For instance, $\frac{1}{2}$ can be rewritten as $\frac{3}{6}$ or $\frac{8}{16}$. Recall that when we *simplify* or *reduce* a fraction, we use the following rule to divide out common factors from the numerator and the denominator.

$$\frac{ac}{bc} = \frac{a}{b}, \qquad \text{where } b \neq 0, c \neq 0$$

Using this rule results in a fraction with a smaller denominator. We can use this same rule in reverse to rename a fraction with a larger denominator.

$$\frac{a}{b} = \frac{ac}{bc}, \qquad \text{where } b \neq 0, c \neq 0$$

This version of the rule allows us to multiply the numerator and the denominator of a fraction by any nonzero number. For example,

$$\frac{3}{7} = \frac{3 \cdot 5}{7 \cdot 5} = \frac{15}{35}$$

Sometimes we need to rename a fraction in this way so that it has a specified denominator.

Exploring the Concept

Renaming Fractions

Compare the process of rewriting the following fractions with new denominators.

(i) $\dfrac{4}{5} = \dfrac{}{15}$

(ii) $\dfrac{3}{x + 2} = \dfrac{}{x^2 - 4}$

We factor each new denominator so that the original denominator is one of the factors.

(i) $\dfrac{4}{5} = \dfrac{}{5 \cdot 3}$

(ii) $\dfrac{3}{x + 2} = \dfrac{}{(x + 2)(x - 2)}$

In (i), the factor 3 is missing in the original denominator. In (ii), the factor $x - 2$ is missing in the original denominator. In each we multiply both the numerator and the denominator of the original fraction by the missing factor to obtain an equivalent fraction with the desired denominator.

(i) $\dfrac{4}{5} = \dfrac{4 \cdot 3}{5 \cdot 3} = \dfrac{12}{15}$
(ii) $\dfrac{3}{x + 2} = \dfrac{3 \cdot (x - 2)}{(x + 2)(x - 2)} = \dfrac{3x - 6}{x^2 - 4}$

A key step in renaming a fraction is determining the factors in the new denominator that are missing in the original denominator. To do this, it may be necessary to factor one or both of the denominators.

EXAMPLE 1

Renaming Rational Expressions

Rewrite each rational expression with the specified denominator.

(a) $\dfrac{5}{3x} = \dfrac{}{12x^3}$
(b) $\dfrac{x}{3 - 5x} = \dfrac{}{5x - 3}$

(c) $\dfrac{2}{x + 3} = \dfrac{}{x^2 + 9x + 18}$
(d) $\dfrac{x + 4}{x^2 - 4x} = \dfrac{}{x^3 - 16x}$

Solution

(a) Because $12x^3 = 3x \cdot 4x^2$, multiply the numerator and the denominator by the factor $4x^2$.

$$\frac{5}{3x} = \frac{5 \cdot 4x^2}{3x \cdot 4x^2} = \frac{20x^2}{12x^3}$$

(b) Because $5x - 3 = -1(3 - 5x)$, multiply the numerator and the denominator by the factor -1.

$$\frac{x}{3 - 5x} = \frac{-1 \cdot x}{-1(3 - 5x)} = \frac{-x}{5x - 3}$$

(c) Because $x^2 + 9x + 18 = (x + 3)(x + 6)$, multiply by $x + 6$.

$$\frac{2}{x + 3} = \frac{2(x + 6)}{(x + 3)(x + 6)} = \frac{2x + 12}{x^2 + 9x + 18}$$

(d) Factor each denominator.

$x^2 - 4x = x(x - 4)$ Original denominator
$x^3 - 16x = x(x^2 - 16) = x(x + 4)(x - 4)$ Specified denominator

The original denominator is missing the factor $(x + 4)$.

$$\frac{x + 4}{x^2 - 4x} = \frac{x + 4}{x(x - 4)}$$

$$= \frac{(x + 4)(x + 4)}{x(x - 4)(x + 4)} \qquad \text{Multiply by } x + 4.$$

$$= \frac{(x + 4)^2}{x(x^2 - 16)} = \frac{x^2 + 8x + 16}{x^3 - 16x}$$

Least Common Multiples

A **common multiple** of two numbers is a number that is divisible by both numbers. For example, 20, 40, 60, . . . are all common multiples of 4 and 5. Of these infinitely many common multiples, 20 is the smallest and is called the **least common multiple (LCM)** of 4 and 5. Once again, factoring is useful in determining the LCM of two or more numbers or expressions. Suppose that we need to know the LCM of 6 and 8.

$$6 = 2 \cdot 3$$
$$8 = 2 \cdot 2 \cdot 2 = 2^3$$
$$\text{LCM: } 2^3 \cdot 3 = 24$$

Note that the LCM consists of each factor that appears in either factorization, and the exponent on each factor of the LCM is the largest one that appears on that factor in either factorization.

EXAMPLE 2

Determining Least Common Multiples

Determine the LCM of the given numbers or expressions.

(a) 10, 12, 15

(b) $2x^3, x^2, 3x$

(c) $x + 2, x^2 + 4x + 4$

(d) $x^2 - 9, x^2 - 2x - 3$

LEARNING TIP

Try not to confuse LCM with GCF. For a GCF, a factor must appear in *all* expressions, and we select the one with the *smallest* exponent. For a LCM, a factor must appear in *at least one* expression, and we select the one with the *largest* exponent.

Solution

(a) $10 = 2 \cdot 5 \qquad 12 = 2^2 \cdot 3 \qquad 15 = 3 \cdot 5$

   LCM: $2^2 \cdot 3 \cdot 5 = 60$

(b) $2x^3, x^2, 3x$

   The LCM consists of the factors 2, 3, and x, and the highest exponent on x is 3. Thus the LCM is $2 \cdot 3 \cdot x^3$, or $6x^3$.

(c) First expression: $x + 2$

   Second expression: $x^2 + 4x + 4 = (x + 2)^2$

   LCM: $(x + 2)^2$

(d) First expression: $x^2 - 9 = (x + 3)(x - 3)$

   Second expression: $x^2 - 2x - 3 = (x - 3)(x + 1)$

   LCM: $(x + 3)(x - 3)(x + 1)$

   Note that we often leave more complicated LCMs in factored form.

Least Common Denominators

In Section 8.3 we learned how to add and subtract rational expressions with like denominators. If the expressions have unlike denominators, we must first rename the fractions so that they have the same (common) denominator. To minimize the work involved, we usually determine the **least common denominator (LCD)**, which is the LCM of the denominators.

Note: In practice, LCM and LCD have similar meanings. We say LCM when we refer to expressions in general, but we say LCD when we refer specifically to expressions that are denominators of fractions.

An LCD is constructed in the same manner as an LCM. The following summarizes the procedure.

Determining a Least Common Denominator

1. Factor each denominator completely.
2. Include in the LCD each factor that appears in at least one denominator.
3. For each factor, use the largest exponent that appears on that factor in any denominator.

EXAMPLE 3 | **Determining a Least Common Denominator**

Determine the LCD of the given fractions.

(a) $\dfrac{2}{3x}, \dfrac{5}{x^2+2x}$

(b) $\dfrac{y}{1-3y}, \dfrac{y+1}{3y-1}$

(c) $\dfrac{t+3}{t^2+10t+25}, \dfrac{t}{t^2-25}$

(d) $\dfrac{x-2}{3x^3+6x^2}, \dfrac{x-1}{x^2+3x+2}, \dfrac{1}{9x}$

Solution

Think About It

We can obtain a common multiple of two expressions simply by multiplying the expressions. If the result is also the *least* common multiple, what do you know about the expressions?

(a) $3x = 3 \cdot x$
$x^2 + 2x = x(x+2)$
LCD: $3x(x+2)$

(b) Because $1 - 3y = -1(3y-1)$, either $1-3y$ or $3y-1$ can be used as the LCD.

(c) $t^2 + 10t + 25 = (t+5)^2$
$t^2 - 25 = (t+5)(t-5)$
LCD: $(t+5)^2(t-5)$

(d) $3x^3 + 6x^2 = 3x^2(x+2)$
$x^2 + 3x + 2 = (x+2)(x+1)$
$9x = 3^2 \cdot x$
LCD: $3^2 \cdot x^2 \cdot (x+2)(x+1) = 9x^2(x+2)(x+1)$

Once the LCD of two or more fractions is determined, we can compare the factors in each original denominator to the factors in the LCD. Then, as we did at the beginning of this section, we can rename each original fraction by multiplying the numerator and the denominator by any missing factors.

EXAMPLE 4 | **Renaming Rational Expressions with a Least Common Denominator**

Determine the LCD for each pair of rational expressions. Then rename each expression as an equivalent expression with that LCD.

(a) $\dfrac{5}{4x^2y}, \dfrac{7}{6xy^3}$

(b) $\dfrac{2x}{x^2+4x+3}, \dfrac{x-3}{x^2-x-2}$

Solution

(a) $4x^2y = 2^2 \cdot x^2 \cdot y$ Factor each denominator.

$6xy^3 = 2 \cdot 3 \cdot x \cdot y^3$

LCD: $2^2 \cdot 3 \cdot x^2 \cdot y^3 = 12x^2y^3$ Construct the LCD.

First fraction: $\dfrac{5}{4x^2y} = \dfrac{5 \cdot 3y^2}{4x^2y \cdot 3y^2} = \dfrac{15y^2}{12x^2y^3}$

Second fraction: $\dfrac{7}{6xy^3} = \dfrac{7 \cdot 2x}{6xy^3 \cdot 2x} = \dfrac{14x}{12x^2y^3}$

(b) $x^2 + 4x + 3 = (x + 3)(x + 1)$ Factor each denominator.

$x^2 - x - 2 = (x + 1)(x - 2)$

LCD: $(x + 3)(x + 1)(x - 2)$ Construct the LCD.

First fraction: $\dfrac{2x}{x^2 + 4x + 3} = \dfrac{2x}{(x + 3)(x + 1)}$

$= \dfrac{2x \cdot (x - 2)}{(x + 3)(x + 1)(x - 2)}$

$= \dfrac{2x^2 - 4x}{(x + 3)(x + 1)(x - 2)}$

Second fraction: $\dfrac{x - 3}{x^2 - x - 2} = \dfrac{x - 3}{(x + 1)(x - 2)}$

$= \dfrac{(x - 3)(x + 3)}{(x + 1)(x - 2)(x + 3)}$

$= \dfrac{x^2 - 9}{(x + 1)(x - 2)(x + 3)}$

Quick Reference 8.4

Renaming Fractions
- The following rule allows us to rename a fraction by multiplying the numerator and the denominator by any nonzero number.

$$\frac{a}{b} = \frac{ac}{bc}, \qquad \text{where } b \neq 0, c \neq 0$$

- To rename a fraction so that it has a specified denominator, factor the original denominator and the specified denominator. Determine which factors in the specified denominator are missing in the original denominator and multiply accordingly.

Least Common Multiples
- A **common multiple** of two numbers is a number that is divisible by both numbers. Of the infinitely many common multiples two numbers have, the smallest is called the **least common multiple (LCM)**.
- To determine the LCM of two or more numbers or expressions, first factor each number or expression. The LCM consists of each factor that appears in any of the factorizations, and the exponent on each factor of the LCM is the largest one that appears on that factor in any of the factorizations.

Least Common Denominators

- The **least common denominator (LCD)** of two or more fractions is the LCM of the denominators.
- To determine the LCD of two or more fractions, use the following procedure.
  1. Factor each denominator completely.
  2. Include in the LCD each factor that appears in at least one denominator.
  3. For each factor, use the largest exponent that appears on that factor in any denominator.
- Once the LCD of two or more fractions is determined, we can compare the factors in each original denominator to the factors in the LCD. Then we can rename each original fraction by multiplying the numerator and the denominator by any missing factors.

Speaking the Language 8.4

1. The process of writing $\frac{2}{3}$ as $\frac{10}{15}$ is called _____ the fraction.
2. A number that is divisible by two or more numbers is called a(n) _____ of the numbers.
3. The _____ denominator of two or more fractions is the smallest number that is divisible by the denominators.
4. Determining the LCD of fractions often begins with _____ the denominators.

Exercises 8.4

Concepts and Skills

1. When we rename a fraction so that it has a specified denominator, it may be useful to factor both the original and the specified denominators. Why?

2. What is the difference between an LCM and an LCD?

In Exercises 3–22, rewrite the rational expression with the specified denominator.

3. $\dfrac{3y}{2x} = \dfrac{}{6xy}$

4. $\dfrac{-b}{5a} = \dfrac{}{45ab}$

5. $\dfrac{-5}{12y^2} = \dfrac{}{60y^5}$

6. $\dfrac{3}{2xy} = \dfrac{}{10x^2y}$

7. $\dfrac{4}{3x} = \dfrac{}{9x^2 + 6x}$

8. $\dfrac{5}{6y} = \dfrac{}{12y^2 + 6y}$

9. $\dfrac{2x - 1}{x + 1} = \dfrac{}{3x^2 + 3x}$

10. $\dfrac{x + 3}{2x - 5} = \dfrac{}{6x^2 - 15x}$

11. $\dfrac{y}{y - 3} = \dfrac{}{3 - y}$

12. $\dfrac{5}{7 - x} = \dfrac{}{x^2 - 7x}$

13. $\dfrac{-6}{x - 3} = \dfrac{}{(x - 3)^2}$

14. $\dfrac{-x}{x + 8} = \dfrac{}{2(x + 8)^2}$

15. $\dfrac{x - 5}{4x + 12} = \dfrac{}{8x^2 + 24x}$

16. $\dfrac{3}{x^2 - x} = \dfrac{}{2x^3 - 2x^2}$

17. $\dfrac{-3}{x + 5} = \dfrac{}{x^2 + 4x - 5}$

18. $\dfrac{7}{x - 7} = \dfrac{}{x^2 - 9x + 14}$

19. $\dfrac{y - 6}{y - 4} = \dfrac{}{y^2 + 2y - 24}$

20. $\dfrac{t + 4}{t + 5} = \dfrac{}{t^2 + 9t + 20}$

21. $\dfrac{2x}{6x - 15} = \dfrac{}{12x^2 - 75}$

22. $\dfrac{4}{2t^2 - t} = \dfrac{}{4t^3 - t}$

23. Consider $3x^5$ and $9x^2$. Compare the process of determining the GCF with that of determining the LCM.

24. Explain how to determine the LCD of two fractions whose denominators are different prime polynomials.

In Exercises 25–40, determine the LCM of the given expressions.

25. $24x$, $8y$

26. $15x$, 27

27. $5x^3$, $10x^2$

28. $3a^5b^2$, $2a^6b^4$

29. b, $b + 2$

30. $x + 5$, $x - 4$

31. $8x$, $4x - 8$

32. $6x^2 + 10x$, $3x^4$

33. $2x - 7$, $7 - 2x$

34. $36t^2 - 25$, $5 - 6t$

35. $x^2 - 49, \quad x - 7$

36. $x + 5, \quad x^2 + 7x + 10$

37. $x^2 - 64, \quad 3x - 24$

38. $x^2 + 3x - 4, \quad x^2 - 16$

39. $x^2 - 6x + 5, \quad x^2 + 4x - 5$

40. $y^2 - 8y + 16, \quad y^2 - 9y + 20$

In Exercises 41–56, determine the LCD of the given fractions.

41. $\dfrac{a}{35}, \quad \dfrac{a+1}{42y}$

42. $\dfrac{x+1}{8a}, \quad \dfrac{x}{10b}$

43. $\dfrac{x+1}{x^4y^3}, \quad \dfrac{x-1}{2xy^2}$

44. $\dfrac{2t+1}{9t^3}, \quad \dfrac{t+1}{6t^5}$

45. $\dfrac{x}{x-7}, \quad \dfrac{x+8}{x-8}$

46. $\dfrac{b}{a}, \quad \dfrac{a}{a-b}$

47. $\dfrac{2}{5x^2}, \quad \dfrac{x+3}{x^2-2x}$

48. $\dfrac{y-1}{15y}, \quad \dfrac{y}{10y+5}$

49. $\dfrac{1}{1-2x}, \quad \dfrac{x}{4x^2-1}$

50. $\dfrac{1}{x-3}, \quad \dfrac{-1}{3-x}$

51. $\dfrac{x-4}{x^2-5x-6}, \quad \dfrac{x}{x-6}$

52. $\dfrac{x-4}{x^2-16}, \quad \dfrac{x}{x+4}$

53. $\dfrac{x}{x^2-1}, \quad \dfrac{x+4}{x^2+2x-3}$

54. $\dfrac{x}{x^2-1}, \quad \dfrac{x+2}{8x-8}$

55. $\dfrac{x^2}{x^2+6x+9}, \quad \dfrac{x^2-2}{x^2+2x-3}$

56. $\dfrac{x+1}{x^2+5x+6}, \quad \dfrac{x+3}{x^2-2x-8}$

In Exercises 57–72, determine the LCD for each pair of rational expressions. Then rename each expression as an equivalent expression with that LCD.

57. $\dfrac{a}{3b^2}, \quad \dfrac{b}{a}$

58. $5, \quad \dfrac{3}{x}$

59. $\dfrac{2}{x}, \quad \dfrac{x}{x+4}$

60. $\dfrac{7}{y-2}, \quad \dfrac{3}{y}$

61. $2x+5, \quad \dfrac{x}{x+3}$

62. $x-2, \quad \dfrac{3}{2x-1}$

63. $\dfrac{x}{x-2}, \quad \dfrac{x+2}{x+1}$

64. $\dfrac{2}{x+3}, \quad \dfrac{2x-1}{x-4}$

65. $\dfrac{x}{2x+6}, \quad \dfrac{x+1}{5x+15}$

66. $\dfrac{x+2}{x^2+2x}, \quad \dfrac{x-3}{x+2}$

67. $\dfrac{3}{(x+3)^2}, \quad \dfrac{2}{x(x+3)}$

68. $\dfrac{x}{4(x-2)}, \quad \dfrac{3}{(x-2)^2}$

69. $\dfrac{x+3}{2x^2-9x+10}, \quad \dfrac{3}{x-2}$

70. $\dfrac{2x}{x+6}, \quad \dfrac{2x-1}{x^2+3x-18}$

71. $\dfrac{x}{4x^2+4x+1}, \quad \dfrac{x+2}{2x^2+3x+1}$

72. $\dfrac{x+1}{3x^2-4x+1}, \quad \dfrac{x-3}{x^2-1}$

🌐 Real-Life Applications

73. At an airport baggage claim, two automated baggage handlers are at work. One handler discharges baggage every 3 seconds, and the other discharges baggage every 4 seconds. Assuming that the two baggage handlers begin at the same time, how many times after that will baggage be discharged simultaneously in a period of 2 minutes?

74. Each of two parallel rows of squares contains 100 squares. In the top row a marker is placed on every third square. In the bottom row a marker is placed on every fifth square. On what squares will there be markers directly across from each other? What do we call such numbers?

75. Two cars start at the same point and travel around a circular loop in opposite directions. Car A makes a complete circuit every 4 minutes, whereas car B takes 7 minutes. What is the least amount of time that will pass before both cars are simultaneously at the starting point? What do we call this number?

76. Two grandfather clocks chime every 15 minutes. However, one clock is running slow and loses 4 minutes every hour. If the clocks are started at exactly noon, at what time will the clocks next chime together?

Challenge

In Exercises 77 and 78, determine the LCM of the given numbers.

77. $14, 33, 78$

78. $36, 48, 162$

In Exercises 79 and 80, suppose that the given polynomials are denominators of fractions. Determine the LCD of the fractions.

79. $x^3 + 1, \quad x^2 + 2x + 1, \quad x^3 - x^2 + x$

80. $(x+2)^2 - y^2, \quad x^2 + 2x + xy, \quad -y^2 + 2y + xy$

In Exercises 81 and 82, determine the LCD for each pair of rational expressions. Then write each expression as an equivalent expression with that LCD.

81. $\dfrac{2}{a^4-b^4}, \quad \dfrac{1}{a^3b+ab^3}$

82. $\dfrac{x-3y}{8x^2-26xy+15y^2}, \quad \dfrac{3x+2y}{6x^2-13xy-5y^2}$

8.5 | Addition and Subtraction (Unlike Denominators)

To add or subtract two rational expressions with unlike denominators, we first use the procedures discussed in Section 8.4 to rename the expressions so that the denominators are the same.

Adding and Subtracting Rational Expressions with Unlike Denominators

1. Determine the LCD of the rational expressions.
2. Rename each expression with that LCD.
3. Add or subtract the numerators and retain the denominator.
4. Simplify the result, if possible.

EXAMPLE **1**

Adding and Subtracting Rational Expressions

Perform the indicated operation.

(a) $\dfrac{y}{2x} + \dfrac{x}{3y} = \dfrac{y \cdot 3y}{2x \cdot 3y} + \dfrac{x \cdot 2x}{3y \cdot 2x}$ The LCD is $6xy$.
Multiply by missing factors.

$= \dfrac{3y^2}{6xy} + \dfrac{2x^2}{6xy}$

$= \dfrac{3y^2 + 2x^2}{6xy}$ Add the numerators.

(b) $\dfrac{3}{x} - \dfrac{2}{x+1} = \dfrac{3(x+1)}{x(x+1)} - \dfrac{2x}{x(x+1)}$ The LCD is $x(x+1)$.
Multiply by missing factors.

$= \dfrac{3x+3}{x(x+1)} - \dfrac{2x}{x(x+1)}$ Remove the parentheses.

$= \dfrac{3x+3-2x}{x(x+1)}$ Subtract the numerators.

$= \dfrac{x+3}{x(x+1)}$ Combine like terms.

LEARNING TIP

After you have renamed the fractions with a common denominator, you will be tempted to divide out common factors, which would ruin your good work. Simplifying comes *after* you have added or subtracted.

(c) $\dfrac{7}{3x-5} + \dfrac{4}{5-3x}$

Because $-1(5-3x) = 3x-5$, multiply the numerator and the denominator of the second fraction by -1.

$\dfrac{7}{3x-5} + \dfrac{4}{5-3x} = \dfrac{7}{3x-5} + \dfrac{-1 \cdot 4}{-1(5-3x)}$

$= \dfrac{7}{3x-5} + \dfrac{-4}{3x-5}$ $-(b-a) = a-b$

$= \dfrac{3}{3x-5}$ Add the numerators.

(d) $x + 1 + \dfrac{x + 3}{2x - 3}$

Write the first fraction with a denominator of 1.

$= \dfrac{x + 1}{1} + \dfrac{x + 3}{2x - 3}$

The LCD is $2x - 3$.

$= \dfrac{(x + 1)(2x - 3)}{1(2x - 3)} + \dfrac{x + 3}{2x - 3}$

Multiply by $2x - 3$.

$= \dfrac{2x^2 - x - 3}{2x - 3} + \dfrac{x + 3}{2x - 3}$

Multiply the binomials.

$= \dfrac{2x^2 - x - 3 + x + 3}{2x - 3}$

Add the numerators.

$= \dfrac{2x^2}{2x - 3}$

Combine like terms.

(e) $\dfrac{x}{x - 4} - \dfrac{3}{x - 5}$

The LCD is $(x - 5)(x - 4)$.

$= \dfrac{x(x - 5)}{(x - 4)(x - 5)} - \dfrac{3(x - 4)}{(x - 5)(x - 4)}$

Multiply by missing factors.

$= \dfrac{x(x - 5) - 3(x - 4)}{(x - 5)(x - 4)}$

Subtract the numerators.

$= \dfrac{x^2 - 5x - 3x + 12}{(x - 5)(x - 4)}$

Remove the parentheses.

$= \dfrac{x^2 - 8x + 12}{(x - 5)(x - 4)}$

Combine like terms.

$= \dfrac{(x - 6)(x - 2)}{(x - 5)(x - 4)}$

Factor in an attempt to simplify.

(f) $\dfrac{x + 4}{x - 3} - \dfrac{x + 1}{x + 2}$

The LCD is $(x - 3)(x + 2)$.

$= \dfrac{(x + 4)(x + 2)}{(x - 3)(x + 2)} - \dfrac{(x + 1)(x - 3)}{(x + 2)(x - 3)}$

Multiply by missing factors.

$= \dfrac{x^2 + 6x + 8}{(x - 3)(x + 2)} - \dfrac{x^2 - 2x - 3}{(x + 2)(x - 3)}$

Multiply the binomials.

$= \dfrac{(x^2 + 6x + 8) - (x^2 - 2x - 3)}{(x + 2)(x - 3)}$

Subtract the numerators.

$= \dfrac{x^2 + 6x + 8 - x^2 + 2x + 3}{(x + 2)(x - 3)}$

Remove the parentheses.

$= \dfrac{8x + 11}{(x + 2)(x - 3)}$

Combine like terms.

Note that none of the results in this example could be simplified.

In each part of Example 1 the LCD was simply the product of the denominators. In Example 2 we illustrate sums and differences in which factoring is needed to determine the LCD.

EXAMPLE 2 **Adding and Subtracting Rational Expressions**

Perform the indicated operation.

(a) $\dfrac{2}{ab^2} + \dfrac{5}{3a^2b}$

$\qquad\qquad\qquad\qquad\qquad\qquad$ The LCD is $3a^2b^2$.

$\quad = \dfrac{2 \cdot 3a}{ab^2 \cdot 3a} + \dfrac{5 \cdot b}{3a^2b \cdot b}$

$\qquad\qquad\qquad\qquad\qquad\qquad$ Multiply by missing factors.

$\quad = \dfrac{6a}{3a^2b^2} + \dfrac{5b}{3a^2b^2}$

$\qquad\qquad\qquad\qquad\qquad\qquad$ Add the numerators.

$\quad = \dfrac{6a + 5b}{3a^2b^2}$

$\qquad\qquad\qquad\qquad\qquad\qquad$ The expression cannot be simplified.

(b) $\dfrac{x+1}{x^2+5x} + \dfrac{8}{x^2-25}$

$\quad = \dfrac{x+1}{x(x+5)} + \dfrac{8}{(x+5)(x-5)}$

$\qquad\qquad\qquad\qquad\qquad\qquad$ Factor the denominators.

$\quad = \dfrac{(x+1)(x-5)}{x(x+5)(x-5)} + \dfrac{8x}{x(x+5)(x-5)}$

$\qquad\qquad\qquad\qquad\qquad\qquad$ The LCD is $x(x+5)(x-5)$.

$\quad = \dfrac{(x+1)(x-5) + 8x}{x(x+5)(x-5)}$

$\qquad\qquad\qquad\qquad\qquad\qquad$ Add the numerators.

$\quad = \dfrac{x^2 - 4x - 5 + 8x}{x(x+5)(x-5)}$

$\qquad\qquad\qquad\qquad\qquad\qquad$ Multiply the binomials.

$\quad = \dfrac{x^2 + 4x - 5}{x(x+5)(x-5)}$

$\qquad\qquad\qquad\qquad\qquad\qquad$ Combine like terms.

$\quad = \dfrac{(x+5)(x-1)}{x(x+5)(x-5)}$

$\qquad\qquad\qquad\qquad\qquad\qquad$ Factor the numerator.

$\quad = \dfrac{x-1}{x(x-5)}$

$\qquad\qquad\qquad\qquad\qquad\qquad$ Divide out the common factor.

(c) $\dfrac{x-1}{x+2} - \dfrac{4}{x^2+4x+4}$

$\quad = \dfrac{x-1}{x+2} - \dfrac{4}{(x+2)^2}$

$\qquad\qquad\qquad\qquad\qquad\qquad$ Factor the denominators.

$\quad = \dfrac{(x-1)(x+2)}{(x+2)(x+2)} - \dfrac{4}{(x+2)^2}$

$\qquad\qquad\qquad\qquad\qquad\qquad$ The LCD is $(x+2)^2$.

$\quad = \dfrac{(x-1)(x+2) - 4}{(x+2)^2}$

$\qquad\qquad\qquad\qquad\qquad\qquad$ Subtract the numerators.

$\quad = \dfrac{x^2 + x - 2 - 4}{(x+2)^2}$

$\qquad\qquad\qquad\qquad\qquad\qquad$ Multiply the binomials.

$\quad = \dfrac{x^2 + x - 6}{(x+2)^2}$

$\qquad\qquad\qquad\qquad\qquad\qquad$ Combine like terms.

$\quad = \dfrac{(x+3)(x-2)}{(x+2)^2}$

$\qquad\qquad\qquad\qquad\qquad\qquad$ The expression cannot be simplified.

(d) $\dfrac{2}{x^2 + 6x + 8} - \dfrac{1}{x^2 + 7x + 12}$

$= \dfrac{2}{(x+4)(x+2)} - \dfrac{1}{(x+3)(x+4)}$ Factor the denominators.

$= \dfrac{2(x+3)}{(x+4)(x+2)(x+3)} - \dfrac{1(x+2)}{(x+3)(x+4)(x+2)}$ The LCD is $(x+3)(x+4)(x+2)$.

$= \dfrac{2(x+3) - 1(x+2)}{(x+3)(x+4)(x+2)}$ Subtract the numerators.

$= \dfrac{2x + 6 - x - 2}{(x+3)(x+4)(x+2)}$ Remove the parentheses.

$= \dfrac{x+4}{(x+3)(x+4)(x+2)}$ Combine like terms.

$= \dfrac{1}{(x+3)(x+2)}$ Divide out the common factor.

Think About It

For the sum $\dfrac{a^{-2}}{b^{-1}} + \dfrac{a^{-1}b}{a}$, what is the LCD? Describe an approach to adding these terms that avoids finding the LCD.

(e) $\dfrac{x^2 - 1}{x^2 - 2x - 15} - \dfrac{x+2}{x+3}$

$= \dfrac{x^2 - 1}{(x-5)(x+3)} - \dfrac{x+2}{x+3}$ Factor the denominators.

$= \dfrac{x^2 - 1}{(x-5)(x+3)} - \dfrac{(x+2)(x-5)}{(x+3)(x-5)}$ The LCD is $(x-5)(x+3)$.

$= \dfrac{x^2 - 1 - (x+2)(x-5)}{(x+3)(x-5)}$ Subtract the numerators.

$= \dfrac{x^2 - 1 - (x^2 - 3x - 10)}{(x+3)(x-5)}$ Multiply the binomials.

$= \dfrac{x^2 - 1 - x^2 + 3x + 10}{(x+3)(x-5)}$ Remove the parentheses.

$= \dfrac{3x + 9}{(x+3)(x-5)}$ Combine like terms.

$= \dfrac{3(x+3)}{(x+3)(x-5)}$ Factor.

$= \dfrac{3}{x-5}$ Divide out the common factor.

Quick Reference 8.5

- Use the following procedure to add or subtract rational expressions with unlike denominators.

  1. Determine the LCD of the rational expressions.

  2. Rename each expression with the LCD.

  3. Add or subtract the numerators and retain the denominator.

  4. Simplify the result, if possible.

Speaking the Language 8.5

1. To add or subtract rational expressions with unlike denominators, we begin by determining the ▓▓▓▓▓ .

2. In $\dfrac{3}{a} - \dfrac{x-4}{a}$, after we combine the fractions, the minus sign applies to the quantity ▓▓▓▓▓ .

3. A simple way to obtain a common denominator in $\dfrac{3}{x} + \dfrac{5}{-x}$ is to ▓▓▓▓▓ the numerator and denominator of either fraction by ▓▓▓▓▓ .

4. If adding two rational expressions results in $\dfrac{2x+4}{x+2}$, the next step is to ▓▓▓▓▓ .

Exercises 8.5

Concepts and Skills

1. If the LCD of two rational expressions is the product of their denominators, what do you know about the denominators?

2. If the denominators of two rational expressions are $x + 5$ and $(x + 5)^2$, which of the following is the LCD? Explain your choice.

   (i) $x + 5$ (ii) $(x + 5)^2$ (iii) $(x + 5)^3$

In Exercises 3 and 4, determine the unknown numerators that make the equations true.

3. $\dfrac{5x-1}{3x-8} + \dfrac{8x-6}{8-3x} = \dfrac{5x-1}{3x-8} + \dfrac{\rule{1.2em}{0.6em}}{3x-8} = \dfrac{\rule{1.2em}{0.6em}}{3x-8}$

4. $\dfrac{x-3}{5x-4} - \dfrac{7x-6}{4-5x} = \dfrac{x-3}{5x-4} + \dfrac{\rule{1.2em}{0.6em}}{5x-4} = \dfrac{\rule{1.2em}{0.6em}}{5x-4}$

In Exercises 5–12, perform the indicated operations.

5. $\dfrac{x}{8} + \dfrac{9}{-8}$

6. $\dfrac{x}{2} - \dfrac{1}{-2}$

7. $\dfrac{-9}{x} - \dfrac{5}{-x}$

8. $\dfrac{1}{y} + \dfrac{4}{-y}$

9. $\dfrac{x^2}{x-9} - \dfrac{-81}{9-x}$

10. $\dfrac{x^2}{x-7} + \dfrac{49}{7-x}$

11. $\dfrac{x-6}{x^2-36} + \dfrac{6-x}{36-x^2}$

12. $\dfrac{x-5}{x^2-25} - \dfrac{x-5}{25-x^2}$

In Exercises 13 and 14, determine the unknown numerators that make the equations true.

13. $\dfrac{1}{x-1} + \dfrac{1}{x} = \dfrac{\rule{1.2em}{0.6em}}{x(x-1)} + \dfrac{\rule{1.2em}{0.6em}}{x(x-1)} = \dfrac{\rule{1.2em}{0.6em}}{x(x-1)}$

14. $\dfrac{x+1}{x-2} - \dfrac{x}{x+1}$

$= \dfrac{\rule{1.2em}{0.6em}}{(x-2)(x+1)} - \dfrac{\rule{1.2em}{0.6em}}{(x-2)(x+1)}$

$= \dfrac{\rule{1.2em}{0.6em}}{(x-2)(x+1)}$

In Exercises 15–34, perform the indicated operations.

15. $x + \dfrac{5}{x}$

16. $\dfrac{4}{y} - y$

17. $8 - \dfrac{7}{x}$

18. $\dfrac{4}{x^2} - 1$

19. $\dfrac{2x+3}{4} - \dfrac{x-5}{7}$

20. $\dfrac{x+3}{5} + \dfrac{x-7}{6}$

21. $2 + \dfrac{8}{x-5}$

22. $y - \dfrac{y^2}{y-2}$

23. $\dfrac{3}{x} + \dfrac{4}{x+7}$

24. $\dfrac{5}{x-2} - \dfrac{3}{x}$

25. $\dfrac{x}{x-2} - \dfrac{x-2}{x}$

26. $\dfrac{9}{x} + \dfrac{5}{x-5}$

27. $\dfrac{8}{x+5} + \dfrac{2}{x-3}$

28. $\dfrac{x}{x-3} - \dfrac{2}{x+7}$

29. $\dfrac{x}{3x-2} - \dfrac{3}{x+4}$

30. $\dfrac{2x}{x+3} + \dfrac{2}{2x-1}$

31. $\dfrac{x+1}{x+5} - \dfrac{x-2}{x-4}$

32. $\dfrac{x+1}{x-4} + \dfrac{x-2}{x+1}$

33. $\dfrac{y+3}{3-y} + \dfrac{y-3}{3+y}$

34. $\dfrac{x-4}{x+4} - \dfrac{x+4}{x-4}$

35. Consider the sum $\dfrac{1}{x} + \dfrac{1}{x+3}$. The LCD is $x(x+3)$, so we rename the fractions as follows.

$$\frac{x+3}{x(x+3)} + \frac{x}{x(x+3)}$$

Is the next step to divide out the common factors? Explain.

36. When adding or subtracting rational expressions, why is it best to leave the LCD in factored form until after the addition or subtraction has been performed?

In Exercises 37–50, perform the indicated operations.

37. $\dfrac{8}{x} + \dfrac{2}{x^2}$

38. $\dfrac{7}{2x} - \dfrac{8}{x^2}$

39. $\dfrac{4}{x+2} - \dfrac{3}{5x+10}$

40. $\dfrac{3}{6x-3} + \dfrac{x}{2x-1}$

41. $\dfrac{x}{x+5} - \dfrac{50}{x^2-25}$

42. $\dfrac{x}{x+2} - \dfrac{8}{x^2-4}$

43. $\dfrac{5}{x+9} - \dfrac{x}{(x+9)^2}$

44. $\dfrac{6}{x-5} - \dfrac{x}{(x-5)^2}$

45. $\dfrac{3}{x+2} + \dfrac{5}{x^2+7x+10}$

46. $\dfrac{8}{x^2-x-6} - \dfrac{2}{x-3}$

47. $\dfrac{4}{x+3} - \dfrac{2}{x^2+4x+3}$

48. $\dfrac{x-6}{x^2-9x+20} + \dfrac{1}{x-5}$

49. $\dfrac{2}{x} - \dfrac{3}{x^2-9x}$

50. $\dfrac{x}{x^2+4x} + \dfrac{3}{x}$

In Exercises 51–72, perform the indicated operations.

51. $\dfrac{4}{5x} - \dfrac{3}{2x}$

52. $\dfrac{4}{5x^2} + \dfrac{3}{2x}$

53. $\dfrac{x+2}{15} + \dfrac{3x-5}{6}$

54. $\dfrac{5x+2}{10} - \dfrac{x-5}{4}$

55. $\dfrac{x}{2x+4} - \dfrac{2}{x^2+2x}$

56. $\dfrac{x}{5x-x^2} + \dfrac{3}{20-4x}$

57. $\dfrac{x}{6x+18} + \dfrac{5}{4x+12}$

58. $\dfrac{3y}{10y-20} - \dfrac{2}{15y-30}$

59. $\dfrac{4}{2y^2+3y} - \dfrac{2y-5}{6y+9}$

60. $\dfrac{3y+2}{12y-16} - \dfrac{2}{3y^2-4y}$

61. $\dfrac{x}{x^2-5x} - \dfrac{1}{x+7}$

62. $\dfrac{2}{x-2} + \dfrac{x}{x^2+3x}$

63. $\dfrac{3}{x^2+9x+18} + \dfrac{1}{x^2-36}$

64. $\dfrac{x}{4x^2-1} - \dfrac{5}{2x^2+5x-3}$

65. $\dfrac{x}{x^2+9x+20} - \dfrac{15}{x^2+13x+40}$

66. $\dfrac{x}{x^2+5x+4} + \dfrac{2}{x^2+8x+7}$

67. $\dfrac{x}{x^2+8x+15} + \dfrac{15}{x^2+4x-5}$

68. $\dfrac{x}{x^2+4x+3} - \dfrac{2}{x^2-2x-3}$

69. $\dfrac{x^2+2x}{x^2+2x-3} - \dfrac{4x-1}{4x-4}$

70. $\dfrac{2x^2-3}{x^2+x-6} - \dfrac{2x-3}{x-2}$

71. $\dfrac{2x+6}{x^2+2x} - \dfrac{x+5}{x^2+x-2}$

72. $\dfrac{2x+1}{x^2-2x} + \dfrac{2-x}{x^2-4x+4}$

Real-Life Applications

73. A pollution treatment facility receives x thousand gallons of water per day. The water contains $\dfrac{2x}{x+1}$ contaminants, and the facility is able to remove $\dfrac{x}{x+2}$ of these particles before discharging the water. Write a simplified rational expression to describe the amount of contaminants in the discharged water.

74. At an athletic club the number (in hundreds) of new members per month t months after its opening is $\dfrac{t}{t+1}$, and the number (in hundreds) of nonrenewals per month is $\dfrac{t}{t+2}$. Write a simplified rational expression for the difference in new members per month and nonrenewals per month.

75. At a certain college the total alumni contributions to the athletic program are given by $\dfrac{20x}{x+1}$, and the contributions to the academic program are given by $\dfrac{5x}{x-1}$, where x is the number of alumni. Write a simplified rational expression for the total contributions by alumni to the college.

76. When the price p of admission exceeds \$1.00, the number of children attending a summer concert series is $\dfrac{30p}{p^2 - 1}$, and the number of adults attending is $\dfrac{50p}{p + 1}$. Write a simplified rational expression to describe the total attendance.

Modeling with Real Data

77. Sales of sport-utility vehicles (SUVs) have risen in recent years. The bar graph shows the SUV market share in the United States. (Source: Ward's Automotive Reports.)

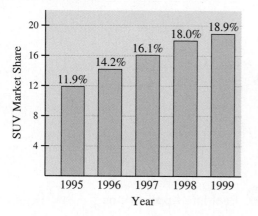

The market share can be modeled by

$$f(x) = 34 - \frac{179}{x + 3},$$

where x is the number of years since 1990.

(a) Write the expression for f as a single rational expression. Call the result $g(x)$.

(b) Evaluate and compare $f(15)$ and $g(15)$. What do the results represent?

78. Refer to the models in Exercise 77.

(a) What is the restricted value for the models?

(b) According to part (a), for what year do the models provide no defined data for the SUV market share?

Challenge

In Exercises 79–82, perform the indicated operations.

79. $\dfrac{3}{x + 4} + \dfrac{6}{x} + \dfrac{12}{x^2 + 4x}$

80. $\dfrac{4}{x - 7} + \dfrac{3}{x} + \dfrac{21}{x^2 - 7x}$

81. $\dfrac{4}{x^2 - 2x - 3} + \dfrac{2}{x^2 + 3x + 2} - \dfrac{5}{x^2 - x - 6}$

82. $\dfrac{2}{x^2 - 4x + 3} + \dfrac{4}{x^2 + x - 2} - \dfrac{5}{x^2 - x - 6}$

In Exercises 83 and 84, use the Order of Operations to perform the indicated operations.

83. $\dfrac{a + 3}{a - 2} - \left[\dfrac{6a}{a^2 - 4} + \dfrac{a - 1}{a + 2} \right]$

84. $\left[\dfrac{1 - y}{y + 2} - \dfrac{9y}{y^2 - 2y - 8} \right] + \dfrac{y + 3}{y - 4}$

8.6 Complex Fractions

Order of Operations • Multiplying by the LCD

Order of Operations

A **complex fraction** is a fraction that contains a fraction in the numerator, the denominator, or both.

$$\dfrac{\dfrac{3}{4}}{5} \qquad \dfrac{a}{3 + \dfrac{5}{a}} \qquad \dfrac{\dfrac{2}{x}}{x + \dfrac{4}{x}} \qquad \dfrac{3 + \dfrac{1}{y}}{3 - \dfrac{1}{y}}$$

To simplify a complex fraction, we eliminate all fractions in the numerator and the denominator.

There are two common methods for simplifying a complex fraction. Because a fraction bar indicates division, one method is to write the complex fraction with a division symbol. Then we carry out the division following the normal Order of Operations. Here is an example from arithmetic.

$$\frac{\dfrac{3}{4}}{\dfrac{12}{7}} = \frac{3}{4} \div \frac{12}{7} = \frac{3}{4} \cdot \frac{7}{12} = \frac{3}{4} \cdot \frac{7}{3 \cdot 4} = \frac{7}{16}$$

EXAMPLE **1**   **Writing Complex Fractions with a Division Symbol**

Use a division symbol to rewrite each complex fraction. Then simplify the result.

(a) $\dfrac{\dfrac{x+3}{5}}{\dfrac{x^2-9}{10}} = \dfrac{x+3}{5} \div \dfrac{x^2-9}{10}$ Rewrite the fraction with a division symbol.

$\qquad = \dfrac{x+3}{5} \cdot \dfrac{10}{x^2-9}$ Multiply by the reciprocal of the divisor.

$\qquad = \dfrac{x+3}{5} \cdot \dfrac{5 \cdot 2}{(x+3)(x-3)}$ Factor.

$\qquad = \dfrac{2}{x-3}$ Divide out common factors.

(b) $\dfrac{\dfrac{3t-5}{2}}{5-3t} = \dfrac{3t-5}{2} \div (5-3t)$ Rewrite the fraction with a division symbol.

$\qquad = \dfrac{3t-5}{2} \cdot \dfrac{1}{5-3t}$ Multiply by the reciprocal of the divisor.

$\qquad = \dfrac{-1(5-3t)}{2} \cdot \dfrac{1}{5-3t}$ Factor out -1.

$\qquad = -\dfrac{1}{2}$ Divide out the common factor.

If the numerator or denominator of a complex fraction contains a sum or difference, we can apply the Order of Operations to simplify the fraction. The following example from arithmetic illustrates a method for simplifying such complex fractions.

$$\frac{\dfrac{1}{2}+\dfrac{2}{3}}{2-\dfrac{1}{3}} = \frac{\dfrac{3}{6}+\dfrac{4}{6}}{\dfrac{6}{3}-\dfrac{1}{3}} = \frac{\dfrac{7}{6}}{\dfrac{5}{3}} = \frac{7}{6} \div \frac{5}{3} = \frac{7}{6} \cdot \frac{3}{5} = \frac{7}{10}$$

EXAMPLE **2**

Simplifying Complex Fractions with the Order of Operations

Simplify the complex fraction $\dfrac{1 + \dfrac{2}{x}}{\dfrac{1}{x}}$.

Solution

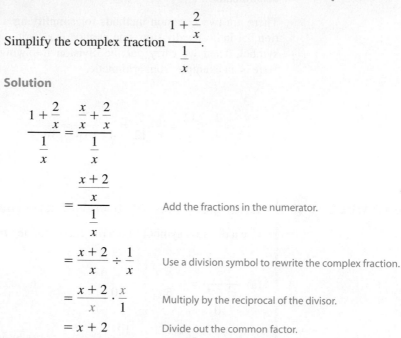

$$\frac{1 + \dfrac{2}{x}}{\dfrac{1}{x}} = \frac{\dfrac{x}{x} + \dfrac{2}{x}}{\dfrac{1}{x}}$$

$$= \frac{\dfrac{x + 2}{x}}{\dfrac{1}{x}} \qquad \text{Add the fractions in the numerator.}$$

$$= \frac{x + 2}{x} \div \frac{1}{x} \qquad \text{Use a division symbol to rewrite the complex fraction.}$$

$$= \frac{x + 2}{x} \cdot \frac{x}{1} \qquad \text{Multiply by the reciprocal of the divisor.}$$

$$= x + 2 \qquad \text{Divide out the common factor.}$$

The following summarizes the Order of Operations method for simplifying a complex fraction.

> **Simplifying a Complex Fraction with the Order of Operations**
>
> 1. Perform any addition or subtraction that is indicated in the numerator or the denominator.
> 2. Write the resulting complex fraction with a division symbol and perform the division.

Multiplying by the LCD

A second method for simplifying a complex fraction uses the fact that we can multiply the numerator and the denominator of a fraction by the same nonzero quantity. Here we multiply the numerator and the denominator by the LCD of every fraction that appears in the complex fraction.

Note: Unlike the Order of Operations method, which may use one LCD for the numerator and a different LCD for the denominator, this second method requires just one LCD, which is the LCD of every fraction in the numerator and the denominator.

For a comparison of methods, we repeat Example 2.

EXAMPLE **3**

Simplifying a Complex Fraction by Multiplying by the LCD

Simplify the given complex fraction.

$$\dfrac{1 + \dfrac{2}{x}}{\dfrac{1}{x}} = \dfrac{x \cdot \left(1 + \dfrac{2}{x}\right)}{x \cdot \left(\dfrac{1}{x}\right)}$$

Multiply the numerator and denominator by the LCD x.

$$= \dfrac{x \cdot 1 + \dfrac{x}{1} \cdot \dfrac{2}{x}}{\dfrac{x}{1} \cdot \dfrac{1}{x}}$$

Distributive Property

$$= \dfrac{x + 2}{1}$$

Perform each multiplication.

$$= x + 2$$

Note that the effect of the Distributive Property is to multiply every term in the numerator and the denominator by the LCD.

The following summarizes the second method we use for simplifying a complex fraction.

Think About It

You could write

$$(x^{-1} + y^{-1}) \div \dfrac{1}{x^{-1} - y^{-1}}$$

as a complex fraction with positive exponents and then simplify the expression. Show a faster method that does not involve a complex fraction.

> **Simplifying a Complex Fraction by Multiplying by the LCD**
>
> 1. Determine the LCD of every fraction in the complex fraction.
> 2. Multiply the numerator and the denominator of the complex fraction by the LCD.
> 3. Simplify, if possible.

EXAMPLE 4

Simplifying a Complex Fraction

Simplify the given complex fractions.

(a) $\dfrac{1 + \dfrac{5}{x}}{1 - \dfrac{25}{x^2}} = \dfrac{x^2 \cdot 1 + \dfrac{x^2}{1} \cdot \dfrac{5}{x}}{x^2 \cdot 1 - \dfrac{x^2}{1} \cdot \dfrac{25}{x^2}}$

The LCD of every fraction is x^2.
Multiply every term by the LCD.

$$= \dfrac{x^2 + 5x}{x^2 - 25}$$

Perform each multiplication.

$$= \dfrac{x(x + 5)}{(x + 5)(x - 5)}$$

Factor.

LEARNING TIP

Write the LCD beside each term of the complex fraction and then simplify the product. Attempting to do this step mentally may lead to errors.

$$= \dfrac{x}{x - 5}$$

Divide out the common factor $x + 5$.

(b) $\dfrac{3 + \dfrac{2}{x + 2}}{\dfrac{x}{x + 2} - 1} = \dfrac{3(x + 2) + (x + 2) \cdot \dfrac{2}{x + 2}}{(x + 2) \cdot \dfrac{x}{x + 2} - 1(x + 2)}$

The LCD is $x + 2$.
Multiply every term by the LCD.

$$= \dfrac{3x + 6 + 2}{x - x - 2}$$

Perform each multiplication.

$$= \dfrac{3x + 8}{-2} \quad \text{or} \quad -\dfrac{3x + 8}{2}$$

Combine like terms.

Quick Reference 8.6

Order of Operations

• A **complex fraction** is a fraction that contains a fraction in the numerator, the denominator, or both. To simplify a complex fraction, we eliminate all fractions in the numerator and the denominator.

• To simplify a complex fraction with the Order of Operations, use the following steps.

1. Perform any addition or subtraction that is indicated in the numerator or the denominator.

2. Write the resulting complex fraction with a division symbol and perform the indicated division.

Multiplying by the LCD

• A second method for simplifying a complex fraction is to multiply the numerator and the denominator by the LCD. The following is the procedure.

1. Determine the LCD of every fraction in the complex fraction.

2. Multiply the numerator and the denominator of the complex fraction by that LCD.

3. Simplify, if possible.

Speaking the Language 8.6

1. A fraction whose numerator is $\frac{2}{3}$ and whose denominator is 25 is called a(n) _____ .

2. Eliminating all the fractions in the numerator and the denominator of a complex fraction is called _____ the fraction.

3. To simplify a complex fraction with the Order of Operations, we begin by performing all indicated operations in the _____ and the _____ .

4. We can simplify a complex fraction by multiplying the numerator and the denominator by the _____ of every fraction in the numerator and the denominator.

Exercises 8.6

Concepts and Skills

1. Consider the following complex fraction.

$$\frac{1 + \dfrac{1}{x}}{1 - \dfrac{1}{x}}$$

Although both methods described in this section are applicable to simplifying this fraction, which method is clearly the more efficient? Why?

2. If we simplify the complex fraction in Exercise 1, we obtain $\dfrac{x + 1}{x - 1}$. When $x = 0$, the value of this expression is -1. Is it therefore correct to say that the value of the expression in Exercise 1 is -1 when $x = 0$? Why?

In Exercises 3–16, use a division symbol to rewrite the given complex fraction. Then simplify the result.

3. $\dfrac{\dfrac{3}{a}}{\dfrac{2}{b}}$

4. $\dfrac{\dfrac{x}{5}}{\dfrac{2}{y}}$

5. $\dfrac{\dfrac{3}{t^4}}{\dfrac{18}{t}}$

6. $\dfrac{\dfrac{z^5}{15}}{\dfrac{z^2}{24}}$

7. $\dfrac{\dfrac{a}{a-3}}{\dfrac{a^2}{3}}$

8. $\dfrac{\dfrac{2x+3}{8x}}{\dfrac{3x+2}{4x^2}}$

9. $\dfrac{\dfrac{y+5}{y-7}}{\dfrac{y-5}{7-y}}$

10. $\dfrac{\dfrac{x-2}{x+3}}{\dfrac{2-x}{x-3}}$

11. $\dfrac{\dfrac{x^2}{x^2-9}}{\dfrac{2x}{x+3}}$

12. $\dfrac{\dfrac{12x}{2x-1}}{\dfrac{8x^5}{4x^2-1}}$

13. $\dfrac{\dfrac{3x+6}{x+2}}{x^2+9}$

14. $\dfrac{\dfrac{6x-15}{x^2+2}}{\dfrac{2x-5}{}}$

15. $\dfrac{\dfrac{9x^2}{6x+6}}{\dfrac{63x^3}{x^2-1}}$

16. $\dfrac{\dfrac{8x}{x+5}}{\dfrac{16x^2}{x^2+x-20}}$

In Exercises 17–40, simplify the complex fraction.

17. $\dfrac{1+\dfrac{1}{2}}{2-\dfrac{3}{4}}$

18. $\dfrac{\dfrac{2}{3}+1}{\dfrac{1}{2}+5}$

19. $\dfrac{\dfrac{3}{4}-\dfrac{1}{3}}{\dfrac{5}{6}+\dfrac{1}{2}}$

20. $\dfrac{\dfrac{5}{4}-\dfrac{1}{6}}{\dfrac{3}{8}+\dfrac{5}{12}}$

21. $\dfrac{\dfrac{3}{x}-\dfrac{5}{9}}{\dfrac{4}{3}-\dfrac{4}{x}}$

22. $\dfrac{\dfrac{2}{5}+\dfrac{5}{t}}{\dfrac{3}{t}-\dfrac{7}{10}}$

23. $\dfrac{4-\dfrac{1}{x^2}}{\dfrac{2}{x}-\dfrac{1}{x^2}}$

24. $\dfrac{49x-\dfrac{16}{x}}{7-\dfrac{4}{x}}$

25. $\dfrac{\dfrac{a}{b}}{\dfrac{2}{a}+b}$

26. $\dfrac{2x+\dfrac{3}{y}}{\dfrac{2x}{y}}$

27. $\dfrac{\dfrac{1}{x}-4}{\dfrac{2}{x}+3}$

28. $\dfrac{\dfrac{3}{a}-5}{\dfrac{5}{b}+6}$

29. $\dfrac{1}{\dfrac{2}{x}+y}$

30. $\dfrac{1}{\dfrac{1}{x}-\dfrac{1}{y}}$

31. $\dfrac{\dfrac{1}{x}+\dfrac{1}{y}}{\dfrac{1}{x}+\dfrac{1}{xy}}$

32. $\dfrac{\dfrac{1}{b}+\dfrac{2}{a}}{\dfrac{1}{a}+\dfrac{4}{ab}}$

33. $\dfrac{4-\dfrac{5}{x}}{\dfrac{4}{x}-\dfrac{5}{x^2}}$

34. $\dfrac{\dfrac{3}{x}+\dfrac{4}{x^2}}{6+\dfrac{8}{x}}$

35. $\dfrac{\dfrac{1}{xy}}{\dfrac{1}{x}+\dfrac{1}{y}}$

36. $\dfrac{\dfrac{2}{a}-\dfrac{1}{b}}{\dfrac{2}{ab}}$

37. $\dfrac{a+b}{\dfrac{1}{a}+\dfrac{1}{b}}$

38. $\dfrac{\dfrac{1}{x^2}-\dfrac{1}{y^2}}{x+y}$

39. $\dfrac{\dfrac{a}{3}+\dfrac{b}{2}}{\dfrac{a}{3}-\dfrac{b}{2}}$

40. $\dfrac{\dfrac{x}{3}-\dfrac{3y}{2}}{\dfrac{x}{2}+\dfrac{y}{6}}$

In Exercises 41–46, simplify the complex fraction.

41. $\dfrac{1+\dfrac{2}{x+4}}{\dfrac{3}{x+4}}$

42. $\dfrac{\dfrac{5}{x-3}}{\dfrac{1}{x-3}-2}$

43. $\dfrac{2+\dfrac{7}{a-2}}{2-\dfrac{1}{a-2}}$

44. $\dfrac{\dfrac{2}{x+1}-3}{3+\dfrac{2}{x+1}}$

45. $\dfrac{\dfrac{y^2}{y+2}-y}{\dfrac{y}{y+2}}$

46. $\dfrac{\dfrac{x}{x-1}}{x+\dfrac{x^2}{x-1}}$

Real-Life Applications

47. In a certain forest area, the number of spotted owls remaining after t trees have been cut is modeled by the following complex fraction.

$$\dfrac{1+\dfrac{20}{t}}{\dfrac{1}{t}+1}$$

(a) Simplify this complex fraction.

(b) According to the result in part (a), how many spotted owls were there initially? Is this a valid use of the simplified expression? Why?

48. During a two-week voter registration drive in a small town, the ratio of registered Democrats to the total number of registered voters after x days was modeled by the following complex fraction.

$$\dfrac{\dfrac{1}{2}+\dfrac{20}{x^2}}{\dfrac{50}{x^2}+1}$$

(a) Simplify this complex fraction.

(b) Compare the percentage of Democrats at the end of the registration drive to the percentage of Democrats at the end of the first day.

49. On a test, the number of algebra problems a student had tried after t minutes was $\dfrac{5t}{3}+\dfrac{5}{2}$. The number answered correctly was $\dfrac{3t}{2}$. Write and simplify a complex fraction to describe the fraction of correct answers.

50. After t months, a city sanitation engineer determines that the amount of recyclable glass collected is given by $t+\dfrac{2}{t}$ and that the amount of recyclable plastic collected is $\dfrac{5}{t}+2t$. Write and simplify a complex fraction that describes the ratio of glass to plastic.

Data Analysis: *Wildfires*

Wildfires are among the most dangerous to control. The following table provides data on wildfires for selected years. (Source: National Interagency Fire Center.)

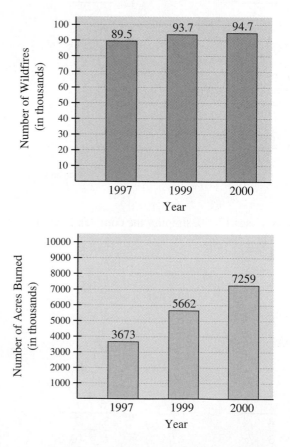

With x representing the number of years since 1990, the number (in thousands) of wildfires can be modeled by the expression $W(x)$, and the number (in thousands) of acres burned by $A(x)$.

$$W(x) = 1.8x + 77.2$$

$$A(x) = 1167x - 4580$$

51. Write an expression $r(x)$ to describe the average number of acres burned per wildfire.

52. By evaluating $r(9)$ and $r(10)$, determine whether the average number of acres burned per wildfire increased or decreased during the two-year period.

53. For which of the years given in the table is the average predicted by $r(x)$ most accurate?

54. Produce the graph of $r(x)$. Assuming that the model is valid for future years, do you infer from the graph that the average number of acres burned per wildfire will increase, decrease, or level off?

Challenge

In Exercises 55–60, simplify the complex fraction.

55. $\dfrac{x^{-1} + 1}{x^{-2} - 1}$

56. $\dfrac{\dfrac{1}{x+h} - \dfrac{1}{x}}{h}$

57. $\dfrac{\dfrac{2}{x} + \dfrac{3}{x+1}}{\dfrac{1}{x+1} - \dfrac{2}{x}}$

58. $\dfrac{1 + \dfrac{1}{x} - \dfrac{12}{x^2}}{1 + \dfrac{6}{x} + \dfrac{8}{x^2}}$

59. $\dfrac{\dfrac{c}{16} - \dfrac{1}{c}}{\dfrac{1}{2} + \dfrac{c+6}{c}}$

60. $\dfrac{x - 5 + \dfrac{7}{x+3}}{x^2 - 7x + 12}$

61. Let $f(x) = \dfrac{x}{1+x}$. Evaluate $f\left(\dfrac{2}{t}\right)$.

8.7 Equations with Rational Expressions

The Graphing Method   •   The Algebraic Method

The Graphing Method

In Chapter 2 we solved equations having fractions as coefficients but no variables in the denominators. In this section we consider equations with rational expressions, which are sometimes called *rational equations*. The following equations are examples.

$$x = \frac{10}{x} + 3 \qquad \frac{1}{10} = \frac{1}{x} - \frac{2}{2x+1} \qquad \frac{12}{x^2 - 9} + 1 = \frac{2}{x - 3}$$

As always, we can estimate the solution(s) of an equation with rational expressions by graphing the two sides of the equation and tracing to the point(s) of intersection.

EXAMPLE 1 **Estimating Solutions of Rational Equations**

Estimate the solutions of each equation.

(a) $3 - \dfrac{4}{x} = 2$

(b) $x = \dfrac{10}{x} + 3$

Solution

(a) Produce the graphs of $y_1 = 3 - \dfrac{4}{x}$ and $y_2 = 2$. (See Fig. 8.6.)

Figure 8.6

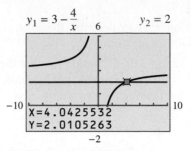

From the graphs, the solution appears to be about 4.

(b) Produce the graphs of $y_1 = \dfrac{10}{x} + 3$ and $y_2 = x$. (See Figs. 8.7 and 8.8.)

Figure 8.7   **Figure 8.8**

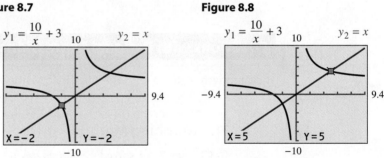

The graphs suggest two solutions: -2 and 5.

The Algebraic Method

The graphing method provides estimates of solutions, but algebraic methods are needed to obtain exact solutions. In our previous experience with rational equations, we usually began by multiplying both sides of the equation by the LCD to clear the fractions. Example 2 reviews this procedure.

EXAMPLE 2

Clearing Fractions in an Equation

Solve

$$\frac{x}{4} + \frac{1}{2} = \frac{x+9}{6}$$

Solution

$$\frac{x}{4} + \frac{1}{2} = \frac{x+9}{6}$$ The LCD is 12.

$$12 \cdot \frac{x}{4} + 12 \cdot \frac{1}{2} = 12 \cdot \frac{x+9}{6}$$ Multiply both sides by the LCD.

$$3x + 6 = 2(x + 9)$$

$$3x + 6 = 2x + 18$$ Distributive Property

$$3x = 2x + 12$$ Subtract 6 from both sides.

$$x = 12$$ Subtract 2x from both sides.

The solution is 12.

This same technique can be used to clear fractions in rational equations that have variables in the denominators. In Example 3 we revisit the equations in Example 1 and verify the estimated solutions algebraically.

EXAMPLE **3**

Solving Rational Equations Algebraically

Solve each equation.

(a) $$3 - \frac{4}{x} = 2$$ The restricted value is 0.

$$3 \cdot x - x \cdot \frac{4}{x} = 2 \cdot x$$ Multiply both sides by the LCD, x.

$$3x - 4 = 2x$$ The resulting equation is a linear equation.

$$x - 4 = 0$$ Subtract 2x from both sides.

$$x = 4$$ Add 4 to both sides.

The solution is 4, as we estimated in Example 1.

(b) $$x = \frac{10}{x} + 3$$ The restricted value is 0.

$$x \cdot x = x \cdot \frac{10}{x} + 3 \cdot x$$ Multiply both sides by the LCD, x.

$$x^2 = 10 + 3x$$ The resulting equation is a quadratic equation.

$$x^2 - 3x - 10 = 0$$ Write the equation in standard form.

$$(x - 5)(x + 2) = 0$$ Factor.

$$x - 5 = 0 \quad \text{or} \quad x + 2 = 0$$ Set each factor equal to 0.

$$x = 5 \quad \text{or} \quad x = -2$$

As we predicted from the graphs in Example 1, the solutions are -2 and 5.

Example 4 illustrates rational equations in which binomials appear in the denominators.

EXAMPLE **4**   **Rational Equations Leading to Linear Equations**

Solve each equation.

(a) $\dfrac{7}{2(x-1)} + \dfrac{1}{2} = \dfrac{2}{x-1}$

(b) $\dfrac{3}{x-8} = \dfrac{2}{x-5}$

(c) $\dfrac{t}{t-2} - \dfrac{1}{t+2} = 1$

Solution

(a) We multiply both sides by the LCD, which is $2(x-1)$.

$$\dfrac{7}{2(x-1)} + \dfrac{1}{2} = \dfrac{2}{x-1} \qquad \text{The restricted value is 1.}$$

$$2(x-1) \cdot \dfrac{7}{2(x-1)} + 2(x-1) \cdot \dfrac{1}{2} = 2(x-1) \cdot \dfrac{2}{x-1}$$

$$7 + x - 1 = 4 \qquad \text{The resulting equation is linear.}$$

$$x + 6 = 4$$

$$x = -2$$

The solution is -2. The solution should be verified by substitution.

(b) Multiply both sides by the LCD, which is $(x-8)(x-5)$, or cross-multiply as shown.

$$\dfrac{3}{x-8} = \dfrac{2}{x-5} \qquad \text{The restricted values are 8 and 5.}$$

$$3(x-5) = 2(x-8) \qquad \text{Cross-multiply.}$$

$$3x - 15 = 2x - 16 \qquad \text{Distributive Property}$$

$$3x = 2x - 1$$

$$x = -1$$

The solution is -1.

(c) Multiply both sides by the LCD, $(t+2)(t-2)$.

$$\dfrac{t}{t-2} - \dfrac{1}{t+2} = 1 \qquad \text{The restricted values are 2 and } -2.$$

$$(t+2)(t-2) \cdot \dfrac{t}{t-2} - (t+2)(t-2) \cdot \dfrac{1}{t+2} = 1(t+2)(t-2)$$

$$t(t+2) - 1(t-2) = (t+2)(t-2)$$

Distributive Property $\qquad t^2 + 2t - t + 2 = t^2 - 4$

$$t^2 + t + 2 = t^2 - 4$$

Subtract t^2 from both sides. $\qquad t + 2 = -4$

$$t = -6$$

The solution is -6.

When a rational expression has a variable in its denominator, we must be careful not to replace that variable with numbers that would cause the denominator to be 0. Recall that such numbers are called *restricted values*.

EXAMPLE **5**

Apparent Solutions That Are Restricted Values

Solve the equation $\dfrac{2x^2 + x + 2}{x - 1} - x - \dfrac{2x + 3}{x - 1} = 0$.

Solution

From the graph in Figure 8.9, we estimate that the solution of the equation is -1.

Figure 8.9

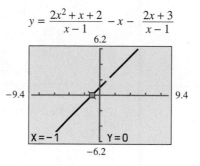

$$y = \frac{2x^2 + x + 2}{x - 1} - x - \frac{2x + 3}{x - 1}$$

In Fig. 8.9, note the "hole" in the graph for $x = 1$. No point with an x-coordinate of 1 can represent a solution because 1 is a restricted value.

To solve algebraically, multiply by the LCD, $x - 1$.

$$\frac{2x^2 + x + 2}{x - 1} - x - \frac{2x + 3}{x - 1} = 0$$

$$(x - 1) \cdot \frac{2x^2 + x + 2}{x - 1} - x(x - 1) - (x - 1) \cdot \frac{2x + 3}{x - 1} = 0(x - 1)$$

$$2x^2 + x + 2 - x^2 + x - 2x - 3 = 0 \qquad \text{Simplify.}$$

$$x^2 - 1 = 0 \qquad \begin{array}{l}\text{Quadratic equation}\\\text{in standard form.}\end{array}$$

$$(x + 1)(x - 1) = 0 \qquad \text{Factor.}$$

$$x + 1 = 0 \quad \text{or} \quad x - 1 = 0 \qquad \begin{array}{l}\text{Set each factor}\\\text{equal to 0.}\end{array}$$

$$x = -1 \quad \text{or} \quad x = 1$$

With the algebraic method, there appear to be two solutions, but the graphing method reveals only one. When we look at the original equation, we realize that 1 is a restricted value and cannot be a solution. The only solution is -1.

As we saw in Example 5, an apparent solution of an equation may be a restricted value and must be disqualified. Such values are called **extraneous solutions.** An important step in solving rational equations is to check for extraneous solutions.

EXAMPLE **6**

Rational Equations Leading to Quadratic Equations

Solve each equation.

(a) $\dfrac{1}{10} = \dfrac{1}{x} - \dfrac{2}{2x + 1}$

(b) $\dfrac{12}{x^2 - 9} + 1 = \dfrac{2}{x - 3}$

LEARNING TIP

To help you remember to check proposed solutions for restricted values, begin by writing the restricted values beside the equation, as in Example 6.

Solution

(a)
$$\frac{1}{10} = \frac{1}{x} - \frac{2}{2x + 1}$$

The restricted values are 0 and $-\dfrac{1}{2}$.

$$10x(2x + 1) \cdot \frac{1}{10} = 10x(2x + 1) \cdot \frac{1}{x} - 10x(2x + 1) \cdot \frac{2}{2x + 1}$$

$$x(2x + 1) = 10(2x + 1) - 20x$$

$$2x^2 + x = 20x + 10 - 20x \qquad \text{Distributive Property}$$

$$2x^2 + x = 10 \qquad \text{The result is a quadratic equation.}$$

$$2x^2 + x - 10 = 0 \qquad \text{Write the equation in standard form.}$$

$$(2x + 5)(x - 2) = 0 \qquad \text{Factor.}$$

$$2x + 5 = 0 \qquad \text{or} \qquad x - 2 = 0 \qquad \text{Set each factor equal to 0.}$$

$$x = -\frac{5}{2} \qquad \text{or} \qquad x = 2 \qquad \text{Solve for } x.$$

Neither value is restricted. The solutions are $-\frac{5}{2}$ and 2.

Think About It

The equation $\dfrac{x^2}{(x - 4)^2} = -1$ has no solution. Is this because 4 is an extraneous solution? How can you tell without actually solving the equation?

(b)
$$\frac{12}{x^2 - 9} + 1 = \frac{2}{x - 3} \qquad \text{Factor the denominator.}$$

$$\frac{12}{(x + 3)(x - 3)} + 1 = \frac{2}{x - 3} \qquad \text{The restricted values are } -3 \text{ and } 3.$$

$$(x + 3)(x - 3) \cdot \frac{12}{(x + 3)(x - 3)} + 1(x + 3)(x - 3)$$
$$= (x + 3)(x - 3) \cdot \frac{2}{x - 3}$$

$$12 + x^2 - 9 = 2(x + 3) \qquad \text{Multiply by } (x + 3)(x - 3).$$

$$x^2 + 3 = 2x + 6 \qquad \text{The result is a quadratic equation.}$$

$$x^2 - 2x - 3 = 0 \qquad \text{Write the equation in standard form.}$$

$$(x - 3)(x + 1) = 0 \qquad \text{Factor.}$$

$$x - 3 = 0 \qquad \text{or} \qquad x + 1 = 0 \qquad \text{Set each factor equal to 0.}$$

$$x = 3 \qquad \text{or} \qquad x = -1$$

Because 3 is a restricted value, it is an extraneous solution. The only solution is -1.

Quick Reference 8.7

The Graphing Method
- Equations with rational expressions are sometimes called *rational equations*.
- By graphing the two sides of a rational equation and tracing to the point(s) of intersection, we can estimate the solution(s) of the equation.

The Algebraic Method
- We clear the fractions in a rational equation by multiplying both sides by the LCD of all the fractions in the equation.
- Clearing fractions in a rational equation may lead to a linear equation, a quadratic equation, or some other type of equation. The resulting equation is solved algebraically with methods appropriate to the type of equation.
- An **extraneous solution** of a rational equation is an apparent solution that is a restricted value and must therefore be disqualified.

Speaking the Language 8.7

1. A rational equation is an equation that contains one or more ▨▨▨▨ expressions.

2. If we graph the left and right sides of a rational equation, a solution of the equation is represented by a(n) ▨▨▨▨ .

3. To solve a rational equation algebraically, we ▨▨▨▨ by multiplying both sides by the LCD.

4. If 3 is an apparent solution of a rational equation, but 3 is a restricted value, then we call 3 a(n) ▨▨▨▨ solution.

Exercises 8.7

Concepts and Skills

1. Suppose that when you use the graphing method to estimate the solution of a rational equation, you see just one point of intersection. But, when you solve the equation algebraically, you obtain two apparent solutions. Explain how this can happen.

2. Explain why the following equation cannot have any extraneous solutions.

$$\frac{8}{x^2 + 1} = \frac{x - 3}{5}$$

In Exercises 3–6, determine whether the number in each part is a solution of the given equation.

3. $7 = \dfrac{4}{x} + \dfrac{3}{x}$

   (a) 0   (b) -1 (c) 1

4. $1 - \dfrac{8}{x^2} = \dfrac{7}{x}$

   (a) 8   (b) -1 (c) 1

5. $\dfrac{3}{x - 3} = \dfrac{2}{2x - 5}$

   (a) $\frac{5}{2}$ (b) $\frac{9}{4}$ (c) 0

6. $\dfrac{7}{x + 8} = \dfrac{8}{x}$

   (a) -64 (b) -8 (c) 6

In Exercises 7–14, use the graphing method to estimate the solution(s) of the equation.

7. $4 = 3 + \dfrac{8}{x}$ **8.** $\dfrac{5}{x + 7} = 1$

9. $\dfrac{1}{x+2} = -1$

10. $3 = \dfrac{4x+1}{x-3}$

11. $x + \dfrac{20}{x} = 12$

12. $x = 5 + \dfrac{24}{x}$

13. $\dfrac{x}{4} = \dfrac{x+6}{x+2}$

14. $4 = \dfrac{(x-2)^2}{x+1}$

In Exercises 15–20, state the restricted values.

15. $\dfrac{1}{x} + \dfrac{4}{x-2} = 5$

16. $\dfrac{x}{4} + \dfrac{2}{x+5} = x$

17. $\dfrac{2}{x^2} + \dfrac{1}{x+1} = 2x$

18. $\dfrac{x+1}{4x} - 2 = \dfrac{3}{(x-2)^2}$

19. $\dfrac{2x+1}{x^2-5x-6} = 1 + \dfrac{1}{x}$

20. $\dfrac{x+1}{x^2-1} = \dfrac{x}{2-x}$

In Exercises 21–34, solve the equation.

21. $\dfrac{t}{3} + \dfrac{1}{2} = -\dfrac{1}{2}$

22. $\dfrac{x}{3} - \dfrac{x}{4} = 1$

23. $\dfrac{7}{x} = \dfrac{4}{x} + \dfrac{15}{8}$

24. $\dfrac{1}{5} = \dfrac{1}{x} - \dfrac{7}{5x}$

25. $\dfrac{x-4}{3} = \dfrac{x}{4}$

26. $\dfrac{7t}{10} = \dfrac{t+2}{2}$

27. $\dfrac{3}{t+3} = 3 + \dfrac{3}{t+3}$

28. $1 + \dfrac{10}{x-5} = \dfrac{2x}{x-5}$

29. $\dfrac{-2}{3x+2} = \dfrac{3}{1-2x}$

30. $\dfrac{4}{z-4} = \dfrac{8}{z+4}$

31. $\dfrac{1}{x} + \dfrac{1}{x-1} = \dfrac{2}{x+2}$

32. $\dfrac{2}{x-1} = \dfrac{1}{x+3} + \dfrac{1}{x}$

33. $\dfrac{4}{t} - \dfrac{3}{t-2} = \dfrac{9}{2t-t^2}$

34. $\dfrac{1}{t} - \dfrac{3}{2-t} = \dfrac{6t+4}{t^2-2t}$

In Exercises 35–48, solve the equation.

35. $14x + \dfrac{6}{x} = 25$

36. $4 = \dfrac{40}{x^2} + \dfrac{27}{x}$

37. $\dfrac{x}{x+2} = \dfrac{4}{x} - \dfrac{2}{x+2}$

38. $\dfrac{x}{x+6} + \dfrac{6}{x+6} = \dfrac{2}{x}$

39. $\dfrac{3t}{t+1} = \dfrac{2}{t-1}$

40. $\dfrac{x-3}{1-x} = \dfrac{2x-1}{x+1}$

41. $\dfrac{8}{t} + \dfrac{3}{t-1} = 3$

42. $\dfrac{1}{x} + \dfrac{2}{x-2} = -1$

43. $\dfrac{x+2}{x-12} = \dfrac{1}{x} + \dfrac{14}{x^2-12x}$

44. $\dfrac{12}{t^2-t} - \dfrac{t}{t-1} = \dfrac{1}{t-1}$

45. $\dfrac{x+8}{x+2} + \dfrac{12}{x^2+2x} = \dfrac{2}{x}$

46. $\dfrac{x+1}{3-x} + \dfrac{12}{x^2+3x} + \dfrac{5}{x} = 0$

47. $\dfrac{3}{x^2+4x+3} = \dfrac{2}{x+1} - 1$

48. $\dfrac{x}{x+4} - 2 = \dfrac{11}{x^2-16}$

49. In each of the following problems, we need to know the LCD. Explain the difference in how the LCD is used.

   (i) Add $\dfrac{1}{x} + \dfrac{1}{5}$.   (ii) Solve $\dfrac{1}{x} + \dfrac{1}{5} = 1$.

50. In the following equation we can clear fractions by factoring the denominators, determining the LCD, and then multiplying both sides by the LCD.

$$\dfrac{1}{2x^2+x-3} = \dfrac{1}{x^2+x-2}$$

  (a) What is an easier method for clearing fractions?

  (b) What is the danger in using that method?

In Exercises 51–66, solve the equation.

51. $\dfrac{4}{t} - 1 = \dfrac{1}{3}$

52. $\dfrac{1}{6y} = \dfrac{1}{4} - \dfrac{1}{3y}$

53. $\dfrac{5}{3} + \dfrac{y}{6} = \dfrac{4}{y}$

54. $\dfrac{x}{2} + \dfrac{3}{x} = \dfrac{5}{2}$

55. $\dfrac{x+3}{x+4} + \dfrac{x+5}{x+4} = 1$

56. $\dfrac{x-10}{x-9} - \dfrac{8-x}{x-9} = 1$

57. $x = \dfrac{4-11x}{3x}$

58. $2x = \dfrac{3x+12}{x-1}$

59. $x - \dfrac{6}{x-3} = \dfrac{2x}{x-3}$

60. $x + \dfrac{8}{x-6} = \dfrac{3x}{x-6}$

61. $\dfrac{7}{2x-5} + 2 = \dfrac{3}{5-2x}$

62. $\dfrac{8}{t-3} + \dfrac{5t}{3-t} = 4$

63. $\dfrac{3}{x+1} - \dfrac{4}{2x-1} = \dfrac{5}{2x^2+x-1}$

64. $\dfrac{5}{x+2} + \dfrac{1}{x+3} = \dfrac{-1}{x^2+5x+6}$

65. $\dfrac{9}{(x-2)^2} = 8 + \dfrac{1}{x-2}$

66. $\dfrac{1}{x+3} + \dfrac{21}{(x+3)^2} = 2$

In Exercises 67–82, solve the equation or simplify the expression.

67. $\dfrac{5}{4} - \dfrac{4}{3} = \dfrac{7}{x}$

68. $\dfrac{5}{2} - \dfrac{15}{4} = \dfrac{5}{x}$

69. $\dfrac{x^2 - 3x - 40}{3x + 15} \cdot \dfrac{9x^2 + 72x}{x^2 - 64}$

70. $\dfrac{4x}{8x^3 - 32x} \cdot (x^2 + x - 6)$

71. $\dfrac{x + 4}{5} = \dfrac{5}{x + 4}$

72. $\dfrac{3x}{x + 2} = \dfrac{2}{2x - 1}$

73. $\dfrac{8}{x} + \dfrac{2}{x^2}$

74. $\dfrac{5x + 28}{x + 3} + \dfrac{2x - 7}{x + 3}$

75. $\dfrac{5x - 20}{x + 4} - \dfrac{8x - 8}{x + 4}$

76. $\dfrac{t}{1 - t^2} - \dfrac{t + 3}{t^2 - t}$

77. $\dfrac{2}{t - 5} + \dfrac{3}{t + 5} = \dfrac{10}{t^2 - 25}$

78. $\dfrac{4}{t + 4} - \dfrac{3}{t - 3} = \dfrac{1}{t^2 + t - 12}$

79. $\dfrac{x - 12}{-16} \div \dfrac{12 - x}{8}$

80. $\dfrac{3x^3 + 6x^2}{2x^2 + x - 6} \div (6x^2 - 15x)$

81. $9 + \dfrac{3}{x + 1} = \dfrac{-4}{x^2 + x}$

82. $1 + \dfrac{5}{x(x - 2)} = \dfrac{-8}{x - 2}$

Data Analysis: *Olympic Swimming*

The Olympic winning times in the 100-meter men's and women's freestyle events in swimming have decreased steadily since 1968.

Year	Women's Times	Men's Times
1968	60.0	52.2
1972	58.6	51.2
1976	55.7	50.0
1980	54.8	50.4
1984	55.9	49.8
1988	54.9	48.6
1992	54.6	49.0
1996	54.5	48.7
2000	53.8	48.3

(Source: American Committee for the Olympic Games.)

The winning times (in seconds) shown in the table can be modeled by the following expressions, where x is the number of years since 1956.

Women: $60.4 - 0.16x$

Men: $52.9 - 0.11x$

83. The formula relating rate r, time t, and distance d is $d = rt$. Solve this formula for r.

84. Use the formula in Exercise 83 and the given model expressions for time to write rational expressions for the men's and women's rates. (Remember that the distance d is 100 meters in both cases.)

85. Write a rational equation describing the fact that in some unknown year the difference of the men's rate and the women's rate was 0.21 meter per second.

86. Describe the first step you would take to solve the equation in Exercise 85.

Challenge

In Exercises 87–89, solve the rational equation.

87. $\dfrac{3}{2x - 1} - \dfrac{5}{8x^2 + 2x - 3} = \dfrac{4}{4x + 3}$

88. $\dfrac{x}{x - 3} - \dfrac{1}{2x - 1} = \dfrac{5x}{2x^2 - 7x + 3}$

89. $\dfrac{x}{x - 4} = \dfrac{4}{x + 2} + \dfrac{24}{x^2 - 2x - 8}$

In Exercises 90–92, solve the equation.

90. $\dfrac{x}{6} + \dfrac{1}{2} = \dfrac{2}{3x} + \dfrac{2}{x^2}$

91. $\dfrac{x^2 - 2}{2} = \dfrac{2x^2 - 3}{x^2 + 1}$

92. $\dfrac{1}{x - 5} - \dfrac{1}{x + 5} = \dfrac{10}{x^2 - 25}$

93. The fact that the following equation has no solution is revealed in different ways depending on the solution method.

$$\dfrac{1}{x - 1} = \dfrac{2}{x - 1}$$

What is your conclusion if you

(a) cross-multiply?

(b) multiply both sides by the LCD?

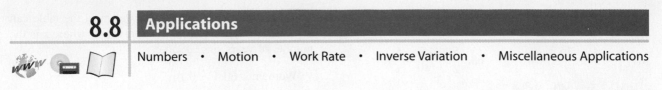

Numbers · Motion · Work Rate · Inverse Variation · Miscellaneous Applications

Some applications lead to equations with rational expressions. The general problem-solving strategy discussed earlier continues to apply to such problems.

As always, some of the examples and exercises in this applications section are organized by category as a matter of convenience. Their purpose is to illustrate ways of thinking about and organizing information so that relationships can be expressed in mathematical language. The process is just as important as the end result.

Numbers

Because the reciprocal of a number x is $\dfrac{1}{x}$, problems involving reciprocals often lead to rational equations.

EXAMPLE 1

Numbers and Their Reciprocals

One number is 3 times another. The sum of their reciprocals is $\frac{4}{9}$. What are the numbers?

Solution

If $x =$ one number, then $3x =$ the other number. Their reciprocals are $\dfrac{1}{x}$ and $\dfrac{1}{3x}$.

$$\frac{1}{x} + \frac{1}{3x} = \frac{4}{9} \qquad \text{The sum of the reciprocals is } \tfrac{4}{9}.$$

$$9x \cdot \frac{1}{x} + 9x \cdot \frac{1}{3x} = 9x \cdot \frac{4}{9} \qquad \text{Multiply by the LCD } 9x.$$

$$9 + 3 = 4x \qquad \text{The resulting equation is linear.}$$

$$4x = 12 \qquad \text{Solve for } x.$$

$$x = 3$$

The numbers are 3 and 9. It is easy to check that $\frac{1}{3} + \frac{1}{9} = \frac{4}{9}$.

EXAMPLE 2

Numbers and Their Reciprocals

The difference of a number and 12 times its reciprocal is 4. What is the number?

Solution

If $x =$ the number, then the reciprocal is $\dfrac{1}{x}$. Also, 12 times the reciprocal is $12 \cdot \dfrac{1}{x} = \dfrac{12}{x}$.

$$x - \frac{12}{x} = 4 \qquad \text{The difference is 4.}$$

$$x \cdot x - x \cdot \frac{12}{x} = 4x \qquad \text{Multiply by the LCD } x.$$

$$x^2 - 12 = 4x \qquad \text{The resulting equation is quadratic.}$$
$$x^2 - 4x - 12 = 0 \qquad \text{Write the equation in standard form.}$$
$$(x - 6)(x + 2) = 0 \qquad \text{Factor.}$$
$$x - 6 = 0 \quad \text{or} \quad x + 2 = 0 \qquad \text{Set each factor equal to 0.}$$
$$x = 6 \quad \text{or} \quad x = -2 \qquad \text{Solve for } x.$$

Two numbers satisfy the conditions: -2 and 6.

Motion

In previous chapters we encountered problems involving distance d, rate r, and time t. The formula $d = rt$ can be solved for r or t to obtain $r = \dfrac{d}{t}$ or $t = \dfrac{d}{r}$. Using these formulas may lead to equations with rational expressions.

EXAMPLE 3

Motion

A trucker drove 48 miles on icy roads and 160 miles on clear roads. If the driver's average speed on the clear roads was twice the speed on the icy roads and the total driving time was 4 hours, what were the respective speeds on each part of the trip?

Solution

If $r = $ the speed on the icy roads, then

$2r = $ the speed on the clear roads.

LEARNING TIP

Drawing a diagram or making a table helps organize information and develop a problem-solving strategy.

	Distance	Rate	Time
Icy Roads	48	r	$\dfrac{48}{r}$
Clear Roads	160	$2r$	$\dfrac{160}{2r}$

$$\frac{48}{r} + \frac{160}{2r} = 4 \qquad \text{The total time was 4 hours.}$$

$$2r \cdot \frac{48}{r} + 2r \cdot \frac{160}{2r} = 2r \cdot 4 \qquad \text{Multiply both sides by } 2r.$$

$$96 + 160 = 8r \qquad \text{The resulting equation is linear.}$$

$$8r = 256 \qquad \text{Solve for } r.$$

$$r = 32$$

The speeds were 32 mph on the icy roads and 64 mph on the clear roads.

EXAMPLE 4

Motion

On a two-day bicycle trip, a rider averaged 4 mph less on the second day than on the first day. If the biker rode 56 miles the first day and 30 miles the second day and the total riding time was 7 hours, how many hours did the biker ride each day?

Solution

If t = the number of hours the biker rode the first day, then

$7 - t$ = the number of hours the biker rode the second day.

	Distance	**Rate**	**Time**
Day 1	56	$\dfrac{56}{t}$	t
Day 2	30	$\dfrac{30}{7-t}$	$7 - t$

$$\frac{30}{7-t} = \frac{56}{t} - 4 \qquad \text{The speed on day 2 was 4 less than the speed on day 1.}$$

$$t(7-t) \cdot \frac{30}{7-t} = t(7-t) \cdot \frac{56}{t} - 4t(7-t) \qquad \text{The LCD is } t(7-t).$$

$$30t = 56(7-t) - 4t(7-t) \qquad \text{Simplify.}$$

$$30t = 392 - 56t - 28t + 4t^2 \qquad \text{The equation is quadratic.}$$

$$4t^2 - 114t + 392 = 0 \qquad \text{Write in standard form.}$$

$$2t^2 - 57t + 196 = 0 \qquad \text{Divide both sides by 2.}$$

$$(2t - 49)(t - 4) = 0 \qquad \text{Factor.}$$

$$2t - 49 = 0 \qquad \text{or} \qquad t - 4 = 0 \qquad \text{Set each factor equal to 0.}$$

$$t = 24.5 \qquad \text{or} \qquad t = 4 \qquad \text{Solve for } t.$$

The total time was 7, so 24.5 is not a solution. The biker rode 4 hours the first day and 3 hours the second day.

Work Rate

If a task can be performed in 4 hours, then $\frac{1}{4}$ of the task can be completed in 1 hour. We say that $\frac{1}{4}$ of the task per hour is the *work rate*. In general, if x hours are needed to complete a task, the work rate is $\dfrac{1}{x}$.

EXAMPLE **5**

Work Rate

A charitable organization provides lunches for the homeless. An experienced volunteer takes 5 hours to prepare all the lunches, and a new worker takes 7 hours to do the same job. On a certain day, both workers began working at the same time, but the new worker left after 2 hours. How much longer did the experienced worker take to complete the job?

Solution

Let t = the amount of additional time the experienced worker took to complete the job. Then the new worker worked 2 hours, and the experienced worker worked $t + 2$ hours. In the table, we multiply the number of hours worked by the work rate to determine the part of the job each person completed.

	Number of Hours to Complete Job	Work Rate	Hours Worked	Part of Job Completed
Experienced Worker	5	$\dfrac{1}{5}$	$t + 2$	$\dfrac{t+2}{5}$
New Worker	7	$\dfrac{1}{7}$	2	$\dfrac{2}{7}$

$$\dfrac{2}{7} + \dfrac{t+2}{5} = 1 \qquad \text{The total of the parts is 1 complete job.}$$

$$35 \cdot \dfrac{2}{7} + 35 \cdot \dfrac{t+2}{5} = 1 \cdot 35 \qquad \text{The LCD is 35.}$$

$$10 + 7(t + 2) = 35 \qquad \text{Simplify.}$$

$$10 + 7t + 14 = 35 \qquad \text{Distributive Property}$$

$$7t + 24 = 35 \qquad \text{The equation is linear.}$$

$$7t = 11 \qquad \text{Solve for } t.$$

$$t = \dfrac{11}{7} \approx 1.57$$

The experienced worker took about $1\frac{1}{2}$ hours to complete the job.

EXAMPLE 6

Work Rate

At a refugee camp, one water desalinization facility takes 5 hours less time than another to process water and fill a water tank. If both facilities are in operation, the same tank can be filled in 6 hours. How long does each facility working alone take to fill the tank?

Solution

Let $t = $ the number of hours required for the slower facility to fill the tank. Then $t - 5 = $ the number of hours required for the faster facility to fill the tank.

	Number of Hours to Fill Tank	Work Rate
Faster Machine	$t - 5$	$\dfrac{1}{t-5}$
Slower Machine	t	$\dfrac{1}{t}$
Together	6	$\dfrac{1}{6}$

The sum of the individual work rates is equal to the combined work rate.

$$\frac{1}{t-5} + \frac{1}{t} = \frac{1}{6}$$ The LCD is $6t(t-5)$.

$$6t(t-5) \cdot \frac{1}{t-5} + 6t(t-5) \cdot \frac{1}{t} = 6t(t-5) \cdot \frac{1}{6}$$

$6t + 6(t-5) = t(t-5)$	Simplify.
$6t + 6t - 30 = t^2 - 5t$	Distributive Property
$12t - 30 = t^2 - 5t$	The equation is quadratic.
$t^2 - 17t + 30 = 0$	Write the equation in standard form.
$(t-15)(t-2) = 0$	Factor.
$t - 15 = 0 \quad$ or $\quad t - 2 = 0$	Set each factor equal to 0.
$t = 15 \quad$ or $\quad t = 2$	Solve for t.

We select $t = 15$ because $t = 2$ leads to a negative time for the faster facility. Working alone, the slower facility takes 15 hours to fill the tank, and the faster facility takes 10 hours.

Inverse Variation

Recall that when y *varies directly* with x, there exists a constant k such that $y = kx$. We now define another kind of variation.

> **Definition of Inverse Variation**
>
> A quantity y **varies inversely** with x if there exists a constant k such that $y = \frac{k}{x}$. The constant k is called the **constant of variation.**

EXAMPLE 7

Inverse Variation

The amount of time required to load all the furnishings in a house into a moving van varies inversely with the number of workers. If two people take 17 hours to load the furnishings, how long will five people take?

Solution

Think About It

Assume that x and y represent positive numbers. For an inverse variation, if $k > 0$, then y decreases as x increases. Describe the values of y relative to x for $k < 0$.

Let $t = $ the amount of time required and $w = $ the number of workers.

$t = \dfrac{k}{w}$	t varies inversely with w.
$17 = \dfrac{k}{2}$	Two people load the van in 17 hours.
$k = 34$	The constant of variation is 34.
$t = \dfrac{34}{w}$	Replace k with 34.
$t = \dfrac{34}{5} = 6.8$	Determine t for five people.

Five people take 6.8 hours to load the van.

Miscellaneous Applications

EXAMPLE   8

Gasoline Mileage

A motorist purchased 25 gallons of gasoline for her car and motorcycle. She rode the motorcycle 320 miles and drove the car 360 miles. The car averaged 8 miles per gallon less than the motorcycle. How many gallons of gasoline did she put in each vehicle?

Solution

If x = the number of gallons of gasoline for the car, then

$25 - x$ = the number of gallons of gasoline for the motorcycle.

	Miles Driven	Number of Gallons	Miles per Gallon
Car	360	x	$\dfrac{360}{x}$
Motorcycle	320	$25 - x$	$\dfrac{320}{25 - x}$

$$\frac{360}{x} = \frac{320}{25 - x} - 8$$

The car's gas mileage is 8 less than the motorcycle's gas mileage.

$$x(25 - x) \cdot \frac{360}{x} = x(25 - x) \cdot \frac{320}{25 - x} - x(25 - x) \cdot 8$$

LCD Is $x(25 - x)$.

$$360(25 - x) = 320x - 8x(25 - x)$$ Simplify.

$$9000 - 360x = 320x - 200x + 8x^2$$ Distributive Property

$$8x^2 + 480x - 9000 = 0$$ Write the equation in standard form.

$$x^2 + 60x - 1125 = 0$$ Divide both sides by 8.

$$(x + 75)(x - 15) = 0$$ Factor.

$$x + 75 = 0 \quad \text{or} \quad x - 15 = 0$$ Zero Factor Property

$$x = -75 \quad \text{or} \quad x = 15$$

The number of gallons must be positive. The motorist put 15 gallons of gasoline in the car and 10 gallons in the motorcycle.

Exercises 8.8

Concepts and Skills

Numbers

In Exercises 1–10, determine the described number or numbers.

1. One number is 3 times another. The sum of their reciprocals is $\frac{7}{6}$.

2. The reciprocal of a number plus 3 times the reciprocal of twice that number is 10.

3. The difference between a number and 5 times its reciprocal is 4.

4. A number is equal to the difference between 7 times its reciprocal and 6.

5. The difference of the reciprocal of a number and the reciprocal of twice the number is 5.

6. One-fifth of the reciprocal of a number is 3 less than the reciprocal of the number.

7. The difference between the reciprocal of a number and the reciprocal of 2 more than the number is $\frac{1}{4}$.

8. Five times the reciprocal of a number is 1 more than 4 times the reciprocal of 3 more than the number.

9. One number is 3 more than another. The difference of their reciprocals is $\frac{3}{4}$.

10. One number is 1 more than another. The sum of their reciprocals is $\frac{8}{3}$.

Real-Life Applications

Motion

11. The time required to canoe upstream 5 miles is the same as the time required to canoe downstream 9 miles. If the speed of the current is 1 mph, what is the speed of the canoe?

12. A boat traveled upstream 40 miles in the same time it took to travel downstream 60 miles. If the speed of the current was 2 mph, what was the boat's speed?

13. A car traveled twice as fast on clear roads as it did on icy roads. If the car took a total of 5 hours to travel 63 miles on icy roads and 84 miles on clear roads, what was the car's speed on the clear roads?

14. Two people were at opposite ends of an island. One was 2 miles from a lighthouse, and the other was 3 miles from the lighthouse. They both walked to the lighthouse at the same rate, but one of them took 20 minutes longer than the other. How long did it take each person to arrive at the lighthouse?

15. A person took half an hour longer to bicycle 30 miles than to jog 9 miles. If the person jogged 9 mph slower than she rode, what was her jogging rate? (*Hint*: Let $t =$ jogging time and assume that the jogging rate is less than 10 mph.)

16. A camper frightened by a bear ran out of the woods to the park ranger's office. After he had run 10 miles,  he walked another 10 miles at a rate that was 5 mph slower than he ran. If the camper took 3 hours to reach the office, how fast did he run?

17. A small plane flew 360 miles with the wind and then 400 miles against the wind. If the wind speed was 20 mph and the total flying time was 8 hours, what was the speed of the plane?

18. At noon a person rented a canoe at a livery and headed 3 miles upstream against a current that was 2 mph. She then turned around and paddled downstream to a point that was 1 mile past the livery. If she arrived at 2:10 P.M., how fast did she paddle?

19. On the first leg of a bike ride, a cyclist's average speed was 5 mph less than her speed on the second leg. She took a half-hour longer to ride 25 miles at the slower rate than to ride 30 miles at the faster rate. What was her rate on the first leg?

20. A sales representative drove from Chicago to Detroit, a trip of 260 miles. He averaged 40 mph for part of the trip and 55 mph for the remainder. If the total driving time was 5 hours, how long did he travel at 55 mph?

Work Rate

21. One construction crew took 10 hours to repair a substation damaged by lightning. If another crew had helped, the same job would have been completed in 6 hours. How long would the second crew working alone have taken to complete the job?

22. Working together, two warehouse workers can load a truck in 6 hours. Working alone, the slower worker would take 5 hours longer than the faster one to complete the job. How long would the faster worker take working alone?

23. One person takes 3 times as long as another to assemble new student information packets for an orientation session. If the two people work together, they can assemble the packets in $1\frac{1}{2}$ hours. If the slower worker is on vacation, how long does the faster one take to do the job?

24. Working together, the math lab coordinator and a student assistant can clean up the math lab at the end of the day in 20 minutes. Working alone, the coordinator takes twice as long as the student to do the same job. How long does the student take to do the job?

25. With one pump a city can fill a water purification pond in 10 hours. With another pump the same pond can be filled in 15 hours. How long do both pumps together take to fill the pond?

26. Two people are restocking the shelves in a grocery store. Working alone, one person takes 2 hours to complete the job, and the other takes 5 hours. How long will they take working together?

27. One copier takes 25 minutes longer than another to reproduce orientation information sheets for all freshmen at a college. Both copiers together take 30 min-

utes to make the copies. How long does each copier take working alone? (*Hint*: Let t = faster copier's time; $750 = 15 \cdot 50$.)

28. Starting at 8 A.M. and working together, two park employees can prepare a soccer field by 9:20 A.M. Working alone, one employee can complete the job in 2 hours less time than the other. How long does each employee take working alone?

29. A student can complete 5 more algebra problems than physics problems in 1 hour. The amount of time the student needs to complete a 45-problem physics assignment is 1 hour more than the time needed to complete a 40-problem algebra assignment. How many algebra problems can the student complete in 1 hour?

30. A caterer takes 1 hour longer to prepare 75 stuffed mushrooms than to prepare 75 pineapple and cream cheese sandwiches. The caterer can prepare 20 more sandwiches than stuffed mushrooms in an hour. How many sandwiches can the caterer make per hour? (*Hint*: Let r = number of stuffed mushrooms that can be prepared per hour; $1500 = 30 \cdot 50$.)

31. A new worker requires 5 hours to complete a data entry job. When he and a more experienced worker do the job together, they can complete it in 2 hours. How long would the experienced person take working alone?

32. A concrete finisher worked from noon until 6 P.M. Then he took a one-hour break and called his partner for help. The two men went back to work at 7 P.M. and finished the job at 9 P.M. If the concrete finisher works twice as fast as his partner, at what time would he have finished the job by himself if he had worked without a break?

Inverse Variation

In Exercises 33–36, assume that m varies inversely with n. Determine the constant of variation for the given values of m and n.

33. $m = 4$ when $n = 5$

34. $m = -3$ when $n = -8$

35. $n = -3.2$ when $m = 5.1$

36. $n = 6.3$ when $m = -0.4$

Exercises 37–44 are inverse variation problems.

37. Two people take 9 hours to lay tile on a lobby floor. How long would 3 people take?

38. The volume of a gas varies inversely with the pressure. If the volume is 60 cubic centimeters (cc) when the pressure is 9 kg/cc, what is the volume if the pressure is 12 kg/cc?

39. The per-person cost of renting a beach-front cottage for a week varies inversely with the number of people sharing the cost. If the per-person cost is $252 for 4 people, how many people share the cottage if the per-person cost is $168?

40. The time required to drive from Duluth to Fargo varies inversely with the speed of the car. If the trip takes 6.25 hours at 40 mph, what speed is required to make the trip in 4 hours?

41. The demand for snow tires varies inversely with the average January temperature. If Tire World sells 2300 snow tires in Burlington, where the average temperature is 20°F, how many snow tires can Tire World expect to sell at its store in Nashville, where the average temperature is 46°F?

42. A baseball pitcher finds that his earned run average (ERA) and his "strikeouts per game" average are inversely related. If his ERA was 3.30 last year when he averaged 8 strikeouts per game, what must his average strikeouts per game be to lower his ERA to 2.64?

43. A political fund-raiser can attract 200 people for a $1000-per-plate dinner. If the number of people attending is inversely related to the price per plate, how many people will attend if the price is $800 per plate?

44. An art dealer sells prints. When 30 copies of a print are available, they can be sold for $10 each. If availability and price are inversely related, how much will it cost to buy a print when only 2 copies are available?

Miscellaneous Applications

45. The unit price of packaged garden salad is 5 times the unit price of bird feed. If salad and bird feed were purchased for $2 and $4, respectively, and the total weight of the purchase was 22 pounds, what was the unit price of the bird feed?

46. A shopper bought bananas for $1.80 and pears for $4.50. He bought 3 more pounds of pears than of bananas and purchased at least 5 pounds of bananas. If the difference in the unit prices was 20 cents and the unit price of pears is greater than that of bananas, how many pounds of pears did the shopper buy?

47. A person made $56 for a certain number of hours of work. A second person made $54 but worked 1 hour more than the first person. If the difference in their hourly wages is $1 per hour, how long did the second person work?

48. A person bought some ABC stock for $1500 and some XYZ stock for $1000. The price per share of XYZ stock was $5 more than that of ABC stock. If a combined total of 150 shares was purchased, what was the price per share of XYZ stock?

Data Analysis: *Two-Parent and One-Parent Families*

The table shows the number (in millions) of two-parent and one-parent family groups for selected years.

Year	Two-Parent	One-Parent
1980	25.2	6.9
1990	24.9	9.7
1995	25.6	11.5
2000	25.7	12.4

(Source: U.S. Bureau of the Census.)

The data can be modeled by the following linear equations, where x represents the number of years since 1980.

Two-parent: $y = 0.03x + 25$

One-parent: $y = 0.28x + 6.9$

49. What trends do the models indicate?

50. Write a rational expression to model the ratio of the number of two-parent family groups to the number of one-parent family groups.

51. Write an equation to indicate that the ratio of two-parent families to one-parent families was 2 to 1.

52. Solve the equation in Exercise 51 and interpret the solution.

Challenge

53. A car traveled 10 mph faster than a motorcycle. The motorcycle took 30 minutes longer to travel 120 miles than the car took to travel 125 miles. How fast did the car travel?

54. A police car was 4 miles from an accident scene, and an ambulance was 6 miles away. The ambulance took 3 minutes longer than the police car to arrive at the scene. If the police car traveled 3 mph faster than the ambulance, how many minutes did the ambulance take to arrive at the scene?

55. Workers at a bakery take twice as long to make 24 dozen cookies as to make 15 dozen donuts. If the workers make 2 dozen more donuts per hour than dozens of cookies, how many dozens of each can they make in an hour?

56. At a paint factory, emptying a full mixing tank takes 2 minutes longer than filling an empty tank. If paint is flowing into and out of the tank at the same time, 24 minutes are required to fill an empty tank. What are the rates of flow of the paint into and out of the tank?

57. After a farmer had baled hay for 3 hours, his neighbor arrived with his baler and helped for 2 hours. When the neighbor left, the farmer still had $\frac{5}{24}$ of the job left to do. If the farmer can work the entire field in 4 hours less than the neighbor, how long would he have taken to hay the entire field if his neighbor had not helped?

Chapter 8 Review Exercises

Section 8.1

1. In each part, evaluate $r(x) = \dfrac{x+6}{x^2-5x}$ as indicated.

   (a) $r(3)$ (b) $r(5)$

2. Determine the restricted values for

$$\frac{2x}{x^2-2x-8}$$

3. Which of the following is the correct way to simplify $\dfrac{3x}{3x+3y}$?

  (i) Divide out $3x$ from the numerator and denominator.

  (ii) Factor out 3 in the denominator and then divide out 3 in the numerator and denominator.

In Exercises 4–9, simplify.

4. $\dfrac{18x+2x^2}{27+3x}$

5. $\dfrac{x^2-4}{(x-2)^2}$

6. $\dfrac{x^2+11x+28}{x^2-x-20}$ **7.** $\dfrac{a-4b}{16b^2-a^2}$

8. $\dfrac{4w^2-9w+5}{2+7w-9w^2}$ **9.** $\dfrac{y+3}{y^2+9}$

In Exercises 10 and 11, determine whether the two given expressions are equivalent.

10. $\dfrac{-(x-y)}{x-y}, \; -1 \quad (x \neq y)$

11. $\dfrac{x+2}{x-2}, \; -1 \quad (x \neq 2)$

12. In each part, determine whether the expression is equal to 1, -1, or neither.

  (a) $\dfrac{x-y}{y-x}$

  (b) $\dfrac{x+y}{y-x}$

  (c) $\dfrac{y+x}{x+y}$

Section 8.2

13. To multiply $\dfrac{x+3}{2x}$ and $\dfrac{5x^2}{2x+6}$, is it better to

  (i) divide out common factors and then multiply or

  (ii) multiply and then simplify the result?

  Explain.

14. Determine the product in Exercise 13.

15. Multiply $\dfrac{x^2-x-12}{x^2+7x+12} \cdot \dfrac{x+4}{x^2-16}$.

16. Multiply $\dfrac{x-2}{3x} \cdot \dfrac{9x^2}{8-4x}$.

17. Write $3 \div (x+2)$ as a product.

18. Divide $\dfrac{x^2+4x+4}{x^2+3x+2} \div \dfrac{x+2}{x+1}$.

19. For what values of x is the following quotient undefined?

$$\frac{x}{x-2} \div \frac{x-3}{x+5}$$

20. Divide $\dfrac{x^2-3x}{x+2} \div \dfrac{12-4x}{x^2+2x}$.

Section 8.3

21. Which of the following is equivalent to -1?

(i) $\dfrac{x}{x-y} - \dfrac{y}{x-y}$

(ii) $\dfrac{x}{x-y} + \dfrac{y}{y-x}$

In Exercises 22–26, perform the indicated operation.

22. $\dfrac{x^2+x-5}{x-1} + \dfrac{4-x^2}{x-1}$

23. $\dfrac{2x-5}{x^2+4x-21} + \dfrac{12-x}{x^2+4x-21}$

24. $\dfrac{8-5x}{x^2-9} + \dfrac{6x-5}{x^2-9}$

25. $\dfrac{x^2-3x}{x^2+2x} - \dfrac{x^2-4x-2}{x^2+2x}$

26. $\dfrac{2-y^2}{y+1} - \dfrac{y^2}{y+1}$

27. Determine the unknown numerator that makes the following equation true.

$$\dfrac{3x^2-5x+1}{x+3} - \dfrac{\rule{1cm}{0.3cm}}{x+3} = \dfrac{2x^2-7x+4}{x+3}$$

28. Simplify $\dfrac{6z+3}{4z} - \dfrac{5z-1}{4z} + \dfrac{3z-4}{4z}$.

Section 8.4

In Exercises 29 and 30, rewrite the rational expression with the specified denominator.

29. $\dfrac{2x+1}{2x^2-x} = \dfrac{\rule{1cm}{0.3cm}}{14x^2-7x}$

30. $\dfrac{y}{3-y} = \dfrac{\rule{1cm}{0.3cm}}{y^2-4y+3}$

In Exercises 31 and 32, determine the LCM of the given expressions.

31. x^2-9, $\quad 3x+9$

32. $x^2+7x+10$, $\quad x^2+3x-10$

In Exercises 33 and 34, determine the LCD of the given fractions.

33. $\dfrac{2}{3x^2}$, $\quad \dfrac{x-1}{x^2+3x}$

34. $\dfrac{x}{x^2-16}$, $\quad \dfrac{x+2}{x^2-7x+12}$

In Exercises 35 and 36, determine the LCD for the given pair of rational expressions. Then rename each expression as an equivalent expression with that LCD.

35. $\dfrac{y}{6y-15}$, $\quad \dfrac{3y}{4y-10}$

36. $\dfrac{x+2}{9x^2-1}$, $\quad \dfrac{2x}{3x^2+5x-2}$

Section 8.5

In Exercises 37–44, add or subtract as indicated.

37. $\dfrac{x}{x+3} + \dfrac{4}{x}$

38. $\dfrac{x}{x-8} - \dfrac{3}{x+2}$

39. $\dfrac{x^2}{x-5} - \dfrac{-25}{5-x}$

40. $\dfrac{2x-1}{4x-5} + \dfrac{x+1}{5-4x}$

41. $\dfrac{2}{x+3} + \dfrac{5}{x^2-9}$

42. $\dfrac{4}{t^2-7t+10} - \dfrac{2}{t-5}$

43. $\dfrac{5}{x^2+2x-24} - \dfrac{1}{x^2+9x+18}$

44. $\dfrac{4}{x^2-9} + \dfrac{3}{2x^2+9x+9}$

Section 8.6

Exercise 45 refers to the fraction in part (a), and Exercise 46 refers to the fraction in part (b). In each exercise, identify the correct method(s) for simplifying the fraction. If a given method is incorrect, explain why.

(a) $\dfrac{\dfrac{x}{5}}{\dfrac{3}{y}}$

(b) $\dfrac{\dfrac{1}{x} - \dfrac{1}{y}}{\dfrac{1}{xy} + \dfrac{1}{x}}$

45. (i) Perform the division after rewriting the expression as $\dfrac{x}{5} \div \dfrac{3}{y}$.

(ii) Multiply the numerator and the denominator by the LCD, $5y$.

46. (i) Perform the division after rewriting the expression as $\dfrac{1}{x} - \dfrac{1}{y} \div \dfrac{1}{xy} + \dfrac{1}{x}$.

(ii) Multiply the numerator and the denominator by the LCD, xy.

In Exercises 47–50, simplify the complex fraction.

47. $\dfrac{\dfrac{3}{a} + \dfrac{5}{b^2}}{\dfrac{4}{b} - \dfrac{1}{a^2}}$

48. $\dfrac{1 - \dfrac{1}{x - 1}}{\dfrac{1}{x - 1}}$

49. $\dfrac{2 - \dfrac{b}{a - b}}{2a - 3b}$

50. $\dfrac{\dfrac{12}{8x - 4}}{\dfrac{18}{14x - 7}}$

Section 8.7

51. Suppose that you intend to use the graphing method to estimate the solutions of

$$\frac{x}{x - 3} = \frac{x}{x^2 + 2}$$

Even before you graph, how do you know that the graphs of the left and right sides do not intersect at $x = 3$?

52. Use the graphing method to estimate the solutions of $x - 2 = \dfrac{24}{x}$.

In Exercises 53–56, solve.

53. $\dfrac{5}{6} + \dfrac{2}{2x - 1} = \dfrac{5}{6x - 3}$

54. $\dfrac{x + 6}{x + 7} = \dfrac{2x - 2}{4x + 1}$

55. $\dfrac{2x + 20}{x + 3} - \dfrac{8 - 2x}{x + 3} = 1$

56. $\dfrac{x - 3}{x - 7} = \dfrac{2}{x} + \dfrac{28}{x^2 - 7x}$

Section 8.8

57. The difference of a negative number and 8 times its reciprocal is 2. What is the number?

58. A commuter lives 13 miles from her office. She drives the first 10 miles in twice the time that it takes her to drive the remaining 3 miles, for which she reduces her speed by 20 mph. How many minutes does she take to travel the first leg of the trip?

59. Working together, a carpet layer and his nephew can lay the carpet in a certain room in $2\frac{2}{5}$ hours. Working alone, the carpet layer could complete the job 2 hours sooner than his less industrious nephew. How long would the nephew take working alone?

60. For a day labor pool, the hourly wages paid vary inversely with the number of workers that are available. If the hourly wage is \$5.25 when 20 workers are in the pool, how many fewer workers must there be for the wages to increase by \$1.75 per hour?

Chapter 8 Test

1. For what value(s) is $\dfrac{x - 4}{-x^2 + 8x - 15}$ not defined?

2. Simplify $\dfrac{x^2 + 3x - 4}{x^2 - 16}$.

3. Simplify $\dfrac{x^2 - y^2}{(y - x)^2}$.

In Questions 4–7, perform the indicated operations.

4. $\dfrac{(x + 3)^2}{x^2 + x - 2} \cdot \dfrac{x^2 - 6x + 5}{x^2 - 2x - 15}$

5. $\dfrac{2 - x}{x + 2} \cdot \dfrac{x^2 + 2x}{4x - 8}$

6. $\dfrac{6x - 18}{2x} \div (x^2 - 9)$

7. $\dfrac{1 - 3x}{x^2 + 6x} + \dfrac{4x + 5}{x^2 + 6x}$

8. Which of the following sequences of keystrokes is correct for entering

$$\dfrac{x}{x + 3} - \dfrac{x - 2}{x}$$

in your calculator? Why?

(i) X ÷ (X + 3) − X − 2 ÷ X

(ii) X ÷ X + 3 − (X − 2) ÷ X

9. Subtract $\dfrac{2x}{y} - \dfrac{x + 4}{y}$.

10. Simplify $\dfrac{x + 3}{x - 1} - \dfrac{x + 5}{x - 1} + \dfrac{x + 1}{x - 1}$.

11. Consider the monomials $3x^2$, $5x^3$, and $2x^4$. Explain the difference in how the exponents are used to determine the LCM and the GCF of the expressions.

12. Determine the LCM of $3x^2$ and $x^3 - x^2$.

13. Determine the LCD of $\dfrac{2x}{x^2 - 4}$ and $\dfrac{5}{x^2 + 6x + 8}$.

14. Rewrite $\dfrac{x}{2x + 4}$ and $\dfrac{x - 1}{x^2 + 2x}$ with a LCD.

In Questions 15–17, add or subtract as indicated.

15. $\dfrac{x}{x - 8} + \dfrac{3}{x + 5}$

16. $\dfrac{y}{y^2 - 1} - \dfrac{4}{y - 1}$

17. $\dfrac{2}{x^2 + 2x + 1} - \dfrac{x}{x^2 + 3x + 2}$

18. Simplify $\dfrac{\dfrac{8x + 64}{2x^3}}{\dfrac{x^2 - 64}{2x^2 - 16x}}$.

19. Simplify $\dfrac{x + \dfrac{1}{y}}{\dfrac{1}{x} - y}$.

20. Solve $\dfrac{4}{x^2} = 3 - \dfrac{1}{x}$.

21. Solve $\dfrac{x + 7}{x + 2} + \dfrac{10}{x^2 + 2x} = \dfrac{2}{x}$.

22. The sum of a number, its reciprocal, and 4 times its reciprocal is 6. What is the number?

23. Working alone, the owner of a gift store can assemble an order of gift baskets in 3 hours less time than her part-time employee. The owner works on the order from 10 A.M. until noon and then leaves the store. The part-time employee continues working on the order from 1 P.M. until 5 P.M., and at that time the job is 90% done. How long would the owner have taken to do the job working alone?

Cumulative Test, Chapters 6–8

1. Simplify.

(a) $(x^2 y)(-x^4 y^2)$

(b) $\dfrac{12a^5}{9a^2}$

(c) $\left(\dfrac{2a}{3b^2}\right)^3$

2. Consider $2x^2 - 5x + 3$.

(a) What are the coefficients?

(b) Evaluate the polynomial for $x = -2$.

3. Simplify $(x^2 + 3) - (2x - 8) + (x^2 + 7x - 4)$.

4. Multiply $(x - 1)(x^2 + 3x - 5)$.

5. Multiply.

(a) $(2x + 1)^2$

(b) $(2x + 1)(2x - 1)$

6. Determine the quotient and remainder.

$$(2x^3 - 3x^2 - 2x - 9) \div (x - 3)$$

7. (i) Use your calculator to evaluate -2^{-2}.

(ii) Now store -2 in X on your calculator and enter X^(−2) on the home screen.

Explain the difference in the two results.

8. Simplify with positive exponents. Assume that x and y are not 0.

$$(2xy^{-3})^{-2}(x^{-2}y^2)$$

9. In 1993 American businesses sent 6 billion e-mail messages. If these messages averaged 50 words in length, use scientific notation to represent the total number of words sent that year by e-mail.

▼ 10. Explain why $x^2 + 2x - xy + 2y$ cannot be factored with the grouping method.

In Questions 11–15, factor the given polynomial completely.

11. (a) $6x^3y^2 - 9x^2y$

    (b) $2ax + ay - 4bx - 2by$

12. (a) $x^2 - 64$

    (b) $x^3 - 64$

13. (a) $2x^2 - 22x + 36$

    (b) $x^4 + 2x^2 - 15$

14. $3x^2 - 5x - 22$

15. $x^5 - x$

16. Solve $x(x - 2) = 35$.

17. The longer leg of a right triangle is 4 inches more than twice the length of the shorter leg. The hypotenuse is 4 inches less than 3 times the length of the shorter leg. Determine the length of the longer leg.

18. Simplify $\dfrac{x^2 + 7x - 18}{x^2 - 2x}$.

19. Divide $\dfrac{x^2 - 25}{x^2 + 5x} \div \dfrac{x^2 - 2x - 15}{2x}$.

20. Subtract $\dfrac{5x + 8}{2x + 6} - \dfrac{5 + 4x}{2x + 6}$.

21. Determine the LCM of $2x^2$ and $4x^2 - 6x$.

22. Add $\dfrac{2}{x^2 + 5x} + \dfrac{x}{x^2 + 4x - 5}$.

23. Simplify $\dfrac{\dfrac{1}{x + 1}}{\dfrac{1}{x} + 1}$.

24. Solve $\dfrac{x - 8}{x - 2} + \dfrac{12}{x^2 - 2x} = \dfrac{2}{x}$.

25. A commuter takes a train for 30 miles and then drives the remaining 20 miles to his home. He averages 20 mph more in his car than on the train and spends 36 minutes longer on the train than in the car. What is the average speed of the train?

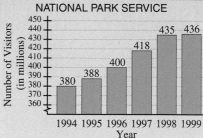

NATIONAL PARK SERVICE

As indicated in the accompanying bar graph, the National Park Service system attracted an increasing number of visitors during the period 1994–1999. To model these data, we can use an expression with a **rational exponent** that is defined in terms of **radicals.**

This model can be used, for example, to project revenues from proposed entrance fee schedules. (For more on this real data application, see Exercises 89–92 at the end of Section 9.7.)

(Source: National Park Service.)

Chapter Snapshot

In this chapter, we consider square roots and higher roots. We learn how to simplify radicals and how to perform the basic operations with them. Then we turn to methods for solving radical equations and use right triangle problems as a specific type of application. Finally, we extend our knowledge of exponents to rational exponents and their properties.

Warm-Up Skills

The following exercises review concepts and skills that you will need in Chapter 9.

1. Determine the square root.

   (a) $\sqrt{81}$ (b) $\sqrt{16}$

In Exercises 2–5, use the properties of exponents to simplify the given expression.

2. (a) $(3x)^2$ (b) $(5ab^3)^2$

3. (a) $x^7 \cdot x^4$ (b) $(6x^5)(3x^3)$

4. (a) $\dfrac{45x^3}{5x^7}$ (b) $\dfrac{24x^9}{3x^3}$

5. (a) $\left(\dfrac{2x^2}{y}\right)^3$ (b) $\left(\dfrac{a^3}{b^9}\right)^2$

6. Show how to use the Distributive Property to add $2x + 7x$.

In Exercises 7 and 8, multiply.

7. $(x + 6)(x - 6)$ 8. $(x + 3)^2$

9. Factor $25 - y^2$.

In Exercises 10 and 11, solve the given equation.

10. $2x - 3 = x + 4$

11. $x^2 - 2x + 1 = 9x - 27$

12. The lengths of the sides of a triangle are 8, 10, and 6. Determine whether the triangle is a right triangle.

9.1 | Radicals

 Square Roots • Evaluating $\sqrt{b^2}$ and $\left(\sqrt{b}\right)^2$ • Higher Roots

Square Roots

Consider the following three problems.

(a) $3^2 = $ ▭   (b) $(-3)^2 = $ ▭   (c) (▭)$^2 = 9$

In (a) we square 3, or raise 3 to the second power, by multiplying 3 by itself. In (b) we multiply -3 by itself. The result in both cases is 9. In (c) we are seeking a number whose square is 9. From the results in parts (a) and (b), we can see that the number is 3 or -3. We say that 3 and -3 are **square roots** of 9.

> **Definition of Square Root**
>
> For real numbers a and b, if $a^2 = b$, then a is a **square root** of b.

If $a^2 = b$, then it is also true that $(-a)^2 = b$. For this reason, each positive number b has two square roots that are opposites of each other, a and $-a$. However, 0 has just one square root because only 0 can be squared to obtain 0. In the real number system, a negative number has no square root because the square of a real number is never negative.

EXAMPLE 1   **Determining the Square Roots of a Real Number**

Determine the square root(s), if any, of the given number.

(a) 49   (b) $\dfrac{16}{25}$ (c) -36

Solution

(a) Because 7^2 and $(-7)^2$ both equal 49, the two square roots of 49 are -7 and 7.

(b) There are two numbers whose square is $\frac{16}{25}$: $-\frac{4}{5}$ and $\frac{4}{5}$. Thus $-\frac{4}{5}$ and $\frac{4}{5}$ are both square roots of $\frac{16}{25}$.

(c) There is no real number whose square is negative. Thus -36 has no square root.

We use the symbol $\sqrt{}$ to indicate the nonnegative or **principal square root** of a number. Thus, although 36 has two square roots, 6 and -6, we say that $\sqrt{36} = 6$.

> **Definition of Principal Square Root**
>
> For any nonnegative real numbers a and b, if $a^2 = b$, then $\sqrt{b} = a$.

In the expression $\sqrt{b}$, $\sqrt{}$ is called the **radical symbol** and b is called the **radicand**. We refer to $\sqrt{b}$ as a **radical** or a **radical expression**.

Note: Observe the difference between the values of $-\sqrt{9}$ and $\sqrt{-9}$. In the first expression, we take the principal square root of 9 and then affix the negative sign: $-\sqrt{9} = -\left(\sqrt{9}\right) = -3$. The second expression does not represent a real number because there is no real number a such that $a^2 = -9$.

EXAMPLE 2

Evaluating Square Roots

Evaluate the given radicals.

(a) $\sqrt{49} = 7$ $7^2 = 49$

(b) $\sqrt{\dfrac{9}{4}} = \dfrac{3}{2}$ $\left(\dfrac{3}{2}\right)^2 = \dfrac{9}{4}$

(c) $-\sqrt{36} = -6$ First determine $\sqrt{36}$; then affix the negative sign.

(d) $6\sqrt{25} = 6 \cdot \sqrt{25} = 6 \cdot 5 = 30$

Numbers whose square roots are rational numbers are called **perfect squares.** For example, 25, 0, 81, $\frac{4}{9}$, and $\frac{1}{4}$ are perfect squares because their square roots are rational numbers. The square roots of positive real numbers that are not perfect squares are irrational numbers.

EXAMPLE 3

Identifying Square Roots as Rational or Irrational Numbers

Classify each number as rational, irrational, or not a real number.

(a) $\sqrt{6}$ (b) $\sqrt{100}$ (c) $\sqrt{-16}$ (d) $-\sqrt{50}$

Solution

(a) Because 6 is not a perfect square, $\sqrt{6}$ is irrational.

(b) Because $100 = 10^2$, $\sqrt{100}$ is rational.

(c) There is no real number whose square is -16. Thus $\sqrt{-16}$ is not a real number.

(d) Because 50 is not a perfect square, $-\sqrt{50}$ is irrational.

We can use a calculator to estimate the square root of a number that is not a perfect square.

EXAMPLE 4

Approximating Square Roots

Use a calculator to approximate each square root. Round the result to the nearest hundredth.

(a) $\sqrt{45}$ (b) $\sqrt{530}$ (c) $-\sqrt{3.9}$

Solution

```
√(45)
                6.71
√(530)
               23.02
-√(3.9)
               -1.97
```

Evaluating $\sqrt{b^2}$ and $\left(\sqrt{b}\right)^2$

Parentheses and radical symbols are grouping symbols. Thus, to evaluate radical expressions of the form $\sqrt{b^2}$ and $\left(\sqrt{b}\right)^2$, we simply follow the Order of Operations.

Exploring the Concept

Evaluating $\sqrt{b^2}$ and $\left(\sqrt{b}\right)^2$

When we evaluate $\sqrt{9^2}$ and $\left(\sqrt{9}\right)^2$, we perform the indicated operations in a different order.

$$\sqrt{9^2} = \sqrt{81} = 9 \qquad \text{Square 9; then take the square root of the result.}$$

$$\left(\sqrt{9}\right)^2 = (3)^2 = 9 \qquad \text{Take the square root of 9; then square the result.}$$

Keep in mind that when we evaluate radical expressions involving square roots, neither the radicand nor the result can be negative.

$$\sqrt{(-9)^2} = \sqrt{81} = 9 \qquad \text{Square } -9; \text{then take the square root of the result.}$$

$$\left(\sqrt{-9}\right)^2 \text{ cannot be evaluated with real numbers} \qquad \text{The radicand cannot be negative.}$$

When the radicand of a square root radical contains a variable, special care must be taken to ensure that the result is not negative. For example, we are tempted to use the definition of square root to say that $\sqrt{x^2} = x$. However, $\sqrt{x^2} = x$ is true only if x is nonnegative. To ensure that the result is nonnegative, we enclose it in absolute value symbols: $\sqrt{x^2} = |x|$.

All of the preceding observations can be summarized by the following rules.

Evaluating $\left(\sqrt{b}\right)^2$ and $\sqrt{b^2}$

1. If $b \geq 0$, then $\left(\sqrt{b}\right)^2 = b$. If $b < 0$, $\left(\sqrt{b}\right)^2$ cannot be evaluated with real numbers.

2. For any real number b, $\sqrt{b^2} = |b|$.

The first rule follows directly from the definition of square root. The second rule is based on the fact that a principal square root must be nonnegative.

EXAMPLE 5

Evaluating $\left(\sqrt{b}\right)^2$

Evaluate or simplify each expression. Assume that all variables represent positive numbers.

(a) $\left(\sqrt{64}\right)^2 = 64 \qquad$ If $b \geq 0$, then $\left(\sqrt{b}\right)^2 = b$.

(b) $\left(-\sqrt{7}\right)^2 = \left(-1 \cdot \sqrt{7}\right)^2 = (-1)^2 \cdot \left(\sqrt{7}\right)^2 = 1 \cdot 7 = 7$

(c) $\left(\sqrt{3y}\right)^2 = 3y = |3y|$

(d) $\left(-3\sqrt{2}\right)^2 = (-3)^2 \cdot \left(\sqrt{2}\right)^2 = 9 \cdot 2 = 18$

(e) $\left(\sqrt{y^2 + 4}\right)^2 = y^2 + 4$

EXAMPLE 6 | **Evaluating $\sqrt{b^2}$**

Evaluate or simplify each expression.

(a) $\sqrt{10^2} = |10| = 10$

(b) $\sqrt{(-7)^2} = |-7| = 7$

(c) $-\sqrt{(-2)^2} = -|-2| = -2$

(d) $\sqrt{9x^2} = \sqrt{(3x)^2} = |3x|$

To simplify the square root of an exponential expression with an even exponent, write the radicand as the square of a quantity. (As illustrated in Example 7, if we assume that the variable represents a positive number, then absolute value symbols are not necessary.)

EXAMPLE 7 | **Evaluating $\sqrt{b^2}$**

Simplify each expression. Assume that all variables represent positive numbers.

(a) $\sqrt{x^6} = \sqrt{(x^3)^2} = x^3$

(b) $\sqrt{y^{12}} = \sqrt{(y^6)^2} = y^6$

(c) $\sqrt{9w^{20}} = \sqrt{(3w^{10})^2} = 3w^{10}$

(d) $\sqrt{36t^{10}} = \sqrt{(6t^5)^2} = 6t^5$

Note: An exponential expression is a perfect square if the expression can be factored so that the exponent on each factor is an even number. An easy way to evaluate such a square root is to take half of each exponent.

Higher Roots

The concept of square root can be extended to higher roots. Just as finding a square root reverses the process of squaring a number, finding a **cube root** or **third root** reverses the process of cubing a number or raising a number to the third power. For example, because $2^3 = 8$, 2 is a cube root of 8.

> **Definition of nth Root**
>
> For real numbers a and b and any integer $n > 1$, if $a^n = b$, then a is an **nth root** of b.

EXAMPLE 8 | **Determining Higher Roots**

Determine the indicated root.

(a) 3rd root of -27 (b) 4th root of 81

(c) 5th root of 32   (d) 6th root of 1

Solution

(a) The 3rd root of -27 is -3. $(-3)^3 = -27$

(b) The 4th roots of 81 are 3 and -3. $3^4 = (-3)^4 = 81$

(c) The 5th root of 32 is 2. $2^5 = 32$

(d) The 6th roots of 1 are -1 and 1. $(-1)^6 = 1^6 = 1$

Observe that there are *two* even roots for all positive numbers b, but there is just *one* real odd root for any number b. Also, whereas even roots of negative numbers are not defined, an odd root of a negative number is a real number.

We use a radical symbol to write the nth root of a number. For even nth roots, the radical symbol means the *principal* (nonnegative) nth root. For example, we write $\sqrt[3]{8} = 2$, $\sqrt[4]{81} = 3$, and $\sqrt[5]{-32} = -2$.

Definition of $\sqrt[n]{b}$

For nonnegative real numbers a and b and any even integer $n > 1$, if $a^n = b$, then $\sqrt[n]{b} = a$.

For any real numbers a and b and any odd integer $n > 1$, if $a^n = b$, then $\sqrt[n]{b} = a$.

Think About It

According to the definition of $\sqrt[n]{b}$, what must be true about n? Now produce the graph of $y = \sqrt[x]{x}$. Does your calculator abide by the definition? What happens to the value of $\sqrt[x]{x}$ as x becomes larger and larger?

In the radical $\sqrt[n]{b}$, n is called the **index** of the radical. When $n = 2$, the index is usually omitted: $\sqrt[2]{b} = \sqrt{b}$.

EXAMPLE 9

Evaluating nth Roots

Evaluate each radical.

(a) $\sqrt[3]{27} = 3$ $3^3 = 27$

(b) $\sqrt[3]{-8} = -2$ $(-2)^3 = -8$

(c) $-\sqrt[4]{16} = -2$ $2^4 = 16$

(d) $\sqrt[4]{-16}$ is not a real number because it is an even root of a negative number.

(e) $\sqrt[5]{-1} = -1$ $(-1)^5 = -1$

Example 9 illustrates the following summary.

Properties of $\sqrt[n]{b}$

If n is	and b is	then $\sqrt[n]{b}$ is
even	positive	positive
	0	0
	negative	not a real number
odd	positive	positive
	0	0
	negative	negative

LEARNING TIP

The properties of $\sqrt[n]{b}$ are easily summarized. For even roots, neither the radicand nor the result can be negative; for odd roots, there are no restrictions at all.

An expression is a **perfect nth power** if the expression can be factored so that the exponent on each factor is divisible by n. To simplify the nth root of an exponential expression, try to write the radicand as a quantity raised to the nth power.

EXAMPLE 10

Simplifying nth Roots

Simplify each expression. Assume that all variables represent positive numbers.

(a) $\sqrt[3]{x^6} = \sqrt[3]{(x^2)^3} = x^2$

(b) $\sqrt[4]{y^{12}} = \sqrt[4]{(y^3)^4} = y^3$

(c) $\sqrt[5]{x^{10}} = \sqrt[5]{(x^2)^5} = x^2$

Quick Reference 9.1

Square Roots

- For real numbers a and b, if $a^2 = b$, then a is a **square root** of b.

- Positive numbers have two square roots that are opposites. The number 0 has just one square root, 0. Negative numbers have no real number square roots.

- We use the symbol $\sqrt{}$ to indicate the nonnegative or **principal square root** of a number. For any nonnegative real numbers a and b, if $a^2 = b$, then $\sqrt{b} = a$.

- In the expression $\sqrt{b}$, $\sqrt{}$ is called the **radical symbol,** and b is called the **radicand.** We refer to $\sqrt{b}$ as a **radical** or a **radical expression.**

- Numbers whose square roots are rational numbers are called **perfect squares.**

Evaluating $\sqrt{b^2}$ and $\left(\sqrt{b}\right)^2$

- If $b \geq 0$, then $\left(\sqrt{b}\right)^2 = b$. If $b < 0$, then $\left(\sqrt{b}\right)^2$ is undefined.

- For any real number b, $\sqrt{b^2} = |b|$.

- To simplify the square root of an exponential expression with an even exponent, write the radicand as the square of a quantity.

Higher Roots

- For real numbers a and b, if $a^3 = b$, then a is a **cube root** of b.

- In general, for real numbers a and b and any integer $n > 1$, if $a^n = b$, then a is an **nth root** of b.

- For nonnegative real numbers a and b and any even integer $n > 1$, if $a^n = b$, then $\sqrt[n]{b} = a$.

  For any real numbers a and b and any odd integer $n > 1$, if $a^n = b$, then $\sqrt[n]{b} = a$.

- In the radical $\sqrt[n]{b}$, n is called the **index** of the radical.

- An expression is a **perfect nth power** if the expression can be factored so that the exponent on each factor is divisible by n. To simplify the nth root of an exponential expression, try to write the radicand as a quantity raised to the nth power.

Speaking the Language 9.1

1. If $a^2 = b$, then a is called a(n) _____ of b.
2. The symbol $\sqrt{\ }$ is called a(n) _____ symbol and indicates the _____ square root of a number.
3. In $\sqrt{x + 3}$, the expression $x + 3$ is called the _____ .
4. In the expression $\sqrt[n]{x}$, the _____ is n.

Exercises 9.1

Concepts and Skills

1. Describe the numbers in each list and give the next three numbers in the list.
   (a) $1, 4, 9, 16, \ldots$
   (b) $1, 8, 27, 64, \ldots$

2. Explain why $\sqrt{-25}$ is not a real number.

In Exercises 3–6, determine the square roots of each number.

3. 25
4. 81
5. $\frac{9}{49}$
6. $\frac{36}{25}$

In Exercises 7–10, produce the graph of $y = \sqrt{x}$. Trace the graph to estimate the value of the radical.

7. $\sqrt{9}$
8. $\sqrt{36}$
9. $\sqrt{10}$
10. $\sqrt{20}$

In Exercises 11–18, evaluate the radical without your calculator.

11. $\sqrt{25}$
12. $\sqrt{49}$
13. $\sqrt{9}$
14. $\sqrt{1}$
15. $-\sqrt{81}$
16. $-\sqrt{4}$
17. $5\sqrt{36}$
18. $-3\sqrt{16}$

In Exercises 19–24, classify the given number as rational, irrational, or not a real number.

19. $\sqrt{100}$
20. $\sqrt{\dfrac{25}{4}}$
21. $\sqrt{-4}$
22. $\sqrt{-1}$
23. $\sqrt{10}$
24. $\sqrt{18}$

In Exercises 25–30, use a calculator to approximate the value to the nearest hundredth.

25. $\sqrt{250}$
26. $\sqrt{90}$
27. $\sqrt{4.16}$
28. $\sqrt{23.75}$
29. $3\sqrt{7}$
30. $-4\sqrt{20}$

31. Produce the graphs of $y_1 = \sqrt{x^2}$ and $y_2 = x$. Explain why the graphs are not the same.

32. Explain why each statement is false.
   (a) $\sqrt{(-7)^2} = -7$
   (b) $-\sqrt{9} = \sqrt{-9}$

In Exercises 33–40, simplify the expression. Assume that all variables represent positive numbers.

33. $\left(\sqrt{30}\right)^2$
34. $\left(\sqrt{12}\right)^2$
35. $\left(-\sqrt{8}\right)^2$
36. $\left(-\sqrt{11}\right)^2$
37. $\left(5\sqrt{3}\right)^2$
38. $\left(-2\sqrt{7}\right)^2$
39. $\left(\sqrt{2x}\right)^2$
40. $\left(\sqrt{3 + x}\right)^2$

In Exercises 41–48, simplify the expression. Assume that all variables represent any real number.

41. $\sqrt{7^2}$
42. $\sqrt{15^2}$
43. $\sqrt{(-12)^2}$
44. $\sqrt{(-9)^2}$
45. $-\sqrt{(-5)^2}$
46. $-\sqrt{(-16)^2}$
47. $\sqrt{25t^2}$
48. $\sqrt{16y^2}$

In Exercises 49–52, simplify the expression. Assume that all variables represent positive numbers.

49. $\sqrt{(5y)^2}$
50. $\sqrt{(7ab)^2}$
51. $\sqrt{(x + 5)^2}$
52. $\sqrt{(x + 7)^2}$

In Exercises 53–60, simplify the expression. Assume that all variables represent positive numbers.

53. $\sqrt{t^6}$

54. $\sqrt{y^8}$

55. $\sqrt{x^{12}}$

56. $\sqrt{w^{16}}$

57. $\sqrt{64y^8}$

58. $2\sqrt{16t^{16}}$

59. $\sqrt{(8ab)^{10}}$

60. $\sqrt{(9xy)^4}$

61. If n is an integer greater than 1, under what circumstances is $\sqrt[n]{k}$ defined?

62. Compare $\sqrt[3]{-8}$ and $\sqrt[4]{-8}$.

In Exercises 63–70, determine the indicated root(s).

63. 3rd root of -27

64. 3rd root of 64

65. 4th root of 16

66. 6th root of 64

67. 5th root of -32

68. 7th root of -1

69. 6th root of -1

70. 4th root of -81

In Exercises 71–78, evaluate each radical.

71. $\sqrt[3]{27}$

72. $\sqrt[3]{8}$

73. $\sqrt[4]{81}$

74. $\sqrt[5]{-1}$

75. $\sqrt[4]{10,000}$

76. $\sqrt[5]{\dfrac{1}{32}}$

77. $\sqrt[4]{\dfrac{16}{81}}$

78. $\sqrt[5]{32}$

In Exercises 79–84, simplify each radical. Assume that all variables represent positive numbers.

79. $\sqrt[3]{t^{15}}$

80. $\sqrt[3]{x^9}$

81. $\sqrt[6]{y^{12}}$

82. $\sqrt[5]{x^{15}}$

83. $\sqrt[7]{y^{14}}$

84. $\sqrt[4]{x^{20}}$

In Exercises 85 and 86, identify the radical expressions that are equal.

85. $\sqrt[3]{8},\ \sqrt[3]{-8},\ -\sqrt[3]{8},\ -\sqrt[3]{-8}$

86. $\sqrt{5^2},\ \sqrt{-5^2},\ -\sqrt{5^2},\ \sqrt{(-5)^2},\ -\sqrt{(-5)^2}$

🌎 Real-Life Applications

87. A caterer uses the formula $p = \dfrac{29\sqrt{T}}{3}$, where T is the temperature (°F), to predict the percentage p of invited guests who will actually attend an outdoor wedding reception. Produce the graph of the formula and use it to estimate the percentage of invited guests who are expected to attend if the temperature is 80°F.

88. The price (in dollars) of a pottery bowl is given by $p = 100 - 10\sqrt{2n + 7}$, where n is the number of bowls created. Produce the graph of this expression and use it to estimate the price of a bowl if the potter makes 6 bowls.

89. A cylindrical waste water treatment tank has a capacity V of 6100 cubic feet. If the height h of the tank is 3.1 feet, what is the approximate diameter of the tank? $\left(\textit{Hint:} \text{ Use } r = \sqrt{\dfrac{V}{\pi h}}, \text{ where } r \text{ is the radius.} \right)$

90. The productivity rating for an assembly line worker is given by $r = 7.5\sqrt{n + 2}$, where n is the number of hours of training the worker has received. What is the approximate productivity rating of a worker after 10 hours of training?

▨ Modeling with Real Data

91. The graph shows the total number of miles (in millions) traveled daily by all vehicles in Atlanta. (Source: The Atlanta Regional Commission.)

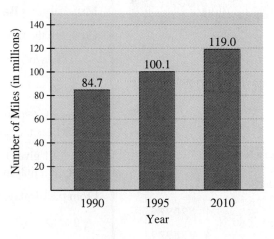

The radical expression $62.5\sqrt[5]{x + 5}$ models the data where x is the number of years since 1990.

(a) Suppose that x is replaced with 12. What does the value of the expression represent?

(b) Use the model expression to estimate the amount of daily travel in 2005.

92. (a) How would you adjust the model expression in Exercise 91 so that x is the number of years since 1985?

(b) Evaluate your new expression for the appropriate value of x to obtain the same result as that in part (b) of Exercise 91.

Challenge

In Exercises 93 and 94, fill in the blank to make the statement true.

93. $\sqrt{} = t^4$

94. $\sqrt[3]{} = -3y$

In Exercises 95 and 96, evaluate the expression.

95. $\sqrt{\sqrt{16}}$

96. $\sqrt{\sqrt{2^3 + 1} + 1}$

In Exercises 97 and 98, produce the graph of the given expression. Trace the graph to estimate the values of x for which the radical is defined.

97. $\sqrt{-x}$

98. $\sqrt{(-x)^2}$

9.2 | Product Rule for Radicals

Products of Radicals • Simplifying with the Product Rule for Radicals • Simplifying Higher-Order Radicals

Products of Radicals

Because radical expressions represent numbers, we want to be able to perform the basic operations with them. We begin with multiplication.

Exploring the Concept

Multiplying Radicals

Compare the following radical expressions.

$$\sqrt{4} \cdot \sqrt{9} = 2 \cdot 3 = 6$$

$$\sqrt{36} = 6$$

Because $\sqrt{4} \cdot \sqrt{9}$ and $\sqrt{36}$ both equal 6, $\sqrt{4} \cdot \sqrt{9} = \sqrt{36}$. Note that the radicand of the result is the product of the radicands of the factors.

Figure 9.1 gives us further numerical evidence that multiplying radicals may involve simply multiplying radicands.

Figure 9.1

```
√(24)
            4.898979
√(2)√(12)
            4.898979
√(3)√(8)
            4.898979
√(4)√(6)
            4.898979
```

From the display, we see that all of the following are true.

$$\sqrt{2}\,\sqrt{12} = \sqrt{24} \qquad \sqrt{3}\,\sqrt{8} = \sqrt{24} \qquad \sqrt{4}\,\sqrt{6} = \sqrt{24}$$

In each case, the result of multiplying two square root radicals is the square root of the product of the radicands.

This conclusion is true for all such products and can be stated symbolically as follows.

The Product Rule for Radicals

For nonnegative real numbers a and b,

$$\sqrt{a}\,\sqrt{b} = \sqrt{ab} \qquad \text{and} \qquad \sqrt{ab} = \sqrt{a}\,\sqrt{b}$$

In words, the result of multiplying two square root radicals is the square root of the product of the radicands.

So that all radicals are defined and results can be expressed without absolute value, we will assume that all variables in this section represent positive numbers.

EXAMPLE **1**

Multiplying with the Product Rule for Radicals

Use the Product Rule for Radicals to multiply the radicals.

(a) $\sqrt{7}\,\sqrt{3} = \sqrt{7\cdot 3} = \sqrt{21}$

(b) $\sqrt{3x}\,\sqrt{2y} = \sqrt{3x\cdot 2y} = \sqrt{6xy}$

(c) $\sqrt{x}\,\sqrt{x+3} = \sqrt{x(x+3)} = \sqrt{x^2+3x}$

Simplifying with the Product Rule for Radicals

In Section 9.1 we found that we could simplify a radical such as $\sqrt{9x^4}$ by writing $\sqrt{9x^4} = \sqrt{(3x^2)^2} = 3x^2$. Another method is to use the Product Rule for Radicals in the form $\sqrt{ab} = \sqrt{a}\,\sqrt{b}$ to factor radicals.

EXAMPLE **2**

Simplifying with the Product Rule for Radicals

Simplify each radical.

(a) $\sqrt{36y^8} = \sqrt{36}\cdot\sqrt{y^8}$ Product Rule for Radicals

$\qquad\quad = 6y^4$ Evaluate each radical.

(b) $\sqrt{81x^6y^{10}} = \sqrt{81}\cdot\sqrt{x^6}\cdot\sqrt{y^{10}}$ Product Rule for Radicals

$\qquad\qquad\ = 9x^3y^5$ Evaluate each radical.

Because the factors of the radicands in Example 2 were all perfect squares, we were able to perform the square root operation and express the result with no square root symbol. Clearly, the square root expression is simplified in such cases.

However, there are three criteria by which we judge whether a square root radical is simplified. In this section, we consider only the first condition.

Simplified Square Root Radicals—Condition 1

The radicand of a simplified square root radical must not contain a perfect square factor.

Earlier, we found three ways of writing $\sqrt{24}$ as a product. One way was $\sqrt{24} = \sqrt{4}\,\sqrt{6}$. Therefore, according to Condition 1, $\sqrt{24}$ is not considered simplified because 4 is a perfect square factor of 24. To simplify $\sqrt{24}$, we use the Product Rule for Radicals to write

$$\sqrt{24} = \sqrt{4\cdot 6} = \sqrt{4}\,\sqrt{6} = 2\sqrt{6}$$

The following summarizes the procedure for satisfying Condition 1 for a simplified square root radical.

Simplifying a Radical When the Radicand Has a Perfect Square Factor

1. Write the radicand as the product of the largest perfect square factor and another factor.
2. Use the Product Rule for Radicals to write the expression as the product of two radicals with the perfect square factor as one radicand.
3. Evaluate the radical with the perfect square radicand.

The next examples illustrate the use of the Product Rule for Radicals for simplifying radical expressions.

EXAMPLE   **3**

Using the Product Rule for Radicals to Simplify a Radical

Simplify each radical.

(a) $\sqrt{45} = \sqrt{9 \cdot 5}$ The largest perfect square factor of 45 is 9.

$\phantom{\sqrt{45}} = \sqrt{9} \cdot \sqrt{5}$ Product Rule for Radicals

$\phantom{\sqrt{45}} = 3\sqrt{5}$

(b) $\sqrt{75} = \sqrt{25 \cdot 3}$ The largest perfect square factor of 75 is 25.

$\phantom{\sqrt{75}} = \sqrt{25} \cdot \sqrt{3}$ Product Rule for Radicals

$\phantom{\sqrt{75}} = 5\sqrt{3}$

(c) Because 21 has no perfect square factor, $\sqrt{21}$ cannot be simplified.

EXAMPLE   **4**

Using the Product Rule for Radicals to Simplify a Radical

Simplify $\sqrt{72}$.

Solution

$\sqrt{72} = \sqrt{36 \cdot 2}$ The largest perfect square factor of 72 is 36.

$\phantom{\sqrt{72}} = \sqrt{36} \cdot \sqrt{2}$ Product Rule for Radicals

$\phantom{\sqrt{72}} = 6\sqrt{2}$

Note that 72 has more than one perfect square factor. Choosing the largest perfect square factor is the most efficient method for simplifying the radical. However, choosing a smaller perfect square factor eventually leads to the correct result. In the following, we choose 9 as the perfect square factor.

$$\sqrt{72} = \sqrt{9 \cdot 8} = \sqrt{9}\,\sqrt{8} = 3\sqrt{8}$$

$$= 3\sqrt{4 \cdot 2} = 3\sqrt{4}\,\sqrt{2} = 3 \cdot 2\sqrt{2} = 6\sqrt{2}$$

Recall that an exponential expression is a perfect square if the exponent is an even number.

EXAMPLE   5

Using the Product Rule for Radicals to Simplify a Radical

Simplify each radical.

(a) $\sqrt{x^7} = \sqrt{x^6 \cdot x}$ The largest perfect square factor of x^7 is x^6.

$\quad\quad = \sqrt{x^6} \cdot \sqrt{x}$ Product Rule for Radicals

$\quad\quad = x^3 \sqrt{x}$

(b) $\sqrt{25x^{11}} = \sqrt{25x^{10} \cdot x}$ The largest perfect square factor of x^{11} is x^{10}.

$\quad\quad\quad = \sqrt{25x^{10}} \cdot \sqrt{x}$ Product Rule for Radicals

$\quad\quad\quad = 5x^5 \sqrt{x}$

(c) $\sqrt{12x^2} = \sqrt{4x^2 \cdot 3}$ The largest perfect square factor of 12 is 4.

$\quad\quad\quad = \sqrt{4x^2} \cdot \sqrt{3}$ Product Rule for Radicals

$\quad\quad\quad = 2x\sqrt{3}$

(d) $\sqrt{18x^{15}} = \sqrt{9x^{14} \cdot 2x}$ The largest perfect square factor is $9x^{14}$.

$\quad\quad\quad = \sqrt{9x^{14}} \cdot \sqrt{2x}$ Product Rule for Radicals

$\quad\quad\quad = 3x^7 \sqrt{2x}$

(e) $\sqrt{28xy^2} = \sqrt{4y^2 \cdot 7x}$ The largest perfect square factor is $4y^2$.

$\quad\quad\quad = \sqrt{4y^2} \cdot \sqrt{7x}$ Product Rule for Radicals

$\quad\quad\quad = 2y \sqrt{7x}$

Sometimes we use the Product Rule for Radicals twice, once to multiply radicals and again to simplify the result.

EXAMPLE   6

Using the Product Rule for Radicals to Multiply and Simplify

Determine the product and then simplify the result.

Think About It

In Example 6, we multiplied the radicals and then simplified the result. In parts (a) and (c), verify that the same result is obtained if we begin by simplifying the radicals and then multiply.

(a) $\sqrt{3} \sqrt{12} = \sqrt{36}$ Multiply the radicands.

$\quad\quad\quad = 6$ Evaluate the radical.

(b) $\sqrt{10} \sqrt{2} = \sqrt{20}$ Multiply the radicands.

$\quad\quad\quad = \sqrt{4 \cdot 5}$ The largest perfect square factor of 20 is 4.

$\quad\quad\quad = \sqrt{4} \sqrt{5}$ Product Rule for Radicals

$\quad\quad\quad = 2\sqrt{5}$

(c) $\sqrt{6x^5} \sqrt{3x^3} = \sqrt{18x^8}$ Multiply the radicands.

$\quad\quad\quad\quad = \sqrt{9x^8 \cdot 2}$ The largest perfect square factor is $9x^8$.

$\quad\quad\quad\quad = \sqrt{9x^8} \sqrt{2}$ Product Rule for Radicals

$\quad\quad\quad\quad = 3x^4 \sqrt{2}$

Simplifying Higher-Order Radicals

The Product Rule for Radicals can be extended to higher roots.

> **The Product Rule for Radicals**
>
> For real numbers a and b and any integer $n > 1$,
> $$\sqrt[n]{a}\,\sqrt[n]{b} = \sqrt[n]{ab} \qquad \text{and} \qquad \sqrt[n]{ab} = \sqrt[n]{a}\,\sqrt[n]{b}$$
> If n is even, then a and b must be nonnegative.

EXAMPLE 7

Simplifying Higher-Order Radicals

Simplify each radical.

(a) $\sqrt[3]{24} = \sqrt[3]{8 \cdot 3}$ The largest perfect cube factor of 24 is 8.

$\qquad\qquad = \sqrt[3]{8} \cdot \sqrt[3]{3}$ Product Rule for Radicals

$\qquad\qquad = 2\sqrt[3]{3}$

(b) $\sqrt[3]{x^5} = \sqrt[3]{x^3 \cdot x^2}$ The largest perfect cube factor of x^5 is x^3.

$\qquad\qquad = \sqrt[3]{x^3} \cdot \sqrt[3]{x^2}$ Product Rule for Radicals

$\qquad\qquad = x \sqrt[3]{x^2}$

(c) $\sqrt[4]{16y^5} = \sqrt[4]{16y^4 \cdot y}$ The largest perfect 4th factor is $16y^4$.

$\qquad\qquad = \sqrt[4]{16y^4} \cdot \sqrt[4]{y}$ Product Rule for Radicals

$\qquad\qquad = 2y \sqrt[4]{y}$

(d) $\sqrt[3]{40x^7} = \sqrt[3]{8x^6 \cdot 5x}$ The largest perfect cube factor is $8x^6$.

$\qquad\qquad = \sqrt[3]{8x^6} \cdot \sqrt[3]{5x}$ Product Rule for Radicals

$\qquad\qquad = 2x^2 \sqrt[3]{5x}$

Quick Reference 9.2

Products of Radicals
- The product of two or more radicals can be found by using the Product Rule for Radicals.

  For nonnegative real numbers a and b,
  $$\sqrt{a}\,\sqrt{b} = \sqrt{ab} \qquad \text{and} \qquad \sqrt{ab} = \sqrt{a}\,\sqrt{b}$$

  In words, the result of multiplying two square root radicals is the square root of the product of the radicands.

Simplifying with the Product Rule for Radicals
- Condition 1 for a simplified square root radical is that the radicand must contain no perfect square factor.

- The Product Rule for Radicals can be used to satisfy Condition 1 as follows.

  1. Write the radicand as the product of the largest perfect square factor and another factor.

  2. Use the Product Rule for Radicals to write the expression as the product of two radicals with the perfect square factor as one radicand.

  3. Evaluate the radical with the perfect square radicand.

- Sometimes we use the Product Rule for Radicals twice, once to multiply radicals and again to simplify the result.

Simplifying Higher-Order Radicals

- The Product Rule for Radicals can be extended to higher roots.

  For real numbers a and b and any integer $n > 1$,

  $$\sqrt[n]{a}\,\sqrt[n]{b} = \sqrt[n]{ab} \qquad \text{and} \qquad \sqrt[n]{ab} = \sqrt[n]{a}\,\sqrt[n]{b}$$

  If n is even, then a and b must be nonnegative.

Speaking the Language 9.2

1. When we multiply two square root radicals, we obtain the ▭▭▭▭ of the product of the ▭▭▭▭ .

2. A condition for a simplified square root radical is that the radicand must contain no ▭▭▭▭ factor.

3. The Product Rule for Radicals can be used to ▭▭▭▭ radicals or to ▭▭▭▭ radicals.

4. The Product Rule for Radicals allows us to write $\sqrt[n]{a}\,\sqrt[n]{b} = \sqrt[n]{ab}$. However, if n is even, then a and b must be ▭▭▭▭ .

Exercises 9.2

Concepts and Skills

1. Both of the following parts illustrate common errors that occur in working with radicals. State the error and correct it.

   (a) $\sqrt{2}\,\sqrt{3} = 6$ (b) $\sqrt{20} = 4\sqrt{5}$

2. According to the Product Rule for Radicals, both of the following are correct. Which one is the correct way to begin simplifying $\sqrt{12}$? Why?

   (i) $\sqrt{12} = \sqrt{2}\,\sqrt{6}$ (ii) $\sqrt{12} = \sqrt{4}\,\sqrt{3}$

In Exercises 3–8, use the Product Rule for Radicals to multiply the radicals.

3. $\sqrt{5}\,\sqrt{7}$ 4. $\sqrt{6}\,\sqrt{11}$

5. $\sqrt{2a}\,\sqrt{5b}$ 6. $\sqrt{6y}\,\sqrt{7x}$

7. $3\sqrt{7} \cdot 2\sqrt{5}$ 8. $-4\sqrt{3} \cdot 5\sqrt{2}$

In Exercises 9–12, use the Product Rule for Radicals to determine the square root.

9. $\sqrt{100x^8}$ 10. $\sqrt{16x^{12}}$

11. $\sqrt{x^6y^2}$ 12. $\sqrt{x^4y^{10}}$

In Exercises 13–18, simplify each radical.

13. $\sqrt{50}$ 14. $\sqrt{63}$

15. $\sqrt{20}$ 16. $\sqrt{98}$

17. $\sqrt{48}$ 18. $\sqrt{80}$

19. If n is a positive even integer, an easy way to simplify $\sqrt{b^n}$ is to take half the exponent. Describe what to do if n is an odd integer, $n > 1$.

20. Which of the following is the most efficient way to begin simplifying $\sqrt[3]{x^7}$? Why?

    (i) $\sqrt[3]{x^7} = \sqrt[3]{x^4 \cdot x^3}$

    (ii) $\sqrt[3]{x^7} = \sqrt[3]{x^5 \cdot x^2}$

    (iii) $\sqrt[3]{x^7} = \sqrt[3]{x^6 \cdot x}$

In Exercises 21–36, simplify each radical.

21. $\sqrt{x^3}$ 22. $\sqrt{x^5}$

23. $\sqrt{x^{11}}$ 24. $\sqrt{y^{17}}$

25. $\sqrt{24x^6}$ 26. $\sqrt{75x^8}$

27. $\sqrt{98y^{10}}$ 28. $\sqrt{44y^{12}}$

29. $\sqrt{25x^7}$ 30. $\sqrt{36y^{11}}$

31. $\sqrt{16y^{15}}$ 32. $\sqrt{64x^9}$

33. $\sqrt{28x^5}$ 34. $\sqrt{45x^7}$

35. $\sqrt{125y^{15}}$ 36. $\sqrt{27w^3}$

In Exercises 37–40, simplify each radical.

37. $\sqrt{x^8 y^3}$ 38. $\sqrt{x^7 y^{10}}$

39. $\sqrt{12x^2 y^7}$ 40. $\sqrt{50x^9 y^4}$

In Exercises 41–50, determine the product and simplify.

41. $\sqrt{2}\sqrt{32}$ 42. $\sqrt{27}\sqrt{3}$

43. $\sqrt{12}\sqrt{5}$ 44. $\sqrt{30}\sqrt{3}$

45. $\sqrt{8x}\sqrt{2x}$ 46. $\sqrt{6x^3}\sqrt{6x}$

47. $\sqrt{a^3}\sqrt{a^2}$ 48. $\sqrt{x^5}\sqrt{x^4}$

49. $\sqrt{2t^3}\sqrt{6t^7}$ 50. $\sqrt{50t^2}\sqrt{2t^5}$

In Exercises 51–60, simplify each radical.

51. $\sqrt[3]{40}$ 52. $\sqrt[3]{54}$

53. $\sqrt[5]{64}$ 54. $\sqrt[4]{48}$

55. $\sqrt[3]{x^7}$ 56. $\sqrt[3]{y^{11}}$

57. $\sqrt[4]{x^7}$ 58. $\sqrt[5]{t^{12}}$

59. $\sqrt[3]{24a^5}$ 60. $\sqrt[3]{64b^{10}}$

In Exercises 61–64, evaluate the radical and write the result in scientific notation.

61. $\sqrt{4.9 \cdot 10^7}$

62. $\sqrt{2.5 \cdot 10^5}$

63. $\sqrt{8.1 \cdot 10^{-3}}$

64. $\sqrt{9 \cdot 10^{-6}}$

In Exercises 65–68, indicate with T or F whether the given statement is true or false.

65. $\sqrt{10 - 3} = \sqrt{10} - \sqrt{3}$

66. $\sqrt{3^2 + 4^2} = 3 + 4$

67. $5\sqrt{2} = \sqrt{50}$

68. $6\sqrt{3} = \sqrt{18}$

69. Consider $\sqrt{16x^{16}}$. Explain the difference in how we treat the coefficient 16 and the exponent 16 as we take the square root.

70. Simplify each of the following radicals and observe the pattern of the results.

    $$\sqrt{4}, \sqrt{40}, \sqrt{400}, \sqrt{4000}, \ldots$$

    On the basis of this pattern, predict the simplified forms of $\sqrt{40,000}$ and $\sqrt{400,000}$.

Real-Life Applications

71. The manager of a small business determined that the yearly cost per employee for fringe benefits is $c = \sqrt{2t + 15}$ and the number of employees is $n = \sqrt{5t}$, where t is the number of years since the company began business. Write an expression for the yearly cost of fringe benefits for the company.

72. The price of electric razors is $p = \sqrt{2a}$, and the number of razors sold is $n = \sqrt{a^2 + 1}$, where a is a measure of advertising during NFL games. Write an expression for the total revenue from the sale of electric razors.

73. An engineer uses the formula $T = \sqrt{20d^3}$ to assign to tires a traction rating T based on the depth d (in inches) of the tread.

    (a) Produce the graph of the formula. Then estimate the traction rating for a tread that is 0.5 inch deep.

    (b) Simplify the formula for T. Then produce the graph of the simplified formula and make the same estimate as in part (a).

74. A ballet company performs in several cities. The average attendance in each city is given by $\sqrt{2a}$, and the number of cities where the company performs is given by $\sqrt{6a^3}$, where a is the level of corporate funding for the tour. Write a simplified expression for the total attendance at all performances for the year.

Data Analysis: *Chicken Consumption*

As shown in the graph the average amount (in pounds) of chicken that a person in the United States eats each year increased during the last two decades. (Source: U.S. Department of Agriculture.)

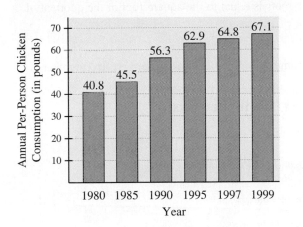

For this period, we can represent the annual per-person chicken consumption $C(t)$ (in pounds) by

$$C(t) = \sqrt{176t + 1184}$$

where t is the number of years since 1980.

75. Produce the graph of $C(t)$ for the 50-year period 1980–2030. What does the shape of the graph suggest about the amount of chicken that will be eaten in the future?

76. Write the model expression in simplified form. (*Hint:* Begin by factoring the radicand, with 16 as a common factor.) What would you expect to be true about the graph of the simplified expression?

77. Produce the graphs of $y = 1.5t + 40.2$ and $C(t)$ in the same coordinate system that you used in Exercise 75. For what years does the linear model appear to be as good as the radical model?

78. Beginning with the last year that you identified in Exercise 77, what is the difference in the trends predicted by the two models?

Challenge

In Exercises 79 and 80, simplify the radical.

79. $\sqrt{2x^2 + 16x + 32}$ **80.** $\sqrt{9x^2 + 9y^2}$

In Exercises 81 and 82, fill in the blank to make the statement true.

81. $\sqrt{\rule{2cm}{0.4pt}} = 2x\sqrt{3x}$ **82.** $\sqrt[3]{\rule{2cm}{0.4pt}} = 2y\sqrt[3]{y}$

In Exercises 83 and 84, simplify the radical.

83. $\sqrt{9t^{2n}}$ **84.** $\sqrt[n]{x^{2n}y^{3n}}$

9.3   Quotient Rule for Radicals

Quotients of Radicals   •   Simplifying with the Quotient Rule for Radicals   •   Rationalizing Denominators   •   Simplifying Higher-Order Radicals

Quotients of Radicals

In Section 9.2, we learned how to multiply radicals. A simple experiment suggests a method for dividing radicals:

$$\frac{\sqrt{36}}{\sqrt{4}} = \frac{6}{2} = 3 \qquad \text{and} \qquad \sqrt{\frac{36}{4}} = \sqrt{9} = 3$$

$$\text{thus} \quad \frac{\sqrt{36}}{\sqrt{4}} = \sqrt{\frac{36}{4}}.$$

This result can be generalized.

The Quotient Rule for Radicals

For nonnegative and real numbers a and b, $b \neq 0$,

$$\frac{\sqrt{a}}{\sqrt{b}} = \sqrt{\frac{a}{b}} \quad \text{and} \quad \sqrt{\frac{a}{b}} = \frac{\sqrt{a}}{\sqrt{b}}$$

In words, a quotient of square roots is equal to the square root of the quotient of the radicands.

In this section we continue to assume that all variables represent positive numbers.

EXAMPLE **1**

Dividing with the Quotient Rule for Radicals

Determine the quotient and simplify.

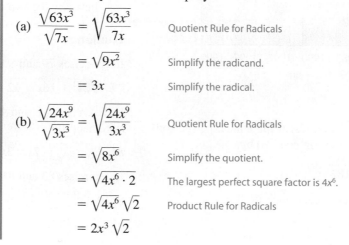

(a) $\dfrac{\sqrt{63x^3}}{\sqrt{7x}} = \sqrt{\dfrac{63x^3}{7x}}$ Quotient Rule for Radicals

$\qquad\qquad = \sqrt{9x^2}$ Simplify the radicand.

$\qquad\qquad = 3x$ Simplify the radical.

LEARNING TIP

For a quotient of radicals, simplifying is often easier if you write the quotient as a single radical and simplify the radicand first. Then simplify the radical.

(b) $\dfrac{\sqrt{24x^9}}{\sqrt{3x^3}} = \sqrt{\dfrac{24x^9}{3x^3}}$ Quotient Rule for Radicals

$\qquad\qquad = \sqrt{8x^6}$ Simplify the quotient.

$\qquad\qquad = \sqrt{4x^6 \cdot 2}$ The largest perfect square factor is $4x^6$.

$\qquad\qquad = \sqrt{4x^6}\,\sqrt{2}$ Product Rule for Radicals

$\qquad\qquad = 2x^3\,\sqrt{2}$

Simplifying with the Quotient Rule for Radicals

In Section 9.2 we stated the first condition for a simplified radical. A second condition is as follows.

Simplified Square Root Radicals—Condition 2

The radicand of a simplified square root radical must not contain a fraction.

Example 2 illustrates the use of the Quotient Rule for Radicals for simplifying radicals.

EXAMPLE **2**

Using the Quotient Rule for Radicals to Simplify a Radical

Simplify the given radicals.

(a) $\sqrt{\dfrac{5}{9}} = \dfrac{\sqrt{5}}{\sqrt{9}}$ Quotient Rule for Radicals

$\qquad\quad = \dfrac{\sqrt{5}}{3}$ Evaluate the denominator.

(b) $\sqrt{\dfrac{18x^2}{49}} = \dfrac{\sqrt{18x^2}}{\sqrt{49}}$ Quotient Rule for Radicals

$= \dfrac{\sqrt{9x^2 \cdot 2}}{\sqrt{49}}$ In the numerator, the largest perfect square factor is $9x^2$.

$= \dfrac{\sqrt{9x^2}\,\sqrt{2}}{\sqrt{49}}$ Product Rule for Radicals

$= \dfrac{3x\,\sqrt{2}}{7}$

It is usually best to simplify the radicand before applying the Product or Quotient Rule for Radicals.

EXAMPLE   3

Simplifying the Square Root of a Quotient

Simplify the given radicals.

(a) $\sqrt{\dfrac{8x^7}{2x}} = \sqrt{4x^6}$ Simplify the radicand.

$= 2x^3$ Evaluate the radical.

(b) $\sqrt{\dfrac{150x^5}{3x^2}} = \sqrt{50x^3}$ Simplify the radicand.

$= \sqrt{25x^2 \cdot 2x}$ The largest perfect square factor is $25x^2$.

$= \sqrt{25x^2}\,\sqrt{2x}$ Product Rule for Radicals

$= 5x\,\sqrt{2x}$

(c) $\sqrt{\dfrac{45x^3}{5x^7}} = \sqrt{\dfrac{9}{x^4}}$ Simplify the radicand.

$= \dfrac{3}{x^2}$

Rationalizing Denominators

The third condition for a simplified radical is as follows.

> **Simplified Square Root Radicals—Condition 3**
>
> A simplified square root radical expression must not contain a radical in the denominator.

Consider the expression $\sqrt{\dfrac{7}{5}}$. Because the radicand contains a fraction, it is not simplified. We can use the Quotient Rule for Radicals to write the expression as $\dfrac{\sqrt{7}}{\sqrt{5}}$.

However, there is now a radical in the denominator. According to Condition 3, the expression is still not simplified.

To eliminate the radical in the denominator, we need to perform an operation that will result in a perfect square radicand in the denominator. Because $\sqrt{5} \cdot \sqrt{5} = \sqrt{25}$, we multiply both the numerator and the denominator by $\sqrt{5}$.

$$\frac{\sqrt{7}}{\sqrt{5}} = \frac{\sqrt{7}}{\sqrt{5}} \cdot \frac{\sqrt{5}}{\sqrt{5}} = \frac{\sqrt{35}}{\sqrt{25}} = \frac{\sqrt{35}}{5}$$

This process is called **rationalizing the denominator** because the denominator is changed to a rational number.

EXAMPLE 4 | **Rationalizing Denominators**

In each expression, rationalize the denominator.

(a) $\sqrt{\dfrac{5}{y}} = \dfrac{\sqrt{5}}{\sqrt{y}}$ Quotient Rule for Radicals

$= \dfrac{\sqrt{5}}{\sqrt{y}} \cdot \dfrac{\sqrt{y}}{\sqrt{y}}$ Multiply the numerator and the denominator by $\sqrt{y}$.

$= \dfrac{\sqrt{5y}}{y}$ $\sqrt{y}\,\sqrt{y} = y$

(b) $\dfrac{10}{\sqrt{5}} = \dfrac{10}{\sqrt{5}} \cdot \dfrac{\sqrt{5}}{\sqrt{5}}$ Multiply the numerator and the denominator by $\sqrt{5}$.

$= \dfrac{10\sqrt{5}}{5}$ $\sqrt{5}\,\sqrt{5} = 5$

$= \dfrac{5 \cdot 2\sqrt{5}}{5}$ Reduce the fraction.

$= 2\sqrt{5}$

(c) $\dfrac{6}{\sqrt{2x}} = \dfrac{6}{\sqrt{2x}} \cdot \dfrac{\sqrt{2x}}{\sqrt{2x}}$ Multiply the numerator and the denominator by $\sqrt{2x}$.

$= \dfrac{6\sqrt{2x}}{2x}$ $\sqrt{2x}\,\sqrt{2x} = 2x$

$= \dfrac{2 \cdot 3\sqrt{2x}}{2x}$ Reduce the fraction.

$= \dfrac{3\sqrt{2x}}{x}$

It is usually best to simplify the numerator and the denominator before rationalizing the denominator.

EXAMPLE 5   **Rationalizing Denominators**

In each expression, rationalize the denominator.

(a) $\sqrt{\dfrac{7}{50}} = \dfrac{\sqrt{7}}{\sqrt{50}}$ Quotient Rule for Radicals

$= \dfrac{\sqrt{7}}{\sqrt{25}\cdot\sqrt{2}}$ Product Rule for Radicals

$= \dfrac{\sqrt{7}}{5\sqrt{2}}$

$= \dfrac{\sqrt{7}}{5\sqrt{2}}\cdot\dfrac{\sqrt{2}}{\sqrt{2}}$ Multiply the numerator and the denominator by $\sqrt{2}$.

$= \dfrac{\sqrt{14}}{5\cdot 2}$ $\sqrt{2}\,\sqrt{2} = 2$

$= \dfrac{\sqrt{14}}{10}$

(b) $\dfrac{3x}{\sqrt{12}} = \dfrac{3x}{\sqrt{4}\cdot\sqrt{3}}$ Product Rule for Radicals

$= \dfrac{3x}{2\sqrt{3}}$

$= \dfrac{3x}{2\sqrt{3}}\cdot\dfrac{\sqrt{3}}{\sqrt{3}}$ Multiply the numerator and the denominator by $\sqrt{3}$.

$= \dfrac{3x\sqrt{3}}{2\cdot 3}$ Reduce the fraction.

$= \dfrac{x\sqrt{3}}{2}$

Simplifying Higher-Order Radicals

The Quotient Rule for Radicals can be stated for higher roots.

The Quotient Rule for Radicals

For real numbers a and b with $b \neq 0$ and any integer $n > 1$,

$$\frac{\sqrt[n]{a}}{\sqrt[n]{b}} = \sqrt[n]{\frac{a}{b}} \quad \text{and} \quad \sqrt[n]{\frac{a}{b}} = \frac{\sqrt[n]{a}}{\sqrt[n]{b}}$$

For even n, a and b must be nonnegative.

EXAMPLE   6   **Simplifying Higher-Order Radicals**

Simplify the given radicals.

(a) $\sqrt[3]{\dfrac{x^2}{8}} = \dfrac{\sqrt[3]{x^2}}{\sqrt[3]{8}}$ Quotient Rule for Radicals

$= \dfrac{\sqrt[3]{x^2}}{2}$

(b) $\sqrt[3]{\dfrac{250}{27}} = \dfrac{\sqrt[3]{250}}{\sqrt[3]{27}}$ Quotient Rule for Radicals

$= \dfrac{\sqrt[3]{125}\,\sqrt[3]{2}}{\sqrt[3]{27}}$ Product Rule for Radicals

Think About It

Show that $\dfrac{5}{\sqrt{5}} = \sqrt{5}$. Is it always

true that $\dfrac{a}{\sqrt[n]{a}} = \sqrt[n]{a}$?

$= \dfrac{5\sqrt[3]{2}}{3}$

(c) $\dfrac{\sqrt[3]{16x^{11}}}{\sqrt[3]{2x^2}} = \sqrt[3]{\dfrac{16x^{11}}{2x^2}}$ Quotient Rule for Radicals

$= \sqrt[3]{8x^9}$ Simplify the radicand.

$= 2x^3$

Quick Reference 9.3

Quotients of Radicals

- The Quotient Rule for Radicals is as follows.

    For nonnegative real numbers a and b, $b \neq 0$,

    $$\dfrac{\sqrt{a}}{\sqrt{b}} = \sqrt{\dfrac{a}{b}} \quad \text{and} \quad \sqrt{\dfrac{a}{b}} = \dfrac{\sqrt{a}}{\sqrt{b}}$$

    In words, a quotient of square roots is equal to the square root of the quotient of the radicands.

Simplifying with the Quotient Rule for Radicals

- Condition 2 for a simplified radical is that the radicand must not contain a fraction.
- We use the Quotient Rule for Radicals to change the square root of a quotient to the quotient of the square roots of the numerator and the denominator.
- It is usually best to simplify the radicand before applying the Product or the Quotient Rule for Radicals.

Rationalizing Denominators

- Condition 3 for a simplified radical is that there must be no radical in the denominator of a fraction.
- If a square root expression has a radical in the denominator and the radicand is a perfect square, we simply take the square root to eliminate the radical. Otherwise, we multiply the numerator and the denominator by the radical in the denominator. This process is called **rationalizing the denominator.**

• It is usually best to simplify the numerator and the denominator before rationalizing the denominator.

Simplifying Higher-Order Radicals

• The Quotient Rule for Radicals can be stated for higher roots.

For real numbers a and b with $b \neq 0$ and any integer $n > 1$,

$$\frac{\sqrt[n]{a}}{\sqrt[n]{b}} = \sqrt[n]{\frac{a}{b}} \quad \text{and} \quad \sqrt[n]{\frac{a}{b}} = \frac{\sqrt[n]{a}}{\sqrt[n]{b}}$$

For even n, a and b must be nonnegative.

Speaking the Language 9.3

1. A quotient of square root radicals is equal to the ▩▩▩▩▩ of the quotient of the ▩▩▩▩▩ .

2. The expression $\sqrt{\dfrac{2}{3}}$ is not considered simplified because a(n) ▩▩▩▩▩ cannot contain a fraction.

3. The radical expression $\dfrac{2}{\sqrt{3}}$ is not considered simplified because it has a(n) ▩▩▩▩▩ in its ▩▩▩▩▩ .

4. Rewriting $\dfrac{2}{\sqrt{3}}$ as $\dfrac{2\sqrt{3}}{3}$ is called ▩▩▩▩▩ the denominator.

Exercises 9.3

Concepts and Skills

1. Which of the following is a correct approach to simplifying the given radical? Which will require the fewest steps?

(i) $\dfrac{\sqrt{36}}{\sqrt{9}} = \sqrt{\dfrac{36}{9}} = \ldots$

(ii) $\dfrac{\sqrt{36}}{\sqrt{9}} = \dfrac{6}{3} = \ldots$

(iii) $\dfrac{\sqrt{36}}{\sqrt{9}} = \dfrac{\sqrt{36}}{\sqrt{9}} \cdot \dfrac{\sqrt{9}}{\sqrt{9}} = \ldots$

2. Why is the following an incorrect application of the Quotient Rule for Radicals?

$$\frac{\sqrt{-12}}{\sqrt{-4}} = \sqrt{\frac{-12}{-4}} = \sqrt{3}$$

In Exercises 3–10, determine the quotient and simplify.

3. $\dfrac{\sqrt{125}}{\sqrt{5}}$

4. $\dfrac{\sqrt{98}}{\sqrt{2}}$

5. $\dfrac{\sqrt{20x^5}}{\sqrt{5x}}$

6. $\dfrac{\sqrt{27x^{12}}}{\sqrt{3x^2}}$

7. $\dfrac{\sqrt{48x}}{\sqrt{3x^3}}$

8. $\dfrac{\sqrt{5y^9}}{\sqrt{125y^3}}$

9. $\dfrac{\sqrt{90y^7}}{\sqrt{2y}}$

10. $\dfrac{\sqrt{72a^3}}{\sqrt{3a}}$

In Exercises 11–20, simplify the given radical.

11. $\sqrt{\dfrac{81}{x^{10}}}$

12. $\sqrt{\dfrac{y^4}{9}}$

13. $\sqrt{\dfrac{y^{12}}{x^{10}}}$

14.  $\sqrt{\dfrac{a^{16}}{b^{18}}}$

15. $\sqrt{\dfrac{x^7}{25}}$

16. $\sqrt{\dfrac{y^3}{49}}$

17. $\sqrt{\dfrac{2}{x^4}}$

18. $\sqrt{\dfrac{7}{y^6}}$

19. $\sqrt{\dfrac{11y}{x^6}}$

20. $\sqrt{\dfrac{7x}{y^4}}$

21. (a) Why is $\sqrt{\dfrac{2}{3}}$ not considered simplified?

(b) We can use the Quotient Rule for Radicals to write $\sqrt{\dfrac{2}{3}} = \dfrac{\sqrt{2}}{\sqrt{3}}$. Now is the expression simplified? Why?

22. One way to rationalize the denominator of $\dfrac{\sqrt{6x^2}}{\sqrt{2x}}$ is to multiply the numerator and the denominator by $\sqrt{2x}$ and then simplify the result. Describe an easier method.

In Exercises 23–32, simplify the given radical.

23. $\sqrt{\dfrac{75a^9}{3a}}$

24. $\sqrt{\dfrac{128x^{13}}{2x^3}}$

25. $\sqrt{\dfrac{2y^7}{98y}}$

26. $\sqrt{\dfrac{162x^3}{8x^{11}}}$

27. $\sqrt{\dfrac{x^6 y^{11}}{x^2 y}}$

28. $\sqrt{\dfrac{a^3 b^3}{ab^5}}$

29. $\sqrt{\dfrac{60x^5}{125y^8}}$

30. $\sqrt{\dfrac{90y^9}{8x^2}}$

31. $\sqrt{\dfrac{6y^3}{50}}$

32. $\sqrt{\dfrac{45a^2}{16a^2 b^4}}$

In Exercises 33–42, simplify.

33. $\dfrac{4}{\sqrt{5}}$

34. $\dfrac{6}{\sqrt{7}}$

35. $\sqrt{\dfrac{5}{6}}$

36. $\sqrt{\dfrac{7}{10}}$

37. $\dfrac{3}{\sqrt{15}}$

38. $\dfrac{7}{\sqrt{7}}$

39. $\dfrac{3}{\sqrt{x}}$

40. $\dfrac{4}{\sqrt{t}}$

41. $\dfrac{9}{\sqrt{3y}}$

42. $\dfrac{10}{\sqrt{5w}}$

In Exercises 43–58, simplify.

43. $\sqrt{\dfrac{5}{24}}$

44. $\dfrac{3}{\sqrt{8}}$

45. $\sqrt{\dfrac{y^2}{5}}$

46. $\sqrt{\dfrac{x^4}{7}}$

47. $\dfrac{10t}{\sqrt{75}}$

48. $\dfrac{14}{\sqrt{28y}}$

49. $\sqrt{\dfrac{x^3}{5}}$

50. $\sqrt{\dfrac{x^7}{3}}$

51. $\dfrac{\sqrt{5}}{\sqrt{4x}}$

52. $\dfrac{\sqrt{10}}{\sqrt{49y}}$

53. $\sqrt{\dfrac{3x^3}{5x^5}}$

54. $\sqrt{\dfrac{8t}{3t^3}}$

55. $\sqrt{\dfrac{5x^4}{60x}}$

56. $\sqrt{\dfrac{90x}{56x^3}}$

57. $\dfrac{\sqrt{48x^3 y}}{\sqrt{2x^4 y^7}}$

58. $\dfrac{\sqrt{7x^3 y^7}}{\sqrt{21x^4 y}}$

In Exercises 59–68, simplify.

59. $\sqrt[3]{\dfrac{w^2}{27}}$

60. $\sqrt[3]{\dfrac{t}{-8}}$

61. $\sqrt[4]{\dfrac{x^3}{16}}$

62. $\sqrt[5]{\dfrac{t^2}{32}}$

63. $\sqrt[3]{\dfrac{54}{125}}$

64. $\sqrt[3]{\dfrac{40}{27}}$

65. $\sqrt[3]{\dfrac{x^5}{64}}$

66. $\sqrt[3]{\dfrac{x^7}{8}}$

67. $\dfrac{\sqrt[4]{48x^{11}}}{\sqrt[4]{3x^3}}$

68. $\dfrac{\sqrt[3]{40x^7}}{\sqrt[3]{5x^2}}$

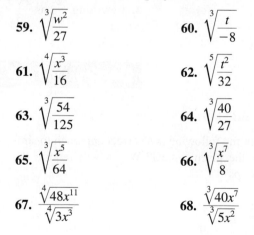 **Real-Life Applications**

69. A certain allergy medication causes drowsiness, which can impair a person's ability to operate certain machinery safely. The level of impairment is measured by $I = \dfrac{300}{\sqrt{5t}}$, where t is the number of hours elapsed after taking the medication. Write the formula in simplified radical form.

70. A fast-food company mounts a major advertising campaign for its Combo Mega Meals. The marketing manager finds that the number of Combo Mega Meals sold is $\sqrt{2a}$ and the revenue is $7\sqrt{a} + 2$, where a is the amount (in thousands of dollars) spent on TV advertising. Write a simplified expression for the average price of a Combo Mega Meal.

71. A county manager estimates that y years after a major industry relocates to the county, the tax revenue from businesses will be $\sqrt{75y^5}$ and the tax revenue from individuals will be $\sqrt{3y^2}$. Write a simplified radical expression for the ratio of tax revenue from businesses to tax revenue from individuals.

72. A city adopts a long-range plan to improve its police department. Funding the plan will require expenditures of $\sqrt{60t^3}$ thousand dollars per year, where t is the number of years from the time the plan is implemented. If the population of the city is predicted to be $\sqrt{3t}$, write a simplified radical expression for the cost per citizen to pay for the plan.

Modeling with Real Data

73. The average number of members of a household in the United States declined from 3.37 in 1950 to 2.62 in 1998. (Source: U.S. Bureau of the Census.) Suppose that the radical expression $\dfrac{5.2}{\sqrt{0.5x + 10}} + 1.73$ is determined to be the most appropriate model of the data, where x is the number of years since 1950.

   (a) Evaluate the expression to estimate the household size in 2010.

   (b) Write the model expression with a rationalized denominator.

   (c) Use the expression in part (b) to estimate the household size in 2010 and compare the answer to that in part (a).

74. Refer to the data in Exercise 73. An equivalent model for the data is $\dfrac{52}{5\sqrt{2x + 40}} + 1.73$.

   (a) Write the expression with a rationalized denominator.

   (b) Use the model to estimate the average number of members of a U.S. household in 2008.

Data Analysis: *Environmental Industries*

The table shows the trend in revenue for environmental industries such as water treatment and solid and hazardous waste disposal.

Year	Environmental Industry Revenue (billions of dollars)
1980	59.0
1990	150.3
1995	179.5
1997	186.1
1998	191.5

(Source: Environmental Business International.)

The radical expression $\dfrac{46x}{\sqrt{x}}$ models the data, where x is the number of years since 1980.

75. Write the model expression as a simplified radical expression.

76. Evaluate the original expression to show that the model is reasonable for the years 1990–1998, but not for 1980.

77. Another model for the data is $\dfrac{39.4(x + 1)}{\sqrt{x + 1}} + 19.5$, where x is the number of years since 1980. Write the model as a simplified radical expression.

78. Why is the model expression in Exercise 77 a better model than the initial model?

Challenge

In Exercises 79–84, rationalize the denominator.

79. $\dfrac{2}{\sqrt[3]{9}}$

80. $\sqrt[3]{\dfrac{7}{2}}$

81. $\dfrac{1}{\sqrt[3]{t^2}}$

82. $\dfrac{9}{\sqrt[4]{w}}$

83. $\sqrt[4]{\dfrac{2}{27}}$

84. $\sqrt[5]{\dfrac{5}{16}}$

In Exercises 85 and 86, simplify.

85. $\dfrac{\sqrt{x^{3n}}}{\sqrt{x^n}}$

86. $\dfrac{1}{\sqrt{x^{2n-1}}}$

9.4 | Operations with Radicals

Addition and Subtraction · Multiplying with the Distributive Property · Rationalizing Binomial Denominators

Addition and Subtraction

Square root radicals that have the same radicand are called **like radicals.** Just as we used the Distributive Property to combine like terms, we also use the Distributive Property to combine like radicals.

$$2x + 7x = (2 + 7)x = 9x \qquad \text{Combining like terms}$$

$$2\sqrt{5} + 7\sqrt{5} = (2 + 7)\sqrt{5} = 9\sqrt{5} \qquad \text{Combining like radicals}$$

Note that when we combine like radicals, we add or subtract the coefficients and retain the radical.

In this section we continue to assume that all variables represent positive numbers.

EXAMPLE 1

LEARNING TIP

In order for us to add or subtract radical expressions, the terms must be identical except for their numerical factors.

Adding and Subtracting Radicals

Combine the like radicals.

(a) $8\sqrt{2} + \sqrt{2} = 8\sqrt{2} + 1\sqrt{2}$ The coefficient of the second term is 1.

       $= (8 + 1)\sqrt{2}$ Like radicals; Distributive Property

       $= 9\sqrt{2}$

(b) $4\sqrt{10} - 7\sqrt{10} = (4 - 7)\sqrt{10}$ Like radicals; Distributive Property

       $= -3\sqrt{10}$

(c) $7\sqrt{x} - 5\sqrt{x} = 2\sqrt{x}$ Combine the coefficients and retain the radical.

(d) $3\sqrt{x} - 5\sqrt{3} - \sqrt{x} + 6\sqrt{3}$

       $= 3\sqrt{x} - 1\sqrt{x} - 5\sqrt{3} + 6\sqrt{3}$ Identify like radicals.

       $= 2\sqrt{x} + \sqrt{3}$ Combine the coefficients.

Note: Radicals are like radicals only if the radicands and indices are the same. Thus neither $3\sqrt{2} + 2\sqrt{3}$ nor $\sqrt[3]{7} + \sqrt{7}$ can be simplified.

Only like radicals can be combined. However, it is sometimes possible to simplify one or both radicals so that they become like radicals and can then be combined.

EXAMPLE 2

Adding and Subtracting Radicals

Simplify and then combine the like radicals.

(a) $\sqrt{18} + \sqrt{50} = \sqrt{9}\sqrt{2} + \sqrt{25}\sqrt{2}$ Product Rule for Radicals

       $= 3\sqrt{2} + 5\sqrt{2}$ Now the radicals are like radicals.

       $= 8\sqrt{2}$ Combine the coefficients.

(b) $\sqrt{45} + \sqrt{75} = \sqrt{9}\sqrt{5} + \sqrt{25}\sqrt{3}$ Product Rule for Radicals

$\qquad\qquad\qquad = 3\sqrt{5} + 5\sqrt{3}$

Because the radicands are not the same, the expression cannot be simplified further.

(c) $5\sqrt{24} - 2\sqrt{54} = 5\sqrt{4}\sqrt{6} - 2\sqrt{9}\sqrt{6}$ Product Rule for Radicals

$\qquad\qquad\qquad = 5 \cdot 2\sqrt{6} - 2 \cdot 3\sqrt{6}$

$\qquad\qquad\qquad = 10\sqrt{6} - 6\sqrt{6}$ The radicands are the same.

$\qquad\qquad\qquad = 4\sqrt{6}$ Combine the coefficients.

(d) $x\sqrt{9x} - \sqrt{16x^3} = x\sqrt{9}\sqrt{x} - \sqrt{16x^2}\sqrt{x}$ Product Rule for Radicals

$\qquad\qquad\qquad = 3x\sqrt{x} - 4x\sqrt{x}$

$\qquad\qquad\qquad = (3 - 4)x\sqrt{x}$ Distributive Property

$\qquad\qquad\qquad = -x\sqrt{x}$

Note: In part (d) of Example 2, the two terms can be combined because both the radicals and the variable factors are identical. Had the expression been $3x\sqrt{x} - 4\sqrt{x}$, the Distributive Property could have been used to factor the expression as $(3x - 4)\sqrt{x}$ but not to combine the terms.

Multiplying with the Distributive Property

We use both the Product Rule for Radicals and the Distributive Property to multiply radical expressions that have more than one term.

EXAMPLE   3

Multiplying Radical Expressions

Multiply the given radical expressions.

(a) $\sqrt{5}(3 - \sqrt{2}) = 3\sqrt{5} - \sqrt{5}\sqrt{2}$ Distributive Property

$\qquad\qquad\qquad = 3\sqrt{5} - \sqrt{10}$ Product Rule for Radicals

(b) $(\sqrt{2} + 2\sqrt{5})(3\sqrt{2} - \sqrt{5})$

$\qquad = 3\sqrt{2}\sqrt{2} - \sqrt{2}\sqrt{5} + 6\sqrt{2}\sqrt{5} - 2\sqrt{5}\sqrt{5}$ FOIL method

$\qquad = 3 \cdot 2 - \sqrt{10} + 6\sqrt{10} - 2 \cdot 5$ Product Rule for Radicals

$\qquad = 6 - \sqrt{10} + 6\sqrt{10} - 10$ and $(\sqrt{b})^2 = b$

$\qquad = -4 + 5\sqrt{10}$ Combine like terms.

(c) $(6 - \sqrt{3})(6 + \sqrt{3}) = 6^2 - (\sqrt{3})^2$ $(A + B)(A - B) = A^2 - B^2$

$\qquad\qquad\qquad = 36 - 3$ $(\sqrt{b})^2 = b$

$\qquad\qquad\qquad = 33$

(d) $(\sqrt{7} + \sqrt{6})^2$

$\qquad = (\sqrt{7})^2 + 2\sqrt{7}\sqrt{6} + (\sqrt{6})^2$ $(A + B)^2 = A^2 + 2AB + B^2$

$\qquad = 7 + 2\sqrt{42} + 6$ Product Rule for Radicals

$\qquad = 13 + 2\sqrt{42}$

Rationalizing Binomial Denominators

In part (c) of Example 3, we used the special product

$$(A + B)(A - B) = A^2 - B^2$$

The expressions $A + B$ and $A - B$ are called **conjugates.** In particular, the expressions $6 - \sqrt{3}$ and $6 + \sqrt{3}$ are square root conjugates. As was shown, their product, 33, does not include a radical.

In general, the product of square root conjugates is an expression that does not contain a radical. We can take advantage of this fact in rationalizing the denominator of a quotient with a binomial in the denominator. We simply multiply the numerator and the denominator by the conjugate of the denominator. This results in an expression with no radical in the denominator.

EXAMPLE **4**

Rationalizing a Binomial Denominator

Rationalize the denominator.

(a) $\dfrac{\sqrt{3}}{\sqrt{3} - 2} = \dfrac{\sqrt{3}}{\sqrt{3} - 2} \cdot \dfrac{\sqrt{3} + 2}{\sqrt{3} + 2}$

The conjugate of $\sqrt{3} - 2$ is $\sqrt{3} + 2$. Multiply the numerator and the denominator by $\sqrt{3} + 2$.

$= \dfrac{\sqrt{3}(\sqrt{3} + 2)}{(\sqrt{3})^2 - 2^2}$

$(A + B)(A - B) = A^2 - B^2$

$= \dfrac{(\sqrt{3})^2 + 2\sqrt{3}}{(\sqrt{3})^2 - 2^2}$

Distributive Property

$= \dfrac{3 + 2\sqrt{3}}{3 - 4}$

$(\sqrt{b})^2 = b$

$= \dfrac{3 + 2\sqrt{3}}{-1}$

Simplify.

$= -3 - 2\sqrt{3}$

(b) $\dfrac{4}{\sqrt{3} + \sqrt{5}} = \dfrac{4}{\sqrt{3} + \sqrt{5}} \cdot \dfrac{\sqrt{3} - \sqrt{5}}{\sqrt{3} - \sqrt{5}}$

The conjugate of $\sqrt{3} + \sqrt{5}$ is $\sqrt{3} - \sqrt{5}$. Multiply the numerator and the denominator by $\sqrt{3} - \sqrt{5}$.

$= \dfrac{4(\sqrt{3} - \sqrt{5})}{(\sqrt{3})^2 - (\sqrt{5})^2}$

$(A + B)(A - B) = A^2 - B^2$

$= \dfrac{4(\sqrt{3} - \sqrt{5})}{3 - 5}$

$(\sqrt{b})^2 = b$

$= \dfrac{4(\sqrt{3} - \sqrt{5})}{-2}$

$= -2(\sqrt{3} - \sqrt{5})$

Reduce the fraction.

$= -2\sqrt{3} + 2\sqrt{5}$

Distributive Property

Think About It

Explain why the expression $\dfrac{1}{\sqrt{x}+2}$

has no restricted value. Now rationalize the denominator. Does the resulting expression have a restricted value? Are the two expressions equivalent?

(c) $\dfrac{\sqrt{3}+1}{\sqrt{3}-1} = \dfrac{\sqrt{3}+1}{\sqrt{3}-1} \cdot \dfrac{\sqrt{3}+1}{\sqrt{3}+1}$

The conjugate of $\sqrt{3}-1$ is $\sqrt{3}+1$. Multiply the numerator and the denominator by $\sqrt{3}+1$.

$= \dfrac{\left(\sqrt{3}\right)^2 + 2\cdot\sqrt{3}\cdot 1 + 1^2}{\left(\sqrt{3}\right)^2 - 1^2}$

$(A+B)^2 = A^2 + 2AB + B^2$

$= \dfrac{3 + 2\sqrt{3} + 1}{3 - 1}$

$\left(\sqrt{b}\right)^2 = b$

$= \dfrac{4 + 2\sqrt{3}}{2}$

$= \dfrac{2\left(2 + \sqrt{3}\right)}{2}$

Factor the numerator.

$= 2 + \sqrt{3}$

Divide out the common factor.

Quick Reference 9.4

Addition and Subtraction

- Square root radicals that have the same radicand are called **like radicals.**
- When we combine like radicals, we add or subtract the coefficients and retain the radical. It is sometimes possible to simplify one or both radicals so that they become like radicals and can be combined.

Multiplying with the Distributive Property

- We use the Distributive Property to multiply radical expressions with more than one term.

Rationalizing Binomial Denominators

- The expressions $A + B$ and $A - B$ are called **conjugates.**
- The product of square root conjugates is an expression that does not contain a radical.
- To rationalize the denominator of a quotient with a binomial in the denominator, multiply the numerator and the denominator by the conjugate of the denominator.

Speaking the Language 9.4

1. Because $2\sqrt{3x}$ and $-5\sqrt{3x}$ have the same radicands, the radicals are called _____ .

2. To combine like radicals, we add or subtract the _____ and retain the _____ .

3. To simplify $\sqrt{2}\left(5 - \sqrt{3}\right)$, we use the _____ Property.

4. Expressions such as $5 - \sqrt{2}$ and $5 + \sqrt{2}$ are called square root _____ .

Exercises 9.4

Concepts and Skills

1. Use the Distributive Property to explain why the terms of $3\sqrt{2} + 2\sqrt{3}$ cannot be combined.

2. For the expression $\sqrt{48} - \sqrt{27}$, what must be done before the subtraction can be performed?

In Exercises 3–12, simplify.

3. $4\sqrt{7} + 3\sqrt{7}$

4. $-2\sqrt{10} + \sqrt{10}$

5. $\sqrt{15} - 3\sqrt{15}$

6. $-\sqrt{3} - 4\sqrt{3}$

7. $3\sqrt{2} - 4\sqrt{3} + \sqrt{3} - 2\sqrt{2}$

8. $2\sqrt{11} + \sqrt{17} - 5\sqrt{17} - 3\sqrt{11}$

9. $5\sqrt{x} - 3\sqrt{x}$

10. $-7\sqrt{2y} + 3\sqrt{2y}$

11. $t\sqrt{7t} + 5t\sqrt{7t}$

12. $-5y\sqrt{6y} - 2y\sqrt{6y}$

In Exercises 13–24, simplify.

13. $\sqrt{12} - \sqrt{3}$

14. $\sqrt{18} + \sqrt{2}$

15. $\sqrt{27} + \sqrt{75}$

16. $\sqrt{98} - \sqrt{50}$

17. $3\sqrt{48} + 2\sqrt{75}$

18. $2\sqrt{45} - \sqrt{80}$

19. $3\sqrt{2t} - \sqrt{98t}$

20. $\sqrt{8x^3} + x\sqrt{32x}$

21. $b\sqrt{90b} + 8\sqrt{10b^3}$

22. $2t\sqrt{27t} - \sqrt{3t^3}$

23. $2\sqrt{72} - \sqrt{300} + \sqrt{200}$

24. $2\sqrt{27t} - \sqrt{108t} + 3\sqrt{75t}$

25. Explain why each of the following is false. Then write the correct answer.

    (a) $\left(3\sqrt{x}\right)^2 = 3x$

    (b) $\left(\sqrt{x} + \sqrt{5}\right)^2 = x + 5$

26. Suppose that you wish to rationalize the denominator of the expression $\dfrac{5}{1 + \sqrt{3}}$. Explain why you cannot accomplish your goal by multiplying the numerator and denominator by $\sqrt{3}$.

In Exercises 27–40, multiply.

27. $\sqrt{5}\left(2 - \sqrt{3}\right)$

28. $\sqrt{3}\left(\sqrt{5} + \sqrt{10}\right)$

29. $\sqrt{2}\left(\sqrt{8} + \sqrt{7}\right)$

30. $\sqrt{3}\left(2\sqrt{15} - 4\sqrt{3}\right)$

31. $\left(2 - \sqrt{6}\right)\left(3 - 2\sqrt{6}\right)$

32. $\left(\sqrt{5} + 3\right)\left(\sqrt{5} - 4\right)$

33. $\left(\sqrt{5} + 2\sqrt{2}\right)\left(3\sqrt{5} + \sqrt{2}\right)$

34. $\left(2\sqrt{6} + \sqrt{7}\right)\left(\sqrt{6} - \sqrt{7}\right)$

35. $\left(\sqrt{x} + 7\right)\left(\sqrt{x} - 2\right)$

36. $\left(\sqrt{x} + y\right)\left(2\sqrt{x} + y\right)$

37. $\left(\sqrt{5} + 2\right)^2$

38. $\left(\sqrt{6} - \sqrt{7}\right)^2$

39. $\left(3 + \sqrt{x}\right)^2$

40. $\left(\sqrt{x} + 2\sqrt{y}\right)^2$

In Exercises 41–46, write the conjugate of the given expression. Then determine the product of the expression and its conjugate.

41. $2\sqrt{3} + 3$

42. $5 - 2\sqrt{7}$

43. $\sqrt{10} - \sqrt{5}$

44. $\sqrt{2} + 2\sqrt{3}$

45. $\sqrt{x} - 4$

46. $\sqrt{3} + y$

In Exercises 47–50, simplify the quotient.

47. $\dfrac{12 - \sqrt{162}}{3}$

48. $\dfrac{\sqrt{36} + \sqrt{48}}{2}$

49. $\dfrac{10 + \sqrt{50}}{15}$

50. $\dfrac{10 - \sqrt{32}}{6}$

In Exercises 51–64, rationalize the denominator.

51. $\dfrac{1}{\sqrt{3} + 5}$

52. $\dfrac{2}{\sqrt{7} - 2}$

53. $\dfrac{14}{3 - \sqrt{2}}$

54. $\dfrac{11}{5 + \sqrt{3}}$

55. $\dfrac{3}{2\sqrt{7} - \sqrt{5}}$

56. $\dfrac{6}{2\sqrt{3} - \sqrt{5}}$

57. $\dfrac{\sqrt{7}}{2\sqrt{5} - \sqrt{7}}$

58. $\dfrac{\sqrt{2}}{\sqrt{6} + 1}$

59. $\dfrac{\sqrt{2}}{\sqrt{6} - \sqrt{2}}$

60. $\dfrac{\sqrt{5}}{\sqrt{10} + \sqrt{15}}$

61. $\dfrac{y}{1 - \sqrt{y}}$

62. $\dfrac{\sqrt{x}}{\sqrt{x} + 2}$

63. $\dfrac{\sqrt{11} + 1}{\sqrt{11} - 1}$

64. $\dfrac{3 - 2\sqrt{6}}{3 + 2\sqrt{6}}$

In Exercises 65–68, fill in the blank to make the statement true.

65. $\underline{\qquad}\left(5 + \sqrt{2}\right) = 5\sqrt{3} + \sqrt{6}$

66. $\underline{\qquad}\left(\sqrt{2} - 7\right) = 2 - 7\sqrt{2}$

67. $\underline{\qquad}\left(\sqrt{2x} - 1\right) = x\sqrt{2} - \sqrt{x}$

68. $\underline{\qquad}\left(\sqrt{t} - \sqrt{2}\right) = 2\sqrt{t} - t\sqrt{2}$

In Exercises 69–74, the lengths of the sides of a triangle are given. Determine whether the triangle is a right triangle.

69. $2\sqrt{3}, 4, 2\sqrt{7}$

70. $5\sqrt{3}, 3\sqrt{5}, 2\sqrt{30}$

71. $\sqrt{7}, \sqrt{7}, 7\sqrt{2}$

72. $5, 5\sqrt{3}, 10$

73. $\sqrt{6}, 2\sqrt{3}, 4$

74. $\sqrt{10}, 5\sqrt{2}, 30$

Geometric Models

75. Determine the area of a rectangle whose length is $\sqrt{2} + \sqrt{15}$ and whose width is $\sqrt{3}$.

76. Determine the perimeter of a rectangle whose length is $5 + 2\sqrt{75}$ and whose width is $1 + 3\sqrt{12}$.

77. The length of a side of a square is $\sqrt{3} + 1$. The length of a rectangle is the same as the length of a side of the square, and the width is 3 less than the length. Determine the total area of the two figures.

78. Determine the total area of the figure.

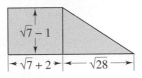

Data Analysis: *The Family Budget*

From 1950 to 1999, the percent of family income that was spent for food decreased from 30.2% to 15.2%. During the same period, the percent spent for clothing decreased from 12.3% to 6.4%. (Source: U.S. Bureau of Economic Analysis.)

If t represents the number of years since 1950, the percents can be modeled by the following expressions.

Food: $F(t) = \dfrac{113}{\sqrt{t+15}} + 1$

Clothing: $C(t) = \dfrac{44}{\sqrt{t+15}} + 0.9$

79. Use the model function F to estimate the percent of income that will be spent for food in 2005.

80. Use a table of values for the model function C to estimate the year in which the percent of income that will be spent for clothing will be below 6.1%.

81. Write equivalent expressions for $F(t)$ and $C(t)$ by rationalizing the denominators.

82. Use the expressions in Exercise 81 to write a simplified expression $T(t)$ for the percent of income spent for food and clothing.

Challenge

In Exercises 83–87, perform the indicated operations.

83. $\sqrt{\dfrac{1}{5}} + \sqrt{20}$

84. $\dfrac{3}{\sqrt{3}} - 2\sqrt{3}$

85. $\dfrac{\sqrt{2}}{\sqrt{3}} + \dfrac{\sqrt{3}}{\sqrt{2}}$

86. $\dfrac{1}{2+\sqrt{2}} - \dfrac{1}{2-\sqrt{2}}$

87. $\left(\sqrt[3]{2} + 1\right)\left(\sqrt[3]{4} - \sqrt[3]{2} + 1\right)$

In Exercises 88 and 89, simplify the expression. Assume that all variables represent positive numbers.

88. $\sqrt{9x^3 + 18x^2} - \sqrt{25x^3 + 50x^2}$

89. $\sqrt{12x^2 + 12x + 3} + \sqrt{3x^2 + 18x + 27}$

9.5   Equations with Radicals

Radical Equations • Real-Life Applications

Radical Equations

An equation that contains a radical with a variable in the radicand is called a **radical equation.** The following are examples of radical equations.

$$\sqrt{x+1} = 3 \qquad \sqrt{x^2+2} = 2 - x$$

$$\sqrt{1-x} = \sqrt{x+1} \qquad \sqrt[3]{x+1} = 4$$

When we solve equations with fractions, we usually begin by eliminating the fractions. Similarly, when we solve radical equations, we first eliminate the radicals. To accomplish this, we need an additional property of equality.

> **Squaring Property of Equality**
>
> For real numbers a and b, if $a = b$ then $a^2 = b^2$.
>
> In words, if two quantities are equal, then their squares are equal.

As we use the Squaring Property of Equality, it is important to realize that squaring both sides of an equation does not necessarily produce an equivalent equation. For instance, the equation $x = 3$ has only one solution, 3, but squaring both sides gives the equation $x^2 = 9$, which has two solutions, -3 and 3. Recall that a solution of a new equation that is not a solution of the original equation is called an *extraneous solution*.

Note: The Squaring Property of Equality is used extensively in the algebraic methods for solving radical equations. Because extraneous solutions can be introduced with this method, it is essential to check all solutions. We will review a variety of methods for doing this.

As with other equations, we can estimate the solution(s) of a radical equation by graphing each side of the equation. In the examples that follow, we illustrate both the graphing method and algebraic methods.

EXAMPLE **1**

Solving a Radical Equation

Solve the equation.

(a) $\sqrt{2x + 1} = 3$

(b) $\sqrt{2x - 3} = \sqrt{x + 4}$

Solution

(a) First we estimate the solution(s) of the equation by graphing. Produce the graphs of $y_1 = \sqrt{2x + 1}$ and $y_2 = 3$. (See Fig. 9.2.)

Figure 9.2

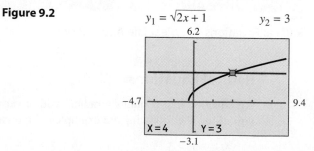

Trace to the point of intersection. The solution appears to be 4.

To solve the equation algebraically, we begin by squaring both sides of the equation to eliminate the radical.

$$\sqrt{2x + 1} = 3$$

$$\left(\sqrt{2x + 1}\right)^2 = 3^2 \qquad \text{Square both sides of the equation.}$$

$$2x + 1 = 9 \qquad \text{Recall that } \left(\sqrt{b}\right)^2 = b.$$

$$2x = 8 \qquad \text{Solve for } x.$$

$$x = 4$$

The solution 4 can be verified with a calculator. (See Fig. 9.3.)

Figure 9.3

```
Y₁(4)
                    3
Y₂(4)
                    3
```

Figure 9.4

```
7→X
                    7
√(2X-3)
         3.31662479
√(X+4)
         3.31662479
```

(b) $\sqrt{2x - 3} = \sqrt{x + 4}$

$$\left(\sqrt{2x - 3}\right)^2 = \left(\sqrt{x + 4}\right)^2 \qquad \text{Square both sides of the equation.}$$

$$2x - 3 = x + 4 \qquad \text{Both radicals are eliminated.}$$

$$x = 7 \qquad \text{Solve for } x.$$

Use a calculator to verify that 7 is the solution. (See Fig. 9.4.)

If possible, isolate the radical term before squaring both sides of an equation.

EXAMPLE **2**

Solving a Radical Equation

Solve $3 + \sqrt{x + 6} = 5$.

Solution

$$3 + \sqrt{x + 6} = 5$$

$$\sqrt{x + 6} = 2 \qquad \text{Subtract 3 from both sides to isolate the radical.}$$

$$\left(\sqrt{x + 6}\right)^2 = 2^2 \qquad \text{Square both sides of the equation.}$$

$$x + 6 = 4 \qquad \text{Solve for } x.$$

$$x = -2$$

This time we verify the solution by substitution.

$$3 + \sqrt{x + 6} = 5 \qquad \text{Original equation}$$

$$3 + \sqrt{-2 + 6} = 5 \qquad \text{Replace } x \text{ with } -2.$$

$$3 + \sqrt{4} = 5 \qquad \text{Evaluate.}$$

$$3 + 2 = 5$$

$$5 = 5 \qquad \text{True}$$

EXAMPLE   **3**

A Radical Equation with No Solution

Solve $\sqrt{x} + 6 = 2$.

Solution

Produce the graphs of $y_1 = \sqrt{x} + 6$ and $y_2 = 2$. (See Fig. 9.5.) Because the graphs do not intersect, the equation appears to have no solution.

Figure 9.5

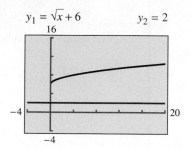

This can be verified algebraically.

$$\sqrt{x} + 6 = 2$$
$$\sqrt{x} = -4 \qquad \text{Subtract 6 from both sides to isolate the radical.}$$
$$\left(\sqrt{x}\right)^2 = (-4)^2 \qquad \text{Square both sides to eliminate the radical.}$$
$$x = 16$$

Although 16 appears to be a solution, substitution reveals that 16 is an extraneous solution.

$$\sqrt{x} + 6 = 2 \qquad \text{Original equation}$$
$$\sqrt{16} + 6 = 2 \qquad \text{Replace } x \text{ with 16 in the original equation.}$$
$$4 + 6 = 2 \qquad \text{Simplify.}$$
$$10 = 2 \qquad \text{False.}$$

The equation has no solution.

EXAMPLE   **4**

Solving a Radical Equation

Solve the equation.

(a) $x = \sqrt{x^2 + x + 5}$ (b) $\sqrt{x^2 + 2} = 2 - x$

Solution

(a) $\quad x = \sqrt{x^2 + x + 5}$
$$x^2 = \left(\sqrt{x^2 + x + 5}\right)^2 \qquad \text{Square both sides of the equation.}$$
$$x^2 = x^2 + x + 5 \qquad \text{Subtract } x^2 \text{ from both sides.}$$
$$0 = x + 5 \qquad \text{Solve for } x.$$
$$x = -5$$

Figure 9.6

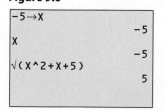

When we use a calculator to check the proposed solution, we find that the left and right sides have different values when $x = -5$. (See Fig. 9.6.) Thus -5 is an extraneous solution, and the equation has no solution.

Figure 9.7

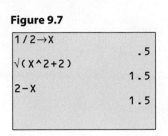

(b)
$$\sqrt{x^2 + 2} = 2 - x$$
$$\left(\sqrt{x^2 + 2}\right)^2 = (2 - x)^2 \qquad \text{Square both sides of the equation.}$$
$$x^2 + 2 = 4 - 4x + x^2 \qquad \text{Use } (A - B)^2 = A^2 - 2AB + B^2.$$
$$2 = 4 - 4x \qquad \text{Subtract } x^2 \text{ from both sides.}$$
$$-2 = -4x \qquad \text{Solve for } x.$$
$$x = \frac{1}{2}$$

Use a calculator to check the proposed solution. (See Fig. 9.7.) The solution is $\frac{1}{2}$.

Sometimes eliminating radicals from an equation results in a quadratic equation.

EXAMPLE **5**

Solving a Radical Equation

Solve the equation.

(a) $x - 1 = 3\sqrt{x - 3}$

(b) $x - 2 = \sqrt{2x - 1}$

Solution

(a)
$$x - 1 = 3\sqrt{x - 3}$$
$$(x - 1)^2 = \left(3\sqrt{x - 3}\right)^2 \qquad \text{Square both sides of the equation.}$$
$$x^2 - 2x + 1 = 9(x - 3) \qquad \text{Use } (A - B)^2 = A^2 - 2AB + B^2.$$
$$x^2 - 2x + 1 = 9x - 27 \qquad \text{Distributive Property}$$
$$x^2 - 11x + 28 = 0 \qquad \text{Write the quadratic equation in standard form.}$$
$$(x - 4)(x - 7) = 0 \qquad \text{Factor.}$$
$$x - 4 = 0 \quad \text{or} \quad x - 7 = 0 \qquad \text{Set each factor equal to 0.}$$
$$x = 4 \quad \text{or} \quad x = 7 \qquad \text{Solve for } x.$$

Both solutions can be verified with a calculator. The equation has two solutions: 4 and 7.

Think About It

Solving
$\sqrt{x} + \sqrt{x + 1} - \sqrt{x + 2} - \sqrt{x + 3} = 0$
algebraically requires many
solution steps, and the result is $0 = 1$.
How could you have predicted,
without solving or graphing, that
the equation would have
no solution?

(b) To estimate the solution of the equation, produce the graphs of $y_1 = x - 2$ and $y_2 = \sqrt{2x - 1}$. (See Fig. 9.8.)

Figure 9.8

$$y_1 = x - 2 \qquad y_2 = \sqrt{2x - 1}$$

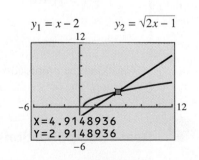

The equation appears to have one solution: approximately 5.

Now solve algebraically.

$$x - 2 = \sqrt{2x - 1}$$
$$(x - 2)^2 = \left(\sqrt{2x - 1}\right)^2 \qquad \text{Square both sides of the equation.}$$
$$x^2 - 4x + 4 = 2x - 1 \qquad \text{Use } (A - B)^2 = A^2 - 2AB + B^2.$$
$$x^2 - 6x + 5 = 0 \qquad \text{Write the quadratic equation in standard form.}$$
$$(x - 5)(x - 1) = 0 \qquad \text{Factor.}$$
$$x - 5 = 0 \quad \text{or} \quad x - 1 = 0 \qquad \text{Set each factor equal to 0.}$$
$$x = 5 \quad \text{or} \quad x = 1 \qquad \text{Solve for } x.$$

When we use a calculator to check the proposed solutions, we find that 1 is an extraneous solution. (See Fig. 9.9.)

Note that the algebraic method produced two apparent solutions, but the graphing method helped us to anticipate that one would be extraneous. The only solution is 5.

Sometimes it is necessary to square both sides of an equation more than one time in order to eliminate all radicals.

Figure 9.9

```
Y₁(5)
              3
Y₂(5)
              3
Y₁(1)
             -1
Y₂(1)
              1
```

EXAMPLE | 6

Solving a Radical Equation

Solve $\sqrt{1 - x} = \sqrt{x} + 1$.

Solution

$$\sqrt{1 - x} = \sqrt{x} + 1$$
$$\left(\sqrt{1 - x}\right)^2 = \left(\sqrt{x} + 1\right)^2 \qquad \text{Square both sides to eliminate the radical on the left.}$$
$$1 - x = x + 2\sqrt{x} + 1 \qquad \text{Use } (A + B)^2 = A^2 + 2AB + B^2.$$
$$-2x = 2\sqrt{x} \qquad \text{Isolate the remaining radical.}$$
$$-x = \sqrt{x} \qquad \text{Divide both sides by 2.}$$
$$(-x)^2 = \left(\sqrt{x}\right)^2 \qquad \text{Square both sides again.}$$
$$x^2 = x \qquad \text{Simplify.}$$
$$x^2 - x = 0 \qquad \text{Write the equation in standard form.}$$
$$x(x - 1) = 0 \qquad \text{Factor.}$$
$$x = 0 \quad \text{or} \quad x - 1 = 0 \qquad \text{Set each factor equal to 0.}$$
$$x = 0 \quad \text{or} \quad x = 1 \qquad \text{Solve for } x.$$

LEARNING TIP

After you eliminate radicals, the resulting equation may be linear, quadratic, or some other type equation. Stop and think about what type of equation you have obtained so that you can select a method for solving it.

Checking the proposed solutions with a calculator, we find that 1 is an extraneous solution. The only solution is 0.

Real-Life Applications

Information given in an application problem can sometimes be modeled with a radical equation.

EXAMPLE 7

Language Development in Chimpanzees

A researcher is studying the development of languages by examining the ability of a chimpanzee to understand symbolic commands. The researcher found that the number of correct responses that the chimpanzee made in a 15-minute session after x weeks of training is given by $9 + \sqrt{x + 4}$. How many weeks of training are required for the chimpanzee to give 17 correct responses in a 15-minute session?

Solution

$$9 + \sqrt{x + 4} = 17 \qquad \text{The number of correct responses is 17.}$$
$$\sqrt{x + 4} = 8 \qquad \text{Subtract 9 from both sides to isolate the radical.}$$
$$\left(\sqrt{x + 4}\right)^2 = 8^2 \qquad \text{Square both sides of the equation.}$$
$$x + 4 = 64 \qquad \text{Solve for } x.$$
$$x = 60$$

Verify the solution by substitution. After 60 weeks of training, the expected number of correct responses is 17.

Quick Reference 9.5

Radical Equations

- An equation that contains a radical with a variable in the radicand is called a **radical equation.**

- The Squaring Property of Equality is used to eliminate radicals from an equation.

    For real numbers a and b, if $a = b$, then $a^2 = b^2$.

- The use of the Squaring Property of Equality can introduce extraneous solutions. Therefore, checking the solutions of a radical equation is an essential step in the solving process.

- By graphing the left and right sides of a radical equation, we can predict the number of solutions and estimate what the solutions are.

- It is usually best, if possible, to isolate the radical before squaring both sides of the equation. Sometimes it is necessary to square both sides twice in order to eliminate all the radicals.

- Eliminating radicals from a radical equation can lead to a linear equation, a quadratic equation, or a higher-degree equation.

Speaking the Language 9.5

1. An equation such as $\sqrt{x + 3} = 5$ is called a(n) ▓▓▓▓▓▓ .

2. In words, the Squaring Property of Equality states that if two numbers are equal, then their ▓▓▓▓▓▓ are equal.

3. Checking solutions of a radical equation is essential because the solving process might introduce ▓▓▓▓▓▓ solutions.

4. To solve an equation such as $\sqrt{2x - 1} - 4 = 12$, a good first step is to ▓▓▓▓▓▓ the radical.

Exercises 9.5

Concepts and Skills

1. Tell whether the following statements are true or false for all real numbers a and b. Justify your answer in each case.

   (a) If $a = b$, then $a^2 = b^2$.

   (b) If $a^2 = b^2$, then $a = b$.

2. Explain how to determine the solution of the equation $\sqrt{x + 7} = -3$ by inspection.

In Exercises 3–12, use the graphing method to estimate the solution(s) of the equation.

3. $\sqrt{4x - 15} = 3$

4. $\sqrt{x^2 + 3} = 3x - 1$

5. $\sqrt{5x - 4} = \sqrt{20 - 3x}$

6. $3\sqrt{x + 13} = x + 9$

7. $\sqrt{4x + 13} - x = x - 1$

8. $2x + 1 = x + \sqrt{22 - 2x}$

9. $\sqrt{6x + 7} = x + 2$

10. $\sqrt{5x + 1} - 2x = 1 - x$

11. $\sqrt{5x - 1} = \sqrt{3x + 3}$

12. $\sqrt{x + 7} = 1 + \sqrt{2x}$

In Exercises 13–28, solve the equation.

13. $\sqrt{x - 5} = 2$

14. $\sqrt{7 - x} = 4$

15. $2\sqrt{x} = \sqrt{x + 15}$

16. $3\sqrt{x - 2} = \sqrt{3x}$

17. $\sqrt{x + 3} = 7$

18. $\sqrt{2x - 1} = 3$

19. $\sqrt{3x + 1} - 7 = -3$

20. $\sqrt{5x - 4} - 3 = 3$

21. $\sqrt{x + 1} + 8 = 6$

22. $\sqrt{x + 5} = 1$

23. $\sqrt{2x + 1} = \sqrt{10 - x}$

24. $\sqrt{7x - 4} = \sqrt{x + 2}$

25. $\sqrt{x - 7} = \sqrt{2 - 2x}$

26. $\sqrt{2x + 1} = \sqrt{3x + 2}$

27. $\sqrt{2x - 24} - \sqrt{x + 8} = 0$

28. $\sqrt{11x - 8} - \sqrt{4x - 1} = 0$

In Exercises 29–36, solve the equation.

29. $x = \sqrt{x^2 - 2x + 12}$

30. $x = \sqrt{x^2 + 5x}$

31. $\sqrt{x^2 + 3x + 3} = x$

32. $x - 3 = \sqrt{x^2 - x + 4}$

33. $\sqrt{x^2 + 7} = x + 1$

34. $\sqrt{x^2 + 1} = 1 - x$

35. $\sqrt{x^2 + 3x} = x + 1$

36. $2 - x = \sqrt{x^2 - 2x}$

37. Explain why $\sqrt{x}\sqrt{x + 3} = 2$ has only one solution whereas $\sqrt{x^2 + 3x} = 2$ has two solutions.

38. Explain why $x + 1 = 2$ has only one solution whereas $\sqrt{(x + 1)^2} = 2$ has two solutions.

In Exercises 39–48, solve the equation.

39. $\sqrt{x + 3} = x + 3$

40. $t - 1 = \sqrt{2t - 2}$

41. $x = -2 + 3\sqrt{x}$

42. $x - 4\sqrt{x} = -3$

43. $t = \sqrt{4t - 3}$

44. $4 = \sqrt{10x - x^2}$

45. $x - \sqrt{3x + 1} = -1$

46. $x = 4 + \sqrt{2x - 8}$

47. $x = 3\sqrt{x - 1} - 1$

48. $x - 4\sqrt{x - 4} = 1$

In Exercises 49–60, solve the equation.

49. $x - 3 = \sqrt{x - 1}$

50. $\sqrt{15 + x} = x + 3$

51. $2\sqrt{x} + x = 8$

52. $x = 4\sqrt{x} + 5$

53. $7\sqrt{x} + x = -10$

54. $x + 3 + 4\sqrt{x} = 0$

55. $x = \sqrt{10 - 3x}$

56. $\sqrt{x}\sqrt{x - 8} = 3$

57. $\sqrt{x - 2} + 4 = x$

58. $5 = x - \sqrt{x + 1}$

59. $\sqrt{1 - 3x} + 1 = x$

60. $2x + 3\sqrt{2x - 3} = 1$

In Exercises 61–68, solve the equation.

61. $\sqrt{x + 2} = \sqrt{x + 8}$

62. $\sqrt{x + 12} - \sqrt{x} = 2$

63. $\sqrt{2t + 5} = \sqrt{2t} + 1$

64. $\sqrt{4t + 27} - \sqrt{4t} = 3$

65. $\sqrt{3t + 1} = \sqrt{1 + 5t}$

66. $\sqrt{2x - 1} = \sqrt{x + 7}$

67. $\sqrt{2x + 1} = \sqrt{x + 7}$

68. $\sqrt{9 - 4t} = 2\sqrt{t} - 3$

In Exercises 69–72, determine the number.

69. Three times the square root of a number is equal to the square root of 16 more than the number.

70. The square root of 3 less than a number is 5 less than the number.

71. The square root of 1 more than twice a number is 7 less than the number.

72. A number is equal to 4 more than the square root of twice the number.

Real-Life Applications

The distance d (in miles) to the horizon from a height h (in feet) is given by $d = 1.5\sqrt{h}$. In Exercises 73 and 74, use this formula to answer the questions.

73. Balloon Adventures advertises that on a clear day a person taking a hot air balloon ride in Chattanooga can see Atlanta 100 miles away. How high must the balloon ascend for this claim to be true?

74. How high should the lookout on a ship be to allow an observer to see another ship 15 miles away?

75. The cost of producing n wooden candle holders is $20\sqrt{2n + 7}$ dollars. How many candle holders can be produced for $340?

76. The population of bison in an area t months after protective measures are enacted is $50t - 300\sqrt{t}$. After how many months is the population 350?

77. An industrial psychologist determines that a person's ability to perform a task without errors diminishes with the number x of continuous hours that the person has worked at the task. The efficiency rating for this task is given by $\sqrt{x - 2} - x + 12$. After how many hours will the person's efficiency rating drop to 8?

78. The price of a print is $80 - 10\sqrt{n}$ dollars, where n is the number of prints produced. If the price of a print is $30, how many prints were produced?

79. A college admissions director estimates that the college's enrollment (in thousands) of students over 25 years of age is given by $\sqrt{2n + 9}$ and the enrollment of students age 25 and younger is given by $\sqrt{n + 15}$, where n is the number of years since 1995. In what year will the enrollment of the two age groups be the same?

80. The traffic count (in thousands per month) on a road is predicted to be $3\sqrt{x - 1} + 24$, where x is the number of months since the road opened. In how many months after the road opens will the traffic count be 30,000 cars per month?

Modeling with Real Data

81. From 45.1 million in 1970, the number of married couples in the United States rose to 55.3 million in 1998. (Source: U.S. Bureau of the Census.) Suppose that the radical expression $2.3\sqrt{x + 1} + 43.1$ is determined to be an appropriate model of the number of married couples (in millions), where x represents the number of years since 1970.

(a) Use the model to write an equation that you can use to determine the year in which the number of married couples will reach 56.9 million.

(b) Solve the equation in part (a) and interpret the solution.

82. Refer to the model in Exercise 81.

(a) What does the solution of the following equation represent?

$$2.3\sqrt{x + 1} + 43.1 = 56.4$$

(b) Solve the equation in part (a) and interpret the solution.

Challenge

In Exercises 83–86, solve the equation.

83. $\sqrt[3]{x} + 5 = 3$

84. $\sqrt[5]{x + 1} = 2$

85. $\sqrt[4]{1 - x} = \sqrt[4]{3x - 7}$

86. $\sqrt[5]{x^2 + 2x - 4} = -1$

In Exercises 87 and 88, solve the formula for the indicated variable.

87. $r = \sqrt{\dfrac{3V}{\pi h}}$; h

88. $v = \sqrt{2pt}$; p

9.6 | Applications with Right Triangles

Lengths of Sides   •   Real-Life Applications

Lengths of Sides

Think About It

Refer to Fig. 9.10, and consider the ratios $\frac{a}{c}$ and $\frac{b}{c}$. Show that the sum of the squares of these ratios is 1 for any right triangle.

Recall that the Pythagorean Theorem relates the lengths a and b of the legs of a right triangle to the length c of the hypotenuse by the formula $a^2 + b^2 = c^2$. (See Fig. 9.10.)

Figure 9.10

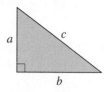

Because this formula involves squares of lengths, the actual lengths are often expressed as square root radicals.

When we use the Pythagorean Theorem to determine the length of a side of a right triangle, we use only the principal (positive) square root because length is positive.

EXAMPLE 1

Determining the Length of the Hypotenuse of a Right Triangle

Determine the length of the hypotenuse of the right triangle in Figure 9.11.

Figure 9.11

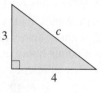

LEARNING TIP

A critical first step in Pythagorean Theorem problems is identifying the hypotenuse and the legs. The hypotenuse is the longest side, and the legs are perpendicular to each other.

Solution

$$c^2 = a^2 + b^2 \qquad \text{Pythagorean Formula}$$
$$c^2 = 3^2 + 4^2 \qquad \text{The lengths of the legs are 3 and 4.}$$
$$c^2 = 9 + 16$$
$$c^2 = 25$$
$$c = \sqrt{25} \qquad \text{Note that because } c \text{ must be positive, } c \text{ is the principal square root of 25.}$$
$$c = 5$$

The length of the hypotenuse is 5.

EXAMPLE 2

Determining the Length of a Leg of a Right Triangle

For each right triangle, determine the length x.

(a) **Figure 9.12**

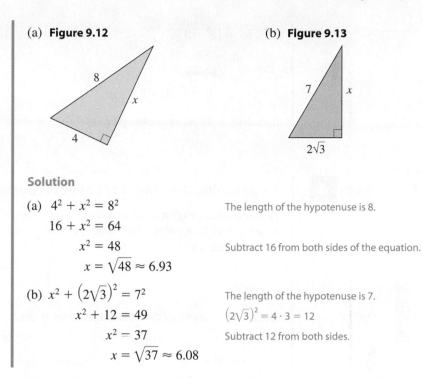

(b) **Figure 9.13**

Solution

(a) $4^2 + x^2 = 8^2$ The length of the hypotenuse is 8.

   $16 + x^2 = 64$

   $x^2 = 48$ Subtract 16 from both sides of the equation.

   $x = \sqrt{48} \approx 6.93$

(b) $x^2 + \left(2\sqrt{3}\right)^2 = 7^2$ The length of the hypotenuse is 7.

   $x^2 + 12 = 49$ $\left(2\sqrt{3}\right)^2 = 4 \cdot 3 = 12$

   $x^2 = 37$ Subtract 12 from both sides.

   $x = \sqrt{37} \approx 6.08$

Real-Life Applications

In applications involving right triangles, a sketch is usually helpful. Be sure to identify the location of the right angle and of the hypotenuse.

EXAMPLE   3

Determining the Height of a Boat Deck

One end of a 17-foot ramp is attached to the deck of a boat. The base of the ramp rests on a pier 15 feet from the boat. (See Fig. 9.14.) How high is the deck above the level of the pier?

Figure 9.14

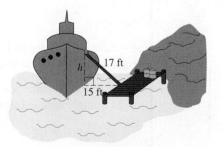

Solution

Let $h =$ the height of the deck above the pier. Note that the ramp is the hypotenuse of the triangle with legs 15 and h.

$$h^2 + 15^2 = 17^2 \qquad \text{Use the Pythagorean Formula with } a = h, b = 15, c = 17.$$
$$h^2 + 225 = 289$$
$$h^2 = 64$$
$$h = \sqrt{64} = 8$$

The deck is 8 feet above the pier.

EXAMPLE **4**

Determining the Length of a Support Beam

A construction engineer must install a support beam from the top of a 10-foot-high retainer wall to a point 4 feet from the base of the wall. (See Fig. 9.15.) How long must the beam be?

Figure 9.15

Solution

Let L = the length of the beam. The beam is the hypotenuse of a right triangle with legs 4 and 10.

$$L^2 = 4^2 + 10^2 \qquad \text{Use the Pythagorean Formula with } a = 4, b = 10, c = L.$$
$$L^2 = 16 + 100$$
$$L^2 = 116$$
$$L = \sqrt{116} \approx 10.77$$

The beam is approximately 10.77 feet long.

EXAMPLE **5**

Determining the Distance Between Two Vehicles

A car leaves Sylvan Grove and travels south at 60 mph. At the same time, a truck leaves the same point and travels east at 56 mph. (See Fig. 9.16.) How far apart are the car and the truck after 90 minutes?

Figure 9.16

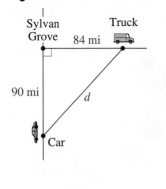

Solution

Because 90 minutes is 1.5 hours, the car traveled 60(1.5) or 90 miles and the truck traveled 56(1.5) or 84 miles.

Let d = the distance between the car and truck. Note that d is the hypotenuse of a right triangle with legs 90 and 84.

$$d^2 = 90^2 + 84^2$$
$$d^2 = 8100 + 7056$$
$$d^2 = 15{,}156$$
$$d = \sqrt{15{,}156} \approx 123.11$$

The car and the truck are approximately 123.11 miles apart.

Quick Reference 9.6

Lengths of Sides

- If the legs of a right triangle have lengths a and b and the hypotenuse has length c, then the Pythagorean Theorem states that $a^2 + b^2 = c^2$.
- If the lengths of any two sides of a right triangle are known, then the length of the third side can be determined from the Pythagorean Theorem.
- Calculations of lengths often involve square root radicals. We use principal square roots because length is a positive measure.

Speaking the Language 9.6

1. The longest side of a right triangle is called the ▨▨▨▨ .

2. The relationship among the sides of a right triangle is given by the ▨▨▨▨ Theorem.

3. If square roots are involved in calculating the lengths of the sides of a right triangle, we use ▨▨▨▨ square roots because length is positive.

4. The sides that form the right angle of a right triangle are called ▨▨▨▨ .

Exercises 9.6

Concepts and Skills

1. If a triangle is a right triangle, how can you identify the side that is the hypotenuse?

2. Explain the difference between the answers to the following two problems.

   (a) What is a number whose square is 25?

   (b) If the square of the length of a hypotenuse is 25, how long is the hypotenuse?

In Exercises 3–12, determine the length of the indicated side. Express decimal answers to the nearest hundredth.

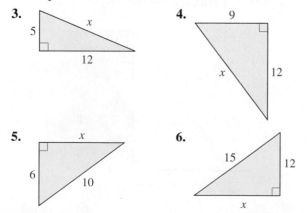

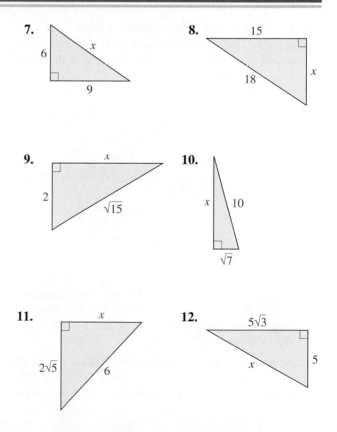

In Exercises 13–22, a and b represent the lengths of the legs of a right triangle, and c represents the length of the hypotenuse. Determine the length of the side that is not given. Express decimal answers to the nearest hundredth.

13. $a = 8, b = 15$ **14.** $a = 15, c = 25$

15. $b = 12, c = 13$ **16.** $a = 10, b = 24$

17. $b = 5, c = 11$ **18.** $b = 9, c = 12$

19. $a = 6\sqrt{2}, c = 10$ **20.** $a = 1, c = \sqrt{10}$

21. $a = \sqrt{7}, b = \sqrt{7}$ **22.** $b = 2, c = 4\sqrt{3}$

🌐 Real-Life Applications

In Exercises 23–44, express all decimal answers to the nearest hundredth.

23. A builder must construct a ramp from a loading platform at a warehouse to a point 12 feet from the building. How long is the ramp if the loading platform is 5 feet above the ground?

24. A guy wire is connected to the top of a 24-foot pole. The other end of the wire is attached to an anchor in the ground 10 feet from the base of the pole. How long is the wire?

25. To determine the distance across a pond, a surveyor measured the distances from a point A to two points B and C, which are on opposite sides of the pond. How wide is the pond?

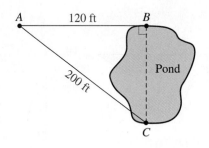

26. The top of a 17-foot ladder rests against the ledge of a window. The base of the ladder is 8 feet from the base of the wall. How high is the window ledge above the ground?

27. At a county fair, a hot air balloon taking people up for a view of the surrounding countryside was tethered to a point on the ground by 200 feet of rope.

If the rope was fully extended and the air current caused the balloon to drift 70 feet from the launch site, how high above the ground was the balloon?

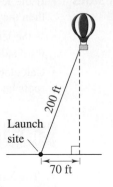

28. A person is stranded on an island 3 miles from a fishing village. The distance from the village to a point on the shore directly across from the island is 2 miles. How far would the person have to swim to get to that point on the shore?

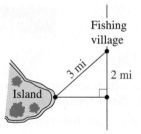

29. A platoon of Army Rangers must position a cable from the ledge of a cliff 60 feet above a 30-foot-wide river so that the other Rangers can slide down the cable from the cliff to a point on the ground that is 10 feet away from the river on the other side. How long must the cable be?

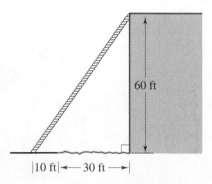

30. Two cars left Santa Fe at the same time. One traveled east at 62 mph, and the other traveled north at 54 mph. After 1 hour how far apart were the cars?

31. Two joggers left the same point at the same time. One went south at 8 mph, and the other went west at 6 mph. How far apart were they after 2 hours?

32. A person rode a bicycle west at 12 mph for two hours and then walked south at 4 mph for 1.5 hours. How far was the person from the starting point at the end of 3.5 hours?

33. A trucker left San Antonio and drove north at 65 mph for 1 hour and then east at 60 mph for two hours. How far from San Antonio was the trucker after 3 hours?

34. A boat left port and traveled south at 30 mph. One hour later a helicopter left the same point and traveled west at 70 mph. What was the distance between the boat and the helicopter 1 hour after the helicopter took off?

35. A small plane left Topeka and flew west at 120 mph. Another plane left the same point 30 minutes later and flew north at 110 mph. How far apart were the planes 1.5 hours after the first one took off?

36. How long is a diagonal path through a 10-meter by 8-meter rectangular garden?

37. A rectangular fence gate is 5 feet by 12 feet. How long is its diagonal support?

38. A square table is 5 feet on each side. How far is it from the center of the table to any corner of the table?

39. As part of a display, the director of a history museum plans to erect a tepee with a diameter of 8 feet and with 8-foot support poles. How high is the tepee at the center?

Figure for 39   **Figure for 40**

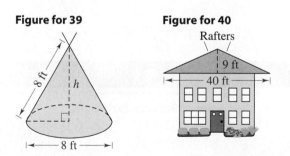

40. The gable end of a house spans 40 feet, and the center of the roof is 9 feet above the bottom of the rafters. How long are the rafters?

41. A third baseman caught a ground ball on the foul line 10 feet behind third base and threw the ball to first base. How far did he throw the ball? (The distance between bases is 90 feet.)

42. A pitcher fielded a bunt on the baseline exactly halfway between home plate and first base. How far was the throw to put out the lead runner at second base? (The distance between bases is 90 feet.)

43. The size of a TV screen is usually given by the length of the diagonal of the screen. What are the dimensions of a square TV screen with a 30-inch diagonal?

44. An airplane is approaching an airport for landing. The plane's altitude is 1200 feet, and the distance from a point on the ground directly beneath it to the airport is 1600 feet. What is the flying distance from the plane to the airport?

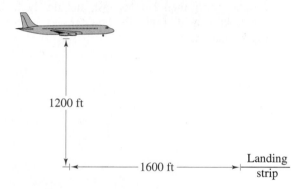

45. Plot the points $A(1, 2)$, $B(6, 5)$, and $C(6, 2)$, and connect the points with line segments. How do you know that the resulting triangle is a right triangle?

46. For the right triangle described in Exercise 45, explain how to determine the lengths of the legs. How can this information be used to determine the distance between points A and B?

In Exercises 47–52, determine the distance between the two given points of a plane. (*Hint*: Plot the points in the coordinate plane. Sketch a right triangle such that the line segment connecting the points is the hypotenuse. Then use the Pythagorean Theorem.)

47. $(-2, 1)$, $(2, 4)$ **48.** $(1, -8)$, $(6, 4)$
49. $(2, 10)$, $(10, 4)$ **50.** $(-6, -1)$, $(-3, -5)$
51. $(3, -2)$, $(7, -8)$ **52.** $(2, 5)$, $(5, 2)$

Challenge

53. Two guy wires attached to the top of a 24-foot pole are anchored at ground level. One guy wire is 25 feet long, and the other is 26 feet long. How far apart are the wires at the ground?

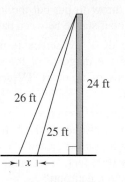

54. A person left Lancaster and drove 3 miles east, then 5 miles north, then 2 miles east, and finally 1 mile south. At the end of this trip, how far was the person from Lancaster?

55. A square with sides 10 feet long is inscribed in a circle. (The corners of the square are points of the circle.) What is the total area inside the circle but outside the square?

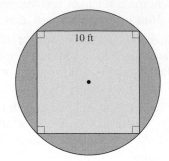

56. A truck pulling a 14-foot-wide mobile home enters an 18-foot-wide underpass constructed with a semicircular arch that begins 8 feet above the roadway. How high can the mobile home be to drive through the underpass?

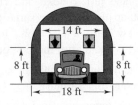

57. (a) An isosceles right triangle has a hypotenuse with a length of 12 feet. How long are the legs of the triangle?

   (b) Suppose that an isosceles right triangle has a hypotenuse of length c. Write a formula for the length of the legs.

9.7 | Rational Exponents

Definition of $b^{1/n}$ • Definition of $b^{m/n}$ • Properties of Rational Exponents

Definition of $b^{1/n}$

We began our discussion of exponents in Chapter 1 by defining natural number exponents. Then we extended the definition to include all integer exponents. In this section we extend the definition of exponent to include all rational number exponents. We begin with an exponent in the form $1/n$, where n is a positive integer.

| **Exploring the Concept** | **Interpreting Exponents in the Form $1/n$** |

Figure 9.17 shows the calculations of $25^{1/2}$, $27^{1/3}$, and $32^{1/5}$. For comparison purposes, we have placed the evaluations of related radical expressions next to the calculator display.

Figure 9.17

```
25^(1/2)
                    5
27^(1/3)
                    3
32^(1/5)
                    2
```

(i) $\sqrt{25} = 5$

(ii) $\sqrt[3]{27} = 3$

(iii) $\sqrt[5]{32} = 2$

From these calculations, we can conclude the following.

$$25^{1/2} = \sqrt[2]{25} \qquad 27^{1/3} = \sqrt[3]{27} \qquad 32^{1/5} = \sqrt[5]{32}$$

Note that in each case the denominator of the exponent is the same as the index of the radical. Thus the general relationship appears to be $b^{1/n} = \sqrt[n]{b}$.

This conclusion is consistent with the properties of exponents discussed in Chapter 6. For example, consider the expression $9^{1/2}$. For the Power to a Power Rule to hold for rational exponents, the following must be true.

$$(9^{1/2})^2 = 9^{2 \cdot (1/2)} = 9^1 = 9$$

Because squaring $9^{1/2}$ results in 9, $9^{1/2}$ must be a square root of 9. We shall agree that $9^{1/2}$ is the *principal* square root of 9; that is, $9^{1/2} = \sqrt{9}$.

These observations lead to the formal definition of the exponent $1/n$.

Definition of $b^{1/n}$

For any integer $n > 1$ and any real number b for which $\sqrt[n]{b}$ is defined, $b^{1/n} = \sqrt[n]{b}$.

Note: In the definition of $b^{1/n}$, if n is even, b must be nonnegative.

In this section, we shall assume that all variables represent positive numbers so that all expressions are defined.

EXAMPLE   **1**   **Evaluating Expressions with Rational Exponents**

Write each expression in radical form and evaluate.

(a) $49^{1/2} = \sqrt{49} = 7$

(b) $(-4)^{1/2}$ is not a real number because $\sqrt{-4}$ is undefined.

(c) $(-8)^{1/3} = \sqrt[3]{-8} = -2$

(d) $16^{1/4} = \sqrt[4]{16} = 2$

Some calculators do not have a special key for higher-order radicals. However, by writing the radical with a rational exponent, we can use a calculator to evaluate higher-order radicals.

EXAMPLE 2

Using a Calculator to Evaluate Higher-Order Radicals

Use a calculator to evaluate each radical. Round the answer to the nearest hundredth.

(a) $\sqrt[4]{20}$ (b) $\sqrt[5]{320}$ (c) $\sqrt[7]{-85}$

Solution

We write each radical as an exponential expression and then use a calculator to evaluate the exponential expression. (See Fig. 9.18.)

Figure 9.18

```
20^(1/4)
            2.11
320^(1/5)
            3.17
(-85)^(1/7)
           -1.89
```

(a) $\sqrt[4]{20} = 20^{1/4}$ The index of the radical is the same as the denominator of the exponent.

(b) $\sqrt[5]{320} = 320^{1/5}$

(c) $\sqrt[7]{-85} = (-85)^{1/7}$

Definition of $b^{m/n}$

Think About It

The quantity $b^{1/n}$ is defined for integers $n > 1$. Determine whether your calculator will compute $\sqrt[0.5]{9}$. Write the expression with a rational exponent and explain why the calculator's result makes sense.

We can extend the definition of $b^{1/n}$ to include rational exponents with a numerator other than 1.

Consider $8^{2/3}$. To make the definition consistent with the Power to a Power Rule, we need $8^{2/3} = (8^{1/3})^2 = \left(\sqrt[3]{8}\right)^2$. This suggests the following definition of $b^{m/n}$.

Definition of a Rational Exponent

For positive integers m and n and any real number b for which $\sqrt[n]{b}$ is defined,

$b^{m/n} = (b^{1/n})^m = \left(\sqrt[n]{b}\right)^m$ and $b^{m/n} = (b^m)^{1/n} = \sqrt[n]{b^m}$.

Note: The definition of a rational exponent assumes that the fraction m/n cannot be reduced.

EXAMPLE 3

Evaluating Expressions with Rational Exponents

Evaluate each expression.

(a) $25^{3/2} = (25^{1/2})^3 = \left(\sqrt{25}\right)^3 = 5^3 = 125$

(b) $8^{5/3} = \left(\sqrt[3]{8}\right)^5 = 2^5 = 32$

(c) $-16^{3/4} = -\left(\sqrt[4]{16}\right)^3 = -2^3 = -8$

(d) $(-32)^{2/5} = \left(\sqrt[5]{-32}\right)^2 = (-2)^2 = 4$

We define a negative rational exponent in the same way as we defined a negative integer exponent.

Definition of a Negative Rational Exponent

For positive integers m and n and any nonzero real number b for which $\sqrt[n]{b}$ is defined, $b^{-m/n} = \dfrac{1}{b^{m/n}}$.

EXAMPLE **4**

Evaluating Expressions with Negative Rational Exponents

Evaluate each expression.

(a) $25^{-1/2} = \dfrac{1}{25^{1/2}} = \dfrac{1}{\sqrt{25}} = \dfrac{1}{5}$

(b) $16^{-3/4} = \dfrac{1}{16^{3/4}} = \dfrac{1}{\left(\sqrt[4]{16}\right)^3} = \dfrac{1}{2^3} = \dfrac{1}{8}$

(c) $8^{-4/3} = \dfrac{1}{8^{4/3}} = \dfrac{1}{\left(\sqrt[3]{8}\right)^4} = \dfrac{1}{2^4} = \dfrac{1}{16}$

(d) $(-8)^{-2/3} = \dfrac{1}{(-8)^{2/3}} = \dfrac{1}{\left(\sqrt[3]{-8}\right)^2} = \dfrac{1}{(-2)^2} = \dfrac{1}{4}$

Note: You can use your calculator to evaluate expressions with rational exponents. Be sure to enclose the rational exponent in parentheses.

Properties of Rational Exponents

All properties of exponents stated in Section 6.1 hold for rational exponents.

EXAMPLE **5**

Simplifying Expressions with Rational Exponents

Simplify each expression.

(a) $5^{1/3} \cdot 5^{4/3} = 5^{1/3+4/3} = 5^{5/3}$ Product Rule for Exponents

(b) $y^{-4/5} \cdot y^{2/5} = y^{-4/5+2/5} = y^{-2/5} = \dfrac{1}{y^{2/5}}$

(c) $\dfrac{x^{3/4}}{x^{-1/4}} = x^{3/4-(-1/4)} = x^{3/4+1/4} = x^1 = x$ Quotient Rule for Exponents

(d) $\dfrac{y}{y^{2/3}} = y^{1-2/3} = y^{3/3-2/3} = y^{1/3}$

(e) $(8^{1/6})^2 = 8^{2 \cdot 1/6} = 8^{1/3} = \sqrt[3]{8} = 2$ Power to a Power Rule

(f) $(y^2)^{-3/4} = y^{2(-3/4)} = y^{-3/2} = \dfrac{1}{y^{3/2}}$

EXAMPLE **6**   **Simplifying Expressions with Rational Exponents**

Simplify each expression.

(a) $(x^3 y^{-1/2})^4 = x^{3 \cdot 4} y^{4(-1/2)} = x^{12} y^{-2} = \dfrac{x^{12}}{y^2}$ Product to a Power Rule

(b) $\left(\dfrac{x^3}{y^9}\right)^{-2/3} = \dfrac{x^{3(-2/3)}}{y^{9(-2/3)}} = \dfrac{x^{-2}}{y^{-6}} = \dfrac{y^6}{x^2}$ Quotient to a Power Rule

Quick Reference 9.7

Definition of $b^{1/n}$
- We define $b^{1/n}$ as follows.

  For any integer $n > 1$ and any real number b for which $\sqrt[n]{b}$ is defined, $b^{1/n} = \sqrt[n]{b}$. If n is even, b must be nonnegative.

- By writing higher-order radicals with rational exponents, we can use a calculator to perform the evaluations.

Definition of $b^{m/n}$
- We define $b^{m/n}$ as follows.

  For positive integers m and n and any real number b for which $\sqrt[n]{b}$ is defined,
  $$b^{m/n} = (b^{1/n})^m = \left(\sqrt[n]{b}\right)^m \qquad \text{and} \qquad b^{m/n} = (b^m)^{1/n} = \sqrt[n]{b^m}$$

- A negative rational exponent is defined as follows.

  For positive integers m and n and any nonzero real number b for which $\sqrt[n]{b}$ is defined,
  $$b^{-m/n} = \frac{1}{b^{m/n}}$$

Properties of Rational Exponents
- All properties of exponents stated in Section 6.1 hold for rational exponents.

Speaking the Language 9.7

1. Rational exponents are defined in terms of ▭▭▭▭ expressions.
2. When we write the expression $b^{m/n}$ as a radical, n is a(n) ▭▭▭▭ and m is a(n) ▭▭▭▭ .
3. The expression $b^{-m/n}$ is equivalent to the ▭▭▭▭ of $b^{m/n}$.
4. When we write $(-4)^{-3/2}$ with a positive exponent, the sign of the ▭▭▭▭ changes, but the sign of the ▭▭▭▭ does not change.

Exercises 9.7

Concepts and Skills

1. Explain why $(-25)^{1/2}$ is not a real number.

2. You can probably compute square roots and cube roots directly with your calculator. Describe a way to use a rational exponent to compute a fourth root on your calculator.

In Exercises 3–12, evaluate the expression.

3. $100^{1/2}$

4. $81^{1/2}$

5. $8^{1/3}$

6. $(-64)^{1/3}$

7. $(-1)^{1/6}$

8. $(-16)^{1/2}$

9. $(-32)^{1/5}$

10. $1^{1/5}$

11. $16^{1/4}$

12. $81^{1/4}$

In Exercises 13–18, use a calculator to evaluate the expression. Round the answer to the nearest hundredth.

13. $\sqrt[4]{96}$

14. $\sqrt[5]{30}$

15. $\sqrt[3]{-30}$

16. $\sqrt[7]{-250}$

17. $\sqrt[6]{795}$

18. $\sqrt[6]{1556}$

19. In the following list, identify the expression that is not a real number and explain why it is not. Then evaluate the other expressions.

   (i) $-4^{3/2}$

   (ii) $4^{-3/2}$

   (iii) $-4^{-3/2}$

   (iv) $(-4)^{-3/2}$

20. Evaluate $(-8)^{-1/3}$ and $\dfrac{1}{8^{1/3}}$. Explain why the results are not equal.

In Exercises 21–34, evaluate the expression.

21. $9^{3/2}$

22. $4^{5/2}$

23. $27^{2/3}$

24. $8^{4/3}$

25. $(-32)^{2/5}$

26. $(-27)^{2/3}$

27. $(-64)^{2/3}$

28. $(-32)^{3/5}$

29. $81^{3/4}$

30. $16^{5/4}$

31. $\left(\dfrac{16}{25}\right)^{3/2}$

32. $\left(\dfrac{1}{8}\right)^{4/3}$

33. $64^{4/3}$

34. $1000^{4/3}$

In Exercises 35–44, evaluate the expression.

35. $16^{-1/2}$

36. $16^{-1/4}$

37. $8^{-2/3}$

38. $32^{-3/5}$

39. $(-8)^{-5/3}$

40. $(-27)^{-2/3}$

41. $\left(\dfrac{27}{8}\right)^{-4/3}$

42. $\left(\dfrac{64}{125}\right)^{-2/3}$

43. $4^{-5/2}$

44. $1000^{-2/3}$

In Exercises 45–50, use a calculator to evaluate the expression. Round the answer to the nearest hundredth.

45. $50^{2/3}$

46. $318^{2/5}$

47. $2^{-5/2}$

48. $7^{-3/7}$

49. $(-55)^{4/5}$

50. $(-130)^{4/3}$

In Exercises 51–68, simplify the expression. Write the result with a positive exponent.

51. $7^{3/5} \cdot 7^{2/5}$

52. $x^{1/3} \cdot x^{2/3}$

53. $a^{-1/2} \cdot a^{3/2}$

54. $7^{-3/2} \cdot 7^{-1/2}$

55. $y^{1/2} \cdot y^{1/3}$

56. $x^{1/2} \cdot x$

57. $(5^3)^{2/3}$

58. $(2^5)^{3/5}$

59. $(y^6)^{1/3}$

60. $(x^{-4})^{3/2}$

61. $(a^{2/3})^{6/5}$

62. $(z^{-3/5})^{10/3}$

63. $\dfrac{10^{4/3}}{10^{1/3}}$

64. $\dfrac{6^{3/5}}{6^{8/5}}$

65. $\dfrac{x}{x^{1/2}}$

66. $\dfrac{y^{1/3}}{y^{1/2}}$

67. $\dfrac{z^{3/2}}{z^{-1/2}}$

68. $\dfrac{y^{-6/5}}{y^{1/5}}$

In Exercises 69–80, simplify the expression. Write the results with positive exponents.

69. $(36x^{10})^{1/2}$

70. $(8y^{12})^{1/3}$

71. $(9x^{-6})^{-1/2}$

72. $(x^{-4}y^{-8})^{-1/4}$

73. $(y^{3/4}z^{-1/2})^8$

74. $(a^{-2/3}b^{5/4})^{12}$

75. $\left(\dfrac{y^3}{8}\right)^{1/3}$

76. $\left(\dfrac{16}{x^{12}}\right)^{1/4}$

77. $\left(\dfrac{a^{-3}}{b^{-9}}\right)^{2/3}$

78. $\left(\dfrac{x^{-4}}{y^2}\right)^{-3/2}$

79. $\left(\dfrac{y^{2/3}}{x^{-1/3}}\right)^6$

80. $\left(\dfrac{a^{3/4}}{b^{-1/2}}\right)^8$

🌐 Real-Life Applications

The speed of a car v (in mph) can be estimated by the formula $v = 3.5d^{1/2}$, where d is the stopping distance (in feet) for a car on wet pavement. In Exercises 81 and 82, use this formula to answer the questions.

81. Following an accident, police find that the car involved left skid marks 230 feet long. What was the approximate speed of the car?

82. A driver applied the brakes 100 feet from an intersection. What was the approximate speed of the car if it came to a stop at the intersection?

In Exercises 83 and 84, produce and trace the graph of $v = 3.5d^{1/2}$ to estimate the answers to the questions.

83. If a car moving at 45 mph makes a sudden stop, how many feet will the car go before it stops?

84. A driver traveling at 33 mph sees a stopped school bus 100 feet ahead and brakes. How far is the car from the bus when the car comes to a stop?

The time t (in seconds) that it takes an object to fall d feet is given by $t = \dfrac{d^{1/2}}{4}$. In Exercises 85 and 86, use this formula to answer the questions.

85. A person on a sidewalk sees a worker drop a bucket of paint from a scaffold 450 feet directly above. How many seconds does the person on the sidewalk have to avoid being hit by the bucket?

86. The observation deck of the Seattle Space Needle is 520 feet above the ground. If a tourist dropped a camera from the observation deck, how many seconds would the camera take to hit the ground?

In Exercises 87 and 88, produce and trace the graph of $t = \dfrac{d^{1/2}}{4}$ to estimate the answer to the questions.

87. To estimate the height of a bridge above the water, a person drops a rock from the bridge. If the rock takes 2 seconds to hit the water, how high is the bridge?

88. A diver jumps from a cliff. If the diver takes 2.3 seconds to reach the water, how high is the cliff?

Data Analysis: *Visitors to the National Park Service System*

The bar graph shows the number of visitors (in millions) to the National Park Service system during the period 1994–1999. (Source: National Park Service.)

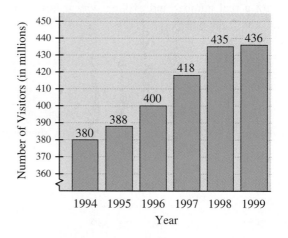

The number of visitors (in millions) to the National Park Service system can be modeled by the following equation, where x is the number of years since 1990.

$$V = 284x^{1/5}$$

89. Write the model equation in radical form. According to the model, how many visitors are predicted in 2005?

90. In 1992, 360 million people visited the park system. Produce the graph of V and trace it to estimate the year in which the model projects a 30% increase in the number of visitors over the 1992 level.

91. Suppose that the National Park Service decided to charge an entrance fee of $5 per vehicle at all system sites. If a survey determined that each car carried an average of 2.8 people, write an expression R to describe the expected revenue that would be generated.

92. Use a table of values for R to estimate the year in which the projected revenue would reach $900 million.

Challenge

In Exercises 93–96, write each radical as an expression with rational exponents. Then use the properties of exponents to simplify the expression and write the result as a radical. Assume that $x > 0$.

93. $\sqrt[4]{x^2}$

94. $\left(\sqrt[4]{x}\right)^{10}$

95. $\sqrt[12]{9^3 x^6}$

96. $\sqrt{\sqrt[3]{x}}$

In Exercises 97 and 98, solve the equation.

97. $y^{-1/2} = 7$

98. $(x - 4)^{1/3} = 4$

Chapter 9 Review Exercises

In all radical expressions, assume that variables represent positive numbers unless otherwise indicated.

Section 9.1

1. (a) Determine the square roots of 16.

   (b) Evaluate $\sqrt{16}$.

2. Identify the numbers in the list that are not rational numbers and explain why they are not.

   (i) $-\sqrt{9}$ (ii) $\sqrt{-9}$

   (iii) $\sqrt{\dfrac{25}{49}}$ (iv) $\sqrt{15}$

3. Evaluate each expression to the nearest hundredth.

   (a) $\sqrt{17} - \sqrt{5}$ (b) $\sqrt{17 - 5}$

4. Simplify.

   (a) $\left(3\sqrt{5}\right)^2$ (b) $\left(\sqrt{x^2}\right)^2$

5. Simplify.

   (a) $\sqrt{(-2)^2}$ (b) $\sqrt{(x + 3)^2}$

6. (a) Determine the cube root of -8.

   (b) Evaluate $\sqrt[3]{-8}$.

7. Evaluate.

   (a) $-\sqrt[5]{32}$ (b) $\sqrt[5]{-32}$

8. Simplify.

   (a) $\sqrt[3]{x^{12}}$ (b) $\sqrt[4]{(3y)^8}$

9. Simplify $\sqrt{x^2}$, where x is any real number.

10. Suppose that a is any real number and $\sqrt[n]{a^n} = a$. What do you know about n? Why?

Section 9.2

11. Multiply $\sqrt{2x}\,\sqrt{3y}$.

12. Simplify $\sqrt{4x^8 y^4}$.

13. Simplify.

   (a) $\sqrt{18}$ (b) $3\sqrt{28}$

14. Simplify.

   (a) $\sqrt{w^{21}}$ (b) $\sqrt{20x^{12}}$

15. Simplify $\sqrt{50a^7 b^{11}}$.

16. Multiply and simplify $\sqrt{3x^5}\,\sqrt{6x^3}$.

17. Simplify.

   (a) $\sqrt[3]{x^8}$ (b) $\sqrt[4]{16x^5}$

18. Simplify $\sqrt{2xy^2}\,\sqrt{10x^2 y}$.

Section 9.3

In Exercises 19–28, simplify.

19. $\dfrac{\sqrt{2x^5}}{\sqrt{32x}}$ **20.** $\sqrt{\dfrac{z^{12}}{16}}$

21. $\sqrt{\dfrac{3y}{49}}$ **22.** $\sqrt{\dfrac{8x^5}{32}}$

23. $\sqrt{\dfrac{48ab^3}{49a^{11}b^3}}$

24. (a) $\sqrt{\dfrac{3}{5}}$ (b) $\sqrt{\dfrac{7}{x}}$

25. $\sqrt{\dfrac{24x^2}{5x^5}}$

26. (a) $\sqrt{\dfrac{a^2}{3}}$ (b) $\sqrt{\dfrac{25y^2}{x}}$

27. $\sqrt[3]{\dfrac{16x^4}{27}}$ **28.** $\dfrac{\sqrt[4]{32x^9}}{\sqrt[4]{2x}}$

Section 9.4

In Exercises 29–32, simplify.

29. $\sqrt{7} - 4\sqrt{7}$ **30.** $\sqrt{20} + \sqrt{45}$

31. $\sqrt{72} + 3\sqrt{32} - \sqrt{48}$

32. $\sqrt{8x} + \sqrt{50x}$

In Exercises 33–36, multiply and simplify.

33. $\sqrt{3}(\sqrt{5} + \sqrt{2})$

34. $\sqrt{5}(\sqrt{10} - 2\sqrt{5})$

35. $(\sqrt{7} - 2)(3\sqrt{2} + \sqrt{7})$

36. $(\sqrt{y} + 3x)^2$

37. Multiply $4 - \sqrt{2}$ by its conjugate.

38. Rationalize the denominator: $\dfrac{\sqrt{2}}{4 + \sqrt{7}}$.

Section 9.5

39. Use the graphing method to estimate the solution(s) of $\sqrt{2x - 3} = x - 5$.

40. If we solve the equation $\sqrt{x} = -2$ algebraically, we obtain $x = 4$. What do we call 4? Why?

In Exercises 41–46, solve the equation.

41. $\sqrt{3x - 4} = \sqrt{x + 6}$

42. $\sqrt{x^2 - 4} = x - 2$

43. $\sqrt{6x + 19} - x = 2$

44. $x = \sqrt{6x - 21} + 2$

45. $\sqrt{x + 5} + \sqrt{x} = 5$

46. $\sqrt{4 - x} = \sqrt{3x} + 2$

47. The square root of 1 more than a number is 1 less than the number. What is the number?

48. The width (in feet) of a rectangle is the square root of a number. The rectangle is 3 feet longer than it is wide, and its area is 4 square feet. How long is the rectangle?

Section 9.6

49. What is the length of the hypotenuse of a right triangle with legs of 4 inches and $5\sqrt{2}$ inches? Round your answer to the nearest hundredth.

50. One cyclist left Milton and traveled north 20 miles. She then rode east to Cannondale, a distance of 30 miles. A second cyclist rode directly from Milton to Cannondale. What was the difference in mileage for the two cyclists?

51. A ham radio operator has a 30-foot antenna with a support wire running from the top of the antenna to an anchor in the ground. If the support wire is 31 feet long, how far is the anchor from the bottom of the antenna?

52. A football field, not including the end zones, is 100 yards long and 160 feet wide. If a player can run the length of the field in 10 seconds, how long would it take him to run diagonally across the field?

Section 9.7

53. Evaluate.

(a) $\left(\dfrac{9}{16}\right)^{1/2}$ (b) $-81^{1/4}$

54. Evaluate the expression to the nearest hundredth.

(a) $\sqrt[6]{27}$ (b) $\sqrt[5]{-80}$

55. Evaluate.

(a) $64^{5/3}$ (b) $(-16)^{3/4}$

56. Evaluate the expression to the nearest hundredth.

(a) $27^{-2/3}$ (b) $(-32)^{-3/5}$

In Exercises 57–60, simplify the expression and express the result with a positive exponent. Assume that all variables represent positive numbers.

57. $25^{3/4} \cdot 25^{-1/4}$

58. $(x^{1/2})^{-4/3}$

59. $\dfrac{x^{1/3}}{x^{1/4}}$

60. $(x^{-3/7}y^{2/7})^{14}$

61. How would you rewrite $2x^{1/2} = 15$ in order to solve the equation with known methods?

Chapter 9 Test

In all radical expressions, assume that variables represent positive numbers unless otherwise indicated.

1. Evaluate.

   (a) $\sqrt{(-3)^2}$

   (b) $\left(\sqrt{-3}\right)^2$

2. Simplify $\left(2\sqrt{x}\right)^2$.

3. For what values of n is $\sqrt[n]{-10}$ a real number? Explain.

4. Simplify.

   (a) $3\sqrt{32}$ (b) $\sqrt{x^9}$

5. Multiply and simplify $\sqrt{2a^3}\sqrt{6a^5}$.

6. Simplify $\sqrt[3]{27x^5}$.

7. Simplify.

   (a) $\sqrt{\dfrac{28x^3}{7}}$ (b) $\sqrt{\dfrac{2}{7}}$

8. Simplify $\sqrt{\dfrac{28x^6}{49}}$.

9. Simplify $\dfrac{\sqrt[3]{16x}}{\sqrt[3]{2x^4}}$.

10. Combine like terms: $\sqrt{18} - \sqrt{2} + \sqrt{50}$.

11. Multiply and simplify: $\left(\sqrt{3} - \sqrt{5}\right)\left(\sqrt{3} + 2\sqrt{5}\right)$.

12. Rationalize the denominator: $\dfrac{x}{x + \sqrt{3}}$.

13. Use the graphing method to estimate to the nearest tenth the solution(s) of $\sqrt{3x + 2} = 6 - x$.

14. Explain how you can tell without solving that the equation $\sqrt{x - 2} = -1$ has no solution.

15. Solve $\sqrt{x^2 + 1} = x - 1$.

16. Solve $2\sqrt{x + 6} - x = 3$.

17. The lengths of the two legs of a right triangle are 5 inches and 11 inches. To the nearest tenth, how long is the hypotenuse?

18. A confused baseball player hits the ball and runs to third base. Once there, he realizes his mistake and runs directly to first base. If the bases are 90 feet apart, what was the total distance that he ran?

19. Explain how to write $x^{2/3}$ in radical form.

20. Evaluate the expression.

   (a) $\left(\dfrac{4}{25}\right)^{1/2}$ (b) $(-8)^{4/3}$

21. Evaluate the expression to the nearest tenth.

   (a) $\sqrt[5]{26}$ (b) $11^{-2/5}$

22. Simplify.

   (a) $3^{1/2} \cdot 3^{-3/2}$ (b) $\left(x^{-1/3}y^{2/3}\right)^6$

The accompanying bar graph shows that the number of commuter rail passengers declined from 1990 to 1992, but then ridership began to increase. These data can be modeled by a quadratic equation whose graph is a ***parabola.***

Graphing and algebraic methods can be used to estimate the years for which given levels of ridership are projected. Corporations and investors rely on such information for long-range planning. (For more on this real data problem, see Exercises 79–82 in Section 10.3.)

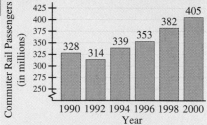

(Source: American Public Transit Association.)

Chapter 10

Quadratic Equations and Functions

Chapter Snapshot

In this final chapter, we introduce additional algebraic methods for solving quadratic equations: the Square Root Property, the method of completing the square, and the Quadratic Formula. We also expand our number system by defining imaginary numbers. We learn how to perform operations with these numbers and how to solve quadratic equations with imaginary number solutions. After a discussion of parabolas—the graphs of quadratic equations—we conclude with a brief treatment of functions, a topic that is central to further study of mathematics.

Warm-Up Skills

The following exercises review concepts and skills that you will need in Chapter 10.

In Exercises 1 and 2, factor the trinomial.

1. $x^2 + 8x + 16$
2. $x^2 - 6x + 9$

In Exercises 3–6, solve the equation.

3. $x^2 + 10x + 25 = 0$
4. $x^2 + 21 = 10x$

5. $x^2 - 9 = 0$
6. $8 + 7x - x^2 = 0$

In Exercises 7 and 8, evaluate the following expression for the given values of the variables.

$$\frac{-b + \sqrt{b^2 - 4ac}}{2a}$$

7. $a = 3, b = 2, c = -5$
8. $a = 9, b = -6, c = 1$

9. Explain how to determine the intercepts of the graph of a linear equation in two variables.

10. Determine the x- and y-intercepts of the graph of the equation $x + 2y = 10$.

11. What is the restricted value of $\dfrac{1}{x + 2}$?

12. Evaluate the polynomial $p(x) = 2x^2 - 3x + 5$ as indicated.

    (a) $p(4)$ (b) $p(0)$ (c) $p(-1)$

10.1 | Special Methods

Graphing and Factoring Methods · The Square Root Property

Graphing and Factoring Methods

Recall that a **quadratic equation** is an equation that can be written in the *standard form* $ax^2 + bx + c = 0$, where a, b, and c are real numbers with $a \neq 0$. We call ax^2 the *quadratic term*, bx the *linear term*, and c the *constant term*. The number a is called the *leading coefficient*.

We first encountered quadratic equations in Chapter 7, where we estimated solutions by graphing and solved algebraically by factoring. We begin with a review of those methods.

Note: Later in this chapter, we will find that a quadratic equation can have solutions that are not real numbers. Until then, our discussion of solutions will always be in reference to *real number solutions*.

EXAMPLE 1

Solving Quadratic Equations by Graphing and Factoring

Estimate the solution(s) of each equation by graphing. Then use the factoring method, if possible, to determine the solution algebraically.

(a) $x^2 - 11x + 24 = 0$ (b) $4x^2 + 20x + 25 = 0$

(c) $2 - 3x - x^2 = 0$ (d) $x^2 - x + 2 = 0$

Solution

(a) To estimate the solutions, produce the graph of $x^2 - 11x + 24$ and trace to the points where the y-coordinate is 0—that is, to the x-intercepts.

From the graph in Figure 10.1, we see that the equation has two solutions. By tracing the graph we estimate that the solutions are 8 and 3.

Figure 10.1

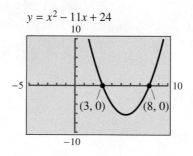

$y = x^2 - 11x + 24$

We can also solve the equation algebraically by factoring.

$x^2 - 11x + 24 = 0$	The equation is in standard form.
$(x - 8)(x - 3) = 0$	Factor.
$x - 8 = 0$ or $x - 3 = 0$	Zero Factor Property
$x = 8$ or $x = 3$	Solve for x.

As we estimated, the two solutions are 8 and 3.

Figure 10.2

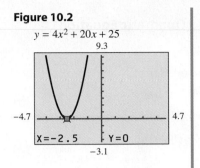

$y = 4x^2 + 20x + 25$

(b) Produce the graph of $4x^2 + 20x + 25$. (See Fig. 10.2.)

Because the graph appears to have only one x-intercept, we estimate that the equation has only one solution, -2.5. Now solve the equation by factoring.

$4x^2 + 20x + 25 = 0$	The equation is in standard form.
$(2x + 5)^2 = 0$	The trinomial is a perfect square.
$2x + 5 = 0$	Zero Factor Property
$2x = -5$	
$x = -2.5$	

The equation has only one solution, -2.5.

(c) Produce the graph of $2 - 3x - x^2$.

From the graph in Figure 10.3 we estimate that the equation has two solutions, -3.54 and 0.54. The equation cannot be solved by factoring.

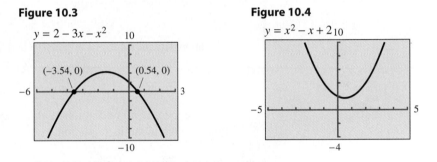

Figure 10.3

$y = 2 - 3x - x^2$

Figure 10.4

$y = x^2 - x + 2$

(d) Produce the graph of $x^2 - x + 2$. (See Fig. 10.4.)

From the graph, we see that the equation has no solution.

Example 1 illustrates the fact that in the real number system, a quadratic equation can have two solutions, one solution, or no solution.

Note: Our inability to factor has no bearing on the number of solutions a quadratic equation may have. For instance, in parts (c) and (d) of Example 1, we were unable to factor. Whereas the equation in part (d) has no solution, the equation in part (c) has two solutions.

The Square Root Property

In Section 9.1 we stated that a is a square root of a nonnegative number b if $a^2 = b$. If we state this definition in terms of radicals, we have the Square Root Property.

> **Square Root Property**
>
> For any nonnegative real number B,
>
> $$\text{if } A^2 = B, \text{ then } A = \sqrt{B} \quad \text{or} \quad A = -\sqrt{B}.$$

We can use the Square Root Property to solve any quadratic equation that can be written in the form $A^2 = B$, even though the equation cannot be solved by factoring.

EXAMPLE 2

Using the Square Root Property to Solve a Quadratic Equation

Use the Square Root Property to solve each equation.

(a) $-5x^2 + 45 = 0$ Isolate the variable term.

$$-5x^2 = -45 \qquad \text{Subtract 45 from both sides.}$$

$$x^2 = 9 \qquad \text{Divide both sides by } -5.$$

$x = \sqrt{9}$ or $x = -\sqrt{9}$ Square Root Property

$x = 3$ or $x = -3$

The solutions are 3 and -3.

LEARNING TIP

A common error is to write $y^2 = 10$ and then $y = \sqrt{10}$. Remember that the Square Root Property provides for two solutions.

(b) $3y^2 - 21 = 9$

$$3y^2 = 30 \qquad \text{Add 21 to both sides of the equation.}$$

$$y^2 = 10 \qquad \text{Divide both sides by 3.}$$

$y = \sqrt{10}$ or $y = -\sqrt{10}$ Square Root Property

$y \approx 3.16$ or $y \approx -3.16$

The approximate solutions are -3.16 and 3.16.

(c) $x^2 + 16 = 0$

$$x^2 = -16$$

Because there is no real number whose square is a negative number, the equation has no real number solution.

Note: For convenience we often combine the positive and negative symbols to write the two square roots of a number. For example, the solutions in part (a) of Example 2 can be written ± 3, which means $+3$ or -3.

When we write an equation in the form $A^2 = B$, the squared term may be a binomial.

EXAMPLE 3

Using the Square Root Property to Solve a Quadratic Equation

Use the Square Root Property to solve each equation.

Think About It

Are the equations $(7 - 2x)^2 = (2x - 7)^2$ and $(2x + 1)^2 = (3 + 2x)^2$ quadratic equations? How many solutions does each equation have?

(a) $(x + 4)^2 = 49$ The equation is in the form $A^2 = B$.

$x + 4 = \sqrt{49}$ or $x + 4 = -\sqrt{49}$ Square Root Property

$x + 4 = 7$ or $x + 4 = -7$ Solve each equation for x.

$x = 3$ or $x = -11$

The solutions are 3 and -11.

(b) $(t - 1)^2 = 15$ The equation is in the form $A^2 = B$.

$t - 1 = \sqrt{15}$ or $t - 1 = -\sqrt{15}$ Square Root Property

$t = 1 + \sqrt{15}$ or $t = 1 - \sqrt{15}$ Solve for t.

$t \approx 4.87$ or $t \approx -2.87$

The approximate solutions are 4.87 and -2.87.

To write the equation in the form $A^2 = B$, it may be necessary to factor.

EXAMPLE   **4** | **Using the Square Root Property to Solve a Quadratic Equation**

Use the Square Root Property to solve $x^2 - 8x + 16 = 100$.

Solution

The left side of the equation is a perfect square trinomial. Begin by factoring the trinomial to write the equation in the form $A^2 = B$.

$x^2 - 8x + 16 = 100$	Factor the left side of the equation.
$(x - 4)^2 = 100$	The equation is in the form $A^2 = B$.
$x - 4 = \sqrt{100}$ or $x - 4 = -\sqrt{100}$	Square Root Property
$x - 4 = 10$ or $x - 4 = -10$	Solve for x.
$x = 14$ or $x = -6$	

The solutions are 14 and -6.

Quick Reference 10.1

Graphing and Factoring Methods

- A **quadratic equation** is an equation that can be written in the *standard form* $ax^2 + bx + c = 0$, where a, b, and c are real numbers with $a \neq 0$.

- One way to estimate the solution(s) of a quadratic equation with the graphing method is to write the equation in standard form and graph the quadratic expression. The number of x-intercepts is the number of solutions that the equation has. The x-coordinates of the x-intercepts are the solutions.

- The previously discussed algebraic method for solving a quadratic equation is to write the equation in standard form, factor the quadratic expression, and apply the Zero Factor Property.

- In the real number system, a quadratic equation can have two solutions, one solution, or no solution.

The Square Root Property

- The Square Root Property is as follows.

  For any nonnegative real number B,

  if $A^2 = B$, then $A = \sqrt{B}$ or $A = -\sqrt{B}$.

- If a quadratic equation can be written in the form $A^2 = B$, then the Square Root Property can be applied to solve the equation.

Speaking the Language 10.1

1. An equation such as $2x^2 + 3x - 5 = 0$ is called a(n) ＿＿＿＿ equation.

2. If we graph the expression $x^2 - 2x - 15$, then the solutions of $x^2 - 2x - 15 = 0$ are represented by the ＿＿＿＿ of the graph.

3. The basis for the factoring method for solving a quadratic equation is the ＿＿＿＿ Property.

4. The ＿＿＿＿ Property guarantees that if $x^2 = 9$, then $x = 3$ or $x = -3$.

Exercises 10.1

Concepts and Skills

1. When we use graphing to estimate solutions of a quadratic equation in standard form, to what point(s) of the graph do we trace? Why?

2. To use the Zero Factor Property to solve a quadratic equation in standard form, we must factor the quadratic expression. Why?

In Exercises 3–10, use the graphing method to estimate the solution(s) of the equation.

3. $x^2 + 6x + 9 = 0$ 4. $x^2 - 8x + 16 = 0$

5. $3 - 4x - x^2 = 0$ 6. $4 + x - x^2 = 0$

7. $x^2 - 5x - 6 = 0$ 8. $x^2 - 6x + 8 = 0$

9. $-x^2 - 8x = 18$ 10. $-x^2 = 6x + 10$

In Exercises 11–24, use the factoring method to solve the equation.

11. $x^2 + 6x + 5 = 0$ 12. $x^2 - 11x + 24 = 0$

13. $49 = 14x - x^2$ 14. $9x^2 = 12x - 4$

15. $6x^2 - 3x = 0$ 16. $6x - 8x^2 = 0$

17. $x^2 - x = 12$ 18. $x^2 + x = 20$

19. $25 - 10y + y^2 = 0$

20. $64x^2 + 48x + 9 = 0$

21. $t^2 + 4t = 60$

22. $y^2 + 60 = 16y$

23. $12x^2 + 15 = 28x$ 24. $8t^2 = 15 + 37t$

25. Describe two algebraic methods for solving the equation $x^2 = 49$. Which method is easier?

26. We say that $x^2 - 5 = 0$ cannot be solved by factoring because $x^2 - 5$ is prime. Does this mean that the equation has no solution? Why? (Give two reasons to support your answer.)

In Exercises 27–38, solve the equation by using the Square Root Property.

27. $x^2 = 49$ 28. $36 - x^2 = 0$

29. $4t^2 - 25 = 0$ 30. $2y^2 = 72$

31. $x^2 + 25 = 0$ 32. $3 + t^2 = 0$

33. $14 - 4z^2 = 0$ 34. $x^2 - 5 = 13$

35. $\frac{1}{5}x^2 = 25$ 36. $\frac{1}{7}y^2 = 49$

37. $7 = -w^2$ 38. $-2 - 5y^2 = 0$

In Exercises 39–50, use the Square Root Property to solve the equation.

39. $(x + 5)^2 = 9$ 40. $(x - 8)^2 = 16$

41. $(w - 9)^2 = 12$ 42. $(3x - 1)^2 = 64$

43. $(x - 6)^2 + 25 = 0$ 44. $(2x - 1)^2 + 10 = 0$

45. $(3 - x)^2 = \frac{49}{16}$ 46. $(4 - x)^2 = \frac{36}{25}$

47. $(3x - 4)^2 = 36$ 48. $(4x - 1)^2 = 27$

49. $9(w + 3)^2 - 4 = 0$ 50. $16(w - 4)^2 - 9 = 0$

51. Which of the following equations is easier to solve with the Square Root Property? Why?

  (i) $x^2 + 4x + 4 = 1$

  (ii) $x^2 + 4x + 3 = 1$

52. Suppose that you use the Square Root Property to solve the following equations.

  (i) $(x - 1)^2 = 0$

  (ii) $(x - 1)^2 = 9$

Why is there a difference in the number of solutions of the two equations?

In Exercises 53–58, use the Square Root Property to solve the equation.

53. $x^2 - 18x + 81 = 25$

54. $x^2 + 8x + 16 = 64$

55. $4x^2 + 20x + 25 = 81$ 56. $9 - 24x + 16x^2 = 1$

57. $x^2 + 6x + 9 = 10$ 58. $16x^2 - 8x + 1 = 8$

In Exercises 59–62, solve by using the Square Root Property.

59. $\frac{3}{x^2} - 4 = 5$ 60. $1 + \frac{5}{x^2} = 26$

61. $\frac{3}{y^2} = \frac{1}{3}$ 62. $\frac{4}{5} = \frac{5}{t^2}$

In Exercises 63–68, solve the equation.

63. $x(x + 12) = 12(3 + x)$

64. $x^2 + 6x = 6(3 + x)$

65. $2(2x^2 + 3) = 3(x^2 + 5)$

66. $3(2x^2 + 3) = 5(x^2 + 2)$

67. $3(x + 1)(x - 1) = 2x^2 + 5$

68. $4(x^2 + 1) = 3(x^2 + 6) + 1$

69. Three times the reciprocal of a positive number is the same as the ratio of 4 less than the number to 4. Determine the number.

70. A negative number is equal to the ratio of 3 to 2 more than the number. Determine the number.

Geometric Models

71. The length of a rectangle is 3 times its width. If the width is doubled and the length is decreased by 4, the area is increased by 3. What is the length of the original rectangle?

72. The length of a rectangle is 1 more than its width. If the width is reduced to half the original width and the length is increased by 6, the area is decreased by 12. What is the width of the original rectangle?

73. The width of a picture is 2 inches less than its length. A 2-inch-wide mat is placed around the picture. The combined area of the picture and the mat is 8 square inches more than twice the area of the picture. What is the width of the picture?

74. The height of a triangular sail is 2 feet more than its base. If the height is increased to twice the base, then the area is increased by 12 square feet. What is the height of the original sail?

75. The length of the base of a triangle is twice its height. What is the length of the base if the area is 25 square feet?

76. The length of the base of a parallelogram is 3 times its height, and the area is 21 square inches. Determine the height.

77. A triangular pennant is hung from the ceiling of a gymnasium. If the area of the pennant is 60 square inches and the height of the triangle is 6 times the length of its base, how far from the ceiling does the pennant hang?

78. The radius of a circular pond is 7 feet. The area of the walk surrounding the pond is 51π square feet. How wide is the walk?

Modeling with Real Data

79. The table shows the percentage of adults whose parents provided some monetary assistance in buying a house.

Age	Percentage
18–29	7%
30–44	15%
45–59	15%
60 and over	11%

(Source: Roper Starch Worldwide.)

The polynomial $-0.0165(x^2 - 94x + 1248)$ models the percentage of those age x who received assistance from their parents.

(a) Estimate the solutions of $x^2 - 94x + 1248 = 0$.

(b) Solve the equation algebraically and interpret the solutions.

80. Refer to the model in Exercise 79.

(a) Suppose that 9.9% of those of a certain age received assistance. Write an equation to determine that age.

(b) Solve the equation in part (a).

Data Analysis: *Computer Viruses*

When a computer virus strikes, both individuals and businesses incur significant costs. The accompanying bar graph shows the increasing cost during the 1990s. (Source: National Computer Security Association.)

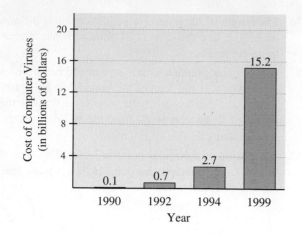

The costs C (in billions of dollars) associated with computer viruses can be estimated by $C(t) = 0.18t^2 + 0.1$ where t is the number of years since 1990.

81. Use the model to write an equation that you can use to determine the year after 1990 in which the cost will be $26 billion. Then use a graph to estimate the solution of the equation.

82. Use the Square Root Property to solve the equation in Exercise 81.

83. Solving with the Square Root Property gives two solutions. Why do we reject one?

84. The cost of computer viruses is ultimately passed on to the consumer. Assume that the population of the United States is 280 million and that the model is valid. What is the estimated per-person cost of computer viruses in 2002?

Challenge

In Exercises 85 and 86, solve the formula for the indicated variable.

85. $V = \dfrac{1}{3}\pi r^2 h$; r

86. $A = P(1 + r)^2$; r

In Exercises 87 and 88, solve for y.

87. $x^2 + y^2 = 49$

88. $\dfrac{x^2}{16} - \dfrac{y^2}{4} = 1$

In Exercises 89 and 90, write a quadratic equation that has the given pair of numbers as solutions. Express your answer in the form $x^2 + bx + c = 0$.

89. $-3, 4$

90. $1, -\dfrac{5}{2}$

10.2 | Completing the Square

Perfect Square Trinomials • Solving Quadratic Equations

Perfect Square Trinomials

In Section 10.1 we used the Square Root Property to solve any equation in the form $A^2 = B$. As we will learn in this section, it is possible to write any quadratic equation in the form $A^2 = B$. The process is called **completing the square.**

If two terms of a trinomial are $x^2 + bx$, completing the square involves determining the constant term c such that $x^2 + bx + c$ is a perfect square trinomial. The technique is based on the special products for the square of a binomial.

$$(A + B)^2 = A^2 + 2AB + B^2$$

$$(A - B)^2 = A^2 - 2AB + B^2$$

Exploring the Concept

Determining the Constant Term of a Perfect Square Trinomial

Suppose that we want to determine the value of c so that $x^2 + 12x + c$ is a perfect square trinomial. We can compare this trinomial with the corresponding model for a perfect square trinomial: $x^2 + 2Bx + B^2$.

$$x^2 + 2Bx + B^2$$

$$x^2 + 12x + \rule{1cm}{0.4em}$$

For the coefficients of x to be equal, we must have $2B = 12$ and so $B = 6$. Replacing B with 6 in the model, we obtain the perfect square trinomial $x^2 + 12x + 36$.

In general, for $x^2 + bx + c$ to be a perfect square trinomial, the constant term c must be the square of $\frac{1}{2}b$.

Determining the Constant Term of a Perfect Square Trinomial

If two terms of a perfect square trinomial are $x^2 + bx$, then the complete trinomial is $x^2 + bx + \left(\dfrac{b}{2}\right)^2$.

Note: This method of determining the constant term of a perfect square trinomial applies only if the leading coefficient is 1.

EXAMPLE **1**

Determining the Constant Term of a Perfect Square Trinomial

Write the perfect square trinomial whose first two terms are given.

(a) $x^2 - 14x + \rule{1cm}{0.4em}$ (b) $x^2 + 9x + \rule{1cm}{0.4em}$

Solution

(a) Because $\left(-\dfrac{14}{2}\right)^2 = 49$, the trinomial is $x^2 - 14x + 49$.

(b) The constant term is $\left(\dfrac{9}{2}\right)^2$ or $\dfrac{81}{4}$. The trinomial is $x^2 + 9x + \dfrac{81}{4}$.

Solving Quadratic Equations

By completing the square, we can prepare any quadratic equation so that the Square Root Property can be applied.

Exploring the Concept

Solving an Equation by Completing the Square

Suppose that we wish to use the Square Root Property to solve the quadratic equation $x^2 + 6x = 7$. This means that we must write the equation in the form $A^2 = B$. We can do this by writing the left side as a perfect square trinomial—that is, by completing the square.

LEARNING TIP

In order for us to solve a quadratic equation by factoring, one side must be 0. For us to complete the square, one side must be a perfect square trinomial. Try to associate forms and methods.

$x^2 + 6x \qquad = 7$		$\frac{1}{2} \cdot 6 = 3$ and $3^2 = 9$
$x^2 + 6x + 9 = 7 + 9$		Add 9 to both sides.
$(x + 3)^2 = 16$		Factor the perfect square trinomial.
$x + 3 = -4$ or	$x + 3 = 4$	Square Root Property
$x = -7$ or	$x = 1$	

The method of completing the square can be used to solve any quadratic equation $ax^2 + bx + c = 0$. However, if the leading coefficient a is a nonzero number other than 1, it is necessary to divide both sides of the equation by a before completing the square. The following summarizes the procedure.

Solving a Quadratic Equation by Completing the Square

1. If necessary, divide both sides by the leading coefficient to obtain a coefficient of 1 for the quadratic term.
2. Write the equation in the form $x^2 + bx = k$.
3. Add $\left(\dfrac{b}{2}\right)^2$ to both sides to complete the square.
4. Factor the resulting perfect square trinomial.
5. Solve with the Square Root Property.

EXAMPLE 2 **Solving a Quadratic Equation by Completing the Square**

Solve each equation by completing the square.

(a) $2x = 35 - x^2$

(b) $x^2 - 10x + 15 = 0$

Solution

(a) Write the equation in the form $ax^2 + bx = c$.

$$2x = 35 - x^2 \qquad \text{Add } x^2 \text{ to both sides.}$$
$$x^2 + 2x = 35 \qquad \text{Complete the square on the left side.}$$
$$x^2 + 2x + 1 = 35 + 1 \qquad \text{Add } \left(\tfrac{1}{2} \cdot 2\right)^2 = 1 \text{ to both sides.}$$
$$(x + 1)^2 = 36 \qquad \text{Factor.}$$
$$x + 1 = -6 \quad \text{or} \quad x + 1 = 6 \qquad \text{Square Root Property}$$
$$x = -7 \quad \text{or} \quad x = 5 \qquad \text{Solve for } x.$$

The solutions are -7 and 5. (Note that the equation could have been solved by factoring, which is usually an easier method.)

(b)
$$x^2 - 10x + 15 = 0 \qquad \text{Subtract 15 from both sides.}$$
$$x^2 - 10x = -15 \qquad \text{Complete the square on the left side.}$$
$$x^2 - 10x + 25 = -15 + 25 \qquad \left(\frac{b}{2}\right)^2 = \left(\frac{-10}{2}\right)^2 = 25$$
$$(x - 5)^2 = 10 \qquad \text{Factor.}$$
$$x - 5 = -\sqrt{10} \quad \text{or} \quad x - 5 = \sqrt{10} \qquad \text{Square Root Property}$$
$$x = 5 - \sqrt{10} \quad \text{or} \quad x = 5 + \sqrt{10} \qquad \text{Solve for } x.$$
$$x \approx 1.84 \quad \text{or} \quad x \approx 8.16$$

The solutions are approximately 1.84 and 8.16.

The method of completing the square will detect quadratic equations that have no solution.

EXAMPLE   3

Solving a Quadratic Equation by Completing the Square

Solve $x^2 + 8x + 20 = 0$ by completing the square.

Solution

$$x^2 + 8x + 20 = 0 \qquad \text{Write the constant on the right side.}$$

$$x^2 + 8x = -20 \qquad \text{Complete the square.}$$

$$x^2 + 8x + 16 = -20 + 16 \qquad \text{Add } \left(\frac{8}{2}\right)^2 = 16 \text{ to both sides.}$$

$$(x + 4)^2 = -4 \qquad \text{Factor.}$$

Because there is no real number whose square is -4, the equation has no real number solution.

For $ax^2 + bx + c = 0$, if the leading coefficient a is not 1, we divide both sides of the equation by a.

EXAMPLE   4

Solving a Quadratic Equation by Completing the Square

Solve each equation.

Think About It

For each of the following equations, which solving method (Square Root Property, completing the square, or factoring) is applicable? In each case, which method is easiest?

(i) $x^2 + 6x = 16$
(ii) $(x + 6)^2 = 16$
(iii) $x^2 + 4x = 16$

(a) $3x^2 + 15x - 24 = 0$

$$3x^2 + 15x \phantom{+ \frac{25}{4}} = 24 \qquad \text{Move the constant to the right side.}$$

$$x^2 + 5x \phantom{+ \frac{25}{4}} = 8 \qquad \text{Divide both sides of the equation by 3.}$$

$$x^2 + 5x + \frac{25}{4} = 8 + \frac{25}{4} \qquad \text{Add } \left(\frac{5}{2}\right)^2 \text{ or } \frac{25}{4} \text{ to both sides.}$$

$$\left(x + \frac{5}{2}\right)^2 = \frac{57}{4} \qquad \text{Factor.}$$

$$x + \frac{5}{2} = -\frac{\sqrt{57}}{2} \qquad \text{or} \qquad x + \frac{5}{2} = \frac{\sqrt{57}}{2} \qquad \text{Square Root Property}$$

$$x = -\frac{5}{2} - \frac{\sqrt{57}}{2} \qquad \text{or} \qquad x = -\frac{5}{2} + \frac{\sqrt{57}}{2} \qquad \text{Solve for } x.$$

$$x \approx -6.27 \qquad \text{or} \qquad x \approx 1.27$$

The solutions are approximately -6.27 and 1.27.

(b) $ x(2x - 3) = 1$

$$2x^2 - 3x = 1 \qquad \text{Distributive Property}$$

$$x^2 - \frac{3}{2}x = \frac{1}{2} \qquad \text{Divide both sides by 2.}$$

$$x^2 - \frac{3}{2}x + \frac{9}{16} = \frac{1}{2} + \frac{9}{16} \qquad \text{Add } \left(\frac{1}{2} \cdot \frac{-3}{2}\right)^2 = \left(-\frac{3}{4}\right)^2 = \frac{9}{16} \text{ to both sides.}$$

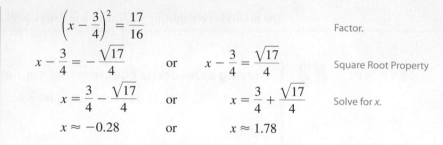

$$\left(x - \frac{3}{4}\right)^2 = \frac{17}{16} \qquad\qquad \text{Factor.}$$

$$x - \frac{3}{4} = -\frac{\sqrt{17}}{4} \quad \text{or} \quad x - \frac{3}{4} = \frac{\sqrt{17}}{4} \qquad \text{Square Root Property}$$

$$x = \frac{3}{4} - \frac{\sqrt{17}}{4} \quad \text{or} \quad x = \frac{3}{4} + \frac{\sqrt{17}}{4} \qquad \text{Solve for } x.$$

$$x \approx -0.28 \quad\quad\quad \text{or} \quad\quad\quad x \approx 1.78$$

The solutions are approximately -0.28 and 1.78.

Quick Reference 10.2

Perfect Square Trinomials

- The process of writing a quadratic equation in the form $A^2 = B$ is called **completing the square.**
- If two terms of a perfect square trinomial are $x^2 + bx$, then the complete trinomial is $x^2 + bx + \left(\dfrac{b}{2}\right)^2$. In words, we can determine the constant term of a perfect square trinomial by squaring half the coefficient of the middle term.

Solving Quadratic Equations

- The method of completing the square can be used to solve any quadratic equation. The following are the steps.
  1. If necessary, divide both sides by the leading coefficient to obtain a coefficient of 1 for the quadratic term.
  2. Write the equation in the form $x^2 + bx = k$.
  3. Add $\left(\dfrac{b}{2}\right)^2$ to both sides to complete the square.
  4. Factor the resulting perfect square trinomial.
  5. Solve with the Square Root Property.

Speaking the Language 10.2

1. A method called ▨▨▨▨▨ can be used to write any quadratic equation in the form $A^2 = B$.

2. The constant term of a perfect square trinomial whose leading coefficient is 1 is the square of ▨▨▨▨▨ of the middle term.

3. To use the method of completing the square to solve $2x^2 - 3x = -5$, we begin by ▨▨▨▨▨ .

4. Once we have completed the square, we use the ▨▨▨▨▨ Property to solve the resulting equation.

Exercises 10.2

Concepts and Skills

1. Describe how to determine a constant c such that $x^2 - 9x + c$ is a perfect square trinomial.

2. Consider $2x^2 + 6x = 0$. Suppose we add $\left(\frac{1}{2} \cdot 6\right)^2 = 9$ to both sides. Will the resulting expression on the left be a perfect square trinomial? Why?

In Exercises 3–10, write the perfect square trinomial whose first two terms are given.

3. $x^2 + 16x +$ ▨

4. $x^2 - 4x +$ ▨

5. $x^2 - 24x +$ ▨

6. $x^2 + 18x +$ ▨

7. $x^2 - 5x +$ ▨

8. $x^2 + x +$ ▨

9. $x^2 + \dfrac{6}{5}x +$ ▨

10. $x^2 - \dfrac{1}{3}x +$ ▨

In Exercises 11–18, solve the equation by completing the square and then by factoring.

11. $x^2 - 4x - 21 = 0$

12. $x^2 + 2x - 3 = 0$

13. $x^2 + 10x + 16 = 0$

14. $x^2 - 8x + 15 = 0$

15. $x^2 + 5x = 6$

16. $x^2 = 3x + 28$

17. $x^2 = 9x - 8$

18. $x^2 + 7x + 6 = 0$

In Exercises 19–32, solve the equation by completing the square.

19. $x^2 - 6x = 16$

20. $x^2 + 12x + 35 = 0$

21. $x^2 + 6x = 2$

22. $x^2 - 4x + 1 = 0$

23. $x^2 = 4x + 2$

24. $x^2 + 16x + 30 = 0$

25. $x^2 + 4x + 7 = 0$

26. $x^2 + 2 = 2x$

27. $x^2 - 5x - 5 = 0$

28. $x^2 - 9x = 4$

29. $x^2 + 5x + 2 = 0$

30. $x = 5 - x^2$

31. $x(x + 7) = -3$

32. $x(x + 8) = -15$

33. The equation $x^2 - 4x = 0$ can be solved by adding 4 to both sides to complete the square. Describe another method. Which method would you choose?

34. The equation $x^2 + 2x - 323 = 0$ can be solved by factoring. Why is the method of completing the square easier?

In Exercises 35–40, solve the equation by completing the square and then by factoring.

35. $6x^2 + x - 2 = 0$

36. $5x^2 + 6x - 8 = 0$

37. $2x^2 + x = 6$

38. $3x^2 + 2 = 7x$

39. $4x^2 + 7x + 3 = 0$

40. $6x^2 + x = 15$

In Exercises 41–52, solve the equation by completing the square.

41. $4x^2 - 4x - 24 = 0$

42. $5x^2 + 30x + 40 = 0$

43. $6x^2 - 3x = 2$

44. $4x^2 + 6x = 3$

45. $4x^2 + 4x + 5 = 0$

46. $3x^2 + 6x + 4 = 0$

47. $2x^2 = 4 - 3x$

48. $4x^2 + 1 = 10x$

49. $6x^2 + 9x + 8 = 0$

50. $4x^2 = 2x - 1$

51. $3x^2 + x - 1 = 0$

52. $5x^2 - 4x - 10 = 0$

53. The product of a negative number and 1 less than half the number is 60. What is the number?

54. The square of 1 more than $\frac{1}{3}$ of a positive number is 6 more than twice the number. What is the number?

55. The product of the measures of two supplementary angles is 7200. What is the measure of the smaller angle?

56. The sum of the squares of the measures of two complementary angles is 4500. What is the measure of the larger angle?

Real-Life Applications

57. In order to combat crime in a 180-square-block section of town, a city plans to install an extra street light at the corner of each block along the perimeter of the section. If the rectangular section is 3 blocks longer than it is wide, how many lights must be installed?

58. A rectangular piece of metal is 6 inches longer than it is wide. The material is made into a box by cutting out 1-inch square corners and folding the sides up. If the volume of the box is 280 cubic inches, what is the length of the original piece of metal?

59. A motorboat traveled downstream 80 miles and then returned 60 miles upstream. If the speed of the current was 8 mph and the entire trip took 5 hours, what was the speed of the boat in still water? (Round your answer to the nearest whole number.)

60. A person drove 30 miles to the airport and returned by the same route. The car's average speed during the trip to the airport was 10 mph faster than on the return trip. If the entire trip took 1 hour, what was the car's speed during the return trip? (Round your answer to the nearest whole number.)

61. Two cars left Abilene at the same time. The northbound car traveled 10 mph faster than the westbound car. After 1 hour the cars were 90 miles apart. What was the speed of the northbound car? (Round your answer to the nearest whole number.)

62. From Buffalo, Casper is due south, and Gillette is due east. The distance from Buffalo to Casper is 50 miles more than the distance from Buffalo to Gillette, and the distance from Gillette to Casper is 70 miles more than the distance from Buffalo to Gillette. To the nearest mile, what is the distance from Gillette to Casper?

Modeling with Real Data

63. Americans are not noted for saving money. The table shows the median net financial assets of families, by age group, in the United States.

Age	Median Net Assets
35–44	$ 700
45–54	2,600
55–64	6,880
65–74	10,000
75 and over	8,275

(Source: Analysis of SIPP Data.)

The polynomial $-0.012(x^2 - 136x + 3783)$ models the median net assets (in thousands of dollars), where x represents age.

(a) Write an equation to estimate the age at which the net worth is expected to be $4800.

(b) Solve the equation in part (a) to find the age.

64. Refer to the model in Exercise 63.

(a) What does the following equation model?

$$-0.012(x^2 - 136x + 3783) = 12$$

(b) Solve the equation in part (a) and interpret the result.

Data Analysis: *Unmarried Mothers*

In 1985, 22 of every 100 births involved an unmarried mother. By 1998, this statistic had risen to 32.8 of every 100 births. (Source: U.S. Center for Health Statistics.) The number B of births to unmarried mothers per 100 births can be approximated by the quadratic equation $B = -0.035(x^2 - 88x + 960)$, where x is the number of years since 1960.

65. Write an equation to describe the year when 34 of every 100 births would involve an unmarried mother.

66. Use the graphing method to estimate the solutions of the equation in Exercise 65 to the nearest whole numbers.

67. Determine the solutions of the equation in Exercise 65 by completing the square. Round your solutions to the nearest integer.

68. What do the solutions in Exercise 67 represent?

Challenge

In Exercises 69–71, determine the value(s) of k such that the trinomial will be a perfect square trinomial.

69. $4x^2 + 3x + k$ **70.** $x^2 + kx + 25$

71. $kx^2 - 8x + 1$

In Exercises 72 and 73, complete the square to write the equation in the form $y - k = (x - h)^2$, where k and h are constants.

72. $x^2 - 4x + 5 - y = 0$ **73.** $x^2 + 6x - y + 11 = 0$

10.3 | The Quadratic Formula

Derivation and Use • The Discriminant

Derivation and Use

Each time we solve a quadratic equation by completing the square, we follow the same steps. When a certain routine is applied repeatedly to solving similar problems, mathematicians try to generalize the routine to make the process more efficient. The result is usually a formula that eliminates the need to perform the same steps over and over.

In this section, we generalize the method of completing the square for solving a quadratic equation $ax^2 + bx + c = 0$. The formula that we derive can be used to obtain the solutions of any quadratic equation.

Exploring the Concept

Derivation of the Quadratic Formula

Consider the general quadratic equation in standard form.

$$ax^2 + bx + c = 0$$

$ax^2 + bx \quad\quad = -c$	Move the constant term to the right side.
$x^2 + \dfrac{b}{a}x \quad\quad = -\dfrac{c}{a}$	Leading coefficient must be 1. Divide both sides by a.
$x^2 + \dfrac{b}{a}x + \dfrac{b^2}{4a^2} = -\dfrac{c}{a} + \dfrac{b^2}{4a^2}$	To complete the square, add $\left(\dfrac{1}{2}\cdot\dfrac{b}{a}\right)^2 = \dfrac{b^2}{4a^2}$ to both sides.
$x^2 + \dfrac{b}{a}x + \dfrac{b^2}{4a^2} = \dfrac{-4ac}{4a^2} + \dfrac{b^2}{4a^2}$	The LCD on the right side is $4a^2$.
$x^2 + \dfrac{b}{a}x + \dfrac{b^2}{4a^2} = \dfrac{b^2 - 4ac}{4a^2}$	Combine the fractions on the right side.
$\left(x + \dfrac{b}{2a}\right)^2 = \dfrac{b^2 - 4ac}{4a^2}$	Factor and write in exponential form.
$x + \dfrac{b}{2a} = \pm\dfrac{\sqrt{b^2 - 4ac}}{2a}$	Square Root Property
$x = \dfrac{-b}{2a} \pm \dfrac{\sqrt{b^2 - 4ac}}{2a}$	Subtract $\dfrac{b}{2a}$ to solve for x.
$x = \dfrac{-b \pm \sqrt{b^2 - 4ac}}{2a}$	Combine the fractions.

This formula is called the **Quadratic Formula.**

The Quadratic Formula

For real numbers a, b, and c, with $a \neq 0$, the solutions of the quadratic equation $ax^2 + bx + c = 0$ are given by the **Quadratic Formula**

$$x = \frac{-b \pm \sqrt{b^2 - 4ac}}{2a}$$

Note: In the Quadratic Formula, the $\pm$ sign means that the solutions are

$$x = \frac{-b + \sqrt{b^2 - 4ac}}{2a} \quad\quad \text{and} \quad\quad x = \frac{-b - \sqrt{b^2 - 4ac}}{2a}$$

Also, note that a quadratic equation must be written in standard form before the Quadratic Formula can be used to solve it.

EXAMPLE **1**

Writing a Quadratic Equation in Standard Form

Write each equation in the standard form $ax^2 + bx + c = 0$ and identify the values of a, b, and c.

(a) $2x^2 = 5x - 2$ (b) $x - x^2 = 0$

Solution

(a) $2x^2 = 5x - 2$

$\quad 2x^2 - 5x + 2 = 0$ One side of the equation must be 0.

$\quad a = 2, \quad b = -5, \quad c = 2$

(b) $x - x^2 = 0$

$\quad -x^2 + x = 0$ Write the left side in descending order.

$\quad a = -1, \quad b = 1, \quad c = 0$

EXAMPLE **2**

Using the Quadratic Formula: Two Real Number Solutions

Use the Quadratic Formula to solve each equation.

(a) $3x^2 + 2x - 5 = 0$ (b) $2x^2 = 4x + 1$ (c) $x(x + 5) = -3$

Solution

(a) $3x^2 + 2x - 5 = 0$

The equation is written in standard form with $a = 3$, $b = 2$, and $c = -5$.

$$x = \frac{-b \pm \sqrt{b^2 - 4ac}}{2a}$$

$$= \frac{-2 \pm \sqrt{2^2 - 4(3)(-5)}}{2(3)}$$ Note that $-b = -2$.

$$= \frac{-2 \pm \sqrt{4 + 60}}{6}$$ Simplify the radicand.

$$= \frac{-2 \pm \sqrt{64}}{6}$$

$$= \frac{-2 \pm 8}{6}$$ Evaluate the radical expression.

$$x = \frac{-2 - 8}{6} \quad \text{or} \quad x = \frac{-2 + 8}{6}$$ Use the $-$ sign for one solution and the $+$ sign for the second solution.

$$x = \frac{-10}{6} \quad \text{or} \quad x = \frac{6}{6}$$ Simplify.

$$x = \frac{-5}{3} \quad \text{or} \quad x = 1$$

There are two solutions: $-\frac{5}{3}$ and 1.

(b) $2x^2 = 4x + 1$

$2x^2 - 4x - 1 = 0$ Write the equation in standard form.

$x = \dfrac{-b \pm \sqrt{b^2 - 4ac}}{2a}$

$= \dfrac{-(-4) \pm \sqrt{(-4)^2 - 4(2)(-1)}}{2(2)}$ Let $a = 2$, $b = -4$, and $c = -1$.

$= \dfrac{4 \pm \sqrt{16 + 8}}{4}$ Simplify the radicand.

$= \dfrac{4 \pm \sqrt{24}}{4} = \dfrac{4 \pm \sqrt{4 \cdot 6}}{4} = \dfrac{4 \pm 2\sqrt{6}}{4} = \dfrac{2 \pm \sqrt{6}}{2}$

$x = \dfrac{2 - \sqrt{6}}{2}$ or $x = \dfrac{2 + \sqrt{6}}{2}$ Use the $-$ sign for one solution and the $+$ sign for the other solution.

$x \approx -0.22$ or $x \approx 2.22$ Use a calculator to evaluate each expression.

The two solutions are approximately -0.22 and 2.22.

LEARNING TIP

Each time you solve a quadratic equation, write the formula. This will help you remember the formula and avoid errors as you substitute.

(c) $x(x + 5) = -3$

$x^2 + 5x = -3$ Distributive Property

$x^2 + 5x + 3 = 0$ Write the equation in standard form.

$x = \dfrac{-b \pm \sqrt{b^2 - 4ac}}{2a}$

$= \dfrac{-5 \pm \sqrt{5^2 - 4(1)(3)}}{2(1)}$ Let $a = 1$, $b = 5$, and $c = 3$.

$= \dfrac{-5 \pm \sqrt{25 - 12}}{2}$ Simplify the radicand.

$= \dfrac{-5 \pm \sqrt{13}}{2}$ Use the $-$ sign for one solution and the $+$ sign for the other solution.

$x = \dfrac{-5 - \sqrt{13}}{2}$ or $x = \dfrac{-5 + \sqrt{13}}{2}$

$x \approx -4.30$ or $x \approx -0.70$ Use a calculator to evaluate each expression.

The two solutions are approximately -4.30 and -0.70.

Note: The Quadratic Formula can be used to solve all quadratic equations, including those for which the factoring method can be used. In Example 2, the factoring method can be used in part (a) but not in parts (b) and (c). However, the Quadratic Formula applied in all parts.

EXAMPLE 3

Using the Quadratic Formula: One Solution

Use the Quadratic Formula to solve $9x^2 - 6x + 1 = 0$.

Solution

$$x = \frac{-(-6) \pm \sqrt{(-6)^2 - 4(9)(1)}}{2(9)}$$ Let $a = 9, b = -6,$ and $c = 1$.

$$= \frac{6 \pm \sqrt{36 - 36}}{18}$$ Simplify the radicand.

$$= \frac{6 \pm \sqrt{0}}{18}$$ Note that the radicand is zero.

$$= \frac{6 \pm 0}{18}$$ Adding or subtracting 0 gives the same solution.

$$= \frac{6}{18} = \frac{1}{3}$$

The only solution is $\frac{1}{3}$.

EXAMPLE 4

Using the Quadratic Formula: No Real Number Solution

Use the Quadratic Formula to solve $x^2 + 2x + 3 = 0$.

Solution

$$x = \frac{-2 \pm \sqrt{2^2 - 4(1)(3)}}{2(1)}$$ Let $a = 1, b = 2,$ and $c = 3$.

$$= \frac{-2 \pm \sqrt{4 - 12}}{2}$$ Simplify the radicand.

$$= \frac{-2 \pm \sqrt{-8}}{2}$$ Note that the radicand is negative.

Because the square root of a negative number is not defined in the real number system, the equation has no real number solution.

The Discriminant

In our previous examples we used the Quadratic Formula to solve quadratic equations and obtained two solutions, one solution, or no solution. Examination of these examples will reveal that the number of solutions is determined by the quantity $b^2 - 4ac$, which is the radicand in the Quadratic Formula and is called the **discriminant.** If the value of $b^2 - 4ac$ is positive, as in Example 2, the square root of that quantity is added to and subtracted from $-b$ to obtain two solutions. If $b^2 - 4ac$ is 0, as in Example 3, the formula reduces to $\frac{-b}{2a}$, and we obtain only one solution. Finally, if $b^2 - 4ac$ is negative, as in Example 4, there is no real number solution because the square root of a negative number is not defined in the real number system.

> **The Discriminant and the Number of Solutions of a Quadratic Equation**
>
> For the quadratic equation $ax^2 + bx + c = 0$,
>
> 1. if $b^2 - 4ac > 0$, the equation has two solutions.
> 2. if $b^2 - 4ac = 0$, the equation has one solution.
> 3. if $b^2 - 4ac < 0$, the equation has no real number solution.

EXAMPLE 5

Using the Discriminant to Determine the Number of Solutions of a Quadratic Equation

Think About It

For a quadratic equation with integer coefficients, we can use the discriminant to determine whether the equation can be solved by factoring. By inspecting the examples in this section, decide what must be true about the discriminant in order for the equation to be solved by factoring.

Without solving the equation, use the discriminant to determine the number of solutions of each equation.

(a) $25x^2 - 10x + 1 = 0$ (b) $x^2 + 3 = 0$ (c) $3x^2 - 2x = 4$

Solution

(a) $25x^2 - 10x + 1 = 0$

Let $a = 25$, $b = -10$, and $c = 1$.

$$b^2 - 4ac = (-10)^2 - 4(25)(1) = 100 - 100 = 0$$

Because the discriminant is 0, the equation has only one solution.

(b) $x^2 + 3 = 0$

Substitute $a = 1$, $b = 0$, and $c = 3$.

$$b^2 - 4ac = 0^2 - 4(1)(3) = 0 - 12 = -12$$

The discriminant is negative, so the equation has no real number solution.

(c) $3x^2 - 2x = 4$

$3x^2 - 2x - 4 = 0$ Write the equation in standard form.

Substitute $a = 3$, $b = -2$, and $c = -4$.

$$b^2 - 4ac = (-2)^2 - 4(3)(-4) = 4 + 48 = 52$$

Because the discriminant is positive, the equation has two solutions.

When you use the Quadratic Formula, consider calculating the discriminant first. If, for example, the discriminant is negative, you can conclude that the equation has no solution without performing unnecessary arithmetic.

Quick Reference 10.3

Derivation and Use

- The Quadratic Formula was derived by applying the method of completing the square to the general quadratic equation $ax^2 + bx + c = 0$.

- For real numbers a, b, and c, with $a \neq 0$, the solutions of the quadratic equation $ax^2 + bx + c = 0$ are given by the **Quadratic Formula**

$$x = \frac{-b \pm \sqrt{b^2 - 4ac}}{2a}$$

- In order for us to use the Quadratic Formula, the quadratic equation must be written in standard form.
- While all previously discussed methods for solving a quadratic equation still apply, the Quadratic Formula can be used to solve any quadratic equation.

The Discriminant
- The quantity $b^2 - 4ac$, which is the radicand in the Quadratic Formula, is called the **discriminant.** The discriminant determines the number of solutions of a quadratic equation $ax^2 + bx + c = 0$.
  1. If $b^2 - 4ac > 0$, the equation has two solutions.
  2. If $b^2 - 4ac = 0$, the equation has one solution.
  3. If $b^2 - 4ac < 0$, the equation has no real number solution.

Speaking the Language 10.3

1. Solving the general quadratic equation $ax^2 + bx + c = 0$ by completing the square leads to the ▨▨▨▨▨ Formula.
2. Before applying the Quadratic Formula, we must write the quadratic equation in ▨▨▨▨▨ form.
3. The radicand in the Quadratic Formula is called the ▨▨▨▨▨.
4. A quadratic equation has no real number solution if the discriminant is ▨▨▨▨▨.

Exercises 10.3

Concepts and Skills

1. What method for solving a quadratic equation was generalized to derive the Quadratic Formula?

2. What step must be taken before the Quadratic Formula can be used to solve $5x = 8 - 3x^2$? Will it then be necessary to divide both sides of the equation by the coefficient of x^2? Why?

In Exercises 3–6, write the equation in standard form. Then state the values of a, b, and c.

3. $2x^2 - 3x - 1 = 0$ 4. $6x - x^2 - 2 = 0$

5. $3x^2 + 2 = 0$ 6. $8x^2 = 2x$

In Exercises 7–12, solve the equation by factoring. Then solve it again with the Quadratic Formula.

7. $x^2 - 6x + 5 = 0$ 8. $x^2 + 14x + 49 = 0$

9. $x^2 + 7x = 0$ 10. $3x^2 - 2x - 1 = 0$

11. $x(3x + 4) = 4$ 12. $2(x^2 - 1) = 3x$

In Exercises 13–26, use the Quadratic Formula to solve the equation.

13. $x^2 + 7x + 6 = 0$ 14. $10 + 3x - x^2 = 0$

15. $x^2 - 6x + 7 = 0$ 16. $x^2 + 2x - 4 = 0$

17. $x^2 + 3x + 6 = 0$ 18. $2x^2 - 5x + 4 = 0$

19. $3 + 2x - 3x^2 = 0$ 20. $4x^2 - 6x - 9 = 0$

21. $15 - x^2 = 0$ 22. $6x^2 - 10x = 0$

23. $12x^2 + 30x = 0$ 24. $9 - 49x^2 = 0$

25. $3x(x - 1) = 2$ 26. $(x + 5)(x - 1) = 1$

27. Suppose that the graph of $ax^2 + bx + c$ intersects the x-axis at only one point. What do you know about the discriminant?

28. Both of the following equations can be solved with the factoring method. For which equation might the Quadratic Formula be a better method? Why?

    (i) $x^2 - 9 = 0$

    (ii) $x^2 + 3x - 270 = 0$

In Exercises 29–48, solve the equation by any method.

29. $10x^2 + x - 3 = 0$

30. $3x^2 - 4x = 0$

31. $3x^2 - 2x + 1 = 0$

32. $x^2 = 4x - 5$

33. $x^2 + 6x = -7$

34. $3 + 7x - 3x^2 = 0$

35. $x^2 + 12x = -35$ **36.** $x^2 + 9x - 81 = 9x$

37. $14 - x^2 = 5x$ **38.** $3x^2 = 7x + 6$

39. $x^2 + 18x + 81 = 0$ **40.** $9x^2 - 6x + 1 = 0$

41. $4x = 11 + \dfrac{3}{x}$ **42.** $\dfrac{4}{x^2} + \dfrac{3}{x} = 1$

43. $x + \dfrac{7}{x} = 9$ **44.** $\dfrac{7}{x} + 1 = -\dfrac{1}{x^2}$

45. $3x(x + 3) = x - 2$ **46.** $(3x - 1)^2 = 10$

47. $(x - 4)(x - 1) = -2$ **48.** $(3x + 1)(x - 1) = -5$

49. Describe the graph of the expression $ax^2 + bx + c$ if the discriminant is negative.

50. Explain why $ax^2 + bx + c = 0$ has only one solution if the discriminant is 0.

In Exercises 51–62, use the discriminant to determine the number of solutions of the equation.

51. $x^2 + 72 = 17x$ **52.** $3x^2 = 2x$

53. $x^2 - 20x + 100 = 0$ **54.** $1 + 2x + x^2 = 0$

55. $x^2 - 1 = -11$ **56.** $x^2 - 6x + 10 = 0$

57. $4x + 2 + x^2 = 0$ **58.** $x^2 = 4(2x - 1)$

59. $16 - 24x + 9x^2 = 0$ **60.** $4x^2 + 20x + 25 = 0$

61. $5x(x + 1) = -2$ **62.** $1 = 2x(1 - 3x)$

Geometric Models

63. For insect control, a border of marigolds is planted around a vegetable garden. The length of the vegetable garden is 10 feet more than its width. The length of the garden including the marigolds is 5 feet less than twice the width of the vegetable garden, and the width is 5 feet more than the width of the vegetable garden. The area of the marigolds is 1025 square feet. What is the length of the vegetable garden?

64. The area of a walk around a circular pond is 16π square meters. The radius of the larger circle is 1 meter less than twice the radius of the pond. What is the radius of the pond?

65. It takes 9 square feet of material to upholster the rectangular seat of a bench. The length of the bench is 1 foot more than twice its width. If there is a 3-inch allowance on all sides for folding the material under, what is the width of the bench?

66. The area of a picture frame is 64 square inches. The length of the picture in the frame is 2 inches more than its width. The length of the frame is 1 inch more than twice the width of the picture, and the width of the frame is 1 inch less than twice the width of the picture. What is the width of the picture?

67. Together, two rooms require 261 square feet of carpet. One room is square, and the second room is rectangular. The second room is 3 feet longer and 1 foot narrower than the square room. What is the length of the second room?

68. One leg of a right triangle is 2 inches shorter than half the other leg, and the hypotenuse is 26 inches. How long is the shorter leg?

69. The hypotenuse of a right triangle is 4 feet longer than its longer leg, and its shorter leg is $\frac{3}{4}$ of the length of the longer leg. How long is the hypotenuse?

70. Two angles are complementary. The measure of one angle is 20° less than the square of the measure of the other angle. What is the measure of the larger angle?

Real-Life Applications

71. An investor owns 200 shares of stock, part in Welltech and part in NGN Bank. The value of the Welltech stock is $1250, and the value of the NGN stock is $4500. If the price per share of NGN is $5 more than the price per share of Welltech, how many shares of Welltech does the investor own?

72. Two pitchers pitched a total of 17 games. The first pitcher allowed 25 runs, and the second pitcher allowed 21 runs. The average number of runs per game for the first pitcher was $\frac{1}{2}$ less than that for the second pitcher. How many games did the first pitcher pitch?

73. A hospital emergency room treated 150 cases on a holiday afternoon. The total charge for those who were treated and released was $36,000, and the total charge for those who were admitted was $22,500. If the average charge for those who were treated and released was $450 less than the average charge for those who were admitted, how many patients were admitted?

74. At a deli a person bought a total of 10 pounds of coleslaw and potato salad. The coleslaw costs $9, and the potato salad costs $21. If the potato salad costs $1.25 per pound more than the coleslaw, what was the price per pound of the coleslaw?

75. Working together, an electrician and his apprentice wired a garage in 1.8 hours. Working alone, the electrician could have completed the job 1.5 hours sooner than his apprentice. How long would the apprentice have taken working alone?

76. Working together along a conveyor belt, two employees can fill a box of candy in 20 seconds less time than it takes the faster employee alone to fill the box. If the slower employee takes 75 seconds to fill a box, how long does the faster employee take?

Modeling with Real Data

77. The table gives the likelihood that a person will buy life insurance within the next year.

Age	Likelihood
18–24	15%
25–34	26%
35–44	24%
45–54	10%
55–64	7%
65 and older	0%

(Source: Opinion Research.)

The polynomial $y = -0.029x^2 + 2.02x - 12.2$ models the likelihood that a person of age x will buy life insurance within the next year.

(a) Estimate the percentage of people of age 40 who are expected to buy life insurance.

(b) At what other age does a person have approximately the same probability of buying insurance as a person of age 40?

78. Refer to the model in Exercise 77.

(a) Write an equation to determine two ages for which the probability of buying insurance is approximately 20%.

(b) Solve the equation in part (a). (Round the answers to the nearest whole number.)

Data Analysis: *Commuter Rail Passengers*

The accompanying bar graph shows that the number of commuter rail passengers declined from 1990 to 1992, but then ridership began to increase. (Source: American Public Transit Association.)

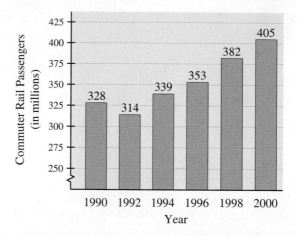

The number of passengers can be modeled with the polynomial $P(t) = 1.1t^2 - 2t + 322$, where t is the number of years since 1990.

79. According to the actual data, by how much did commuter rail travel increase from 1992 to 2000?

80. Write an equation to describe the year in which the increase in ridership from 2000 to that year is the same as the increase in Exercise 79.

81. Solve the equation in Exercise 80 and interpret the solution.

82. On the basis of a comparison of your answers in Exercises 79 and 81, what conclusion can you draw?

Challenge

In Exercises 83 and 84, solve the equation.

83. $x^2 - 4\sqrt{2}x - 4 = 0$

84. $\sqrt{2}x^2 + \sqrt{5}x + \sqrt{8} = 0$

85. Solve $y^2 + 6y + (5 - x) = 0$ for y. (*Hint*: Use the Quadratic Formula with $c = 5 - x$.)

In Exercises 86 and 87, determine the values of k such that (a) the equation has two unequal real number solutions and (b) the equation has no real number solutions.

86. $x^2 + 6x + k = 0$

87. $kx^2 - 3x + 3 = 0$

10.4 Complex Numbers

Definitions • Operations with Complex Numbers • Complex Number Solutions of Quadratic Equations

Definitions

We have seen that some quadratic equations have no real number solution. For instance, when we attempt to solve $x^2 + 1 = 0$, we need to determine a number x such that $x^2 = -1$. However, no real number squared is -1. In order for all quadratic equations to have solutions, we must define a new set of numbers.

Note: To divide 3 by 5, it was necessary to invent rational numbers. To subtract 5 from 3, it was necessary to invent negative numbers. It is not at all unusual for mathematicians to define sets of numbers when they are needed.

We begin our construction of the new set of numbers by defining the **imaginary number i.**

Definition of the Imaginary Number i

The **imaginary number i** has the following properties:

$$i = \sqrt{-1} \qquad \text{and} \qquad i^2 = -1$$

Now we write other numbers, such as $\sqrt{-3}$, in terms of i.

Definition of $\sqrt{-n}$

For any positive real number n, $\sqrt{-n} = i\sqrt{n}$.

Note: Because the definition of $\sqrt{-n}$ is for a *positive* real number n, the radicand $-n$ is *negative*.

EXAMPLE | **1** | **Using the Definition of $\sqrt{-n}$**

Write each number in terms of i.

(a) $\sqrt{-10} = i\sqrt{10}$

(b) $\sqrt{-9} = i\sqrt{9} = 3i$

(c) $\sqrt{-12} = i\sqrt{12} = i\sqrt{4 \cdot 3} = 2i\sqrt{3}$

In addition to the number i, numbers that are multiples of i and numbers that are the sum of a real number and a multiple of i also are called imaginary numbers. We call the expanded number system that includes i the **complex number system.**

> **Definition of a Complex Number**
>
> A **complex number** can be written in the *standard form* $a + bi$, where a and b are real numbers. The number a is called the **real part,** and the number b is called the **imaginary part.**

EXAMPLE 2

Complex Numbers

Each of the following numbers is a complex number.

Number	Standard Form	Real Part	Imaginary Part
$3 - 7i$	$3 + (-7)i$	3	-7
4	$4 + 0i$	4	0
$5i$	$0 + 5i$	0	5

Note: For the complex number $a + bi$, if $b = 0$, the complex number is also a real number, and if $b \neq 0$, the complex number is an imaginary number. The set of complex numbers includes all real numbers and all imaginary numbers.

The set of real numbers and the set of imaginary numbers are subsets of the set of complex numbers. A complex number is either real or imaginary, but not both.

In Figure 10.5, we see that 3, $\sqrt{7}$, $4 + 0i$, $-\sqrt{5}$, and $\frac{7}{8}$ are real numbers. The numbers i, $0 - 3i$, $3 - 7i$, $\sqrt{-5}$, and $\frac{1}{2}i$ are imaginary numbers. Every number in the figure is a complex number.

Figure 10.5

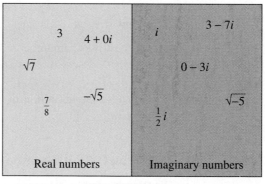

Complex Numbers

Operations with Complex Numbers

LEARNING TIP

If you think of *i* as a variable, you will see that the process of adding, subtracting, and multiplying complex numbers is the same as for binomials.

It can be shown that the associative, commutative, and distributive properties hold for complex numbers. The patterns we use for adding, subtracting, and multiplying complex numbers are similar to those that we used for polynomials.

To add or subtract complex numbers, we use the Distributive Property to combine the real parts and the imaginary parts.

EXAMPLE ▌3▐   **Addition and Subtraction of Complex Numbers**

Perform the indicated operations.

(a) $(3 - 8i) + (2 + 6i) = (3 + 2) + (-8 + 6)i = 5 - 2i$

(b) $(4 + 3i) - (2 - 5i) = 4 + 3i - 2 + 5i = 2 + 8i$

(c) $i - (-3 + 7i) = i + 3 - 7i = 3 - 6i$

We also use the Distributive Property to multiply complex numbers. Again, the methods are the same as for polynomials. However, remember that $i^2 = -1$.

EXAMPLE ▌4▐   **Multiplying Complex Numbers**

Perform the indicated operations.

(a) $2i(3 - 8i) = 6i - 16i^2$ Distributive Property

$\qquad\qquad = 6i - 16(-1)$ $i^2 = -1$

$\qquad\qquad = 16 + 6i$ Write the result in the form $a + bi$.

(b) $(3i)^2 = 9i^2 = 9(-1) = -9$ The result is a real number, but it is also the complex number $-9 + 0i$.

Think About It

A method for simplifying i^n, where $n > 4$, is to begin by dividing the exponent by 4. Describe the rest of the process.

(c) $(2 + 5i)(1 - 3i) = 2 - 6i + 5i - 15i^2$ FOIL method

$\qquad\qquad\qquad = 2 - i - 15(-1)$ $i^2 = -1$

$\qquad\qquad\qquad = 17 - i$

(d) $(4 + 3i)(4 - 3i) = 4^2 - (3i)^2$ $(A + B)(A - B) = A^2 - B^2$

$\qquad\qquad\qquad = 16 - (-9)$ $(3i)^2 = 9i^2 = -9$

$\qquad\qquad\qquad = 16 + 9$

$\qquad\qquad\qquad = 25$

Pairs of complex numbers, such as $4 + 3i$ and $4 - 3i$ in part (d) of Example 4, are called **complex conjugates.**

> **Definition of Complex Conjugates**
>
> The numbers $a + bi$ and $a - bi$ are **complex conjugates.**

In Example 4, we found that $(4 + 3i)(4 - 3i) = 4^2 + 3^2 = 25$. In general, the product of complex conjugates is the sum of the squares of the real and imaginary parts, and the result is always a real number.

> **The Product of Complex Conjugates**
>
> The product $(a + bi)(a - bi)$ is the real number $a^2 + b^2$.

The rule for dividing a complex number $a + bi$ by a nonzero real number c is similar to that used for dividing a binomial by a monomial.

$$\frac{a + bi}{c} = \frac{a}{c} + \frac{b}{c}i$$

In words, divide the real part and the imaginary part by the number c.

EXAMPLE 5

Dividing a Complex Number by a Real Number

Divide $3 + 8i$ by 2.

Solution

$$\frac{3 + 8i}{2} = \frac{3}{2} + \frac{8}{2}i = \frac{3}{2} + 4i$$

Dividing a complex number by an imaginary number is similar to rationalizing the denominator. Because the product of complex conjugates is a real number, we multiply the numerator and the denominator by the conjugate of the divisor.

$$\frac{a + bi}{c + di} = \frac{a + bi}{c + di} \cdot \frac{c - di}{c - di}$$

EXAMPLE 6

Dividing Complex Numbers

Determine each quotient.

(a) $\dfrac{1 + 4i}{2 - 3i} = \dfrac{1 + 4i}{2 - 3i} \cdot \dfrac{2 + 3i}{2 + 3i}$ Multiply the numerator and denominator by $2 + 3i$, which is the conjugate of the denominator.

$$= \frac{2 + 3i + 8i + 12i^2}{2^2 + 3^2}$$ FOIL method and product of complex conjugates

$$= \frac{2 + 11i + 12(-1)}{4 + 9}$$ $i^2 = -1$

$$= \frac{-10 + 11i}{13}$$ Simplify.

$$= -\frac{10}{13} + \frac{11}{13}i$$ Standard form

(b) $\dfrac{5 - i}{2i} = \dfrac{5 - i}{2i} \cdot \dfrac{-2i}{-2i}$ Multiply the numerator and denominator by $-2i$, which is the conjugate of the denominator.

$$= \frac{-10i + 2i^2}{-4i^2}$$ Distributive Property

$$= \frac{-10i + 2(-1)}{-4(-1)}$$ $i^2 = -1$

$$= \frac{-2 - 10i}{4}$$ Simplify.

$$= -\frac{2}{4} - \frac{10}{4}i$$ $\dfrac{a + bi}{c} = \dfrac{a}{c} + \dfrac{b}{c}i$

$$= -\frac{1}{2} - \frac{5}{2}i$$ Standard form

Complex Number Solutions of Quadratic Equations

As we noted at the beginning of this section, the equation $x^2 = -1$ has no real number solution. However, i^2 and $(-i)^2$ are both equal to -1. Thus, if we consider solutions in the complex number system, the equation has two solutions: $-i$ and i.

We can extend the Square Root Property to include complex numbers.

Square Root Property

For complex numbers A and B, if $A^2 = B$, then $A = \pm\sqrt{B}$.

Note: In the real number system the Square Root Property requires B to be nonnegative. In the complex number system this requirement can be removed.

EXAMPLE 7

Using the Square Root Property to Determine Complex Number Solutions

Determine all complex number solutions of each equation.

(a) $x^2 + 25 = 0$

(b) $(x - 5)^2 = -9$

Solution

(a) $x^2 + 25 = 0$

$$x^2 = -25 \qquad \text{Isolate } x^2.$$

$$x = \pm\sqrt{-25} \qquad \text{Square Root Property}$$

$$x = \pm 5i \qquad \text{Definition of } \sqrt{-n}.$$

The complex number solutions are $5i$ and $-5i$.

(b) $(x - 5)^2 = -9$

$$x - 5 = \pm\sqrt{-9} \qquad \text{Square Root Property}$$

$$x - 5 = \pm 3i \qquad \text{Definition of } \sqrt{-n}.$$

$$x = 5 \pm 3i \qquad \text{Solve for } x.$$

The complex number solutions are $5 + 3i$ and $5 - 3i$.

We can also use the Quadratic Formula to obtain complex number solutions of a quadratic equation.

EXAMPLE 8

Determining Complex Number Solutions with the Quadratic Formula

Use the Quadratic Formula to solve each equation.

(a) $x^2 + 2x + 5 = 0$

(b) $x^2 - 4x + 7 = 0$

Solution

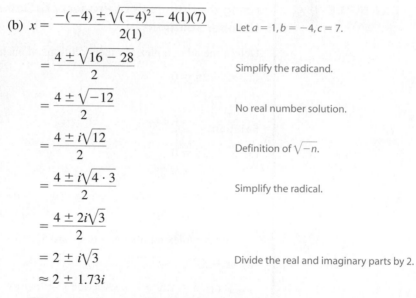

(a) $x = \dfrac{-b \pm \sqrt{b^2 - 4ac}}{2a}$

$= \dfrac{-2 \pm \sqrt{2^2 - 4(1)(5)}}{2(1)}$ Let $a = 1, b = 2, c = 5$.

$= \dfrac{-2 \pm \sqrt{4 - 20}}{2}$ Simplify the radicand.

$= \dfrac{-2 \pm \sqrt{-16}}{2}$ The radicand is negative, so no real number solution exists.

$= \dfrac{-2 \pm 4i}{2}$ Definition of $\sqrt{-n}$.

$= -1 \pm 2i$ Divide the real and imaginary parts by 2.

The complex number solutions are $-1 + 2i$ and $-1 - 2i$.

(b) $x = \dfrac{-(-4) \pm \sqrt{(-4)^2 - 4(1)(7)}}{2(1)}$ Let $a = 1, b = -4, c = 7$.

$= \dfrac{4 \pm \sqrt{16 - 28}}{2}$ Simplify the radicand.

$= \dfrac{4 \pm \sqrt{-12}}{2}$ No real number solution.

$= \dfrac{4 \pm i\sqrt{12}}{2}$ Definition of $\sqrt{-n}$.

$= \dfrac{4 \pm i\sqrt{4 \cdot 3}}{2}$ Simplify the radical.

$= \dfrac{4 \pm 2i\sqrt{3}}{2}$

$= 2 \pm i\sqrt{3}$ Divide the real and imaginary parts by 2.

$\approx 2 \pm 1.73i$

The approximate complex number solutions are $2 + 1.73i$ and $2 - 1.73i$.

Quick Reference 10.4

Definitions   • The **imaginary number i** has the following properties:

$$i = \sqrt{-1} \qquad \text{and} \qquad i^2 = -1$$

• For any positive real number n, $\sqrt{-n} = i\sqrt{n}$.

• A **complex number** can be written in the *standard form $a + bi$*, where a and b are real numbers. The number a is called the **real part,** and the number b is called the **imaginary part.**

• The expanded number system that includes i is called the **complex number system.** If $b = 0$, the complex number is also a real number, and if $b \neq 0$, the complex number is an imaginary number.

Operations with Complex Numbers

- The associative, commutative, and distributive properties hold for complex numbers.
- To add or subtract complex numbers, we use the Distributive Property to combine the real parts and the imaginary parts.
- To multiply complex numbers, use the Distributive Property in a manner similar to the way it is used to multiply polynomials.
- The numbers $a + bi$ and $a - bi$ are **complex conjugates.** Their product is $a^2 + b^2$, which is a real number.
- To divide a complex number $a + bi$ by a nonzero real number c, use the following rule.

$$\frac{a + bi}{c} = \frac{a}{c} + \frac{b}{c}i$$

  In words, divide the real part and the imaginary part by the number c.
- To divide a complex number by an imaginary number, multiply the numerator and the denominator of the quotient by the conjugate of the divisor.

Complex Number Solutions of Quadratic Equations

- We can extend the Square Root Property to include complex numbers. For complex numbers A and B, if $A^2 = B$, then $A = \pm\sqrt{B}$.
- The Quadratic Formula can be used to obtain complex number solutions of a quadratic equation.

Speaking the Language 10.4

1. The number defined as $\sqrt{-1}$ is called the ▩▩▩ number i.
2. A number of the form $a + bi$, where a and b are real numbers, is called a(n) ▩▩▩ number.
3. The numbers $a + bi$ and $a - bi$ are called ▩▩▩, and their product is a(n) ▩▩▩ number.
4. We use the ▩▩▩ Property to add, subtract, and multiply complex numbers.

Exercises 10.4

Concepts and Skills

1. What is the difference between an imaginary number and a complex number?
2. If a and b are real numbers, how do the following products differ?
   (i) $(a + b)(a - b)$
   (ii) $(a + bi)(a - bi)$

In Exercises 3–12, write the number in terms of i and with a simplified radical.

3. $\sqrt{-15}$
4. $\sqrt{-7}$
5. $\sqrt{-49}$
6. $\sqrt{-16}$
7. $\sqrt{-144}$
8. $\sqrt{-100}$
9. $\sqrt{-20}$
10. $\sqrt{-54}$
11. $\sqrt{-75}$
12. $\sqrt{-44}$

In Exercises 13–20, identify (a) the real part and (b) the imaginary part of the complex number.

13. $5 + 7i$
14. $-2 + 5i$
15. $10i$
16. $-8i$
17. 12
18. 5
19. $15i - 7$
20. $i - 2$

In Exercises 21–32, add or subtract the complex numbers.

21. $2i + (3 + 4i)$ **22.** $-6 + (2 - 5i)$

23. $(3 - 2i) + (4 + i)$ **24.** $(1 + i) + (3 + 2i)$

25. $(7 + 3i) - (-2 + 5i)$

26. $(-1 + 4i) - (1 - 2i)$

27. $5 - (3 - 2i)$ **28.** $-3i - (4 - 3i)$

29. $(6 + i) + (6 - i)$

30. $(1 + 3i) + (1 - 2i)$

31. $(-3 + 2i) - (-1 - 4i)$

32. $(7 + i) - (6 - 5i)$

In Exercises 33–44, multiply the complex numbers.

33. $i(2 + 3i)$ **34.** $3i(-1 + 5i)$

35. $-2i(6 - i)$ **36.** $-i(-2 - 7i)$

37. $(5i)^2$ **38.** $(-2i)^2$

39. $(5 - 4i)^2$ **40.** $(1 + 4i)^2$

41. $(3 - 2i)(2 - i)$ **42.** $(7 + i)(11 - i)$

43. $(2 - i)(1 + 2i)$ **44.** $(1 - 6i)(1 - 2i)$

In Exercises 45–50, (a) write the conjugate of the complex number and (b) determine the product of the number and its conjugate.

45. $5 - 3i$ **46.** $1 + 4i$

47. $7 + 2i$ **48.** $-3 - 2i$

49. $-1 + i$ **50.** $4 - 3i$

▼ 51. Consider the following two expressions.

(i) $\dfrac{1}{1 + i}$ (ii) $\dfrac{1}{1 + \sqrt{3}}$

Compare the process of finding the quotient in (i) to the process of rationalizing the denominator in (ii).

▼ 52. Explain why $x^2 + 4 = 0$ has no real number solution but has two complex number solutions.

In Exercises 53–62, divide the complex numbers.

53. $\dfrac{-12 + 15i}{3}$ **54.** $\dfrac{8 - 12i}{-2}$

55. $\dfrac{3 + 2i}{i}$ **56.** $\dfrac{1 - 4i}{-i}$

57. $\dfrac{3i}{1 - 2i}$ **58.** $\dfrac{2i}{3 + i}$

59. $\dfrac{1 + 3i}{1 - 3i}$ **60.** $\dfrac{2 - i}{2 + i}$

61. $\dfrac{5}{2 + i}$ **62.** $\dfrac{2}{1 + 3i}$

▼ 63. What have you observed about the imaginary solutions of a quadratic equation?

▼ 64. If applying the Quadratic Formula to solving a quadratic equation results in imaginary number solutions, what must be true about the discriminant?

In Exercises 65–74, use the Square Root Property to determine the complex solutions of the equation.

65. $y^2 + 4 = 0$ **66.** $y^2 + 25 = 0$

67. $t^2 + 5 = 0$ **68.** $10 + x^2 = 0$

69. $(x + 5)^2 = -16$ **70.** $(x - 2)^2 = -81$

71. $(3x - 1)^2 = -9$ **72.** $(2x + 1)^2 = -36$

73. $(x + 2)^2 + 18 = 0$ **74.** $(x - 5)^2 + 24 = 0$

In Exercises 75–90, use the Quadratic Formula to determine the complex solutions of the equation.

75. $x^2 - 2x + 5 = 0$ **76.** $x^2 + 6x + 10 = 0$

77. $x^2 + 10x + 26 = 0$ **78.** $x^2 - 4x + 20 = 0$

79. $4 + 3x + x^2 = 0$ **80.** $4 - x + x^2 = 0$

81. $x^2 - 3x + 9 = 0$ **82.** $x^2 - 5x + 9 = 0$

83. $x^2 + 4x = -7$ **84.** $x^2 = 2x - 8$

85. $4x^2 - 4x + 5 = 0$

86. $2x^2 + 6x + 5 = 0$

87. $9x^2 + 12x + 5 = 0$

88. $8 + 2x + 5x^2 = 0$

89. $4x^2 = -2x - 1$

90. $9x^2 = 6x - 17$

In Exercises 91–94, determine the number.

91. The product of a number and 4 more than the number is -8.

92. Twice the product of a number and 1 less than the number is -1.

93. Ten more than the square of a number is 6 times the number.

94. The difference between 4 times a number and 5 is the square of the number.

Challenge

In Exercises 95 and 96, solve the equation.

95. $x^2 + 2ix - 5 = 0$ **96.** $x^2 + ix + 6 = 0$

In Exercises 97–100, evaluate the expression.

97. i^3 **98.** i^5

99. i^4 **100.** i^6

10.5 | Graphs of Quadratic Equations

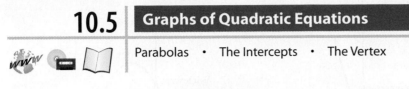

Parabolas • The Intercepts • The Vertex

Parabolas

A **quadratic equation in two variables** can be written in the form $y = ax^2 + bx + c$, where $a \neq 0$. For any x-value, we can determine the corresponding y-value and form the ordered pair (x, y). Each pair can then be represented by a point in the coordinate plane.

Exploring the Concept

The Graph of a Quadratic Expression

Consider $y = x^2 + 2x - 3$. The following table shows a partial list of ordered pairs that satisfy the equation, and Figure 10.6 shows the points that represent the pairs.

Figure 10.6

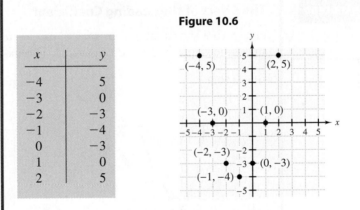

x	y
-4	5
-3	0
-2	-3
-1	-4
0	-3
1	0
2	5

Rather than plot individual points, we can use a calculator to graph the expression $x^2 + 2x - 3$. (See Fig. 10.7.) We have done this before to estimate solutions of quadratic equations.

Figure 10.7

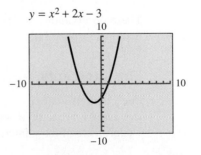

$y = x^2 + 2x - 3$

The familiar U-shaped graph of a quadratic expression is called a **parabola.** Recall that the points where a graph intersects the x- and y-axes are called the x- and y-intercepts. The lowest point of a parabola (or the highest point, if the parabola opens downward) is called the **vertex.** (See Fig. 10.8.)

Figure 10.8

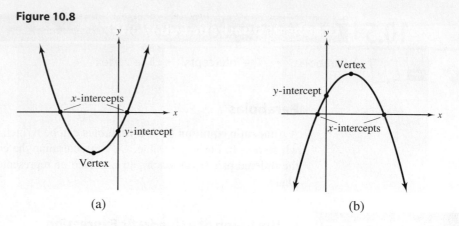

(a)   (b)

By experimenting with the graphs of quadratic equations in the simple form $y = ax^2$, we can observe the effect of the leading coefficient a.

Exploring the Concept

The Effect of the Leading Coefficient

Figure 10.9 shows the graphs of $y = ax^2$ for $a = \frac{1}{3}$, 1, and 3, and Figure 10.10 shows the graphs of $y = ax^2$ for $a = -\frac{1}{3}$, -1, and -3.

Figure 10.9   **Figure 10.10**

$y = 1x^2$
$y = 3x^2$ $y = \frac{1}{3}x^2$

$y = -1x^2$
$y = -3x^2$ $y = -\frac{1}{3}x^2$

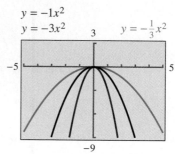

Observe that all the parabolas in Figure 10.9 open upward and that $a > 0$. Similarly, all the parabolas in Figure 10.10 open downward and $a < 0$.

These observations are true in general: The graph of $y = ax^2 + bx + c$ is a parabola that opens upward if $a > 0$ and downward if $a < 0$.

The Intercepts

The graph of $y = ax^2 + bx + c$ is a parabola that has exactly one y-intercept. Recall that a y-intercept of a graph is a point whose x-coordinate is 0. To determine the y-coordinate of the y-intercept of a parabola, we replace x with 0.

$$y = ax^2 + bx + c$$
$$y = a(0)^2 + b(0) + c = c$$

Thus the y-intercept is $(0, c)$.

To determine the x-intercepts of a parabola, we replace y with 0. Thus the equation $y = ax^2 + bx + c$ becomes $0 = ax^2 + bx + c$, a quadratic equation that we can solve with a variety of methods. Because the equation can have two solutions, one solution, or no solution, a parabola can have two x-intercepts, one x-intercept, or no x-intercept.

EXAMPLE **1**   **Determining the Orientation and Intercepts of a Parabola**

From the given equation, predict the orientation of the graph. Then produce the graph of the equation and estimate the intercepts. Finally, determine the intercepts algebraically.

(a) $y = 8 + 7x - x^2$ (b) $y = x^2 - 8x + 16$ (c) $y = -x^2 + 2x - 2$

Solution

(a) $y = 8 + 7x - x^2$

Figure 10.11

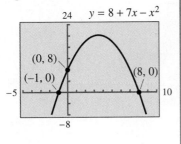

Because $a = -1$, the parabola opens downward. (See Fig. 10.11.)

From the graph, the y-intercept appears to be $(0, 8)$, and the x-intercepts appear to be $(-1, 0)$ and $(8, 0)$. To verify the y-intercept we note that $c = 8$, so the y-intercept is $(0, 8)$.

To determine the x-intercepts algebraically, we replace y with 0 and solve the resulting equation.

$$8 + 7x - x^2 = 0 \qquad \text{Let } y = 0.$$
$$(8 - x)(1 + x) = 0 \qquad \text{Factor.}$$
$$8 - x = 0 \quad \text{or} \quad 1 + x = 0 \qquad \text{Set each factor equal to 0.}$$
$$x = 8 \quad \text{or} \quad x = -1 \qquad \text{Solve for } x.$$

The two x-intercepts are $(8, 0)$ and $(-1, 0)$.

(b) $y = x^2 - 8x + 16$

Figure 10.12

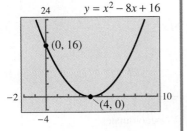

Because $a = +1$, the parabola opens upward. We estimate the intercepts to be $(0, 16)$ and $(4, 0)$. (See Fig 10.12.)

The y-intercept is $(0, 16)$ because $c = 16$.

To determine the one x-intercept, we replace y with 0 and solve the quadratic equation.

$$x^2 - 8x + 16 = 0 \qquad \text{Let } y = 0.$$
$$(x - 4)^2 = 0 \qquad \text{The trinomial is a perfect square.}$$
$$x - 4 = 0 \qquad \text{Square Root Property}$$
$$x = 4 \qquad \text{Solve for } x.$$

The x-intercept is $(4, 0)$.

(c) $y = -x^2 + 2x - 2$

Figure 10.13

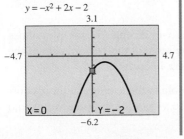

Again, $a = -1$, so the parabola opens downward. (See Fig. 10.13.) The y-intercept is $(0, -2)$, but there appears to be no x-intercept. To verify this, evaluate the discriminant.

$$b^2 - 4ac = 2^2 - 4(-1)(-2) = 4 - 8 = -4$$

Because the discriminant is negative, the equation $-x^2 + 2x - 2 = 0$ has no real number solution, and the parabola has no x-intercept.

The Vertex

We have seen that a parabola opens upward and its vertex is the lowest point if $a > 0$. We also observed that a parabola opens downward and its vertex is the highest point if $a < 0$.

One way to determine the vertex of a parabola is to produce a graph and trace to the lowest or highest point.

EXAMPLE 2

Estimating a Vertex by Graphing

Produce a graph of $y = x^2 + 2x - 3$ and estimate the coordinates of the vertex.

Solution

Figure 10.14

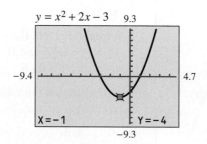

The estimated vertex is $(-1, -4)$. (See Fig. 10.14.)

We now examine the relationship between the vertex of a parabola and the x-intercepts.

Exploring the Concept

The Coordinates of the Vertex

Figure 10.15 shows the graph of $y = x^2 - 10x + 16$.

Figure 10.15

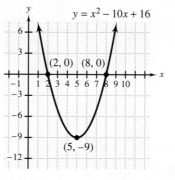

Figure 10.16

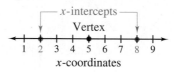

Observe the x-coordinates of the vertex and the x-intercepts. If we plot those x-coordinates on a number line (see Fig. 10.16.), we see that the x-coordinate of the vertex lies midway between the x-coordinates of the x-intercepts.

In general, the x-coordinate of the vertex of a parabola is the average of the x-coordinates of the x-intercepts. In other words, if h is the x-coordinate of the vertex and

x_1 and x_2 are the x-coordinates of the x-intercepts, then $h = \dfrac{x_1 + x_2}{2}$. We can use the Quadratic Formula to determine x_1 and x_2.

$$x_1 = \frac{-b + \sqrt{b^2 - 4ac}}{2a} \qquad \text{and} \qquad x_2 = \frac{-b - \sqrt{b^2 - 4ac}}{2a}$$

If we replace x_1 and x_2 with these expressions in the formula for h, it can be shown that $h = -\dfrac{b}{2a}$. (See Exercise 91 for a derivation of this formula.)

The following summary describes how this fact can be used to determine the vertex of a parabola algebraically.

Determining the Vertex of a Parabola Algebraically

For a parabola whose equation is $y = ax^2 + bx + c$, the following steps can be used to determine the vertex $V(h, k)$.

1. Determine the first coordinate with the formula $h = -\dfrac{b}{2a}$.

2. Determine the second coordinate k by evaluating the expression $ax^2 + bx + c$ for $x = h$.

LEARNING TIP

For a linear equation, the slope and the intercepts are important features of the graph. For a quadratic equation, the intercepts and the vertex are important features.

As always, it is wise to estimate the vertex of a parabola graphically and to determine the vertex algebraically.

EXAMPLE 3

Think About It

By examining the graphs of $y = (x - h)^2$ and $y = x^2 + k$ for various values of h and k, describe the graph of an equation in the form $y = (x - h)^2 + k$.

The Vertex of a Parabola

Consider the parabola whose equation is $y = 2x^2 + 8x - 3$.

(a) Estimate the vertex $V(h, k)$ graphically.

(b) Determine the vertex algebraically.

Solution

(a) The estimated vertex is $(-2, -11)$. (See Fig. 10.17.)

Figure 10.17

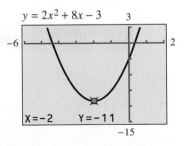

$y = 2x^2 + 8x - 3$

X=-2   Y=-11

Figure 10.18

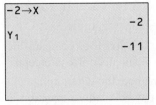

-2→X

Y₁

-2

-11

(b) The first coordinate of the vertex is $h = -\dfrac{b}{2a} = -\dfrac{8}{2 \cdot 2} = -2$. The second coordinate k is found by evaluating the expression $2x^2 + 8x - 3$ for $x = -2$. (See Fig. 10.18.) The vertex is $V(-2, -11)$.

Quick Reference 10.5

Parabolas
- A **quadratic equation in two variables** can be written in the form
$$y = ax^2 + bx + c, \quad a \neq 0$$
- The graph of a quadratic equation is called a **parabola.** The points where a graph intersects the x- and y-axes are called the x- and y-intercepts. The highest or lowest point of a parabola is called the **vertex.**
- For a parabola whose equation is $y = ax^2 + bx + c$, the parabola opens upward if $a > 0$ and downward if $a < 0$.

The Intercepts
- The parabola whose equation is $y = ax^2 + bx + c$ has exactly one y-intercept $(0, c)$.
- The x-intercept(s), if any, are found by letting $y = 0$ and then solving the equation $ax^2 + bx + c = 0$. The parabola may have two x-intercepts, one x-intercept, or no x-intercept.

The Vertex
- We can estimate the coordinates of the vertex of a parabola by producing the graph and tracing to the highest or lowest point.
- For a parabola whose equation is $y = ax^2 + bx + c$, the following steps can be used to determine the vertex $V(h, k)$ algebraically.
  1. Determine the first coordinate with the formula $h = -\dfrac{b}{2a}$.
  2. Determine the second coordinate k by evaluating the expression $ax^2 + bx + c$ for $x = h$.

Speaking the Language 10.5

1. An equation in the form $y = ax^2 + bx + c$ is called a(n) ▢▢▢▢▢ equation in ▢▢▢▢▢ .
2. The graph of $y = ax^2 + bx + c$ is a(n) ▢▢▢▢▢ .
3. The ▢▢▢▢▢ of a parabola is the highest or lowest point of the graph.
4. The graph of $y = ax^2 + bx + c$ has exactly one ▢▢▢▢▢ , and its coordinates are $(0, c)$.

Exercises 10.5

Concepts and Skills

1. For what condition is the vertex of the graph of the equation $y = ax^2 + bx + c$ the highest point of the graph? the lowest point?

2. Suppose that the graph of $y = ax^2 + bx + c$ is a parabola whose vertex is $V(-2, 5)$. What is the algebraic interpretation of the second coordinate, 5?

In Exercises 3–10, state whether the parabola opens upward or downward.

3. $y = x^2 - 3x$
4. $y = -5 + x^2$
5. $y = 7 - x^2$
6. $y = 4 + 3x - x^2$
7. $y = 2x^2 - 3x - 5$
8. $y = 2 - 5x - 3x^2$
9. $y = 1 + 2x - x^2$
10. $y = 4 - x + 3x^2$

11. For the graph of $y = ax^2 + bx + c$, compare the number of y-intercepts to the number of x-intercepts.

12. Describe an algebraic method for determining the x-intercepts, if any, of a parabola whose equation is $y = ax^2 + bx + c$.

In Exercises 13–28, use the graph to estimate the intercepts of the parabola. Then determine the intercepts algebraically.

13. $y = x^2 + 2x - 3$
14. $y = x^2 - 5x + 4$
15. $y = 9 - 6x + x^2$
16. $y = -x^2 + 2x - 1$
17. $y = 2x - 2x^2$
18. $y = -15x + 3x^2$
19. $y = x^2 - 3x + 5$
20. $y = 1 + x + 2x^2$
21. $y = -4x^2 - 4x - 1$
22. $y = 4x^2 + 12x + 9$
23. $y = 12 + 5x - 2x^2$
24. $y = 1 - 9x^2$
25. $y = -x^2 + x - 2$
26. $y = 3x - 5 - 2x^2$
27. $y = x^2 - 25$
28. $y = 3x^2 - 4x + 1$

In Exercises 29–36, use a graph to estimate the vertex. Then determine the vertex algebraically.

29. $y = x^2 + 6x$
30. $y = x^2 - 4x$
31. $y = x^2 + 2x - 2$
32. $y = x^2 - 4x - 1$
33. $y = x^2 - 6x + 4$
34. $y = x^2 + 10x + 15$
35. $y = 2x - x^2$
36. $y = -6x - x^2$

In Exercises 37–48, describe the graph by identifying the vertex and the orientation of the parabola.

37. $y = x^2 + 6x - 7$
38. $y = x^2 - 4x + 5$
39. $y = -x^2 + 2x - 1$
40. $y = x^2 + 6x + 9$
41. $y = x^2 + 4$
42. $y = -x^2 + 2$
43. $y = -x^2 - 3x + 2$
44. $y = -x^2 + 5x + 6$
45. $y = 2x^2 - 12x$
46. $y = -3x^2 + 18x$
47. $y = -3x^2 + 6x + 1$
48. $y = 2x^2 + 12x + 11$

49. Suppose that the vertex of a parabola is $V(2, 5)$. Describe the number of x-intercepts if
(a) $a > 0$.
(b) $a < 0$.

50. Consider the parabola $y = ax^2 + bx + c$. Explain how the discriminant can indicate the number of x-intercepts that the parabola has.

In Exercises 51–58, use the given value of a and the given vertex to determine the number of x-intercepts of the parabola.

51. $a = -2$, $V(3, 4)$
52. $a = 1$, $V(-2, -5)$
53. $a = 3$, $V(-1, 0)$
54. $a = -1$, $V(3, 0)$
55. $a = 2$, $V(-2, 4)$
56. $a = -4$, $V(0, -2)$
57. $a = -1$, $V(0, -4)$
58. $a = 4$, $V(4, -3)$

In Exercises 59–66, determine the number of x-intercepts of the parabola. (*Hint*: Consider the discriminant.)

59. $y = 6x - 2x^2$
60. $y = 6x^2 + x - 2$
61. $y = x^2 + 10x + 25$
62. $y = -4x^2 + 4x - 1$

63. $y = 2x^2 + x - 2$
64. $y = 1 + 4x - x^2$
65. $y = 3 + x + x^2$
66. $y = -x^2 + 2x - 4$

In Exercises 67–74, use the given information to determine the number of x-intercepts.

67. The vertex is $(2, 3)$, and the parabola opens upward.
68. The vertex is $(5, -7)$, and the parabola opens upward.
69. The vertex is $(-4, 5)$, and the parabola opens downward.
70. The vertex is $(-2, -6)$, and the parabola opens downward.
71. The vertex is $(2, 0)$.
72. The vertex is $(-6, 0)$.
73. The vertex is $(4, 2)$, and the y-intercept is $(0, 5)$.
74. The vertex is $(-5, 3)$, and the y-intercept is $(0, -2)$.

In Exercises 75–78, the graph of $y = ax^2 + bx + c$ is given. From the following list, select the statements that are true.

  (i) $a > 0$
  (ii) $a < 0$
  (iii) The equation $ax^2 + bx + c = 0$ has two real number solutions.
  (iv) The equation $ax^2 + bx + c = 0$ has one real number solution.
  (v) The equation $ax^2 + bx + c = 0$ has no real number solution.
  (vi) The graph contains $(3, 0)$.
  (vii) The graph contains the point $(0, -9)$.
  (viii) The first coordinate of the vertex is 3.
  (ix) The vertex is also the x-intercept.
  (x) The vertex is also the y-intercept.

75. **76.**

77. **78.**

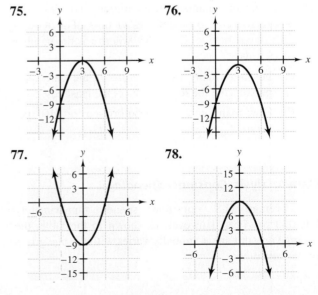

In Exercises 79 and 80, use a graph to approximate the answer to each question.

79. Consider the product of a number and 6 less than the number.

   (a) For what number is the product the smallest? What is that product?

   (b) For what numbers is the product 7?

   (c) For what number is the product -10?

80. Consider the product of a number and the difference of 8 and the number.

   (a) For what number is the product the largest? What is that product?

   (b) For what numbers is the product -9?

   (c) For what number is the product 20?

Real-Life Applications

In Exercises 81 and 82, use a graph to approximate the answer to each question.

81. The height h in feet of a ball t seconds after it is thrown is given by $h = -16t^2 + 96t$.

   (a) After how many seconds does the ball reach its highest point? How high is the ball at that time?

   (b) How high is the ball after 1 second?

   (c) After how many seconds does the ball hit the ground?

82. A volunteer organization determines that the daily per-person cost c of housing victims of a flood is given by $c = x^2 - 12x + 56$, where x is the number of victims.

   (a) For how many flood victims is the daily per-person cost the least? What is the least cost?

   (b) How many victims can be housed for $101 per person?

Data Analysis: *Consumer Spending*

The average U.S. consumer spends approximately $30,000 per year. However, as shown in the table, the average amount spent annually varies with the age group of the consumer.

Age	Amount Spent
24 and under	$17,258
25–34	29,554
35–44	37,196
45–54	37,427
55–64	31,704
65–74	22,862
75 and over	17,794

(Source: Bureau of Labor Statistics.)

The amount A spent per year can be estimated by the equation $A = -24.25x^2 + 2428x - 23{,}100$, where x is the person's age.

83. Use a graph to estimate the vertex. Then determine the vertex algebraically. Round results to the nearest whole number.

84. What does the vertex represent?

85. At what ages do consumers spend the average of $30,000?

86. What do the x-intercepts represent in this case?

Challenge

In Exercises 87–90, use the graphing method to estimate the solution of the system of equations. Then use the substitution method to solve the system algebraically.

87. $y = x^2$
    $y = 2 - x$

88. $y = x + 3$
    $y = x^2 + 3x - 5$

89. $y = 4 - x^2$
    $y = x^2 - 4$

90. $y = 3 - x^2$
    $y = x^2 + 5x$

91. Consider the parabola in the figure.

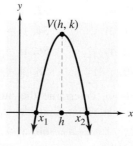

We observed that h lies midway between the x-coordinates of the x-intercepts. In other words, h is the average of x_1 and x_2.

$$h = \frac{x_1 + x_2}{2}$$

In this formula, replace x_1 and x_2 with the values we obtain by using the Quadratic Formula.

$$x_1 = \frac{-b + \sqrt{b^2 - 4ac}}{2a}$$

$$x_2 = \frac{-b - \sqrt{b^2 - 4ac}}{2a}$$

Then simplify the resulting expression to show that $h = \frac{-b}{2a}$.

10.6 | Functions

 Definitions • Functions Defined by a Rule • Graphs of Functions • Domain and Range of a Function

Definitions

An important concept in mathematics is that of a *function*. Although the concept of functions is developed more fully in later courses, we give a brief introduction in this section.

A person's pay often depends on the number of hours worked. The following table shows some typical numbers of hours worked and the corresponding wages.

Hours	10	20	30	40
Wages	60	120	180	240

In this case, the person's pay is a function of time. The information in the table can be written as a set of ordered pairs. We usually give the set a letter name such as f, g, h, or any other letter. This particular set of ordered pairs is an example of a function.

$$f = \{(10, 60), (20, 120), (30, 180), (40, 240)\}$$

An essential characteristic of a function is that no two different ordered pairs have the same first coordinate.

> **Definition of a Function**
>
> A **function** is a set of ordered pairs in which no two different ordered pairs have the same first coordinate. The set of all first coordinates is the **domain** of the function, and the set of all second coordinates is the **range** of the function.

EXAMPLE **1**

Identifying Functions

Determine whether the given set of ordered pairs is a function. If the set is a function, give the domain and the range.

(a) $f = \{(0, 1), (2, 4), (5, 6), (8, 7), (9, 6)\}$

(b) $g = \{(3, 5), (-3, 5), (2, 4), (3, 7)\}$

Solution

(a) No two pairs have the same first coordinate, so the set f represents a function. The domain is $\{0, 2, 5, 8, 9\}$, and the range is $\{1, 4, 6, 7\}$.

(b) Two pairs, $(3, 5)$ and $(3, 7)$, have the same first coordinate. Therefore, the set g does not represent a function.

Functions Defined by a Rule

Rather than listing the ordered pairs, we often describe a function with a rule in the form of an equation. For example, the equation $y = 2x$ assigns twice the value of x to y. When a function is described by a rule, we usually use function notation. In this case, we write $f(x) = 2x$ rather than $y = 2x$.

Note: Recall that the symbol $f(x)$ is read "f of x." We can use $f(x)$ and y interchangeably; that is, $y = f(x)$. Thus the value of y is a function of (depends on) the value of x.

The names given to certain categories of functions reflect the type of expression used in the rule. Thus the function $f(x) = 2x$ is called a *linear function*. Similarly, $g(x) = 7$ is called a *constant function*, and $h(x) = x^2 + 4x - 5$ is called a *quadratic function*. When we describe a function with a rule, we often think of the definition of a function in terms of the rule rather than in terms of ordered pairs.

Alternative Definition of a Function

A function $y = f(x)$ is a rule that assigns any permissible value of the variable x to a unique value y. The domain is the set of all permissible values of x.

EXAMPLE **2**

Determining Whether an Equation Represents a Function

Determine whether the given equation represents a function.

(a) $y = x^2$ (b) $x = y^2$

LEARNING TIP

To decide whether an equation defines a function, ask yourself, "For a value of x, is more than one value of y possible?" If the answer is yes, then the equation does not define a function.

Solution

(a) For each real number x, x^2 is unique. Thus $y = x^2$ represents a quadratic function. The function may be written with function notation: $f(x) = x^2$.

(b) For $x = 9$, there are two values for y: -3 and 3. The equation $x = y^2$ does not represent a function.

Graphs of Functions

We produce the graph of a function in exactly the same way that we graph expressions. For example, to graph $f(x) = x^2 + 3$, we simply enter $x^2 + 3$ in a calculator and graph. We can also sketch the graph by plotting the pairs (x, y), where $y = x^2 + 3$. Note again that we often use y and $f(x)$ interchangeably when we work with functions and their graphs.

An easy way to determine whether an equation represents a function is to examine the graph of the equation.

Exploring the Concept

The Vertical Line Test

Figure 10.19 shows a graph with a vertical line intersecting two points of the graph. The two points of intersection have the same first coordinate. This graph does not represent a function because a function cannot have two ordered pairs with the same first coordinate. Figure 10.20 shows a graph for which it is not possible to draw a vertical line that would intersect the graph more than once. This graph represents a function.

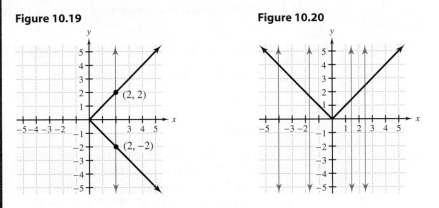

Figure 10.19

Figure 10.20

This simple test for determining whether a graph represents a function is called the **Vertical Line Test**.

The Vertical Line Test

If it is possible to draw a vertical line that intersects a graph at more than one point, the graph does not represent a function. If it is not possible to draw such a line, the graph does represent a function.

EXAMPLE **3**

Determining Whether a Graph Represents a Function

Determine whether the given graph represents a function.

Think About It

Suppose that it is possible to draw a horizontal line that intersects a graph at more than one point. Could the graph represent a function? Can you visualize such a graph that does not represent a function?

Figure 10.21

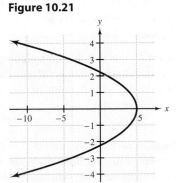

Figure 10.22

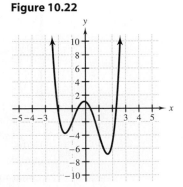

Solution

(a) It is possible to draw a vertical line that intersects the graph at more than one point. Therefore, the graph does not represent a function.

(b) The graph represents a function because no vertical line can be drawn that would intersect the graph at more than one point.

Domain and Range of a Function

A graph can be used to estimate the domain and the range of a function.

EXAMPLE 4

Estimating the Domain and the Range of a Function

Use a graph of the given function to estimate the domain and the range of the function.

(a) $f(x) = x + 4$ (b) $g(x) = x^2 - 4$

Solution

(a) From the graph in Figure 10.23, we see that the line extends forever both horizontally and vertically. Thus any number is a possible first coordinate, and any number is a possible second coordinate. The domain and the range of this linear function are the set of all real numbers **R.**

Our conclusion can be verified by noting that any real number is a permissible replacement for x; and the resulting value of $x + 4$ can also be any real number.

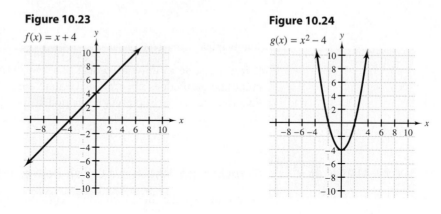

Figure 10.23

$f(x) = x + 4$

Figure 10.24

$g(x) = x^2 - 4$

(b) In Figure 10.24 the parabola continues to widen horizontally, so the domain is **R.** However, no y-coordinate of the graph is less than -4. Thus the range is the set of all y-values such that $y \geq -4$.

Again, we can verify our conclusion by noting that x can be replaced with any real number. Also, from our study of parabolas, we know that the vertex is $V(0, -4)$, which is the lowest point of the graph. Therefore, every point has a y-coordinate that is at least -4.

Note: Officially, domains and ranges are sets of numbers. However, sometimes we can describe the numbers with an inequality, as we did in Example 4(b). In the remaining examples and exercises, we will use such inequalities to describe domains and ranges.

When we determine the domain of a function, we often look for numbers that are not in the domain. Remember that an even root radicand must be nonnegative, and a denominator cannot be 0.

EXAMPLE **5**

Determining the Domain of a Function

Determine the domain of each function.

(a) $h(x) = \sqrt{x - 3}$ (b) $F(x) = \dfrac{1}{x + 2}$

Solution

(a) Because the radicand must be nonnegative, $x - 3 \geq 0$ or $x \geq 3$. The domain is $x \geq 3$.

(b) We must eliminate any value of x that makes the denominator 0.

$x + 2 = 0$ Set the denominator equal to 0.

$x = -2$ Solve for x.

Because -2 makes the denominator 0, this value is not in the domain. The domain is **R**, $x \neq 2$.

Note: In Example 5, h is a *radical function* and F is a *rational function*.

The topic of functions plays a central role in the study of mathematics. This brief introduction provides you with the basics that you will need in later courses.

Quick Reference 10.6

Definitions

• A **function** is a set of ordered pairs in which no two different ordered pairs have the same first coordinate. The set of all first coordinates is the **domain** of the function, and the set of all second coordinates is the **range** of the function.

Functions Defined by a Rule

• A function $y = f(x)$ can be defined alternatively as a rule that assigns any permissible value of the variable x to a unique value y. The domain is the set of all permissible values of x.

Graphs of Functions

• We produce the graph of a function in the same way that we graph expressions.

• An easy way to determine whether a graph represents a function is to use the **Vertical Line Test.**

If it is possible to draw a vertical line that intersects a graph at more than one point, the graph does not represent a function. If it is not possible to draw such a line, the graph does represent a function.

Domain and Range of a Function

• A graph can be used to estimate the domain and the range of a function.

• When we determine the domain of a function, we often look for numbers that are not in the domain. For example, radicands of even roots must be nonnegative, and denominators cannot be 0.

Speaking the Language 10.6

1. Suppose that a set of ordered pairs has no pairs in which the first coordinates are the same. Such a set is called a(n) ▨▨▨▨ .

2. For the function $\{(2, 3), (5, -1), (-4, 0)\}$, the set $\{2, 5, -4\}$ is called the ▨▨▨▨ , and the set $\{3, -1, 0\}$ is called the ▨▨▨▨ .

3. A method for determining whether a graph represents a function is called the ▨▨▨▨ .

4. We read the symbol $g(x)$ as ▨▨▨▨ .

Exercises 10.6

Concepts and Skills

1. The set $A = \{(3, 5), (7, 5)\}$ has two pairs with the same second coordinate. Does this mean that A is not a function? Why?

2. Consider the function $f(x) = x + 1$. What are the algebraic and graphical meanings of $f(3)$?

In Exercises 3–8, determine whether the given set is a function. If the set is a function, give the domain and the range.

3. $\{(3, 2), (5, 1), (7, 2), (4, -3)\}$

4. $\{(1, 3), (2, 7), (4, 3)\}$

5. $\{(3, 2), (-1, 1), (0, 0), (3, 4)\}$

6. $\{(5, 1), (3, 2), (3, 0), (2, 2)\}$

7. $\{(1, 2), (2, 3), (3, 4), (4, 5)\}$

8. $\{(-1, 1), (-2, 2), (-3, 3), (-4, 4)\}$

In Exercises 9–14, write the set of ordered pairs and determine whether the set is a function. If the set is a function, give the domain and the range.

9. The first coordinate is a positive integer less than 5, and the second coordinate is 3 less than the first.

10. The first coordinate is a negative integer larger than -4, and the second coordinate is twice the first.

11. The first coordinate is 4, and the second coordinate is a nonnegative integer less than or equal to 5.

12. The first coordinate is 0, and the second coordinate is a negative integer greater than or equal to -6.

13. The first coordinate is an integer between -2 and 2, and the second coordinate is 4.

14. The first coordinate is an integer between -3 and 4, and the second coordinate is 0.

In Exercises 15–28, determine whether the equation represents a function.

15. $x = |y|$

16. $y = |x|$

17. $y = x^2 + 5$

18. $y^2 = x + 1$

19. $y = \sqrt{x}$

20. $x = \sqrt{y}$

21. $x^2 + y^2 = 4$

22. $x^2 - y^2 = 16$

23. $y = x^3$

24. $x = \sqrt[3]{y}$

25. $x^2 = y^2$

26. $|x| = |y|$

27. $y = \dfrac{1}{x}$

28. $x = 2y + 7$

29. Explain why $y = x + 1$ is a function but $y \le x + 1$ is not a function.

30. Suppose that the graphs of $x = 3$ and some second equation are drawn in the same coordinate system. If the graphs have two points of intersection, how do you know that the second equation does not represent a function?

In Exercises 31–42, determine whether the graph represents a function.

31.

32.

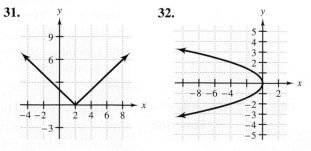

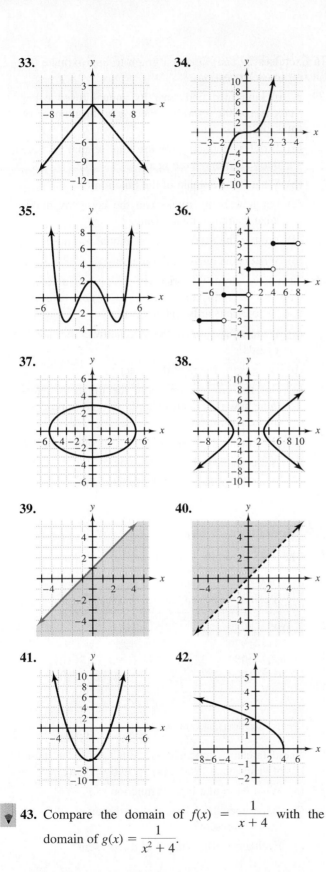

33.

34.

35.

36.

37.

38.

39.

40.

41.

42.

43. Compare the domain of $f(x) = \dfrac{1}{x+4}$ with the domain of $g(x) = \dfrac{1}{x^2+4}$.

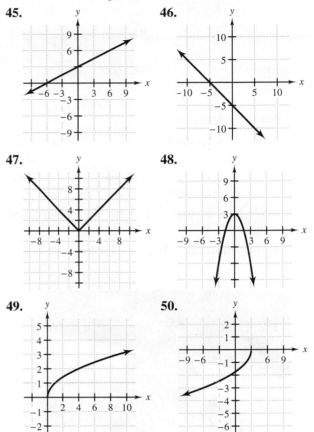

44. Describe, in general, the range of a quadratic function $f(x) = ax^2 + bx + c$ if $a > 0$.

In Exercises 45–50, use the given graph to estimate the domain and the range of the function.

45.

46.

47.

48.

49.

50.

In Exercises 51–58, produce the graph of the function. Then use the graph to estimate the domain and the range.

51. $f(x) = x - 8$ **52.** $f(x) = 10 - x$

53. $h(x) = 9 - x^2$ **54.** $g(x) = x^2 - 5$

55. $f(x) = \sqrt{9 - x}$ **56.** $h(x) = \sqrt{x + 7}$

57. $f(x) = x^2 + 4x + 3$ **58.** $g(x) = 3 + 2x - x^2$

In Exercises 59–76, determine the domain of the given function.

59. $f(x) = 1 - 2x$ **60.** $f(x) = 3x$

61. $g(x) = \dfrac{3}{x - 1}$ **62.** $h(x) = \dfrac{5}{x + 4}$

63. $h(x) = \sqrt{x + 7}$ **64.** $f(x) = \sqrt{x - 2}$

65. $f(x) = x^2 + 3$ **66.** $g(x) = 4 - x^2$

67. $f(x) = \sqrt{3 - x}$

68. $f(x) = \sqrt{1 - x}$

69. $g(x) = \dfrac{x + 5}{x}$

70. $g(x) = \dfrac{x - 2}{x + 2}$

71. $F(x) = \dfrac{x}{x^2 + 1}$

72. $G(x) = \dfrac{1 - x}{x^2 + 3}$

73. $f(x) = \sqrt{5 + x^2}$

74. $h(x) = \dfrac{2x}{\sqrt{x^2 + 2}}$

75. $h(x) = \sqrt[3]{x - 1}$

76. $g(x) = \sqrt[3]{x + 8}$

🌐 Real-Life Applications

77. For at most 6 people to rent a condo at a ski resort, the per-person cost $c(n)$ is given by $c(n) = \dfrac{420}{n}$, where n is the number of people.

(a) What are the domain and the range of the given function?

(b) Write the function as a set of ordered pairs.

78. For a pizza with up to 4 toppings, the price $p(t)$ is $p(t) = 8 + 1.5t$, where t is the number of toppings.

(a) What are the domain and the range of the given function?

(b) Write the function as a set of ordered pairs.

In Exercises 79 and 80, use a graph to approximate the answer to each question.

79. A trucking firm finds that its annual revenue per truck is $10,000 - 500x$, where x is the number of trucks in use.

(a) Write a function f for the total annual revenue derived from operating x trucks.

(b) What is the domain of the function f?

(c) For how many trucks can the company maximize its total annual revenue?

(d) What is the maximum total annual revenue?

(e) What is the range of the function f?

80. At a local store, the price (in dollars) of a Rocky Colavito baseball card is $30 - x$, where x is the number of such cards that are available.

(a) Write a function for the income from the sale of x cards.

(b) What is the range of this function?

(c) Use the information about the range to determine how many cards should be sold to produce the greatest income.

(d) How many items are sold if the income is $176?

(e) Is it possible to have an income of $250?

(f) What is the domain of the function?

📊 Modeling with Real Data

81. The table shows the trend in survival rates for cancer.

Period	Cancer Survival Rates
1955–1969	40%
1974–1976	49%
1980–1982	51%
1989–1995	59%
1998–1999	60%

(Source: American Cancer Society.)

The data can be modeled as a function of the number of years since 1960: $S(x) = 39.6 + 0.56x$.

(a) What particular type of function is S?

(b) What does the y-intercept of the graph of function S represent?

(c) Evaluate $S(30)$. What does it represent?

82. Refer to the model in Exercise 81.

    (a) What is the slope of the graph of the model function?

    (b) According to the model, what is the survival rate predicted to be in 2005?

Challenge

In Exercises 83–86, use a graph to estimate the domain and the range of the function.

83. $f(x) = \dfrac{|x|}{x}$ **84.** $g(x) = |x| + x$

85. $f(x) = \dfrac{1}{4}x^4 - \dfrac{1}{3}x^3 - 6x^2$

86. $f(x) = \sqrt{x^2}$

In Exercises 87 and 88, write a linear function that satisfies the conditions.

87. $f(0) = 3$ and $f(1) = 5$

88. $f(1) = -1$ and $f(-2) = 5$

Chapter 10 Review Exercises

Section 10.1

1. Use the graphing method to estimate the solution(s) of $x^2 - 5x = 24$.

2. Use the factoring method to solve $6x^2 - x - 2 = 0$.

3. Use the Square Root Property to solve the equation $x^2 - 28 = 0$.

4. Solve $(3x - 2)^2 = 16$.

5. Use the Square Root Property to solve the equation $36 - 12x + x^2 = 9$.

6. Use the Square Root Property to solve the equation $x(x - 4) = 9 - 4x$.

7. Solve $(x - 1)^2 = 4(x + 2)$.

8. Solve $\dfrac{16}{x^2} - 5 = -1$.

9. The product of two consecutive odd integers is 9 more than twice the smaller integer. What is the larger positive integer?

10. The length of a rectangle is 4 inches greater than the width. If a 1-inch strip is cut off each side, the area of the remaining rectangle is 12 square inches more than half the original area. How long is the original rectangle?

Section 10.2

11. Explain how to write a perfect square trinomial whose first two terms are x^2 and $10x$.

In Exercises 12–16, determine the real number solutions by completing the square.

12. $x^2 + 2x - 63 = 0$ **13.** $x^2 - 8x = 4$

14. $x(x - 5) = 2$ **15.** $x^2 + 34 = 10x$

16. $2x^2 - 4x = 1$

17. The product of the measures of two complementary angles is 1625. What is the measure of the smaller angle?

18. Twice the square of an integer is 3 less than 7 times the integer. What is the number?

Section 10.3

19. Write $-x = 5 - 2x^2$ in standard form and state the values of a, b, and c.

20. In the Quadratic Formula, what is $b^2 - 4ac$ called? What does its value tell us about the number of solutions of $ax^2 + bx + c = 0$?

In Exercises 21 and 22, solve the equation by using the Quadratic Formula.

21. $4x - 1 - 2x^2 = 0$ **22.** $x^2 + 4 = x$

In Exercises 23–28, use any method to determine the real number solutions of the equation.

23. $6x^2 + 7x - 10 = 0$ **24.** $2 = 3x(x - 2)$

25. $(x - 1)(x - 3) = 6$ **26.** $6x^2 = x - 3$

27. $9 - 24x = -16x^2$ **28.** $\dfrac{4}{x} = 2 + \dfrac{3}{2x^2}$

29. Use the discriminant to determine the number of real number solutions of $6x^2 = 6x - 1$.

30. A circular hole with a radius of 2 inches is cut out of the center of a circular piece of metal. (See the figure.) The area of the resulting ring is 96π square inches. How wide is the ring?

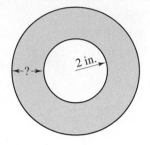

Section 10.4

31. Write $\sqrt{-11}$ in terms of i.

32. Identify (a) the real part and (b) the imaginary part of the complex number $-6 + 4i$.

In Exercises 33–38, perform the indicated operations.

33. $-2i + (4 + 3i)$

34. $(2 - 3i) - (5 + 4i)$

35. $-7i(3 + 4i)$

36. $(3 + i)(8 - 2i)$

37. $\dfrac{-4 + 6i}{2}$

38. $\dfrac{1 + 2i}{2 + 3i}$

In Exercises 39 and 40, determine the complex number solutions.

39. $x - 3 = x^2$

40. $x(2x + 1) = -1$

Section 10.5

41. Describe the orientation of the parabola whose equation is $y = 6 + 4x - x^2$.

42. Use a graph to estimate the vertex of the parabola whose equation is $y = 0.4x^2 - 6x + 3$.

In Exercises 43 and 44, determine the vertex of the parabola algebraically.

43. $y = -x^2 + 4$

44. $y = x^2 + 6x - 11$

45. Suppose that the graph of $y = ax^2 + bx + c$ opens downward and has a vertex $V(-2, 7)$. What are (a) the geometric interpretation and (b) the algebraic interpretation of V?

46. If the graph of $y = ax^2 + bx + c$ has only one x-intercept, what is true about the discriminant?

In Exercises 47 and 48, use a graph to estimate the intercepts. Then determine the intercepts algebraically.

47. $y = x^2 - 8x + 7$

48. $y = x^2 + x + 3$

49. If the vertex of a parabola is $V(2, -4)$ and the y-intercept is $(0, 1)$, how many x-intercepts does the parabola have?

50. A company's revenue $r(x)$ is estimated by the function $r(x) = x(8 - x)$, where x is the number of sales representatives employed. For how many sales representatives is the revenue at a maximum?

Section 10.6

51. Which of the following sets, if either, is a function? Why?

(i) $\{(2, 3), (3, 3), (4, 3)\}$

(ii) $\{(-1, 0), (-1, 1)\}$

52. Consider a set of ordered pairs in which the first coordinate is a whole number less than 3 and the second coordinate is a natural number less than 2.

(a) Is this set a function?

(b) If so, state the domain and the range.

53. Which of the following equations, if either, represents a function?

(i) $y = \sqrt{x + 1}$

(ii) $y^2 = x + 1$

54. Suppose that $f(x) = ax^2 + bx + c$. For what points of the graph of f is $f(x) = 0$?

In Exercises 55 and 56, determine the domain of the given function.

55. $f(x) = \sqrt{4 - x}$

56. $g(x) = \dfrac{x + 1}{x - 2}$

In Exercises 57 and 58, use a graph to estimate the range of the given function.

57. $h(x) = \sqrt{x + 2}$

58. $f(x) = 4 - 6x - x^2$

59. Which of the following graphs, if either, represents a function?

(i) A parabola with two x-intercepts

(ii) A circle with its center at the origin

60. The sum of the base and height of a parallelogram is 24 inches.

(a) Write a function for the area of the parallelogram.

(b) Use the function to determine the greatest possible area of the parallelogram.

(c) What is the domain of the function?

Chapter 10 Test

1. Use a graph to estimate the solutions of the equation $x^2 + 4x - 2 = 0$.

2. Solve the equation in Question 1 by completing the square.

3. Use the Square Root Property to solve the equation $x^2 - 6x + 9 = 16$.

4. The product of the measures of two supplementary angles is 6875. Use the method of completing the square to determine the measure of the smaller angle.

5. Solve $-2x^2 - 3x + 1 = 0$ by using the Quadratic Formula.

6. The perimeter of a rectangle is 38 inches, and the area is 88 square inches. Use the Quadratic Formula to determine the length of the rectangle.

7. Use the discriminant to determine the number of real number solutions of $3x^2 + 5x = -4$.

8. Multiply $\sqrt{-4}\,\sqrt{-9}$.

9. Simplify $\dfrac{3 - i}{3 + i}$.

10. Determine the complex number solutions of the equation $5x^2 + x + 2 = 0$.

11. What name do we give to the graph of a quadratic expression?

12. Consider $y = -x^2 - 4x + 12$. Explain how to determine

    (a) the orientation of the graph.

    (b) the intercepts of the graph.

13. Determine the vertex of the parabola whose equation is $y = 2x^2 + 4x - 3$.

14. Which of the following represents a function?

    (i) $\{(1, 2), (2, 3), (3, 3)\}$

    (ii) $y = \sqrt{x + 1}$

    (iii) $y = \dfrac{1}{x + 1}$

15. For each part in Question 14 that is a function, determine the domain and the range.

16. Explain how to determine the range of the function $y = ax^2 + bx + c$ if the vertex is $(-5, 4)$ and a < 0.

17. A rectangular town park is 300 feet long and 200 feet wide. When a sidewalk of uniform width was installed around the entire perimeter of the park, the grassy area was reduced to 56,064 square feet. How wide is the sidewalk?

18. A farmer's daily egg production $p(t)$ is modeled by $p(t) = -t^2 + 150t - 4800$, where t is the constant temperature of the henhouse.

    (a) At what temperature does the model predict maximum egg production?

    (b) How many eggs will be produced at that temperature?

Cumulative Test, Chapters 9–10

For Questions 1–6, assume that $x > 0$.

1. Simplify $\sqrt{(3x)^2}$.

2. Simplify $\sqrt{48x^3}$.

3. Simplify $\dfrac{\sqrt{5x^2}}{\sqrt{20x^6}}$.

4. Rationalize the denominator and simplify $\dfrac{\sqrt{7}}{\sqrt{7x}}$.

5. Add $\sqrt{12x} + \sqrt{75x}$.

6. Multiply and simplify $\sqrt{2x^5}\,\sqrt{6x^4}$.

7. Solve $x - \sqrt{5x - 1} = -1$.

8. Solve $\sqrt{x^2 + 12} = -2x$.

9. One part of a modern sculpture is a right triangle whose hypotenuse is 25 inches long. The difference of the lengths of the legs is 5 inches. How long is the shorter leg?

10. Simplify $(-8)^{-2/3}$.

11. Use the Square Root Property to solve the equation $x(x - 5) = 16 - 5x$.

12. Suppose that you are solving an equation of the form $ax^2 + bx + c = 0$.

    (a) If $c = 0$, why is the factoring method the best method for solving?

    (b) If $b = 0$, why is the use of the Square Root Property a good option?

13. Solve $x^2 + 4 = 6x$ by completing the square.

14. Use the Quadratic Formula to solve the equation $6x^2 - 5x - 2 = 0$.

15. Determine the product of $2i - 3$ and its conjugate.

16. Determine the complex number solutions of the equation $x^2 + x + 1 = 0$.

17. The equation $x^2 + 2x + 1 = 0$ has just one solution. Is it possible for an equation with an imaginary number solution to have just one solution? Why?

18. Consider the graph of $y = x^2 - 8x + 9$.

    (a) Describe the orientation of the graph.

    (b) What is the vertex?

    (c) What are the intercepts?

19. Give two values of x for which the set of ordered pairs $\{(-2, 3), (5, -1), (x, 8)\}$ would not be a function. Explain your answer.

20. The difference of two numbers is 1. The numbers vary inversely; that is, their product is k, which is the constant of proportionality. What is the smallest possible value of k?

Chapter 1

Speaking the Language

Section 1.1 *(page 10)*

1. integers   **2.** rational number   **3.** terminating, repeating
4. coordinate

Section 1.2 *(page 19)*

1. factor, divisor   **2.** base, exponent
3. opposite, absolute value   **4.** grouping symbols

Section 1.3 *(page 27)*

1. associative   **2.** reciprocal (or multiplicative inverse)
3. Distributive   **4.** factoring

Section 1.4 *(page 33)*

1. negative   **2.** absolute value   **3.** left, right
4. subtracting

Section 1.5 *(page 40)*

1. equivalent   **2.** reducing (or simplifying)   **3.** prime
4. common denominator

Section 1.6 *(page 48)*

1. sum   **2.** opposites (or additive inverses)   **3.** numerators
4. difference, inequality

Section 1.7 *(page 56)*

1. factors   **2.** like, unlike   **3.** opposite (or additive inverse)
4. odd

Section 1.8 *(page 63)*

1. dividend, divisor   **2.** multiplication
3. reciprocal (or multiplicative inverse)   **4.** 0, undefined

Chapter 2

Warm-Up Skills *(page 71)*

1. $3x - 15$ **2.** $4(x + y)$ **3.** $-12y$ **4.** 17   **5.** -1
6.

$$\begin{array}{c} \frac{5}{2} \\ \leftarrow\!\!\blacklozenge\!\!+\!\!+\!\!+\!\!\blacklozenge\!\!+\!\!|\!\!\blacklozenge\!\!+\!\!|\!\!+\!\!\blacklozenge\!\!\rightarrow \\ {\scriptstyle -4\,-3\,-2\,-1\ \ 0\ \ 1\ \ 2\ \ 3\ \ 4\ \ 5} \end{array}$$

7. $<$ **8.** $=$ **9.** $>$

10. Use the definition of subtraction to write the expression as $x + (-10)$, and use the Commutative Property of Addition to write $-10 + x$.

11. $-1 \cdot a$ **12.** $\left(6 \cdot \frac{5}{6}\right)(y + 3)$

Speaking the Language

Section 2.1 *(page 79)*

1. term, coefficient   **2.** equivalent   **3.** like, simplifying
4. changing the sign

Section 2.2 *(page 89)*

1. origin   **2.** ordered   **3.** quadrants   **4.** y-axis

Section 2.3 *(page 100)*

1. expression   **2.** variable, expression   **3.** tracing
4. store

Section 2.4 *(page 111)*

1. equation, solution   **2.** linear   **3.** conditional
4. point of intersection

Section 2.5 *(page 119)*

1. solutions   **2.** isolate   **3.** Addition
4. multiply, reciprocal

Section 2.6 *(page 126)*

1. Distributive   **2.** Addition, Multiplication   **3.** clearing
4. identity

Section 2.7 *(page 135)*

1. $\geq, \leq$ **2.** linear inequality in one variable   **3.** Ø, **R**
4. double, between

Section 2.8 *(page 144)*

1. positive   **2.** reverse   **3.** true, **R**
4. the variable in the middle

Chapter 3

Warm-Up Skills *(page 151)*

1. -3 **2.** $-\frac{2}{3}$ **3.** -21 **4.** 7
5. -4 **6.** 45   **7.** 86   **8.** $1.5x + 65$

9. For both, we begin by dividing both sides by -3. For the inequality, we must also reverse the inequality symbol.

10. $\frac{5}{2}$ **11.** 850   **12.** 50

Speaking the Language

Section 3.1   *(page 156)*

1. ratio   **2.** proportion   **3.** Cross-Products   **4.** similar

Section 3.2   *(page 166)*

1. percent   **2.** times   **3.** two, left, two, right
4. the original number

Section 3.3   *(page 174)*

1. formula   **2.** hypotenuse   **3.** Pythagorean Theorem
4. solving

Section 3.4   *(page 183)*

1. equality   **2.** supplementary, complementary
3. consecutive even, consecutive odd   **4.** $y - x$

Chapter 4

Warm-Up Skills   *(page 209)*

1. $(8, 3)$
2. If (a, b) is a point of the x-axis, then b is 0. If (a, b) is a point of the y-axis, then $a = 0$.
3. $y = 2x - 12$ **4.** $y = 2x + 5$ **5.** $y = -\frac{3}{2}x + 6$
6. (a) 0   (b) 0   **7.** (a) 0   (b) $\frac{1}{5}$ **8.** $(1, -2)$
9. parallel, perpendicular   **10.** $-\frac{3}{2}$
11. unchanged, reversed   **12.** $y > -4$

Speaking the Language

Section 4.1   *(page 215)*

1. linear, variables   **2.** ordered pair   **3.** line   **4.** $2x + 1$

Section 4.2   *(page 224)*

1. x-intercept   **2.** x-coordinate   **3.** second, constant
4. vertical, horizontal

Section 4.3   *(page 233)*

1. falls   **2.** run, rise   **3.** slope   **4.** undefined, 0

Section 4.4   *(page 240)*

1. slope–intercept   **2.** slope   **3.** slope   **4.** varies directly

Section 4.5   *(page 248)*

1. parallel   **2.** coincide   **3.** perpendicular
4. quadrilateral, parallelogram, trapezoid

Section 4.6   *(page 257)*

1. point–slope   **2.** slope–intercept   **3.** slope
4. vertical, horizontal

Section 4.7   *(page 266)*

1. linear inequality in two variables   **2.** boundary
3. half-plane   **4.** test

Chapter 5

Warm-Up Skills   *(page 273)*

1. 12   **2.** Ø   **3.** **R**   **4.** neither   **5.** parallel
6. neither
7. (a) $(3, 1)$ (b) $(2, -2)$ **8.** $\frac{7}{4}y + 5$
9. Both graphs include the region below the boundary line $y = x + 3$. However, the graph of $y \leq x + 3$ also includes the boundary line.
10. $5n + 10d = 345$, where $n =$ the number of nickels and $d =$ the number of dimes
11. $48x + 62y = 306$, where $x =$ the time at 48 mph and $y =$ the time at 62 mph
12. 2

Speaking the Language

Section 5.1   *(page 280)*

1. system   **2.** point of intersection   **3.** consistent
4. dependent, parallel

Section 5.2   *(page 290)*

1. addition, elimination   **2.** opposites   **3.** inconsistent
4. unique, no, infinitely many

Section 5.3   *(page 297)*

1. substitution   **2.** x **3.** dependent   **4.** parallel

Section 5.5   *(page 313)*

1. system of inequalities   **2.** solution   **3.** intersection
4. nonnegative

Chapter 6

Warm-Up Skills   *(page 321)*

1. (a) 32   (b) -125 **2.** (a) 1   (b) $\frac{1}{27}$
3. (a) 5000   (b) $\frac{1}{2}$ **4.** -8 **5.** 13   **6.** 15
7. $x^2 - 3x + 1$ **8.** $18 - 15x$ **9.** $x + 4y + 1$
10. $-6x + 5$ **11.** $4x^2 + 1$ **12.** $3x - 2$

Speaking the Language

Section 6.1   *(page 328)*

1. base, add   **2.** numerator, denominator   **3.** bases
4. zero

Section 6.2   *(page 336)*

1. monomials   **2.** degree   **3.** function, polynomial
4. descending, trinomial

Section 6.3   *(page 342)*

1. combining, Distributive   **2.** missing   **3.** change the sign
4. grouping symbols

Section 6.4   *(page 349)*

1. Distributive   **2.** each term of the other polynomial
3. like terms   **4.** Associative

Section 6.5   *(page 356)*

1. outer, inner   **2.** difference of two squares
3. square, twice, square   **4.** Distributive

Section 6.6   *(page 364)*

1. subtracting   **2.** each term, monomial
3. remainder, divisor   **4.** divisor, remainder, dividend

Section 6.7   *(page 371)*

1. reciprocal   **2.** factor, term   **3.** negative   **4.** positive

Section 6.8   *(page 378)*

1. scientific notation   **2.** right   **3.** first nonzero digit
4. negative

Chapter 7

Warm-Up Skills   *(page 385)*

1. $-2y^3 + 2y^2 - 8y$ **2.** $a^4b^3 + a^3b^2 - ab^2$
3. $49y^2 - 25$ **4.** $45 - 20a^2$ **5.** $16x^2 + 8xy + y^2$
6. $x^2 + 5x - 14$ **7.** $3a^2 - 5ab - 2b^2$ **8.** $x^3 - 8$
9. (a) $-5, -2$ **(b)** $2, 4$ **10. (a)** $-6, 2$ **(b)** $-3, 5$
11. $\frac{7}{2}$ **12.** 0

Speaking the Language

Section 7.1   *(page 392)*

1. factoring   **2.** greatest common factor   **3.** polynomials
4. grouping method

Section 7.2   *(page 399)*

1. sum, difference   **2.** perfect squares, twice
3. sum of two cubes, $A + B$ **4.** greatest common factor

Section 7.3   *(page 407)*

1. quadratic   **2.** like, different
3. linear term, leading coefficient   **4.** multiplying

Section 7.4   *(page 413)*

1. linear (or middle)   **2.** common factor
3. linear (or middle), grouping   **4.** prime

Section 7.6   *(page 423)*

1. quadratic   **2.** Zero Factor   **3.** standard   **4.** product

Chapter 8

Warm-Up Skills   *(page 439)*

1. (a) $\frac{14}{55}$ **(b)** $\frac{6}{5}$ **2. (a)** $\frac{13}{12}$ **(b)** $-\frac{11}{15}$ **3.** $\frac{3}{11}$
4. -8, undefined, $\frac{1}{3}$ **5.** $5x + 10$
6. We solve a linear equation by using the properties of
equations to isolate the variable. We solve a quadratic
equation by factoring and applying the Zero Factor
Property.
7. $-3, 7$ **8.** 12 **9. (a)** $(x + 3)^2$ **(b)** $(x + 5)(x - 5)$
10. (a) $3x(x - 3)$ **(b)** $(x + 5)(x - 2)$
11. $-w(w - 1); w(-w + 1)$
12. (a) $\dfrac{3}{4}$ **(b)** $\dfrac{4y^4}{5x^2}$

Speaking the Language

Section 8.1   *(page 446)*

1. rational   **2.** restricted   **3.** common factor
4. break, branches

Section 8.2   *(page 452)*

1. fractions   **2.** divide out   **3.** reciprocal, divisor
4. numerators, denominators

Section 8.3   *(page 458)*

1. common denominator   **2.** numerators, denominator
3. simplify   **4.** groups

Section 8.4   *(page 466)*

1. renaming   **2.** common multiple   **3.** least common
4. factoring

Section 8.5   *(page 472)*

1. LCD   **2.** $x - 4$ **3.** multiply, -1 **4.** simplify

Section 8.6   *(page 478)*

1. complex fraction   **2.** simplifying
3. numerator, denominator   **4.** LCD

Section 8.7   *(page 487)*

1. rational   **2.** point of intersection   **3.** clear the fractions
4. extraneous

Chapter 9

Warm-Up Skills   *(page 505)*

1. (a) 9   (b) 4   **2.** (a) $9x^2$ (b) $25a^2b^6$

3. (a) x^{11} (b) $18x^8$ **4.** (a) $\dfrac{9}{x^4}$ (b) $8x^6$

5. (a) $\dfrac{8x^6}{y^3}$ (b) $\dfrac{a^6}{b^{18}}$ **6.** $2x + 7x = (2 + 7)x = 9x$

7. $x^3 - 36$ **8.** $x^2 + 6x + 9$ **9.** $(5 + y)(5 - y)$

10. 7   **11.** 4, 7   **12.** Yes

Speaking the Language

Section 9.1   *(page 512)*

1. square root   **2.** radical, principal   **3.** radicand
4. index

Section 9.2   *(page 519)*

1. square root, radicands   **2.** perfect square
3. multiply, simplify   **4.** nonnegative

Section 9.3   *(page 527)*

1. square root, radicands   **2.** radicand
3. radical, denominator   **4.** rationalizing

Section 9.4   *(page 533)*

1. like radicals   **2.** coefficients, radical   **3.** Distributive
4. conjugates

Section 9.5   *(page 541)*

1. radical equation   **2.** squares   **3.** extraneous   **4.** isolate

Section 9.6   *(page 547)*

1. hypotenuse   **2.** Pythagorean   **3.** principal   **4.** legs

Section 9.7   *(page 554)*

1. radical   **2.** index, exponent   **3.** reciprocal
4. exponent, base

Chapter 10

Warm-Up Skills   *(page 561)*

1. $(x + 4)^2$ **2.** $(x - 3)^2$ **3.** -5 **4.** 3, 7   **5.** $-3, 3$
6. $-1, 8$ **7.** 1   **8.** $\frac{1}{3}$

9. To determine the x-intercept, replace y with 0 and solve
for x. To determine the y-intercept, replace x with 0 and
solve for y.

10. $(0, 5), (10, 0)$ **11.** -2 **12.** (a) 25   (b) 5   (c) 10

Speaking the Language

Section 10.1   *(page 565)*

1. quadratic   **2.** x-intercepts   **3.** Zero Factor
4. Square Root

Section 10.2   *(page 572)*

1. completing the square   **2.** half the coefficient
3. dividing both sides by 2   **4.** Square Root

Section 10.3   *(page 580)*

1. Quadratic   **2.** standard   **3.** discriminant   **4.** negative

Section 10.4   *(page 589)*

1. imaginary   **2.** complex   **3.** complex conjugates, real
4. Distributive

Section 10.5   *(page 596)*

1. quadratic, two variables   **2.** parabola   **3.** vertex
4. y-intercept

Section 10.6   *(page 604)*

1. function   **2.** domain, range   **3.** Vertical Line Test
4. g of x

Answers to Exercises

Chapter 1

Section 1.1 *(page 10)*

1. The set of whole numbers includes the natural numbers and 0.

3. $\{0, 1, 2, 3\}$ **5.** $\{\ldots, -1, 0, 1, 2, 3\}$

7. $\{8, 10, 12, 14\}$ **9.** $\{0, 1, 2, \ldots\}$

11. $\{-8, -7, -6, -5\}$

13. Any integer n can be written as $\dfrac{n}{1}$.

15. True **17.** True **19.** False **21.** False **23.** True

25. Zero **27.** Whole numbers

29. The number 0.35 is a terminating decimal, whereas $0.\overline{35}$ is a repeating decimal.

31. 1.4, terminating **33.** $0.8\overline{3}$, repeating

35. $-0.\overline{45}$, repeating

37. (a) $5, \frac{35}{5}$ (b) $0, 5, \frac{35}{5}$ (c) $-3, 0, 5, \frac{35}{5}$
(d) $-3, \frac{-2}{3}, 0, 5, \frac{35}{5}$ (e) None

39. (a) $1, \sqrt{9}$ (b) $1, \sqrt{9}$ (c) $1, \sqrt{9}$
(d) $1, \sqrt{9}, \frac{4}{3}$ (e) $-\sqrt{7}$

41. (a) $532, \frac{18}{6}$ (b) $532, \frac{18}{6}$ (c) $532, \frac{18}{6}$
(d) $-2\frac{1}{3}, 532, -0.35, \frac{18}{6}, 1.\overline{14}$ (e) $\sqrt{5}$

43.

45.

47.

49.

51. $5 > 0, 0 < 5$ **53.** $-12 \geq -20, -20 \leq -12$

55. $24 > -17, -17 < 24$ **57.** $n \geq -5, -5 \leq n$

59. $y \leq 4, 4 \geq y$ **61.** False **63.** True **65.** True

67. False **69.** $<$ **71.** $=$ **73.** $>$ **75.** $<$ **77.** $>$

79. $\{\ldots, -9, -8, -7\}$ **81.** $\{0, 1\}$ **83.** $\{0\}$ **85.** $=$

87. $9.1 < 13.1, 13.1 < 13.3, 13.3 > 10.8, 10.8 > 9.6$

89. December–January

91. (a) False (b) True (c) True (d) False

93. $\frac{9}{5}$ **95.** $\frac{2}{3}$ **97.** (a) $0.\overline{1}, 0.\overline{2}$ (b) $0.\overline{3}, 0.\overline{4}$ (c) $\frac{7}{9}$

Section 1.2 *(page 19)*

1. Addends are the numbers in a sum; factors are the numbers in a product.

3. (c) **5.** (b) **7.** (g) **9.** (f) **11.** $2(5), 2 \cdot 5, (2)(5)$

13. A natural number exponent indicates repeated multiplication.

15. $6 \cdot 6 \cdot 6 \cdot 6$ **17.** $t \cdot t \cdot t \cdot t \cdot t$ **19.** 5^4 **21.** y^5

23. x^{14} **25.** 64 **27.** 32 **29.** 1024 **31.** 759,375

33. 9 **35.** 7 **37.** 26

39. The $-$ symbol is used for subtraction, as in $6 - 2$; for negative numbers, as in -4; and for opposites, as in $-t$.

41. 13 **43.** -8 **45.** $-7, 7$ **47.** $2, 2$ **49.** $\frac{5}{8}, \frac{5}{8}$

51.

53.

55. $-8, 8$ **57.** $-3, 3$ **59.** $-6, 2$ **61.** $>$ **63.** $=$

65. $=$ **67.** $<$ **69.** 13 **71.** 11 **73.** 19 **75.** 16

77. 6 **79.** 1 **81.** 2 **83.** 1 **85.** 1

87. $59 + 370 = 429$ **89.** $87 \cdot 236 = 20,532$ **91.** 0

93. 2 **95.** 4 **97.** Griffey: 48; McGwire: 65

99. Debates, newspaper stories, TV news

101. No, a person may have chosen more than one category.

103. $a^2 = x$ **105.** $7(2 + 3 \cdot 1)$ or $7(2 + 3) \cdot 1$

107. $(4^2 - 1) \cdot 3$

Section 1.3 *(page 27)*

1. In a sum the addends can be added in any order. In a product the factors can be multiplied in any order.

3. $73 + (92 + 8) = 173$; Associative Property of Addition

5. $(5 \cdot 4) \cdot 7 = 140$; Associative Property of Multiplication

7. Associative Property of Addition

9. Property of Additive Inverses

11. Additive Identity Property **13.** Distributive Property

15. Commutative Property of Multiplication

17. For addition the number 0 is the additive identity.

19. $1 \cdot c$ **21.** $7x$ **23.** $5(x + y)$ **25.** x **27.** $4(x + 6)$

29. (a) Associative Property of Addition
(b) Commutative Property of Addition
(c) Associative Property of Addition
(d) Distributive Property
(e) Commutative Property of Multiplication

31. (a) Associative Property of Multiplication
(b) Commutative Property of Multiplication
(c) Associative Property of Multiplication

33. (d)   **35.** (e)   **37.** (b)

39. $n \cdot 11 = n(10 + 1) = 10n + n$ **41.** $2x + 6$

43. $4y - 12$ **45.** $2x + xy$ **47.** $ax - 5a$ **49.** $4(x + y)$

51. $7(y - z)$ **53.** $a(x + y)$ **55.** $3(x + 2)$ **57.** $8(a - 2)$

59. $15 \cdot 100 + 15 \cdot 2$ **61.** $7 \cdot 100 - 7 \cdot 2$

63. $17(98 + 2)$ **65.** $31(13 - 3)$

67. $8 \cdot 140 + 8 \cdot 132$ and $8(140 + 132)$

69. (a) 15%   (b) 27%

71. $73(5 + 3 + 2)$ **73.** $15(100 - 2)$

75. $\frac{3}{4} \cdot 8$; Associative Property of Multiplication

77. -2; Distributive Property

Section 1.4   *(page 33)*

1. Add the absolute values and use the sign of the addends.

3. 3   **5.** -4 **7.** -10 **9.** -8 **11.** 0   **13.** 2

15. 8   **17.** 8   **19.** -7 **21.** 20   **23.** -55 **25.** 8

27. 0   **29.** -22 **31.** -1 **33.** -11 **35.** -17

37. -6 **39.** 1   **41.** -9 **43.** 0   **45.** -6.8

47. -51.57 **49.** -10.31 **51.** 9   **53.** -4 **55.** 2

57. 0   **59.** -45 **61.** 3   **63.** -18.3 **65.** -7

67. 0   **69.** -0.222

71. The Commutative Property of Addition allows us to reorder, and the Associative Property of Addition allows us to regroup.

73. (a) Commutative Property of Addition

(b) Associative Property of Addition

75. (a) Commutative Property of Addition

(b) Associative Property of Addition

(c) Property of Additive Inverses

(d) Additive Identity Property

77. 1   **79.** 1   **81.** 1   **83.** 8   **85.** -4

87. $-964 + 351 = -613$ **89.** $-9 + 5 = -4$

91. $(-5 + 2) + 12 = 9$ **93.** -7 **95.** -12

97. 2-yard gain   **99.** 212 pounds   **101.** Positive

103. Negative   **105.** 81

Section 1.5   *(page 41)*

1. Equivalent fractions are fractions with the same decimal value.

3. $\frac{3}{18}, \frac{5}{30}, \frac{7}{42}$ **5.** $-\frac{4}{14}, -\frac{10}{35}, -\frac{16}{56}$ **7.** 9   **9.** 21

11. 25   **13.** 12   **15.** 28

17. A fraction can be reduced if there is a common factor in the numerator and denominator.

19. $\frac{2}{3}$ **21.** $\frac{1}{8}$ **23.** $-\frac{5}{2}$ **25.** $\frac{12}{35}$ **27.** $-\frac{9}{2}$

29. Although 8 is a common denominator, 4 is the least common denominator.

31. $\frac{9}{5}$ **33.** 1   **35.** $\frac{2}{7}$ **37.** $\frac{1}{2}$ **39.** $\frac{5}{6}$ **41.** $\frac{1}{2}$ **43.** $\frac{19}{7}$

45. $\frac{41}{36}$ **47.** $\frac{49}{90}$ **49.** $\frac{1}{2}$ **51.** $\frac{2}{3}$ **53.** $\frac{7}{10}$ **55.** $\frac{7}{8}$ **57.** $5\frac{7}{12}$

59. $6\frac{1}{3}$ **61.** $\frac{-5}{3}$ **63.** $\frac{5}{8}$ **65.** $\frac{31}{7}$ **67.** $\frac{11}{15}$ **69.** $\frac{15}{16}$

71. $1\frac{1}{8} + \frac{3}{8} = 1\frac{1}{2}$ **73.** $2\frac{2}{3} + \frac{-5}{6} = 1\frac{5}{6}$ **75.** $26\frac{1}{8}$

77. (a) $\frac{11}{20}$ (b) $\frac{13}{100}$

79. $\frac{142}{100,000}, \frac{50}{100,000}, \frac{34}{100,000}, \frac{25}{100,000}, \frac{71}{50,000}, \frac{1}{2,000}, \frac{17}{50,000}, \frac{1}{4,000}$

81. $\frac{251}{100,000}$; In these four occupations there are 251 fatalities per 100,000 workers.

83. $\frac{10}{a}$ **85.** (ii); If $b = 4$ and $d = 6$, the LCD is 12, not 24.

Section 1.6   *(page 48)*

1. Change the minus sign to a plus sign and change 8 to -8.

3. -11 **5.** -10 **7.** 13   **9.** 6   **11.** -13 **13.** 52

15. 18   **17.** -24 **19.** -3 **21.** 48   **23.** -20 **25.** -3

27. Taking the opposite of a difference changes the order in which the numbers are subtracted.

29. $7 - (-5)$; 12   **31.** $-6 - 4$; -10 **33.** $9 - a$

35. -10 **37.** 13   **39.** -8 **41.** -14 **43.** 0

45. $-17 - (-18)$; 1   **47.** $-1 - 9$; -10 **49.** $-\frac{1}{10}$

51. $-\frac{14}{3}$ **53.** $\frac{17}{16}$ **55.** -34 **57.** 74.23   **59.** 5

61. 6   **63.** -5 **65.** 43   **67.** 5   **69.** $-\frac{22}{15}$ **71.** -21

73. -32 **75.** 32   **77.** 59.76   **79.** -80 **81.** -15

83. (a) Definition of Subtraction

(b) Associative Property of Addition

85. (a) Property of the Opposite of a Difference

(b) Definition of Subtraction

(c) Associative Property of Addition

(d) Property of Additive Inverses

(e) Additive Identity Property

87. -5 **89.** -24 **91.** -7 **93.** 1   **95.** 30,398 feet

97. 1.9 inches below average

99. $+11,931$; -2175; -1259; -1695; $+4142$; $+11,908$

101. 2.73 miles   **103.** (a) Negative   (b) Positive

105. $3 - (7 - 2)$ **107.** $2(3 - 2) - 5$

Section 1.7   *(page 56)*

1. In both cases multiply the absolute values of the numbers. For like signs the product is positive; for unlike signs the product is negative.

3. -28 **5.** -10 **7.** 12   **9.** 12   **11.** -100

13. -44 **15.** -7 **17.** 0   **19.** 6.21   **21.** -4.83

23. -4 **25.** -1

27. Because there is an even number of negative factors, the product is positive.

29. -24 **31.** 12 **33.** -36 **35.** -243 **37.** 256

39. -1

41. The Multiplication Property of -1 states that multiplying a number by -1 results in the opposite of the number.

43. $-\frac{1}{5}$ **45.** $\frac{10}{3}$ **47.** $\frac{85}{2}$ **49.** 28.512 **51.** -20

53. -40 **55.** $\frac{4}{21}$ **57.** 72 **59.** 96 **61.** 0 **63.** -3

65. 36 **67.** 1 **69.** 2 **71.** $29(0.63); 18.27$

73. $-\frac{3}{4} \cdot 32; -24$

75. (a) Commutative Property of Multiplication

(b) Associative Property of Multiplication

77. (a) Associative Property of Multiplication

(b) Multiplication Property of -1

79. $2, 3$ **81.** $-3, 4$ **83.** $-2, -6$ **85.** $-8, 1$

87. -3 **89.** 6 **91.** $\$51.95$ **93.** 75

95. (a)
Brazil	1.235
China	0.305
Germany	1.030
Mexico	1.892
Russia	0.055
United States	5.638
Zimbabwe	0.041

(b) United States

97. Negative **99.** Positive **101.** $-5(6 - 8) = 10$

Section 1.8 *(page 63)*

1. We call -12 the dividend, 2 the divisor, and -6 the quotient.

3. -2 **5.** -3 **7.** 8 **9.** 8 **11.** 16 **13.** -3

15. -4 **17.** Undefined **19.** 0 **21.** $\frac{1}{2}$ **23.** 8

25. -4 **27.** -1 **29.** 2.19 **31.** -0.51

33. $-\frac{x}{y}, \frac{-x}{y}$; Property of Signs in Quotients **35.** $\frac{5}{8}$

37. $\frac{3}{5}$ **39.** $-\frac{y}{5}, \frac{-y}{5}$ **41.** Invert the divisor and multiply.

43. (a) -5 (b) $\frac{1}{5}$ **45.** (a) $\frac{3}{4}$ (b) $-\frac{4}{3}$

47. (a) 0 (b) None **49.** $-\frac{5}{6}$ **51.** $\frac{15}{2}$ **53.** $-\frac{3}{32}$

55. $-\frac{9}{4}$ **57.** -28 **59.** $\frac{35}{2}$ **61.** -3 **63.** -19.34

65. $-\frac{1}{3}$ **67.** 4 **69.** -9 **71.** Undefined **73.** -3

75. -2 **77.** -2 **79.** $-16.4 \div (-4.1); 4$

81. $-10 \div (-20); \frac{1}{2}$

83. (a) Multiplication Rule for Fractions

(b) Distributive Property

(c) Multiplication Rule for Fractions

85. (a) Multiplication Rule for Fractions

(b) Associative Property of Multiplication

(c) Property of Multiplicative Inverses

(d) Multiplicative Identity Property

87. 14 **89.** -40 **91.** $-4\frac{7}{8}$

93. (a) $23,660$ (b) $\frac{40}{91}, \frac{600}{1183}, \frac{45}{2366}, \frac{20}{1183}, \frac{1}{91}, \frac{15}{2366}$

95.
Army	71,154
Navy	50,038
Marines	9,782
Air Force	65,513

97.
Army	$\frac{61}{71}$
Navy	$\frac{21}{25}$
Marines	$\frac{9}{10}$
Air Force	$\frac{9}{11}$

99. Negative **101.** Positive **103.** $\frac{5}{6}$

Chapter Review Exercises *(page 66)*

1. $\{-3, -2, -1, 0\}$

3. (a) -0.625, terminating (b) $0.1\overline{6}$, repeating

5. (a) $\{0, 1, 2, 3\}$ (b) $\{\ldots, -1, 0, 1, 2\}$

7. (a) True (b) True (c) False (d) True

9. (a) Multiplication (b) Division

(c) Subtraction (d) Addition

11. b^4 **13.** 46 **15.** -10 **17.** (a) $=$ (b) $>$ (c) $<$

19. Property of Multiplicative Inverses

21. Commutative Property of Addition

23. Property of Additive Inverses

25. $6x + 12$ **27.** Additive identity, 0 **29.** -19 **31.** 9.3

33. -1 **35.** -1 **37.** -3 **39.** (ii) **41.** $-\frac{25}{40}$ **43.** 24

45. $\frac{2}{45}$ **47.** The number n must contain a factor of 2 or 3.

49. Change minus to plus and change -7 to 7.

51. 35 **53.** -9 **55.** 7 **57.** 1 **59.** -28 **61.** 84

63. -6 **65.** -32 **67.** 10 **69.** -7 **71.** 4

73. $-\frac{1}{12}$ **75.** 9 **77.** 1

Chapter Test *(page 68)*

1. [*1.1*] (a) True (b) True

(c) False. Because π is an irrational number, $\frac{\pi}{3}$ is also irrational.

(d) True

2. [*1.2*]

3. [*1.1*] $n \le 7, 7 \ge n$ **4.** [*1.2*] $2^4 = 16$

5. [*1.2*] (a) 22.2 (b) -4 (c) 4

6. [*1.2*] (a) Divisor (b) Addend (c) Factor

7. [*1.3*] (a) Associative Property of Addition

(b) Multiplication Property of -1

(c) Property of Additive Inverses

(d) Commutative Property of Multiplication

8. [*1.3*] **(a)** $4(x + 3)$ **(b)** $3a + 6$

9. [*1.3*] The Property of Multiplicative Inverses states that the product of a nonzero number and its reciprocal is always 1.

10. [*1.7*] 21   **11.** [*1.4*] -1 **12.** [*1.6*] -7

13. [*1.8*] -5 **14.** [*1.5*] $-\frac{17}{60}$ **15.** [*1.8*] -16

16. [*1.8*] -2 **17.** [*1.5*] $\frac{9}{2}$ **18.** [*1.6*] $k - (-3)$ or $k + 3$

19. [*1.7*] $-8, 3$ **20.** [*1.4*] 258

Chapter 2

Section 2.1   *(page 79)*

1. A numerical expression is an algebraic expression that does not contain a variable.

3. -1 **5.** -6 **7.** -9 **9.** 29   **11.** 10   **13.** 9

15. 6   **17.** $\frac{1}{4}$ **19.** -11 **21.** 5.92   **23.** -1.12

25. There are two terms in $2x + 1$ but only one term in $2(x + 1)$. In $2(x + 1)$, both 2 and $x + 1$ are factors.

27. $3y, x, -8$; $3, 1, -8$ **29.** $x^2, 3x, -6$; $1, 3, -6$

31. $7b, -c, \dfrac{3a}{7}$; $7, -1, \dfrac{3}{7}$ **33.** $(3 + 8)x = 11x$

35. $(-1 + 3)y = 2y$ **37.** $-8x$ **39.** $3t^2$ **41.** $-x + 3$

43. $6x - 7$ **45.** $-m - n$ **47.** $b + 1$ **49.** $-4a^3 + a^2$

51. $-ab + 8ac$ **53.** $-\dfrac{2}{3}a + \dfrac{7}{6}b$ **55.** $3.3x - 3.9y$

57. Remove parentheses and combine like terms.

59. $-15x$ **61.** $-y$ **63.** $\frac{3}{4}x$ **65.** $15b + 20$

67. $6x - 15$ **69.** $6x + 2y$ **71.** $3x - 3y + 15$

73. $-3x - 12$ **75.** $-x + 4$ **77.** $-3 - n$

79. $-2x - 5y + 3$ **81.** $2x - 2$ **83.** $-3x + 8$ **85.** $7y$

87. $a + 12$ **89.** $3a + 2$ **91.** $-15x$ **93.** $16 - x$

95. $-10x - 26$ **97.** 84   **99.** \$229   **101.** $5n - 8$

103. 0.28   **105.** 59.60

Section 2.2   *(page 89)*

1. The number lines are called the x-axis and the y-axis. Their point of intersection is called the origin.

3. I   **5.** III   **7.** II   **9.** IV   **11.** x-axis   **13.** y-axis

15. $A(5, 0), B(-7, 2), C(2, 6), D(0, -4), E(-3, -5), F(4, -4)$

17. $A(0, 6), B(-5, 0), C(7, 3), D(-6, -3), E(-2, 8), F(3, -3)$

19. The order of the coordinates is different.

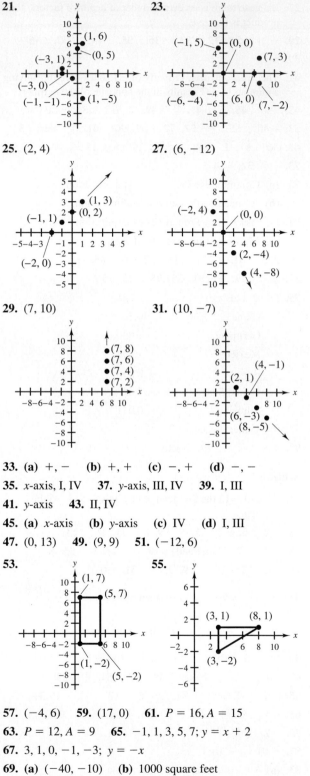

21.

23.

25. $(2, 4)$ **27.** $(6, -12)$

29. $(7, 10)$ **31.** $(10, -7)$

33. (a) $+, -$ **(b)** $+, +$ **(c)** $-, +$ **(d)** $-, -$

35. x-axis, I, IV   **37.** y-axis, III, IV   **39.** I, III

41. y-axis   **43.** II, IV

45. (a) x-axis   **(b)** y-axis   **(c)** IV   **(d)** I, III

47. $(0, 13)$ **49.** $(9, 9)$ **51.** $(-12, 6)$

53.

55.

57. $(-4, 6)$ **59.** $(17, 0)$ **61.** $P = 16, A = 15$

63. $P = 12, A = 9$ **65.** $-1, 1, 3, 5, 7$; $y = x + 2$

67. $3, 1, 0, -1, -3$; $y = -x$

69. (a) $(-40, -10)$ **(b)** 1000 square feet

71. (a) I   **(b)** $(0, 25), (30, 17), (70, 11)$

73. Sales/marketing and written communication

75. The points are associated with the areas in which more than 50% of the companies offer training.

77. $(2, 3)$ **79.** $(-4, 2)$ **81.** $(7, 5)$

Section 2.3   *(page 100)*

1. It is not necessary to enter the expression each time it is to be evaluated.

3. $-7, -6, -2, 5$ **5.** $23.8, 0.7, -5.6, -18.2$

7. $24, -21, 0, 24$ **9.** $0, 14, 20, 36$ **11.** $9, 1, 0, 17$

13. $0.4, 1, 0.52, 0.51$

15.

$x - 3$	-5	-2	1	4	7
	$(-2, -5)$	$(1, -2)$	$(4, 1)$	$(7, 4)$	$(10, 7)$

17.

$2x + 5$	11	19	5	-3	-5
	$(3, 11)$	$(7, 19)$	$(0, 5)$	$(-4, -3)$	$(-5, -5)$

19.

$x^2 - 3x - 4$	-6	0	14	14	36
	$(1, -6)$	$(4, 0)$	$(6, 14)$	$(-3, 14)$	$(-5, 36)$

21. You obtain an error message for $x = 1$ because division by 0 is not defined.

23. 4   **25.** 1   **27.** $-6, 0, 4$ **29.** $-5, 2, 9$

31. $17, -1, -16$

33. The first coordinate, 3, is the value of the variable; the second coordinate, 7, is the value of the expression for $x = 3$.

35.

x	2	4	0	5	-1	-2
$x - 4$	-2	0	-4	1	-5	-6

37.

x	-32	-8	-12	16	6	28
$\frac{1}{4}x + 3$	-5	1	0	7	4.5	10

39. (a) 3   **(b)** 9   **(c)** -5 **(d)** -10

41. (a) -1 **(b)** 2   **(c)** 5   **(d)** 8

43. (a) -23 **(b)** 6   **(c)** 20   **(d)** 33

45. (a) $-2, 11$ **(b)** $-1, 10$ **(c)** $1, 8$ **(d)** $2, 7$

47. (a) $-12, 22$ **(b)** $-3, 13$ **(c)** 5   **(d)** $1, 9$

49. 1   **51.** 2   **53.** 3 pizzas   **55.** 1999

57.

1965	0	675.3
1970	5	1943.3
1975	10	3211.3
1980	15	4479.3
1985	20	5747.3
1990	25	7015.3
1995	30	8283.3
2000	35	9551.3

59. 1985   **61.** 1 more than twice x

63. Twice the sum of x and 1   **65. (a)** 30.75   **(b)** 30.75

Section 2.4   *(page 111)*

1. An equation has an $=$ symbol but an expression does not.

3. Expression   **5.** Equation   **7.** Equation   **9.** Expression

11. A solution of an equation is a value of the variable that makes the equation true.

13. Yes   **15.** No   **17.** No   **19.** Yes   **21.** Yes

23. Yes   **25.** Yes   **27.** No   **29.** 14   **31.** 12   **33.** 4

35. -14 **37.** -32 **39.** -20

41. Graph each side of the equation and trace to the point of intersection. The x-coordinate of that point is the estimated solution.

43. -1.25 **45.** 0.14 **47.** -1.71 **49.** -7 **51.** -12

53. -12 **55.** -20 **57.** 8   **59.** -9 **61.** 15   **63.** -9

65. 15

67. (a) The graphs of the two sides of the equation do not intersect. The solution set is $\varnothing$.

  (b) The graphs of the two sides of the equation coincide. The solution set is **R**.

69. 0   **71.** $\varnothing$ **73.** R   **75.** $\varnothing$ **77.** R   **79.** 8

81. 24   **83.** $\varnothing$ **85.** R   **87. (a)** 2004   **(b)** 2006

89. 110   **91.** $-6, 30$ **93.** $\varnothing$

Section 2.5   *(page 119)*

1. Equivalent equations are equations that have exactly the same solutions.

3. 9   **5.** 3.5   **7.** (i)   **9.** (ii)   **11.** 4   **13.** 2   **15.** 14

17. -18.7 **19.** 0   **21.** -1 **23.** 5   **25.** 2   **27.** -6

29. 15   **31.** 16   **33.** 17   **35.** 3   **37.** 2   **39.** 0

41. When we add 3 to both sides of $x - 3 = 12$, the left side becomes $x - 3 + 3 = x + 0 = x$. When we divide both sides of $3x = 12$ by 3, the left side becomes $\frac{3x}{3} = x$.

  In both cases, the variable is isolated.

43. (ii)   **45.** (i)

47. All of the given steps are correct. In each case the result is $1x = 20$ because

   (i) $\frac{4}{3}$ and $\frac{3}{4}$ are reciprocals whose product is 1.

   (ii) $\frac{3}{4}$ divided by $\frac{3}{4}$ is 1.

   (iii) multiplying by 4 and dividing by 3 is equivalent to multiplying by $\frac{4}{3}$.

49. 4   **51.** -3 **53.** 3   **55.** 0   **57.** -36 **59.** 20

61. 12   **63.** 0   **65.** 5   **67.** -8 **69.** 0   **71.** 7

73. 5   **75.** 3   **77.** $\frac{-1}{8}$ **79.** Distributive Property

81. Multiplication Property of Equations

83. Addition Property of Equations

85. 9   **87.** $145x - 263 = 2200$ **89.** $x = b - a$

91. $x = \dfrac{a + b}{5}$ **93.** -4 **95.** 3   **97.** 5

Section 2.6 *(page 126)*

1. Dividing both sides by 2 would create fractions and make the equation more difficult to solve. The best first step is to add 5 to both sides.

3. -2 **5.** $\frac{2}{3}$ **7.** 3   **9.** -1 **11.** -7 **13.** -5

15. 0   **17.** 4   **19.** 0   **21.** -3 **23.** 3   **25.** $\frac{10}{3}$ **27.** 3

29. $-\frac{2}{3}$ **31.** 2   **33.** -2 **35.** -9 **37.** -3 **39.** 3

41. The number n is the LCD of all fractions in the equation.

43. $\frac{7}{2}$ **45.** -1 **47.** -19 **49.** 4   **51.** 16   **53.** 3

55. 7   **57.** -1.2 **59.** -4.3 **61.** Identity, **R**

63. Contradiction, Ø   **65.** Conditional, 0

67. Conditional, 0   **69.** Contradiction, Ø   **71.** -8

73. 0   **75.** **R**   **77.** Ø   **79.** Ø   **81.** **R**

83. **(a)** (i) The cost at a private college is $150,000.

     (ii) The cost at a public college is double the 1997 cost.

   (b) (i) 18.61 (2009)

     (ii) 20.87 (2011)

85. Men, 10.22 seconds; women, 11.16 seconds

87. 2289; men and women, 5.8 seconds

89. 6   **91.** **R**   **93.** $x = \dfrac{c - b}{a}$ **95.** -3

Section 2.7 *(page 136)*

1. The value of $x + 3$ is less than or equal to 7.

3. 9, 11   **5.** $-4, -2, -1$ **7.** 2, 5

9. The symbols [and] indicate that an endpoint is included. The symbols (and) indicate that an endpoint is not included.

11.

13.

15.

17.

19.

21.

23. $n \geq 0$ **25.** $x \leq 3$ **27.** $n < 0$ **29.** $x \leq 6$

31. $x < -3$ **33.** $x \geq 5$ **35.** $-4 < x \leq 7$

37. The point of intersection represents a solution if the inequality symbol is $\leq$ or $\geq$.

39. $x > -15$

41. $x \leq 0$

43. $x > -17$ **45.** $x \geq -5$ **47.** $x < -3$ **49.** $x \leq 7$

51. $x < 0$ **53.** $x < 2$ **55.** $x \geq -8$ **57.** $x < 12$

59. If the inequality is $y_1 > y_2$ or $y_1 \geq y_2$, the solution set is **R**. If the inequality is $y_1 < y_2$ or $y_1 \leq y_2$, the solution set is Ø.

61. **R**   **63.** (a) **R**   (b) Ø

65. $-11 \leq x \leq 23$

67. $-19 < x \leq 13$

69. $-13 < x \leq 9$

71. $-6 < x < 8$

73. $x \leq 9$ **75.** Ø   **77.** $x > -9$ **79.** **R**   **81.** $x < 0$

83. Ø   **85.** $x + 6 \geq 3; x \geq -3$ **87.** $8 - x > -2; x < 10$

89. **(a)** $-1.01x + 27.5 \geq 20$ **(b)** $x \leq 7.4$; before 1988

91. **(a)** $-0.83x + 23.9 > -1.01x + 27.5$

   (b) $x > 20$; after 2000

93. Ø   **95.** (a) Ø   (b) **R**   **97.** **R**   **99.** Ø

Section 2.8 *(page 144)*

1. The properties allow us to add the same number to both sides of an equation or an inequality.

3. $>$ **5.** $>$ **7.** $>$ **9.** $<$ **11.** $x < -4$ **13.** $x \geq -5$

15. $x \geq 0$ **17.** $x < -8$ **19.** $x > 4$ **21.** $x > -3$

23. $x \leq 0$ **25.** $x \leq 16$ **27.** $x \geq -5$

29. No. Because we multiplied both sides by a negative number, the inequality symbol must be reversed: $32x > 89$.

31. $x > -3$ **33.** $x \geq -2$ **35.** $x \geq -4$ **37.** $x \leq -7$

39. $x < 2$ **41.** $x \leq -1$ **43.** $x < -3$ **45.** $y > \frac{3}{2}$

47. $t \geq -3$ **49.** $x \geq -\frac{1}{12}$ **51.** $t < \frac{9}{10}$ **53.** $t \geq 4$

55. If the resulting inequality is true, the solution set is **R**. If the resulting inequality is false, the solution set is Ø.

57. Ø   **59.** R   **61.** Ø   **63.** $2 < x < 9$

65. $-7 < x \leq -2$ **67.** $0 \leq y \leq 6$ **69.** $-2 < x < 0$

71. $0 \leq x \leq 2$ **73.** $x \leq \frac{3}{2}$ **75.** Ø   **77.** R

79. $x > -4$ **81.** $x > -3$ **83.** Ø   **85.** $y \geq 1.5$

87. $0.57x + 27.4 > 65$; 2006 and all years after 2006

89. $x < 40$ **91.** $x < b - a$ **93.** $x \geq \dfrac{-b}{a}$

95. $x < -2$ **97.** $7 < x < 10$

Chapter Review Exercises   *(page 146)*

1. (a) 31   **(b)** 0   **3.** -8 **5. (a)** $-2xy + 5x$ **(b)** $7c^2$

7. $-6a + 3b - 12$ **9.** $-13x + 2y - 4$

11. $(4, -1)$ **13.** Q

15. The set of all points with a y-coordinate of 0 is the x-axis.

17. Origin   **19.** $-11, -5, 3, 11$ **21.** 18

23. The y-coordinate will be -3 because the y-coordinate is the value of the expression when $x = 5$.

25. (ii)   **27.** Yes   **29.** Yes   **31.** $(-2, 20)$

33. (a) The lines are parallel.   **(b)** The lines coincide.

35. (iii)   **37.** (ii)   **39.** 7   **41.** 8   **43.** 18   **45.** $\frac{15}{2}$

47. -4 **49.** -1 **51.** -9 **53.** R   **55.** Ø   **57.** (iii)

59. The point of intersection represents a solution for $\leq$ or $\geq$ inequalities.

61. (a) $x \leq 2$ **(b)** $-1 < x \leq 3$

63. (a) $8 + x > x - 6$ **(b)** $8 + x < x - 6$

65. (iii)   **67.** $x \geq 11$ **69.** $x \geq -20$ **71.** $x > -2$

73. $-4 \leq x < 3$ **75.** R

Chapter Test   *(page 149)*

1. [2.1] 1   **2.** [2.1] $a + 2b$ **3.** [2.1] $7x + 13$

4. [2.1] Because the expression can be written as $\frac{1}{2}(x - 9)$, the coefficient is $\frac{1}{2}$.

5. [2.2] II

6. [2.2] The set of all points whose x-coordinates are 0 is the y-axis.

7. [2.3] The y-coordinate of a point of the graph of an expression corresponds to the value of the expression. If x is replaced with -4, the value of the expression $3x - 2$ is -14. Thus if the x-coordinate is -4, the y-coordinate is -14.

8. [2.3] 5   **9.** [2.4] 10   **10.** [2.4] **(a)** 16   **(b)** 5

11. [2.4] **(a)** Identity   **(b)** Contradiction   **12.** [2.4] None

13. [2.6] 21   **14.** [2.6] -2 **15.** [2.6] -16 **16.** [2.6] **R**

17. [2.6] Ø   **18.** [2.6] 6   **19.** [2.7] $x - 3 \geq -2$

20. [2.7] $x > 2$

21. [2.7] **(a)** Ø   **(b)** R; The graph of $9 - x$ is always above the graph of $-11 - x$. Thus the solution set for part (a) is Ø and for part (b) is **R**.

22. [2.8] $x > -2$ **23.** [2.8] $x \geq -1$

24. [2.8] $-15 \leq x < 6$ **25.** [2.8] $5 \leq x < 7$

Chapter 3

Section 3.1   *(page 157)*

1. A ratio is a quotient used to compare two numbers.

3. $\frac{3}{2}$ **5.** $\frac{10}{13}$ **7.** $\frac{1}{200}$ **9.** $\frac{173}{42}$ **11.** $\frac{11}{13}$ **13.** 8

15. (a) $\frac{4}{15}$ **(b)** $\frac{10}{7}$ **(c)** $\frac{1}{2}$ **17. (a)** $\frac{4}{5}$ **(b)** 2   **(c)** $\frac{18}{23}$

19. $\frac{7}{5}$ **21.** $\frac{3}{4}$ **23.** $\frac{3}{4}$ **25.** $\frac{4}{5}$

27. The unit price is the price charged for each unit, such as weight, volume, or length, in which the item is sold.

29. 6.88¢   **31.** 8¢   **33.** $18.62   **35.** 13¢

37. 64 ounces   **39.** 11 ounces

41. A proportion is an equation stating that two ratios are equal.

43. 28   **45.** $\frac{9}{4}$ **47.** $\frac{5}{2}$ **49.** $\frac{1}{3}$ **51.** 7   **53.** -18

55. -17 **57.** 3.75   **59.** 10, 12   **61.** 57¢

63. $8.\overline{8}$ ounces   **65.** 82.5 calories   **67.** 15 miles

69. (a) 6.6 feet   **(b)** 7 inches   **71.** 20 pounds

73. 792 votes   **75.** 80   **77.** 14,000   **79.** 63, 105

81. 31.5 yards   **83.** 10.5 gallons   **85.** 2380   **87.** 962

89. 9   **91.** 48-ounce; with coupon, 16-ounce   **93.** 200

Section 3.2   *(page 166)*

1. Percent means per 100.   **3. (a)** 0.28   **(b)** $\frac{7}{25}$

5. (a) 2.35   **(b)** $\frac{47}{20}$ **7. (a)** 0.172   **(b)** $\frac{43}{250}$

9. (a) 0.0575   **(b)** $\frac{23}{400}$ **11.** 37.5%   **13.** 55%

15. 690%   **17.** 0.02%   **19.** 87.5%   **21.** $46\frac{2}{3}\%$

23. 175%   **25.** 78.96   **27.** 180   **29.** 68%   **31.** 300

33. 414   **35.** 8%   **37.** 608   **39.** 125%   **41.** 6000

43. The student's result, 1.2, is the amount of increase. The answer is found by adding this amount to 10.

45. 770.5   **47.** 360   **49.** 119.6   **51.** 1592

53. (a) 25%   **(b)** 20%   **55. (a)** 10%   **(b)** $9.\overline{09}\%$

57. Statement (ii) is true. In statement (i), the percentage is correct, but we don't know that the club has 100 members. Statement (iii) is not true because 40 is not 60% of 70.

59. 25%   **61.** 16.1 inches   **63.** 3065   **65.** No

67. $8869.50   **69.** 11,450   **71. (a)** 4%   **(b)** 96%

73. (a) United States, Japan, France, Germany

(b) 28 million

75. 100%

77. No. We know only percentages, not the actual number of home buyers.

79. (i) **81.** $120,000

Section 3.3 *(page 175)*

1. A formula is an equation that serves as instructions for calculating a certain quantity.

3. $175 **5.** $P = 26$ feet; $A = 30$ square feet

7. $C \approx 20.11$ meters; $A \approx 32.17$ square meters

9. (a) 50 **(b)** 89.6 **11. (a)** 165 **(b)** 64

13. (a) 30 **(b)** 25.2 **15. (a)** 37.515 **(b)** 48

17. (a) 78.54 **(b)** 43.01

19. The hypotenuse is the longest side of a right triangle. It is the side opposite the right angle.

21. No **23.** Yes **25.** Yes **27.** No **29.** $66.\overline{6}$ km/hr

31. (a) 84 feet **(b)** 156 feet **(c)** 0 feet

33. Yes **35.** The procedure is exactly the same.

37. $s = \dfrac{P}{4}$ **39.** $L = \dfrac{A}{W}$ **41.** $P = \dfrac{I}{rt}$ **43.** $d = \dfrac{C}{\pi}$

45. $F = \dfrac{9}{5}C + 32$ **47.** $b = \dfrac{2A}{h}$ **49.** $b = 2A - a$

51. $y = -x + 8$ **53.** $y = 5x - 7$ **55.** $y = -\dfrac{7}{2}x + 5$

57. $y = \dfrac{3}{4}x - 1$ **59.** $y = 2x - 5$ **61.** $y = \dfrac{15}{8}x - \dfrac{5}{4}$

63. (a) 10 feet **(b)** 12 feet **(c)** 15.87 feet

65. (a) $102.96 **(b)** 15%

67. (a) 88° **(b)** 63° **(c)** 21° **(d)** 9° **(e)** 79° **(f)** 64°

69. $y = \dfrac{c - ax}{b}$ **71.** $y = mx + b$ **73.** $m = \dfrac{y - a}{x - b}$

75. $P = \dfrac{A}{1 + rt}$ **77.** $I = \dfrac{E}{R + r}$

Section 3.4 *(page 183)*

1. If the information contains a verb, the final form of the translation will be an equation.

3. $x - (-7)$ **5.** $\dfrac{x}{2}$ **7.** $2x - 2$ **9.** $\dfrac{1}{4}x = 8$

11. $6x - 1 = 11$ **13.** $2x - 1 = x + 5$ **15.** $3x - 4$

17. $x - 5 = -8$ **19.** $2 - 3x$ **21.** $2x + 5 = 9$

23. $x - \frac{1}{2}$ **25.** $3W$ **27.** $x, x + 8$ **29.** $x, x + 7$

31. $x, \dfrac{1}{2}x + 5$

33. The angles are supplementary angles because the sum of their measures is $x + (180 - x) = 180$.

35. $180 - x$ **37.** $130 - m$ **39.** $x, 90 - x$ **41.** $x, 90 - x$

43. $n + 1$ **45.** $x + 1, x + 3$ **47.** $3x + 3$ **49.** $2x + 2$

51. $0.3x$ **53.** $0.92q$ **55.** $18d$ **57.** $60t$ **59.** $2500y$

61. $1.25x$ **63.** $0.66x$ **65.** $1.05x$

67. $3x - 40$, $x =$ number of miles driven in the afternoon

69. $4x - 14$, $x =$ length **71.** $35 + 0.3x$, $x =$ number of miles

73. $400 - 0.01x$, $x =$ amount invested at 7%

75. $300 - 5n$, $n =$ number of nickels **77.** $20 - 0.15x$

79. $30 - 0.3x$, $x =$ number of pounds of grapes

81. $320 - 2x$, $x =$ number of hours at $6 per hour

83. (a) $\frac{1}{3}x + 16$ **(b)** $\frac{1}{3}x + 16 = 20$

85. $x^2 - 49$ **87.** $(n + 3)^2$ **89.** $5.5x + 1.25y$

Section 3.5 *(page 191)*

1. The equation may not be correct. The answer should be checked against the wording of the problem.

3. 10 **5.** 3 **7.** At most 11 **9.** 14 **11.** 12

13. Less than 9, less than 16 **15.** 8, 28 **17.** $-1, -6$

19. $-15, -17$ **21.** 32, 34, 36 **23.** 0, 1, 2

25. 27, 29, 31 **27.** 5 feet, 20 feet, 10 feet

29. 20 yards, 35 yards, 45 yards

31. Because the sum of the measures of the three angles is 180° and the measure of the right angle is 90°, the sum of the measures of the other two angles is 90°. The other two angles are the same size, so each has a measure of 45°.

33. 30°, 60°, 90° **35.** 40°, 60°, 80° **37.** 11 meters, 4 meters

39. Length no more than 20 feet, width no more than 15 feet

41. 60° **43.** 50° **45.** 30°, 150° **47.** 20°, 70°

49. 428 **51.** 8 **53.** 9 feet, 12 feet **55.** 18

57. 8 years **59.** 16 fiction, 48 nonfiction

61. 18 inches **63.** 60 minutes

65. Decaffeinated, 24 pounds; Regular, 48 pounds; Premium, 18 pounds

67. $700 **69.** 3 feet, 2 feet, 2 feet **71.** Less than 65° F

73. $0.57 billion **75.** 1998 and 2001

77. Any two consecutive even integers

79. Any three consecutive odd integers

Section 3.6 *(page 200)*

1. The unit price of the mixture must be between $1.20 and $1.80.

3. 19 dimes, 9 quarters **5.** 28 pennies, 6 nickels, 8 dimes

7. 10 fives, 5 tens, 3 twenties

9. First class, 1480; second class, 540

11. 130 tapes, 260 CD's   **13.** $3700 at 6%, $4800 at 7%

15. $5200 at 9%, $4300 at 12%   **17.** $2364, $118.20 tax

19. 14 ml   **21.** 26 gallons

23. (a) If the cars travel in opposite directions, then the distance between them is the sum of their distances: $55t + 60t$ or $115t$.

(b) If the cars travel in the same direction, then the distance between them is the positive difference of their distances: $60t - 55t$ or $5t$.

25. 63 miles, 84 miles   **27.** 4 mph, 6 mph

29. Walking, 4 mph; jogging, 6 mph

31. Brick house, 40 years; stucco house, 20 years

33. Shortstop, 20; left fielder, 5   **35.** corn, 600 acres

37. 24 seconds   **39.** $2400 at 8%, $4600 at 11%

41. $112,500   **43.** $2600 at 15%, $1100 at 18%

45. 4 pages per minute, 6 pages per minute   **47.** $9200

49. Faculty lot, 240; student lot, 720

51. $5700 at 6%, $500 at 14%   **53.** $120   **55.** $56,500

57. 75 shares Coca-Cola; 60 shares AT&T

59. $820 at 18%; $660 at 21%   **61.** 2.5 hours

63. 0.2 quadrillion BTUs   **65.** 23.1 quadrillion BTUs

67. 8 pennies, 6 nickels, 11 dimes, 7 quarters   **69.** 3 years

Chapter Review Exercises   *(page 203)*

1. (a) $\frac{43}{57}$ **(b)** $\frac{57}{100}$ **3.** 15¢ per ounce or $2.40 per pound

5. 1824   **7.** 4000   **9.** 1350 mg   **11.** All

13. (a) $\frac{47}{100}$, 0.47   **(b)** $\frac{11}{10}$, 1.1   **(c)** $\frac{1}{5000}$, 0.0002

15. 25.6   **17.** 150   **19.** 20%   **21.** 39.6°F

23. From 16 to 20 is a 25% increase; from 20 to 16 is a 20% decrease.

25. $21.60   **27. (a)** Yes   **(b)** No   **29.** $b_2 = \dfrac{2A}{h} - b_1$

31. $b = 3Q - a - c$ **33.** $y = -3x + 2$ **35.** 59°F

37. (a) $2x - 5$ **(b)** $2x = x + 4$ **39.** $10 - c$

41. $x - 2, x - 1$

43. (a) $4L - 6$, where L is the length

(b) $0.70p$, where p is the pre-discount price

45. 6   **47.** 1   **49.** 100°   **51.** 40°   **53.** 15   **55.** $9000

57. $\frac{1}{3}$ quart   **59.** 6

Chapter Test   *(page 206)*

1. [3.1] 4.5¢ per ounce   **2.** [3.1] 160

3. [3.1] $AB = 10, DE = 6$ **4.** [3.2] 80

5. [3.2] 16   **6.** [3.2] 10   **7.** [3.2] $473.33

8. [3.3] **(a)** $b = \dfrac{2A}{h}$ **(b)** 14 inches

9. [3.3] $P = 2L + 2W$ **10.** [3.3] $y = \frac{5}{4}x - 5$

11. [3.3] The triangle is a right triangle if the square of the longest side equals the sum of the squares of the other two sides. Because $4.25^2 = 2^2 + 3.75^2$, the triangle is a right triangle.

12. [3.4] $23 - w$ **13.** [3.4] Complement

14. [3.4] The first integer is $n - 2$, and the third integer is $n + 2$. Therefore, the sum is $(n - 2) + n + (n + 2) = 3n$.

15. [3.5] 15   **16.** [3.5] 8 feet, 32 feet, 20 feet   **17.** [3.5] 57°

18. [3.6] 22 pounds at $3.40/lb, 28 pounds at $6.10/lb

19. [3.6] 1:00 P.M.   **20.** [3.6] 23.5 cubic centimeters

Cumulative Test: Chapters 1–3   *(page 207)*

1. [1.1] Rational numbers have terminating or repeating decimal names. Irrational numbers have decimal names that do not terminate and do not repeat.

2. [1.2] **(a)** $=$ **(b)** $<$ **(c)** $>$

3. [1.3] Associative Property of Multiplication   **4.** [1.4] -2

5. [1.6] $\frac{20}{3}$ **6.** [1.8] -6 **7.** [2.1] -7

8. [2.1] $-3x + 5a + 2$ **9.** [2.2] $+, -$

10. [2.3] The y-coordinate is the value of the expression $1 - 5x$ for $x = 3$. Thus, $y = 1 - 5(3) = -14$.

11. [2.2] **(a)** Identity   **(b)** Contradiction

12. [2.5] $-\frac{1}{3}$

13. [2.7] $x > -2$

14. [3.1] 24,000   **15.** [3.2] $492,000

16. [3.3] $y = \dfrac{1}{10}x - 4$ **17.** [3.3] $3L - 8$ **18.** [3.5] 800

19. [3.5] 28°, 62°, 90°   **20.** [3.6] 13 minutes

Chapter 4

Section 4.1   *(page 215)*

1. If both were 0, the equation would not have a variable term.

3. $x + y = 24$, $x = $ number of women, $y = $ number of men

5. $x + y = 180$, $x = $ measure of one angle, $y = $ measure of the other angle

7. $x + y = 12$, $x = $ length of one piece, $y = $ length of the other piece

9. $x + y = 12$, $x = $ number of pounds of Brazilian coffee, $y = $ number of pounds of Turkish coffee

11. Yes   **13.** No   **15.** Yes   **17.** No

19. The value of x must be written first: $(1, 3)$.

21. $(5, -15), (0, -5), (-6, 7)$ **23.** $(6, -9), (2, -1), (-4, 11)$

25. $(0, 2), (3, 6), \left(1, \frac{10}{3}\right)$

27. Solve the equation for y; then enter $y = 3x - 4$.

29.

x	4	-2	-3
y	-4	2	3

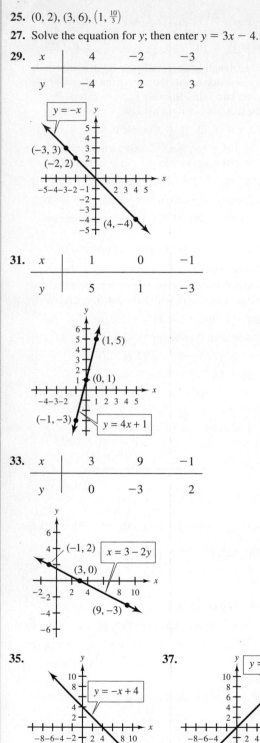

31.

x	1	0	-1
y	5	1	-3

33.

x	3	9	-1
y	0	-3	2

35. **37.**

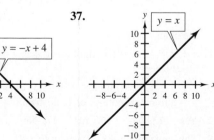

39. **41.**

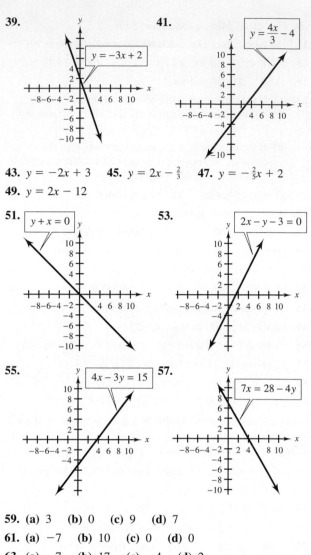

43. $y = -2x + 3$ **45.** $y = 2x - \frac{2}{3}$ **47.** $y = -\frac{2}{5}x + 2$

49. $y = 2x - 12$

51. **53.**

55. **57.**

59. (a) 3   (b) 0   (c) 9   (d) 7

61. (a) -7 (b) 10   (c) 0   (d) 0

63. (a) -7 (b) 17   (c) -4 (d) 2

65. (ii)   **67.** (ii); $a = -15$; $b = 1$

69. $y = x + 3$ **71.** $y = 2x$ **73.** $x + y = 7$ **75.** $y = -x$

77. $y = 1.2x + 5$; \$15.80   **79.** $y = 2x + 3$; 11 men

81. (a)

x	11	**19**	23	**27**
y	**3.5**	7.5	**9.5**	11.5

(b) 1997

83. (a) According to the model, the average refund in 1992 was \$1006.

(b) According to the model, the average refund in 1997 was \$1441.

85. 3   **87.** -4 **89.** 10

Section 4.2   *(page 224)*

1. The points where the graph intersects the x- and y-axes.

3. $(-7, 0), (0, 21)$ **5.** $(10, 0), (0, 15)$

7. $(-18, 0), (0, -9)$ **9.** $(8, 0), (0, 20)$

11. Replace x with 0 and solve for y. Write the equation in the form $y = ax + b$. The y-intercept is $(0, b)$.

13. $(0, 0)$ **15.** $(2, 0), (0, -4)$ **17.** $(7, 0), (0, -3)$

19. $\left(\frac{3}{2}, 0\right), (0, 4)$ **21.** $y = -\frac{3}{2}x + 9; (0, 9)$

23. $y = \frac{1}{3}x - 6; (0, -6)$ **25.** $y = 5; (0, 5)$

27. $y = 2x + 10; (0, 10)$ **29.** $y = -\frac{3}{4}x + 3; (0, 3)$

31. (i), (iii)   **33.** (i), (ii)   **35.** (iii)   **37.** (ii)

39. The graph is a horizontal line with y-intercept $(0, b)$ and no x-intercept.

41.

x	4	0	-7
y	-2	-2	-2

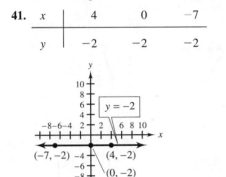

43.

x	2	2	2
y	4	0	-7

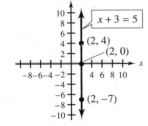

45.   **47.**

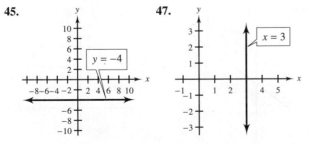

49.   **51.**

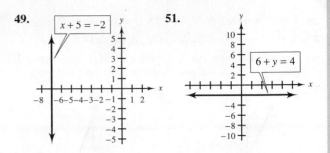

53. Vertical   **55.** Neither   **57.** Horizontal

59. Horizontal   **61.** $x = 7$ **63.** $2y - 6 = 8$

65. $y = 6$ **67.** $x = -3$ **69.** $x = 4$ **71.** $y = 7$

73. $y = -x + 22$

(a) $(22, 0), (0, 22)$

(b) The y-intercept corresponds to "no windows have been washed," and the x-intercept corresponds to "all windows washed."

75. $y = 1.25x + 5$

(a) \$12.50

(b) The y-intercept $(0, 5)$ corresponds to the basic charge for rental.

77. The y-intercept $(0, 37)$ represents the number of subscribers in 1995.

79. The y-intercept $(0, 10.5)$ represents the number of subscribers in 1990.

81. $x = -2$ **83.** $y = x - 4$ **85.** $(-4, 0), (0, 2)$

Section 4.3   *(page 234)*

1. The rise is greater than the run.   **3.** 2   **5.** $-\frac{5}{3}$

7. Undefined   **9.** 0   **11.** Any two points may be used.

13. 2   **15.** 1   **17.** -1 **19.** 0   **21.** Undefined   **23.** $\frac{1}{6}$

25. -1 **27.** $\frac{1}{5}$ **29.** $-\frac{2}{5}$ **31.** $\frac{5}{2}$ **33.** -8.19 **35.** $\frac{2}{3}$

37. No, the ratio must be $\frac{3}{5}$.   **39.** $-\frac{1}{2}$ **41.** Undefined

43. $\frac{1}{3}$ **45.** $\frac{3}{4}$

47. For a zero slope the rise is 0. For an undefined slope the run is 0.

49. 0   **51.** Undefined   **53.** Undefined   **55.** 0   **57.** 0

59. Undefined   **61.** 0   **63.** Negative   **65.** 5   **67.** 0

69. 3   **71.** (a) $\frac{7}{20}$ (b) $\frac{7}{20}$ **73.** 1; 3   **75.** $\frac{1}{2}$

77. 0 or negative

Section 4.4   *(page 241)*

1. The coefficient of x is the slope, and the constant term is the y-coordinate of the y-intercept.

3. $y = -3x + 4; -3; (0, 4)$ **5.** $y = 4x; 4; (0, 0)$

7. $y = 2x - 4; 2; (0, -4)$ **9.** $y = -5; 0; (0, -5)$

11. $y = -\frac{2}{3}x + 4$; $-\frac{2}{3}$; $(0, 4)$ **13.** $y = \frac{3}{2}x - 3$; $\frac{3}{2}$; $(0, -3)$

15. $y = -\frac{3}{4}x + \frac{5}{2}$; $-\frac{3}{4}$; $(0, \frac{5}{2})$ **17.** (i), (ii)   **19.** (ii), (iii)

21. Because the slope can be written as $\frac{-2}{1}$ or $\frac{2}{-1}$, move down 2 and right 1 from the given point or move up 2 and left 1.

23.   **25.**   **27.**   **29.**   **31.**   **33.**   **35.**   **37.**   **39.**   **41.**

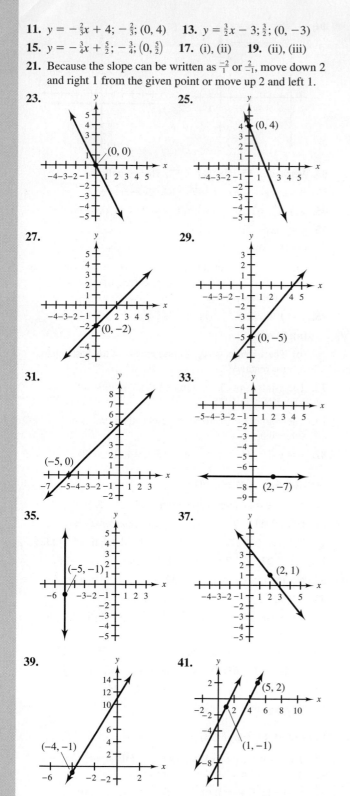

43.

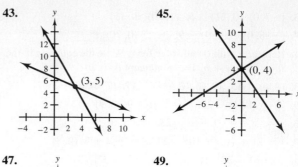

45.

47.

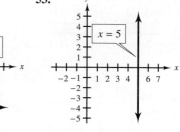

49.

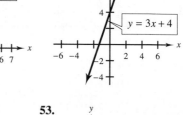

51.

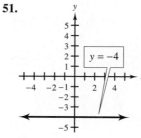

53.

55.

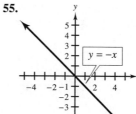

57.

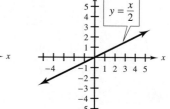

59.

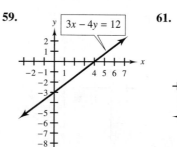

61.

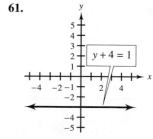

63.

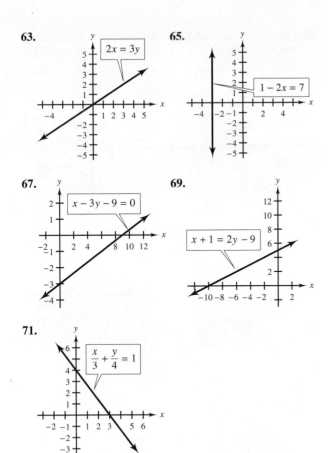

65.

67.

69.

71.

73. 0;

75. Undefined;

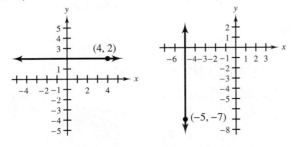

77. The slope is k and the graph contains the origin.

79. Yes   **81.** Yes   **83.** No   **85.** 12   **87.** -12

89. 2820 pounds   **91.** $\frac{3}{4}$ cup

93. (a) $y = \frac{12}{25}x - 4$

    (b) $\frac{12}{25}$; The positive slope indicates that the percentage increases with increasing salaries.

95. $y = \frac{83}{50}x + \frac{27}{50}$; $\frac{83}{50}$; Because the constant term is not 0, it is not a direct variation.

97.

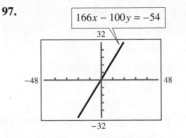

101. A, B, or $C = 0$ **103.** $A = 0$

105. A and B have the same sign.

Section 4.5   *(page 249)*

1. For nonvertical lines, write the equations in the form $y = mx + b$ and compare the slopes and y-intercepts. The lines are parallel if the slopes are the same and the y-intercepts are different.

3. Parallel   **5.** Perpendicular   **7.** Neither

9. Perpendicular   **11.** Parallel   **13.** Perpendicular

15. Neither   **17.** Perpendicular   **19.** Parallel

21. Perpendicular   **23.** Neither

25. The slope of a vertical line is not defined.

27. (a) 1   (b) -1 **29.** (a) $-\frac{1}{4}$ (b) 4

31. (a) 0   (b) Undefined   **33.** (a) $\frac{3}{2}$ (b) $-\frac{2}{3}$

35. (a) Undefined   (b) 0   **37.** (a) $\frac{3}{5}$ (b) $-\frac{5}{3}$

39. Parallel   **41.** Neither   **43.** Perpendicular

45. Perpendicular   **47.** Neither   **49.** Undefined

51. Undefined   **53.** 0   **55.** 2   **57.** $(2, 4)$

63. From 1995 to 1998, mail increased by an average of 5.7 billion pieces per year.

65. 1995: 206,857; 1998: 217,680

67. -3;   **69.** $-\frac{3}{2}$;

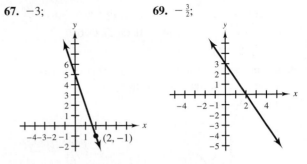

71. It is possible for two lines in space not to intersect and yet not to be parallel.

Section 4.6   *(page 257)*

1. The slope and y-intercept are needed.

3. $y = -x + 6$ **5.** $y = \frac{5}{2}x - 2$ **7.** $y = 2x + \frac{3}{4}$

9. $y = -2x + 3$ **11.** $y = \frac{2}{3}x + 1$ **13.** $y = 3$

15. $y = -3x - 12$ **17.** $y = x - 5$ **19.** $y = -\frac{2}{3}x + 6$

21. $y = -7$ **23.** $y = 2x + 6$

25. The slope and any point of the line are required.

27. $2x - y = 1$ **29.** $5x + 2y = 7$ **31.** $18x - 6y = 5$

33. $5x - 20y = -2$ **35.** $y = -2x + 4$ **37.** $y = -\frac{5}{2}x - 5$

39. $y = -2x + 15$ **41.** $y = 5$ **43.** $y = -\frac{3}{2}x + 6$

45. $x = -8$ **47.** $y = \frac{3}{2}x + \frac{1}{2}$ **49.** $y = \frac{7}{5}x + 7$

51. The slope of a vertical line is not defined.

53. $y = 7$ **55.** $x = 4$

57. We need the slope and a point of the line.

59. (a) $y = x - 8$ (b) $y = -x$

61. (a) $y = \frac{3}{4}x + \frac{9}{4}$ (b) $y = -\frac{4}{3}x + \frac{13}{3}$

63. (a) $y = \frac{5}{3}x - \frac{35}{3}$ (b) $y = -\frac{3}{5}x - \frac{13}{5}$

65. (a) $y = -\frac{1}{2}x + \frac{7}{2}$ (b) $y = 2x - 4$

67. (a) $y = \frac{1}{3}x$ (b) $y = -3x$

69. (a) $y = 2$ (b) $x = -5$ **71.** (a) $x = 3$ (b) $y = 1$

73. $y = -3x - 5$ **75.** $y = x + 7$ **77.** $y = \frac{1}{2}x + 5$

79. $x = -3$ **81.** $x = 1$ **83.** $y = 2x + 7$ **85.** $y = -\frac{3}{4}x$

87. $y = 120x + 500$, $x =$ number of months since July 1,
$y =$ amount deposited in the account

89. $y = -\frac{3141}{7}x + 8000$, $x =$ number of months since logging
began, $y =$ number of trees remaining

91. (a) $(0, 126), (3, 142)$ (b) $m = \frac{16}{3}, (0, 126)$
  (c) $y = \frac{16}{3}x + 126$

93. $(0, 37), (2, 327)$; $m = 145, (0, 37)$

95. $(7, 37), (9, 327)$; $m = 145$ **97.** -10 **99.** -3

Section 4.7 *(page 266)*

1. A solution is a pair (x, y) that satisfies the inequality.

3. $(-2, 3), (6, -8)$ **5.** $(4, -3), (8, 2), (-6, 0)$

7. $(4, 3), (1, -4)$

9. It is not correct. The inequality can be rewritten as
$y > 2x - 3$. Shade above the line.

11. (b) **13.** (a)

15.

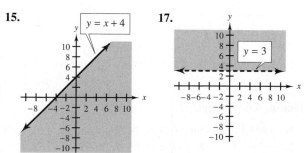

17.

19. For $\leq$ or $\geq$ use a solid line, and for $<$ or $>$ use a dashed
line.

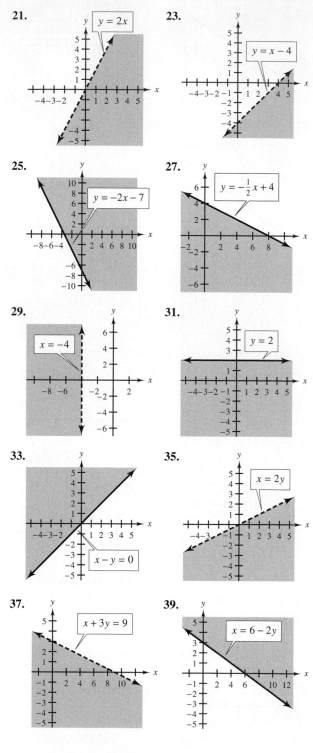

41.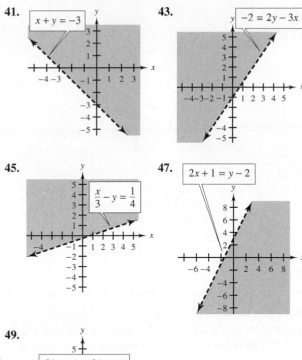

43.

45.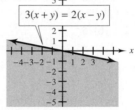

47.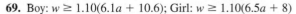

49.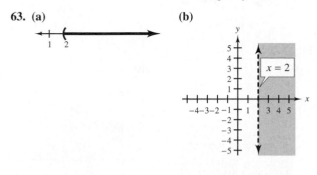

51. $y \geq x + 2$ **53.** $x + y < 10$ **55.** $y - 2x > 0$

57. $y \leq \frac{1}{2}x$ **59.** $y - (x - 3) \geq 0$

61. We must know whether the inequality is being considered as a one-variable or a two-variable inequality.

63. (a)

(b)

65. (a)

(b)

67. Boy: $w > 6.1a + 10.6$; Girl: $w > 6.5a + 8$

69. Boy: $w \geq 1.10(6.1a + 10.6)$; Girl: $w \geq 1.10(6.5a + 8)$

71.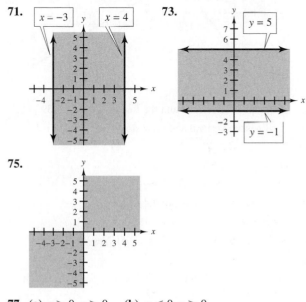

73.

75.

77. (a) $x > 0, y > 0$ **(b)** $x < 0, y > 0$
  (c) $x < 0, y < 0$ **(d)** $x > 0, y < 0$

Chapter Review Exercises *(page 268)*

1. Equations (ii) and (iii) cannot be written in the form $Ax + By = C$.

3. (i), (ii), (iv)

5.

x	0	3	7
y	-4	-1	3

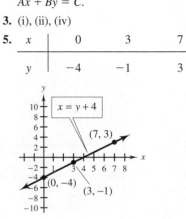

7. $a = 3; b = -7$ **9.** (ii), (iii) **11.** $(0, 9), (-18, 0)$

13. $y = \frac{2}{5}x - 4; (0, -4)$ **15.** The graph is a vertical line.

17. (a) Vertical **(b)** Neither **(c)** Horizontal

19. $y = 2x + 50; (0, 50); (-25, 0)$ **21.** (iii)

23. Any two points of the line can be used to determine the slope.

25. $-\frac{2}{3}$ **27. (a)** -2 **(b)** 0 **29.** $13, 5$

31. $y = \frac{4}{5}x - 1$; slope is $\frac{4}{5}$, y-intercept is $(0, -1)$)

33.

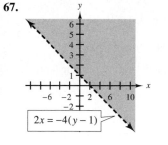

35.

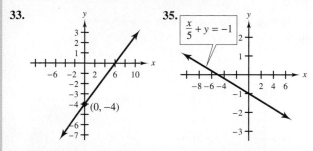

37. The slope is undefined and the line is vertical. **39.** 112

41. The lines are vertical and their slopes are not defined.

43. Parallel **45. (a)** $-\frac{3}{4}$ **(b)** $\frac{4}{3}$ **47.** $-\frac{3}{2}$

49. The triangle is a right triangle

51. $y = \frac{3}{5}x - 2$ **53.** $y = -\frac{1}{2}x - \frac{1}{2}$

55. The slope of the line is not defined.

57. $y = x + 9$ **59.** $y = -3x + 16$

61. (a) No **(b)** Yes **(c)** Yes **(d)** Yes

63. Above

65. Any point other than a point of the line can be used as a test point.

67.

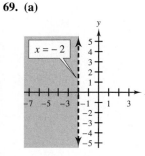

69. (a)

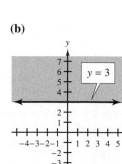

(b)

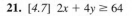

Chapter Test *(page 271)*

1. [*4.1*] (i), (iii) **2.** [*4.1*]

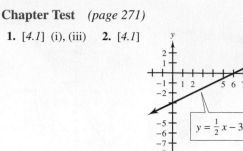

3. [*4.3*] The graph of equation (ii) is different from the other two graphs. All three graphs have the same y-intercept, $(0, 1)$. However, the slope in (ii) is $-\frac{2}{3}$, whereas the slope in (i) and (iii) is $+\frac{2}{3}$.

4. [*4.2*] $(3, 0), (0, -2)$ **5.** [*4.2*] $y = -\frac{1}{2}x - \frac{5}{2}; \left(0, -\frac{5}{2}\right)$

6. [*4.2*] **(a)** $y = -7$ **(b)** $x = 8$ **7.** [*4.3*] -4

8. [*4.3*] $-\frac{5}{2}$

9. [*4.3*] **(a)** Undefined **(b)** Negative **(c)** Positive **(d)** 0

10. [*4.4*] $y = \frac{3}{7}x - 2; \frac{3}{7}; (0, -2)$

11. [*4.4*] Plot the point $(-2, -5)$. From that point move up 2 units and right 5 units to arrive at another point of the line. Draw a line through these two points.

12. [*4.4*] 15 **13.** [*4.5*] **(a)** $-\frac{1}{5}$ **(b)** 5

14. [*4.5*] Parallel **16.** [*4.6*] $y = \frac{3}{5}x + 5$

17. [*4.6*] $4x - 3y = 0$ **18.** [*4.6*] $y = -\frac{1}{2}x - 4$

19. [*4.7*]

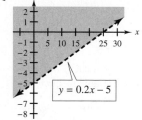

20. [*4.7*]

(a) **(b)**

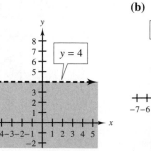

21. [*4.7*] $2x + 4y \geq 64$

Chapter 5

Section 5.1 *(page 280)*

1. A solution of a system of equations is a pair of numbers that satisfies both equations.

3. $(-2, -5)$ 5. $(5, 3), (-5, -1)$ 7. $a = 3, b = -5$

9. $a = -3, c = 3$ 11. $b = -1, c = 8$

13. No, two lines cannot intersect at exactly two points.

15. $(0, 0)$ 17. $(12, -9)$ 19. $(9, 11)$ 21. $(12, -7)$

23. $(4, -2)$ 25. $(-8, -3)$

27. (a) The lines intersect at one point.

   (b) The lines coincide.   (c) The lines are parallel.

   (d) Not possible

29. 1   31. Infinitely many   33. 0   35. 1

37. Inconsistent   39. Consistent   41. Consistent

43. Dependent   45. Independent   47. Independent

49. $(5, 4)$ 51. No solution   53. $(-5, 3)$

55. Infinitely many solutions   57. $\left(\frac{32}{3}, 10\right)$ 59. $(11, 16)$

61. $(2.8, -1.6)$ 63. $(1.8, -1.4)$ 65. No solution

67. Infinitely many solutions   69. $\left(\frac{1}{3}, -\frac{1}{4}\right)$ 71. $\left(-\frac{9}{8}, \frac{7}{2}\right)$

73. 26, 43   75. $84°, 96°$

77. (a) Male: $y = 0.22x + 7.18$ Female: $y = 0.38x + 3.44$

   (b) $(23.38, 12.32)$

79. (a) Cancellations: $y = -95x + 3034$
      Ticketing: $y = 64x + 624$

   (b) Because the slopes of the lines are unequal, the lines must intersect. The point of intersection represents the solution of the system.

81. $(8, -4)$ 83. No solution

85. $(-6, 9), (12, 21), (12, -18)$; Yes

Section 5.2 *(page 290)*

1. The coefficients of one of the variables are opposites.

3. $(5, -2)$ 5. $(-7, -1)$ 7. No solution   9. $(-3, 8)$

11. $(3, 2)$ 13. $\left(-1, \frac{5}{3}\right)$ 15. $(5, 3)$ 17. $(-10, -8)$

19. When the equations are added, at least one variable is eliminated.

21. $(5, -13)$ 23. $\left(\frac{1}{3}, 1\right)$ 25. No solution   27. $(1, 1)$

29. Infinitely many solutions   31. $(0, -1)$ 33. $(1, -2)$

35. $(-1, -3)$ 37. Infinitely many solutions   39. $(3, -2)$

41. No solution   43. $(4, -2)$

45. Multiply by the LCD to clear fractions.

47. $(2, 3)$ 49. No solution   51. $(10, 13)$ 53. $\left(\frac{3}{4}, \frac{1}{2}\right)$

55. No solution   57. $(1, -1)$ 59. $(0, 1)$ 61. $(0, 0)$

63. $(2, 6)$ 65. $(0, -2)$ 67. 12, 7   69. $24°, 66°$

71. (a) Men: $y = 9.5x + 25.3$ Women: $y = 10.8x + 19.6$

   (b) $(4.4, 67.0)$

73. The slope of the first equation indicates that the percentage of male online shoppers is decreasing. The slope of the second equation indicates that the percentage of female online shoppers is increasing.

75. $(11.5, 50)$ 77. $x = \dfrac{a + b}{2}, y = \dfrac{a - b}{2}$

79. -5 81. -4

Section 5.3 *(page 298)*

1. (a) Multiply the second equation by -3 and add the equations.

   (b) Solve the second equation for y.

3. $(1, 1)$ 5. $(15, 7)$ 7. $(5, 3)$ 9. $(-4, 1)$

11. $(-20, -4)$ 13. $(1, 2)$ 15. $(0, 4)$ 17. $(8, -5)$

19. $(-4, -3)$ 21. $(0, -1)$ 23. $(-1, 2)$

25. The graphs coincide, and the system has infinitely many solutions.

27. $(3, 6)$ 29. No solution   31. Infinitely many solutions

33. $(-10, 3)$ 35. Infinitely many solutions   37. $(0, 0)$

39. No solution   41. $6, -10$ 43. 10 feet, 22 feet

45. Energy: $y = -0.13x + 2.7$
    Environment: $y = 0.07x + 1.4$
    General science: $y = 0.24x + 2.4$

47. $(6.5, 1.8)$; The solution indicates that in 1997 the amounts spent in the two categories were the same.

49. (a) 3   (b) 3   51. $(7, 3, -2)$

53. $(-3, 6), (3, 6), (-3, -1)$
    The triangle is not an isosceles triangle.

Section 5.4 *(page 306)*

1. The product of the number of items and the unit value is the total value of the items.

3. 29 nickels, 14 quarters   5. 23 fives, 8 twenties

7. 12 ounces basil, 8 ounces oregano

9. Wind: 20 mph; Plane: 180 mph

11. Current: 2 mph; Canoe: 5 mph

13. Freeway: 186 miles; Secondary roads: 120 miles

15. The volume of the solution is needed. To determine the amount of acid, multiply the concentration (0.2) by the volume of the solution.

17. $2650 at 4%, $5600 at 5.5%

19. Clothing items: $3700; Other items: $6250

21. 75% solution: 10 liters; 45% solution: 5 liters

23. Car: 6 years; Truck: 3 years

25. Wheat: 160 acres; Corn: 40 acres

27. Homes: 450; Apartments: 150   **29.** 36 cars, 12 trucks

31. Apples: 7.5 pounds; Pears: 7.5 pounds

33. Icy roads: 20 mph; Wet roads: 48 mph

35. $\frac{3}{7}$ **37.** 5 feet, 10 feet   **39.** 560 cans, 282 bottles

41. 25% solution: 40 ml; 40% solution: 20 ml

43. 28, 17   **45.** 20 attempts, 10 made

47. 8 inches, 12 inches

49. Northbound biker: 10 mph; Southbound biker: 15 mph

51. Upstairs: 42 square yards; Basement: 30 square yards

53. 3200 Republicans, 1600 Democrats

55. 45% alloy: 6 pounds; 20% alloy: 24 pounds

57. 516 for, 344 against   **59.** Taco: 80¢; Cola: 75¢

61. (a) Under 10 acres: $y = -2.9x + 183$

  1000–1999 acres: $y = -0.1x + 102$

  (b) 2016

63. New York: $y = 9.38x + 47.5$

  Detroit: $y = 1.66x + 41.1$

  Boston: $y = 8.82x + 36.3$

65. (0.7, 44.2); According to the model, the payrolls for Boston and Detroit were the same in 1995.

67. 5 miles, 7 miles

69. Dog: 12 ft/sec; Cat: 15 ft/sec

Section 5.5   *(page 314)*

1. No, the pair must satisfy both inequalities.

3. (i)   **5.** (i), (ii)

7.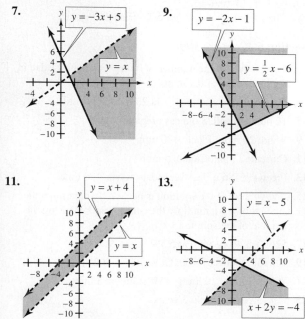

9.

11.

13.

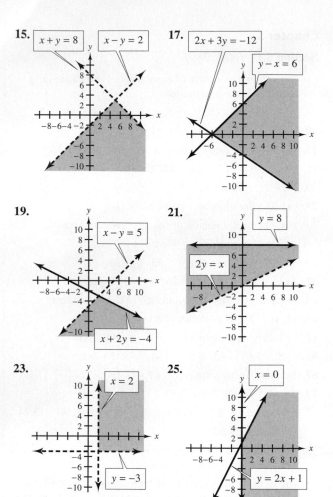

15. $x + y = 8$ $x - y = 2$

17. $2x + 3y = -12$ $y - x = 6$

19. $x - y = 5$ $x + 2y = -4$

21. $y = 8$ $2y = x$

23. $x = 2$ $y = -3$

25. $x = 0$ $y = 2x + 1$

27. The boundary lines are parallel, and the half-planes are in opposite directions; that is, they do not intersect.

29. ∅

31. 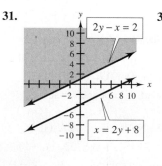 $2y - x = 2$ $x = 2y + 8$

33. 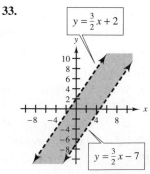 $y = \frac{3}{2}x + 2$ $y = \frac{3}{2}x - 7$

35. 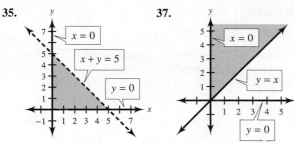 **37.**

39. $y \le -x + 5, x \ge 0, y \ge 0$

41.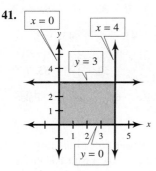

43. Points of the line $y = x - 5$ **45.** Ø

47. 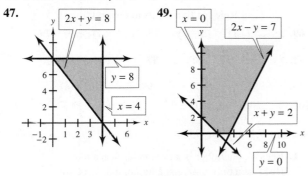 **49.**

43. It is easier to produce the graphs of boundary lines and to determine which half-plane to shade.

45. (ii)

47. 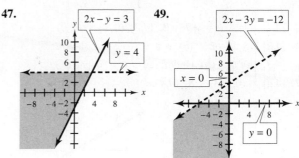 **49.**

Chapter Test *(page 318)*

1. [5.1] $a = -3, b = 2$ **2.** [5.1] $(-3, 4)$

3. [5.1] **(a)** The lines coincide.

   (b) The lines are parallel.

   (c) The lines intersect at one point.

4. [5.1] The slopes are different. **5.** [5.2] $(-6, -7)$

6. [5.2] $(0, -4)$ **7.** [5.2] No solution

8. [5.2] Infinitely many solutions

9. [5.3] Solve one equation for one variable.

10. [5.3] $(-2, 4)$

11. [5.3] If $a = -2$, the system has infinitely many solutions. Otherwise, the system has no solution.

12. [5.3] $(2, -7)$ **13.** [5.4] $18, -1$

14. [5.4] 15 feet, 22 feet **15.** [5.4] 14 nickels

16. [5.4] Plane: 400 mph **17.** [5.4] $6000

18. [5.4] 2 years

19. [5.5]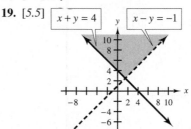

20. [5.5]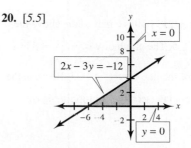

Chapter Review Exercises *(page 315)*

1. A solution is a pair of numbers that satisfies each equation.

3. (ii) **5.** $(0, -5)$ **7.** **(a)** 0 **(b)** 1 **(c)** Infinitely many

9. $(8, -15)$ **11.** $(2, -5)$ **13.** $(4, -5)$ **15.** $(4, -3)$

17. Ø

19. The sum $x + y = 3$ does not give any information about the number of solutions of the original system.

21. It is not easy to solve for a variable in either equation.

23. $(-6, 7)$ **25.** $(3, -2)$

27. Infinitely many solutions, dependent equations

29. No, the resulting equation has two variables.

31. $16, -9$ **33.** $11°, 79°$

35. 40 newer cards, 15 older cards

37. Bus: 30 minutes; Car: 40 minutes

39. $51,000 **41.** 8000 fiction, 6000 nonfiction

Cumulative Test: Chapters 4–5   (page 319)

1. [4.1]

x	5	6	10
y	-2	-1	3

2. [4.1] $a = 4, b = -7$ **3.** [4.2] $(0, 4), (-3, 0)$

4. [4.2] **(a)** $x = 2$ **(b)** $y = -5$ **5.** [4.3] $\frac{-9}{10}$

6. [4.3] **(a)** 0   **(b)** Undefined   **(c)** Negative

7. [4.4] $\frac{5}{3}, (0, -3)$ **8.** [4.4]

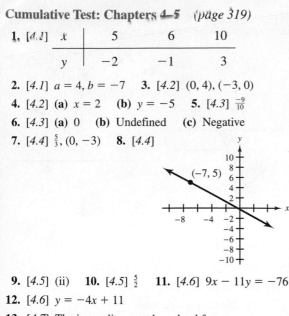

9. [4.5] (ii)   **10.** [4.5] $\frac{5}{2}$ **11.** [4.6] $9x - 11y = -76$

12. [4.6] $y = -4x + 11$

13. [4.7] The inequality must be solved for y.

14. [4.7]

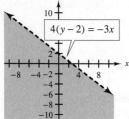

$4(y - 2) = -3x$

15. [5.1] 0   **16.** [5.1] $(-7, -12)$ **17.** [5.2] $(-3, -1)$

18. [5.2] $(8, 15)$ **19.** [5.3] $(-4, 9)$ **20.** [5.4] $89°$

21. [5.4] 1.5 miles   **22.** [5.4] 5 ounces

23. [5.5]

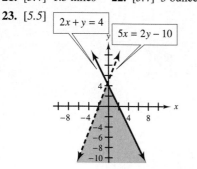

$2x + y = 4$ $5x = 2y - 10$

24. [5.5] The graph of $x + 2y > 10$ is a half-plane above the line $y = -\frac{1}{2}x + 5$. The graph of $y < -\frac{1}{2}x - 3$ is a half-plane below the line $y = -\frac{1}{2}x - 3$. Because the half-planes do not intersect, the solution set is empty.

Chapter 6

Section 6.1   (page 328)

1. Because the bases are the same, we multiply $x^2 \cdot x^3$ by adding the exponents. Because they are not like terms, $x^2 + x^3$ cannot be simplified.

3. $-16, 16, 16$ **5.** $49, -49, 49$ **7.** 7^4 **9.** y^8

11. x^{10} **13.** $(3x)^{14}$ **15.** $-4x^9$ **17.** $8x^{12}$ **19.** $-10x^5y^7$

21. $-30x^8$ **23.** x^8y^5 **25.** **(a)** $2x^4$ **(b)** $-35x^8$

27. **(a)** Not possible   **(b)** $64y^9$

29. **(a)** $-2n$ **(b)** $-18n^3$ **31.** 5^8 **33.** a^{15} **35.** $-x^{28}$

37. x^4y^4 **39.** $64y^3$ **41.** $\frac{a^5}{b^5}$ **43.** $\frac{-32}{x^5}$ **45.** $9x^8$

47. $-32y^{15}$ **49.** $\frac{a^{20}}{b^{15}}$ **51.** $\frac{27n^3}{125m^3}$ **53.** $\frac{64x^6}{y^2}$ **55.** n^{14}

57. $3x^{10}$ **59.** $72x^{16}$ **61.** False   **63.** True   **65.** False

67. Use the Quotient Rule for Exponents to simplify $\frac{x^5}{x^3} = x^2$ and then use the Power to a Power Rule to simplify $(x^2)^2 = x^4$. Use the Quotient to a Power Rule and the Power to a Power Rule to write $\frac{x^{10}}{x^6}$ and then use the Quotient Rule for Exponents to obtain x^4.

69. x^3 **71.** t^4 **73.** x^4y **75.** $\frac{4}{3}c^2d$ **77.** a^{12} **79.** a

81. $8y^3$ **83.** $\frac{t^3}{27}$ **85.** 1   **87.** 1   **89.** 3   **91.** $\frac{y^2}{x^2}$

93. x^3 **95.** $9t^2$ **97.** $6d^3$ **99.** $8x^3$ **101.** **(a)** $\frac{1}{8}x^2$

103. 5   **105.** 3   **107.** 3^{2n+1} **109.** y^{2n^2}

Section 6.2   (page 336)

1. Because a number is a monomial, 1 is a monomial. Because a polynomial is a monomial or the sum of monomials, 1 is also a polynomial.

3. No   **5.** Yes   **7.** Yes   **9.** No

11. Binomial; $-y, -7$; $-1, -7$ **13.** Monomial; $-a^2b$; -1

15. Trinomial; $6, -t, -t^2$; $6, -1, -1$

17. Polynomial; $2x^2y, 5yz, -7z, 1$; $2, 5, -7, 1$

19. $-1, 1$ **21.** $1, 5$ **23.** $5, 2$ **25.** $2, 6$ **27.** 1

29. 3   **31.** 6   **33.** 3   **35.** **(a)** $-y^2 + 16$ **(b)** $16 - y^2$

37. **(a)** $-x^2 + 3x + 4$ **(b)** $4 + 3x - x^2$

39. **(a)** $-3t^3 + t^2 - t + 2$ **(b)** $2 - t + t^2 - 3t^3$

41. **(a)** $-3x^5 + 8x^4 - 6x^2 + x$ **(b)** $x - 6x^2 + 8x^4 - 3x^5$

43. $G(3)$ is the value of the polynomial when the variable is replaced with 3.

45. $5, -1$ **47.** $6, 6$ **49.** $-4, 0$ **51.** $5(3) + 4$

53. $4 - (2)^2$ **55.** $f(2)$ **57.** $-(-2)^2 + 2(-2) + 5$

59. $P(3) = 3; P(-2) = 8$ **61.** $F(-3) = 3; F(4) = 17$

63. $P(-2) = -3; P(3) = -3$

65. (a) -2 (b) 9 (c) 0 (d) 7

67. (a) 12 (b) -27 (c) 1 (d) 4

69. $Q(-3) = 0; Q(0) = 9$ **71.** $P(-1) = 2; P(1) = 2$

73. $P(6) = -16; P(1) = -\frac{1}{6}$ **75.** 0 **77.** -11 **79.** -5

81. -53 **83.** $x^3 - x^2y$ **85.** 1292

87. (a) 2

   (b) $S(12) = 264.1$; credit card spending between Thanksgiving and Christmas is projected to be $264.1 billion in 2002.

89. $P_1(21) = 37$ $P_1(55) = 25$

   $P_1(30) = 34$ $P_1(70) = 20$

91. $P_2(x)$ **93.** $4t + 13$ **95.** $18a^3 + 6a$

97. $-3x^{3n} + 5x^{3n-1} + x^n$

Section 6.3 *(page 343)*

1. To add or subtract polynomials, we remove parentheses and combine like terms. That is, we simplify.

3. $8x - 1$ **5.** $x^2 + 2x - 5$ **7.** $6x^2 + 2x + 3$

9. $2y^3 + y^2 - 2y + 1$ **11.** $4x^2 - xy - y^2$ **13.** $7x + 1$

15. $x^2 + 7x$ **17.** $-w + 2$ **19.** $-2x^2 + 2x + 9$

21. $4x^3 + x^2 - 5x - 8$ **23.** $x^2 - 6x + 6$ **25.** 6

27. $-3x^3 + 2x + 8$ **29.** $y^2 + 6y$ **31.** $-b^2 + 8b + 3$

33. $3x - 1$ **35.** $-3x^2 - 8x + 10$ **37.** $4y^2 + y + 6$

39. $-x^3 + x^2 + 8x + 16$ **41.** $-4xy - 3x^2$

43. $2x^2y + 2x + 3$ **45.** $1 - 12x$ **47.** $x^2 - 5x + 2$

49. $-2t^2 - 7t - 6$ **51.** $2x^2 + 6x + 6$ **53.** $6x^3 - 9$

55. $3x + 7$ **57.** $2x^2 + 3x + 13$ **59.** $14x^2 + 5x + 11$

61. $2w^3 - 4w^2 + w + 1$ **63.** $5x^2 - x + 17$

65. $4t^2 + 5t - 8$ **67.** $z^3 - 6z^2 + 2z + 9$

69. Adding $-x^2$ to both sides of the equation eliminates the second-degree term. Then the remaining variable can be isolated.

71. -3 **73.** 0 **75.** 2 **77.** $-2x + 1$ **79.** $1 - x$

81. $a = 1, b = 2, c = 1$ **83.** $2y^2 + 2y + 4$

85. $-2x^3 - 2x + 230$ **87.** $x^3 - x^2 + 3x - 5$

89. 2 seconds

91. (a) $F(x) = 0.26x + 5.5$ (b) $M(x) = 0.1x + 1.2$

93. $5x + 1$ **95.** $11y^2 + 9$ **97.** $-3a$

Section 6.4 *(page 349)*

1. The product of binomials may have two terms, such as $(x + 3)(x - 3) = x^2 - 9$, or it may have four terms, such as $(x + 3)(a - 2) = ax - 2x + 3a - 6$.

3. $-15x^2$ **5.** $-10y^3$ **7.** $3x^3y^5$ **9.** $12x^3 - 21x^2$

11. $-15x^3 - 5x^2 + 10x$ **13.** $-4x^6 + 12x^4 - 8x^3$

15. $18x^3 + 24x^2 - 30x$ **17.** $x^3y^2 + 3x^2y^3 + xy^4$

19. $x^4y^2 + x^3y + 2x^4$ **21.** $x^2 + 4x - 21$

23. $y^2 + 7y + 6$ **25.** $2x^2 - 5x - 3$ **27.** $6a^2 - 7a - 5$

29. $x^2 + 8x + 16$ **31.** $x^3 + 4x^2 - x - 10$ **33.** $x^3 - 27$

35. $3y^3 - 13y^2 + 6y - 8$ **37.** $2x^3 - 3x^2 - 5x + 3$

39. $x^4 + 3x^3 - 3x^2 - 9x + 4$ **41.** $3t^2 + 8t - 35$

43. $x^3 - 6x^2 + 15x - 14$

45. The result is the same. The Associative Property of Multiplication allows us to group any two factors.

47. $2x^3 - 6x^2 - 36x$ **49.** $x^3 + 2x^2 - 5x - 6$

51. $8x^3 + 2x^2 - 7x - 3$ **53.** $2y^3 + 4y^2 + 2y$

55. $-3x^5 - 3x^4 + 6x^3$ **57.** $x^3 + 3x^2 + 3x + 1$

59. $x^2 - 4$ **61.** $-x^2 + 8x + 6$ **63.** $2x^2 - 8$

65. $-x^2 + 8x + 40$ **67.** $-x^2 + 5x + 5$

69. $4x^2 + 8x - 8$ **71.** $2x^4 + 5x^3 - 3x^2$ **73.** 5

75. 6 **77.** $2x^3 + 7x^2 + x - 1$ **79.** $24x + 228$

81. 12 inches, 15 inches **83.** 10 feet, 14 feet

85. 1 mile, 3 miles

87. (a) Total expenditures on home video games

   (b) $T(x) = 2.64x^2 + 275.46x + 2546.1$

89. 1698.65

91. Both numbers approximate the total monthly cable TV revenue (in millions of dollars) in 1997. The table value is $1700.02 million.

93. $3x^{2n+2} + 2x^{n+2} - 5x^2$

95. $x^5 + 3x^4 - 7x^3 + 17x^2 - 29x + 5$ **97.** $a = -2, b = 13$

Section 6.5 *(page 356)*

1. Because (ii) is the product of binomials, the FOIL method applies. Because (i) is the product of a binomial and a monomial, FOIL does not apply.

3. $x^2 + 7x + 10$ **5.** $y^2 - 4y - 21$ **7.** $16 - 6x - x^2$

9. $6x^2 - 5x - 6$ **11.** $4x^2 - 9x + 2$ **13.** $2x^2 + 3x - 35$

15. $8a^2 - 26a + 15$ **17.** $x^2 - 2xy - 3y^2$

19. $20x^2 + 9xy - 20y^2$ **21.** $x^4 + 2x^2 - 15$

23. $a^4 + 6a^2 + 8$ **25.** $a^2b^2 + 5ab + 6$

27. In (i) the terms are identical, $2x$ and $2x$. In (ii) the terms are opposites, $2x$ and $-2x$.

29. $x^2 - 49$ **31.** $9 - y^2$ **33.** $9x^2 - 25$ **35.** $36 - 25y^2$

37. $9x^2 - 49y^2$ **39.** $4a^2 - b^2$ **41.** $x^4 - 9$

43. $x^2 + 4x + 4$ **45.** $81 - 18x + x^2$ **47.** $4x^2 + 4x + 1$

49. $16 - 56x + 49x^2$ **51.** $a^2 - 6ab + 9b^2$

53. $x^6 + 4x^3 + 4$ **55.** $y^2 - 5y + 6$ **57.** $c^2 - 25$

59. $9 + 6b + b^2$ **61.** $16 - 49b^2$ **63.** $6y^2 + 19y + 15$

65. $9x^2 + 30x + 25$ **67.** $2a^2 - 11ab - 6b^2$

69. $y^2 - 36x^2$ **71.** $3x + 49$ **73.** $12x - 36$

75. $9y^3 + 12y^2 + 4y$ **77.** $x^4 - 16$ **79.** $-8x + 26$

81. $2x^2 - 3x + 25$ **83.** $-\frac{4}{3}$ **85.** $-\frac{1}{2}$ **87.** 6   **89.** 3, 4

91. $-\frac{1}{2}$ **93.** $9x^2 - 49$ **95.** $8x + 16$

97. 5 feet, 11 feet   **99.** 5.5 feet square   **101.** 63 feet

103. $p(x) = -0.003x^2 + 0.115x + 1.174$

105. In 2000 the per-mile tax was 2.06¢. The per-mile tax can be approximated by evaluating $p(25)$.

107. $x^4 - 81$ **109.** $4x + 2$ **111.** $t^2 + 7t + 9$

113. 25

Section 6.6   *(page 365)*

1. A is the dividend, B is the divisor, and C is the quotient.

3. $\frac{2}{3}x$ **5.** $\frac{1}{4}y$ **7.** $5x - 8$ **9.** $y^4 + 3y^2 - 1$

11. $\frac{1}{2}x^2 - \frac{3}{2}x + 1$ **13.** $3ab^2 - 4a^2b$ **15.** $\frac{1}{2}x^2 + 3x - 1$

17. $\frac{3}{x} - 2$ **19.** $x^2 + x - \frac{1}{3}$ **21.** $3x - 1 + \frac{1}{x}$

23. First, divide x^3 by x and place the result x^2 above the $3x^2$. Second, multiply x^2 by $x - 4$ and place the result $x^3 - 4x^2$ below $x^3 + 3x^2$.

25. $Q = x + 3, R = 0$ **27.** $Q = -x + 3, R = 0$

29. $Q = 2x + 5, R = 0$ **31.** $Q = x + 3, R = 10$

33. $Q = 2x + 5, R = 20$ **35.** $Q = 3x - 5, R = -2$

37. $Q = 2x - 1, R = 4$ **39.** $Q = 2x - 3, R = 0$

41. $Q = -3x - 1, R = 0$ **43.** $Q = x + 2, R = 8$

45. $Q = x - 2, R = -3$ **47.** $Q = 2x + 1, R = -3$

49. Include the missing terms with 0 coefficients: $x^3 + 0x^2 + 0x - 5$.

51. $Q = 2x + 5, R = 0$ **53.** $Q = x + 5, R = 20$

55. $Q = 3x + 4, R = 4$ **57.** $Q = x^2 + 3x - 1, R = 0$

59. $Q = 2x^2 + x - 2, R = 0$ **61.** $Q = 2x^2 + x + 3, R = 8$

63. $Q = 2x^2 - x - 1, R = 0$

65. $Q = x^3 + x^2 + x + 1, R = 0$

67. $Q = x^3 - 3x^2 + 9x - 7, R = 11$ **69.** $3x + 4$

71. $x + 3 + \dfrac{10}{x - 7}$ **73.** $2x - 1 + \dfrac{5}{x + 3}$

75. $x^2 - x - 2 + \dfrac{1}{x - 2}$ **77.** $3x^2 + 4x + 4 + \dfrac{3}{x - 1}$

79. $x^3 - 5x^2 + 20x - 108 + \dfrac{392}{x + 4}$ **81.** $3t + 1$

83. $x^2 + 4$ **87.** $Q(x) = 27x - 1039, R = 193{,}390$

89. The value, 215, is the estimated number of volunteers per country at the 2000 Olympics.

91. $Q = 1, R = 4$ **93.** $2x^2 + x - 15$ **95.** -10

97. $x + 3$

Section 6.7   *(page 371)*

1. If the exponent in the denominator is larger than the exponent in the numerator, applying the Quotient Rule for Exponents results in a negative exponent.

3. $\frac{1}{36}$ **5.** $-\frac{1}{8}$ **7.** $\frac{9}{16}$ **9.** $\frac{8}{15}$ **11.** $\frac{1}{5}$ **13.** 2   **15.** $-\frac{1}{9}$

17. 81   **19.** $\dfrac{1}{3}$ **21.** $\dfrac{9}{16}$ **23.** $\dfrac{1}{x}$ **25.** $\dfrac{10}{3}$ **27.** $\dfrac{1}{(-12)^9}$

29. x^5 **31.** y^{-5} **33.** $\dfrac{1}{a^{-6}}$ **35.** $-7x^{-4}$ **37.** $\dfrac{5}{x^4}$

39. $\dfrac{x^5}{3}$ **41.** $\dfrac{1}{7t}$ **43.** $\dfrac{-2y}{x^2}$ **45.** $\dfrac{7}{ab^6}$ **47.** $\dfrac{x}{3}$

49. $\dfrac{1}{x^2y^3}$ **51.** $\dfrac{1}{a^3b^2}$ **53.** $\dfrac{4b}{5a^3}$

55. Apply the Quotient to a Power Rule and then the Power to a Power Rule to obtain $\dfrac{\left(x^{-2}\right)^{-1}}{\left(y^{-3}\right)^{-1}} = \dfrac{x^2}{y^3}$.

57. $\dfrac{1}{a^7}$ **59.** $\dfrac{1}{y^9}$ **61.** $\dfrac{-8}{y^4}$ **63.** $35a^2$ **65.** y^8 **67.** $\dfrac{1}{7^6}$

69. x^6 **71.** $\dfrac{1}{a^9}$ **73.** $\dfrac{1}{2x^5}$ **75.** $\dfrac{2}{3x^5}$ **77.** x^{10} **79.** $\dfrac{a^2}{3}$

81. $\dfrac{y^4}{x^6}$ **83.** a^2b **85.** $a^{12}b^6$ **87.** $\dfrac{y^3}{x^2z^2}$ **89.** $18x^6$

91. $\dfrac{y^3}{x^2}$ **93.** y^4 **95.** $n = 3, m = -1$

97. (a) $\dfrac{260}{x}$ (b) 17.3¢ per minute

99. $H_1(10) = 44$; the result is close to 43, the actual number of hours.

101. $H_2(x) = 0.068x^3 - 2.5x^2 + 33.1x - 105$

103. $\dfrac{8x^3}{3y^5}$ **105.** $3x^6y^7$ **107.** $\dfrac{1}{14}$

Section 6.8   *(page 379)*

1. Each time we multiply by 10, the decimal point is moved one place to the right.

3. 2900   **5.** 0.04   **7.** 639,000   **9.** 0.0067

11. 70,000   **13.** 0.00000756   **15.** 4,285,000,000

17. 0.00000000025534   **19.** $5.6 \cdot 10^3$ **21.** $3.07 \cdot 10^5$

23. $5.2 \cdot 10^7$ **25.** $2 \cdot 10^9$ **27.** $3 \cdot 10^{-2}$ **29.** $4.74 \cdot 10^{-5}$

31. $8 \cdot 10^{-8}$ **33.** $3.4569 \cdot 10^{-1}$ **35.** $9.0 \cdot 10^{19}$

37. $9.3 \cdot 10^{-10}$ **39.** $4.0 \cdot 10^{-4}$ **41.** $3.1 \cdot 10^{10}$

43. $4.7 \cdot 10^5$ **45.** $2.9 \cdot 10^{-4}$ **47.** $7 \cdot 10^3$ **49.** 524

51. 4.73   **53.** $6 \cdot 10^3$ **55.** $2.4 \cdot 10^{13}$ **57.** $3 \cdot 10^4$

59. $5 \cdot 10^{-12}$ **61.** $2.7 \cdot 10^{16}$ **63.** $1.25 \cdot 10^{-10}$
65. $1.1 \cdot 10^9$ **67.** $1.3 \cdot 10^7$ **69.** $7.4 \cdot 10^{47}$ **71.** $1.98 \cdot 10^{11}$
73. $1 \cdot 10^{-12}$ **75.** $7.0 \cdot 10^{-4}$ **77.** $3.68 \cdot 10^9$ **79.** $1 \cdot 10^{20}$
81. $9.3 \cdot 10^{-5}$ **83.** $1.4 \cdot 10^7$ **85.** $1.14 \cdot 10^{-6}$
87. $2.5 \cdot 10^{-4}$ inches **89.** $6.70 \cdot 10^8$ miles **91.** $\$3.35 \cdot 10^9$
93. $4.35 \cdot 10^{10}$ **95.** $\$5.25 \cdot 10^5, \$2.20 \cdot 10^6$
97. $\$1.76 \cdot 10^8$ **99.** 6027 years

Chapter Review Exercises *(page 381)*

1. (a) Base $-4, (-4)^2 = 16$ (b) Base $4, -4^2 = -16$

3. y^{12} **5.** $\dfrac{3}{2}x^4$ **7.** y^8 **9.** $\dfrac{2}{3}x$ **11.** (i), (iii)

13. In (a) the degree is 3, the value of the exponent. In (b) the degree is 5, the sum of the exponents. In (c) the degree is 0 because 9 can be written as $9x^0$, and the exponent on the variable is 0.

15. $2x^5 + 4x^3 - x^2 + 6$ **17.** 5 **19.** 1
21. $3t^4 + t^3 + 2t^2 - 3t - 1$ **23.** $9x^2y + 15xy^2 - 9xy$
25. $-x^2 + 5x + 6$ **27.** $2x^2 - 5x + 16$
29. $-x^2 + 9x + 8$ **31.** $-6x^3y^5$ **33.** $40x^2 + x - 6$
35. $3x^3 + 9x^2 + 6x$ **37.** $2x^2 + 10$ **39.** $\frac{2}{3}$
41. $x^2 + 2x - 3$ **43.** $16 - a^2$ **45.** $4a^2 + 12ab + 9b^2$
47. $-13x^2 + 4x$ **49.** -2

51. We stop dividing when the remainder is 0 or when the degree of the remainder is less than the degree of the divisor.

53. $3a^3 - 2a^2 + 4$ **55.** $x^2 + 2x$ **57.** $Q = x^2 + 2x, R = 2$

59. $x^2 - x + 3 + \dfrac{1}{x+3}$ **61.** $\dfrac{1}{9}$ **63.** $\dfrac{-5}{x^2}$ **65.** $\dfrac{y^3}{x^2}$

67. $-\dfrac{x^6}{8}$ **69.** $\dfrac{1}{x^6y^4}$

71. If $n < 0$, then $-n$ is a positive number. Therefore, b^{-n} already has a positive exponent.

73. 0.00028 **75.** $6.7 \cdot 10^{-5}$ **77.** $1.9 \cdot 10^{-7}$
79. $4.8 \cdot 10^{21}$ **81.** $4 \cdot 10^{12}$ dollars

Chapter Test *(page 383)*

1. [6.1] $-8x^5y^6$ **2.** [6.1] $32x^5y^{10}$ **3.** [6.1] $25x^2y^2$
4. [6.2] $x^3 + 2x^2 + 8x - 7$; 1, 2, 8, -7; Degree 3
5. [6.2] -2 **6.** [6.2] -4 **7.** [6.3] $4x^3 + x^2 + x - 5$
8. [6.3] $x^3 - 2x^2 + 11x - 7$ **9.** [6.3] 5
10. [6.4] The FOIL method applies to products of binomials.
11. [6.4] $-6x^5y^9$ **12.** [6.4] $-2a^3b + 6a^2b^2 - 2ab^3$
13. [6.4] $2x^3 + 7x^2 - 5x - 4$ **14.** [6.5] $2x^2 + 7x - 15$
15. [6.5] $4x^2 - 20x + 25$ **16.** [6.5] $p^2q^2 - 49$
17. [6.6] $Q = 2x^5 - 3x^3 + 5x^2, R = 0$

18. [6.6] $Q = x^2 - 3x + 2, R = -6$
19. [6.6] $Q = x^3 + x^2 + 2x + 2, R = 3$

20. [6.7] $-\dfrac{1}{9}$ **21.** [6.7] $\dfrac{a^{12}}{b^8}$ **22.** [6.7] y^3

23. [6.8] The number n is a real number such that $1 \le n < 10$, and p is any integer.
24. [6.8] (a) $5.9 \cdot 10^{-4}$ (b) $6.84 \cdot 10^5$
25. [6.8] $1.5 \cdot 10^{-4}$ **26.** [6.8] $5.5 \cdot 10^{-5}$

Chapter 7

Section 7.1 *(page 392)*

1. To factor a number completely, write the number as a product of prime numbers.
3. 3 **5.** 1 **7.** 3 **9.** $3y$ **11.** xy^3 **13.** $3x$
15. (a) All are correct factorizations of the polynomial.
(b) Parts (ii) and (iii) are complete factorizations.
17. 8 **19.** $3x + 1$ **21.** 5 **23.** $7(x + 3)$
25. $8(4x + 3y - 7)$ **27.** $6a(3a - 4)$ **29.** Prime
31. $3x(x^2 - 3x + 1)$ **33.** $5x^2y(2x - y^2)$
35. $8a^3b^4(3a + 2)$ **37.** $xy(x + y - 5)$ **39.** $8(3y + 1)$
41. $7(3a + 5b - 1)$ **43.** $4x(3x + 1)$
45. $a^2b^3(ab - b + 1)$ **47.** $4y + 3z$ **49.** $2x - 1$
51. $3(x - 5); -3(5 - x)$ **53.** $4(3 - x); -4(-3 + x)$
55. $2(-x^2 + x + 5); -2(x^2 - x - 5)$
57. The expression is the sum of the terms $x(a + b)$ and $y(a + b)$, not a product.
59. $(x - 8)(x + 5)$ **61.** $(y + 3)(3y - 2)$
63. $(x - 4)(x + 1)$ **65.** $(x - 3)(x^2 + 2)$
67. $(a + 2)(x + 5)$ **69.** $(y - 1)(2x + 3)$
71. $(x + 3)(y - 1)$ **73.** Prime **75.** $(t^2 + 3)(t + 1)$
77. $(2x^2 - 1)(x + 1)$ **79.** $(x + y)(z + 1)$
81. $(a - 1)(x + 1)$ **83.** $(x + 2y)(z - 3)$
85. $3(a + b)(a - 2)$ **87.** $3t(t - 2)(t^2 + 5)$ **89.** NF
91. NF **93.** CF **95.** F **97.** $x^2 + 15x; x(x + 15)$
99. $3x^2(x + 2); x + 2$ **101.** $C(x) = 2(x + 12)$
103. The common factor is the annual increase in the percentage of people who use computers in their jobs.
105. $(x + 3)(x - y + z)$ **107.** $2x - 5$ **109.** $\frac{1}{4}(3x - 5)$

Section 7.2 *(page 399)*

1. In (i), 10 is not a perfect square, and in (iii), the expression is a sum, not a difference.
3. $(x + 5)(x - 5)$ **5.** $(10 + a)(10 - a)$
7. $(8 + 7w)(8 - 7w)$ **9.** $(a + 2b)(a - 2b)$
11. $(1 + x)(1 - x)$ **13.** $\left(x + \frac{1}{2}\right)\left(x - \frac{1}{2}\right)$

15. $(x^2 + 10)(x^2 - 10)$ **17.** $2(x + 2)(x - 2)$

19. $4(x + 2)(x - 2)$ **21.** $y^5(10 + y)(10 - y)$

23. $x(x + y)(x - y)$ **25.** $(x^2 + 9)(x + 3)(x - 3)$

27. $(x^4 + 1)(x^2 + 1)(x + 1)(x - 1)$

29. The constant term is -9. The constant term of a perfect square trinomial must be positive.

31. $(x + 3)^2$ **33.** $(1 - 2y)^2$ **35.** $(x + 4y)^2$

37. $(7x - 3y)^2$ **39.** $(y^2 + 3)^2$ **41.** $(x^3 - 2)^2$

43. $5(x - 3)^2$ **45.** $-x(2x + 7)^2$

47. Yes. First factor out the common factor x. The result $x^3 + 1$ is the sum of two cubes.

49. $(x + 7)(x^2 - 7x + 49)$ **51.** $(z + 1)(z^2 - z + 1)$

53. $5(1 + y)(1 - y + y^2)$ **55.** $(11 - t)(121 + 11t + t^2)$

57. $(4 - y)(16 + 4y + y^2)$ **59.** $(x - 2y)(x^2 + 2xy + 4y^2)$

61. $(5 + x)(25 - 5x + x^2)$ **63.** $(4y + z)(16y^2 - 4yz + z^2)$

65. $(2a - b)(4a^2 + 2ab + b^2)$ **67.** $9(x - 2)(x^2 + 2x + 4)$

69. $27x^6(1 + x)(1 - x + x^2)$ **71.** $(x - 9)^2$

73. $\left(\dfrac{3}{5}t + 1\right)\left(\dfrac{3}{5}t - 1\right)$

75. $2mn(m + 5n)(m^2 - 5mn + 25n^2)$ **77.** $4a^2(a + 1)^2$

79. Prime   **81.** $2(6y + 1)(6y - 1)$ **83.** $(10x - 3y)^2$

85. $3y(y + 4x)(y - 4x)$ **87.** $(2a - 3)(4a^2 + 6a + 9)$

89. $(7x + 9y^3)(7x - 9y^3)$ **91.** $(x^2 - 3y)(x^4 + 3x^2y + 9y^2)$

93. $x^2 - 36$; $(x + 6)(x - 6)$ **95.** $(5 - 3x)(25 + 15x + 9x^2)$

97. The factor $25x + 585$ represents the weekly salary.

99. Weekly: $B(x) = 52(36x + 549)$;
Monthly: $B(x) = 12(156x + 2379)$

101. $(x + 2)^2(x - 2)^2$ **103.** $(x + 3 + y)(x + 3 - y)$

105. $(x - y + 3)^2$

Section 7.3   *(page 407)*

1. Both m and n are positive.   **3.** $(x + 3)(x + 1)$

5. $(y + 3)(y - 1)$ **7.** $(a + 3)(a + 2)$ **9.** $(x + 3)(x + 2)$

11. $(x - 2)(x - 6)$ **13.** $(y - 4)(y - 5)$ **15.** $(x + 4)(x + 3)$

17. The numbers m and n have opposite signs, and the one with the larger absolute value is negative.

19. $(x + 2)(x - 1)$ **21.** $(y - 5)(y + 2)$

23. $(y + 7)(y - 2)$ **25.** $(x - 9)(x + 1)$

27. $(x + 8)(x + 1)$ **29.** $(z - 9)(z + 5)$ **31.** $(x + 2)(x + 7)$

33. The leading coefficient is -1. A possible first step is to factor out the -1. Writing it in ascending order makes it unnecessary to factor out a common factor.

35. $(x + 3y)(x + 2y)$ **37.** $(x + 7y)(x - 2y)$

39. $(2 - xy)(1 - xy)$ **41.** $(x^2 + 2)(x^2 + 7)$

43. $(x^3 - 5)(x^3 - 2)$ **45.** $-1(x + 5)(x - 1)$

47. $-1(t - 8)(t + 4)$ **49.** $-1(x - 7)(x - 5)$

51. $5(x + 3)(x + 4)$ **53.** $4y^7(y + 1)^2$ **55.** $x^2(x^2 + 3x + 1)$

57. $-3x^2(x - 6)(x + 1)$ **59.** $(x - 10)(x - 8)$

61. $(x + 4)(x + 7)$ **63.** $(w + 10)(w - 4)$

65. $(c - 7)(c + 5)$ **67.** Prime   **69.** $-2(x - 7)(x + 6)$

71. $x^2(x - 12)(x + 4)$ **73.** $(x - 12y)(x - y)$

75. $x(x - 4)(x - 1)$ **77.** $-3(x^2 - 4x + 12)$

79. $(x - 4)(x - 1)$ **81.** $-1(w + 17)(w - 2)$

83. $(x + 4y)(x - y)$ **85.** $x^5(x + 10)(x - 8)$

87. $xy(x^2 + 7x - 10)$ **89.** $(a + 12)(a - 3)$

91. $(a - 4x)(a + 3x)$ **93.** Prime

95. $(x + 8)(x + 1)$ **97.** $-16(t - 4)(t + 1)$

99. **(a)** $0.6(-t + 129)$ **(b)** 68.4%

101. $(x + 3)(x - 7 - w)$ **103.** $\frac{1}{3}(x - 3)(x + 1)$

105. $-5, 5, -7, 7$

Section 7.4   *(page 414)*

1. Because 11 and 37 are prime, there are few possible factors and the trial-and-check method is a good choice for (ii). In (i), because 12 and 36 have several factors, there are many possible factors to try.

3. $(3x + 2)(x + 3)$ **5.** $(2x + 3)(x - 4)$

7. $(x - 3)(6x - 5)$ **9.** $(2x + 1)(x + 1)$

11. $(4x + 3)(3x - 2)$ **13.** $(12x + 5)(x - 1)$

15. $(8y - 5)(y - 5)$ **17.** $(2x + 3)(2x + 5)$

19. $(11x + 10)(x - 1)$ **21.** $(3x + y)(2x + 3y)$

23. $(7a - 3b)(2a + 3b)$ **25.** $(7x + 2z)(x - 5z)$

27. First factor out the common factor 2. Then use either the trial-and-check method or the grouping method to factor the trinomial.

29. $(2x + 5)(x + 1)$ **31.** $(4x - 3)^2$ **33.** $(5y - 1)(y - 2)$

35. **(a)** $(7 + 2x)(2 - x)$ **(b)** $-1(2x + 7)(x - 2)$

37. **(a)** $(6 + x)(7 - x)$ **(b)** $-1(x + 6)(x - 7)$

39. **(a)** $(9 - x)(4 + x)$ **(b)** $-1(x - 9)(x + 4)$

41. $5(2x - 3)(x - 2)$ **43.** $-7(4x + 3)(3x - 2)$

45. $10(t - 2)(t + 1)$

47. No. For the polynomial $x^2 + 4x + 3$, $ac = 3$ is prime, and the polynomial can be factored as $(x + 3)(x + 1)$.

49. $(3x + 2)(x + 1)$ **51.** $(4x - 3)(x - 2)$

53. $(2x + 3)(x - 1)$ **55.** Prime   **57.** $2x(3x - 4)(x + 2)$

59. $(5t - 11)(t - 1)$ **61.** $(5b + 4)(2b - 3)$

63. $(5x + 3)(x + 1)$ **65.** $(3x - 1)(x + 2)$

67. $(6x - y)(2x - 3y)$ **69.** $(2a + 5)(2a - 3)$

71. $(10y + 7)(y - 1)$ **73.** $w^4(2w + 5)(w - 4)$

75. $(4x + 5)(x - 1)$ **77.** Prime   **79.** $(7z + 9)(z - 2)$

81. $(4t + 3)(7t - 2)$ **83.** $(2x - 3)(x - 7)$

85. $(3a - w)(a + 3w)$ **87.** $(5x + 6)(x + 3)$

89. $(7x - 3)(x + 1)$ **91.** $(3a + 2b)(a - b)$

93. $(4x + 9)(2x - 5)$ **95.** $(2a - 3)(a - 8)$

97. $(2x + 1)(x + 5)$ **99.** $(2x + 1)(x + 4)$; Base: $2x + 1$

101. (a) $P(x) = 0.01(x^2 - 78x + 6631)$

   (b) The only possible factorizations are
   $(x - 1)(x - 6631)$ and $(x - 19)(x - 349)$. Neither
   results in the middle term, $-78x$, when the factors
   are multiplied.

   (c) About 53%

103. $\frac{1}{25}(5x + 2)(10x - 1)$ **105.** $(2x^5 + 1)(x^5 - 4)$

107. $(2x + 1)(2x - 1)(3x + 2)(3x - 2)$

Section 7.5 *(page 418)*

1. $(1 + x)(1 - x)$ **3.** $(2 + 3x)^2$ **5.** $4y^2(y + 2)(y + 1)$

7. $(2x + 5)(4x - 1)$ **9.** $(y + 5)(y + 9)$

11. $-4(y - 3)(y + 1)$ **13.** $(6x + 5)(x + 1)$

15. $x^2(2x + 3y)(6x - 5y)$ **17.** $(y - 8)(y - 7)$

19. $(a + 1)(x + 7)$ **21.** Prime   **23.** $x^3(x^2y + 2xy^2 + 1)$

25. $(ab + 5)^2$ **27.** $(b + 6)(b - 5)$ **29.** $(5x + 6)(x - 1)$

31. $(8x + 7)(x - 2)$ **33.** $(a - 2b)(c + 2)$

35. $(t - 8)(t + 3)$ **37.** Prime   **39.** $(8t^5 + 1)(8t^5 - 1)$

41. $(3 - y)(9 + 3y + y^2)$ **43.** $(t + 1)(t - 1)(t - 5)$

45. $-3x^2(2x - 1)(x - 1)$ **47.** Prime

49. $3x(2x - 5)(x + 1)$ **51.** $c(-2x + c - 5)$

53. $(a^2 - b)(c^2 + d)$ **55.** $(x - 4b)(x - 5b)$

57. $2x(x - 3)(x^2 + 3x + 9)$ **59.** $(4b + 3c)(2b + c)$

61. $3a(a + 4)(a - 2)$ **63.** $(4y - 5z)(2y + 3z)$

65. $8x(x + 2)(x - 2)$ **67.** $(x^3 + 2)(x + 1)$

69. $(x - 12y)(x + 2y)$ **71.** $x(x + 6y)^2$

73. $b^2(a^2 - 8ab + 24b^2)$ **75.** $2x^2(x - 8)(x + 3)$

77. $a(2a - 1)(4a^2 + 2a + 1)$ **79.** $-6xyz^4(x^3z + 4y^4)$

81. $(a + 11b)(a - 2b)$ **83.** $(4x + 7z)(3x - 7z)$

85. $(4y^2 + 1)(2y + 1)(2y - 1)$ **87.** $(10x - 9)^2$

89. $yz^2(4y + z)(3y - z)$ **91.** $2x(x^2 - 3x + 18)$

93. $(4a + b)(16a^2 - 4ab + b^2)$

Section 7.6 *(page 424)*

1. For $a = 0$, the equation becomes $bx + c = 0$, which for
   $b \neq 0$ is a linear equation.

3. (a) 2   (b) 2   (c) 1   (d) 0

5. 10, 25   **7.** No solution   **9.** $-10, 30$

11. (a) 2   (b) No solution   (c) $-2, 6$ (d) $-4, 8$

13. One side of the equation must be 0, and the other side
   must be written in factored form.

15. $-5, 7$ **17.** $-\frac{7}{3}, 5$ **19.** $\frac{1}{2}, 8$ **21.** $-3, 1$ **23.** $0, \frac{1}{4}$

25. $-4, -1, 3$ **27.** $-9, 0, 2$ **29.** $-4, -2$ **31.** $-7, 3$

33. $0, 1$ **35.** $-8, 8$ **37.** -5 **39.** $-9, 3$ **41.** $-6, 1$

43. $-4, 5$ **45.** $-4, 0$ **47.** $\frac{5}{2}$ **49.** $-\frac{7}{2}, \frac{7}{2}$ **51.** $0, \frac{5}{2}$

53. $-\frac{1}{2}, 4$ **55.** $-4, -\frac{1}{4}$ **57.** $-\frac{1}{4}, \frac{1}{2}$ **59.** $\frac{5}{6}, 2$ **61.** $-2, \frac{4}{9}$

63. (a) There are 3 factors: 2, x, and $x - 1$.

   (b) No, the factor 2 can never be 0.

65. $-7, 4$ **67.** $-7, 1$ **69.** $1, 2$ **71.** $-6, -3$

73. $-\frac{9}{2}, 5$ **75.** $-\frac{3}{2}, -1$ **77.** $-\frac{1}{5}, \frac{3}{2}$ **79.** $-7, 0, 4$

81. $-1, 0, 1$ **83.** $-9, 0, 2$ **85.** $-\frac{4}{3}, 0, \frac{4}{3}$ **87.** $-2, 2$

89. $-\frac{1}{3}, \frac{1}{3}, 3$ **91.** $-1.4, 2.4$ **93.** 1.4 **95.** $2.1, -0.7$

97. (a) $80, -35$

   (b) Because age cannot be negative, -35 is not
   meaningful.

   (c) No one of age 80 received assistance.

99. $(36x + 1132)(0.11x + 3.80)$

101. $(36x + 1132)(0.11x + 3.80) = 10,000$; The solution is
   approximately 17, which represents 2007.

103. $-3, -\frac{5}{2}, -\frac{4}{3}, 2$ **105.** $-9, -3, 3$ **107.** $-3, -1, 1, 3$

109. $-3b$

Section 7.7 *(page 430)*

1. 7 and 9   **3.** -4 and -3 **5.** -5 and 6 or -6 and 5

7. -7 or 3   **9.** -8 or 3   **11.** $-\frac{1}{2}$ or $\frac{7}{3}$ **13.** $-\frac{2}{3}$ or 4

15. -3 or $\frac{1}{2}$ **17.** 7, 10   **19.** 7, 11   **21.** 7, 12

23. 3, 7, 11   **25.** 2, 3, 8   **27.** 5 feet, 8 feet

29. 6 meters, 7 meters   **31.** 8   **33.** 10 feet   **35.** 3, 4, 5

37. 9, 12, 15   **39.** 15 inches, 20 inches, 25 inches

41. 5 yards, 12 yards, 13 yards   **43.** 5 feet, 12 feet, 13 feet

45. 12 feet, 15 feet   **47.** 6 inches   **49.** 20 mph

51. 3 yards, 4 yards   **53.** 25   **55.** 10   **57.** 30

59. 50   **61.** 13 mg   **63.** \$5

65. (a) $0.3x^2 - 3.6x + 36.2 = 36.2$ (b) 2002

67. $x^2 + 22x + 588 = 1428$ **69.** 2010   **71.** 8, 9

73. 2 inches   **75.** 25 feet, 65 feet

Chapter Review Exercises *(page 434)*

1. $5 \cdot 5 \cdot 3 \cdot 11$ **3.** (a) $5y^2$ (b) $3xz$

5. (a) $9x^2(3x^2 - 5x - 1)$ (b) $4x(xy - 1)$

7. $(a^2 + 7)(b - 1)$ **9.** $x(x^2 - 6)(x + 3)$

11. (a) $(4a + 3)(4a - 3)$ (b) $(pq + 2)(pq - 2)$

13. $(b^2 + 4)(b + 2)(b - 2)$

15. (a) $(x - 6)^2$ (b) $2a(x + 4)^2$

17. (a) $(x + 3)(x^2 - 3x + 9)$ (b) $3(1 + 2a)(1 - 2a + 4a^2)$

19. $(x + 1)(x - 1)^3$

21. (a) $(c + 7)(c + 3)$ (b) $(x + 9)(x + 2)$

23. $(y - 6)(y + 5)$ **25.** $(x^2 + 5)(x^2 - 3)$

27. (a) $3(y - 8z)(y + 4z)$ (b) Prime

29. $(x + 1)^3(x - 1)^2(x + 2)$

31. (a) $(3x + 1)(2x + 5)$ (b) $(2x - 3)(x - 4)$

33. (a) $(3a + 7b)(a + 2b)$ (b) $(2x - y)(x - 9y)$

35. $-1(3x + 8)(x - 3)$

37. (a) Prime (b) $2a(3a + 4)(8a + 7)$

39. $(2x + 1)(x - 1)(2x - 1)^2$ **41.** $4(x + 2)(x - 2)$

43. $(t^2 + 4)(t + 2)(t - 2)$ **45.** $(a^2 + 1)(a + 1)$

47. $(11x - 3)(x + 2)$ **49.** $2x(x + 3)^2$

51. (a) $-10, 30$ (b) 10 (c) No solution

53. $-0.4, 0, \frac{1}{2}$ **55.** $-4, 7$ **57.** $-4, 9$ **59.** $-8, 0, 5$

61. 9 **63.** 5 **65.** 13 feet **67.** 15 inches **69.** 15 feet

Chapter Test *(page 437)*

1. [7.1] $2x^4y^2$ **2.** [7.1] $(x - 8)^2$

3. [7.1] $(x - 1)(x^4 - 2)$

4. [7.2] Difference of two squares; $9(x + 2)(x - 2)$

5. [7.2] Sum of two cubes; $(y + 1)(y^2 - y + 1)$

6. [7.2] Difference of two cubes; $8(a - 2)(a^2 + 2a + 4)$

7. [7.2] Perfect square trinomial; $(2x - 5)^2$

8. [7.3] $(x + 8)(x - 4)$ **9.** [7.3] $(a - 5b)(a + 2b)$

10. [7.3] $-1(x - 5)(x - 3)$ **11.** [7.4] $c(c + 7)(c - 2)$

12. [7.4] Prime **13.** [7.4] $(2x + 3)(6x - 5)$

14. [7.4] $2(3a^2 - ab - 3b^2)$ **15.** [7.4] $8x + 3$

16. [7.5] $(x^2 + 6)(x + 1)(x - 1)$

17. [7.5] $(x - 1)(x + 2)(x - 2)$

18. [7.5] $5(3x + 1)^2$ **19.** [7.5] $(a^2 + 4)(a + 2)(a - 2)$

20. [7.6] The quadratic and linear terms cannot be combined, so the variable cannot be isolated.

21. [7.6] $0, 7$ **22.** [7.6] $-6, 8$ **23.** [7.6] $-\frac{1}{2}, \frac{5}{3}$

24. [7.6] $-1, 1, 4$

25. [7.6] The expression $ax^2 + bx + c$ is a perfect square trinomial.

26. [7.7] 1.5 inches **27.** [7.7] 12 feet **28.** [7.7] 3 hours

Chapter 8

Section 8.1 *(page 447)*

1. Because the expression is not defined for -2, the calculator gives an error message.

3. $0, \frac{1}{2}$, undefined **5.** $-\frac{13}{10}$, undefined, 5 **7.** $\frac{3}{4}$ **9.** $-\frac{1}{9}$

11. -2 **13.** $0, 4$ **15.** None **17.** $-5, 5$

19. Numerator: $4x, -3$; Denominator: $x, -3$

21. Numerator: $4, x - 4$; Denominator: $3, x - 4$

23. In the numerator x is a term, not a factor. Only factors can be divided out.

25. $\frac{y}{3}$ **27.** Cannot simplify further **29.** $\frac{2x + 1}{3}$

31. $\frac{x + 3}{x - 3}$ **33.** $\frac{3}{x - 6}$ **35.** $\frac{w + 4}{w - 3}$

37. The expressions are opposites because each is -1 times the other.

39. -1 **41.** -1 **43.** Cannot simplify further

45. $-\frac{1}{2 + 3y}$ **47.** -4 **49.** $-\frac{3x}{2}$ **51.** $-\frac{2z}{z + 2}$

53. $-\frac{2(x + 1)}{3}$ **55.** $\frac{2}{3}$ **57.** $\frac{x - 4}{x + 5}$ **59.** $-(x + 3)$

61. $x - 4y$ **63.** $y + 3$ **65.** -5

67. (a) 1 (b) Neither (c) -1 **69.** No **71.** Yes

73. No **75.** $3 + x$ **77.** $\frac{3x + 4}{5}$ 0

79. The coefficient of x gives the slope. Because in each case the coefficient of x is positive, the life expectancy is increasing for both men and women.

81. $r(100) \approx 1.09$; approximately 9% **83.** $-(1 - x)^2$

85. $x - 2$ **87.** $\frac{1}{x - y - 2}$ **89.** $x - 4$

Section 8.2 *(page 452)*

1. Multiplying first may create expressions that are difficult to factor. If we divide out common factors first, then the remaining factors are easier to multiply.

3. $\frac{7x^2}{4}$ **5.** $\frac{-3(2x + 1)}{2x - 1}$ **7.** $\frac{y - 2}{4}$ **9.** $\frac{4}{x}$ **11.** $\frac{w + 6}{3w^4}$

13. $2(x - 3)$ **15.** -1 **17.** $\frac{y}{8}$ **19.** $\frac{1}{5}$

21. Change the operation to multiplication and take the reciprocal of the divisor.

23. $\frac{1}{x}$ **25.** $a + 3$ **27.** $\frac{x - 1}{x + 2}$ **29.** $\frac{5a^2}{2b}$ **31.** $\frac{2}{x}$ **33.** $\frac{1}{2}$

35. $-\frac{1}{3x}$ **37.** $-\frac{7}{4(2x - 5y)}$ **39.** $\frac{h + 1}{6h}$ **41.** 1

43. $\frac{x}{(2x - 3)(2x - 5)}$ **45.** $\frac{d + 3}{2(d + 2)}$ **47.** $-\frac{3}{4z}$

49. $-\frac{1}{ab}$ **51.** $\frac{x(x + 2)}{y(x - 3)}$ **53.** $\frac{a + 1}{4a^2}$ **55.** $\frac{1}{3}$

57. $\frac{x + 6}{x - 8}$ **59.** $\frac{x + 4}{x + 8}$ **61.** $4x$ **63.** $\frac{2x + 1}{5}$

65. (a) No, the denominator cannot be 0.   (b) 77,429

67. $1 - x$ **69.** 1   **71.** x^2

Section 8.3   *(page 458)*

1. In order for us to add fractions, the denominators must be the same. Reducing $\frac{3}{9}$ first would result in fractions with different denominators.

3. $\dfrac{9}{2x}$ **5.** $\dfrac{5x - 5}{x}$ **7.** -4 **9.** $\dfrac{2x - 3}{2x + 3}$ **11.** y

13. $-y - 1$ **15.** $\dfrac{x + 5}{x - 2}$ **17.** $\dfrac{4}{x + 2}$ **19.** $\dfrac{1}{x}$ **21.** $2t^2$

23. $\dfrac{4}{x}$ **25.** -1 **27.** 4   **29.** -2 **31.** $\dfrac{3}{x - 6}$

33. $x + 1$ **35.** $\dfrac{1}{2x - 5}$

37. Insert parentheses to write the numerator as $x - (2a + 3)$.

39. $\dfrac{9x - 5}{x}$ **41.** $\dfrac{2x - 7}{5}$ **43.** $\dfrac{7z - 3}{3z^2}$ **45.** $\dfrac{-3x + 5y}{y}$

47. 1   **49.** -3 **51.** 1   **53.** 2   **55.** $\dfrac{2}{x + 2}$ **57.** $\dfrac{x + 2}{x - 2}$

59. $\dfrac{x}{x + 2}$ **61.** $x + 8$ **63.** $x^2 + 2x$ **65.** 6   **67.** $\dfrac{3z - 1}{3z}$

69. (a) C_2 (b) 8,000   (c) x

71. $t - 4$ **73.** (a) 1   (b) 1

75. $\dfrac{1}{x + 2}$ **77.** $\dfrac{2x - 3}{x - 5}$ **79.** $\dfrac{x + 1}{2x}$

Section 8.4   *(page 466)*

1. When both denominators are in factored form, it is easier to determine the factors that are in the new denominator but not in the original denominator.

3. $\dfrac{9y^2}{6xy}$ **5.** $\dfrac{-25y^3}{60y^5}$ **7.** $\dfrac{12x + 8}{9x^2 + 6x}$ **9.** $\dfrac{6x^2 - 3x}{3x^2 + 3x}$

11. $\dfrac{-y}{3 - y}$ **13.** $\dfrac{-6x + 18}{(x - 3)^2}$ **15.** $\dfrac{2x^2 - 10x}{8x^2 + 24x}$

17. $\dfrac{-3x + 3}{x^2 + 4x - 5}$ **19.** $\dfrac{y^2 - 36}{y^2 + 2y - 24}$ **21.** $\dfrac{4x^2 + 10x}{12x^2 - 75}$

23. The GCF consists of each factor that appears in both terms. For a repeated factor, use the smallest exponent that appears on the factor. The LCM consists of each factor that appears in at least one term. For a repeated factor, use the largest exponent that appears on the factor.

25. $24xy$ **27.** $10x^3$ **29.** $b(b + 2)$ **31.** $8x(x - 2)$

33. $2x - 7$ **35.** $(x + 7)(x - 7)$ **37.** $3(x + 8)(x - 8)$

39. $(x + 5)(x - 5)(x - 1)$ **41.** $210y$ **43.** $2x^4y^3$

45. $(x - 7)(x - 8)$ **47.** $5x^2(x - 2)$ **49.** $(2x + 1)(2x - 1)$

51. $(x - 6)(x + 1)$ **53.** $(x + 1)(x - 1)(x + 3)$

55. $(x + 3)^2(x - 1)$ **57.** $\dfrac{a^2}{3ab^2}, \dfrac{3b^3}{3ab^2}$

59. $\dfrac{2x + 8}{x(x + 4)}, \dfrac{x^2}{x(x + 4)}$ **61.** $\dfrac{2x^2 + 11x + 15}{x + 3}, \dfrac{x}{x + 3}$

63. $\dfrac{x^2 + x}{(x - 2)(x + 1)}, \dfrac{x^2 - 4}{(x - 2)(x + 1)}$

65. $\dfrac{5x}{10(x + 3)}, \dfrac{2x + 2}{10(x + 3)}$ **67.** $\dfrac{3x}{x(x + 3)^2}, \dfrac{2x + 6}{x(x + 3)^2}$

69. $\dfrac{x + 3}{(2x - 5)(x - 2)}, \dfrac{6x - 15}{(2x - 5)(x - 2)}$

71. $\dfrac{x^2 + x}{(2x + 1)^2(x + 1)}, \dfrac{2x^2 + 5x + 2}{(2x + 1)^2(x + 1)}$ **73.** 10

75. 28 minutes   The number is the LCM of 4 and 7.

77. 6006   **79.** $x(x + 1)^2(x^2 - x + 1)$

81. $\dfrac{2ab}{ab(a^2 + b^2)(a + b)(a - b)}, \dfrac{a^2 - b^2}{ab(a^2 + b^2)(a + b)(a - b)}$

Section 8.5   *(page 472)*

1. The denominators have no common factors.

3. $-8x + 6, -3x + 5$ **5.** $\dfrac{x - 9}{8}$ **7.** $-\dfrac{4}{x}$ **9.** $x + 9$

11. $\dfrac{2}{x + 6}$ **13.** $x, x - 1, 2x - 1$ **15.** $\dfrac{x^2 + 5}{x}$ **17.** $\dfrac{8x - 7}{x}$

19. $\dfrac{10x + 41}{28}$ **21.** $\dfrac{2x - 2}{x - 5}$ **23.** $\dfrac{7x + 21}{x(x + 7)}$ **25.** $\dfrac{4x - 4}{x(x - 2)}$

27. $\dfrac{10x - 14}{(x + 5)(x - 3)}$ **29.** $\dfrac{x^2 - 5x + 6}{(3x - 2)(x + 4)}$

31. $\dfrac{-6x + 6}{(x + 5)(x - 4)}$ **33.** $\dfrac{12y}{(3 + y)(3 - y)}$

35. The fractions have the same denominator. The next step is to add the numerators.

37. $\dfrac{8x + 2}{x^2}$ **39.** $\dfrac{17}{5(x + 2)}$ **41.** $\dfrac{x - 10}{x - 5}$ **43.** $\dfrac{4x + 45}{(x + 9)^2}$

45. $\dfrac{3x + 20}{(x + 5)(x + 2)}$ **47.** $\dfrac{4x + 2}{(x + 3)(x + 1)}$ **49.** $\dfrac{2x - 21}{x(x - 9)}$

51. $-\dfrac{7}{10x}$ **53.** $\dfrac{17x - 21}{30}$ **55.** $\dfrac{x - 2}{2x}$ **57.** $\dfrac{2x + 15}{12(x + 3)}$

59. $\dfrac{4 - y}{3y}$ **61.** $\dfrac{12}{(x - 5)(x + 7)}$ **63.** $\dfrac{4x - 15}{(x + 3)(x + 6)(x - 6)}$

65. $\dfrac{x - 12}{(x + 4)(x + 8)}$ **67.** $\dfrac{x + 9}{(x + 3)(x - 1)}$ **69.** $\dfrac{-3}{4(x + 3)}$

71. $\dfrac{x - 3}{x(x - 1)}$ **73.** $\dfrac{x^2 + 3x}{(x + 1)(x + 2)}$ **75.** $\dfrac{25x^2 - 15x}{(x + 1)(x - 1)}$

77. (a) $g(x) = \dfrac{34x - 77}{x + 3}$

(b) 24.1; in 2005, the SUV market share is projected to be 24.1%.

79. $\dfrac{9}{x}$ **81.** $\dfrac{1}{(x+1)(x+2)}$ **83.** $\dfrac{2}{a-2}$

Section 8.6   (page 478)

1. Simplifying the complex fraction can be accomplished in just one step by multiplying the numerator and the denominator by the LCD x.

3. $\dfrac{3b}{2a}$ **5.** $\dfrac{1}{6t^3}$ **7.** $\dfrac{3}{a(a-3)}$ **9.** $-\dfrac{y+5}{y-5}$

11. $\dfrac{x}{2(x-3)}$ **13.** $3(x^2+9)$ **15.** $\dfrac{x-1}{42x}$ **17.** $\dfrac{6}{5}$

19. $\dfrac{5}{16}$ **21.** $\dfrac{27-5x}{12x-36}$ **23.** $2x+1$ **25.** $\dfrac{a^2}{2b+ab^2}$

27. $\dfrac{1-4x}{2+3x}$ **29.** $\dfrac{x}{2+xy}$ **31.** $\dfrac{y+x}{y+1}$ **33.** x **35.** $\dfrac{1}{y+x}$

37. ab **39.** $\dfrac{2a+3b}{2a-3b}$ **41.** $\dfrac{x+6}{3}$ **43.** $\dfrac{2a+3}{2a-5}$ **45.** -2

47. (a) $\dfrac{t+20}{1+t}$

     (b) 20

     It is not valid because the original expression is not defined if t is 0.

49. $\dfrac{9t}{10t+15}$ **51.** $r(x)=\dfrac{1167x-4580}{1.8x+77.2}$ **53.** 1997

55. $\dfrac{x}{1-x}$ **57.** $\dfrac{5x+2}{-x-2}$ **59.** $\dfrac{c-4}{24}$ **61.** $\dfrac{2}{t+2}$

Section 8.7   (page 487)

1. One of the apparent solutions may be a restricted value.

3. (a) No   **(b)** No   **(c)** Yes

5. (a) No   **(b)** Yes   **(c)** No

7. 8   **9.** -3 **11.** 2, 10   **13.** $-4, 6$ **15.** 0, 2

17. $-1, 0$ **19.** $-1, 0, 6$ **21.** -3 **23.** $\frac{8}{5}$ **25.** 16

27. No solution   **29.** $-\frac{8}{5}$ **31.** $\frac{2}{5}$ **33.** -1 **35.** $\frac{2}{7}, \frac{3}{2}$

37. 4   **39.** $-\frac{1}{3}, 2$ **41.** $\frac{2}{3}, 4$ **43.** $-2, 1$ **45.** -4

47. $-2, 0$

49. In (i), we rewrite each fraction with the LCD and then add the numerators. In (ii), we multiply both sides (each term) of the equation by the LCD to clear the fractions in the equation.

51. 3   **53.** $-12, 2$ **55.** No solution   **57.** $-4, \frac{1}{3}$

59. $-1, 6$ **61.** 0   **63.** 6   **65.** $\frac{7}{8}, 3$ **67.** -84 **69.** $3x$

71. $-9, 1$ **73.** $\dfrac{8x+2}{x^2}$ **75.** -3 **77.** 3   **79.** $\dfrac{1}{2}$ **81.** $-\dfrac{2}{3}$

83. $r=\dfrac{d}{t}$ **85.** $\dfrac{100}{52.9-0.11x}-\dfrac{100}{60.4-0.16x}=0.21$

87. -2 **89.** No solution   **91.** $-2, -1, 1, 2$

93. (a) Cross-multiplication results in the possible solution 1, which is an extraneous solution. Therefore, the equation has no solution.

     (b) Multiplication by the LCD results in the contradiction $1 = 2$. Again, the equation has no solution.

Section 8.8   (page 496)

1. $\frac{8}{7}$ and $\frac{24}{7}$ **3.** -1 or 5   **5.** $\frac{1}{10}$ **7.** -4 or 2

9. -4 and -1 or 1 and 4   **11.** 3.5 mph   **13.** 42 mph

15. 6 mph   **17.** 100 mph   **19.** 10 mph   **21.** 15 hours

23. 2 hours   **25.** 6 hours   **27.** 50 minutes, 75 minutes

29. 20   **31.** $\frac{10}{3}$ hours   **33.** 20   **35.** -16.32

37. 6 hours   **39.** 6   **41.** 1000   **43.** 250

45. 20¢ per pound   **47.** 9 hours

49. The number of one-parent family groups is increasing at a faster rate than the number of two-parent family groups.

51. $\dfrac{0.03x+25}{0.28x+6.9}=2$ **53.** 50 mph

55. 8 dozen cookies, 10 dozen donuts   **57.** 8 hours

Chapter Review Exercises   (page 499)

1. (a) $-\dfrac{3}{2}$ **(b)** Not defined   **3.** (ii)   **5.** $\dfrac{x+2}{x-2}$

7. $-\dfrac{1}{4b+a}$ **9.** Cannot simplify   **11.** No

13. By dividing out common factors and then multiplying, we ensure that the result is already in simplified form.

15. $\dfrac{1}{x+4}$ **17.** $\dfrac{3}{1}\cdot\dfrac{1}{x+2}$ **19.** $-5, 2, 3$ **21.** Neither

23. $\dfrac{1}{x-3}$ **25.** $\dfrac{1}{x}$ **27.** x^2+2x-3 **29.** $\dfrac{14x+7}{14x^2-7x}$

31. $3(x+3)(x-3)$ **33.** $3x^2(x+3)$

35. $\dfrac{2y}{6(2y-5)}, \dfrac{9y}{6(2y-5)}$ **37.** $\dfrac{x^2+4x+12}{x(x+3)}$ **39.** $x+5$

41. $\dfrac{2x-1}{(x+3)(x-3)}$ **43.** $\dfrac{4x+19}{(x+6)(x-4)(x+3)}$

45. Both are correct.   **47.** $\dfrac{3ab^2+5a^2}{4a^2b-b^2}$ **49.** $\dfrac{1}{a-b}$

51. The left side of the equation is not defined if x is 3.

53. $\frac{3}{10}$ **55.** No solution   **57.** -2 **59.** 6 hours

Chapter Test   (page 501)

1. [8.1] 3, 5   **2.** [8.1] $\dfrac{x-1}{x-4}$ **3.** [8.1] $\dfrac{x+y}{x-y}$

4. [8.2] $\dfrac{x+3}{x+2}$ **5.** [8.2] $-\dfrac{x}{4}$ **6.** [8.2] $\dfrac{3}{x(x+3)}$

7. [8.3] $\dfrac{1}{x}$

8. [8.3] Neither sequence is correct. The denominator of the first fraction and the numerator of the second fraction must be enclosed in parentheses.

9. [8.3] $\dfrac{x-4}{y}$ **10.** [8.3] 1

11. [8.3] We use the largest exponent to determine the LCM, whereas the smallest exponent is used to determine the GCF.

12. [8.4] $3x^2(x-1)$ **13.** [8.4] $(x+2)(x-2)(x+4)$

14. [8.4] $\dfrac{x^2}{2x(x+2)}, \dfrac{2x-2}{2x(x+2)}$ **15.** [8.5] $\dfrac{x^2+8x-24}{(x-8)(x+5)}$

16. [8.5] $\dfrac{-3y-4}{(y+1)(y-1)}$ **17.** [8.5] $\dfrac{-x^2+x+4}{(x+1)^2(x+2)}$

18. [8.6] $\dfrac{8}{x^2}$ **19.** [8.6] $\dfrac{x^2y+x}{y-xy^2}$ **20.** [8.7] $-1, \dfrac{4}{3}$

21. [8.7] -3 **22.** [8.8] 1 or 5 **23.** [8.8] 5 hours

Cumulative Test: Chapters 6–8 *(page 502)*

1. [6.1] **(a)** $-x^6y^3$ **(b)** $\dfrac{4a^3}{3}$ **(c)** $\dfrac{8a^3}{27b^6}$

2. [6.2] **(a)** $2, -5, 3$ **(b)** 21 **3.** [6.3] $2x^2+5x+7$

4. [6.4] x^3+2x^2-8x+5

5. [6.5] **(a)** $4x^2+4x+1$ **(b)** $4x^2-1$

6. [6.6] $Q=2x^2+3x+7$ $R=12$

7. [6.7] In (i) the base is 2, and the expression is evaluated as $-(2^{-2})$. In (ii) the base is -2, so the expression is evaluated as $(-2)^{-2}$.

8. [6.7] $\dfrac{y^8}{4x^4}$ **9.** [6.8] $3 \cdot 10^{11}$

10. [7.1] After we write
$$x^2+2x-xy+2y=x(x+2)-y(x-2),$$
the resulting two terms do not have a common factor. This same problem occurs with other groupings.

11. [7.1] **(a)** $3x^2y(2xy-3)$ **(b)** $(a-2b)(2x+y)$

12. [7.2] **(a)** $(x+8)(x-8)$ **(b)** $(x-4)(x^2+4x+16)$

13. [7.3] **(a)** $2(x-9)(x-2)$ **(b)** $(x^2+5)(x^2-3)$

14. [7.4] $(3x-11)(x+2)$ **15.** [7.5] $x(x^2+1)(x+1)(x-1)$

16. [7.6] $-5, 7$ **17.** [7.7] 24 inches **18.** [8.1] $\dfrac{x+9}{x}$

19. [8.2] $\dfrac{2}{x+3}$ **20.** [8.3] $\dfrac{1}{2}$ **21.** [8.4] $2x^2(2x-3)$

22. [8.5] $\dfrac{x^2+2x-2}{x(x+5)(x-1)}$ **23.** [8.6] $\dfrac{x}{(x+1)^2}$ **24.** [8.7] 8

25. [8.8] 30 mph

Chapter 9

Section 9.1 *(page 512)*

1. **(a)** The numbers are the squares of the natural numbers. The next three numbers are 25, 36, and 49.

(b) The numbers are the cubes of the natural numbers. The next three numbers are 125, 216, and 343.

3. $5, -5$ **5.** $\frac{3}{7}, -\frac{3}{7}$ **7.** 3 **9.** 3.16 **11.** 5 **13.** 3

15. -9 **17.** 30 **19.** Rational **21.** Not a real number

23. Irrational **25.** 15.81 **27.** 2.04 **29.** 7.94

31. Because $\sqrt{x^2}$ represents a nonnegative number, $y_1 \geq 0$. In $y_2 = x$, y_2 represents any real number.

33. 30 **35.** 8 **37.** 75 **39.** $2x$ **41.** 7 **43.** 12

45. -5 **47.** $|5t|$ **49.** $5y$ **51.** $x+5$ **53.** t^3 **55.** x^6

57. $8y^4$ **59.** $(8ab)^5$

61. If n is odd, then $\sqrt[n]{k}$ is defined for all values of k. If n is even, then $\sqrt[n]{k}$ is a real number only if $k \geq 0$.

63. -3 **65.** $-2, 2$ **67.** -2 **69.** Not a real number

71. 3 **73.** 3 **75.** 10 **77.** $\frac{2}{3}$ **79.** t^5 **81.** y^2 **83.** y^2

85. $\sqrt[3]{8} = -\sqrt[3]{-8}, \sqrt[3]{-8} = -\sqrt[3]{8}$ **87.** 86.46% **89.** 50 feet

91. **(a)** The number of miles (in millions) traveled daily in 2002.

(b) 113.79 million miles

93. t^8 **95.** 2 **97.** $x \leq 0$

Section 9.2 *(page 519)*

1. **(a)** The radical sign was omitted in the result:
$$\sqrt{2}\sqrt{3} = \sqrt{6}.$$

(b) A radical sign is missing on the 4:
$$\sqrt{20} = \sqrt{4}\sqrt{5} = 2\sqrt{5}.$$

3. $\sqrt{35}$ **5.** $\sqrt{10ab}$ **7.** $6\sqrt{35}$ **9.** $10x^4$ **11.** x^3y

13. $5\sqrt{2}$ **15.** $2\sqrt{5}$ **17.** $4\sqrt{3}$

19. Write $\sqrt{b^n}$ as $\sqrt{b^{n-1}} \cdot \sqrt{b}$. Then take half of the even exponent $n-1$ to obtain $b^{(n-1)/2}\sqrt{b}$.

21. $x\sqrt{x}$ **23.** $x^5\sqrt{x}$ **25.** $2x^3\sqrt{6}$ **27.** $7y^5\sqrt{2}$

29. $5x^3\sqrt{x}$ **31.** $4y^7\sqrt{y}$ **33.** $2x^2\sqrt{7x}$ **35.** $5y^7\sqrt{5y}$

37. $x^4y\sqrt{y}$ **39.** $2xy^3\sqrt{3y}$ **41.** 8 **43.** $2\sqrt{15}$ **45.** $4x$

47. $a^2\sqrt{a}$ **49.** $2t^5\sqrt{3}$ **51.** $2\sqrt[5]{5}$ **53.** $2\sqrt[5]{2}$ **55.** $x^2\sqrt[3]{x}$

57. $x\sqrt[4]{x^3}$ **59.** $2a\sqrt[3]{3a^2}$ **61.** $7 \cdot 10^3$ **63.** $9 \cdot 10^{-2}$

65. F **67.** T

69. To simplify the radical, take the square root of the coefficient and take half of the exponent: $4x^8$.

71. $\sqrt{5t}\sqrt{2t+15} = \sqrt{10t^2+75t}$

73. **(a)** 1.58 **(b)** $T = 2d\sqrt{5d}$, 1.58

75. The graph suggests a steady increase in the amount of chicken that will be eaten.

77. 1980–2000　**79.** $(x + 4)\sqrt{2}$　**81.** $12x^3$　**83.** $3t^n$

Section 9.3　*(page 527)*

1. All three approaches are correct, but (ii) will require the fewest steps.

3. 5　**5.** $2x^2$　**7.** $\dfrac{4}{x}$　**9.** $3y^3\sqrt{5}$　**11.** $\dfrac{9}{x^5}$　**13.** $\dfrac{y^6}{x^5}$

15. $\dfrac{x^3\sqrt{x}}{5}$　**17.** $\dfrac{\sqrt{2}}{x^2}$　**19.** $\dfrac{\sqrt{11y}}{x^3}$

21. (a) The radicand is a fraction. This violates Condition 2 for a simplified square root radical.

(b) The expression is still not simplified because the denominator contains a radical. This violates Condition 3 for a simplified square root radical.

23. $5a^4$　**25.** $\dfrac{y^3}{7}$　**27.** x^2y^5　**29.** $\dfrac{2x^2\sqrt{3x}}{5y^4}$　**31.** $\dfrac{y\sqrt{3y}}{5}$

33. $\dfrac{4\sqrt{5}}{5}$　**35.** $\dfrac{\sqrt{30}}{6}$　**37.** $\dfrac{\sqrt{15}}{5}$　**39.** $\dfrac{3\sqrt{x}}{x}$　**41.** $\dfrac{3\sqrt{3y}}{y}$

43. $\dfrac{\sqrt{30}}{12}$　**45.** $\dfrac{y\sqrt{5}}{5}$　**47.** $\dfrac{2t\sqrt{3}}{3}$　**49.** $\dfrac{x\sqrt{5x}}{5}$　**51.** $\dfrac{\sqrt{5x}}{2x}$

53. $\dfrac{\sqrt{15}}{5x}$　**55.** $\dfrac{x\sqrt{3x}}{6}$　**57.** $\dfrac{2\sqrt{6x}}{xy^3}$　**59.** $\dfrac{\sqrt[3]{w^2}}{3}$　**61.** $\dfrac{\sqrt[4]{x^3}}{2}$

63. $\dfrac{3\sqrt[3]{2}}{5}$　**65.** $\dfrac{x\sqrt[3]{x^2}}{4}$　**67.** $2x^2$　**69.** $I = \dfrac{60\sqrt{5t}}{t}$　**71.** $5y\sqrt{y}$

73. (a) 2.55　**(b)** $\dfrac{5.2\sqrt{0.5x + 10}}{0.5x + 10} + 1.73$　**(c)** 2.55

75. $46\sqrt{x}$　**77.** $39.4\sqrt{x + 1} + 19.5$

79. $\dfrac{2\sqrt[3]{3}}{3}$　**81.** $\dfrac{\sqrt[3]{t}}{t}$　**83.** $\dfrac{\sqrt[4]{6}}{3}$　**85.** x^n

Section 9.4　*(page 534)*

1. Because the two terms do not have a common factor, the Distributive Property does not apply, and the terms cannot be combined.

3. $7\sqrt{7}$　**5.** $-2\sqrt{15}$　**7.** $\sqrt{2} - 3\sqrt{3}$　**9.** $2\sqrt{x}$

11. $6t\sqrt{7t}$　**13.** $\sqrt{3}$　**15.** $8\sqrt{3}$　**17.** $22\sqrt{3}$　**19.** $-4\sqrt{2t}$

21. $11b\sqrt{10b}$　**23.** $22\sqrt{2} - 10\sqrt{3}$

25. (a) The Power of a Product Rule states that each factor must be squared. Thus $\left(3\sqrt{x}\right)^2 = (3)^2\left(\sqrt{x}\right)^2 = 9x$.

(b) The square of a binomial is a special product whose pattern is $(A + B)^2 = A^2 + 2AB + B^2$. Thus $\left(\sqrt{x} + \sqrt{5}\right)^2 = \left(\sqrt{x}\right)^2 + 2 \cdot \sqrt{x} \cdot \sqrt{5} + \left(\sqrt{5}\right)^2 = x + 2\sqrt{5x} + 5$.

27. $2\sqrt{5} - \sqrt{15}$　**29.** $4 + \sqrt{14}$　**31.** $18 - 7\sqrt{6}$

33. $19 + 7\sqrt{10}$　**35.** $x + 5\sqrt{x} - 14$　**37.** $9 + 4\sqrt{5}$

39. $9 + 6\sqrt{x} + x$　**41.** $2\sqrt{3} - 3, 3$　**43.** $\sqrt{10} + \sqrt{5}, 5$

45. $\sqrt{x} + 4, x - 16$　**47.** $4 - 3\sqrt{2}$　**49.** $\dfrac{2 + \sqrt{2}}{3}$

51. $\dfrac{5 - \sqrt{3}}{22}$　**53.** $2\left(3 + \sqrt{2}\right)$　**55.** $\dfrac{3\left(2\sqrt{7} + \sqrt{5}\right)}{23}$

57. $\dfrac{2\sqrt{35} + 7}{13}$　**59.** $\dfrac{\sqrt{3} + 1}{2}$　**61.** $\dfrac{y\left(1 + \sqrt{y}\right)}{1 - y}$

63. $\dfrac{6 + \sqrt{11}}{5}$　**65.** $\sqrt{3}$　**67.** $\sqrt{x}$　**69.** Yes　**71.** No

73. No　**75.** $3\sqrt{5} + \sqrt{6}$　**77.** $5 + \sqrt{3}$　**79.** 14.5%

81. $F(t) = \dfrac{113\sqrt{t + 15}}{t + 15} + 1$; $C(t) = \dfrac{44\sqrt{t + 15}}{t + 15} + 0.9$

83. $\dfrac{11\sqrt{5}}{5}$　**85.** $\dfrac{5\sqrt{6}}{6}$　**87.** 3　**89.** $(3x + 4)\sqrt{3}$

Section 9.5　*(page 542)*

1. (a) True. The Multiplication Property of Equality allows us to multiply both sides by the same nonzero number. Because $a = b$, we multiply the left side by a and the right side by b to obtain $a^2 = b^2$. If a and b are 0, then clearly $a^2 = b^2$.

(b) False. For example, $(3)^2 = (-3)^2$, but $3 \neq -3$.

3. 6　**5.** 3　**7.** 3　**9.** $-1, 3$　**11.** 2　**13.** 9　**15.** 5

17. 46　**19.** 5　**21.** No solution　**23.** 3　**25.** No solution

27. 32　**29.** 6　**31.** No solution　**33.** 3　**35.** 1

37. The solutions of $\sqrt{x^2 + 3x} = 2$ are -4 and 1. The first equation is not satisfied by $x = -4$ because $\sqrt{-4}$ is not a real number.

39. $-2, -3$　**41.** 1, 4　**43.** 1, 3　**45.** 0, 1　**47.** 2, 5

49. 5　**51.** 4　**53.** No solution　**55.** 2　**57.** 6

59. No solution　**61.** 1　**63.** 2　**65.** 0, 3　**67.** 2

69. 2　**71.** 12　**73.** 4444 feet　**75.** 141

77. 6 hours　**79.** 2001

81. (a) $2.3\sqrt{x + 1} + 43.1 = 56.9$　**(b)** 35; the year 2005

83. -8　**85.** No solution　**87.** $h = \dfrac{3V}{\pi r^2}$

Section 9.6　*(page 547)*

1. The hypotenuse is the longest side of the triangle and is opposite the right angle.

3. 13　**5.** 8　**7.** $\sqrt{117} \approx 10.82$　**9.** $\sqrt{11} \approx 3.32$

11. 4　**13.** $c = 17$　**15.** $a = 5$　**17.** $a = 4\sqrt{6} \approx 9.80$

19. $b = 2\sqrt{7} \approx 5.29$　**21.** $c = \sqrt{14} \approx 3.74$　**23.** 13.00 feet

25. 160.00 feet　**27.** 187.35 feet　**29.** 72.11 feet

31. 20.00 miles　**33.** 136.47 miles　**35.** 210.95 miles

37. 13.00 feet　**39.** 6.93 feet　**41.** 134.54 feet

43. 21.21 inches

45. One leg is horizontal and the other is vertical.

47. 5 **49.** 10 **51.** $2\sqrt{13} \approx 7.21$ **53.** 3 feet

55. 57.08 square feet **57.** (a) $6\sqrt{2} \approx 8.49$ feet (b) $\dfrac{c\sqrt{2}}{2}$

Section 9.7 *(page 555)*

1. In the definition of $b^{1/n}$, the expression $\sqrt[n]{b}$ must be defined. In this case, $\sqrt{-25}$ is not a real number.

3. 10 **5.** 2 **7.** Not a real number **9.** -2 **11.** 2

13. 3.13 **15.** -3.11 **17.** 3.04

19. The expression in (iv) is not a real number because the base is negative and n is even.

(i) -8 (ii) $\frac{1}{8}$ (iii) $-\frac{1}{8}$

21. 27 **23.** 9 **25.** 4 **27.** 16 **29.** 27 **31.** $\frac{64}{125}$

33. 256 **35.** $\frac{1}{4}$ **37.** $\frac{1}{4}$ **39.** $-\frac{1}{32}$ **41.** $\frac{16}{81}$ **43.** $\frac{1}{32}$

45. 13.57 **47.** 0.18 **49.** 24.68 **51.** 7 **53.** a

55. $y^{5/6}$ **57.** 25 **59.** y^2 **61.** $a^{4/5}$ **63.** 10 **65.** $x^{1/2}$

67. z^2 **69.** $6x^5$ **71.** $\dfrac{x^3}{3}$ **73.** $\dfrac{y^6}{z^4}$ **75.** $\dfrac{y}{2}$ **77.** $\dfrac{b^6}{a^2}$

79. x^2y^4 **81.** 53 mph **83.** 165 feet **85.** 5.3 seconds

87. 64 feet **89.** $V = 284\sqrt[5]{x}$; 488 million

91. $R = \dfrac{5\left(284x^{1/5}\right)}{2.8}$ **93.** $\sqrt{x}$ **95.** $\sqrt{3x}$ **97.** $\frac{1}{49}$

Chapter Review Exercises *(page 557)*

1. (a) $-4, 4$ (b) 4 **3.** (a) 1.89 (b) 3.46

5. (a) 2 (b) $x + 3$ **7.** (a) -2 (b) -2 **9.** $|x|$

11. $\sqrt{6xy}$ **13.** (a) $3\sqrt{2}$ (b) $6\sqrt{7}$ **15.** $5a^3b^5\sqrt{2ab}$

17. (a) $x^2\sqrt[3]{x^2}$ (b) $2x\sqrt[4]{x}$ **19.** $\dfrac{x^2}{4}$ **21.** $\dfrac{\sqrt{3y}}{7}$ **23.** $\dfrac{4\sqrt{3}}{7a^5}$

25. $\dfrac{2\sqrt{30x}}{5x^2}$ **27.** $\dfrac{2x\sqrt[3]{2x}}{3}$ **29.** $-3\sqrt{7}$ **31.** $18\sqrt{2} - 4\sqrt{3}$

33. $\sqrt{15} + \sqrt{6}$ **35.** $3\sqrt{14} + 7 - 6\sqrt{2} - 2\sqrt{7}$ **37.** 14

39. 8.83 **41.** 5 **43.** 5 **45.** 4 **47.** 3 **49.** 8.12 inches

51. 7.81 feet **53.** (a) $\frac{3}{4}$ (b) -3

55. (a) 1024 (b) Not a real number **57.** 5 **59.** $x^{1/12}$

61. Write the equation as $2\sqrt{x} = 15$ and solve the radical equation.

Chapter Test *(page 559)*

1. [9.1] (a) 3 (b) Not a real number **2.** [9.1] $4x$

3. [9.1] Because the radicand is negative, the expression is a real number only if the index is an odd integer.

4. [9.2] (a) $12\sqrt{2}$ (b) $x^4\sqrt{x}$

5. [9.2] $2a^4\sqrt{3}$ **6.** [9.2] $3x\sqrt[3]{x^2}$

7. [9.3] (a) $2x\sqrt{x}$ (b) $\dfrac{\sqrt{14}}{7}$

8. [9.3] $\dfrac{2x^3\sqrt{7}}{7}$ **9.** [9.3] $\dfrac{2}{x}$ **10.** [9.4] $7\sqrt{2}$

11. [9.4] $\sqrt{15} - 7$ **12.** [9.4] $\dfrac{x\left(x - \sqrt{3}\right)}{x^2 - 3}$ **13.** [9.5] 2.8

14. [9.5] A principal square root is nonnegative. Therefore, this equation is false.

15. [9.5] No solution **16.** [9.5] 3

17. [9.6] 12.1 inches **18.** [9.6] 217.28 feet

19. [9.7] The base x is the radicand, the denominator of the exponent is the index, and the numerator is the power: $x^{2/3} = \left(\sqrt[3]{x}\right)^2$.

20. [9.7] (a) $\frac{2}{5}$ (b) 16 **21.** [9.7] (a) 1.9 (b) 0.4

22. [9.7] (a) $\dfrac{1}{3}$ (b) $\dfrac{y^4}{x^2}$

Chapter 10

Section 10.1 *(page 566)*

1. Because the value of y must be 0, we trace to the x-intercepts.

3. -3 **5.** $-4.65, 0.65$ **7.** $-1, 6$

9. No real number solution **11.** $-5, -1$ **13.** 7

15. $0, \frac{1}{2}$ **17.** $-3, 4$ **19.** 5 **21.** $-10, 6$ **23.** $\frac{5}{6}, \frac{3}{2}$

25. Both the factoring method and the Square Root Property may be used to solve the equation. Applying the Square Root Property is easier.

27. ± 7 **29.** $\pm\frac{5}{2}$ **31.** No real number solution

33. $\pm\dfrac{\sqrt{14}}{2} \approx \pm 1.87$ **35.** $\pm 5\sqrt{5} \approx \pm 11.18$

37. No real number solution **39.** $-8, -2$

41. $9 \pm 2\sqrt{3} \approx 5.54, 12.46$ **43.** No real number solution

45. $\frac{5}{4}, \frac{19}{4}$ **47.** $\frac{-2}{3}, \frac{10}{3}$ **49.** $\frac{-11}{3}, \frac{-7}{3}$

51. Because the left side of (i) is a perfect square trinomial, it is easier to solve with the Square Root Property.

53. $4, 14$ **55.** $-7, 2$ **57.** $-3 \pm \sqrt{10} \approx -6.16, 0.16$

59. $\pm\dfrac{\sqrt{3}}{3} \approx \pm 0.58$ **61.** ± 3 **63.** ± 6 **65.** ± 3

67. $\pm 2\sqrt{2} \approx \pm 2.83$ **69.** 6 **71.** 9 **73.** 8 inches

75. 10 feet **77.** 26.83 inches

79. (a) 16, 78

(b) People of ages 16 and 78 received no assistance.

81. $0.18t^2 + 0.1 = 26$; $t = 12$

83. The solution $t = -12$ corresponds to 1978, but the question is about the year after 1990 when the cost will be $26 billion.

85. $r = \sqrt{\dfrac{3V}{\pi h}}$ **87.** $y = \pm\sqrt{49 - x^2}$ **89.** $x^2 - x - 12 = 0$

Section 10.2   *(page 573)*

1. Take half of -9, the coefficient of x, and square the result.

3. $x^2 + 16x + 64$ **5.** $x^2 - 24x + 144$ **7.** $x^2 - 5x + \frac{25}{4}$

9. $x^2 + \frac{6}{5}x + \frac{9}{25}$ **11.** $-3, 7$ **13.** $-8, -2$ **15.** $-6, 1$

17. $1, 8$ **19.** $-2, 8$ **21.** $-3 \pm \sqrt{11} \approx -6.32, 0.32$

23. $2 \pm \sqrt{6} \approx -0.45, 4.45$ **25.** No real number solution

27. $\dfrac{5 \pm 3\sqrt{5}}{2} \approx -0.85, 5.85$

29. $\dfrac{-5 \pm \sqrt{17}}{2} \approx -4.56, -0.44$

31. $\dfrac{-7 \pm \sqrt{37}}{2} \approx -6.54, -0.46$

33. Because the equation is in standard form and the left side can be factored, the factoring method is easier.

35. $\frac{-2}{3}, \frac{1}{2}$ **37.** $-2, \frac{3}{2}$ **39.** $-1, -\frac{3}{4}$ **41.** $-2, 3$

43. $\dfrac{3 \pm \sqrt{57}}{12} \approx -0.38, 0.88$ **45.** No real number solution

47. $\dfrac{-3 \pm \sqrt{41}}{4} \approx -2.35, 0.85$ **49.** No real number solution

51. $\dfrac{-1 \pm \sqrt{13}}{6} \approx -0.77, 0.43$ **53.** -10 **55.** $60°$

57. 54 **59.** 29 mph   **61.** 68 mph

63. (a) $-0.012(x^2 - 136x + 3783) = 4.8$ (b) $47, 89$

65. $-0.035(x^2 - 88x + 960) = 34$ **67.** $42, 46$

69. $\frac{9}{16}$ **71.** 16 **73.** $y - 2 = (x + 3)^2$

Section 10.3   *(page 580)*

1. The Quadratic Formula is derived by completing the square.

3. $2, -3, -1$ **5.** $3, 0, 2$ **7.** $1, 5$ **9.** $-7, 0$

11. $-2, \frac{2}{3}$ **13.** $-6, -1$ **15.** $3 \pm \sqrt{2} \approx 1.59, 4.41$

17. No real number solution   **19.** $\dfrac{1 \pm \sqrt{10}}{3} \approx -0.72, 1.39$

21. $\pm\sqrt{15} \approx \pm 3.87$ **23.** $-\frac{5}{2}, 0$

25. $\dfrac{3 \pm \sqrt{33}}{6} \approx -0.46, 1.46$

27. Because the graph has only one x-intercept, the discriminant is 0.

29. $-\frac{3}{5}, \frac{1}{2}$ **31.** No real number solution

33. $-3 \pm \sqrt{2} \approx -1.59, -4.41$ **35.** $-7, -5$ **37.** $-7, 2$

39. -9 **41.** $-\frac{1}{4}, 3$ **43.** $\dfrac{9 \pm \sqrt{53}}{2} \approx 0.86, 8.14$

45. $\dfrac{-4 \pm \sqrt{10}}{3} \approx -2.39, -0.28$ **47.** $2, 3$

49. The graph has no x-intercepts.   **51.** 2 **53.** 1 **55.** 0

57. 2 **59.** 1 **61.** 0 **63.** 45 feet   **65.** 1.5 feet

67. 14 feet   **69.** 20 feet   **71.** 50 **73.** 30 **75.** 4.5 hours

77. (a) 22.2% (b) 30 **79.** 91 million

81. 13.5 (2003); The model predicts an increase of 91 million passengers between 2000 and 2003.

83. $2\sqrt{2} \pm 2\sqrt{3} \approx -0.64, 6.29$ **85.** $y = -3 \pm \sqrt{x + 4}$

87. (a) $k < \frac{3}{4}$ (b) $k > \frac{3}{4}$

Section 10.4   *(page 589)*

1. A complex number can be written in the form $a + bi$, where a and b are any real numbers. An imaginary number is a complex number with $b \neq 0$.

3. $i\sqrt{15}$ **5.** $7i$ **7.** $12i$ **9.** $2i\sqrt{5}$ **11.** $5i\sqrt{3}$

13. (a) 5 (b) 7 **15.** (a) 0 (b) 10

17. (a) 12 (b) 0 **19.** (a) -7 (b) 15 **21.** $3 + 6i$

23. $7 - i$ **25.** $9 - 2i$ **27.** $2 + 2i$ **29.** 12

31. $-2 + 6i$ **33.** $-3 + 2i$ **35.** $-2 - 12i$ **37.** -25

39. $9 - 40i$ **41.** $4 - 7i$ **43.** $4 + 3i$

45. (a) $5 + 3i$ (b) 34 **47.** (a) $7 - 2i$ (b) 53

49. (a) $-1 - i$ (b) 2

51. In each we multiply both the numerator and the denominator by the conjugate of the denominator. In (i) multiply by the complex conjugate $1 - i$, and in (ii) multiply by $1 - \sqrt{3}$.

53. $-4 + 5i$ **55.** $2 - 3i$ **57.** $-\frac{6}{5} + \frac{3}{5}i$

59. $-\frac{4}{5} + \frac{3}{5}i$ **61.** $2 - i$

63. The imaginary solutions of a quadratic equation with real coefficients are complex conjugates.

65. $\pm 2i$ **67.** $\pm i\sqrt{5} \approx \pm 2.24i$ **69.** $-5 \pm 4i$

71. $\frac{1}{3} \pm i$ **73.** $-2 \pm 3i\sqrt{2} \approx -2 \pm 4.24i$ **75.** $1 \pm 2i$

77. $-5 \pm i$ **79.** $\dfrac{-3 \pm i\sqrt{7}}{2} \approx -1.5 \pm 1.32i$

81. $\dfrac{3 \pm 3i\sqrt{3}}{2} \approx 1.5 \pm 2.60i$ **83.** $-2 \pm i\sqrt{3} \approx -2 \pm 1.73i$

85. $\frac{1}{2} \pm i$ **87.** $-\frac{2}{3} \pm \frac{1}{3}i$ **89.** $\dfrac{-1 \pm i\sqrt{3}}{4} \approx -0.25 \pm 0.43i$

91. $-2 \pm 2i$ **93.** $3 \pm i$ **95.** $-i \pm 2$ **97.** $-i$ **99.** 1

Section 10.5   *(page 596)*

1. If $a < 0$ the parabola opens downward and the vertex is the highest point. If $a > 0$ the parabola opens upward and the vertex is the lowest point.

3. Upward   **5.** Downward   **7.** Upward   **9.** Downward

11. The graph is a parabola and always has one y-intercept. The graph may have no, one, or two x-intercepts.

13. y-intercept: $(0, -3)$ x-intercepts: $(1, 0), (-3, 0)$

15. y-intercept: $(0, 9)$ x-intercept: $(3, 0)$

17. y-intercept: $(0, 0)$ x-intercepts: $(1, 0), (0, 0)$

19. y-intercept: $(0, 5)$ x-intercepts: none

21. y-intercept: $(0, -1)$ x-intercept: $\left(-\frac{1}{2}, 0\right)$

23. y-intercept: $(0, 12)$ x-intercepts: $\left(-\frac{3}{2}, 0\right), (4, 0)$

25. y-intercept: $(0, -2)$ x-intercepts: none

27. y-intercept: $(0, -25)$ x-intercepts: $(-5, 0), (5, 0)$

29. $(-3, -9)$ **31.** $(-1, -3)$ **33.** $(3, -5)$ **35.** $(1, 1)$

37. $(-3, -16)$; Upward **39.** $(1, 0)$; Downward

41. $(0, 4)$; Upward **43.** $\left(-\frac{3}{2}, \frac{17}{4}\right)$; Downward

45. $(3, -18)$; Upward **47.** $(1, 4)$; Downward

49. (a) The parabola opens upward and has no x-intercepts.

   (b) The parabola opens downward and has two x-intercepts.

51. 2 **53.** 1 **55.** 0 **57.** 0 **59.** 2 **61.** 1 **63.** 2

65. 0 **67.** 0 **69.** 2 **71.** 1 **73.** 0

75. (ii), (iv), (vi), (vii), (viii), (ix) **77.** (i), (iii), (vi), (vii), (x)

79. (a) $3, -9$ **(b)** $7, -1$ **(c)** None

81. (a) 3 seconds, 144 feet **(b)** 80 feet **(c)** 6 seconds

83. $(50, 37675)$ **85.** $32, 68$ **87.** $(1, 1), (-2, 4)$

89. $(-2, 0), (2, 0)$

Section 10.6 *(page 604)*

1. No, a function can have two pairs with the same second coordinate but not the same first coordinate.

3. Function
   Domain = $\{3, 4, 5, 7\}$
   Range = $\{-3, 1, 2\}$

5. Not a function

7. Function
   Domain = $\{1, 2, 3, 4\}$
   Range = $\{2, 3, 4, 5\}$

9. $\{(1, -2), (2, -1), (3, 0), (4, 1)\}$
   Function
   Domain = $\{1, 2, 3, 4\}$
   Range = $\{-2, -1, 0, 1\}$

11. $\{(4, 0), (4, 1), (4, 2), (4, 3), (4, 4), (4, 5)\}$
   Not a function

13. $\{(-1, 4), (0, 4), (1, 4)\}$
   Function
   Domain = $\{-1, 0, 1\}$
   Range = $\{4\}$

15. Not a function **17.** Function **19.** Function

21. Not a function **23.** Function **25.** Not a function

27. Function

29. For $y = x + 1$, each value of x corresponds to exactly one value of y. For the inequality $y \leq x + 1$, for any value of x there are many values of y.

31. Function **33.** Function **35.** Function

37. Not a function **39.** Not a function **41.** Function

43. The domain of f is **R**, $x \neq -4$ and the domain of g is **R**.

45. Domain: **R** Range: **R**

47. Domain: **R** Range: $y \geq 0$

49. Domain: $x \geq 0$ Range: $y \geq 0$

51. Domain: **R** Range: **R**

53. Domain: **R** Range: $y \leq 9$

55. Domain: $x \leq 9$ Range: $y \geq 0$

57. Domain: **R** Range: $y \geq -1$

59. **R** **61.** **R**, $x \neq 1$ **63.** $x \geq -7$ **65.** **R**

67. $x \leq 3$ **69.** **R**, $x \neq 0$ **71.** **R** **73.** **R** **75.** **R**

77. (a) Domain = $\{1, 2, 3, 4, 5, 6\}$
      Range = $\{70, 84, 105, 140, 210, 420\}$

   (b) $\{(1, 420), (2, 210), (3, 140), (4, 105), (5, 84), (6, 70)\}$

79. (a) $f(x) = x(10,000 - 500x)$ **(b)** $[0, 20]$ **(c)** 10

   (d) \$50,000 **(e)** $0 \leq y \leq 50,000$

81. (a) Linear

   (b) The estimated 1960 survival rate of 39.6%

   (c) $S(30) = 56.4$; The estimated 1990 survival rate of 56.4%

83. Domain: **R**, $x \neq 0$ Range: $\{-1, 1\}$

85. Domain: **R** Range: $y \geq -53.33$

87. $f(x) = 2x + 3$

Chapter Review Exercises *(page 607)*

1. $-3, 8$ **3.** $\pm 2\sqrt{7} \approx \pm 5.29$ **5.** $3, 9$ **7.** $-1, 7$ **9.** 5

11. The third term is the square of half the coefficient of x: $\left(\frac{1}{2} \cdot 10\right)^2 = 25$.

13. $4 \pm 2\sqrt{5} \approx -0.47, 8.47$ **15.** No real number solution

17. $25°$ **19.** $2, -1, -5$ **21.** $\dfrac{2 \pm \sqrt{2}}{2} \approx 0.29, 1.71$

23. $-2, \dfrac{5}{6}$ **25.** $2 \pm \sqrt{7} \approx -0.65, 4.65$ **27.** $\frac{3}{4}$ **29.** 2

31. $i\sqrt{11}$ **33.** $4 + i$ **35.** $28 - 21i$ **37.** $-2 + 3i$

39. $\dfrac{1 \pm i\sqrt{11}}{2} \approx 0.5 \pm 1.66i$ **41.** Downward **43.** $(0, 4)$

45. (a) The vertex is the highest point of the graph.

   (b) The y-coordinate of the vertex is the maximum value of y.

47. y-intercept: $(0, 7)$ x-intercepts: $(1, 0), (7, 0)$ **49.** 2

51. No two ordered pairs of a function can have the same first coordinate. Thus (i) is a function and (ii) is not a function.

53. (i)   **55.** $x \le 4$ **57.** $y \ge 0$ **59.** (i)

Chapter Test   *(page 609)*

1. [*10.1*] $-4.45, 0.45$ **2.** $-2 \pm \sqrt{6} \approx -4.45, 0.45$

3. [*10.1*] $-1, 7$ **4.** [*10.2*] $55°$

5. [*10.3*] $\dfrac{-3 \pm \sqrt{17}}{4} \approx -1.78, 0.28$

6. [*10.3*] 11 inches

7. [*10.3*] None   **8.** [*10.4*] -6 **9.** [*10.4*] $\frac{4}{5} - \frac{3}{5}i$

10. [*10.4*] $\dfrac{-1 \pm i\sqrt{39}}{10} \approx -0.1 \pm 0.62i$

11. [*10.5*] Parabola

12. [*10.5*]

   (a) Because $a < 0$, the parabola opens downward.

   (b) The constant term is the y-coordinate of the y-intercept. The y-intercept is $(0, 12)$. To determine the x-intercept, solve $-x^2 - 4x + 12 = 0$ to obtain $x = -6$ and $x = 2$. The x-intercepts are $(-6, 0)$ and $(2, 0)$.

13. [*10.5*] $(-1, -5)$ **14.** [*10.6*] (i), (ii), (iii)

15. [*10.6*]

   (i) Domain: $\{1, 2, 3\}$
      Range: $\{2, 3\}$

   (ii) Domain: $x \ge -1$
       Range: $y \ge 0$

   (iii) Domain: $\mathbf{R}, x \ne -1$
        Range: $\mathbf{R}, y \ne 0$

16. [*10.6*] Because the parabola opens downward, the range is $y \le 4$.

17. [*10.3*] 4 feet   **18.** [*10.6*]   **(a)** $75°$ **(b)** 825

Cumulative Test, Chapters 9–10   *(page 609)*

1. [*9.1*] $3x$ **2.** [*9.2*] $4x\sqrt{3x}$ **3.** [*9.3*] $\dfrac{1}{2x^2}$

4. [*9.3*] $\dfrac{\sqrt{x}}{x}$ **5.** [*9.4*] $7\sqrt{3x}$ **6.** [*9.4*] $2x^4\sqrt{3x}$

7. [*9.5*] $1, 2$ **8.** [*9.5*] -2 **9.** [*9.6*] 15 inches

10. [*9.7*] $\frac{1}{4}$ **11.** [*10.1*] ± 4

12. [*10.1*]

   (a) There is a common factor x.

   (b) It is easy to isolate the x^2 term.

13. [*10.2*] $3 \pm \sqrt{5} \approx 0.76, 5.24$

14. [*10.3*] $\dfrac{5 \pm \sqrt{73}}{12} \approx -0.30, 1.13$

15. [*10.4*] 13

16. [*10.4*] $\dfrac{-1 \pm i\sqrt{3}}{2} \approx -0.5 \pm 0.87i$

17. [*10.4*] No, if an equation has an imaginary solution, then the conjugate is also a solution.

18. [*10.5*]   **(a)** Upward

       (b) $(4, -7)$

       (c) $(0, 9), (6.65, 0), (1.35, 0)$

19. [*10.6*] Either -2 or 5 results in two pairs with the same first coordinate.

20. [*10.6*] $-\frac{1}{4}$

Index of Data Analysis

Index of Real-Life Applications

Index of Modeling with Real Data

Index

Index of Calculator Key Words
for the TI-83 Plus

Key Word	Function	Page
Absolute Value	Calculate an absolute value	16
Add	Perform addition	13
Alpha	Enter a variable other than x	170
Cursor	Activate the general cursor	86
Decimal	Display the decimal window setting	86
Divide	Perform division	5
Evaluate	Evaluate an expression	73
Evaluate Y	Evaluate an expression	334
Exponent	Raise a number to a power	14
Fraction	Report result in fraction form	39
Graph	Display a graph	85
Home	Display the home screen	85
Integer	Display the integer window setting	86
Mode	Set number of decimal places	6
Multiply	Perform multiplication	13
Negative	Enter negative numbers	4
Scientific	Display results in scientific notation	376
Square Root	Calculate a square root	7
Standard	Display the standard window setting	86
Store	Assign a value to a variable	73
Subtract	Perform subtraction	13
Table	Display a table of values	93
Table Set	Configure a table display	93
Trace	Activate the tracing cursor	96
Window	Enter viewing window settings	86
Y Screen	Display the Y screen	85
Y Var	Access a Y variable	94
Zoom In	Magnify a portion of a graph	107